Birkhäuser

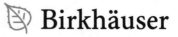

Applied and Numerical Harmonic Analysis

More information about this series at http://www.springer.com/series/4968

Shayne F. D. Waldron

An Introduction to Finite Tight Frames

Birkhäuser

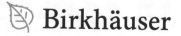

Shayne F. D. Waldron
Department of Mathematics
University of Auckland
Auckland
New Zealand

ISSN 2296-5009 ISSN 2296-5017 (electronic)
Applied and Numerical Harmonic Analysis
ISBN 978-1-4939-7920-2 ISBN 978-0-8176-4815-2 (eBook)
https://doi.org/10.1007/978-0-8176-4815-2

Mathematics Subject Classification (2010): 20C15, 41A10, 42C15, 42C40

Printed on acid-free paper

This book is published under the trade name Birkhäuser
The registered company is Springer Science+Business Media, LLC
The registered company address is: 233 Spring Street, New York, NY 10013, U.S.A.

To my wife Catherine Ann Etheredge.

ANHA Series Preface

The *Applied and Numerical Harmonic Analysis (ANHA)* book series aims to provide the engineering, mathematical, and scientific communities with significant developments in harmonic analysis, ranging from abstract harmonic analysis to basic applications. The title of the series reflects the importance of applications and numerical implementation, but richness and relevance of applications and implementation depend fundamentally on the structure and depth of theoretical underpinnings. Thus, from our point of view, the interleaving of theory and applications and their creative symbiotic evolution is axiomatic.

Harmonic analysis is a wellspring of ideas and applicability that has flourished, developed, and deepened over time within many disciplines and by means of creative cross-fertilization with diverse areas. The intricate and fundamental relationship between harmonic analysis and fields such as signal processing, partial differential equations (PDEs), and image processing is reflected in our state-of-the-art *ANHA* series.

Our vision of modern harmonic analysis includes mathematical areas such as wavelet theory, Banach algebras, classical Fourier analysis, time-frequency analysis, and fractal geometry, as well as the diverse topics that impinge on them.

For example, wavelet theory can be considered an appropriate tool to deal with some basic problems in digital signal processing, speech and image processing, geophysics, pattern recognition, biomedical engineering, and turbulence. These areas implement the latest technology from sampling methods on surfaces to fast algorithms and computer vision methods. The underlying mathematics of wavelet theory depends not only on classical Fourier analysis, but also on ideas from abstract harmonic analysis, including von Neumann algebras and the affine group. This leads to a study of the Heisenberg group and its relationship to Gabor systems, and of the metaplectic group for a meaningful interaction of signal decomposition methods. The unifying influence of wavelet theory in the afore-mentioned topics illustrates the justification for providing a means for centralizing and disseminating information from the broader, but still focused, area of harmonic analysis. This will be a key role of *ANHA*. We intend to publish with the scope and interaction that such a host of issues demands.

Along with our commitment to publish mathematically significant works at the frontiers of harmonic analysis, we have a comparably strong commitment to publish major advances in the following applicable topics in which harmonic analysis plays a substantial role:

Antenna theory	*Prediction theory*
Biomedical signal processing	*Radar applications*
Digital signal processing	*Sampling theory*
Fast algorithms	*Spectral estimation*
Gabor theory and applications	*Speech processing*
Image processing	*Time-frequency and time-scale analysis*
Numerical partial differential equations	*Wavelet theory*

The above point of view for the *ANHA* book series is inspired by the history of Fourier analysis itself, whose tentacles reach into so many fields.

In the last two centuries Fourier analysis has had a major impact on the development of mathematics, on the understanding of many engineering and scientific phenomena, and on the solution of some of the most important problems in mathematics and the sciences. Historically, Fourier series were developed in the analysis of some of the classical PDEs of mathematical physics; these series were used to solve such equations. In order to understand Fourier series and the kinds of solutions they could represent, some of the most basic notions of analysis were defined, e.g., the concept of “function.” Since the coefficients of Fourier series are integrals, it is no surprise that Riemann integrals were conceived to deal with uniqueness properties of trigonometric series. Cantors set theory was also developed because of such uniqueness questions.

A basic problem in Fourier analysis is to show how complicated phenomena, such as sound waves, can be described in terms of elementary harmonics. There are two aspects of this problem: first, to find, or even define properly, the harmonics or spectrum of a given phenomenon, e.g., the spectroscopy problem in optics; second, to determine which phenomena can be constructed from given classes of harmonics, as done, for example, by the mechanical synthesizers in tidal analysis.

Fourier analysis is also the natural setting for many other problems in engineering, mathematics, and the sciences. For example, Wiener’s Tauberian theorem in Fourier analysis not only characterizes the behavior of the prime numbers, but also provides the proper notion of spectrum for phenomena such as white light; this latter process leads to the Fourier analysis associated with correlation functions in filtering and prediction problems, and these problems, in turn, deal naturally with Hardy spaces in the theory of complex variables.

Nowadays, some of the theory of PDEs has given way to the study of Fourier integral operators. Problems in antenna theory are studied in terms of unimodular trigonometric polynomials. Applications of Fourier analysis abound in signal processing, whether with the fast Fourier transform (FFT), or filter design, or the adaptive modeling inherent in time-frequency-scale methods such as wavelet theory. The coherent states of mathematical physics are translated and modulated Fourier transforms, and these are used, in conjunction with the uncertainty principle, for dealing with signal reconstruction in communications theory. We are back to the raison d'être of the *ANHA* series!

College Park, MD, USA John J. Benedetto

Preface

This book gives a unified introduction to the rapidly developing area of *finite tight frames*. Fifteen years ago, the existence of equal-norm tight frames of $n > d$ vectors for $\mathbb{R}^d$ and $\mathbb{C}^d$ was not widely known. Now equal-norm tight frames are known to be common, and those with optimal cross-correlation and symmetry properties are being constructed and classified. The impetus behind these rapid developments are applications to areas as diverse as signal processing, quantum information theory, multivariate orthogonal polynomials and splines, and compressed sensing.

It can be thought of as an extension of the first chapter of Ole Christensen's book: *An introduction to Frames and Riesz Bases* (in this series), which deals mostly with the infinite dimensional case. For finite dimensional Hilbert spaces the technicalities of Riesz bases disappear (though infinite frames are still of interest), and, with some work, usually a nice tight frame can be constructed explicitly. Hence the focus is on finite *tight* frames, which are the most intuitive generalisation of orthonormal bases. In addition to analogues of familiar ideas from the infinite dimensional setting such as group frames, there is a special geometry, e.g., the variational characterisation and the frame force.

The book is structured into chapters, with the first paragraph intended to give a feeling for its content. These give a logical development, while being as independent as possible. For example, one could jump to those giving the important examples of *harmonic frames*, *equiangular frames* and *SICs*, referring back to the motivating chapters on *symmetries* and *group frames* as desired. Similarly, in the text we give forward references to such nice examples. Grey boxes are used to emphasise or paraphrase some key ideas, and may help readers to navigate this book.

Each chapter has *Notes*, which primarily give a brief description of my source material and suggestions for further reading. The interdisciplinary nature of the subject makes it difficult (and often senseless) to give exact attribution for many results, e.g., the discrete form of Naĭmark's theorem, and so I have attempted to do so only in some cases. This *introductory* book is deliberately as short as possible, and so its scope and list of references is far from definitive. If only for this reason, some contributions may not be noted explicitly.

The chapters conclude with *Exercises*, which contain more details and examples. Brief solutions are given, since some are parts of proofs, and in the interests of being both introductory and self-contained. Those marked with a$^{\mathtt{m}}$ are suitable for using a computer algebra package such as `matlab`, `maple`, and `magma`. To get a real sense for finite tight frames the reader is encouraged construct the various examples given in the book numerically or algebraically. In particular, the *harmonic frames*, *MUBs* and *Weyl–Heisenberg SICs* for $d \leq 100$ (or higher), which really are quite remarkable.

This book can be used for an introductory (graduate or undergraduate) course in finite tight frames: chapters 1, 2, 5–8 and 11 together with some examples of interest, e.g., group frames (chapter 10), equiangular tight frames (chapter 12) and SICs (chapter 14). The latter chapters rely heavily on parts of classical algebra, such as graph theory (real equiangular tight frames), character theory (harmonic frames), representation theory (group frames), and Galois theory (SICs). Examples from these could be included in a graduate algebra course, e.g., Theorem 10.8 as an orthogonal decomposition of $\mathbb{F}G$–modules. It could also act as the theoretical background for a course involving applications of tight frames, e.g., some of the topics treated in *Finite frames Theory and Applications* (in this series). It is also suitable for self-study because of the complete proofs and solutions for exercises.

How does one come to write such a book? My initial interest in finite tight frames was accidental. While investigating the eigenstructure of the Bernstein operator, I encountered deficiencies in the existing orthogonal and biorthogonal expansions for the Jacobi polynomials on a triangle. Having heard of frames while a graduate student in Madison, I supposed that a finite tight frame might just be what was required. This eventually proved to be so (see §15) and along the way I found that finite frames are a fascinating and very active area of research.

When I started this book, there was just enough material for a short book on *finite tight* frames, a small part of frame theory, at the time. Since then, there has been an explosion of research on finite tight frames, which continues apace. I have called time on the project: the book contains results from last week (Lucas-Fibonnacci SICs), but not next week (maybe a proof of Zauner's conjecture). There are chapters that could have been written, e.g., ones on erasures and data transmission. I maintain a strong interest in finite tight frames. On my home page, there are various links of relevance to the book (lists of SICs, harmonic frames, typos, etc).

It is my hope that this book conveys the basic theory of finite tight frames, in a friendly way, being mindful of its connections with existing areas and continuing development. The applications given are just the tip of an iceberg: an invitation to use finite tight frames in any area where there is a natural inner product on a finite dimensional space.

Shayne F. D. Waldron
Honeymoon Valley, Far North
Aotearoa (New Zealand)
July 2017

Acknowledgements

There are many people to thank: Catherine for her support; my mentors: Carl de Boor (Orcas island) and Len Bos (Verona); my colleagues at home: Tom ter Elst, Sione Ma'u and Warren Moors, and in Italy: Marco Vianello and Stefano Di Marchi; my students: Irine Peng, Richard Vale, Simon Marshall, Nick Hay, Helen Broome, Tan Do, Jennifer Bramwell, Xiaoyang Li, Paul Harris, Tuan Chien and Daniel Hughes; many in the frame community: John Benedetto and Pete Casazza for their ongoing encouragement with this book, Joe Renes for introducing me to SICs and others for their careful and insightful reading of various chapters of the book; algebraists that I have bothered: Charles Leedham-Green, Mark Lewis, Martin Liebeck, Gabriel Verret, and those that I will, now that I have time.

Contents

1 Introduction ... 1
1.1 Some history ... 2
1.2 Desirable properties ... 3
Notes ... 4
Exercises ... 4

2 Tight frames ... 7
2.1 Normalised tight frames ... 7
2.2 Unitarily equivalent finite tight frames ... 10
2.3 Projective and complex conjugate equivalences ... 11
2.4 The analysis, synthesis and frame operators ... 12
2.5 The Gramian ... 13
2.6 Tight frames as orthogonal projections ... 15
2.7 The construction of tight frames from orthogonal projections ... 17
2.8 Complementary tight frames ... 18
2.9 Partition frames ... 20
2.10 Real and complex tight frames ... 21
2.11 SICs and MUBs ... 22
Notes ... 23
Exercises ... 24

3 Frames ... 31
3.1 Motivation ... 31
3.2 A frame and its dual ... 32
3.3 The canonical tight frame ... 35
3.4 Unitarily equivalent frames ... 40
3.5 Similar frames and orthogonal frames ... 41
3.6 Frames as orthogonal projections ... 43
3.7 Condition numbers and the frame bounds ... 46
3.8 Normalising frames and the distances between them ... 47

3.9 Approximate inverses of the frame operator . . . 53
3.10 Alternate duals . . . 54
3.11 Oblique duals . . . 57
Notes . . . 60
Exercises . . . 60

4 Canonical coordinates for vector spaces and affine spaces . . . 71
4.1 The canonical Gramian of a spanning sequence . . . 72
4.2 The canonical coordinates of a spanning sequence . . . 74
4.3 A characterisation of the canonical coordinates . . . 76
4.4 Properties of the canonical coordinates . . . 78
4.5 The canonical inner product for a spanning sequence . . . 80
4.6 Canonical coordinates for cyclotomic fields . . . 81
4.7 Generalised barycentric coordinates for affine spaces . . . 87
4.8 The Bernstein frame . . . 92
4.9 Properties of the Bernstein frame . . . 93
4.10 The generalised Bernstein operator . . . 94
Notes . . . 98
Exercises . . . 98

5 Combining and decomposing frames . . . 99
5.1 Unions . . . 99
5.2 Direct sums . . . 100
5.3 Lifting . . . 104
5.4 Complements . . . 105
5.5 Sums . . . 106
5.6 Tensor products . . . 108
5.7 Decompositions . . . 110
Notes . . . 110
Exercises . . . 110

6 Variational characterisations of tight frames . . . 113
6.1 Welch bound equality sequences . . . 113
6.2 The variational characterisation . . . 114
6.3 The frame potential . . . 115
6.4 Spherical t-designs and the Waring formula . . . 116
6.5 Other characterisations of spherical t-designs . . . 117
6.6 The existence of cubature formulas . . . 119
6.7 Tight frames of symmetric tensors . . . 120
6.8 Cubature on the real and complex spheres . . . 122
6.9 Spherical (t,t)-designs . . . 126
6.10 Weighted (t,t)-designs . . . 128
6.11 Complex projective t-designs . . . 129
6.12 Isometric embeddings . . . 131

6.13 Weighted $(2,2)$-designs of orthonormal bases ... 132
6.14 The frame force ... 135
6.15 Local and global minimisers of the frame potential ... 136
6.16 The numerical construction of spherical (t,t)-designs ... 139
Notes ... 142
Exercises ... 142

7 The algebraic variety of tight frames ... 151
7.1 The real algebraic variety of normalised tight frames ... 152
7.2 The algebraic variety of equal-norm tight frames ... 153
7.3 The density of rational tight frames ... 155
7.4 The existence of tight frames with given norms ... 157
7.5 The construction of tight frames with given norms ... 158
Notes ... 162
Exercises ... 162

8 Projective unitary equivalence and fusion frames ... 165
8.1 Projective unitary equivalence ... 166
8.2 The m-products ... 167
8.3 The frame graph ... 168
8.4 Characterisation of projective unitary equivalence ... 169
8.5 Reconstruction from the m-products ... 172
8.6 Triple products, equiangular lines, SICs and MUBs ... 174
8.7 Projective similarity and canonical m-products for vector spaces ... 176
8.8 Fusion frames ... 179
8.9 Signed frames ... 181
8.10 Scaling the vectors of a frame to obtain a tight frame ... 182
Notes ... 186
Exercises ... 186

9 Symmetries of tight frames ... 189
9.1 The symmetries of a sequence of vectors ... 190
9.2 The symmetry group of a sequence of vectors ... 191
9.3 The projective symmetry group of a sequence of vectors ... 195
9.4 Symmetries of combinations of frames ... 197
9.5 Maximally symmetric tight frames ... 198
9.6 Algorithms and examples ... 200
9.7 Case study: 16 equiangular lines in $\mathbb{R}^6$ ... 205
9.8 Case study: A spherical $(4,4)$-design of 12 lines in $\mathbb{C}^2$ ... 206
Notes ... 206
Exercises ... 207

10 Group frames ... 209
10.1 Representations and G-frames ... 210
10.2 The frame operator of a G-frame ... 211
10.3 Group matrices and the Gramian of a G-frame ... 213
10.4 Identifying G-frames from their Gramian ... 216
10.5 Irreducible G-frames ... 218
10.6 The vertices of the Platonic solids ... 222
10.7 Irreducible G-frames from vertex-transitive graphs ... 224
10.8 Maschke's theorem and homogeneous G-frames ... 225
10.9 The characterisation of all tight G-frames ... 227
10.10 G-frames of multivariate orthogonal polynomials ... 231
10.11 G-invariant frames ... 234
10.12 Frames invariant under the action of an abelian group ... 236
10.13 The minimal number of generators for a G-invariant frame ... 238
10.14 The coinvariants of a finite reflection group ... 240
Notes ... 241
Exercises ... 241

11 Harmonic frames ... 245
11.1 Introduction ... 245
11.2 Character tables ... 246
11.3 Harmonic frames ... 247
11.4 Harmonic frames with distinct vectors and with real vectors ... 249
11.5 Combining and decomposing harmonic frames ... 250
11.6 Real harmonic frames ... 251
11.7 Unitary equivalence preserving the group structure ... 253
11.8 Noncyclic harmonic frames ... 256
11.9 Unitary equivalence not preserving the group structure ... 257
11.10 The number of cyclic harmonic frames ... 258
11.11 Projective unitary equivalence of harmonic frames ... 260
11.12 The projective symmetry group of a harmonic frame ... 262
Notes ... 263
Exercises ... 263

12 Equiangular and Grassmannian frames ... 265
12.1 Equiangular lines and frames ... 266
12.2 Grassmannian frames ... 268
12.3 Equiangular harmonic frames and difference sets ... 269
12.4 Equiangular tight frames from block designs ... 275
12.5 Tremain equiangular tight frames ... 278
12.6 Equiangular frames and their signature matrices ... 279
12.7 The reduced signature matrix and projective unitary equivalence ... 284
12.8 The relative bound on the number of real equiangular lines ... 287
12.9 The connection with algebraic graph theory ... 289

12.10 Real equiangular tight frames and strongly regular graphs 291
12.11 Conditions for the existence of real equiangular tight frames 295
12.12 A list of real equiangular tight frames 297
12.13 Nontight real equiangular frames from graphs 299
12.14 Spherical two-distance tight frames 304
12.15 Two-distance tight frames and partial difference sets............ 307
12.16 The standard m-distance tight frame 309
12.17 Complex equiangular tight frames 310
12.18 Algebraic equations for tight complex equiangular lines......... 316
12.19 Mutually unbiased bases and s-angular tight frames 318
12.20 Mutually unbiased bases and Hadamard matrices 320
12.21 Examples of MUBs 324
Notes 326
Exercises 326

13 Tight frames generated by nonabelian groups 331
13.1 The identification of the G-matrices with the group algebra 332
13.2 Tight G-frames as idempotents of the group algebra 333
13.3 Characters of nonabelian groups 334
13.4 Central G-frames 336
13.5 The classification of central tight G-frames 337
13.6 The idempotents and the homogeneous decomposition.......... 339
13.7 An illustrative example 340
13.8 The highly symmetric tight frames 341
13.9 The construction of highly symmetric tight frames 342
13.10 Complex polytopes and finite reflection groups 343
13.11 Projective representations 351
13.12 The canonical abstract (error) group 352
13.13 Nice error frames 353
13.14 Tensor products of nice error frames 355
13.15 Computing all nice error frames........................... 356
13.16 Examples of nice error frames 357
Notes 359

14 Weyl–Heisenberg SICs.. 361
14.1 Maximal sets of complex equiangular lines and vectors 362
14.2 SICs... 363
14.3 Quantum measurements 364
14.4 Group covariance 365
14.5 The discrete Heisenberg group and Weyl–Heisenberg SICs 366
14.6 The Hoggar lines 367
14.7 The Clifford group (normaliser of the Heisenberg group)........ 368
14.8 Generators for the Clifford group.......................... 372
14.9 Indexing the Clifford operations........................... 373
14.10 Appleby indexing...................................... 374

14.11 Symplectic unitaries ... 378
14.12 Permutation matrices ... 380
14.13 Calculating a symplectic matrix from its symplectic index ... 382
14.14 The Zauner matrix ... 384
14.15 The Scott–Grassl numerical SICs ... 385
14.16 Symmetries of the Weyl–Heisenberg SICs ... 389
14.17 Monomial representations of the Clifford group ... 391
14.18 The Clifford trace and symplectic unitaries of order 3 ... 392
14.19 Conjugates of the canonical order 3 symplectic unitaries ... 398
14.20 The SIC field of a Weyl–Heisenberg SIC ... 400
14.21 The Galois group of a generic SIC ... 401
14.22 The Galois action on a fiducial projector ... 402
14.23 Constructing exact SICs from numerical SICs ... 404
14.24 The Galois action on a centred fiducial ... 404
14.25 Minimal and maximal SIC fields ... 410
14.26 Ray class fields ... 412
14.27 Equivalent equations for SIC fiducial vectors ... 414
Notes ... 418
Exercises ... 419

15 Tight frames of orthogonal polynomials on the simplex ... 429
15.1 Jacobi polynomials and their Bernstein coefficients ... 430
15.2 The Bernstein–Durrmeyer operator ... 434
15.3 Tight frames of Jacobi polynomials with symmetries ... 435
15.4 The orthogonal polynomials of Appell and Proriol ... 437
Notes ... 438
Exercises ... 439

16 Continuous tight frames for finite dimensional spaces ... 441
16.1 Continuous and discrete frames ... 441
16.2 The analysis and synthesis operators ... 443
16.3 Reproducing kernels ... 444
16.4 Zonal harmonics ... 445
16.5 Homogeneous polynomials ... 448
16.6 Orthogonal polynomials for a radially symmetric weight ... 451
16.7 Functions on the complex sphere ... 461
16.8 *G*-frames for infinite groups ... 463
Notes ... 465
Exercises ... 466

Solutions ... 471

References ... 557

Index ... 569

ANHA Back Matter ... 583

Chapter 1
Introduction

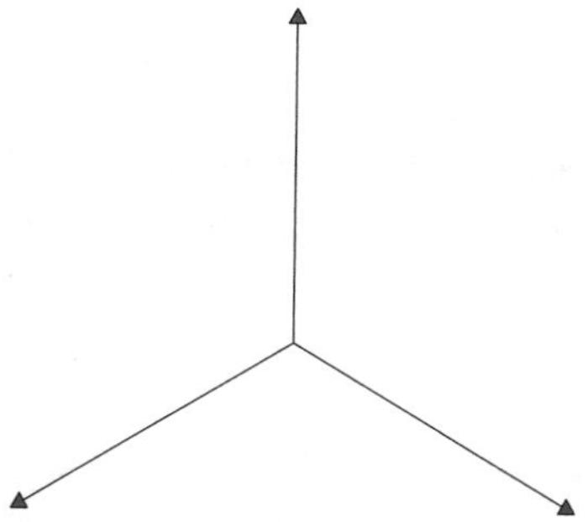

The prototypical example of a *finite tight frame* is three equally spaced unit vectors u_1, u_2, u_3 in $\mathbb{R}^2$, which provide the following *redundant* decomposition

$$f = \frac{2}{3}\sum_{j=1}^{3}\langle f, u_j\rangle u_j, \qquad \forall f \in \mathbb{R}^2. \tag{1.1}$$

Such generalisations of an orthogonal expansion have been used extensively for infinite dimensional function spaces. Most notably, in the area of wavelets, where they allow expansions in terms of functions which have nicer properties, such as good time-frequency localisation and a simple description, than is possible for an orthogonal expansion. Despite the fact that in these methods, ultimately a numerical approximation is computed in a finite dimensional subspace, until recently little attention has been paid to frames for finite dimensional spaces.

Over the last decade, it has become increasingly apparent that tight frames for finite dimensional spaces are useful for similar reasons. They can have *desirable properties*, such as good time-frequency localisation, and share symmetries of the space, which may be impossible for an orthonormal basis. In addition, there are computational advantages of stability and robustness to erasures. There is also a special geometry (different for real and complex spaces) which has no analogue in the infinite dimensional setting.

S.F.D. Waldron, *An Introduction to Finite Tight Frames*, Applied and Numerical Harmonic Analysis, https://doi.org/10.1007/978-0-8176-4815-2_1

1.1 Some history

Like many great ideas in mathematics, frames (which have been crucial to the development of wavelets) have, with hindsight, been around in some form or other for quite a while. For example, in 1937, Schönhardt [Sch37] proved the generalisation of (1.1) to n equally spaced unit vectors $u_1,\ldots,u_n \in \mathbb{R}^2$, i.e.,

$$f = \frac{2}{n}\sum_{j=1}^{n} \langle f, u_j\rangle u_j, \qquad \forall f \in \mathbb{R}^2. \tag{1.2}$$

This idea received some attention, with Brauer and Coxeter [BC40] extending the result to the orbit of any irreducible group of orthogonal matrices (also see [Had40]). They also mention the possibility of extending the result to the orbit of a continuous group, e.g., taking the group of rotations gives the continuous version of (1.2)

$$f = \frac{2}{2\pi}\int_0^{2\pi} \langle f, u_\theta\rangle u_\theta \, d\theta, \qquad \forall f \in \mathbb{R}^2, \qquad u_\theta := \begin{pmatrix}\cos\theta\\ \sin\theta\end{pmatrix}, \tag{1.3}$$

which is an example of a continuous tight frame. I don't doubt that there are even earlier instances.

In 1952, Duffin and Schaeffer set out the modern theory of frames in their seminal paper [DS52], which included the definition in terms of *frame bounds*

$$A\|f\|^2 \le \sum_j |\langle f, f_j\rangle|^2 \le B\|f\|^2, \qquad \forall f.$$

They were interested in Fourier type series for functions in $L_2[-\pi,\pi]$ involving functions $f_j : t \mapsto e^{i\lambda_j t}$, for frequencies $\lambda_j \in \mathbb{R}$, which might not be integers (see Young [You01] for an excellent account).

From the late 1980s came the *wavelet* era (see, e.g., [Dau92], [Kai94]). Here, frames were used to obtain Fourier expansions for $L_2(\mathbb{R}^d)$ in terms of functions with both a simple description and good time-frequency localisation. At the risk of over simplification, this was done by taking a single (wavelet) function and obtaining the others from it by applying the operations of

$$\textit{translation}: \quad (\mathscr{T}_a f)(x) := f(x-a), \quad a \in \mathbb{R}^d, \tag{1.4}$$

$$\textit{modulation}: \quad (\mathscr{M}_b f)(x) := e^{2\pi i b\cdot x} f(x), \quad b \in \mathbb{R}^d, \tag{1.5}$$

$$\textit{dilation}: \quad (\mathscr{D}_c f)(x) := c^{\frac{d}{2}} f(cx), \quad c > 0. \tag{1.6}$$

The parameters a,b,c may be chosen to be either discrete or continuous. The theory has two strands: when the operations form a group (Gabor systems), and when they don't (wavelet systems). In the former case, the group allows a description of the dual frame as the orbit of a single function, and in the latter the method of *multiresolution analysis* yields a suitable function.

1.2 Desirable properties

Surely, three equally spaced unit vectors (aka the **Mercedes–Benz frame**) give the "nicest" possible tight frame of three vectors for $\mathbb{R}^2$. In this example, one can see many desirable properties that one might hope for more generally in a tight frame. To illustrate these properties, many of which do extend to quite general situations, we also consider a *second prototypical example*: the four unit vectors $\Phi := (v, Sv, \Omega v, S\Omega v)$ in $\mathbb{C}^2$, given by

$$v := \frac{1}{\sqrt{6}}\begin{pmatrix}\sqrt{3+\sqrt{3}}\\ e^{\frac{\pi}{4}i}\sqrt{3-\sqrt{3}}\end{pmatrix}, \qquad S := \begin{pmatrix}0 & 1\\ 1 & 0\end{pmatrix}, \quad \Omega := \begin{pmatrix}1 & 0\\ 0 & -1\end{pmatrix}, \tag{1.7}$$

which form a tight frame for $\mathbb{C}^2$, i.e.,

$$f = \frac{2}{4}\sum_{\phi\in\Phi}\langle f,\phi\rangle\phi, \qquad \forall f \in \mathbb{C}^2. \tag{1.8}$$

The operations S and Ω are discrete analogues of translation and modulation, and the tight frame Φ, which will be termed a *SIC*, is a discrete analogue of a Gabor system (Weyl–Heisenberg frame).

In addition to being *tight*, we might hope that a finite frame has some of the following properties.

- *Equal norms.* As in both examples, the vectors in the tight frame have equal norms. Equivalently, the decomposition of the form (1.1) and (1.8) is a sum of one-dimensional projections with equal weightings.
- *Symmetries.* The frame be invariant under some group G of symmetries, such as the three equally spaced vectors (which are invariant under the group of order 3 generated by rotation through $2\pi/3$). Equivalently, the frame is the G-orbit of one or a small number of vectors.
- *Equiangularity.* The (equal norm) vectors in both frames have the *cross-correlation* $|\langle\phi,\psi\rangle|$ constant for all vectors $\phi \neq \psi$. For vectors in $\mathbb{R}^d$, this is equivalent to the vectors having equal angles between each other.
- *Robustness to erasures.* In both examples, each pair of frame vectors spans a 2-dimensional space. Hence if all but two of the coefficients $\langle f,\phi\rangle$ (ϕ a frame vector) are lost, then f can be reconstructed from these values.
- *Stability.* Suppose the three coefficients in (1.1) are perturbed, say to $\langle f,u_j\rangle + a_j$. Then the error in the computed f is $a_1u_1 + a_2u_2 + a_3u_3$, which is bounded by the norm of $a = (a_j)$. This error might even be zero, for a nonzero perturbation (take $a_1 = a_2 = a_3$), a phenomenon which cannot occur for an orthogonal expansion.

A further property, not shared by these examples, is that of *sparseness*, i.e., the frame vectors having many zeros. This is the discrete analogue of having small or compact support, and is of importance in algorithms for compressed sensing and data compression.

Notes

In some sense, the theory of frames is dual to that of compressed sensing (sparse sampling). In compressed sensing, the (sparseness) structure of some vectors allows them to be represented by using fewer vectors than in a basis, while in finite frame theory more vectors than are needed for a basis are used in a redundant expansion which has more desirable properties than would be possible by using just a basis.

Exercises

1.1. Prove (1.1) holds, by using the fact $\langle f,u_j\rangle = u_j^* f$, to write it in the matrix form

$$\frac{2}{3}VV^* = I, \qquad V := [u_1,u_2,u_3], \quad I := \begin{pmatrix} 1 & 0 \\ 0 & 1 \end{pmatrix}.$$

Hint. Any two sets of three equally spaced unit vectors are rotations of each other.

1.2. *Robustness to erasures.*
(a) Suppose one of the coefficients in (1.1) is lost, say $\langle f,u_3\rangle$. Show that $f \in \mathbb{R}^2$ can be reconstructed from the remaining two, by giving an explicit formula for this, i.e., find the basis for $\mathbb{R}^2$ dual to the functionals $f \mapsto \langle f,u_j\rangle$, $j = 1,2$.
(b) Now suppose a coefficient is changed, say $\langle f,u_3\rangle$, can this be detected?
(c) What if two of the coefficients are changed?

1.3. *Equiangularity/equispacing.*
(a) Verify that the three unit vectors in $\mathbb{C}^2$ given by

$$v_1 := \frac{1}{\sqrt{2}}\begin{bmatrix} 1 \\ 1 \end{bmatrix}, \quad v_2 := \frac{1}{\sqrt{2}}\begin{bmatrix} 1 \\ \omega \end{bmatrix}, \quad v_3 := \frac{1}{\sqrt{2}}\begin{bmatrix} 1 \\ \omega^2 \end{bmatrix}, \qquad \omega := e^{\frac{2\pi i}{3}}$$

form a tight frame for $\mathbb{C}^2$, i.e., $f = \frac{2}{3}\sum_{j=1}^3 \langle f,v_j\rangle v_j$, for all $f \in \mathbb{C}^2$.
(b) Show these three vectors and those of Exer. 1.1 are equiangular, i.e.,

$$|\langle v_j,v_k\rangle| = |\langle u_j,u_k\rangle| = \frac{1}{2}, \qquad j \neq k.$$

(c) Show that both these sets of three vectors are equally spaced, with

$$\|v_j - v_k\| = \frac{\sqrt{3}}{\sqrt{2}}, \quad \|u_j - u_k\| = \sqrt{3}, \qquad j \neq k.$$

(d) How can the distances between the real vectors (u_j) be larger than those for the complex vectors (v_j)? Isn't there more space in $\mathbb{C}^2$ than in $\mathbb{R}^2$!

1.4. *Gabor and Wavelet systems.*
(a) Show that translation and modulation satisfy the *commutation relation*

$$\mathcal{T}_a \mathcal{M}_b = e^{-2\pi i a \cdot b} \mathcal{M}_b \mathcal{T}_a.$$

(b) Use this to conclude if A and B are nonzero subgroups of $(\mathbb{R}, +)$, then

$$G := \{c \mathcal{T}_a \mathcal{M}_b : a \in A, b \in B, c \in C\}, \qquad C := \{e^{2\pi i a \cdot b} : a \in A, b \in B\}.$$

is a group, which is abelian only if $\mathbb{R}^d = \mathbb{R}$. The G-orbit $\{\mathcal{T}_a \mathcal{M}_b \phi\}_{a \in A, b \in B}$ (up to scalar multipliers) of a suitable $\phi \in L_2(\mathbb{R}^d)$ is called a *Gabor system/frame*.
(c) For a suitable function $\psi \in L_2(\mathbb{R})$, a *wavelet system/frame* is given by the functions

$$\psi_{jk} := 2^{\frac{j}{2}} \psi(2^j \cdot - k) = \mathcal{D}_{2^j} \mathcal{T}_k \psi, \qquad j, k \in \mathbb{Z}.$$

Show $\{\mathcal{D}_{2^j} \mathcal{T}_k\}_{j,k \in \mathbb{Z}}$ is *not* a group, and so wavelet systems are not group orbits.

Chapter 2
Tight frames

Decompositions like those in our two prototypical examples, i.e.,

$$f = \sum_{j \in J} \langle f, f_j \rangle f_j, \qquad \forall f \in \mathscr{H}, \tag{2.1}$$

will come from what is called a tight frame $(f_j)_{j \in J}$. The basic ingredients are

- $\mathscr{H}$ – a real or complex Hilbert space (for us usually finite dimensional)
- J – an index set (often with a group structure)
- $(f_j)_{j \in J}$ – a sequence (set, or multiset) of vectors in $\mathscr{H}$
- $\sum_{j \in J}$ – a sum (for us usually finite, but sometimes continuous)

The emphasis here is on the possible *redundancy* (over completeness) of the vectors (f_j) in the expansion, i.e., the case when (2.1) is not an orthogonal expansion.

In the first instance, you are encouraged to consider $\mathscr{H}$ as $\mathbb{R}^d$ or $\mathbb{C}^d$, with the usual (Euclidean) inner product, and to think in familiar matrix terms.

2.1 Normalised tight frames

The *polarisation identity* (see Exer. 2.1) implies that (2.1) is equivalent to

$$\|f\|^2 = \sum_{j \in J} |\langle f, f_j \rangle|^2, \qquad \forall f \in \mathscr{H},$$

which explains the following definition.

Definition 2.1. A countable sequence $(f_j)_{j \in J}$ in a Hilbert space $\mathscr{H}$ is said to be a **tight frame** (for $\mathscr{H}$) if there exists a (**frame bound**) $A > 0$, such that

$$A\|f\|^2 = \sum_{j \in J} |\langle f, f_j \rangle|^2, \qquad \forall f \in \mathscr{H}. \tag{2.2}$$

Further, $(f_j)_{j \in J}$ is **normalised** if $A = 1$, and **finite** if J is finite.

S.F.D. Waldron, *An Introduction to Finite Tight Frames*, Applied and Numerical Harmonic Analysis, https://doi.org/10.1007/978-0-8176-4815-2_2

The *Bessel identity* (2.2) is equivalent (see Exer. 2.2) to either of the identities

$$\textit{Parseval}: \qquad f = \frac{1}{A}\sum_{j\in J}\langle f, f_j\rangle f_j, \quad \forall f \in \mathscr{H}, \tag{2.3}$$

$$\textit{Plancherel}: \qquad \langle f, g\rangle = \frac{1}{A}\sum_{j\in J}\langle f, f_j\rangle\langle f_j, g\rangle, \quad \forall f, g \in \mathscr{H}. \tag{2.4}$$

For $A=1$ (2.2) says that $f \mapsto (\langle f, f_j\rangle)_{j\in J}$ is an *isometry*, and so normalised tight frames for $\mathscr{H}$ are equivalent to isometries $\mathscr{H} \to \ell_2(J)$. The maps taking normalised tight frames to normalised tight frames are the *partial isometries* (Exer. 2.7).

We prefer the term *normalised tight frame*[1] to *Parseval frame* (which is also used), as it emphasizes the fact the frame bound $A > 0$ is simply a normalising factor, i.e., if (f_j) is a tight frame, then $(f_j/\sqrt{A})$ is the unique positive scalar multiple of it which is a normalised tight frame. We will soon see that this *normalised* version of a tight frame is convenient in many situations.

We have defined a tight frame to be a sequence, which is standard, but not universal. By contrast with a basis (which can be a set or a sequence), a tight frame *can have repeated vectors*. At times, e.g., when the vectors in a frame are all distinct or the indexing is unimportant, it can be convenient to think of them as a (multi)set. We will not labour this point, making statements such as the set $\{f_j\}_{j\in J}$ is a tight frame, without further explanation.

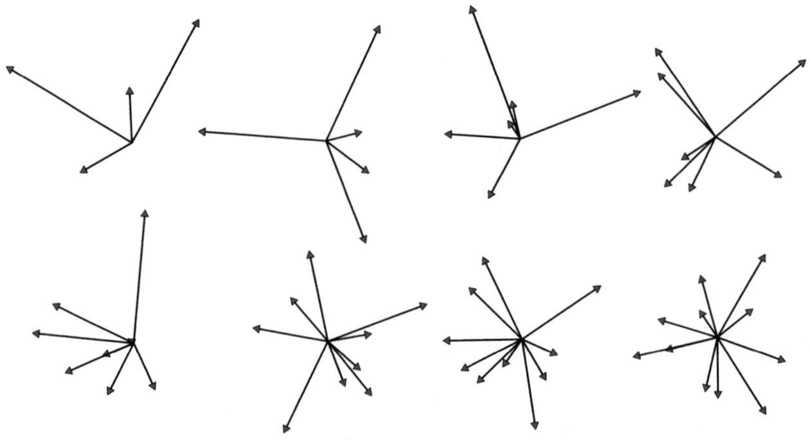

Fig. 2.1: Examples of normalised tight frames of $n = 4, 5, \ldots, 11$ vectors for $\mathbb{R}^2$.

[1] This term dates back to [HL00]. Just to confuse matters, the term *normalised tight frame* has also been used for a tight frame with $\|f_j\| = 1$, $\forall j \in J$ (we call these *unit-norm* tight frames).

Example 2.1. (Exer. 2.4) An orthonormal basis is a normalised tight frame. These are the only normalised tight frames in which all the vectors have unit length (all vectors in a normalised tight frame have length ≤ 1).

Example 2.2. (Exer. 2.5) The unitary image of a normalised tight frame is again a normalised tight frame, and the only invertible linear maps which map a normalised tight frame to a normalised tight frame are the unitary maps.

Example 2.3. (Exer. 2.6) The orthogonal projection of a normalised tight frame is again a normalised tight frame (for its span). In particular, if U is an $n \times n$ unitary matrix, then the columns of any $d \times n$ submatrix is a normalised tight frame of n vectors for $\mathbb{C}^d$. This is effectively the projection of the orthonormal basis for $\mathbb{C}^n$ given by the columns of U onto the d-dimensional subspace of vectors which are zero in some fixed $n-d$ coordinates.

We say that $(f_j)_{j\in J}$ is an **equal-norm** tight frame if $\|f_j\| = \|f_k\|$, $\forall j,k \in J$, and is a **unit-norm** tight frame if $\|f_j\| = 1$, $\forall j \in J$.

Example 2.4. (Exer. 2.8) Equal-norm tight frames of n vectors for $\mathbb{F}^d$ can be obtained from an $n \times n$ unitary matrix U with entries of constant modulus, by taking the columns of any $d \times n$ submatrix. Examples of such U include the **Hadamard matrices** (real entries) and the **Fourier (transform) matrix**

$$U = F_n = \frac{1}{\sqrt{n}} \begin{bmatrix} 1 & 1 & 1 & \cdots & 1 \\ 1 & \omega & \omega^2 & \cdots & \omega^{n-1} \\ 1 & \omega^2 & \omega^4 & \cdots & \omega^{2(n-1)} \\ \vdots & \vdots & \vdots & & \vdots \\ 1 & \omega^{n-1} & \omega^{2(n-1)} & \cdots & \omega^{(n-1)(n-1)} \end{bmatrix}, \qquad \omega := e^{\frac{2\pi i}{n}}. \tag{2.5}$$

Equal-norm tight frames which come from the Fourier matrix in this way will be known as *harmonic* tight frames (see Chapter 11).

In §2.6, we will show that every normalised tight frame can be obtained as the orthogonal projection of an orthonormal basis (in a larger space).

Example 2.5. (Exer. 2.9) *The tight frames for* $\mathbb{R}^2$. A sequence of vectors $(v_j)_{j=1}^n$, $v_j = (x_j, y_j) \in \mathbb{R}^2$ is a tight frame for $\mathbb{R}^2$ if and only if the **diagram vectors** which are defined by $w_j := (x_j + iy_j)^2 \in \mathbb{C}$, $1 \leq j \leq n$, sum to zero (in $\mathbb{C}$).

In applications, the interest in (2.3) is usually that it gives a decomposition of the identity into a weighted sum of projections (see Chapter 8), i.e.,

$$I = \sum_{j\in J} c_j P_j, \qquad c_j := \frac{\|f_j\|^2}{A}, \quad P_j f := \frac{\langle f, f_j \rangle}{\langle f_j, f_j \rangle} f_j, \tag{2.6}$$

where the particular (unit modulus) scalar multiple of f_j that is used to define the orthogonal projection P_j is unimportant. The pair (P_j),(c_j) above is a **fusion frame** (see §8.8). When taking this point of view, we will use the epithet *projective*.

2.2 Unitarily equivalent finite tight frames

Before giving any further concrete examples of finite tight frames, we define an equivalence, under which any set of three equally spaced vectors with the same norm in $\mathbb{R}^2$ would be considered *equivalent*.

Definition 2.2. We say that two normalised tight frames $(f_j)_{j\in J}$ for $\mathscr{H}$ and $(g_j)_{j\in J}$ for $\mathscr{K}$, with the same index set J, are **(unitarily) equivalent** if there is a unitary transformation $U : \mathscr{H} \to \mathscr{K}$, such that $g_j = Uf_j, \forall j \in J$.

Since unitary transformations preserve inner products, unitarily equivalent tight frames have the same inner products (angles) between their vectors. Furthermore, these inner products uniquely determine the equivalence classes (see §2.5).

This equivalence is *dependent on the indexing*, which is appropriate when set J has some natural (e.g., group) structure. The normalised tight frames of two vectors $(e_1,0)$ and $(0,e_1)$ for the one-dimensional space $\mathscr{H} = \text{span}\{e_1\}$ are not equivalent, since there is no unitary map $e_1 \mapsto 0$ (or $0 \mapsto e_1$). For such cases, where it is useful to consider these as equivalent, we extend our definition of equivalence as follows.

Definition 2.3. We say that two finite normalised tight frames $(f_j)_{j\in J}$ for $\mathscr{H}$ and $(g_j)_{j\in K}$ for $\mathscr{K}$ are **(unitarily) equivalent up to reordering** if there is a bijection $\sigma : J \to K$ for which $(f_j)_{j\in J}$ and $(g_{\sigma j})_{j\in J}$ are unitarily equivalent.

We will say that tight frames are unitarily equivalent (up to reordering) if after normalisation they are, in which case we say they are equal up to unitary equivalence (and reordering).

Example 2.6. Let u_1, u_2, u_3 be equally spaced unit vectors in $\mathbb{R}^2$, and R_θ be rotation through an angle θ. Then each of the sets of six vectors

$$\{u_1,u_2,u_3,R_\theta u_1,R_\theta u_2,R_\theta u_3\}, \qquad 0<\theta\le\frac{\pi}{3}$$

forms a tight frame. Since unitary maps preserve angles, *none* of these are unitarily equivalent (up to reordering).

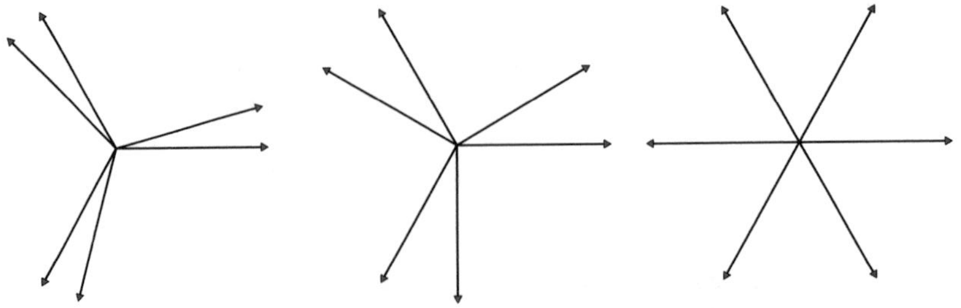

Fig. 2.2: The unitarily inequivalent tight frames obtained for $\theta = \frac{\pi}{12}, \frac{\pi}{6}, \frac{\pi}{3}$.

2.3 Projective and complex conjugate equivalences

There are other *equivalences* which appear in the frame literature. Most notably, the normalised tight frames of Definition 2.2 are **projectively (unitarily) equivalent** if

$$g_j = \alpha_j U f_j, \qquad \forall j \in J,$$

where U is unitary, and $|\alpha_j| = 1$, $\forall j$.

All tight frames $(\alpha_j f_j)$, $|\alpha_j| = 1$, $\forall j$, obtained from a given tight frame (f_j), are projectively unitarily equivalent, but are *not* unitarily equivalent, in general.

Example 2.7. For tight frames of n nonzero vectors in $\mathbb{R}^2$ the equivalence classes for *projective equivalence up to reordering* are in 1–1 correspondence with convex polygons with n sides (see Exer. 2.10).

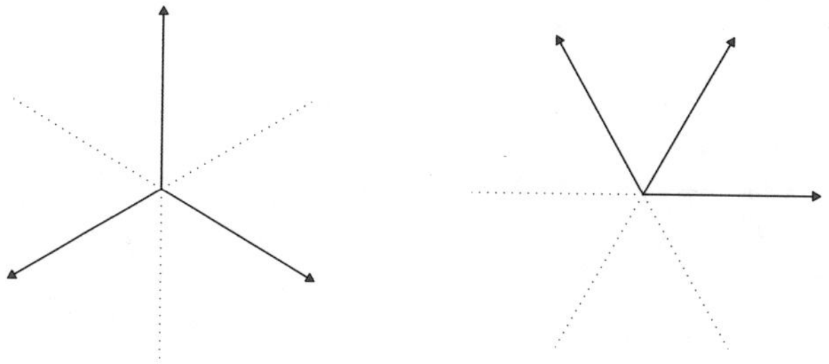

Fig. 2.3: Equal–norm tight frames of three vectors for $\mathbb{R}^2$ which are projectively equivalent, but are *not* unitarily equivalent.

The **complex conjugation map** on $\mathbb{C}^d$ is the *antilinear* map

$$\mathbb{C}^d \to \mathbb{C}^d : v = (v_j) \mapsto \overline{v} := (\overline{v_j}).$$

Since $\langle \overline{v}, \overline{w} \rangle = \overline{\langle v, w \rangle}$, this maps a tight frame (f_j) for $\mathbb{C}^d$ to a tight frame $(\overline{f_j})$, and these are said to be **complex conjugate** (or **anti**) **equivalent**. The conjugation map $C : \mathscr{H} \to \overline{\mathscr{H}}$ extends these ideas to $\mathscr{H}$ (see Exer. 2.11).

These basic types of equivalences can be combined, in the obvious way, to obtain others, e.g., the tight frames $(f_j)_{j\in J}$ and $(g_k)_{k\in K}$ for $\mathscr{H}$ and $\mathscr{K}$ would be *anti projectively unitarily equivalent up to reordering* if

$$\overline{g_{\sigma j}} = \alpha_j U f_j, \qquad \forall j \in J,$$

for $\sigma : J \to K$ a bijection, $|\alpha_j| = 1$, $\forall j$, and $U : \mathscr{H} \to \mathscr{K}$ unitary.

2.4 The analysis, synthesis and frame operators

The Parseval identity (2.3) consists of an extraction of "coordinates"

$$c_j = \langle f, f_j \rangle, \qquad j \in J$$

for the vector f (analysis), and a reconstruction of f from these (synthesis). Many important properties of a tight frame follow from this factorisation.

For simplicity of presentation, we suppose J is finite, write $\mathbb{F}$ for $\mathbb{R}$ or $\mathbb{C}$, $\ell_2(J)$ for $\mathbb{F}^J$, with the usual inner product, and $I = I_{\mathscr{H}}$ for the identity on $\mathscr{H}$.

Definition 2.4. For a finite sequence $(f_j)_{j\in J}$ in $\mathscr{H}$ the **synthesis operator (reconstruction operator** or **pre-frame operator**) is the linear map

$$V := [f_j]_{j\in J} : \ell_2(J) \to \mathscr{H} : a \mapsto \sum_{j\in J} a_j f_j,$$

and its dual is the **analysis operator** (or **frame transform operator**)

$$V^* : \mathscr{H} \to \ell_2(J) : f \mapsto (\langle f, f_j \rangle)_{j\in J}.$$

It is convenient to make little distinction between the sequence $(f_j)_{j\in J}$ and the linear map $V = [f_j]_{j\in J}$, which we will say has ***j*-th column** f_j.

The product $S := VV^* : \mathscr{H} \to \mathscr{H}$ is known as the **frame operator**. A simple calculation (see Exer. 2.12) shows the trace of S and S^2 are given by

$$\operatorname{trace}(S) = \|S^{\frac{1}{2}}\|_F^2 = \sum_{j\in J} \|f_j\|^2, \qquad \operatorname{trace}(S^2) = \|S\|_F^2 = \sum_{j\in J}\sum_{k\in J} |\langle f_j, f_k \rangle|^2. \quad (2.7)$$

Proposition 2.1. *A finite sequence $(f_j)_{f\in J}$ in $\mathscr{H}$ is a tight frame for $\mathscr{H}$ (with frame bound A) if and only if*

$$S = VV^* = AI_{\mathscr{H}}, \qquad V := [f_j]_{f\in J}. \quad (2.8)$$

In particular, a tight frame satisfies

$$\sum_{j\in J} \|f_j\|^2 = dA, \qquad d := \dim(\mathscr{H}), \quad (2.9)$$

and

$$\sum_{j\in J}\sum_{k\in J} |\langle f_j, f_k \rangle|^2 = \frac{1}{d}\Big(\sum_{j\in J} \langle f_j, f_j \rangle\Big)^2. \quad (2.10)$$

Proof. Since

$$Sf = VV^* f = \sum_j \langle f, f_j \rangle f_j, \qquad \forall f \in \mathscr{H},$$

the Parseval identity (2.3) implies the condition (2.2) is equivalent to (2.8). Taking the trace of (2.8) and its square gives

$$\sum_j \|f_j\|^2 = \operatorname{trace}(S) = \operatorname{trace}(AI_{\mathscr{H}}) = dA,$$

$$\sum_j \sum_k |\langle f_j, f_k \rangle|^2 = \operatorname{trace}(S^2) = \operatorname{trace}(A^2 I_{\mathscr{H}}) = \frac{1}{d}(Ad)^2 = \frac{1}{d}\Big(\sum_j \langle f_j, f_j \rangle\Big)^2,$$

which are (2.9) and (2.10). □

The equations (2.3) and (2.8) will be referred to as the **Parseval identity**, (2.9) as the **trace formula**, and (2.10) as the **variational formula**.

For $\mathscr{H} = \mathbb{F}^d$ and $|J| = n$, V is a $d \times n$ matrix, and the condition (2.8) says that the columns of V are orthogonal and of length $\sqrt{A}$, i.e., $V/\sqrt{A}$ is a *coisometry*, equivalently, $V^*/\sqrt{A}$ is an *isometry*.

In §6.2, we show that the variational formula characterises tight frames for finite dimensional spaces. There is *no* infinite dimensional counterpart for this result.

2.5 The Gramian

Unitary equivalence has the advantage (over projective unitary equivalence) that it preserves the inner product between vectors, and hence the Gramian matrix. Indeed, we will show that the Gramian characterises the equivalence class.

Definition 2.5. For a finite sequence of n vectors $(f_j)_{j\in J}$ in $\mathscr{H}$, the **Gramian**[2] or **Gram matrix** is the $n \times n$ Hermitian matrix

$$\operatorname{Gram}((f_j)_{j\in J}) := [\langle f_k, f_j \rangle]_{j,k\in J}.$$

This is the matrix representing the linear map $V^*V : \ell_2(J) \to \ell_2(J)$ with respect to the standard orthonormal basis $\{e_j\}_{j\in J}$.

The possible Gramian matrices are precisely the orthogonal projections:

Theorem 2.1. *An $n \times n$ matrix $P = [p_{jk}]_{j,k\in J}$ is the Gramian matrix of a normalised tight frame $(f_j)_{j\in J}$ for the space $\mathscr{H} := \operatorname{span}\{f_j\}_{j\in J}$ if and only if it is an orthogonal projection matrix, i.e., $P = P^* = P^2$. Moreover,*

$$d = \dim(\mathscr{H}) = \operatorname{rank}(P) = \operatorname{trace}(P) = \sum_{j\in J} \|f_j\|^2. \tag{2.11}$$

Proof. ($\Longrightarrow$) Let $\Phi = (f_j)_{j\in J}$ be a normalised tight frame, and $P = \operatorname{Gram}(\Phi)$. Take $f = f_\ell$ in (2.3) to get $f_\ell = \sum_{j\in J} \langle f_\ell, f_j \rangle f_j$, and take the inner product of this with f_k to obtain

[2] Note the (j,k)-entry of the Gramian is $\langle f_k, f_j \rangle = f_j^* f_k$ (so it factors V^*V), *not* $\langle f_j, f_k \rangle$, which is sometimes used to define the Gramian.

$$\langle f_k, f_\ell \rangle = \sum_{j\in J} \langle f_j, f_\ell \rangle \langle f_k, f_j \rangle \iff p_{\ell k} = \sum_{j\in J} p_{\ell j} p_{jk} \iff P = P^2.$$

But P is Hermitian, since $\overline{p_{jk}} = \overline{\langle \phi_k, \phi_j \rangle} = \langle \phi_j, \phi_k \rangle = p_{kj}$, and so is an orthogonal projection.

($\Longleftarrow$) Suppose that P is an $n \times n$ matrix, such that $P = P^* = P^2$. The columns of P are $f_j := Pe_j$, $j \in J$, where $\{e_j\}_{j\in J}$ is the standard orthonormal basis of $\ell_2(J)$. Fix $f \in \mathscr{H} := \operatorname{span}\{f_j\}_{j=1}^n \subset \ell_2(J)$. Then $f = Pf$, so that

$$f = P\Big(\sum_{j\in J} \langle Pf, e_j \rangle e_j\Big) = \sum_{j\in J} \langle f, Pe_j \rangle Pe_j = \sum_{j\in J} \langle f, f_j \rangle f_j,$$

i.e., $(f_j)_{j=1}^n$ is a normalised tight frame for $\mathscr{H}$, with Gramian P.

Finally, taking the trace of P gives (2.11). □

The condition that $P = \operatorname{Gram}(\Phi)$ be an orthogonal projection is equivalent to it having exactly d nonzero eigenvalues all equal to 1 (see Exer. 2.17).

Corollary 2.1. *(Characterisation of unitary equivalence) Normalised tight frames are unitarily equivalent if and only if their Gramians are equal.*

Proof. Let $\Phi = (f_j)_{j\in J}$, $\Psi = (g_j)_{j\in J}$ be normalised tight frames for $\mathscr{H}$ and $\mathscr{K}$.

($\Longrightarrow$) If Φ and Ψ are unitarily equivalent, i.e., $g_j = Uf_j$, $\forall j$, for some unitary $U : \mathscr{H} \to \mathscr{K}$, then their Gramians are equal since

$$\langle g_j, g_k \rangle = \langle Uf_j, Uf_k \rangle = \langle f_j, f_k \rangle.$$

($\Longleftarrow$) Suppose the Gramians of Φ and Ψ are equal, i.e., $\langle g_j, g_k \rangle = \langle f_j, f_k \rangle$, $\forall j, k$. Then, by Exer. 2.19, there is a unitary $U : \mathscr{H} \to \mathscr{K}$ with $g_j = Uf_j$, $\forall j$. Hence Φ and Ψ are unitarily equivalent. □

In other words:

The properties of a tight frame (up to unitary equivalence) are determined by its Gramian.

Example 2.8. Equal-norm tight frames of three vectors for $\mathbb{C}^2$ are given by

$$\Phi := \left(\begin{bmatrix} 1 \\ 1 \end{bmatrix}, \begin{bmatrix} \omega \\ \omega^2 \end{bmatrix}, \begin{bmatrix} \omega^2 \\ \omega \end{bmatrix} \right), \qquad \Psi := \left(\begin{bmatrix} 1 \\ 1 \end{bmatrix}, \begin{bmatrix} 1 \\ \omega \end{bmatrix}, \begin{bmatrix} 1 \\ \omega^2 \end{bmatrix} \right), \qquad \omega := e^{\frac{2\pi i}{3}}.$$

These harmonic frames are not unitarily equivalent since their Gramians

$$\operatorname{Gram}(\Phi) = \begin{pmatrix} 2 & -1 & -1 \\ -1 & 2 & -1 \\ -1 & -1 & 2 \end{pmatrix}, \qquad \operatorname{Gram}(\Psi) = \begin{pmatrix} 2 & 1+\omega & 1+\omega^2 \\ 1+\omega^2 & 2 & 1+\omega \\ 1+\omega & 1+\omega^2 & 2 \end{pmatrix}$$

are different. They are however projectively unitarily equivalent (see Exer. 2.21).

2.6 Tight frames as orthogonal projections

We have seen (Exer. 2.6) that the orthogonal projection of an orthonormal basis is a normalised tight frame (for its span). The converse is also true.

Theorem 2.2. *(Naĭmark) Every finite normalised tight frame $\Phi = (f_j)_{j\in J}$ for $\mathscr{H}$ is the orthogonal projection of an orthonormal basis for $\ell_2(J)$. Indeed, the orthogonal projection $P = \mathrm{Gram}(\Phi)$ of the standard orthonormal basis $(e_j)_{j\in J}$ (onto the column space of the Gramian) is unitarily equivalent to Φ via $f_j \mapsto Pe_j$, i.e.,*

$$\langle Pe_j, Pe_k\rangle_{\ell_2(J)} = \langle f_j, f_k\rangle_{\mathscr{H}}, \qquad \forall j,k \in J.$$

Proof. By Theorem 2.1, $P = \mathrm{Gram}(\Phi)$ is an orthogonal projection, and so

$$\langle Pe_j, Pe_k\rangle = \langle Pe_j, e_k\rangle = (k,j)\text{-entry of } P = \langle f_j, f_k\rangle.$$

□

When Ψ and Φ are unitarily equivalent, then we will say that Ψ is a **copy** of Φ. With this terminology, Naĭmark's theorem says:

A canonical copy of a tight frame Φ is given by the columns of $\mathrm{Gram}(\Phi)$.

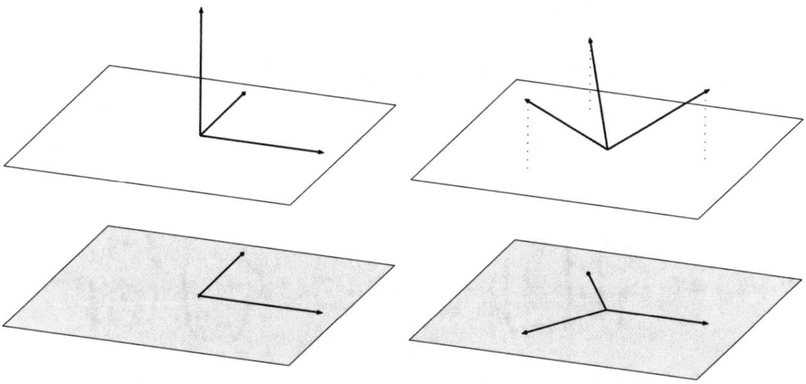

Fig. 2.4: The normalised tight frames $\{0, e_1, e_2\}$ and three equally spaced vectors obtained as the orthogonal projection of an orthonormal basis for $\mathbb{R}^3$ onto $\mathbb{R}^2$.

This is one of those often rediscovered theorems, which can be considered as a special case of Naĭmark's theorem (see [AG63] and Exer. 2.26). Hadwiger [Had40] showed that $(f_j)_{j=1}^n$ in $\mathbb{R}^d$ is a **coordinate star** (normalised tight frame) if and only if it is a **Pohlke normal star** (projection of an orthonormal basis). In signal processing, this method of obtaining tight frames is called *seeding*.

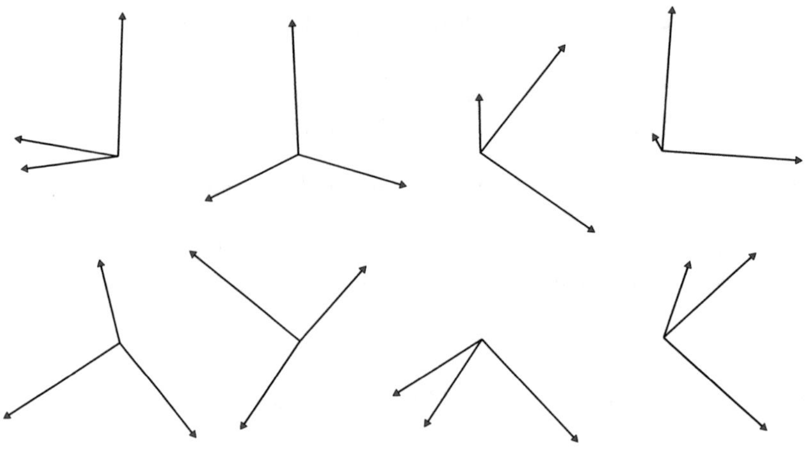

Fig. 2.5: Examples of normalised tight frames of three vectors for $\mathbb{R}^2$ obtained as the orthogonal projection of an orthonormal basis for $\mathbb{R}^3$.

Example 2.9. The Gramian of the three equally spaced vectors in $\mathbb{R}^2$ is

$$P = V^*V = \begin{bmatrix} \frac{2}{3} & -\frac{1}{3} & -\frac{1}{3} \\ -\frac{1}{3} & \frac{2}{3} & -\frac{1}{3} \\ -\frac{1}{3} & -\frac{1}{3} & \frac{2}{3} \end{bmatrix}, \qquad V = [v_1, v_2, v_3] = \sqrt{\frac{2}{3}} \begin{bmatrix} 1 & -\frac{1}{2} & -\frac{1}{2} \\ 0 & \frac{\sqrt{3}}{2} & -\frac{\sqrt{3}}{2} \end{bmatrix}. \quad (2.12)$$

The (particular choice of) vectors are normalised $\|v_j\| = \sqrt{\frac{2}{3}}$ (see Exer. 2.15), so that (v_j) is a normalised tight frame, and hence P is an orthogonal projection.

The columns (Pe_j) of P give a canonical copy of this normalised tight frame (up to unitary equivalence), e.g.,

$$\langle Pe_1, Pe_1 \rangle = \langle \begin{pmatrix} \frac{2}{3} \\ -\frac{1}{3} \\ -\frac{1}{3} \end{pmatrix}, \begin{pmatrix} \frac{2}{3} \\ -\frac{1}{3} \\ -\frac{1}{3} \end{pmatrix} \rangle = \frac{2}{3}, \quad \langle Pe_1, Pe_2 \rangle = \langle \begin{pmatrix} \frac{2}{3} \\ -\frac{1}{3} \\ -\frac{1}{3} \end{pmatrix}, \begin{pmatrix} -\frac{1}{3} \\ \frac{2}{3} \\ -\frac{1}{3} \end{pmatrix} \rangle = -\frac{1}{3}.$$

Example 2.10. A **cross** in $\mathbb{R}^n$ is the set obtained by taking an orthonormal basis and its negatives $\{\pm e_1, \ldots, \pm e_n\}$, and the orthogonal projection of a cross onto a d-dimensional subspace V is a called a **eutactic star** (see Coxeter [Cox73]). In view of Theorem 2.2, a eutactic star is precisely a tight frame of the form $\{\pm a_1, \ldots, \pm a_n\}$ for V, i.e., the union of a tight frame $\{a_1, \ldots, a_n\}$ and the equivalent frame obtained by taking its negative. When the vectors a_i all have the same length, one obtains a so-called **normalised** eutactic star. Since equal-norm tight frames always exist (see Chapters 7 and 11), so do normalised eutactic stars in $\mathbb{R}^d$ for every $n \geq d$.

2.7 The construction of tight frames from orthogonal projections

The Gramian of a normalised tight frame $\Phi = (v_j)_{j=1}^n$ for a d-dimensional space is an orthogonal projection P. By Theorem 2.2, the columns (Pe_j) of P give a canonical copy of the frame (up to unitary equivalence) as a d-dimensional subspace of $\mathbb{F}^n$. To obtain a copy of Φ in $\mathbb{F}^d$, we consider the rows of $P = \text{Gram}(\Phi)$.

Theorem 2.3. *(Row construction). Let $P \in \mathbb{C}^{n\times n}$ be an orthogonal projection matrix of rank d. The columns of $V = [v_1,\ldots,v_n] \in \mathbb{C}^{d\times n}$ are a normalised tight frame for $\mathbb{C}^d$ with Gramian P if and only if the rows of V are an orthonormal basis for the row space of P. In particular, such a V can always be obtained by applying the Gram–Schmidt process to the rows of P.*

Proof. ($\Longrightarrow$) Suppose the columns of V are a normalised tight frame for $\mathbb{C}^d$ with Gramian P, i.e., $VV^* = I$ (the rows of V are orthonormal) and $P = V^*V$. Then

$$\text{row}(P) = \text{row}(V^*V) \subset \text{row}(V) = \text{row}(VV^*V) \subset \text{row}(V^*V) = \text{row}(P),$$

so that $\text{row}(P) = \text{row}(V)$, and the rows of V are an orthonormal basis for $\text{row}(P)$.

($\Longleftarrow$) Suppose the rows of V are an orthonormal basis for $\text{row}(P)$. Then $VV^* = I$, and we have

$$(V^*V)^2 = V^*(VV^*)V = V^*V,$$

so that V^*V is an orthogonal projection matrix with the same row space (and hence column space) as P. Thus, $V^*V = P$. □

In other words:

A frame $V = [v_1,\ldots,v_n]$ is a copy of a normalised tight frame Φ if and only if the rows of V are an orthonormal basis for the row space of $\text{Gram}(\Phi)$.

Example 2.11. For the three equally spaced vectors of Example 2.9, applying the Gram–Schmidt process to the first two rows of the Gramian P gives

$$P = \begin{bmatrix} \frac{2}{3} & -\frac{1}{3} & -\frac{1}{3} \\ -\frac{1}{3} & \frac{2}{3} & -\frac{1}{3} \\ -\frac{1}{3} & -\frac{1}{3} & \frac{2}{3} \end{bmatrix}, \quad \longrightarrow \quad V = [v_1, v_2, v_3] = \sqrt{\frac{2}{3}} \begin{bmatrix} 1 & -\frac{1}{2} & -\frac{1}{2} \\ 0 & \frac{\sqrt{3}}{2} & -\frac{\sqrt{3}}{2} \end{bmatrix}. \tag{2.13}$$

Example 2.12. For Φ a tight frame with frame bound A, the matrix $P = \frac{1}{A}\text{Gram}(\Phi)$ is an orthogonal projection, and so a copy of Φ is given by an orthogonal basis for the row space of $Q = \text{Gram}(\Phi)$ consisting of vectors of length $\sqrt{A}$. For the four equally spaced unit vectors of (1.7), we have $A = 2$, and applying the Gram–Schmidt process to the first two rows of the Gramian Q gives

$$Q=\begin{bmatrix} 1 & \frac{1}{\sqrt{3}} & \frac{1}{\sqrt{3}} & -\frac{i}{\sqrt{3}} \\ \frac{1}{\sqrt{3}} & 1 & -\frac{i}{\sqrt{3}} & \frac{1}{\sqrt{3}} \\ \frac{1}{\sqrt{3}} & \frac{i}{\sqrt{3}} & 1 & -\frac{1}{\sqrt{3}} \\ \frac{i}{\sqrt{3}} & \frac{1}{\sqrt{3}} & -\frac{1}{\sqrt{3}} & 1 \end{bmatrix} \longrightarrow V=\begin{bmatrix} 1 & \frac{1}{\sqrt{3}} & \frac{1}{\sqrt{3}} & -\frac{i}{\sqrt{3}} \\ 0 & \frac{\sqrt{2}}{\sqrt{3}} & -\frac{i}{\sqrt{2}}-\frac{1}{\sqrt{2}\sqrt{3}} & \frac{1}{\sqrt{2}}+\frac{i}{\sqrt{2}\sqrt{3}} \end{bmatrix}.$$

Here $\mathrm{col}(Q) \neq \mathrm{row}(Q)$, and applying Gram–Schmidt to the columns of Q (instead of the rows) does not give a copy of Φ.

2.8 Complementary tight frames

Tight frames are determined (up to unitary equivalence) by their Gramian matrix P, which is an orthogonal projection matrix (when the frame is normalised), and all orthogonal projection matrices correspond to normalised tight frames. Thus there is normalised tight frame with Gramian given by the complementary projection $I-P$.

Definition 2.6. Given a finite normalised tight frame Φ with Gramian P, we call any normalised tight frame with Gramian $I-P$ its **complement**. More generally, we say that two tight frames are **complements** of each other, if after normalisation the sum of their Gramians is the identity I.

The complement of a finite tight frame is *unique* up to unitary equivalence (and normalisation), the complement of the complement is the frame itself, and a tight frame is equiangular (or equal-norm) if and only if its complement is.

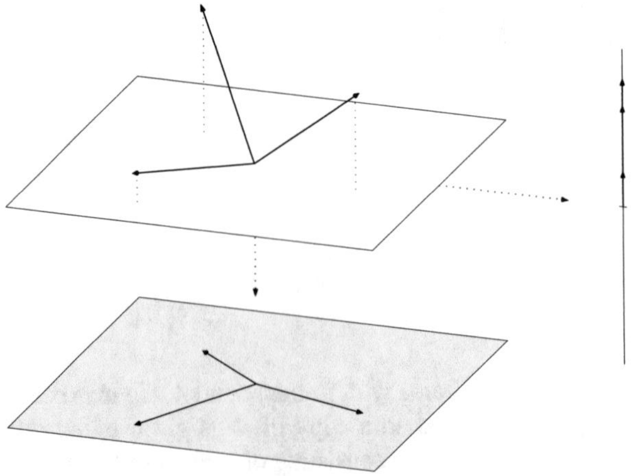

Fig. 2.6: Projecting an orthonormal basis for $\mathbb{R}^3$ onto a two-dimensional subspace and its orthogonal complement, thereby obtaining a normalised tight frame of three vectors for $\mathbb{R}^2$ and the complementary tight frame of three vectors for $\mathbb{R}$.

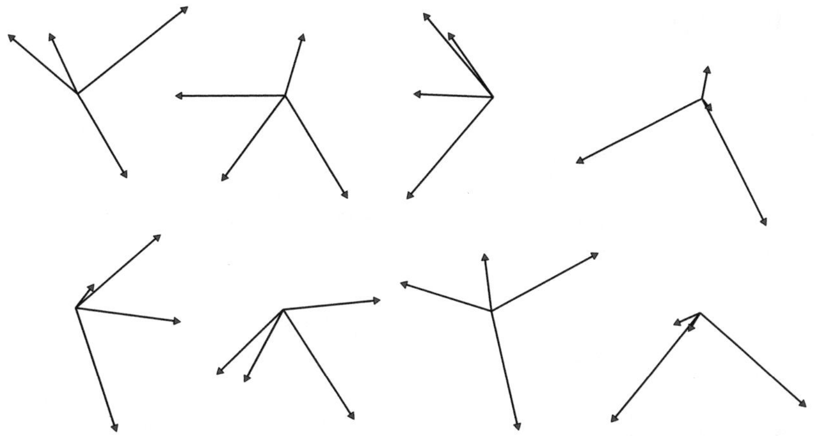

Fig. 2.7: Examples of normalised tight frames of four vectors for $\mathbb{R}^2$ and the complementary frames for $\mathbb{R}^2$ (below). A tight frame and its complement can never be unitarily equivalent.

In view of (2.11), the complement of a tight frame of n vectors for a space of dimension d is a tight frame of n vectors for a space of dimension $n-d$.

A tight frame and its complement can never be unitarily or projectively unitarily equivalent (Exer. 2.23), though they do have the same symmetries (see §9.2, §9.3).

Example 2.13. The complement of an orthonormal basis for $\mathbb{C}^d$ is the frame for the zero vector space given by d zero vectors.

Example 2.14. By (2.12) the Gramian of the complementary frame to the three equally spaces vectors in $\mathbb{R}^2$ is

$$Q = I - P = \begin{bmatrix} 1 & 0 & 0 \\ 0 & 1 & 0 \\ 0 & 0 & 1 \end{bmatrix} - \begin{bmatrix} \frac{2}{3} & -\frac{1}{3} & -\frac{1}{3} \\ -\frac{1}{3} & \frac{2}{3} & -\frac{1}{3} \\ -\frac{1}{3} & -\frac{1}{3} & \frac{2}{3} \end{bmatrix} = \begin{bmatrix} \frac{1}{3} & \frac{1}{3} & \frac{1}{3} \\ \frac{1}{3} & \frac{1}{3} & \frac{1}{3} \\ \frac{1}{3} & \frac{1}{3} & \frac{1}{3} \end{bmatrix}.$$

By the row construction (Theorem 2.3), this is the Gramian of normalised tight frame $\{\frac{1}{\sqrt{3}}, \frac{1}{\sqrt{3}}, \frac{1}{\sqrt{3}}\}$ of three repeated vectors for $\mathbb{R}$. The Gramian of an arbitrary normalised tight frame of three vectors for $\mathbb{C}^2$ is considered in Exer. 2.22.

We call the tight frame of $n = d+1$ vectors for $\mathbb{R}^d$ which is the complement of $\{\frac{1}{\sqrt{d+1}}, \ldots, \frac{1}{\sqrt{d+1}}\}$, the **vertices of the (regular) simplex** in $\mathbb{R}^d$. This has Gramian

$$P = [p_{jk}], \qquad p_{jk} := \begin{cases} \frac{-1}{d+1}, & j \neq k; \\ \frac{d}{d+1}, & j = k. \end{cases} \tag{2.14}$$

To find a copy in $\mathbb{R}^d$, one can apply the method of Theorem 2.3 to the Gramian. This example can be generalised to obtain *partition frames*.

2.9 Partition frames

Definition 2.7. Let $\alpha = (\alpha_1, \ldots, \alpha_k) \in \mathbb{Z}^k$ be a partition of n, i.e.,

$$n = \alpha_1 + \cdots + \alpha_k, \qquad 1 \le \alpha_1 \le \alpha_2 \le \cdots \le \alpha_k.$$

The **α-partition frame** for $\mathbb{R}^d$, $d = n - k$, is the complement of the normalised tight frame of n vectors for $\mathbb{R}^k$ given by

$$\Big(\underbrace{\frac{e_1}{\sqrt{\alpha_1}}, \ldots, \frac{e_1}{\sqrt{\alpha_1}}}_{\alpha_1 \text{ times}}, \ldots, \underbrace{\frac{e_k}{\sqrt{\alpha_k}}, \ldots, \frac{e_k}{\sqrt{\alpha_k}}}_{\alpha_k \text{ times}}\Big). \tag{2.15}$$

It is said to be **proper** if $\alpha_j \ge 2$, $\forall j$.

The Gramian of the α-partition frame is the block diagonal $n \times n$ matrix

$$P = \begin{bmatrix} B_1 & & & & \\ & \ddots & & & \\ & & B_j & & \\ & & & \ddots & \\ & & & & B_k \end{bmatrix}, \qquad B_j := \begin{bmatrix} \frac{\alpha_j-1}{\alpha_j} & \frac{-1}{\alpha_j} & \frac{-1}{\alpha_j} & \cdots & \frac{-1}{\alpha_j} \\ \frac{-1}{\alpha_j} & \frac{\alpha_j-1}{\alpha_j} & \frac{-1}{\alpha_j} & \cdots & \frac{-1}{\alpha_j} \\ \frac{-1}{\alpha_j} & \frac{-1}{\alpha_j} & \frac{\alpha_j-1}{\alpha_j} & & \frac{-1}{\alpha_j} \\ \vdots & \vdots & & \ddots & \frac{-1}{\alpha_j} \\ \frac{-1}{\alpha_j} & \frac{-1}{\alpha_j} & \frac{-1}{\alpha_j} & \frac{-1}{\alpha_j} & \frac{\alpha_j-1}{\alpha_j} \end{bmatrix} \tag{2.16}$$

where the above B_j is a $\alpha_j \times \alpha_j$ orthogonal projection matrix of rank $\alpha_j - 1$. Since each normalised tight frame is unitarily equivalent to the columns of its Gramian, it follows that the vectors in a proper partition frame are distinct and nonzero. If $\alpha_j = 1$, then the corresponding partition frame vector is zero.

Example 2.15. (Simplex) For $n = d + 1$, the trivial partition $\alpha = (d + 1)$ gives the vertices of the simplex in $\mathbb{R}^d$.

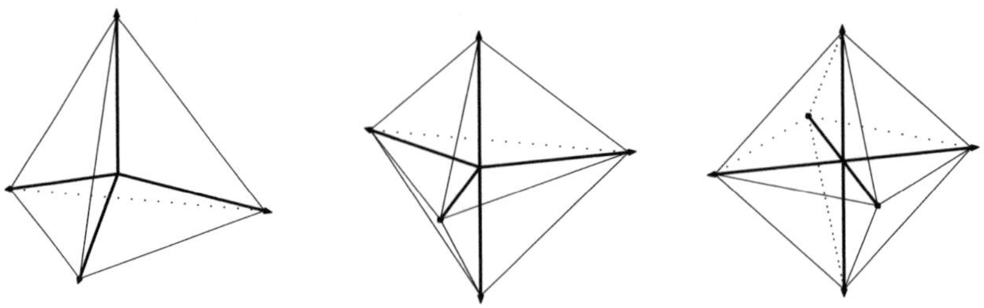

Fig. 2.8: The proper α-partition frames in $\mathbb{R}^3$ for $\alpha = (4), (2,3)$ and $(2,2,2)$, respectively. These are the vertices of the tetrahedron, trigonal bipyramid and octagon. In four dimensions, the possible choices for α are (5), $(2,4)$, $(3,3)$ and $(2,2,3)$.

Table 2.1: The proper partition frames in $\mathbb{R}^2$ and $\mathbb{R}^3$. Here $|G|$ is the order of their symmetry group $G=\mathrm{Sym}(\Phi)$ (see Chapter 9, Exer. 9.5).

Partition	n	Description of partition frame	$\|G\|$
(3)	3	three equally spaced vectors in $\mathbb{R}^2$	6
(2,2)	4	four equally spaced vectors in $\mathbb{R}^2$	8
(4)	4	vertices of the tetrahedron in $\mathbb{R}^3$	24
(2,3)	5	vertices of the trigonal bipyramid in $\mathbb{R}^3$	12
(2,2,2)	6	vertices of the octahedron in $\mathbb{R}^3$	48

2.10 Real and complex tight frames

A tight frame for a real Hilbert space is a tight frame for its complexification (see Exer. 2.28). We will call frames that come in this way *real* tight frames.

Definition 2.8. We say that a tight frame $(f_j)_{j\in J}$ is **real** if its Gramian is a real matrix, and otherwise it is **complex**.

By Theorem 2.3 (row construction), a tight frame for a space $\mathscr{H}$ of dimension d is real if and only if there is a unitary matrix $U:\mathscr{H}\to\mathbb{F}^d$ for which $Uf_j\in\mathbb{R}^d$, $\forall j$. Moreover, a frame is complex if and only if its complementary frame is.

Example 2.16. The vertices of a simplex (or partition frame) are a *real* frame, by definition. The second prototypical example (1.7) and the example (a) of Exer. 1.3 are *complex* frames, since their Gramians (after normalisation) are

$$\begin{bmatrix} \frac{1}{2} & \frac{1}{2\sqrt{3}} & \frac{1}{2\sqrt{3}} & \frac{-i}{2\sqrt{3}} \\ \frac{1}{2\sqrt{3}} & \frac{1}{2} & \frac{-i}{2\sqrt{3}} & \frac{1}{2\sqrt{3}} \\ \frac{1}{2\sqrt{3}} & \frac{i}{2\sqrt{3}} & \frac{1}{2} & \frac{-1}{2\sqrt{3}} \\ \frac{i}{2\sqrt{3}} & \frac{1}{2\sqrt{3}} & \frac{-1}{2\sqrt{3}} & \frac{1}{2} \end{bmatrix}, \qquad \begin{bmatrix} \frac{2}{3} & \frac{-\omega^2}{3} & -\frac{\omega}{3} \\ \frac{-\omega}{3} & \frac{2}{3} & -\frac{\omega^2}{3} \\ \frac{-\omega^2}{3} & -\frac{\omega}{3} & \frac{2}{3} \end{bmatrix}. \tag{2.17}$$

There are intrinsic differences between the classes of real and complex frames, e.g., see Exer. 1.3, §6.8, §12.1, §12.10 and §12.17.

The real algebraic variety of real and complex normalised finite tight frames (and unit-norm tight frames) is considered in Chapter 7.

Remark 2.1. One could extend the Definition 2.8 to other fields, e.g., say that the three equally spaced unit vectors in $\mathbb{R}^2$ are a *rational* tight frame, since their Gramian has rational entries. In this case, the columns of the Gramian give a copy of this frame in a rational inner product space (Example 2.9), but the row construction (Example 2.11) does not give a copy in $\mathbb{Q}^2$ (with the Euclidean inner product). These ideas are explored in [CFW15].

2.11 SICs and MUBs

There are many interesting and useful examples of *equal-norm* tight frames (v_j), e.g., *group frames* (see §10). Those for which the *cross-correlation* $|\langle v_j, v_k\rangle|$, $j \neq k$, takes a small number of values are of particular interest (especially for applications). We briefly mention two such classes of frames: the *SICs* and the *MUBs*. These are simple to describe, and come with some intriguing conjectures, which are still unproven (despite considerable work on them). Indeed, the construction of SICs (see §14) and maximal sets of MUBs (see §14.2, §8.6) are two central problems in the theory of finite tight frames.

Definition 2.9. A tight frame of d^2 unit vectors (v_j) for $\mathbb{C}^d$ is a **SIC** if

$$|\langle v_j, v_k\rangle|^2 = \frac{1}{d+1}, \qquad j \neq k.$$

SICs can be viewed as maximal sets of complex equiangular lines. It follows from the bounds of Theorem 12.2 on such lines that SICs are *complex frames*. Their origins as quantum measurements and the known constructions are detailed in Chapter 14. The conjecture that SICs exist in every dimension d is known as *Zauner's conjecture* or the *SIC problem*.

Example 2.17. The second prototypical example (1.7) is a SIC for $\mathbb{C}^2$. The case $d = 3$ seems to be an exception for SICs. Here the SICs form a continuous family, while for $d \neq 3$, there is currently only a finite number of SICs for $\mathbb{C}^d$ known.

Definition 2.10. A tight frame consisting of m orthonormal bases $\mathscr{B}_1, \ldots, \mathscr{B}_m$ is said to be a **MUB** (or a set of m **MUBs**) for $\mathbb{C}^d$ if the bases are mutually unbiased, i.e.,

$$|\langle v, w\rangle|^2 = \frac{1}{d}, \qquad v \in \mathscr{B}_j,\ w \in \mathscr{B}_k, \quad j \neq k.$$

Mutually unbiased bases have similar uses in quantum-state determination as SICs do. The maximal number $\mathscr{M}(d)$ of MUBs for $\mathbb{C}^d$ is bounded above by $d+1$ (Proposition 12.12). This bound is attained for d a prime power. Beyond this not much is known [BWB10], e.g., for $d = 6$ (the first d which is not a prime power), it is only known that

$$3 \leq \mathscr{M}(6) \leq 7.$$

The *MUB problem* is to say anything more, e.g., to show that $\mathscr{M}(6) = 3$ (as is commonly believed).

Example 2.18. Three mutually unbiased bases in $\mathbb{C}^2$ are given by

$$\mathscr{B}_1 = \left\{\begin{bmatrix}1\\0\end{bmatrix}, \begin{bmatrix}0\\1\end{bmatrix}\right\}, \quad \mathscr{B}_2 = \left\{\frac{1}{\sqrt{2}}\begin{bmatrix}1\\1\end{bmatrix}, \frac{1}{\sqrt{2}}\begin{bmatrix}1\\-1\end{bmatrix}\right\}, \quad \mathscr{B}_3 = \left\{\frac{1}{\sqrt{2}}\begin{bmatrix}1\\i\end{bmatrix}, \frac{1}{\sqrt{2}}\begin{bmatrix}1\\-i\end{bmatrix}\right\}.$$

The first two are *real* MUBs. The question on how many real MUBs there are and its connection with *association schemes* is of interest (see [LMO10]). Three MUBs for $\mathbb{R}^4$ can be obtained by choosing a subset of the vertices of the 24-*cell* in $\mathbb{R}^4$.

Notes

The key idea (not to be underestimated) of this section is:

Tight frames are best understood via their Gramian.

Indeed, a tight frame $\Phi = (f_j)_{j=1}^n$ is determined up to unitary equivalence (and normalisation) by its Gramian $P = P_\Phi$, which is an orthogonal projection matrix. The columns of P_Φ give a (canonical) copy of Φ, and so the kernel of P_Φ is the space of linear dependencies between the vectors in Φ, i.e.,

$$\ker(P_\Phi) = \{a \in \mathbb{F}^n : Pa = \sum_j a_j P e_j = 0\} = \{a \in \mathbb{F}^n : \sum_j a_j f_j = 0\} =: \operatorname{dep}(\Phi).$$

Since P_Φ is determined by $\ker(P_\Phi)$, this observation allows the theory of tight frames to be extended to any finite dimensional vector space over a subfield of $\mathbb{C}$ which is closed under conjugation (see Chapter 4).

Many notions of *equivalence* of tight frames appear in the literature (see [Bal99], [HL00], [GKK01], [Fic01], [HP04]). Here we use a descriptive terminology (from which all of these can be described). For finite tight frames viewed as sequences of vectors, *unitary equivalence* is the natural equivalence, and when viewed as (weighted) projections (fusion frames), *projective unitary equivalence* is natural. Unitary equivalence is determined by the Gramian (Corollary 2.1), and projective unitary unitary equivalence is determined by certain m-products (see Chapter 8).

It is implicit in the Definition 2.1 of a *tight frame* that $\mathcal{H}$ be separable, i.e., have a countable orthonormal basis. The theory extends, in the obvious way, to nonseparable spaces, with J now an uncountable index set. In these cases, it turns out that all tight frames for $\mathcal{H}$ (with nonzero vectors) have the same infinite cardinality, i.e., the Hilbert dimension of $\mathcal{H}$. By way of contrast, if $\mathcal{H}$ has finite dimension d, then there exist tight frames for $\mathcal{H}$ with any countable cardinality greater than or equal to d.

We will have good reason to consider representations such as (1.3), where the sum $\sum_{j\in J}$ is replaced by a continuous sum (with respect to some measure). This generalisation (see Chapter 16) will be called a *continuous tight frame*, with the special case of Definition 2.1 referred to as a *(discrete) tight frame*.

The book [HKLW07] covers the material of this section. It has a section on frames in $\mathbb{R}^2$ (for tight frames in $\mathbb{R}^3$ see [Fic01]). The popular article [KC07a], [KC07b] advocates the use of tight frames in a number of engineering applications. It outlines *standard terminology* for frames (resulting from an e-mail discussion within the frame community), which we adopt, except for our preference of *normalised tight frame* over *Parseval frame*. In this parlance a *ENPTF* is a *equal-norm Parseval tight frame*, and similarly.

Exercises

2.1. By expanding, or otherwise, verify the *polarisation identity* for an inner product space $\mathcal{H}$, i.e., $\forall f,g \in \mathcal{H}$ that

$$\Re\langle f,g\rangle = \frac{1}{4}(\|f+g\|^2 - \|f-g\|^2),$$
$$\Im\langle f,g\rangle = \frac{1}{4}(\|f+ig\|^2 - \|f-ig\|^2), \qquad \text{(for } \mathcal{H} \text{ complex)}.$$

2.2. Use the polarisation identity to show that the following conditions are equivalent to being a finite tight frame

$$\textit{Parseval:} \qquad f = \frac{1}{A}\sum_{j\in J}\langle f,f_j\rangle f_j, \quad \forall f \in \mathcal{H},$$

$$\textit{Plancherel:} \qquad \langle f,g\rangle = \frac{1}{A}\sum_{j\in J}\langle f,f_j\rangle\langle f_j,g\rangle, \quad \forall f,g \in \mathcal{H}.$$

2.3. *Orthogonal projection formula.*
Let $(f_j)_{j\in J}$ be a finite tight frame (with frame bound A) for a subspace $\mathcal{K} \subset \mathcal{H}$. Show that P the orthogonal projection onto this subspace is given by

$$P = \frac{1}{A}VV^* : f \mapsto \frac{1}{A}\sum_{j\in J}\langle f,f_j\rangle f_j, \qquad V := [f_j]_{j\in J}.$$

2.4. *Orthogonal bases and tight frames.*
(a) Show that an orthogonal basis $(f_j)_{j\in J}$ for $\mathcal{H}$ is a tight frame if and only if all its vectors have the same norm and that it is a normalised tight frame if and only if it is an orthonormal basis.
(b) Show that if (f_j) is a normalised tight frame, then $\|f_j\| \leq 1, \forall j \in J$, and

$$\|f_j\| = 1 \qquad \Longleftrightarrow \qquad f_j \perp f_k, \quad \forall k \neq j.$$

In particular, the only normalised tight frames whose vectors all have unit length are the orthonormal bases.

2.5. *Unitary images of tight frames.*
(a) Show that the image of a tight frame $(f_j)_{j\in J}$ under a unitary map U is a tight frame with the same frame bound.
(b) Show that if (f_j) is a finite normalised tight frame for $\mathcal{H}$, and T is a linear map for which (Tf_j) is also, then T is a unitary map.

2.6. *Projections of normalised tight frames are normalised tight frames.*
A linear map $P : \mathcal{H} \to \mathcal{H}$ on a Hilbert space is an **orthogonal projection** if $P^2 = P$

and $P^* = P$. Show that if $(f_j)_{j\in J}$ is a normalised tight frame for a Hilbert space $\mathscr{K}$ and P is an orthogonal projection onto a subspace $\mathscr{H} \subset \mathscr{K}$, then (Pf_j) is a normalised tight frame for $\mathscr{H}$. (This is obvious in the context of Theorem 2.2.)

2.7. *Partial isometries map normalised tight frames to normalised tight frames.* A linear map $L : \mathscr{H} \to \mathscr{K}$ between Hilbert spaces is an **isometry** if $L^*L = I_{\mathscr{H}}$, i.e., it is norm preserving:

$$\|Lx\| = \|x\|, \qquad \forall x \in \mathscr{H}.$$

It is a **coisometry** if $L^* : \mathscr{K} \to \mathscr{H}$ is an isometry, i.e., L^* is norm preserving.
Let Φ be a finite normalised tight frame for $\mathscr{H}$, and $Q : \mathscr{H} \to \mathscr{K}$ be a linear map. Show that the following are equivalent
(a) Q is a partial isometry, i.e., its restriction to $(\ker Q)^\perp = \operatorname{ran}(Q^*)$ is an isometry.
(b) QQ^* is an orthogonal projection.
(c) Q^*Q is an orthogonal projection.
(d) $Q\Phi$ is a normalised tight frame (for its span).
Remark. Since unitary maps and orthogonal projections are partial isometries, this generalises Exercises 2.5 and 2.6. It appears as a special case in Exer. 3.5.

2.8.^{m} If U is an $n \times n$ unitary matrix (or a scalar multiple of one) with entries of constant modulus, then an *equal-norm tight frame* for $\mathbb{F}^d$ is given by the columns of the $d \times n$ submatrix obtained from it by selecting *any* d of its rows.
(a) When $\mathbb{F} = \mathbb{R}$, such U, with entries ± 1, are called **Hadamard matrices**. Use the `matlab` function `hadamard(n)` (defined for $n, \frac{n}{12}$ or $\frac{n}{20}$ a power of 2) to construct equal-norm tight frames of n vectors in $\mathbb{R}^d$.
Remark: It can be shown that if a Hadamard matrix exists, then $n = 1, 2$ or n is divisible by 4. The **Hadamard conjecture** is that there exists a Hadamard matrix of size $n = 4k$, for every k. The smallest open case (in 2010) is $n = 668$.
(b) Show that the Fourier matrix $F = \frac{1}{\sqrt{n}}[\omega^{jk}]_{0\le j,k<n}$, $\omega = e^{\frac{2\pi i}{n}}$ of (2.5) is unitary and has order 4. Use the `matlab` function `fft(X)` (Discrete Fourier transform) to construct F, and hence equal-norm tight frames of n vectors in $\mathbb{C}^d$.
Remark: It is always possible to obtain a real frame in this way.

2.9. *Tight frames for* $\mathbb{R}^2$.
(a) Show the vectors $(v_j)_{j=1}^n$, $v_j = (x_j, y_j) \in \mathbb{R}^2$ are a tight frame for $\mathbb{R}^2$ if and only if the **diagram vectors** $w_j := (x_j + iy_j)^2 \in \mathbb{C}$ sum to zero (in $\mathbb{C}$).
(b) Show that two tight frames for $\mathbb{R}^2$ are projectively unitarily equivalent if and only if their diagram vectors are scalar multiples of each other.
(c) Show that up to projective unitary equivalence the only equal-norm tight frame of three vectors for $\mathbb{R}^2$ is three equally spaced unit vectors.
(d) Show that all unit-norm tight frames of four vectors for $\mathbb{R}^2$ are the union of two orthonormal bases. This gives a one-parameter family of projectively unitarily inequivalent unit-norm tight frames of four vectors for $\mathbb{R}^2$.
(e) Show the tight frames of five unit vectors for $\mathbb{R}^2$ with diagram vectors

$$\{e^{i\theta}, e^{-i\theta}, e^{i(\theta+\psi)}, e^{-i(\theta+\psi)}, -1\}, \qquad 0 < \theta < \frac{\pi}{2}, \quad \cos 2\theta + \cos(2\theta + 2\psi) = \frac{1}{2}$$

are projectively unitarily inequivalent, and that none is the union of an orthonormal basis and three equally spaced vectors.

2.10. *Projective unitary equivalence in* $\mathbb{R}^2$.
(a) For unit-norm tight frames of n vectors for $\mathbb{R}^2$ show that the equivalence classes for *projective unitary equivalence up to reordering* are in 1–1 correspondence with convex n-gons with sides of unit length (given by a sum of diagram vectors).
(b) What do subsets of orthonormal vectors correspond to on the polygon?
(c) What is the n-gon corresponding to the tight frame for $\mathbb{R}^2$ given by n equally spaced unit vectors?
(d) Does every finite tight frames for $\mathbb{R}^2$ correspond to some convex polygon?

2.11. The **complex conjugate** of $\mathcal{H}$ is the Hilbert space $\overline{\mathcal{H}}$ of all formal complex conjugates with addition, scalar multiplication and inner product given by

$$\overline{v} + \overline{w} = \overline{v + w}, \qquad \alpha\overline{v} = \overline{\overline{\alpha}v}, \qquad \langle \overline{v}, \overline{w} \rangle = \overline{\langle v, w \rangle}. \tag{2.18}$$

(a) Show that the **conjugation map** $C : \mathcal{H} \to \overline{\mathcal{H}} : v \mapsto \overline{v}$ is antilinear.
(b) Suppose that $\Phi = (f_j)$ is a sequence of vectors in $\mathcal{H}$ and $\overline{\Phi} := (\overline{f_j}) \subset \overline{\mathcal{H}}$. Show that the frame operator and Gramian satisfy

$$S_{\overline{\Phi}} = C S_{\Phi} C^{-1}, \qquad \mathrm{Gram}(\overline{\Phi}) = \overline{\mathrm{Gram}(\Phi)}.$$

Hence C maps tight frames to a tight frames (with the same bound A).
(c) Suppose that $\mathcal{H} = V$, with V a subspace of $\mathbb{C}^d$. Show that $\overline{\mathcal{H}}$ is isomorphic to the subspace $\overline{V} := \{\overline{v} : v \in V\}$ of $\mathbb{C}^d$, where $\overline{v} = \overline{(v_j)} := (\overline{v_j})$.

2.12. Show the frame operator S for a sequence of vectors $f_1, \ldots, f_n$ satisfies:
(a) $\mathrm{trace}(S) = \sum_j \|f_j\|^2$.
(b) $\mathrm{trace}(S^2) = (\|S\|_F)^2 = \sum_j \sum_k |\langle f_j, f_k \rangle|^2$.
Hint: The trace operator satisfies $\mathrm{trace}(AB) = \mathrm{trace}(BA)$.

2.13. *Trace formula.* Show that if $(f_j)_{j \in J}$ is a finite normalised tight frame for $\mathcal{H}$ and $L : \mathcal{H} \to \mathcal{H}$ is a linear transformation, then its trace is given by

$$\mathrm{trace}(L) = \sum_{j \in J} \langle L f_j, f_j \rangle.$$

In particular, when L is the identity map, we obtain the *trace formula* (2.9).

2.14. Let $\mathcal{H}$ be have finite dimension $d \geq 1$. Show that
(a) There exists a tight frame $(f_j)_{j=1}^{\infty}$ for $\mathcal{H}$, with infinitely many nonzero vectors.
(b) For any such tight frame, $\|f_j\| \to 0$ as $j \to \infty$.
(c) There are no equal-norm tight frames for $\mathcal{H}$ with infinitely many vectors.
Remark: In contrast, the *continuous* tight frame (u_θ) for $\mathbb{R}^2$ of (1.3) has uncountably many equal-norm vectors.

2.15. *Equal-norms.* Show that if (f_j) is an *equal-norm* tight frame of n vectors (with frame bound A) for a space $\mathscr{H}$ of dimension d, then

$$\|f_j\| = \sqrt{\frac{dA}{n}}, \qquad \forall j.$$

In particular, if (f_j) is unit-norm, i.e., $\|f_j\| = 1$, $\forall j$, then $\frac{1}{A} = \frac{d}{n}$,

2.16. *Equiangularity.* Show that if (f_j) is an *equiangular* tight frame of $n > 1$ vectors (with frame bound A) for a space $\mathscr{H}$ of dimension d, then its Gramian satisfies

$$\langle f_j, f_j \rangle = \frac{dA}{n}, \quad \forall j, \qquad |\langle f_j, f_k \rangle| = \frac{A}{n}\sqrt{\frac{d(n-d)}{n-1}}, \quad j \neq k.$$

2.17. Let $\Phi = (f_j)_{j\in J}$ be a finite sequence of vectors in $\mathscr{H}$, where $d = \dim(\mathscr{H})$, and $V := [f_j]_{j\in J}$. Show Φ is a normalised tight frame for $\mathscr{H}$ if and only if
(a) $\mathrm{Gram}(\Phi) = V^*V$ has exactly d nonzero eigenvalues all equal to 1.
(b) The frame operator $S_\Phi = VV^*$ has all its eigenvalues equal to 1.
(c) The synthesis operator V has d singular values equal to 1.
(d) The analysis operator V^* has d singular values equal to 1.

2.18. *Isometries.* Let $\Phi = (f_j)_{j\in J} \subset \mathscr{H}$, and $V^* : \mathscr{H} \to \ell_2(J) : f \mapsto (\langle f, f_j \rangle)_{j\in J}$ be the analysis operator. Show that the following are equivalent
(a) Φ is a normalised tight frame for $\mathscr{H}$.
(b) V^* is *inner product preserving*, i.e., $\langle V^*f, V^*g \rangle = \langle f, g \rangle$, $\forall f, g \in \mathscr{H}$.
(c) V^* is an *isometry*, i.e., $\|V^*f\| = \|f\|$, $\forall f \in \mathscr{H}$.

2.19. Suppose that $\Phi = (f_j)_{j=1}^n$ and $\Psi = (g_j)_{j=1}^n$ are sequences of vectors, with $\mathscr{H} = \mathrm{span}(\Phi)$ and $\mathscr{K} = \mathrm{span}(\Psi)$. Show there is a unitary map $U : \mathscr{H} \to \mathscr{K}$ with $g_j = Uf_j$, $\forall j$ if and only if $\langle f_j, f_k \rangle = \langle g_j, g_k \rangle$, $\forall j,k$, i.e., $\mathrm{Gram}(\Phi) = \mathrm{Gram}(\Psi)$.

2.20. (a) Express *unitary equivalence up to reordering* in terms of the Gramian.
(b) Express *projective unitary equivalence up to reordering* in terms of the Gramian.
(c) Show that a necessary, but not sufficient, condition for normalised tight frames $(f_j)_{j\in J}$ and $(g_j)_{j\in K}$ to be projectively equivalent up to reordering is that there is a permutation $\sigma : J \to K$ with $|\langle g_{\sigma j}, g_{\sigma k} \rangle| = |\langle f_j, f_k \rangle|$, $\forall j,k \in J$. In particular, the multisets $\{|\langle f_j, f_k \rangle|\}_{j,k\in J}$ and $\{|\langle g_j, f_k \rangle|\}_{j,k\in J}$ must be equal.

2.21. (a) Show that normalised tight frames Φ and Ψ are projectively unitarily equivalent up to reordering if and only if their complements are.
(b) Show that all equal-norm tight frames of $n = d+1$ vectors in $\mathbb{F}^d$ are projectively unitarily equivalent, and hence are equiangular.
(c) For the unitarily inequivalent equal-norm frames of three vectors for $\mathbb{C}^2$ given in Example 2.8, find a unitary matrix U (and scalars α_j) which gives the projective unitary equivalence $g_j = \alpha_j U f_j$, where $\Phi = (f_j)$ and $\Psi = (g_j)$.

2.22. Find all possible normalised tight frames of three vectors for $\mathbb{C}^2$ up to unitary equivalence.

2.23. Show that no tight frame can be unitarily equivalent to its complement. Can a tight frame be projectively unitarily equivalent to its complement?

2.24.^{m} Write a `matlab` function for the complementary tight frame using `null`.

2.25.^{m} (a) By using an inductive argument based on complements, prove that an equal-norm tight frame of n vectors for $\mathbb{F}^d$ can be constructed, for all $n > d$.
(b) Write a function `ENTF(n,d)` to construct such equal-norm tight frames.
(c) Construct an equal-norm tight frame of 8 vectors for $\mathbb{R}^3$.

2.26. *M. A. Naĭmark's theorem.*
An **orthogonal resolution of the identity** for a Hilbert space $\mathscr{H}$ is a one-parameter family $(E_t)_{t\in\mathbb{R}}$ of orthogonal projections on $\mathscr{H}$, for which $t \mapsto E_t$ is left continuous, and

$$\lim_{t\to-\infty} E_t = 0, \quad \lim_{t\to\infty} E_t = I_{\mathscr{H}}, \qquad E_s E_t = E_{\min\{s,t\}}.$$

A **generalised resolution of the identity** is a family $(F_t)_{t\in\mathbb{R}}$, for which the differences $F_t - F_s$, $s < t$ are bounded positive operators, $t \mapsto F_t$ is left continuous, and

$$\lim_{t\to-\infty} F_t = 0, \quad \lim_{t\to\infty} F_t = I_{\mathscr{H}}.$$

Naĭmark's theorem (see, e.g., [AG63]) says that every generalised resolution of the identity for $\mathscr{H}$ is the orthogonal projection onto $\mathscr{H}$ of an orthogonal resolution of the identity for some larger Hilbert space $\mathscr{K} \supset \mathscr{H}$.
(a) Let $(f_j)_{j=1}^n$ be a finite normalised tight frame for which none of the vectors are zero. Show that a generalised resolution of the identity is given by

$$F_t f := \sum_{j\le t} \langle f, f_j\rangle f_j, \qquad \forall f \in \mathscr{H}.$$

(b) By Naĭmark's theorem, there is a Hilbert space $\mathscr{K} \supset \mathscr{H}$, and an orthogonal resolution of the identity (E_t) for $\mathscr{K}$, such that $F_t = PE_t$, where P is the orthogonal projection of $\mathscr{K}$ onto $\mathscr{H}$. Conclude that

$$I_{\mathscr{H}} = \sum_{j=1}^n (F_j - F_{j-1}) = \sum_{j=1}^n PQ_j, \qquad Q_j := E_j - E_{j-1}$$

where Q_j is an orthogonal projection, and $Q_j \perp Q_k$, $k \neq j$.
(c) Show that $\mathscr{K}$ can be taken to be n dimensional.
(d) Prove Naĭmark's theorem for $\mathscr{H} = \mathbb{F}^d$ by taking $V = [f_1, \ldots, f_n]$ which has orthonormal rows, and extend it to obtain a unitary matrix.

2.27. Suppose that $(u_j + iv_j)_{j=1}^n$ is a normalised tight frame of n vectors for $\mathbb{C}^d$, where $u_j, v_j \in \mathbb{R}^d$. Prove that $(u_1, \ldots, u_n, v_1, \ldots, v_n)$ is a normalised tight frame of $2n$ vectors for $\mathbb{R}^d$.

2.28. Show that a tight frame for $\mathbb{R}^d$ is tight frame for $\mathbb{C}^d$.

2.29. *Normalised tight frames and linear mappings.*
Let $(f_j)_{j\in J}$ and $(g_k)_{k\in K}$ be finite normalised tight frames for $\mathscr{H}$ and $\mathscr{K}$. Denote the vector space of all linear maps $\mathscr{H} \to \mathscr{K}$ by $\mathscr{L}(\mathscr{H},\mathscr{K})$.
(a) Show that the *Hilbert–Schmidt* inner product on $\mathscr{L}(\mathscr{H},\mathscr{K})$ satisfies

$$\langle L, M\rangle_{HS} := \operatorname{trace}(M^*L) = \sum_{j\in J}\langle Lf_j, Mf_j\rangle = \sum_{k\in K}\langle M^*g_k, L^*g_k\rangle.$$

Remark: Taking $M = I_{\mathscr{H}}$ gives the trace formula of Exer. 2.13.
(b) Let f_j^* be $\mathscr{H} \to \mathbb{F} : f \mapsto \langle f, f_j\rangle$. Show that $(g_k f_j^*)_{j\in J,k\in K}$ is a normalised tight frame (of rank one maps) for $\mathscr{L}(\mathscr{H},\mathscr{K})$ with the Hilbert–Schmidt inner product.

2.30. *Matrices with respect to a normalised tight frame.*
Normalised tight frames can be used to represent vectors and linear maps in much the same way as orthonormal bases. Suppose that $(f_j)_{j\in J}$ and $(g_k)_{k\in K}$ are finite normalised tight frames for $\mathscr{H}$ and $\mathscr{K}$, and let $V = [f_j]_{j\in J}$, $W = [g_k]_{k\in K}$. Then the **coordinates** x of $f \in \mathscr{H}$ with respect to (f_j), and the **matrix** A representing a linear map $L : \mathscr{H} \to \mathscr{K}$ with respect to (f_j) and (g_k) are

$$x = [f] := V^*f \in \mathbb{F}^J, \qquad A = [L] := W^*LV \in \mathbb{F}^{J\times J}.$$

(a) Show that $[Lf] = Ax$, and f, L can be recovered via $f = Vx$, $L = WAV^*$.
(b) Show that $[\alpha L + \beta M] = \alpha[L] + \beta[M]$, $\alpha, \beta \in \mathbb{F}$, and $[L^*] = [L]^*$.
(c) Show that the composition of linear maps satisfies $[ML] = [M][L]$.
(d) Suppose $L : \mathscr{H} \to \mathscr{H}$, and $W = V$. Show that f is an eigenvector of L for the eigenvalue λ if and only if $Ax = \lambda x$, i.e., eigenvectors of L correspond to the eigenvectors of A that are in the range of V^*.
(e) Show L and A have the same singular values, and hence the same rank.

Chapter 3
Frames

A finite (normalised) tight frame is a spanning set $(f_j)_{j=1}^n$ for $\mathscr{H}$, for which

$$f = \sum_{j=1}^n \langle f, f_j \rangle f_j, \qquad \forall f \in \mathscr{H}.$$

This expansion can be further generalised, by replacing the rank one orthogonal projections $f \mapsto \langle f, f_j \rangle f_j$, by rank one projections $f \mapsto \langle f, g_j \rangle f_j$, to obtain what is called a (nontight) frame expansion.

This elegant theory includes orthogonal and biorthogonal expansions as special cases. It will be mostly used as a route to obtain the so-called *canonical* tight frame. This is a normalised tight frame naturally associated with a given frame, which is as close as possible to it.

3.1 Motivation

Suppose that $(f_j)_{j=1}^n$ spans $\mathscr{H}$, so that each $f \in \mathscr{H}$ can be reconstructed

$$f = \sum_{j=1}^n c_j f_j, \quad \Longleftrightarrow \quad Vc = f, \quad V := [f_j]_{j=1}^n,$$

for *some* coefficients $c_j = c_j(f) \in \mathbb{F}$, which are unique if and only if (f_j) is a basis, i.e., $n = \dim(\mathscr{H})$. However, there is always a *unique* least squares solution (one minimising $\sum_j |c_j|^2$) given by

$$c = V^\dagger f = V^* S^{-1} f \qquad \Longleftrightarrow \qquad c_j = \langle f, S^{-1} f_j \rangle,$$

where $S := VV^*$, and $V^\dagger = V^*(VV^*)^{-1} : \mathscr{H} \to \mathbb{F}^n$ is the **pseudoinverse** of V (see Exercises 3.2 and 3.3). The coefficients $c_j = \langle f, S^{-1} f_j \rangle$ are *linear* functions of f, whose Riesz representers

$$g_j = \tilde{f}_j := S^{-1} f_j$$

will be called the *dual frame*.

S.F.D. Waldron, *An Introduction to Finite Tight Frames*, Applied and Numerical Harmonic Analysis, https://doi.org/10.1007/978-0-8176-4815-2_3

3.2 A frame and its dual

A finite frame $(f_j)_{j\in J}$ for $\mathcal{H}$ is simply a spanning set (see Exer. 3.6), i.e.,

$$V = [f_j]_{j\in J} : \ell_2(J) \to \mathcal{H} \text{ is } onto\ \mathcal{H},$$

which implies that $S = VV^*$ is (boundedly) invertible. For $\mathcal{H}$ infinite dimensional (or J infinite), the following condition ensures that V and S can be defined, such that V is onto and S has a bounded inverse.

Definition 3.1. A countable sequence $(f_j)_{j\in J}$ in a Hilbert space is said to be a **frame** (for $\mathcal{H}$) if there exists (frame bounds) $A, B > 0$, such that

$$A\|f\|^2 \le \sum_{j\in J} |\langle f, f_j\rangle|^2 \le B\|f\|^2, \qquad \forall f \in \mathcal{H}. \tag{3.1}$$

The best possible A, B are called **the (optimal) frame bounds**.

A finite frame for $\mathcal{H}$ is precisely a spanning sequence for $\mathcal{H}$.

Given a frame $\Phi = (f_j)$ for $\mathcal{H}$, we recall the **frame operator** $S = S_\Phi : \mathcal{H} \to \mathcal{H}$ is the *self-adjoint operator* defined by $S := VV^*$, i.e.,

$$Sf := \sum_{j\in J} \langle f, f_j\rangle f_j, \qquad \forall f \in \mathcal{H},$$

where $V := [f_j]_{j\in J}$ is the synthesis operator. The frame bounds (3.1) can be written

$$\langle Af, f\rangle \le \langle Sf, f\rangle \le \langle Bf, f\rangle, \quad \forall f \in \mathcal{H} \quad \iff \quad AI_{\mathcal{H}} \le S \le BI_{\mathcal{H}}. \tag{3.2}$$

From this, it follows that S is *positive*, with a *bounded inverse* satisfying

$$\frac{1}{B} I_{\mathcal{H}} \le S^{-1} \le \frac{1}{A} I_{\mathcal{H}}$$

and the optimal frame bounds are (cf Exer. 3.7)

$$A = A_\Phi := \text{smallest eigenvalue of } S, \qquad B = B_\Phi := \text{largest eigenvalue of } S.$$

We can therefore define the dual frame, which is indeed a frame.

Definition 3.2. Given a frame $\Phi = (f_j)_{j\in J}$ for $\mathcal{H}$, with frame operator S, the **(canonical) dual frame**[1] $\tilde{\Phi} = (g_j)_{j\in J}$ for $\mathcal{H}$ is defined by

$$g_j = \tilde{f}_j := S^{-1} f_j.$$

---

[1] Sometimes the *dual frame* is called the *canonical dual frame* to distinguish it from a so-called *alternate dual frame* (see §3.10).

The synthesis operator for the dual frame $\tilde{\Phi}$ is $W=[\tilde{f}_j]=S^{-1}V$, and so

$$S_{\tilde{\Phi}}=WW^*=(S^{-1}V)V^*S^{-1}=S^{-1}=S_{\Phi}^{-1}. \tag{3.3}$$

Thus the dual frame $\Psi=\tilde{\Phi}$ *is a frame*, with $\tilde{\Psi}=\Phi$, and optimal frame bounds

$$A_{\tilde{\Phi}}=\frac{1}{B_{\Phi}}, \qquad B_{\tilde{\Phi}}=\frac{1}{A_{\Phi}}.$$

The Gramian of a frame and its dual are pseudoinverses of each other (Exer. 3.3)

$$\mathrm{Gram}(\tilde{\Phi})=\mathrm{Gram}(\Phi)^{\dagger}, \qquad \mathrm{Gram}(\Phi)=\mathrm{Gram}(\tilde{\Phi})^{\dagger}. \tag{3.4}$$

Proposition 3.1. *Let $(f_j)_{j\in J}$ be a finite frame for $\mathscr{H}$, i.e., a spanning set, and let $\tilde{f}_j:=S^{-1}f_j$ be the dual frame. Then we have the frame expansion*

$$f=\sum_{j\in J}\langle f,\tilde{f}_j\rangle f_j=\sum_{j\in J}\langle f,f_j\rangle \tilde{f}_j, \qquad \forall f\in\mathscr{H}. \tag{3.5}$$

Moreover, suppose that $f=\sum_j c_jf_j$, for some coefficients c_j, then

$$\sum_{j\in J}|c_j|^2=\sum_{j\in J}|\langle f,\tilde{f}_j\rangle|^2+\sum_{j\in J}|c_j-\langle f,\tilde{f}_j\rangle|^2. \tag{3.6}$$

Proof. Since $[\tilde{f}_j]_{j\in J}=S^{-1}V$, $V:=[f_j]_{j\in J}$, the equation (3.5) can be written

$$I_{\mathscr{H}}=V(S^{-1}V)^*=(S^{-1}V)V^*, \qquad S:=VV^*,$$

which clearly holds. Suppose $f=Vc$, and write c as the least squares solution, plus the error

$$c=V^*S^{-1}f+(c-V^*S^{-1}f).$$

These vectors are orthogonal (see Exer. 3.2) since

$$\langle V^*S^{-1}f,c-V^*S^{-1}f\rangle=\langle S^{-1}f,Vc-VV^*S^{-1}f\rangle=\langle S^{-1}f,f-f\rangle=0,$$

and so, by Pythagoras, $\|c\|^2=\|V^*S^{-1}f\|^2+\|c-V^*S^{-1}f\|^2$, which gives (3.6), since $(V^*S^{-1}f)_j=((S^{-1}V)^*f)_j=\langle f,\tilde{f}_j\rangle$. □

It follows from Proposition 3.1 that given a finite spanning sequence (f_j) for a vector space (over $\mathbb{R}$ or $\mathbb{C}$), for each vector f there are unique coefficients (c_j) with $f=\sum_j c_jf_j$ and $\sum_j|c_j|^2$ minimal. These c_j are linear functions of f and are called the *canonical coordinates* of f (see § 4).

Example 3.1. For (f_j) a basis for $\mathscr{H}$, (3.5) implies that $(\langle\cdot,\tilde{f}_j\rangle)$ is the corresponding *dual basis* (*coordinate functionals*), i.e., (f_j) and $(\tilde{f}_j)$ form a **biorthogonal system**

$$\langle f_j,\tilde{f}_k\rangle=\begin{cases}1, & j=k;\\ 0, & j\neq k.\end{cases}$$

Example 3.2. If (f_j) is a tight frame, then (3.2) reduces to $S = AI_{\mathcal{H}}$, and so the dual frame is given by $\tilde{f}_j = S^{-1} f_j = \frac{1}{A} f_j$.

Example 3.3. Consider the frame $\Phi = (e_1, \alpha e_2, \beta e_2)$, $\alpha^2 + \beta^2 > 0$, for $\mathbb{R}^2$. Then

$$V = \begin{bmatrix} 1 & 0 & 0 \\ 0 & \alpha & \beta \end{bmatrix}, \quad S = \begin{bmatrix} 1 & 0 \\ 0 & \alpha^2+\beta^2 \end{bmatrix}, \quad W = S^{-1}V = \begin{bmatrix} 1 & 0 & 0 \\ 0 & \frac{\alpha}{\alpha^2+\beta^2} & \frac{\beta}{\alpha^2+\beta^2} \end{bmatrix},$$

so the dual frame is $\tilde{\Phi} = (e_1, \frac{\alpha}{\alpha^2+\beta^2} e_2, \frac{\beta}{\alpha^2+\beta^2} e_2)$, and (3.5) gives the expansion

$$x = \begin{pmatrix} x_1 \\ x_2 \end{pmatrix} = x_1 e_1 + \frac{\alpha x_2}{\alpha^2+\beta^2}(\alpha e_2) + \frac{\beta x_2}{\alpha^2+\beta^2}(\beta e_2), \qquad \forall x \in \mathbb{R}^2.$$

Other choices for $\tilde{\Phi}$ that satisfy (3.5) include $(e_1, \frac{1}{\alpha} e_2, 0)$ and $(e_1, 0, \frac{1}{\beta} e_2)$ $(\alpha, \beta \neq 0)$, which give the (orthogonal) expansions

$$x = x_1 e_1 + \frac{x_2}{\alpha}(\alpha e_2) + 0(\beta e_2) = x_1 e_1 + 0(\alpha e_2) + \frac{x_2}{\beta}(\beta e_2), \qquad \forall x \in \mathbb{R}^2.$$

Observe that these coefficients have greater ℓ_2-norm than those for the dual frame.

This illustrates the tendency of a frame to distribute information about a function *evenly* over the coefficients, rather than concentrate it on a few. Indeed, since S is invertible, it follows that $f_j \neq 0$ if and only if $\tilde{f}_j \neq 0$.

Example 3.4. If (f_j) is a normalised tight frame for $\mathcal{H}$ and $T : \mathcal{H} \to \mathcal{H}$ is an invertible linear map, then the frames $(T^* f_j)$ and $(T^{-1} f_j)$ are dual (see Exer. 3.4).

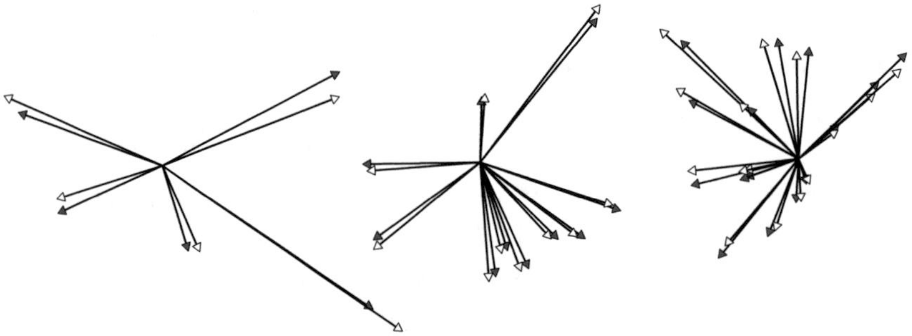

Fig. 3.1: A frame (solid heads) and its dual, for 5, 10 and 15 random vectors in $\mathbb{R}^2$.

Example 3.5. (Exer. 3.5) The image $\Psi = Q\Phi$ of a frame Φ under a linear map $Q : \mathcal{H} \to \mathcal{K}$ is again a frame (for its span), with frame bounds

$$A_\Psi \geq A_\Phi \|Q^\dagger\|^{-2}, \qquad B_\Psi \leq B_\Phi \|Q\|^2.$$

3.3 The canonical tight frame

Given a frame (or its dual), there is a naturally associated *tight frame*, which inherits properties of the original (see, e.g., Theorem 10.1).

Definition 3.3. If $\Phi = (f_j)$ is a frame for $\mathscr{H}$, with frame operator $S = S_\Phi$, then the corresponding **canonical tight frame** $\Phi^{\mathrm{can}} = (f_j^{\mathrm{can}})$ is given by

$$g_j = f_j^{\mathrm{can}} := S^{-\frac{1}{2}} f_j, \qquad \forall j.$$

To find the canonical tight frame, one must calculate $S^{-\frac{1}{2}}$, the positive square root of the inverse of the frame operator S. In practice, this may be a difficult numerical or analytical calculation (see §3.9). If $V = U_1 \Sigma U_2^*$ is a singular value decomposition of $V = [f_j]$, then (see Exer. 3.9) the canonical tight frame is given by

$$[f_j^{\mathrm{can}}] = U_1 \begin{bmatrix} I & 0 \end{bmatrix} U_2^* = U_1 U_2^*. \tag{3.7}$$

Theorem 3.1. *Let $\Phi = (f_j)_{j \in J}$ be a finite frame for $\mathscr{H}$. Then the canonical tight frame $f_j^{\mathrm{can}} := S_\Phi^{-\frac{1}{2}} f_j$ is a normalised tight frame, i.e.,*

$$f = \sum_{j \in J} \langle f, f_j^{\mathrm{can}} \rangle f_j^{\mathrm{can}}, \qquad \forall f \in \mathscr{H}. \tag{3.8}$$

Moreover, $(\tilde{\Phi})^{\mathrm{can}} = \Phi^{\mathrm{can}}$, and the Gramian is the orthogonal projection matrix with the same kernel as the synthesis operator $V = [f_j]$, and satisfies

$$\mathrm{Gram}(\Phi^{\mathrm{can}}) = \mathrm{Gram}(\Phi)\,\mathrm{Gram}(\tilde{\Phi}) = \mathrm{Gram}(\tilde{\Phi})\,\mathrm{Gram}(\Phi). \tag{3.9}$$

Proof. Let $U = [f_j^{\mathrm{can}}]_{j \in J} = S^{-\frac{1}{2}} V$ be the synthesis operator of (f_j^{can}). Then

$$UU^* = (S^{-\frac{1}{2}} V)(V^* S^{-\frac{1}{2}}) = S^{-\frac{1}{2}} S S^{-\frac{1}{2}} = I_{\mathscr{H}},$$

which is (3.8). The other observations follow similarly (see Exer. 3.10). □

Example 3.6. If (f_j) is a basis for $\mathscr{H}$, then the canonical tight frame is an orthonormal basis known as the **symmetric** or **Löwdin orthogonalisation** of (f_j). It was first obtained via a *symmetric* version of the Gram–Schmidt algorithm and is the closest orthonormal basis to (f_j) (see Corollary 3.3).

Example 3.7. If (f_j) is an equal-norm frame for $\mathscr{H}$, then the dual frame and the canonical tight frame need not have vectors of equal length, e.g., the canonical dual of the frame of Example 3.3 is $\Phi^{\mathrm{can}} = (e_1, \frac{\alpha}{\sqrt{\alpha^2+\beta^2}} e_2, \frac{\beta}{\sqrt{\alpha^2+\beta^2}} e_2)$ (take $\alpha = \beta = 1$). They do satisfy (see Exer. 3.13)

$$\langle f_j, \tilde{f}_j \rangle = \langle \tilde{f}_j, f_j \rangle = \|f_j^{\mathrm{can}}\|^2, \ \forall j \quad \Longrightarrow \quad \sum_{j \in J} \langle f_j, \tilde{f}_j \rangle = d = \dim(\mathscr{H}).$$

Example 3.8. (Exer. 3.16) If $\Phi = (f_j)$ is a frame for $\mathscr{H}$, and $Q : \mathscr{H} \to \mathscr{H}$ is an invertible linear map, then $\Psi := Q\Phi = (Qf_j)$ is a frame for $\mathscr{H}$, and

$$\Psi^{\mathrm{can}} = U\Phi^{\mathrm{can}},$$

where U is unitary. Equivalently, $\mathrm{Gram}(\Psi^{\mathrm{can}}) = \mathrm{Gram}(\Phi^{\mathrm{can}})$.

The synthesis operator of the canonical tight frame for $\Phi = (f_j)_{j\in J}$

$$U = [f_j^{\mathrm{can}}]_{j\in J} = S^{-\frac{1}{2}}V : \ell_2(J) \to \mathscr{H}$$

is a *partial isometry*, which appears in the polar decomposition of $V = [f_j]_{j\in J}$.

Corollary 3.1. *(Polar decomposition). Let $\Phi = (f_j)_{j\in J}$ be a finite frame for $\mathscr{H}$, and $U = [f_j^{\mathrm{can}}]_{j\in J}$. Then the polar decomposition of the synthesis operator $V = [f_j]_{j\in J}$ into a product of a partial isometry and a positive operator is*

$$V = UG^{\frac{1}{2}} = S^{\frac{1}{2}}U, \qquad G = \mathrm{Gram}(\Phi) = V^*V, \quad S = S_\Phi = VV^*. \tag{3.10}$$

Proof. Let $V = U_1\Sigma U_2^*$ be a singular value decomposition of V. Then (3.10) is the polar decomposition of $A = V$, where

$$U = U_1U_2^*, \qquad P = (V^*V)^{\frac{1}{2}} = G^{\frac{1}{2}}, \qquad Q = (VV^*)^{\frac{1}{2}} = S^{\frac{1}{2}},$$

and we recognise that U is $[f_j^{\mathrm{can}}]$, by (3.7). □

The Gramian of the canonical tight frame is an orthogonal projection

$$P = \mathrm{Gram}(\Phi^{\mathrm{can}}) = V^*S^{-1}V : \ell_2(J) \to \ell_2(J), \tag{3.11}$$

which gives the coefficients c with $f = \sum_j c_j f_j$ of minimal ℓ_2-norm.

Corollary 3.2. *Let $\Phi = (f_j)_{j\in J}$ be a finite frame for $\mathscr{H}$, and $V = [f_j]_{j\in J}$. If $f = Va$, then the coefficients c of minimal ℓ_2-norm with $f = Vc$ are given by $c = \mathrm{Gram}(\Phi^{\mathrm{can}})a$. In particular, V can be expanded*

$$V = V\,\mathrm{Gram}(\Phi^{\mathrm{can}}), \tag{3.12}$$

where

$$\mathrm{ran}(V^*) = \ker(V)^\perp = \mathrm{ran}(\mathrm{Gram}(\Phi^{\mathrm{can}})). \tag{3.13}$$

Proof. By Proposition 3.1, the coefficients c of minimal ℓ_2-norm are given by

$$c = V^*S^{-1}f = V^*S^{-1}Va = (S^{-\frac{1}{2}}V)^*(S^{-\frac{1}{2}}V)a = \mathrm{Gram}(\Phi^{\mathrm{can}})a. \tag{3.14}$$

Left multiplying $c = V^*S^{-1}Va = \mathrm{Gram}(\Phi^{\mathrm{can}})a$ by V gives

$$VV^*S^{-1}Va = Va = V\,\mathrm{Gram}(\Phi^{\mathrm{can}})a, \quad \forall a \quad \Longrightarrow \quad V = V\,\mathrm{Gram}(\Phi^{\mathrm{can}}).$$

Finally, since S is invertible and V is onto, we obtain (3.13) from (3.11). □

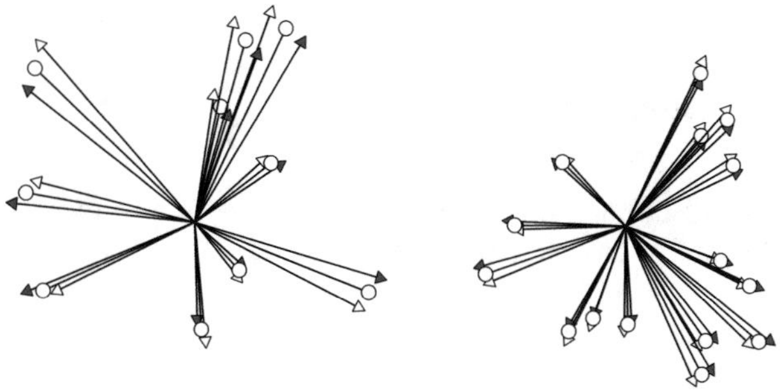

Fig. 3.2: A frame of n randomly chosen unit vectors (solid arrowheads), its dual and the canonical tight frame (circles), for $n=8$ and $n=16$.

Thus, if $V=[f_j]$ is the synthesis operator of a finite frame $\Phi=(f_j)$, then

$$V = V\,\mathrm{Gram}(\Phi^{\mathrm{can}}) = U\,\mathrm{Gram}(\Phi)^{\frac{1}{2}} = S_\Phi^{\frac{1}{2}}U,$$

where $\mathrm{Gram}(\Phi^{\mathrm{can}})$ is an orthogonal projection, $U:=[f_j^{\mathrm{can}}]$ is a coisometry, $\mathrm{Gram}(\Phi)$ is positive semidefinite, and S_Φ is positive definite.

For a given a frame $(f_j)_{j=1}^n$, the canonical tight frame is the *closest* tight frame $(g_j)_{j=1}^n$ in the sense of minimising the least squares error

$$\sum_{j=1}^n \|f_j-g_j\|^2 = \|[f_j]_{j=1}^n-[g_j]_{j=1}^n\|_F^2.$$

Theorem 3.2. *Let $\Phi=(f_j)_{j=1}^n$ be a frame for $\mathscr{H}$, and $\lambda_1,\ldots,\lambda_d$ be the eigenvalues of S_Φ. If $\Psi=(g_j)_{j=1}^n$ is a tight frame for $\mathscr{H}$, with frame bound A, then*

$$\sum_{j=1}^n \|f_j-g_j\|^2 \ge \sum_{k=1}^d (\sqrt{\lambda_k}-\sqrt{A})^2, \qquad (3.15)$$

with equality if and only if $\Psi=\sqrt{A}\Phi^{\mathrm{can}}$.

Proof. With $V=[f_j]_{j=1}^n$, $W=[g_j]_{j=1}^n$, we compute

$$\begin{aligned}\textstyle\sum_j \|f_j-g_j\|^2 &= \mathrm{trace}((V-W)^*(V-W))\\ &= \mathrm{trace}(V^*V)+\mathrm{trace}(W^*W)-\mathrm{trace}(W^*V)-\mathrm{trace}(V^*W)\\ &= \mathrm{trace}(VV^*)+\mathrm{trace}(WW^*)-\mathrm{trace}(W^*V)-\overline{\mathrm{trace}(W^*V)}\\ &= \mathrm{trace}(S_\Phi)+\mathrm{trace}(S_\Psi)-2\Re\,\mathrm{trace}(W^*V)\\ &= \textstyle\sum_k \lambda_k+\sum_k A-2\Re\,\mathrm{trace}(W^*V).\end{aligned}$$

Hence we must prove that

$$\Re(\text{trace}(W^*V)) = \Re\sum_j \langle f_j, g_j\rangle \le \sum_k \sqrt{\lambda_k}\sqrt{A}. \tag{3.16}$$

Let $u_1,\ldots,u_d$ an orthonormal basis of eigenvectors for S_Φ corresponding to the eigenvalues $\lambda_1,\ldots,\lambda_d$. Then (3.16) can be expanded

$$\begin{aligned}\Re\sum_{j=1}^n \langle f_j, g_j\rangle &= \Re\sum_{j=1}^n\sum_{k=1}^d \langle f_j, u_k\rangle\langle u_k, g_j\rangle \\ &= \sum_{k=1}^d \Re\sum_{j=1}^n \langle f_j, u_k\rangle\overline{\langle g_j, u_k\rangle} \le \sum_{k=1}^d \sqrt{\lambda_k}\sqrt{A}.\end{aligned}$$

It would therefore suffice to prove that

$$\Re\sum_{j=1}^n \langle f_j, u_k\rangle\overline{\langle g_j, u_k\rangle} \le \sqrt{\lambda_k}\sqrt{A}, \qquad \forall k$$

which we now do. Firstly,

$$\Re\sum_{j=1}^n \langle f_j, u_k\rangle\overline{\langle g_j, u_k\rangle} \le \Big|\sum_{j=1}^n \langle f_j, u_k\rangle\overline{\langle g_j, u_k\rangle}\Big|,$$

with equality if and only if

$$\sum_{j=1}^n \langle f_j, u_k\rangle\overline{\langle g_j, u_k\rangle} = \Big|\sum_{j=1}^n \langle f_j, u_k\rangle\overline{\langle g_j, u_k\rangle}\Big| \ge 0. \tag{3.17}$$

By the Cauchy–Schwarz inequality,

$$\begin{aligned}\Big|\sum_{j=1}^n \langle f_j, u_k\rangle\overline{\langle g_j, u_k\rangle}\Big| &\le \Big(\sum_j |\langle f_j, u_k\rangle|^2\Big)^{\frac{1}{2}}\Big(\sum_j |\langle g_j, u_k\rangle|^2\Big)^{\frac{1}{2}} \\ &= \sqrt{\langle S_\Phi u_k, u_k\rangle}\sqrt{\langle S_\Psi u_k, u_k\rangle} = \sqrt{\lambda_k}\sqrt{A},\end{aligned}$$

with equality if and only if

$$\langle f_j, u_k\rangle = c_k\langle g_j, u_k\rangle, \quad \forall j, \qquad |c_k A| = \sqrt{\lambda_k}\sqrt{A}. \tag{3.18}$$

Thus we obtain the desired inequality (3.15).

There is equality in (3.15) if and only if (3.17) and (3.18) hold. These together imply $c_k = \sqrt{\lambda_k}/\sqrt{A}$, and hence

$$g_j = \sum_{k=1}^{d} \langle g_j, u_k \rangle u_k = \sum_{k=1}^{d} \frac{\sqrt{A}}{\sqrt{\lambda_k}} \langle f_j, u_k \rangle u_k = \frac{\sqrt{A}}{\sqrt{\lambda_k}} \sum_{k=1}^{d} \langle S_\Phi^{-\frac{1}{2}} f_j, S_\Phi^{\frac{1}{2}} u_k \rangle u_k$$

$$= \frac{\sqrt{A}}{\sqrt{\lambda_k}} \sum_{k=1}^{d} \langle S_\Phi^{-\frac{1}{2}} f_j, \sqrt{\lambda_k} u_k \rangle u_k = \sqrt{A} S_\Phi^{-\frac{1}{2}} f_j = \sqrt{A} f_j^{\mathrm{can}},$$

as claimed. □

Corollary 3.3. *Let $\Phi = (f_j)_{j=1}^n$ be a frame for $\mathscr{H}$, and $\lambda_1, \ldots, \lambda_d$ be the eigenvalues of the frame operator $S = S_\Phi$. Then*

$$\min\{\sum_{j=1}^{n} \|f_j - g_j\|^2 : \Psi = (g_j) \text{ is a tight frame for } \mathscr{H}\} = \sum_{k=1}^{d} \lambda_k - \frac{1}{d}\Big(\sum_{k=1}^{d} \sqrt{\lambda_k}\Big)^2,$$

which is attained if and only if

$$\Psi = \sqrt{A}\Phi^{\mathrm{can}}, \qquad \sqrt{A} := \frac{1}{d}\sum_{k=1}^{d} \sqrt{\lambda_k}.$$

Proof. The minimum of $\sqrt{A} \mapsto \sum_k (\sqrt{\lambda_k} - \sqrt{A})^2$ occurs when its derivative is zero:

$$-2\sum_{k=1}^{d} (\sqrt{\lambda_k} - \sqrt{A}) = 0 \quad \Longrightarrow \quad \sqrt{A} = \frac{1}{d}\sum_{k=1}^{d} \sqrt{\lambda_k},$$

and this minimum value $\sum_k (\sqrt{\lambda_k} - \sqrt{A})^2$ can be simplified as claimed. □

In terms of matrices (see Exer. 3.17), this result says that the minimum

$$\min_{\substack{W \in \mathbb{F}^{d\times n},\, A>0 \\ WW^* = AI}} \|V - W\|_F, \qquad \operatorname{rank}(V) = d$$

is uniquely attained for

$$W = \frac{\operatorname{trace}((VV^*)^{\frac{1}{2}})}{d}(VV^*)^{-\frac{1}{2}} V = \frac{\operatorname{trace}(S^{\frac{1}{2}})}{d} S^{-\frac{1}{2}} V.$$

Let $V = [f_j]$ be a finite frame for $\mathbb{F}^d$. Then there is a unique tight frame closest to it given by

$$\sqrt{A}(f_j^{\mathrm{can}}), \qquad \sqrt{A} = \operatorname{trace}((VV^*)^{\frac{1}{2}})/d.$$

3.4 Unitarily equivalent frames

As for tight frames, we say that finite frames $\Phi = (f_j)_{j\in J}$ and $\Psi = (g_j)_{j\in J}$, with the same index set J, are **unitarily equivalent** if there is a $c > 0$ and a unitary U for which

$$g_j = cUf_j, \qquad \forall j \in J.$$

The dual and canonical tight frames are related as follows (see Exer. 3.16)

$$\tilde{g}_j = \frac{1}{c}U\tilde{f}_j, \quad g_j^{\mathrm{can}} = Uf_j^{\mathrm{can}}, \qquad \forall j \in J.$$

Moreover, the Gramians satisfy $\mathrm{Gram}(\Psi) = c^2\,\mathrm{Gram}(\Phi)$, which gives the following generalisation of Corollary 2.1.

Unitarily equivalent frames are uniquely determined by their Gramians (up to a positive scalar multiplier).

The Gramian of a normalised tight frame is a projection matrix, and vice versa. Any positive semidefinite matrix is the Gramian of some frame.

Theorem 3.3. *An $n \times n$ matrix $M = [m_{jk}]_{j,k\in J}$ is the Gramian matrix of a frame $\Phi = (f_j)_{j\in J}$ for the space $\mathscr{H} := \mathrm{span}\{f_j\}_{j\in J}$ if and only if it is positive semidefinite, i.e., $M = M^*$ and $\langle Mf, f\rangle \geq 0$, $\forall f \in \mathscr{H}$. Moreover,*

$$d = \dim(\mathscr{H}) = \mathrm{rank}(M), \tag{3.19}$$

and a copy of the frame is given by the columns of any matrix L for which $L^*L = M$ (e.g., the positive square root $L = M^{\frac{1}{2}}$, or a Cholesky factor $L = R$).

Proof. Since the Gramian can be factorised as $\mathrm{Gram}(\Phi) = V^*V$, $V = [f_j]_{j\in J}$, it is positive semidefinite: $\langle V^*Vf, f\rangle = \langle Vf, Vf\rangle \geq 0$.

Conversely, suppose that M is a positive semidefinite $n \times n$ matrix. Then M can be factored $M = L^*L$ for some $m \times n$ matrix. For example, let $m = n$, and take $L = R$ from a Cholesky factorisation $M = R^*R$ (which exists, though it may not be numerically stable to calculate), or with $M = UDU^*$ a unitary diagonalisation of M, take the positive square root $L = M^{\frac{1}{2}} := D^{\frac{1}{2}}U^*$. The columns of L are a frame with Gramian M, since

$$\langle Le_j, Le_k\rangle = \langle L^*Le_j, e_k\rangle = \langle Me_j, e_k\rangle = (k,j)\text{-entry of}\,M.$$

□

This result, i.e., that every positive semidefinite matrix is the Gramian of some sequence of vectors (which is a frame for its span) is well known (cf [Ros97]).

Example 3.9. (Exer. 3.19) A sequence of n unit vectors $\Phi = (v_j)_{j=1}^n$ in $\mathbb{R}^d$, $d > 1$, is said to be **θ-isogonal**[2] if

$$\langle v_j, v_k \rangle = a := \cos\theta, \qquad \forall j \neq k.$$

The eigenvalues of the Gramian M of such a sequence are $na - a + 1$ and $1 - a$ (multiplicity $n-1$). Since the Gramian is positive semidefinite, $\frac{-1}{n-1} \leq a \leq 1$, and its rank is either 1 ($a = 1$), $n-1$ ($a = \frac{-1}{n-1}$), or n. Thus the only isogonal configurations up to unitary equivalence are the $d+1$ vertices of the regular simplex in $\mathbb{R}^d$, or a unique set of d vectors in $\mathbb{R}^d$ with $\frac{-1}{n-1} < a < 1$. In the latter case

$$\mathrm{Gram}(\tilde{\Phi}) = M^\dagger = \begin{pmatrix} b & c & \cdots & c \\ c & b & & c \\ \vdots & & \ddots & \vdots \\ c & c & \cdots & b \end{pmatrix}, \qquad \frac{c}{b} = \frac{-a}{na - 2a + 1},$$

and so the dual frame $\tilde{\Phi}$ is isogonal, and it is unitarily equivalent to Φ if and only if $a = 0$, i.e., Φ is an orthonormal basis.

Remark 3.1. As for tight frames, a frame is said to be **real** if its Gramian is real, and to be **complex** otherwise. A frame is real if and only if its dual is, in which case the canonical tight frame is real also. However, if the canonical tight frame is real, then the frame itself need not be real (see Exer. 3.20).

3.5 Similar frames and orthogonal frames

We define a second equivalence relation on frames, for which each equivalence class contains a unique tight frame (up to unitary equivalence).

Definition 3.4. Frames $\Phi = (f_j)_{j\in J}$ and $\Psi = (g_j)_{j\in J}$ for $\mathcal{H}$ and $\mathcal{K}$, with the same index set J, are said to be **similar** if there is an invertible linear map $Q : \mathcal{H} \to \mathcal{K}$ such that

$$g_j = Qf_j, \qquad \forall j \in J.$$

For frames which are not tight, this equivalence relation is weaker than unitary equivalence. Indeed, a frame $\Phi = (f_j)$, its dual frame and canonical tight frame are all similar, since

$$\tilde{f}_j = S^{-1} f_j, \qquad f_j^{\mathrm{can}} = S^{-\frac{1}{2}} f_j,$$

but they are unitarily equivalent if and only if Φ is tight.

Example 3.10. Every basis is similar to an orthonormal basis.

---

[2]Isogonal vectors appear in the structure of soap films and bubbles, see [Mur93].

Theorem 3.4. *There is exactly one tight frame (up to unitary equivalence) in each equivalence class of similar frames, namely the canonical tight frame.*

Proof. Suppose $\Phi = (f_j)$ and $\Psi = (g_j)$ are similar, via Q, and $S = S_\Phi$. Then

$$g_j = Qf_j = QS^{\frac{1}{2}}S^{-\frac{1}{2}}f_j = Tf_j^{\mathrm{can}}, \qquad \forall j,$$

where $T = QS^{\frac{1}{2}}$ is invertible. Hence $\Psi = (Tf_j^{\mathrm{can}})$ is tight if and only if

$$S_\Psi = T[f_j^{\mathrm{can}}][f_j^{\mathrm{can}}]^*T^* = TT^* = c^2 I, \qquad c > 0,$$

i.e., $T = cU$, with U unitary, and Ψ is unitarily equivalent to Φ^{can}. □

In other words, the study of frames up to similarity reduces to the study of tight frames up to unitary equivalence.

Finite frames are similar if and only if their canonical Gramians are equal.

Since orthogonal projections, such as $P = \mathrm{Gram}(\Phi^{\mathrm{can}})$, are uniquely determined by their range (or kernel), Theorem 3.4 implies the equivalence classes of similar frames (or of unitarily equivalent tight frames) of n vectors for a d-dimensional space are in 1–1 correspondence with the d-dimensional subspaces of $\mathbb{F}^n$, i.e., points on the Grassmannian $\mathrm{Gr}_d(\mathbb{F}^n)$.

These considerations lead to many conditions equivalent to similarity.

Proposition 3.2. *(Similarity) Let $\Phi = (f_j)_{j\in J}$ and $\Psi = (g_j)_{j\in J}$ be finite frames with synthesis operators V and W. Then the following are equivalent*
(a) *Φ and Ψ are similar.*
(b) $\mathrm{Gram}(\Phi^{\mathrm{can}}) = \mathrm{Gram}(\Psi^{\mathrm{can}})$*, i.e.,* $V^*(VV^*)^{-1}V = W^*(WW^*)^{-1}W$.
(c) $\mathrm{ran}(V^*) = \mathrm{ran}(W^*)$*, or, equivalently,* $\ker(V) = \ker(W)$.
Furthermore, these imply the equivalent conditions
(d) *WV^* is invertible.*
(e) $\sum_j \langle f, f_j\rangle g_j \neq 0$, $\forall f \neq 0$.

Proof. Recall, from (3.13), that $\mathrm{ran}(V^*) = \mathrm{ran}(\mathrm{Gram}(\Phi^{\mathrm{can}}))$.
(a)$\Longleftrightarrow$(b) Theorem 3.4 and Corollary 2.1.
(b)$\Longleftrightarrow$(c) The orthogonal projections $\mathrm{Gram}(\Phi^{\mathrm{can}})$ and $\mathrm{Gram}(\Psi^{\mathrm{can}})$ (which are determined by their ranges) are equal if and only if $\mathrm{ran}(V^*) = \mathrm{ran}(W^*)$.
(c) $\Longrightarrow$ (d) If $\mathrm{ran}(V^*) = \mathrm{ran}(W^*)$, then $\mathrm{ran}(WV^*) = \mathrm{ran}(WW^*) = \mathrm{ran}(S_\Psi)$, so that WV^* onto, and hence invertible.
(d)$\Longleftrightarrow$(e) This is follows immediately since $WV^*f = \sum_j \langle f, f_j\rangle g_j$. □

At the other extreme, we consider frames which are far from being similar.

Definition 3.5. Frames $\Phi = (f_j)_{j\in J}$ and $\Psi = (g_j)_{j\in J}$ for $\mathcal{H}$ and $\mathcal{K}$, with the same index set J, are said to be **orthogonal** (or **strongly disjoint**) if

$$\mathrm{ran}(V^*) \perp \mathrm{ran}(W^*), \qquad V = [f_j],\ W = [g_j].$$

Orthogonal frames *cannot* be similar (similar frames have $\operatorname{ran}(V^*) = \operatorname{ran}(W^*)$). There are many conditions equivalent to orthogonality (see Lemma 5.1), including

$$\sum_j \langle f, f_j \rangle g_j = 0, \quad \forall f \in \mathcal{H} \quad \iff \quad WV^* = 0. \tag{3.20}$$

This can be written suggestively as

$$\sum_j g_j f_j^* = 0, \qquad f_j^* : \mathcal{H} \to \mathbb{F} : f \mapsto \langle f, f_j \rangle. \tag{3.21}$$

Example 3.11. (1-dimensional frames) For $\mathcal{H} = \mathcal{K} = \mathbb{F}$, the condition (3.21) is orthogonality of vectors. Hence the rows of any unitary matrix, e.g., the Fourier matrix (2.5), are orthogonal frames for $\mathbb{C}$.

Example 3.12. (Projections) Let (f_j) be a normalised tight frame for $\mathcal{H}$, and P, Q be orthogonal projections onto subspaces $\mathcal{H}_1$, $\mathcal{H}_2$, with $\mathcal{H}_1 \perp \mathcal{H}_2$, i.e., $PQ = 0$. Then

$$\Phi = (Pf_j), \qquad \Psi = (Qf_j)$$

are *orthogonal* normalised tight frames for $\mathcal{H}_1$ and $\mathcal{H}_2$, since

$$WV^* = QUU^*P^* = QP = 0, \qquad U = [f_j],\ V = [Pf_j] = PU,\ W = [Qf_j] = QU.$$

In §5.2, we will see that effectively all orthogonal frames appear in this way.

3.6 Frames as orthogonal projections

We have seen (in §2.6) that every finite normalised tight frame is the orthogonal projection of an orthonormal basis (tight frame without redundancy). Thus, it is natural to ask whether every frame and its dual are the projection of a biorthogonal system with the same frame bounds (cf Exer. 3.5).

Let (f_j) and (g_j) be a biorthogonal system for $\mathcal{K}$, i.e., (f_j) be a basis for $\mathcal{K}$, with (g_j) the dual basis uniquely determined by

$$\langle f_j, g_k \rangle = \delta_{jk} \quad \iff \quad VW^* = WV^* = I, \quad V := [f_j],\ W := [g_j],$$

and P be the orthogonal projection onto a subspace $\mathcal{H}$. Then (see Exer. 3.12)

$$f = \sum_j \langle f, Pg_j \rangle Pf_j = \sum_j \langle f, Pf_j \rangle Pg_j, \qquad \forall f \in \mathcal{H},$$

but the frames $\Phi = (Pf_j)$ and $\Psi = (Pg_j)$ for $\mathcal{H}$ need not be *the* (canonical) duals of each other (they are alternate duals in the sense of §3.10). They are dual if and only if the synthesis operators of $\tilde{\Phi}$ and Ψ are equal, i.e.,

$$(PV(PV)^*)^{-1}PV = PW = P(V^*)^{-1},$$

which is equivalent to

$$PS = PSP \iff SP = PSP \iff SP = PS, \qquad S := VV^*.$$

In other words (see Exer. 3.21):

Let (f_j) and (g_j) be a biorthogonal system for $\mathscr{K}$, $S = VV^*$, $V = [f_j]$, and P be the orthogonal projection of $\mathscr{K}$ onto a subspace $\mathscr{H}$. Then (Pf_j) and (Pg_j) are (canonically) dual frames for $\mathscr{H}$ if and only if

$$\mathscr{H} \text{ is an invariant subspace of } S, \text{ i.e., } S\mathscr{H} \subset \mathscr{H}.$$

Therefore, the answer to the above question takes the following form.

Theorem 3.5. *Every finite frame and its dual is the orthogonal projection of a biorthogonal system (onto an invariant subspace of frame operator). Moreover, the two bases making up the biorthogonal system can be taken to have the same frame bounds as their projections, i.e., the frame and its dual.*

Proof. Let $(\phi_j)_{j\in J}$ be a frame for $\mathscr{H}$, with frame bounds A and B. This will be the projection of a basis $(\phi_j + \psi_j)_{j\in J}$ in a larger Hilbert space $\mathscr{H} \oplus \mathscr{L}$.

To see what this might be, consider the condition that $\mathscr{H}$ be invariant under the frame operator $S = S_{(\phi_j+\psi_j)}$ of $(\phi_j + \psi_j)$, i.e.,

$$Sf = \sum_j \langle f, \phi_j + \psi_j \rangle (\phi_j + \psi_j) = \sum_j \langle f, \phi_j \rangle \phi_j + \sum_j \langle f, \phi_j \rangle \psi_j \in \mathscr{H}, \quad \forall f \in \mathscr{H}.$$

This is precisely the condition that the frames be orthogonal (Definition 3.5), i.e.,

$$\sum_j \langle f, \phi_j \rangle \psi_j = 0, \quad \forall f \in \mathscr{H} \iff \sum_j \langle g, \psi_j \rangle \phi_j = 0, \quad \forall g \in \mathscr{L}, \tag{3.22}$$

which implies (cf Theorem 5.1)

$$S_{(\phi_j+\psi_j)}(f+g) = S_{(\phi_j)}(f) + S_{(\psi_j)}(g), \qquad \forall f \in \mathscr{H}, g \in \mathscr{L}.$$

Hence the dual frame to (f_j), $f_j := \phi_j + \psi_j$, is

$$\tilde{f}_j = S^{-1}_{(\phi_j+\psi_j)}(\phi_j + \psi_j) = S^{-1}_{(\phi_j)}\phi_j + S^{-1}_{(\psi_j)}\psi_j = \tilde{\phi}_j + \tilde{\psi}_j,$$

and the orthogonal projections of f_j and $\tilde{f}_j$ onto $\mathscr{H}$ are ϕ_j and $\tilde{\phi}_j$.

Further, (3.22) implies

$$\sum_j |\langle f+g, \phi_j + \psi_j \rangle|^2 = \sum_j |\langle f, \phi_j \rangle|^2 + \sum_j |\langle g, \psi_j \rangle|^2, \qquad \forall f \in \mathscr{H}, g \in \mathscr{L},$$

so that $(\phi_j + \psi_j)$ will have the desired frame bounds (A and B) provided that

$$A\|g\|^2 \le \sum_j |\langle g, \psi_j\rangle|^2 \le B\|g\|^2, \qquad \forall g \in \mathscr{L}. \tag{3.23}$$

Let $V := [\phi_j]$. The bijection $[\phi_j + \psi_j] : \ell_2(J) \to \mathscr{H} \oplus \mathscr{L}$ maps the kernel of V onto $\mathscr{L}$ (by a dimension count). Hence, we take

$$\mathscr{L} := \ker(V) = \operatorname{ran}(V^*)^\perp,$$

and it therefore remains only to find $\psi_j \in \mathscr{L} = \ker(V)$ satisfying (3.22) and (3.23).

Let $Q : \mathscr{L} \to \ell_2(J)$ be positive (self-adjoint), with range $\mathscr{L}$, and eigenvalues between $\sqrt{A}$ and $\sqrt{B}$, e.g., $Qg := \lambda g$, $\forall g$, where $\sqrt{A} \le \lambda \le \sqrt{B}$. Then

$$\psi_j := Q^* e_j,$$

where $(e_j)_{j\in J}$ is the standard basis for $\ell_2(J)$, satisfies

$$\sum_j \langle g, \psi_j\rangle \phi_j = \sum_j \langle g, Q^* e_j\rangle \phi_j = \sum_j \langle Qg, e_j\rangle \phi_j = V(Qg) = 0,$$

$$A\|g\|^2 \le \sum_j |\langle g, \psi_j\rangle|^2 = \sum_j |\langle g, Q^* e_j\rangle|^2 = \sum_j |\langle Qg, e_j\rangle|^2 = \|Qg\|^2 \le B\|g\|^2,$$

which completes the proof. □

The sum $(\phi_j + \psi_j)$ is the motivating example of a *direct sum* (see §5.2). The above argument holds with $\mathscr{L}$ a proper subspace of $\ker(V)$, with the only difference being that $(\phi_j + \psi_j)$ is not a basis. It also extends to frames for infinite dimensional $\mathscr{H}$, where the role of a basis (and dual basis) is replaced by that of a *Riesz basis* (and *dual Riesz basis*). We give the relevant facts (cf [Chr03]).

Definition 3.6. A sequence of vectors (f_j) is a **Riesz basis** for a Hilbert space $\mathscr{H}$ if it is a perturbation of an orthonormal basis (e_j) for $\mathscr{H}$, i.e., there is a bounded invertible operator $T : \mathscr{H} \to \mathscr{H}$, such that $f_j = Te_j$, $\forall j$.

Equivalently, (f_j) is a Riesz basis if it is an unconditional basis with

$$0 < \inf_j \|f_j\| \le \sup_j \|f_j\| < \infty.$$

From the orthogonal expansion, we have

$$f = T(T^{-1}f) = T\sum_j \langle T^{-1}f, e_j\rangle e_j = \sum_j \langle f, (T^{-1})^* e_j\rangle Te_j,$$

and similarly for T replaced by $(T^{-1})^*$. This gives the *biorthogonal expansion*

$$f = \sum_j \langle f, g_j\rangle f_j = \sum_j \langle f, f_j\rangle g_j, \qquad \forall f \in \mathscr{H},$$

where $g_j := (T^*)^{-1} e_j = (T^{-1})^* e_j$ is called the **dual Riesz basis** (cf Example 3.4).

3.7 Condition numbers and the frame bounds

The frame bounds for (f_j) imply, and are equivalent to, a number of similar bounds for the various maps obtainable from $V = [f_j]$ (see Exer. 3.24).

Proposition 3.3. *Let $\Phi = (f_j)_{j\in J}$ be a finite sequence in $\mathcal{H}$, and $V = [f_j]$, $S = VV^*$. The frame bounds (3.1) are equivalent to the inequalities*

$$A\|f\| \le \|Sf\| \le B\|f\|, \qquad \forall f \in \mathcal{H}, \tag{3.24}$$

$$\sqrt{A}\|f\| \le \|V^*f\| \le \sqrt{B}\|f\|, \qquad \forall f \in \mathcal{H}, \tag{3.25}$$

$$A\|c\| \le \|\mathrm{Gram}(\Phi)c\| \le B\|c\|, \qquad \forall c \in \mathrm{ran}(\mathrm{Gram}(\Phi)), \tag{3.26}$$

$$\sqrt{A}\|c\| \le \|Vc\| \le \sqrt{B}\|c\|, \qquad \forall c \in \mathrm{ran}(V^*), \tag{3.27}$$

which are sharp if and only if A and B are the optimal frame bounds, respectively.

The frame expansion (3.5) of a function $f \in \mathcal{H}$ involves the representation

$$f = Vc = \sum_{j\in J} c_j f_j, \qquad c = V^*S^{-1}f \in \mathrm{ran}(V^*).$$

When f is constructed from c in this way, there will be rounding errors which yield a perturbed vector $c+\delta c$. The perturbation will have a component δa in $\mathrm{ran}(V^*)$, and a component $\delta b \in \ker(V) = \mathrm{ran}(V^*)^\perp$. Hence, we will obtain a perturbed function

$$f + \delta f = V(c+\delta c) = V(c+\delta a+\delta b) = Vc + V(\delta a), \qquad \delta a \in \mathrm{ran}(V^*).$$

By (3.27), the relative error in the computed f satisfies

$$\frac{\sqrt{A}\|\delta a\|}{\sqrt{B}\|c\|} \le \frac{\|\delta f\|}{\|f\|} \le \frac{\sqrt{B}\|\delta a\|}{\sqrt{A}\|c\|} \le \frac{\sqrt{B}\|\delta c\|}{\sqrt{A}\|c\|}. \tag{3.28}$$

where A and B are the frame bounds. This estimate motivates the following.

Definition 3.7. The **condition number** (of the frame expansion) of a frame Φ, with frame bounds A and B, is

$$\mathrm{cond}(\Phi) := \sqrt{\frac{B}{A}} \ge 1.$$

We observe that a frame has condition number 1 if and only if it is tight, and that a frame and its dual have the same condition number, since

$$\mathrm{cond}(\tilde{\Phi}) = \frac{1/\sqrt{A}}{1/\sqrt{B}} = \frac{\sqrt{B}}{\sqrt{A}} = \mathrm{cond}(\Phi).$$

In view of (3.28), the relative error in the computed value $f+\delta f$ of f is bounded by the condition number times the relative error in the frame coefficients c. If Φ is a basis, then $\text{cond}(\Phi) = \|V\|\|V^{-1}\|$, the usual condition number of a basis, and V has kernel 0, so that $\delta a = \delta c$, and we obtain the lower estimate on the relative error

$$\frac{1}{\text{cond}(\Phi)} \frac{\|\delta c\|}{\|c\|} \leq \frac{\|\delta f\|}{\|f\|}.$$

In contrast, for a frame that is *not* a basis, a nonzero perturbation $\delta c \in \ker(V)$ in the coefficients leads to *no error* in the constructed f.

Another measure of how errors propagate in calculations with a frame (and its dual) is the **condition number** of the frame operator $S = S_\Phi$, i.e.,

$$\text{cond}(S_\Phi) := \|S\|\|S^{-1}\| = \frac{B}{A} = \text{cond}(\Phi)^2 \geq 1.$$

Thus, either $\text{cond}(\Phi)$ or $\text{cond}(S_\Phi)$, and its distance from 1, can be used as a measure of how well conditioned calculations with the frame Φ are.

Both the *redundancy* and the *tightness* (as measured by the condition number) of a frame make it better conditioned than a basis.

3.8 Normalising frames and the distances between them

Here we consider whether or not there is a natural *normalisation* of a frame, which extends that of a tight frame. Since the canonical tight frame of a tight frame is its normalised version (which is its own dual), we would hope that a normalised frame and its dual are both *close*, in some sense, to the canonical tight frame.

There does seem to be one normalisation which captures *all* the properties one might reasonably hope for, and so we proceed directly to it. We then investigate how close this normalised frame and its dual are to the canonical tight frame for the following metrics

$$\text{dist}(\Phi,\Psi) = \|\text{Gram}(\Phi) - \text{Gram}(\Psi)\| \quad \text{(frames up to unitary equivalence)}$$

$$\text{dist}_B(\Phi, Q\Phi) = \log(\max(\|I-Q\|, \|I-Q^{-1}\|)+1) \qquad \text{(for similar frames)}.$$

If a frame $\Phi = (f_j)$ with frame bounds A and B is multiplied by a scalar $c > 0$, then the resulting frame $\Psi = c\Phi = (cf_j)$ and its dual $\tilde{\Psi} = \frac{1}{c}\tilde{\Phi}$ have frame bounds c^2A, c^2B and $\frac{1}{c^2B}, \frac{1}{c^2A}$, respectively. Thus there is a unique scaling which ensures that a frame and its dual have the same optimal frame bounds, i.e.,

$$c^2A = \frac{1}{c^2B}, \quad c^2B = \frac{1}{c^2A} \quad \Longleftrightarrow \quad c = \frac{1}{\sqrt[4]{AB}}.$$

Definition 3.8. We say that a frame Φ is **normalised** if its optimal frame bounds A and B satisfy

$$AB = 1.$$

We emphasise this is simply a normalising factor:

If $\Phi = (f_j)$ is a frame, with optimal frame bounds A and B, then

$$\frac{1}{\sqrt[4]{AB}}\Phi = (\frac{1}{\sqrt[4]{AB}} f_j)$$

is the unique positive scalar multiple of it which is a normalised frame.

After this normalisation (used in Figures 3.1 and 3.2), the frame bounds are

$$\frac{\sqrt{A}}{\sqrt{B}} = \frac{1}{\operatorname{cond}(\Phi)}, \qquad \frac{\sqrt{B}}{\sqrt{A}} = \operatorname{cond}(\Phi). \tag{3.29}$$

Hence being normalised is equivalent to either of

$$\|S_\Phi\| = \|S_{\tilde{\Phi}}\|, \qquad \|\operatorname{Gram}(\Phi)\| = \|\operatorname{Gram}(\tilde{\Phi})\|.$$

A natural distance on the unitarily equivalent frames (where the non unit scalar multiples are not identified) is given by

$$\operatorname{dist}(\Phi, \Psi) := \|\operatorname{Gram}(\Phi) - \operatorname{Gram}(\Psi)\|,$$

where $\|\cdot\|$ is the induced (spectral) norm. Let $\lambda_1, \ldots, \lambda_d > 0$ be the eigenvalues of the frame operator S_Φ. Then (see Exer. 3.25)

$$\operatorname{dist}(c\Phi, \frac{1}{c}\tilde{\Phi}) = \|S_{c\Phi} - S_{\frac{1}{c}\tilde{\Phi}}\| = \max_{1 \le j \le d} |c^2\lambda_j - \frac{1}{c^2}\lambda_j^{-1}|, \tag{3.30}$$

$$\operatorname{dist}(c\Phi, \Phi^{\mathrm{can}}) = \|S_{c\Phi} - S_{\Phi^{\mathrm{can}}}\| = \max_{1 \le j \le d} |c^2\lambda_j - 1|, \tag{3.31}$$

$$\operatorname{dist}(\frac{1}{c}\tilde{\Phi}, \Phi^{\mathrm{can}}) = \|S_{\frac{1}{c}\tilde{\Phi}} - S_{\Phi^{\mathrm{can}}}\| = \max_{1 \le j \le d} |\frac{1}{c^2}\lambda_j^{-1} - 1|. \tag{3.32}$$

The following inequality shows that (3.30) is minimised by a unique $c > 0$.

Lemma 3.1. *The following inequality holds*

$$\max\{\left|A - \frac{1}{A}\right|, \left|\lambda - \frac{1}{\lambda}\right|, \left|B - \frac{1}{B}\right|\} \ge \frac{B-A}{\sqrt{AB}}, \qquad 0 < A \le \lambda \le B,$$

with equality if and only if $AB = 1$.

Proof. Since $\lambda \mapsto \lambda - \frac{1}{\lambda}$ is increasing for $\lambda > 0$, and zero at $\lambda = 1$, the maximum is either $|A - \frac{1}{A}|$ or $|B - \frac{1}{B}|$, which is bounded below by $\frac{B-A}{\sqrt{AB}}$ (see Exer. 3.26). □

Theorem 3.6. *Let Φ be a finite frame with optimal frame bounds A and B, then*

$$\operatorname{dist}(\Phi,\tilde{\Phi}) = \|\operatorname{Gram}(\Phi)-\operatorname{Gram}(\tilde{\Phi})\| = \|S_\Phi - S_{\tilde{\Phi}}\| \geq \frac{B-A}{\sqrt{AB}},$$

with equality if and only if Φ is normalised.

Proof. Let $A=\lambda_1\leq\cdots\leq\lambda_d=B$ be the eigenvalues of S_Φ. Then by Exer. 3.25 (with $\alpha=1$, $\beta=-1$, $\gamma=0$) and Lemma 3.1, we have

$$\|\operatorname{Gram}(\Phi)-\operatorname{Gram}(\tilde{\Phi})\| = \|S_\Phi - S_{\tilde{\Phi}}\| = \max_{1\leq j\leq d}\left|\lambda_j-\frac{1}{\lambda_j}\right| \geq \frac{B-A}{\sqrt{AB}},$$

with equality if and only if $AB=1$, i.e., when Φ is normalised. □

We note that any function of the ratio $\frac{B}{A}$ of the optimal frame bounds, such as the condition number, and

$$\frac{B-A}{\sqrt{AB}} = \sqrt{\frac{B}{A}}-\sqrt{\frac{A}{B}},$$

is invariant under any scaling of the frame.

We now investigate the minima of the distances (3.31) and (3.32) of $c\Phi$ and its dual $\frac{1}{c}\tilde{\Phi}$, from the canonical tight frame Φ^{can}. They are the same

$$\min_{c>0}\operatorname{dist}(c\Phi,\Phi^{\mathrm{can}}) = \min_{c>0}\operatorname{dist}(\frac{1}{c}\tilde{\Phi},\Phi^{\mathrm{can}}) = \frac{B-A}{A+B}<1,$$

which occur for $c^2=\frac{2}{A+B}$ and $c^2=\frac{A+B}{2AB}$, respectively (see Exer. 3.27). These values for c are equal if and only if $A=B$, i.e., the frame is tight.

Hence, when Φ is not tight, the scaling which makes Φ closest to Φ^{can} does not make $\tilde{\Phi}$ closest to Φ^{can}. Thus we seek to minimise the maximum of these distances. The minimum is $\sqrt{\frac{B}{A}}-1$, which occurs once (see Exer. 3.27), when

$$c^4 = \frac{2}{A+B}\cdot\frac{A+B}{2AB} = \frac{1}{AB} \quad\iff\quad c\Phi \text{ and } \tfrac{1}{c}\tilde{\Phi} \text{ are normalised.}$$

In this case, and only this case, the distances are equal:

Theorem 3.7. *Let Φ be a finite frame with optimal frame bounds A and B, then*

$$\|\operatorname{Gram}(\Phi)-\operatorname{Gram}(\Phi^{\mathrm{can}})\| = \|\operatorname{Gram}(\tilde{\Phi})-\operatorname{Gram}(\Phi^{\mathrm{can}})\|,$$

which is equivalent to

$$\|S_\Phi - I\| = \|S_{\tilde{\Phi}}-I\|,$$

if and only if Φ is normalised. In this case, the common distance is

$$\operatorname{dist}(\Phi,\Phi^{\mathrm{can}}) = \operatorname{dist}(\tilde{\Phi},\Phi^{\mathrm{can}}) = \operatorname{cond}(\Phi)-1 = \sqrt{\frac{B}{A}}-1. \tag{3.33}$$

Proof. Let $A=\lambda_1\le\cdots\le\lambda_d=B$ be the eigenvalues of S_Φ. For $\lambda>0$, the functions $\lambda\mapsto|\lambda-1|$ and $\frac{1}{\lambda}\mapsto|\frac{1}{\lambda}-1|$ decrease to 0 at $\lambda=1$ and then increase, and so take their maxima over $[A,B]$ and $[\frac{1}{B},\frac{1}{A}]$ at an endpoint. Thus (see Exer. 3.25), we obtain

$$\|\mathrm{Gram}(\Phi)-\mathrm{Gram}(\Phi^{\mathrm{can}})\|=\|S_\Phi-I\|=\max_j|\lambda_j-1|=\max\{|A-1|,|B-1|\},$$

$$\|\mathrm{Gram}(\tilde\Phi)-\mathrm{Gram}(\Phi^{\mathrm{can}})\|=\|S_{\tilde\Phi}-I\|=\max_j\Big|\frac{1}{\lambda_j}-1\Big|=\max\Big\{\Big|\frac{1}{A}-1\Big|,\Big|\frac{1}{B}-1\Big|\Big\}.$$

By Exer. 3.28, these are equal if and only if $AB=1$, i.e., Φ is normalised. The frame bounds of Φ after normalisation are given by (3.29), and so the common distance is

$$\max\Big\{\Big|\frac{\sqrt{A}}{\sqrt{B}}-1\Big|,\Big|\frac{\sqrt{B}}{\sqrt{A}}-1\Big|\Big\}=\max\Big\{\frac{\sqrt{B}-\sqrt{A}}{\sqrt{B}},\frac{\sqrt{B}-\sqrt{A}}{\sqrt{A}}\Big\}=\frac{\sqrt{B}}{\sqrt{A}}-1.$$

□

We now summarise our results so far:

If Φ is a finite frame with optimal frame bounds A and B, then the following are equivalent

1. Φ is normalised, i.e., $AB=1$.
2. Φ and $\tilde\Phi$ have the same frame bounds.
3. $\|S_\Phi\|=\|S_{\tilde\Phi}\|$, i.e., $\|\mathrm{Gram}(\Phi)\|=\|\mathrm{Gram}(\tilde\Phi)\|$.
4. $\|S_\Phi-S_{\tilde\Phi}\|=\frac{B-A}{\sqrt{AB}}$, i.e., $\|\mathrm{Gram}(\Phi)-\mathrm{Gram}(\tilde\Phi)\|=\frac{B-A}{\sqrt{AB}}$.
5. $\|S_\Phi-I\|=\|S_{\tilde\Phi}-I\|$.
6. $\|\mathrm{Gram}(\Phi)-\mathrm{Gram}(\Phi^{\mathrm{can}})\|=\|\mathrm{Gram}(\tilde\Phi)-\mathrm{Gram}(\Phi^{\mathrm{can}})\|$.

There is a unique closest tight frame to a normalised frame Φ and its dual $\tilde\Phi$ (individually and simultaneously) amongst all tight frames similar to Φ.

Theorem 3.8. *Let Φ be a finite frame which is normalised, i.e., its optimal frame bounds satisfy $B=\frac{1}{A}$, then*

$$\min_{\substack{\Psi\,\mathrm{tight}\\ \Psi=Q\Phi}}\max\{\mathrm{dist}(\Phi,\Psi),\mathrm{dist}(\tilde\Phi,\Psi)\}=\frac{1-A^2}{2A}.$$

This is attained for $\Psi=\frac{\sqrt{A^2+1}}{\sqrt{2A}}\Phi^{\mathrm{can}}$ (up to unitary equivalence), for which

$$\mathrm{dist}(\Phi,\Psi)=\mathrm{dist}(\tilde\Phi,\Psi)=\frac{1-A^2}{2A}\le\frac{1-A}{A}=\mathrm{dist}(\Phi,\Phi^{\mathrm{can}})=\mathrm{dist}(\tilde\Phi,\Phi^{\mathrm{can}}),$$

with equality in the above if and only if Φ is tight.

Proof. If Ψ is tight and is similar to Φ, then it is similar to Φ^{can}, and so

$$\Psi = cU\Phi^{\text{can}}, \qquad c > 0, \quad U \text{ unitary},$$

(see Exer. 2.5). For this Ψ, $\text{Gram}(\Psi) = c^2\text{Gram}(\Phi^{\text{can}})$. Hence (see Exer. 3.27)

$$\text{dist}(\Phi, \Psi) = \text{dist}(\tilde{\Phi}, \Psi) = \max\{|A - c^2|, \Big|\frac{1}{A} - c^2\Big|\}.$$

This function of c has a minimum value of $\frac{1-A^2}{2A}$, which occurs precisely when

$$c^2 = \frac{A + \frac{1}{A}}{2} = \frac{A^2+1}{2A}$$

(the minimum must occur when $|A - c^2| = |\frac{1}{A} - c^2|$). By (3.33), we have

$$\frac{1-A^2}{2A} \le \text{dist}(\Phi, \Phi^{\text{can}}) = \text{dist}(\tilde{\Phi}, \Phi^{\text{can}}) = \frac{\sqrt{B}}{\sqrt{A}} - 1 = \frac{1-A}{A}.$$

Since $\frac{1-A}{A} - \frac{1-A^2}{2A} = \frac{(A-1)^2}{2A}$, this inequality is sharp, unless Φ is tight. □

Another notion of distance between (certain) frames is based on the theory of perturbations of orthonormal bases (called Riesz bases). We follow [Bal99]. A frame $\Psi = (g_j)_{j\in J}$ is **close** to a frame $\Phi = (f_j)_{j\in J}$ if there exists a $\lambda \ge 0$ such that

$$\|\sum_{j\in J} c_j(g_j - f_j)\| \le \lambda \|\sum_{j\in J} c_j f_j\|, \qquad \forall c \in \ell_2(J),$$

with the infimum over such λ called the **closeness bound**, and denoted by $\text{cl}(\Psi, \Phi)$. We say that Φ and Ψ are **near** if Φ is close to Ψ and Ψ is close to Φ. This is an equivalence relation, and a metric dist_B called the **quadratic distance** can be defined on all frames which are near each other by

$$\text{dist}_B(\Phi, \Psi) := \log(\max\{\text{cl}(\Phi, \Psi), \text{cl}(\Psi, \Phi)\} + 1).$$

Frames Φ and Ψ are near if and only if they are similar (see Exer. 3.30), i.e., $\Phi = Q\Psi$, for some invertible Q, in which case

$$\text{dist}_B(\Phi, \Psi) := \log(\max\{\|Q - I\|, \|Q^{-1} - I\|\} + 1). \tag{3.34}$$

Unlike $\text{dist}(\Phi, \Psi) = \|\text{Gram}(\Phi) - \text{Gram}(\Psi)\|$, dist_B is *scaling invariant*, i.e.,

$$\text{dist}_B(c\Phi, c\Psi) = \text{dist}_B(\Phi, \Psi), \qquad \forall c \ne 0.$$

A frame, its dual and canonical tight frame are all near. By (3.34), we have

$$\text{dist}_B(c\Phi, \frac{1}{c}\tilde{\Phi}) = \log(\max\{\|c^2 S_\Phi - I\|, \|\frac{1}{c^2} S_\Phi^{-1} - I\|\} + 1),$$

$$\text{dist}_B(c\Phi, \Phi^{\text{can}}) = \text{dist}_B(\frac{1}{c}\tilde{\Phi}, \Phi^{\text{can}}) = \log(\max\{\|cS_\Phi^{\frac{1}{2}} - I\|, \|\frac{1}{c} S_\Phi^{-\frac{1}{2}} - I\|\} + 1).$$

By minimising these distances over all $c > 0$, we obtain the following analogue of Theorems 3.6 and 3.7.

Theorem 3.9. *Let Φ be a finite frame with optimal frame bounds A and B, then*

$$\mathrm{dist}_B(\Phi, \tilde{\Phi}) \geq \frac{1}{2}(\log B - \log A), \tag{3.35}$$

$$\mathrm{dist}_B(\Phi, \Phi^{\mathrm{can}}) = \mathrm{dist}_B(\tilde{\Phi}, \Phi^{\mathrm{can}}) \geq \frac{1}{4}(\log B - \log A), \tag{3.36}$$

with equality in each if and only if Φ is normalised.

Proof. Since log is strictly increasing, it suffices to minimise the maxima in these distances. By Exer. 3.25 (and a slight variation of Exer. 3.27), these are

$$\max\{\|c^2 S_\Phi - I\|, \|\frac{1}{c^2} S_\Phi^{-1} - I\|\} = \max\{|c^2 A - 1|, |c^2 B - 1|, \left|\frac{1}{c^2 A} - 1\right|, \left|\frac{1}{c^2 B} - 1\right|\},$$

$$\max\{\|c S_\Phi^{\frac{1}{2}} - I\|, \|\frac{1}{c} S_\Phi^{-\frac{1}{2}} - I\|\} = \max\{|c\sqrt{A} - 1|, |c\sqrt{B} - 1|, \left|\frac{1}{c\sqrt{A}} - 1\right|, \left|\frac{1}{c\sqrt{B}} - 1\right|\}.$$

By Exer. 3.27 (with appropriate changes of variables), the respective minima of these are $\sqrt{B/A} - 1$ and $\sqrt{\sqrt{B}/\sqrt{A}} - 1$, giving the dist_B distances $\frac{1}{2}\log(\frac{B}{A})$ and $\frac{1}{4}\log(\frac{B}{A})$, which are attained if and only if $c^2 = \frac{1}{\sqrt{AB}}$, i.e., $c\Phi$ and $\frac{1}{c}\tilde{\Phi}$ are normalised. □

The analogue of Theorem 3.8 is as follows.

Theorem 3.10. *Let Φ be a finite frame with optimal frame bounds A and B, then*

$$\min_{\substack{\Psi \text{ tight} \\ \Psi = Q\Phi}} \mathrm{dist}_B(\Phi, \Psi) = \frac{1}{4}(\log B - \log A),$$

with equality if $\Psi = \sqrt[4]{AB}\,\Phi^{\mathrm{can}}$ (and possibly other unitarily equivalent frames). Thus, if Φ is normalised, then Φ^{can} is a best tight frame approximation to it.

Proof. Since Ψ is tight and similar to Φ^{can}, we have $\Psi = cU\Phi^{\mathrm{can}}$, where $c > 0$ and U is unitary. Hence

$$\mathrm{dist}_B(\Phi, \Psi) = \log(\max\{\|cUS_\Phi^{-\frac{1}{2}} - I\|, \|(cUS_\Phi^{-\frac{1}{2}})^{-1} - I\|\} + 1).$$

By Exer. 3.23, the maximum above is only made smaller by choosing $U = I$, and so, effectively, we need to minimise $\max\{\|cS_\Phi^{-\frac{1}{2}} - I\|, \|\frac{1}{c}S_\Phi^{\frac{1}{2}} - I\|\}$, i.e.,

$$\max\{\left|\frac{c}{\sqrt{A}} - 1\right|, \left|\frac{c}{\sqrt{B}} - 1\right|, \left|\frac{\sqrt{A}}{c} - 1\right|, \left|\frac{\sqrt{B}}{c} - 1\right|\}.$$

By Exer. 3.27, this is minimised if and only if $c = \sqrt[4]{AB}$, with minimum $(B/A)^{\frac{1}{4}} - 1$, which gives the minimal distance $\mathrm{dist}_B(\Phi, c\Phi^{\mathrm{can}}) = \frac{1}{4}\log(\frac{B}{A})$. □

3.9 Approximate inverses of the frame operator

Calculations with a frame Φ require the inverse of the frame operator S_Φ, e.g., to determine the dual frame. This can be done numerically using standard iterative algorithms for the calculating the inverse based on an approximate inverse.

Definition 3.9. An **approximate (left) inverse** of bounded linear map S on $\mathcal{H}$ is a bounded linear map M on $\mathcal{H}$ for which

$$\|I_{\mathcal{H}} - MS\| < 1.$$

If M is an approximate left inverse of S, then MS is boundedly invertible. Hence if S is invertible, then S^{-1} can be calculated as $S^{-1} = (MS)^{-1}M$ (see Exer. 3.31).

Let A and B be known, but possibly *not* optimal frame bounds for a frame Φ. Then $\frac{2}{A+B}I_{\mathcal{H}}$ is an approximate inverse of $S = S_\Phi$, with

$$\|I - \frac{2}{A+B}S\| \le \frac{B-A}{A+B} < 1.$$

This leads to the standard **fixed point iteration method** for finding the solution of $Sg = h$ (see Exer. 3.32)

$$g_0 := 0, \qquad g_{k+1} := g_k - \frac{2}{A+B}Sg_k + \frac{2}{A+B}h, \qquad k = 0,1,2,\ldots,$$

with g_n converging to g at the rate

$$\|g_n - g\| \le \|S^{-1}h\| \Bigl(\frac{B-A}{A+B}\Bigr)^n \le \frac{\|h\|}{A}\Bigl(\frac{B-A}{A+B}\Bigr)^n.$$

In particular, taking $h = f$ gives a sequence converging to $S^{-1}f$, and the choice $h = Sf$, i.e., $g = f$, leads to the so called **frame algorithm**

$$g_0 := 0, \qquad g_{k+1} := g_k + \frac{2}{A+B}S(f - g_k), \qquad k = 0,1,2,\ldots. \tag{3.37}$$

The frame algorithm, which requires some estimate for the frame bounds, allows f to be reconstructed from the coefficients $\langle f, f_j\rangle$ without calculating the dual frame. It can be accelerated by standard techniques, including the **Chebyshev method** and the **conjugate gradient method** (see [Gro93] for a detailed analysis).

It is also possible to calculate $S^{-\frac{1}{2}}$ (and hence the canonical tight frame) numerically in terms of S, via the absolutely convergent series expansion

$$S^{-\frac{1}{2}} = \sqrt{\frac{2}{A+B}} \sum_{j=0}^{\infty} \frac{(2j)!}{2^{2j}(j!)^2}\Bigl(I - \frac{2}{A+B}S\Bigr)^j$$

(see Exer. 3.33 for details).

3.10 Alternate duals

The notion of a basis and its dual functionals can be further generalised as follows.

Definition 3.10. Finite frames (f_j) and (g_j) for $\mathscr{H}$, with synthesis operators V and W, are said to be **dual** if $VW^* = I$, i.e., they give the reconstruction formula

$$f = \sum_j \langle f, g_j \rangle f_j = \sum_j \langle f, f_j \rangle g_j, \qquad \forall f \in \mathscr{H}. \tag{3.38}$$

Since $(VW^*)^* = WV^*$, being dual does not depend on the order of the frames. The set $\mathscr{A} = \mathscr{A}_\Phi$ of all frames dual to a given finite frame $\Phi = (f_j)_{j\in J}$ for $\mathscr{H}$ is an *affine subspace* of $\mathscr{H}^J$, which contains the *(canonical) dual frame* $\tilde{\Phi}$. The elements of $\mathscr{A}$ are commonly referred to as **dual frames** and those of $\mathscr{A} \setminus \tilde{\Phi}$ as **alternate** (or **noncanonical**) **dual frames** of the frame Φ.

Example 3.13. Let (f_j) and (g_j) be a biorthogonal system for $\mathscr{K}$, and P be the orthogonal projection onto a subspace $\mathscr{H}$ of $\mathscr{K}$, then (Pf_j) and (Pg_j) are dual frames for $\mathscr{H}$. Moreover, (Pg_j) is the canonical dual of (Pf_j) if and only if $\mathscr{H}$ is an invariant subspace of the frame operator for (f_j). See Exer. 3.21 for details.

A finite frame (f_j) is dual to (g_j), where we write $g_j = \tilde{f}_j + h_j$, if and only if

$$[f_j][g_j]^* = [f_j][\tilde{f}_j]^* + [f_j][h_j]^* = I + [f_j][h_j]^* = I \quad \Longleftrightarrow \quad [f_j][h_j]^* = 0,$$

i.e., the frame (h_j) is orthogonal to (f_j), in the sense of Definition 3.5, see (3.20).

Some conditions equivalent to being dual include the following.

Proposition 3.4. *Let $\Phi = (f_j)_{j\in J}$ and $\Psi = (g_j)_{j\in J}$ be finite frames for $\mathscr{H}$, with synthesis operators V and W. Let $S = VV^*$, and $P_\Phi = V^*S^{-1}V$ be the canonical Gramian of Φ. Then the following are equivalent*
(a) *Φ and Ψ are dual.*
(b) *W is a left inverse of V^*, i.e., $WV^* = I$.*
(c) *W^* is a right inverse of V, i.e., $VW^* = I$.*
(d) *$\Psi - \tilde{\Phi}$ is orthogonal to Φ, i.e., $g_j = \tilde{f}_j + h_j$, with (h_j) orthogonal to (f_j).*
(e) $\operatorname{ran}((W - S^{-1}V)^*) \perp \operatorname{ran}(V^*)$.
(f) *$W = S^{-1}V + L(I - P_\Phi)$, where $L : \ell_2(J) \to \mathscr{H}$ is a linear operator.*
(g) *$W^* = V^*S^{-1} + (I - P_\Phi)R$, where $R : \mathscr{H} \to \ell_2(J)$ is a linear operator.*
(h) *$Q = V^*W$ is a projection, i.e., $Q^2 = Q$.*

Proof. (a)$\Longleftrightarrow$(b)$\Longleftrightarrow$(c) Use $VW^*f = \sum_j \langle f, g_j \rangle f_j$ and $(VW^*)^* = (WV^*)$.
(d),(e)$\Longleftrightarrow$(a) As observed ($\Psi - \tilde{\Phi}$ and Φ have synthesis maps $W - S^{-1}V$ and V).
(b)$\Longrightarrow$(f) Suppose that $WV^* = I$, and take $L = W$. Then we have

$$S^{-1}V + L(I - P_\Phi) = S^{-1}V + W - (WV^*)S^{-1}V = W.$$

(f)$\Longrightarrow$(b) We have $WV^* = (S^{-1}V + L(I - V^*S^{-1}V))V^* = I + L(V^* - V^*) = I$.
(f)$\Longleftrightarrow$(g) Take adjoints.

(b)$\Longrightarrow$(h) If $WV^* = I$, then $(V^*W)^2 = V^*(WV^*)W = V^*W$.
(h)$\Longrightarrow$(b) If $(V^*W)^2 = V^*W$, then $V(V^*WV^*W)W^* = V(V^*W)W$. Since VV^* and WW^* are invertible (Φ and Ψ are frames), they cancel to give $WV^* = I$. □

Corollary 3.4. *Let Φ be a frame of n vectors for $\mathcal{H}$, where $d = \dim(\mathcal{H})$. Then the affine subspace $\mathcal{A}$ of all frames dual to Φ has dimension $d(n-d)$. In particular, there exist alternate dual frames if and only if Φ is not a basis.*

Proof. By (3.13), we have that $\ker(I - P_\Phi) = \operatorname{ran}(P_\Phi) = \operatorname{ran}(V^*)$. Thus by (f), the dual frames are in 1–1 correspondence with linear maps $L|_{\mathcal{K}} : \mathcal{K} \to \mathcal{H}$, where $\mathcal{K} = \operatorname{ran}(V^*)^\perp = \ker(V)$, $\dim(\mathcal{K}) = n - d$. The space of linear maps $\mathcal{K} \to \mathcal{H}$ is isomorphic to the subspace $\mathcal{A} - S^{-1}V$, and so $\mathcal{A}$ has dimension $d(n-d)$. □

Let $\Phi = (f_j)$ be a finite frame for $\mathcal{H}$, $V = [f_j]$, with canonical Gramian P_Φ. Then all frames (g_j) that are dual to (f_j) are given by

$$W = [g_j] = [\tilde{f}_j] + L(I - P_\Phi), \tag{3.39}$$

where $L : \ker(V) \to \mathcal{H}$ and $P_\Phi = V^*(VV^*)^{-1}V$.

Example 3.14. (Exer. 3.34) Let $\Phi = (f_1, f_2, f_3)$ be the tight frame of three equally spaced unit vectors in $\mathbb{R}^2$. Then the affine space $\mathcal{A}_\Phi$ of all duals of Φ has dimension $d(n-d) = 2(3-1) = 2$ and is given by

$$(\tilde{f}_1 + w, \tilde{f}_2 + w, \tilde{f}_3 + w), \qquad w \in \mathbb{R}^2.$$

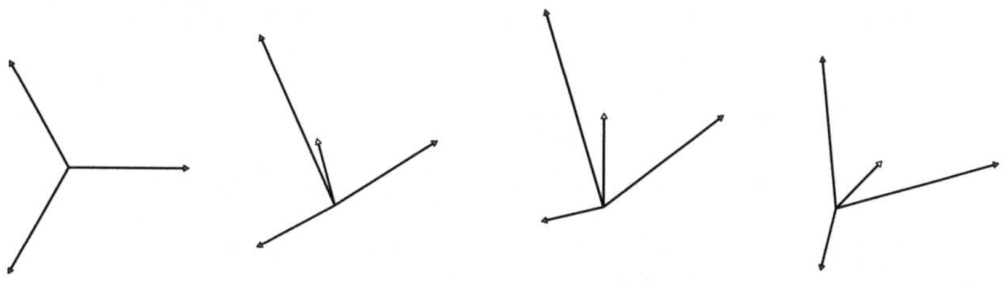

Fig. 3.3: The tight frame of three equally spaced vectors for $\mathbb{R}^2$ and some alternate duals. Here the vector w of Example 3.14 is depicted with a hollow arrowhead.

If (g_j) is dual to a finite frame (f_j) for $\mathcal{H}$, then taking $c_j = \langle f, g_j \rangle$ in (3.6) gives

$$\sum_j |\langle f, g_j \rangle|^2 = \sum_j |\langle f, \tilde{f}_j \rangle|^2 + \sum_j |\langle f, g_j - \tilde{f}_j \rangle|^2, \qquad \forall f \in \mathcal{H}.$$

Thus the *canonical* dual is characterised by minimising $\sum_j |\langle f, g_j \rangle|^2$, $\forall f \in \mathcal{H}$.

In a similar vein, taking the Frobenius norm: $\|A\|_F^2 := \text{trace}(AA^*)$ of the formula (3.39) for a dual frame (g_j) of (f_j), using (3.12), i.e., $V = VP_\Phi$, gives

$$\begin{aligned}\sum_j \|g_j\|^2 = \|W\|_F^2 &= \text{trace}\big((S^{-1}VP_\Phi + L(I-P_\Phi))(P_\Phi V^* S^{-1} + (I-P_\Phi)L^*)\big)\\ &= \|S^{-1}V\|_F^2 + \|L(I-P_\Phi)\|_F^2 = \sum_j \|\tilde{f}_j\|^2 + \|L(I-P_\Phi)\|_F^2.\end{aligned} \tag{3.40}$$

Thus the *canonical* dual frame to (f_j) is the unique dual frame (g_j) of (f_j) which minimises $\sum_j \|g_j\|^2$.

Example 3.15. (Exer. 3.35) Let $\Phi = (f_j)$ be a finite frame for $\mathscr{H}$, $V = [f_j]$, and $Q : \ell_2(J) \to \ell_2(J)$ be an invertible linear map. Then

$$W = [g_j] = (VQ^*QV^*)^{-1}VQ^*Q$$

gives a frame (g_j) dual to (f_j), which uniquely minimises $\|WQ^{-1}\|_F$, $W = [g_j]$. For D an invertible first-order difference operator, the (alternate) dual frame (g_j) which minimises the Sobolev-type norm $\|W(D^r)^*\|_F$ is called the r-th order **Sobolev dual**. These are motivated by $\Sigma\Delta$-quantisation (see [BLPY10]).

The canonical tight frame can also be characterised in terms of similarity:

Proposition 3.5. *No two distinct duals of a finite frame Φ are similar to each other. Thus the canonical dual $\tilde{\Phi}$ is the unique dual of Φ which is similar to Φ.*

Proof. Suppose that (g_j) and (Qg_j) are similar frames, which are duals of $\Phi = (f_j)$. Then

$$Q^* f = \sum_j \langle Q^* f, g_j \rangle f_j = \sum_j \langle f, Qg_j \rangle f_j = f, \qquad \forall f \in \mathscr{H},$$

so that $Q = I$. Hence $\tilde{\Phi} = S^{-1}\Phi$ is the only dual which is similar to Φ. □

For finite frames (f_j) and (g_j) for $\mathscr{H}$, with synthesis operators V and W, the condition $VW^* = I$ for being dual can be weakened as follows:

(f_j) and (g_j) are **approximately dual frames** if $\|VW^* - I\| < 1$,

(f_j) and (g_j) are **pseudodual frames** if VW^* is invertible.

It is easy to verify that these notions do not depend on the frame order, and that the pairs of frames satisfying the different dualities satisfy the inclusions:

$$\text{canonically dual} \subset \text{dual} \subset \text{approximately dual} \subset \text{pseudodual}.$$

Example 3.16. The set of all frames approximately dual to a given finite frame Φ is a convex set. If Φ has optimal frame bounds A and B, then $c\Phi$ is approximately dual to Φ for any scalar $0 < c < \frac{2}{B}$ (see Exer. 3.37).

Example 3.17. If two frames are similar, then their synthesis operators are related $W = QV$, where Q is invertible. Since $VW^* = (VV^*)Q^*$ is invertible, it follows that similar frames (for the same space) are pseudoduals. They may not necessarily approximate duals (see Example 3.16).

3.11 Oblique duals

The dual frame expansion (3.38) can be generalised by allowing the coefficients $c_j = \langle f, g_j \rangle$ to be given by vectors g_j from outside the space $\mathscr{H}$. For example, in signal processing one may try and choose the g_j to be outside some subspace for which measurements are known to be corrupted by noise.

Definition 3.11. Let (f_j) and (g_j) be finite frames for subspaces $\mathscr{V}$ and $\mathscr{W}$ of $\mathscr{H}$, and $V := [f_j]$, $W := [g_j]$. Then (g_j) is an **oblique dual**[3] of (f_j) if $VW^*|_{\mathscr{V}} = I_{\mathscr{V}}$, i.e.,

$$f = \sum_j \langle f, g_j \rangle f_j, \qquad \forall f \in \mathscr{V}.$$

Example 3.18. Let $\mathscr{V}_k(\mu)$ be the space of multivariate orthogonal polynomials of degree k for a measure μ (see §10.10), and Q be the orthogonal projection onto it. Let (f_j) be a frame for $\mathscr{V}_k(\mu)$, and g_j be leading term of $\tilde{f}_j$ (its homogeneous term of degree k), so that (g_j) is a frame for the homogeneous polynomials of degree k. Since $g_j = Q\tilde{f}_j$ and $Qf = f$, $f \in \mathscr{V}_k(\mu)$, we have

$$\sum_j \langle f, g_j \rangle f_j = \sum_j \langle f, Q\tilde{f}_j \rangle f_j = \sum_j \langle f, \tilde{f}_j \rangle f_j = f, \quad \forall f \in \mathscr{V}_k(\mu),$$

i.e., (g_j) is an oblique dual of (f_j).

Example 3.19. Suppose (f_j) is a finite frame for $\mathscr{V}$, and $(\hat{g}_j) \subset \mathscr{V}$ is a dual frame. Let Q be the orthogonal projection onto $\mathscr{V}$. For $h_j \in \mathscr{V}^\perp = \mathscr{H} \ominus \mathscr{V}$, define

$$g_j := \hat{g}_j + h_j, \qquad \mathscr{W} := \operatorname{span}\{g_j\}.$$

Then (g_j) is an oblique dual of (f_j), since $Qg_j = \hat{g}_j$ gives

$$f = \sum_j \langle f, \hat{g}_j \rangle f_j = \sum_j \langle f, Qg_j \rangle f_j = \sum_j \langle Qf, g_j \rangle f_j = \sum_j \langle f, g_j \rangle f_j, \quad \forall f \in \mathscr{V}.$$

Similarly, if $(g_j) \subset \mathscr{W}$ is an oblique dual, then (Qg_j) is a dual frame for (f_j). It may be that $\dim(\mathscr{W}) > \dim(\mathscr{V})$. For example, let $\mathscr{V} = \mathbb{R}^2 \times 0 \subset \mathbb{R}^3$, and (f_j) be the normalised tight frame for $\mathscr{V}$ given by three equally spaced vectors. Then $(\tilde{f}_j + ce_3)$, $c \neq 0$, $e_3 = (0,0,1)$ is an oblique dual where $\mathscr{W} = \mathbb{R}^3$.

---

[3] The term **pseudodual** is also used in the literature, e.g., see [LO04].

Let $\Phi = (f_j)$ be a finite frame for $\mathscr{V}$, $V = [f_j]$, with canonical Gramian P_Φ. Then all frames (g_j) that are oblique duals of (f_j) are given by

$$W = [g_j] = [\tilde{f}_j] + L(I - P_\Phi) + [h_j], \tag{3.41}$$

where $L : \ker(V) \to \mathscr{H}$, $P_\Phi = V^*(VV^*)^{-1}V$, and $h_j \in \mathscr{V}^\perp$.

The condition of being an oblique dual is not symmetric, i.e., if (g_j) is an oblique dual of (f_j), then (f_j) may or may not be an oblique dual of (g_j) (see Example 3.19). For (g_j) a frame for $\mathscr{W}$ to be an oblique dual of a frame (f_j) for $\mathscr{V}$, we must have

$$\mathscr{V} \cap \mathscr{W}^\perp = 0.$$

Otherwise

$$0 \neq f \in \mathscr{V} \cap \mathscr{W}^\perp \quad \Longrightarrow \quad f = \sum_j \langle f, g_j \rangle f_j = 0.$$

Thus a necessary condition to ensure being an oblique dual is symmetric is that

$$\mathscr{V} \cap \mathscr{W}^\perp = \mathscr{W} \cap \mathscr{V}^\perp = 0,$$

which is equivalent to the algebraic direct sums

$$\mathscr{H} = \mathscr{V} \oplus_a \mathscr{W}^\perp = \mathscr{W} \oplus_a \mathscr{V}^\perp,$$

and implies that $\dim(\mathscr{V}) = \dim(\mathscr{W})$ (see Exer. 3.38). We now show this condition is sufficient. For an algebraic direct sum $\mathscr{H} = \mathscr{V} \oplus_a \mathscr{W}^\perp$, the **oblique projection** of $\mathscr{H}$ onto $\mathscr{V}$ along $\mathscr{W}^\perp$ is the linear map $P = P_{\mathscr{V},\mathscr{W}^\perp}$ on $\mathscr{H}$ given by

$$P|_\mathscr{V} = I_\mathscr{V}, \qquad P(\mathscr{W}^\perp) = 0.$$

Proposition 3.6. *Let (f_j) and (g_j) be finite frames for subspaces $\mathscr{V}$ and $\mathscr{W}$ of $\mathscr{H}$, with $\mathscr{V} \cap \mathscr{W}^\perp = \mathscr{W} \cap \mathscr{V}^\perp = 0$. Then the following are equivalent*

1. *The oblique projection $P_{\mathscr{V},\mathscr{W}^\perp}$ of $\mathscr{H}$ onto $\mathscr{V}$ is given by $P_{\mathscr{V},\mathscr{W}^\perp} f = \sum_j \langle f, g_j \rangle f_j$.*
2. *The oblique projection $P_{\mathscr{W},\mathscr{V}^\perp}$ of $\mathscr{H}$ onto $\mathscr{W}$ is given by $P_{\mathscr{W},\mathscr{V}^\perp} f = \sum_j \langle f, f_j \rangle g_j$.*
3. *(g_j) is an oblique dual of (f_j), i.e., $f = \sum_j \langle f, g_j \rangle f_j$, $\forall f \in \mathscr{V}$.*
4. *(f_j) is an oblique dual of (g_j), i.e., $f = \sum_j \langle f, f_j \rangle g_j$, $\forall f \in \mathscr{W}$.*

Proof. The condition $\mathscr{V} \cap \mathscr{W}^\perp = \mathscr{W} \cap \mathscr{V}^\perp = 0$ ensures $P_{\mathscr{V},\mathscr{W}^\perp}$ and $P_{\mathscr{W},\mathscr{V}^\perp}$ are well defined and satisfy $(P_{\mathscr{V},\mathscr{W}^\perp})^* = P_{\mathscr{W},\mathscr{V}^\perp}$ (see Exer. 3.38). Let $V = [f_j]$ and $W = [g_j]$.
1.$\Longleftrightarrow$2. These conditions are $P_{\mathscr{V},\mathscr{W}^\perp} = VW^*$, $P_{\mathscr{W},\mathscr{V}^\perp} = WV^*$, which are equivalent.
1.$\Longrightarrow$3. If $P_{\mathscr{V},\mathscr{W}^\perp} = VW^*$, then $f = VW^* f = \langle f, g_j \rangle f_j$, $\forall f \in \mathscr{V}$.
3.$\Longrightarrow$1. Define P on $\mathscr{H}$ by $Pf := \sum_j \langle f, g_j \rangle f_j$, then $Pf = f$, $\forall f \in \mathscr{V}$ and $Pf = 0$, $\forall f \in \mathscr{W}^\perp$ (since then $\langle f, g_j \rangle = 0$), so that $P = P_{\mathscr{V},\mathscr{W}^\perp}$.
2.$\Longleftrightarrow$4. Interchange (f_j) and (g_j) in the argument for 1.$\Longleftrightarrow$3. □

For a given finite frame for $\mathscr{V}$, there is a *canonical* oblique dual from $\mathscr{W}$.

Theorem 3.11. *Suppose that $\mathscr{V}$ and $\mathscr{W}$ are subspaces of $\mathscr{H}$, with*

$$\mathscr{V} \cap \mathscr{W}^\perp = \mathscr{W} \cap \mathscr{V}^\perp = 0,$$

and (f_j) is a finite frame for $\mathscr{V}$. Then the unique coefficients $c_j = c_j(f)$ satisfying

$$\mathscr{P}_{\mathscr{V},\mathscr{W}^\perp} f = \sum_j c_j f_j, \qquad f \in \mathscr{H} \tag{3.42}$$

which have minimal ℓ_2-norm are given by

$$c = (W^*V)^\dagger W^* f, \qquad V := [f_j], \tag{3.43}$$

where W is the synthesis operator of any frame for $\mathscr{W}$.

Proof. It is easy to verify (take $\Lambda = W^*$ in Exer. 3.40) that $\mathscr{P}_{\mathscr{V},\mathscr{W}^\perp} = V(W^*V)^\dagger W^*$, and so we seek a minimum norm solution c to

$$Vc = V(W^*V)^\dagger W^* f.$$

The unique such c is given by $c = V^\dagger V(W^*V)^\dagger W^* f$. Since $V^\dagger V$ is the orthogonal projector onto the range of V^*, and $\operatorname{ran}((W^*V)^\dagger) = \operatorname{ran}(V^*W) \subset \operatorname{ran}(V^*)$, we can simplify this to $c = (W^*V)^\dagger W^* f$. □

For the $c_j(f) = \langle f, g_j \rangle$ of (3.43), the frame (g_j) for $\mathscr{W}$, which is given by

$$[g_j] = ((W^*V)^\dagger W^*)^* = W(V^*W)^\dagger, \tag{3.44}$$

is called the **canonical oblique dual** of (f_j) in $\mathscr{W}$. As would be hoped, the canonical oblique dual of this (g_j) in $\mathscr{V}$ is (f_j) (see Exer. 3.39).

Example 3.20. If $\mathscr{V} = \mathscr{W} = \mathscr{H}$ and $W = V$, then (3.44) gives (cf Exer. 3.9)

$$[\tilde{f}_j] = V(V^*V)^\dagger = V \operatorname{Gram}(\Phi)^\dagger, \qquad \Phi = (f_j).$$

Example 3.21. In signal processing, (3.42) can be used as follows (see [Eld03]). Let P be the orthogonal projection onto $\mathscr{W}$. Then $c_j(f) = \langle f, Pg_j \rangle = \langle Pf, g_j \rangle$, so that the sampling of a signal f can be done by first projecting it onto the *sampling space* $\mathscr{W}$ (where signals can be accurately measured) and then measuring the projected signal Pf.

Corollary 3.5. *Suppose that $\Phi = (f_j)$ is a finite sequence in $\mathscr{H}$, with $V := [f_j]$, and $S := VV^* : \mathscr{H} \to \mathscr{H}$ (which may not be invertible). Then (f_j) is a finite frame for $\mathscr{V} := \operatorname{span}\{f_j\}$, with the canonical dual frame $(\tilde{f}_j)$ given by*

$$[\tilde{f}_j] = V \operatorname{Gram}(\Phi)^\dagger = S^\dagger V.$$

Proof. Take $\mathscr{W} = \mathscr{V}$ in (3.44), and use $V \operatorname{Gram}(\Phi)^\dagger = S^\dagger V$ (see Exer. 3.11). □

Notes

An early appearance of the canonical tight frame was in Löwdin's well-known work in Quantum Chemistry in the late 1940s, where he constructed "orthonormalised atomic orbitals" from a basis of orbitals (see [Löw70], [AEG80]). This also known as *Schweinler–Wigner orthogonalisation* (see [SW70]). Symmetric Gram–Schmidt methods for infinite dimensional spaces are explored in [FPT02].

The material of this chapter extends to infinite dimensional spaces, see [HL00], [Chr03]. Source material and further reading include: dual frames [Li95], oblique dual frames [Eld03], [CE04], [LO04], approximate dual frames [CL10], [LY09] and optimal dual frames for erasures [LH10].

Exercises

3.1. *Rank one projections*. Let $A_j : \mathscr{H} \to \mathscr{H} : f \mapsto \langle f, g_j \rangle f_j$, where $f_j, g_j \in \mathscr{H}$.
(a) Show that $\operatorname{trace}(A_j) = \langle f_j, g_j \rangle$.
(b) Show that $A = A_j$ is a scalar multiple of a projection, i.e., $A = cP$, where $c \in \mathbb{F}$ and $P^2 = P$, provided $\langle f_j, g_j \rangle \neq 0$.
(c) Show P is orthogonal if and only if f_j and g_j are multiples of each other.
(d) Show that Hilbert–Schmidt (Frobenius) inner product between them is

$$\operatorname{trace}(A_j A_k^*) = \langle f_j, f_k \rangle \langle g_k, g_j \rangle.$$

3.2. *Least squares solution.*
Let $\mathscr{A} := \{c \in \ell_2(J) : Vc = f\}$ be all possible sequences of coefficients from which $f \in \mathscr{H}$ can be reconstructed $f = \sum_j c_j v_j$, where (f_j) spans $\mathscr{H}$.
(a) Show $\mathscr{A}$ is an affine subspace, i.e., $\lambda a + (1-\lambda) b \in \mathscr{A}$, $\forall a, b \in \mathscr{A}$, $\lambda \in \mathbb{R}$.
(b) Show that $c = V^* S^{-1} f \in \mathscr{A}$. Here $c_j = \langle f, S^{-1} f_j \rangle$, where $S = VV^*$.
(c) Since affine subspaces are translates of linear subspaces, (a) and (b) give

$$\mathscr{A} = V^* S^{-1} f + \ker(V).$$

Show that $c = V^* S^{-1} f$ is the unique solution to $f = Vc$ of minimal ℓ_2-norm.

3.3. *Pseudoinverse.* Take the definition of the **pseudoinverse** of $A : \mathscr{H} \to \mathscr{K}$ to be the unique linear map $A^\dagger : \mathscr{K} \to \mathscr{H}$ satisfying

$$AA^\dagger, A^\dagger A \text{ are Hermitian}, \qquad AA^\dagger A = A \text{ and } A^\dagger A A^\dagger = A^\dagger.$$

(a) Show that if A is onto, then $A^\dagger = A^*(AA^*)^{-1}$. In particular, for $V = [f_j]_{j \in J}$ the synthesis operator of a finite spanning set for $\mathscr{H}$

$$V^\dagger = V^*(VV^*)^{-1} = V^* S^{-1}, \qquad S := VV^*.$$

(b) Show that if P is an orthogonal projection P, then $P^\dagger = P$.
(c) Show that the Gramian of a frame and its dual are pseudoinverses, i.e.,

$$\mathrm{Gram}(\tilde{\Phi}) = \mathrm{Gram}(\Phi)^\dagger.$$

(d) Let $\Phi = (f_j)$ be a finite frame with synthesis operator $V = [f_j]$. Show that

$$\mathrm{Gram}(\Phi^{\mathrm{can}}) = V^\dagger V = \mathrm{Gram}(\Phi)\,\mathrm{Gram}(\Phi)^\dagger.$$

3.4. *Perturbation of a normalised tight frame.*
Let $(f_j)_{j\in J}$ be a normalised tight frame for $\mathscr{H}$, and $T : \mathscr{H} \to \mathscr{H}$ be invertible. Show that the frames $(T^* f_j)$ and $(T^{-1} f_j)$ for $\mathscr{H}$ are dual, i.e.,

$$f = \sum_{j\in J} \langle f, T^* f_j\rangle T^{-1} f_j = \sum_{j\in J} \langle f, T^{-1} f_j\rangle T^* f_j, \qquad \forall f \in \mathscr{H}.$$

3.5. If $\Phi = (f_j)$ is a finite frame for $\mathscr{H}$, then its image $\Psi = (Qf_j)$ under a linear map $Q : \mathscr{H} \to \mathscr{K}$ is a frame for its span, with frame bounds satisfying

$$A_\Phi \|Q^\dagger\|^{-2} \le A_\Psi \le B_\Psi \le B_\Phi \|Q\|^2.$$

In particular, if Q is a partial isometry, e.g., an orthogonal projection or unitary map, then

$$A_\Phi \le A_\Psi \le B_\Psi \le B_\Phi,$$

and so partial isometries map tight frames to tight frames (cf Exer. 2.7).

3.6. Let $(f_j)_{j=1}^n$ be a finite sequence in $\mathscr{H}$, with synthesis operator $V = [f_j]_{j=1}^n$. Prove the following are equivalent:
(a) $A := \inf_{f\ne 0} \sum_{j=1}^n \frac{|\langle f, f_j\rangle|^2}{\|f\|^2} > 0$.
(b) (f_j) is a frame for $\mathscr{H}$.
(c) (f_j) spans $\mathscr{H}$.
(d) V is onto.
(e) V^* is 1–1.
(f) $\langle Sf, f\rangle = \langle V^* f, V^* f\rangle = \|V^* f\|^2 = \sum_j |\langle f, f_j\rangle|^2 > 0, \forall f \ne 0$.

3.7. Suppose that (f_j) is a finite frame, and $V = [f_j]$. Let $\{\lambda_j\}$ be the eigenvalues of $S = VV^*$ (so $\sigma_j = \sqrt{\lambda_j}$ are the nonzero singular values of V). Show that
(a) The optimal frame bounds are $A = \min_j \lambda_j$ and $B = \max_j \lambda_j$.
(b) The frame bounds (3.1) are equivalent to $A \le \min_j \lambda_j$ and $B \ge \max_j \lambda_j$.

3.8. Let (f_j) be a finite frame with optimal frame bounds A and B. Show that
(a) $\|f_j\|^2 \le B, \forall j$.
(b) $\|f_j\|^2 = B$ if and only if $f_j \perp \mathrm{span}_{k\ne j} f_k$.
(c) $\|f_j\|^2 < A$ implies $f_j \in \mathrm{span}_{k\ne j} f_k$.

3.9. Let $\Phi = (f_j)$ be a finite frame for $\mathcal{H}$, and

$$V = U_1 \Sigma U_2^*, \qquad \Sigma = \operatorname{diag}(\sigma_1, \sigma_2, \dots)$$

be a singular value decomposition of the synthesis operator $V = [f_j]$. Show that
(a) $[\tilde{f}_j] = V \operatorname{Gram}(\Phi)^\dagger = U_1 \operatorname{diag}(1/\sigma_1, 1/\sigma_2, \dots) U_2^*$.
(b) $[f_j^{\mathrm{can}}] = V(\operatorname{Gram}(\Phi)^\dagger)^{\frac{1}{2}} = U_1 \operatorname{diag}(1, 1, \dots) U_2^*$.

3.10. *Canonical tight frame.* Let $\Phi = (f_j)_{j \in J}$ be a frame for $\mathcal{H}$. Show that
(a) $(\tilde{\Phi})^{\mathrm{can}} = \Phi^{\mathrm{can}}$.
(b) $\operatorname{Gram}(\Phi^{\mathrm{can}}) = \operatorname{Gram}(\Phi) \operatorname{Gram}(\tilde{\Phi}) = \operatorname{Gram}(\tilde{\Phi}) \operatorname{Gram}(\Phi)$.
(c) $f_j^{\mathrm{can}} = \tilde{f}_j^{\mathrm{can}} = S_\Phi^{-\frac{1}{2}} f_j = S_{\tilde{\Phi}}^{\frac{1}{2}} f_j = S_{\tilde{\Phi}}^{-\frac{1}{2}} \tilde{f}_j = S_\Phi^{\frac{1}{2}} \tilde{f}_j$, $\forall j \in J$.

3.11. *Commutativity of the synthesis, frame and Gramian operators.*
Let $\Phi = (f_j)_{j \in J}$ be a finite sequence in $\mathcal{H}$, with synthesis operator $V = [f_j]_{j \in J}$, frame operator $S = VV^*$ and Gramian $G = V^*V$. Show that these satisfy
(a) $S^j V = V G^j$, $j = 1, 2, \dots$.
(b) $S^{\frac{1}{2}} V = V G^{\frac{1}{2}}$.
(c) $S^\dagger V = V G^\dagger$.

3.12. *Orthogonal projection formula* (generalises Exer. 2.3).
Suppose $(f_j)_{j \in J}$ is a finite frame for a subspace $\mathcal{K} \subset \mathcal{H}$, and let $V := [f_j]_{j \in J}$, $W := [\tilde{f}_j]_{j \in J}$. Show the orthogonal projection onto this subspace is given by

$$P = VW^* = WV^* : f \mapsto \sum_{j \in J} \langle f, \tilde{f}_j \rangle f_j = \sum_{j \in J} \langle f, f_j \rangle \tilde{f}_j.$$

3.13. Let (f_j) be finite frame for $\mathcal{H}$. Prove the analogue of the trace formula (2.9)

$$\langle f_j, \tilde{f}_j \rangle = \langle \tilde{f}_j, f_j \rangle = \|f_j^{\mathrm{can}}\|^2, \ \forall j \quad \Longrightarrow \quad \sum_{j \in J} \langle f_j, \tilde{f}_j \rangle = d = \dim(\mathcal{H}).$$

3.14. Show that if $V = [f_1, \dots, f_n]$ maps onto a proper subspace $\mathcal{K}$ of $\mathcal{H}$, so that $S = VV^* : \mathcal{H} \to \mathcal{H}$ is *not* invertible. Then the dual frame and canonical tight frame for (f_j) are given by

$$\tilde{f}_j = S^\dagger f_j, \qquad f_j^{\mathrm{can}} = (S^\dagger)^{\frac{1}{2}} f_j = (S^{\frac{1}{2}})^\dagger f_j.$$

3.15. If the Gramian of a frame $\Phi = (f_j)_{j=1}^n$ can be factored $\operatorname{Gram}(\Phi) = L^* L$, where $L = [v_1, \dots, v_n]$ is an $m \times n$ matrix, then the columns of L give a copy of Φ as a subspace of $\mathbb{F}^m$ (cf Th. 3.3). Show that the dual frame and canonical tight frame of this copy are given by the columns of $(LL^*)^\dagger L$ and $((LL^*)^\dagger)^{\frac{1}{2}} L$.

3.16. Let $\Phi = (f_j)$ be a finite frame for $\mathcal{H}$ with synthesis operator $V = [f_j]$, and $\Psi = (g_j) := Q\Phi = (Qf_j)$, where $Q : \mathcal{H} \to \mathcal{H}$ is invertible. Show that
(a) Ψ is a frame for $\mathcal{H}$.
(b) $\tilde{\Psi} = (Q^*)^{-1}\tilde{\Phi}$.
(c) $\Psi^{\text{can}} = U\Phi^{\text{can}}$, $U := (QVV^*Q^*)^{-\frac{1}{2}}Q(VV^*)^{\frac{1}{2}}$, where U is unitary.
(d) If $Q = cU$, with $c > 0$ and U unitary, then this U is the U of part (c).

3.17. Let V be a $d \times n$ matrix, $n \geq d$, with full rank, i.e., $\text{rank}(V) = d$. Use the singular value decomposition $V = U_1 \Sigma U_2^*$ to show that the minimum

$$\min_{\substack{W \in \mathbb{F}^{d\times n}, A>0 \\ WW^*=AI}} \|V - W\|_F$$

is uniquely attained for

$$W = \frac{1}{d}\text{trace}((VV^*)^{\frac{1}{2}})(VV^*)^{-\frac{1}{2}}V = \frac{1}{d}\text{trace}(S^{\frac{1}{2}})S^{-\frac{1}{2}}V.$$

Remark: This generalises the problem of finding the unitary matrix W which best approximates a square matrix V (cf [HJ90], Problem 3 of §7.4).

3.18. Modify the argument of Theorem 3.2 to show that for any sequences

$$\sum_{j=1}^{n} \|f_j - g_j\|^2 \geq \sum_{k=1}^{d} \lambda_k + \sum_{k=1}^{d} \mu_k - 2\sum_{k_1=1}^{d}\sum_{k_2=1}^{d} \sqrt{\lambda_{k_1}}\sqrt{\mu_{k_2}}|\langle u_{k_1}, v_{k_2}\rangle|,$$

where (λ_k), (μ_k) are the eigenvalues of the frame operators for (f_j), (g_j), and (u_k), (v_k) are corresponding orthonormal bases of eigenvectors.

3.19. *Isogonal configurations* (Example 3.9). Let $\Phi = (u_j)_{j=1}^n$ be a sequence of isogonal of unit vectors in $\mathbb{R}^d$, $d > 1$, i.e., one with Gramian

$$M = \text{Gram}(\Phi) = \begin{cases} 1, & j = k; \\ a, & j \neq k. \end{cases}$$

(a) Show the eigenvalues of M are $na - a + 1$ and $1 - a$ (of multiplicity $n - 1$).
(b) Determine the condition on a that ensures M is positive semidefinite and the corresponding rank of M (which can be 1, $n-1$ or n).
(c) When $\text{rank}(M) = n - 1$, conclude (u_j) are the vertices of the simplex.
(d) When $\text{rank}(M) = n$, show that the dual frame $(\tilde{u}_j)$ is an isogonal configuration.

3.20. *Real and complex frames.* Let Φ be a finite frame.
(a) Show that Φ is real if and only if its dual $\tilde{\Phi}$ is real.
(b) Show that if Φ is real, then canonical tight frame Φ^{can} is real.
(c) If the canonical tight frame is real, then does it follow that Φ and $\tilde{\Phi}$ are?

3.21. Suppose that (f_j) and (g_j) form a biorthogonal system for $\mathscr{K}$, and P is the orthogonal projection onto a subspace $\mathscr{H}$. Let S be the frame operator for (f_j). Show that the dual frame of (Pf_j) (for $\mathscr{H}$) being equal to (Pg_j) is equivalent to
(a) $PS = PSP$.
(b) $SP = PSP$.
(c) $SP = PS$.
(d) $\mathscr{H}$ is invariant under S, i.e., $S\mathscr{H} \subset \mathscr{H}$.

3.22. Let $L : \mathscr{H} \to \mathscr{K}$ be a linear map between finite dimensional Hilbert spaces, with singular values $\sigma_1, \sigma_2, \cdots, \sigma_m$. Show that

$$\|Lx\| \le \big(\max \sigma_j\big)\|x\|, \qquad \forall x \in \mathscr{H},$$

with equality for $x \neq 0$ if and only if $x \in (\ker L)^\perp$ is a right singular vector for the largest singular value (which implies $\|L\| = \max_j \sigma_j$), and

$$\big(\min_{\sigma_j \neq 0} \sigma_j\big)\|x\| \le \|Lx\|, \qquad \forall x \in (\ker L)^\perp,$$

with equality for $x \neq 0$ if and only if x is a right singular vector for the smallest *nonzero* singular value.

3.23. Let L be a self-adjoint invertible map on a finite dimensional Hilbert spaces. Show that if U is unitary, then

$$\|L - U\|, \|LU - I\|, \|UL - I\| \ge \|L - I\|.$$

3.24. Suppose that $\Phi = (f_j)$ is a finite frame for $\mathscr{H}$ with (possibly not optimal) frame bounds A and B. Let $V = [f_j]$. Show that the frame bounds (3.1) are equivalent to any of the bounds

$$\sqrt{A}\|c\| \le \|Vc\| \le \sqrt{B}\|c\|, \qquad \forall c \in \operatorname{ran}(V^*),$$

$$\sqrt{A}\|f\| \le \|V^* f\| \le \sqrt{B}\|f\|, \qquad \forall f \in \mathscr{H},$$

$$A\|f\| \le \|Sf\| \le B\|f\|, \qquad \forall f \in \mathscr{H},$$

$$A\|c\| \le \|\mathrm{Gram}(\Phi)c\| \le B\|c\|, \qquad \forall c \in \operatorname{ran}(\mathrm{Gram}(\Phi)),$$

and these are sharp if and only if A and B are the optimal frame bounds.

3.25. Suppose that $\Phi = (f_j)$ is a finite frame. Let $\lambda_1, \ldots, \lambda_d$ be the eigenvalues of its frame operator S_Φ. Use a singular value decomposition for the synthesis operator $V = [f_j]$, to show that for any scalars $\alpha, \beta, \gamma \in \mathbb{F}$ one has

$$\begin{aligned}\|\alpha S_\Phi + \beta S_{\tilde{\Phi}} + \gamma I\| &= \|\alpha \mathrm{Gram}(\Phi) + \beta \mathrm{Gram}(\tilde{\Phi}) + \gamma \mathrm{Gram}(\Phi^{\mathrm{can}})\| \\ &= \max_{1 \le j \le d} \big|\alpha\lambda_j + \beta\lambda_j^{-1} + \gamma\big|.\end{aligned}$$

In particular, since $S_{c\Phi} = c^2 S_\Phi$ and $S_{\frac{1}{c}\tilde{\Phi}} = \frac{1}{c^2} S_{\tilde{\Phi}}$, we obtain (3.30), (3.31), (3.32), by taking (α, β, γ) to be $(c^2, \frac{1}{c^2}, 0)$, $(c^2, 0, -1)$, $(0, \frac{1}{c^2}, -1)$, respectively. We also have

$$\|S_\Phi\| = \|\mathrm{Gram}(\Phi)\| = \|V\|^2 = \max_j \lambda_j = B_\Phi.$$

3.26. Show that the inequality

$$\max\Bigl\{\Bigl|A - \frac{1}{A}\Bigr|, \Bigl|B - \frac{1}{B}\Bigr|\Bigr\} \geq \frac{B-A}{\sqrt{AB}}, \qquad 0 < A \leq B$$

holds, with equality if and only if $AB = 1$.

3.27. Let Φ be a finite frame. Here we investigate how close $c\Phi$, its dual $\frac{1}{c}\tilde{\Phi}$ and Φ^{can} can be for the metric $\mathrm{dist}(\Phi, \Psi) := \|\mathrm{Gram}(\Phi) - \mathrm{Gram}(\Psi)\|$.
(a) Fix $0 < A \leq B$. Show that

$$\min_{t>0} \max\{|tA - 1|, |tB - 1|\} = \min_{t>0} \max\Bigl\{\Bigl|\frac{1}{tA} - 1\Bigr|, \Bigl|\frac{1}{tB} - 1\Bigr|\Bigr\} = \frac{B-A}{A+B},$$

which are attained if and only if $t = \frac{2}{A+B}$ and $t = \frac{A+B}{2AB}$, respectively.
(b) Let A and B be the optimal frame bounds for Φ. Use (a) to show that

$$\min_{c>0} \mathrm{dist}(c\Phi, \Phi^{\mathrm{can}}) = \min_{c>0} \mathrm{dist}(\frac{1}{c}\tilde{\Phi}, \Phi^{\mathrm{can}}) = \frac{B-A}{A+B} < 1,$$

which are attained for $c^2 = \frac{2}{A+B}$ and $c^2 = \frac{A+B}{2AB}$, respectively.
(c) Fix $0 < A \leq B$. Show that

$$\min_{t>0} \max\Bigl\{|tA - 1|, |tB - 1|, \Bigl|\frac{1}{tA} - 1\Bigr|, \Bigl|\frac{1}{tB} - 1\Bigr|\Bigr\} = \sqrt{\frac{B}{A}} - 1,$$

which is attained if and only if $t = 1/\sqrt{AB}$.
(d) Let A and B be the optimal frame bounds for Φ. Show that

$$\min_{c>0} \max\bigl\{\mathrm{dist}(c\Phi, \Phi^{\mathrm{can}}), \mathrm{dist}(\frac{1}{c}\tilde{\Phi}, \Phi^{\mathrm{can}})\bigr\} = \sqrt{\frac{B}{A}} - 1,$$

which is attained if and only if $c^4 = \frac{1}{AB}$.
(e) Suppose Φ is normalised, i.e., $AB = 1$. Let $c > 0$ and U be unitary. Show that

$$\mathrm{dist}(\Phi, cU\Phi^{\mathrm{can}}) = \mathrm{dist}(\tilde{\Phi}, cU\Phi^{\mathrm{can}}) = \max\Bigl\{|A - c^2|, \Bigl|\frac{1}{A} - c^2\Bigr|\Bigr\},$$

and

$$\min_{c>0} \max\Bigl\{|A - c^2|, \Bigl|\frac{1}{A} - c^2\Bigr|\Bigr\} = \frac{1 - A^2}{2A},$$

with the minimum occurring for $c^2 = \frac{1}{2}(A + \frac{1}{A}) = \frac{A^2+1}{2A}$.

3.28. Let $0 < A \le B$. Show that

$$\max\{|A-1|,|B-1|\} = \max\Big\{\Big|\frac{1}{A}-1\Big|,\Big|\frac{1}{B}-1\Big|\Big\} \iff AB = 1.$$

3.29. Prove the following *one-sided* version of Proposition 3.2.
If $\Phi = (f_j)$ and $\Psi = (g_j)$ are finite frames, with the same index set, and synthesis operators V and W, then the following are equivalent
(a) $\Psi = Q\Phi$ for some linear map Q (possibly not invertible).
(b) $\mathrm{Gram}(\Phi^{\mathrm{can}})\,\mathrm{Gram}(\Psi^{\mathrm{can}}) = \mathrm{Gram}(\Psi^{\mathrm{can}})$.
(c) $\mathrm{ran}(W^*) \subset \mathrm{ran}(V^*)$, or, equivalently, $\ker(V) \subset \ker(W)$.
Furthermore, these imply the equivalent conditions
(d) VW^* is 1–1.
(e) $\sum_j \langle g, g_j\rangle f_j \neq 0$, $\forall g$.

3.30. Suppose that $\Psi = (g_j)_{j\in J}$ is *close* to a finite frame $\Phi = (f_j)_{j\in J}$ for $\mathscr{H}$, i.e.,

$$\|\sum_{j\in J} c_j(g_j - f_j)\| \le \lambda \|\sum_{j\in J} c_j f_j\|, \qquad \forall c \in \ell_2(J).$$

Use Exer. 3.29 to show
(a) $\Psi = Q\Phi$ for some linear map Q.
(b) The *closeness bound* (smallest λ) is $\mathrm{cl}(\Psi,\Phi) = \mathrm{cl}(Q\Phi,\Phi) = \|Q - I\|$.
(c) If $\lambda := \mathrm{cl}(\Psi,\Phi) < 1$, then Φ is close to Ψ, with $\mathrm{cl}(\Phi,\Psi) < \frac{\lambda}{1-\lambda}$. In particular, the Q of (b) is invertible (since $\|Q - I\| < 1$).

3.31. Let $S : X \to Y$ be a bounded linear map between normed linear spaces, and $M : Y \to X$ be an approximate left inverse of S, i.e., a bounded linear map with

$$\|I - MS\| < 1.$$

(a) Show that S is invertible.
(b) Let $k \in X$. Show that the map $F : X \to X$ given by

$$F(g) := (I - MS)g + k$$

is a contraction map with constant $\kappa = \|I - MS\|$, and its fixed point g satisfies

$$MSg = k,$$

i.e., MS is invertible.
(c) Show that $(MS)^{-1}$ is bounded.

3.32. Let A and B be (possibly not optimal) frame bounds for a finite frame Φ for $\mathscr{H}$.
(a) Show that $M := \frac{2}{A+B} I$ is an approximate left inverse for $S = S_\Phi$, i.e.,

$$\|I - \frac{2}{A+B} S\| \le \frac{B-A}{A+B}.$$

(b) Consider the contraction mapping $F:\mathscr{H}\to\mathscr{H}$ given by

$$F(g) := (I-MS)g+Mh,$$

which has fixed point g satisfying $MSg = Mh$, i.e., $g = S^{-1}h$ (see Exer. 3.31). Estimate the error in the fixed point iteration method

$$g_0 := 0, \qquad g_{k+1} := (I-MS)g_k + Mh = g_k - \frac{2}{A+B}Sg_k + \frac{2}{A+B}h,$$

for finding $g = S^{-1}h$ as $\lim g_n$.

3.33. Let A and B be (possibly not optimal) frame bounds for a finite frame Φ, with frame operator $S = S_\Phi$. Show that $S^{-\frac{1}{2}}$ can be calculated via the absolutely convergent series

$$S^{-\frac{1}{2}} = \sqrt{\frac{2}{A+B}}\sum_{j=0}^{\infty}\frac{(2j)!}{2^{2j}(j!)^2}\Big(I-\frac{2}{A+B}S\Big)^j.$$

3.34. Let (f_j) be the tight frame for $\mathbb{R}^2$ given by the three equally spaced vectors $f_j = (\cos\frac{2\pi j}{3}, \sin\frac{2\pi j}{3})$, $0\le j\le 2$ and $(\tilde{f}_j)$, $\tilde{f}_j = \frac{2}{3}f_j$, be the canonical dual frame.
(a) Show that all dual frames (g_j) of (f_j) are given by

$$g_j = \tilde{f}_j + w, \qquad w\in\mathbb{R}^2.$$

(b) From (a), it follows that there is a dual (g_j) of Φ with $g_1\in\mathbb{R}^2$ arbitrary. Show that there is a pseudodual (g_j) of Φ with $g_1, g_2\in\mathbb{R}^2$, $g_1\ne g_2$ arbitrary.

3.35. Let $\Phi = (f_j)_{j\in J}$ be a finite frame for $\mathscr{H}$, with synthesis operator $V = [f_j]$, and $Q:\ell_2(J)\to\ell_2(J)$ be an invertible linear map. Define a norm on the linear maps $\ell_2(J)\to\mathscr{H}$ by

$$\|W\|_Q := \|WQ^{-1}\|_F \quad \text{(Frobenius norm)}.$$

(a) Show that frames with synthesis operators V and W are dual if and only if the frames with synthesis operators VQ^* and $U = WQ^{-1}$ are dual.
(b) Show that there is a unique dual frame (g_j) of Φ minimising $\|W\|_Q$, $W = [g_j]$, given by

$$W = (VQ^*QV^*)^{-1}VQ^*Q.$$

Remark. For $J = \{1,\ldots,n\}$ and $Q^{-1} = (D^*)^r$, $r\ge 1$, where D is the first-order difference operator

$$D = \begin{pmatrix} 1 & -1 & & & \\ & 1 & -1 & & \\ & & \ddots & \ddots & \\ & & & 1 & -1 \\ & & & & 1 \end{pmatrix},$$

the dual frames of part (b) are called r-th order **Sobolev duals** of Φ (see [BLPY10]).

3.36. Let $\Phi = (f_j)$ be a finite frame with frame bounds A and B, and $\hat{\Phi} = (\hat{f}_j)$ be a perturbation satisfying

$$\sum_j |\langle f, \hat{f}_j - f_j \rangle|^2 \leq R\|f\|^2, \qquad \forall f.$$

(a) Show that if $R < A$, then $\hat{\Phi}$ is a frame with bounds $(\sqrt{A}-\sqrt{R})^2$ and $(\sqrt{B}+\sqrt{R})^2$.
(b) Show that if $R < \frac{A}{4}$, then the canonical dual of $\hat{\Phi}$ is an approximate dual of Φ.

3.37. Show that a finite frame Φ with optimal frame bounds A and B is approximately dual to the scalar multiple $c\Phi$ if and only if $0 < c < \frac{2}{B}$, and that $\|VW^*\|$ is minimised by the choice $c = \frac{2}{A+B}$.

3.38. Suppose that $\mathscr{V}$ and $\mathscr{W}$ are finite dimensional subspaces of $\mathscr{H}$.
(a) Show that the following are equivalent
 (i) $\mathscr{V} \cap \mathscr{W}^\perp = \mathscr{W} \cap \mathscr{V}^\perp = 0$.
 (ii) $\mathscr{H} = \mathscr{V} \oplus_a \mathscr{W}^\perp = \mathscr{W} \oplus_a \mathscr{V}^\perp$ (algebraic direct sums).
(b) Show that if $\mathscr{V} \cap \mathscr{W}^\perp = \mathscr{W} \cap \mathscr{V}^\perp = 0$, then the oblique projections satisfy

$$\dim(\mathscr{V}) = \dim(\mathscr{W}), \qquad (P_{\mathscr{V},\mathscr{W}^\perp})^* = P_{\mathscr{W},\mathscr{V}^\perp}.$$

3.39. Let (f_j) be a finite frame for $\mathscr{V}$ and (g_j) be its canonical oblique dual in $\mathscr{W}$. Show that the canonical oblique dual of (g_j) in $\mathscr{V}$ is (f_j).

3.40. Suppose $(f_j)_{j=1}^n$ is a finite sequence in a linear space X, and $(\lambda_k)_{k=1}^m$ is a finite sequence of linear functionals $X \to \mathbb{F}$. Let $V := [f_j] : \mathbb{F}^n \to X$ and $\Lambda' := [\lambda_k] : \mathbb{F}^m \to X'$, i.e., $\Lambda(f) := (\lambda_k(f))_{k=1}^m$, and

$$P := V(\Lambda V)^\dagger \Lambda, \qquad \mathscr{V} := \operatorname{span}(f_j) \subset X, \quad \mathscr{L} := \operatorname{span}(\lambda_k) \subset X'.$$

(a) Show that $P : X \to X$ is a projection, i.e., $P^2 = P$.
(b) Show that $\Lambda(Pf) = \Lambda(f)$, $\forall f \in \mathscr{V} = \operatorname{ran}(V)$.
(c) Show that $\operatorname{rank}(P) = \operatorname{rank}(\Lambda V) = \operatorname{rank}(\Lambda|_{\mathscr{V}}) = \dim(\mathscr{L}|_{\mathscr{V}})$.
(d) Show that if $\dim(\mathscr{V}) \leq \dim(\mathscr{L}|_{\mathscr{V}})$, then P projects onto $\mathscr{V}$, i.e., $PV = V$.
(e) Show that if $\dim(\mathscr{L}) \leq \dim(\mathscr{L}|_{\mathscr{V}})$, then P interpolates $\mathscr{L}$, i.e., $\Lambda P = \Lambda$.
(f) What is the formula for P if X is a Hilbert space, and so $\lambda_k(f) = \langle f, g_k \rangle$.

3.41.[m] *Orthogonal polynomials*. For a suitable (nonnegative) *weight* function w on (a,b), an inner product can be defined on the univariate polynomials by

$$\langle f, g \rangle := \int_a^b f(t)g(t)\,w(t)\,dt.$$

If the Gram–Schmidt algorithm is applied to the first $n+1$ monomials, or any other sequence of polynomials with degrees $0,1,\ldots,n$, then polynomials obtained are (for various weights) the *classical orthogonal polynomials*.

Fix some standard weight, e.g., the *Legendre weight* $w = 1$ $(a = -1, b = 1)$. Investigate the orthonormal basis obtained by taking the canonical tight frame for the monomials (*Löwdin orthogonalisation*). What happens for other bases?

Chapter 4
Canonical coordinates for vector spaces and affine spaces

If $(f_j)_{j\in J}$ is a finite frame for $\mathscr{H}$, i.e., a spanning sequence for $\mathscr{H}$, then each $f \in \mathscr{H}$ can be written

$$f = \sum_j c_j f_j,$$

for some choice of coefficients $c = (c_j)$. The unique coefficients which minimise

$$\|c\|^2 = \sum_j |c_j|^2$$

are $c_j = c_j(f) := \langle f, \tilde{f}_j \rangle$, where $(\tilde{f}_j)$ is the canonical dual frame. We observe that

The linear functionals $f \mapsto c_j(f)$ do not depend on the inner product on $\mathscr{H}$.

In this way, it is possible to extend the frame expansion (and other elements of frame theory) to any finite spanning sequence for a vector space over any subfield of the complex numbers which is closed under conjugation.

The linear functionals $f \mapsto c_j(f)$ will be called the *canonical coordinates* for f with respect to the spanning sequence (f_j) for the vector space $X = \mathrm{span}\{f_j\}$. They depend only on the vector space structure of X, though they can be calculated via the canonical dual frame if X is endowed with an inner product. They generalise the dual basis (the case when the vectors are linearly independent), and are characterised by the fact that the associated Gramian matrix is an *orthogonal projection*. The unique inner product for which a finite spanning sequence is a normalised tight frame is given by the Euclidean inner product between the canonical coordinates of vectors.

The canonical coordinates are ideally suited to situations where there is a natural spanning set for a vector space (which is not a basis), e.g., the n-th roots of unity in cyclotomic field (as vector space over the rationals). In such cases, computations can be done directly with the canonical coordinates, in an efficient and stable way, which preserves the geometry of the spanning sequence.

S.F.D. Waldron, *An Introduction to Finite Tight Frames*, Applied and Numerical Harmonic Analysis, https://doi.org/10.1007/978-0-8176-4815-2_4

4.1 The canonical Gramian of a spanning sequence

Throughout, let X be a finite dimensional vector space over a subfield $\mathbb{F}$ of $\mathbb{C}$. To be able to calculate $\sum_j |c_j|^2$ for a vector $c = (c_j) \in \mathbb{F}^J$, it is necessary that $\mathbb{F}$ be closed under conjugation. Therefore, from now on:

We assume that $\mathbb{F}$ is closed under complex conjugation.

Recall (Theorem 2.2) that if $\Phi = (f_j)_{j\in J}$ is finite normalised tight frame with $P = \mathrm{Gram}(\Phi)$ (an orthogonal projection), then Φ is unitarily equivalent to (Pe_j), the columns of P. In particular, they have the same **(linear) dependencies**, i.e.,

$$\mathrm{dep}(\Phi) := \{c \in \mathbb{F}^J : \sum_j c_j f_j = 0\} = \{c \in \mathbb{F}^J : \sum_j c_j P e_j = Pc = 0\} = \ker(P).$$

Thus P is the orthogonal projection onto $\mathrm{dep}(\Phi)^\perp$. We now show that the orthogonal decomposition of $\mathbb{F}^J$ extends to the case when $\mathbb{F}$ is not $\mathbb{R}$ or $\mathbb{C}$, e.g., $\mathbb{F} = \mathbb{Q}$.

Lemma 4.1. *(Orthogonal projections in $\mathbb{F}^J$). Suppose $\overline{\mathbb{F}} = \mathbb{F}$, and $\mathscr{W}$ is a subspace of $\mathbb{F}^J$ (J finite). Then there is the orthogonal direct sum decomposition*

$$\mathbb{F}^J = \mathscr{W} \oplus \mathscr{W}^\perp, \qquad \mathscr{W}^\perp := \{x \in \mathbb{F}^J : \langle x, a\rangle = 0, \forall a \in \mathscr{W}\},$$

and matrices $Q, P \in \mathbb{F}^{J\times J}$ giving the orthogonal projections onto $\mathscr{W}$ and $\mathscr{W}^\perp$.

Proof. It suffices to show there exists matrix Q giving the orthogonal projection onto $\mathscr{W}$ (then take P to be the complementary orthogonal projection $P = I - Q$).

We can apply Gram–Schmidt orthogonalisation, without normalising, to any spanning set for $\mathscr{W}$ to obtain an orthogonal basis of vectors $\{w_1, \ldots, w_r\}$ in $\mathbb{F}^J$ (it may not be possible to normalise and stay within the field $\mathbb{F}$). The j-th column of the matrix Q is then defined by

$$Qe_j = \sum_{k=1}^{r} \frac{\langle e_j, v_k\rangle}{\langle v_k, v_k\rangle} v_k \in \mathbb{F}^J.$$

By construction, this is the orthogonal projection of e_j on $\mathscr{W}$. □

Definition 4.1. Let $\Phi = (f_j)_{j\in J}$ be a finite sequence in a vector space X over $\mathbb{F}$ (with $\overline{\mathbb{F}} = \mathbb{F}$). The **canonical Gramian** of Φ, denoted by $P_\Phi \in \mathbb{F}^{J\times J}$, is the orthogonal projection onto $\mathrm{dep}(\Phi)^\perp$.

By Lemma 4.1, P_Φ is well defined and can be calculated by

$$P_\Phi = I - \sum_{k=1}^{r} \frac{v_k v_k^*}{\langle v_k, v_k\rangle}, \tag{4.1}$$

where (v_k) is any orthogonal basis for $\mathrm{dep}(\Phi)$. Such a (v_k) can be obtained by applying Gram–Schmidt (without normalising) to any spanning set for $\mathrm{dep}(\Phi)$.

We can generalise Proposition 3.2 as follows.

Proposition 4.1. *(Similarity) Let $\Phi = (f_j)_{j\in J}$ and $\Psi = (g_j)_{j\in J}$ be finite spanning sequences. Then the following are equivalent*

1. *Φ and Ψ are similar, i.e., there is an invertible linear map $Q : f_j \mapsto g_j$.*
2. *$P_\Phi = P_\Psi$.*
3. *$\mathrm{dep}(\Phi) = \mathrm{dep}(\Psi)$.*

In other words:

Every spanning sequence Φ corresponds to a unique normalised tight frame determined by $\mathrm{dep}(\Phi)$ (with Gramian P_Φ).

Example 4.1. (Basis) If $\Phi = (f_j)$ is a basis, then $\mathrm{dep}(\Phi) = 0$ and so $P_\Phi = I$.

Example 4.2. (Simplex) Suppose that $\Phi = (f_j)_{j=1}^n$ has just one dependency

$$f_1 + f_2 + \cdots + f_n = 0.$$

Then $\mathbf{1} = (1,\ldots,1)$ spans $\mathrm{dep}(\Phi)$, so that $P_\Phi = I - \frac{1}{n}\mathbf{1}^*\mathbf{1}$, $d = \dim(X) = n-1$, i.e.,

$$(P_\Phi)_{jk} = \begin{cases} \frac{d}{d+1}, & j = k; \\ \frac{-1}{d+1}, & j \neq k. \end{cases}$$

Example 4.3. (Frames) Let $(f_j)_{j\in J}$ be a finite frame with synthesis map $V = [f_j]$. Then P_Φ can be calculated as above from a spanning sequence for $\mathrm{dep}(\Phi) = \ker(V)$, or by

$$P_\Phi = V^* S^{-1} V, \qquad S := VV^*. \tag{4.2}$$

This follows, since $V^*S^{-1}V$ is clearly an orthogonal projection, and its kernel is $\mathrm{dep}(\Phi) = \ker(V)$ (since V is onto, V^* is 1–1). The (j,k)-entry of P_Φ is

$$\begin{aligned}(P_\Phi)_{jk} &= (S^{-1}Ve_j)^* S^{-1}(Ve_k) = (S^{-1}f_j)^* f_k = \langle f_k, \tilde{f}_j\rangle \\ &= (S^{-\frac{1}{2}} f_j)^* S^{-\frac{1}{2}} f_k = \langle f_k^{\mathrm{can}}, f_j^{\mathrm{can}}\rangle.\end{aligned} \tag{4.3}$$

In particular, P_Φ is the Gramian of the canonical tight frame associated with Φ.

Example 4.4. (Equiangular tight frames) We can use the identification of a spanning sequence Φ with the unique normalised tight frame with Gramian P_Φ to extend elements of frame theory to spanning sequences. For example, we say a spanning sequence Φ is **equiangular** if the tight frame given by the columns of $P_\Phi = [p_{jk}]$ is, i.e., P_Φ has a constant diagonal and $|p_{jk}| = C$, $j \neq k$. In this way, an equiangular tight frame could be given by a spanning sequence for a vector space X (which is not $\mathbb{C}^d$), as in Example 4.2.

4.2 The canonical coordinates of a spanning sequence

We now define the *canonical coordinates* via the canonical Gramian.

Theorem 4.1. *Let $\Phi = (f_j)_{j\in J}$ be a finite spanning sequence for a vector space X over $\mathbb{F}$ (with $\overline{\mathbb{F}} = \mathbb{F}$) and $f = \sum_j a_j f_j \in X$, $a \in \mathbb{F}^J$. Then there are unique coefficients $c = c^\Phi(f) \in \mathbb{F}^J$ of minimal ℓ_2-norm with*

$$f = \sum_j c_j(f) f_j, \tag{4.4}$$

which are given by

$$c^\Phi(f) = P_\Phi a, \tag{4.5}$$

where P_Φ is the canonical Gramian of Φ.

Proof. Let $V = [f_j]$ (the synthesis map for Φ).

Since $I - P_\Phi$ is the orthogonal projection onto $\mathrm{dep}(\Phi) = \ker(V)$, we have

$$V = V\big(P_\Phi + (I - P_\Phi)\big) = VP_\Phi. \tag{4.6}$$

Thus, $f = Va = V(P_\Phi a) = V(c^\Phi(f)) = \sum_j c_j(f) f_j$.

Finally, we show the choice $c = c^\Phi$ has minimal ℓ_2-norm. For any choice of coefficients $c \in \mathbb{F}^J$, we have the orthogonal decomposition

$$c = (c - P_\Phi c) + P_\Phi c \in \mathrm{dep}(\Phi) \oplus \mathrm{dep}(\Phi)^\perp,$$

and so Pythagoras gives

$$\|c\|^2 = \|c - P_\Phi c\|^2 + \|P_\Phi c\|^2.$$

If $f = \sum_j c_j f_j$, then $c^\Phi = P_\Phi c$, and we obtain $\|c\|^2 \geq \|c^\Phi\|^2$ with equality if and only if $c = c^\Phi$. □

It follows from (4.5), that $f \mapsto c_j^\Phi(f)$, $j \in J$, are linear functionals and

$$P_\Phi = [c_j^\Phi(f_k)]_{j,k\in J}. \tag{4.7}$$

Definition 4.2. The **canonical coordinates** (or **canonical dual functionals**) for a finite spanning sequence $\Phi = (f_j)$ for an $\mathbb{F}$-vector space X are the linear functionals $c^\Phi = (c_j^\Phi)$ given by

$$c^\Phi(f) := P_\Phi a, \qquad f = \sum_j a_j f_j,$$

where P_Φ is the canonical Gramian of Φ. We call (4.4) the **canonical expansion**, and (4.6) the **canonical factorisation** of the synthesis map $V = [f_j]$.

By Theorem 4.1, the canonical coordinates c^Φ are well defined. They are the minimal ℓ_2-norm coefficients (c_j) for which $f = \sum_j c_j f_j$, and

The linear map

$$L_\Phi : X \to \operatorname{ran}(P_\Phi) = \operatorname{dep}(\Phi)^\perp : f \mapsto c^\Phi(f)$$

is a vector space isomorphism, with $L_\Phi(f_j) = c^\Phi(f_j) = P_\Phi e_j$. In particular,

$$c^\Phi(f) = 0 \quad \iff \quad f = 0.$$

Example 4.5. (Dual basis) If $\Phi = (f_j)$ is a basis, then $P_\Phi = I$, and (4.7) implies that the canonical coordinates are the dual basis. The canonical coordinates therefore generalise the dual functionals to the case when (f_j) is not a basis.

Example 4.6. (Frames) If $\Phi = (f_j)$ is a frame, then (4.3) and (4.7) give

$$c_j^\Phi(f) = \langle f, \tilde{f}_j \rangle, \qquad \forall f,$$

i.e., $\tilde{f}_j$ is the Riesz representer of the linear functional c_j^Φ.

Example 4.7. (Matrices) We can define *matrices* with respect to spanning sequences in the usual way. The **(canonical) matrix** representing a linear map $L : X \to Y$ with respect to spanning sequences $\Phi = (f_j)_{j=1}^n$ and $\Psi = (g_k)_{k=1}^m$ for X and Y is the $A = A_L \in \mathbb{F}^{m\times n}$ given by

$$j\text{-th column of } A = Ae_j = c^\Psi(Lf_j).$$

i.e., $c^\Psi(Lf) = A(c^\Phi(f))$, $\forall f \in X$. The map $L \mapsto A_L$ is linear, and L can be recovered from $A = A_L$ via

$$L = WAc^\Phi, \qquad W = [g_k].$$

The canonical coordinates for a Φ which spans X can be computed from any spanning sequence (λ_k) for the algebraic dual space X' as follows.

Proposition 4.2. *Let $\Phi = (f_1,\ldots,f_n)$ span X. If $\Lambda = (\lambda_k)_{k=1}^m : X \to \mathbb{F}^m$ is any 1–1 linear map, i.e., $\lambda_1,\ldots,\lambda_m$ span X', then the canonical coordinates for Φ are given by*

$$c^\Phi(f) = (\Lambda V)^+ \Lambda f, \qquad V = [f_1,\ldots,f_n], \tag{4.8}$$

where $(\Lambda V)^+$ is the pseudoinverse of the matrix $\Lambda V \in \mathbb{F}^{m\times n}$.

Proof. Let $a = c(f) \in \mathbb{F}^n$ be a solution to (4.4), i.e., to $Va = f$. Since the linear functionals $\lambda_1,\ldots,\lambda_m$ span X', $Va = f$ is equivalent to $\lambda_k(Va) = \lambda_k(f)$, $\forall k$, i.e.,

$$\Lambda V a = \Lambda f,$$

where $\Lambda V \in \mathbb{F}^{m\times n}$ and $\Lambda f \in \mathbb{F}^n$. This (possibly) underdetermined linear system has a unique minimal ℓ_2-norm (least squares) solution $a = c^\Phi(f)$ given by (4.8). □

4.3 A characterisation of the canonical coordinates

Let $\Phi = (f_j)_{j\in J}$ span a vector space X, and $\Psi = (\lambda_j)_{j\in J}$ be linear functionals on X. We say that Φ and Ψ are **dual** sequences (see Exer. 4.1) if

$$f = \sum_j \lambda_j(f) f_j, \qquad \forall f \in X. \tag{4.9}$$

This implies (f_j) spans X and (λ_j) spans X', and is equivalent to

$$\lambda = \sum_j \lambda(f)\lambda_j, \qquad \forall \lambda \in X'.$$

The matrix $G = [\lambda_j(f_k)]$ is called the **Gramian** of Φ and Ψ. We now show that:

> (λ_j) are the canonical coordinates (canonical dual functionals) for (f_j) if and only if $G = [\lambda_j(f_k)]$ is an orthogonal projection of rank $d = \dim(\text{span}\{f_j\})$.

Recall that the canonical isomorphism between X and its bidual X'' is

$$\hat{}: X \to X'' : f \mapsto \hat{f}, \qquad \hat{f}(\lambda) := \lambda(f), \quad \forall \lambda \in X'.$$

Theorem 4.2. *(Characterisation). Suppose X is an $\mathbb{F}$-vector space (with $\overline{\mathbb{F}} = \mathbb{F}$), $\Phi = (f_j)$ in X and $\Psi = (\lambda_j)$ in X' are dual, i.e., satisfy (4.9), and G is the Gramian*

$$G = \Lambda V = [\lambda_j(f_k)], \qquad V := [f_j] : \mathbb{F}^J \to X, \quad \Lambda = (\lambda_j) : X \to \mathbb{F}^J.$$

Then the following are equivalent

1. $c^\Phi = \Psi$.
2. $c^\Psi = \hat{\Phi}$.
3. $G = G^*$, *i.e., G is an orthogonal projection.*
4. $P_\Phi = G$.
5. $P_\Psi = G^T$.
6. $P_\Psi = \overline{P_\Phi^T}$.
7. $\text{dep}(\Psi) = \overline{\text{dep}(\Phi)}$.

Proof. First, observe that (4.9) can be written as $V\Lambda = I_X$, and so

$$G^2 = \Lambda(V\Lambda V) = \Lambda V = G,$$

i.e., G is a projection of rank $d = \dim(X)$.
1.$\Longrightarrow$2. Suppose that $\Psi = c^\Phi$. Then $\Psi = (c_j^\Phi)$ spans X' and $(\hat{f}_j)$ spans X'', and so Proposition 4.2 gives

$$c^\Psi = (LW)^+ L, \qquad W := [c_j^\Phi] : \mathbb{F}^J \to X', \quad L = (\hat{f}_j) : X' \to \mathbb{F}^J.$$

Now $LW = [\hat{f}_j(c_k^\Phi)] = [c_k^\Phi(f_j)] = P_\Phi^T$ is an orthogonal projection, so $(LW)^+ = LW$, and we obtain

$$\begin{aligned} c_j^\Psi(\lambda) &= (P_\Phi^T L\lambda)_j = \sum_k (P_\Phi^T)_{jk}(L\lambda)_k = \sum_k c_k^\Phi(f_j)\hat{f}_k(\lambda) = \sum_k c_k^\Phi(f_j)\lambda(f_k) \\ &= \lambda\Big(\sum_k c_k^\Phi(f_j)f_k\Big) = \lambda(f_j) = \hat{f}_j(\lambda), \quad \forall \lambda \in X' \quad \Longrightarrow \quad c_j^\Psi = \hat{f}_j. \end{aligned}$$

2.$\Longrightarrow$5. If $(\hat{f}_j)$ are the canonical coordinates for Ψ, then (4.7) gives

$$P_\Psi = [\hat{f}_j(\lambda_k)] = [\lambda_k(f_j)] = G^T.$$

5.$\Longrightarrow$4. If $P_\Psi = G^T$, then G is an orthogonal projection matrix of rank $d = \dim(X)$. Moreover,

$$P_\Phi G = (c^\Phi V)(\Lambda V) = c^\Phi (V\Lambda)V = c^\Phi V = P_\Phi,$$

and so $G = P_\Phi$.
4.$\Longrightarrow$3. Immediate (P_Φ is an orthogonal projection).
3.$\Longrightarrow$7. Suppose $G = G^*$, i.e., $\lambda_j(f_k) = \overline{\lambda_k(f_j)}$, $\forall k$. Then

$$\begin{aligned} a \in \operatorname{dep}(\Psi) \quad &\Longleftrightarrow \quad \sum_j a_j\lambda_j(f_k) = 0, \quad \forall k \quad \Longleftrightarrow \quad \sum_j a_j\overline{\lambda_k(f_j)} = 0, \quad \forall k \\ &\Longleftrightarrow \quad \sum_j \overline{a_j}\lambda_k(f_j) = 0, \quad \forall k \quad \Longleftrightarrow \quad \lambda_k\Big(\sum_j \overline{a_j}f_j\Big) = 0, \quad \forall k \\ &\Longleftrightarrow \quad \sum_j \overline{a_j}f_j = 0 \quad \Longleftrightarrow \quad \overline{a} \in \operatorname{dep}(\Phi). \end{aligned}$$

7.$\Longrightarrow$6. Suppose that $\operatorname{dep}(\Psi) = \overline{\operatorname{dep}(\Phi)}$. Let (v_k) be an orthogonal basis for $\operatorname{dep}(\Phi)$, so $(\overline{v_k})$ is an orthogonal basis for $\operatorname{dep}(\Psi)$, and from (4.1), we obtain

$$P_\Psi = I - \sum_k \frac{\overline{v_k}(\overline{v_k})^*}{\langle \overline{v_k}, \overline{v_k}\rangle} = \overline{I - \sum_k \frac{v_k v_k^*}{\langle v_k, v_k\rangle}} = \overline{P_\Phi} = P_\Phi^T.$$

6.$\Longrightarrow$1. For each $f \in X$, we have the orthogonal decomposition

$$\Psi(f) = (\lambda_j(f)) = c^\Phi(f) + \rho(f) \in \operatorname{ran}(P_\Phi) \oplus \ker(P_\Phi).$$

For $a \in \ker(P_\Phi)$, we have

$$\langle \rho(f), a\rangle = \langle \Psi(f), a\rangle = \sum_j \overline{a_j}\lambda_j(f).$$

Now suppose $P_\Phi^T = P_\Psi$, so that $\overline{a} \in \ker(P_\Phi^T) = \ker(P_\Psi)$, and we obtain

$$\langle \rho(f), a\rangle = 0, \quad \forall a \quad \Longrightarrow \quad \rho(f) = 0,$$

i.e., $\Psi = c^\Phi$. □

Example 4.8. The second equivalence shows that any spanning sequence Ψ for X' is the canonical coordinates for some Φ, i.e., the Φ given by $\hat{\Phi} = c^{\Psi}$. For example, let $X = \Pi_1$ the (three dimensional) space of linear polynomials on $\mathbb{R}^2$ and Ψ be the point evaluations

$$\Psi = \big(\delta_{(0,0)}, \delta_{(1,0)}, \delta_{(0,1)}, \delta_{(a,b)}\big), \qquad \delta_x : \Pi_1 \to \mathbb{R} : f \mapsto f(x).$$

These are the canonical coordinates for Φ consisting of the linear polynomials (with the obvious indexing)

$$\begin{aligned}
f_{(0,0)}(x,y) &= \frac{(ab-1-a-b^2)x+(ab-1-b-a^2)y+1+a^2+b^2}{(a+b)^2+(a-1)^2+(b-1)^2}, \\
f_{(1,0)}(x,y) &= \frac{(2+ab-a+2b^2-2b)x+(a-a^2-2ab)y+ab+a^2-a}{(a+b)^2+(a-1)^2+(b-1)^2}, \\
f_{(0,1)}(x,y) &= \frac{(b-b^2-2ab)x+(2+ab-b+2a^2-2a)y+ab+b^2-b}{(a+b)^2+(a-1)^2+(b-1)^2}, \\
f_{(a,b)}(x,y) &= \frac{(2a+b-1)x+(a+2b-1)y+1-a-b}{(a+b)^2+(a-1)^2+(b-1)^2}.
\end{aligned} \tag{4.10}$$

Note that these polynomials are continuous functions of (a,b).

4.4 Properties of the canonical coordinates

Because $P_\Phi = [c_j^{\Phi}(f_k)]$ is an orthogonal projection, the canonical coordinates share many properties of the coordinates for a basis (where $P_\Phi = I$). We write the sequence obtained by removing the vector f_j from $\Phi = (f_1,\ldots,f_n)$ as

$$\Phi \setminus f_j := (f_1,\ldots,f_{j-1},f_{j+1},\ldots,f_n).$$

Proposition 4.3. *(Properties). The canonical coordinates $c^{\Phi} = (c_j)$ for Φ satisfy*

1. $c_j(f_k) = \overline{c_k(f_j)}$.
2. $|c_j(f_k)| \le 1$ *with* $|c_j(f_k)| = 1$ *if and only if* $k = j$ *and* $f_j \notin \operatorname{span}(\Phi \setminus f_j)$, *in which case* $c_j(f_j) = 1$ *and* $c_j = 0$ *on* $\operatorname{span}(\Phi \setminus f_j)$.
3. $c_j(f_j) \ge 0$, *with* $c_j(f_j) = 0$ *if and only if* $f_j = 0$.
4. $\sum_j c_j(f_j) = d = \dim(X)$.
5. $c_j = \alpha c_k$, $\alpha \in \mathbb{F}$ *if and only if* $f_j = \overline{\alpha} f_k$.

Proof. Let $P = [p_{jk}] = [c_j(f_k)]$. These properties follow from those of orthogonal projection matrices, e.g., $P = P^*$, gives $c_j(f_k) = p_{jk} = \overline{p_{kj}} = \overline{c_k(f_j)}$.

Orthogonal projections P have the property $\|Px\| \le \|x\|$ with equality if and only if $Px = x$, so that

$$|c_j(f_k)| = |p_{jk}| \le \|Pe_k\| \le \|e_k\| = 1,$$

with equality if and only if $Pe_k = e_k$ and $|p_{jk}| = 1$, i.e., $j = k$, $c_j(f_j) = 1$ and $c_j(f_\ell) = 0$, $\ell \neq j$. The condition $Pe_j = e_j$ is equivalent to the j-th column of P not being in the span of the others, and since the columns of P and the vectors of Φ have the same linear dependencies, this is the same as $f_j \notin \operatorname{span}(\Phi \setminus f_j)$.

Similarly,

$$c_j(f_j) = \langle e_j, Pe_j \rangle = \langle Pe_j, Pe_j \rangle = \|Pe_j\|^2 \geq 0,$$

with equality if and only if $Pe_j = 0$, i.e., $c^\Phi(f_j) = 0$, which gives $f_j = 0$.

We have

$$\sum_j c_j(f_j) = \operatorname{trace}(P) = \operatorname{rank}(P) = d.$$

Finally,

$$\begin{aligned} f_j = \overline{\alpha} f_k \quad &\iff \quad c_\ell(f_j) = c_\ell(\overline{\alpha} f_k) = \overline{\alpha} c_\ell(f_k), \quad \forall \ell \\ &\iff \quad c_j(f_\ell) = \alpha c_k(f_\ell), \quad \forall \ell \\ &\iff \quad \sum_\ell c_j(f_\ell) f_\ell = \alpha \sum_\ell c_k(f_\ell) f_\ell \quad \iff \quad f_j = \alpha f_k. \end{aligned}$$

□

Example 4.9. If $\Phi = (f_j)_{j=1}^n$ satisfies $\sum_j f_j = 0$, i.e., $(1, \ldots, 1) \in \operatorname{dep}(\Phi)$, and Q is the orthogonal projection onto $(1, \ldots, 1)^\perp$, then

$$|c_j(f_k)| \leq \|P_\Phi e_k\| \leq \|Q e_k\| = \|e_k - \frac{1}{n} \sum_\ell e_\ell\| = \sqrt{(1 - \frac{1}{n})^2 + \frac{n-1}{n^2}} = \sqrt{1 - \frac{1}{n}}.$$

The canonical coordinates transform naturally under the action of a linear map.

Proposition 4.4. *(Linear maps). Suppose that $\Phi = (f_j)$ spans the $\mathbb{F}$-vector space X, $L : X \to Y$ is an invertible $\mathbb{F}$-linear map, and $\Psi = L\Phi = (Lf_j)$. Then, the canonical coordinates for Φ and Ψ satisfy*

$$c_j^{L\Phi}(Lf) = c_j^\Phi(f), \qquad \forall f \in X. \tag{4.11}$$

Proof. Choose Λ as in (4.8), so that

$$c^\Phi(f) = (\Lambda V)^+ \Lambda f, \qquad V = [f_j].$$

Then, $\Lambda L^{-1} : Y \to \mathbb{F}^m$ is 1–1, and so, with $W = [Lf_j] = LV$, we have

$$c^\Psi(Lf) = (\Lambda L^{-1} W)^+ \Lambda L^{-1}(Lf) = (\Lambda V)^+ \Lambda f = c^\Phi(f).$$

□

The canonical coordinates c^Φ have the same symmetries as Φ (see Example 9.5 of §9.2 for details).

4.5 The canonical inner product for a spanning sequence

We observe that if $\overline{\mathbb{F}} = \mathbb{F}$, then inner products can be defined on $\mathbb{F}$-vector spaces in the usual way. In this case, there is a unique inner product for which a spanning sequence is a normalised tight frame.

Theorem 4.3. *Let $\Phi = (f_j)_{j\in J}$ be a finite spanning sequence for an $\mathbb{F}$-vector space X (with $\overline{\mathbb{F}} = \mathbb{F}$). Then there exists a unique inner product on X for which Φ is a normalised tight frame, namely*

$$\langle f,g\rangle_\Phi := \langle c^\Phi(f), c^\Phi(g)\rangle. \tag{4.12}$$

Proof. The linear map $X \to \mathbb{F}^J : f \mapsto c^\Phi(f)$ is 1–1, and so (4.12) defines an inner product on X. For this,

$$\langle f_k, f_j\rangle_\Phi = \langle P_\Phi e_k, P_\Phi e_j\rangle = \langle P_\Phi e_k, e_j\rangle = (P_\Phi)_{jk},$$

so that $\mathrm{Gram}(\Phi) = P_\Phi$, and Φ is a normalised tight frame (Theorem 2.1).

Conversely, for any inner product $\langle f,g\rangle_X$ on X for which Φ is a normalised tight frame $\mathrm{Gram}(\Phi) = P_\Phi$, and hence (since Φ is a spanning set)

$$\langle f_k, f_j\rangle_\Phi = \langle f_k, f_j\rangle_X, \quad \forall j,k \quad \Longrightarrow \quad \langle f,g\rangle_\Phi = \langle f,g\rangle_X, \quad \forall f,g \in X.$$

This proves the uniqueness. □

Definition 4.3. The **canonical inner product** $\langle\cdot,\cdot\rangle_\Phi$ for a spanning sequence Φ for an $\mathbb{F}$-vector space X is the unique inner product on X for which Φ is a normalised tight frame, i.e., the inner product given by (4.12).

Example 4.10. (Frames) If Φ is a finite frame for $\mathcal{H}$, with frame operator $S = VV^*$, then

$$\langle f,g\rangle_\Phi = \langle S^{-\frac{1}{2}} f, S^{-\frac{1}{2}} g\rangle = \langle f, S^{-1} g\rangle, \qquad \forall f,g \in \mathcal{H}. \tag{4.13}$$

This follows from $c_j^\Phi(f) = \langle f, S^{-1} f_j\rangle = \langle S^{-\frac{1}{2}} f, S^{-\frac{1}{2}} f_j\rangle$ (see Exer. 4.2).

In summary:

Whenever there is a natural spanning sequence for a vector space, it can be viewed as a normalised tight frame (in a unique way), and computations can be done directly with it, in an efficient and stable way. This avoids the need to obtain a basis by thinning which may destroy the inherent geometry.

We now illustrate this principle for the cyclotomic fields (as $\mathbb{Q}$-vector spaces).

4.6 Canonical coordinates for cyclotomic fields

Let ω be the primitive n-th root of unity

$$\omega := e^{\frac{2\pi i}{n}}.$$

The **cyclotomic field** $\mathbb{Q}(\omega)$ is a $\mathbb{Q}$-vector space of dimension $d = \varphi(n)$, where φ is the *Euler phi* (*totient*) function. The number of primitive n-th roots is $\varphi(n)$, but they do not form a basis for $\mathbb{Q}(\omega)$ in general, e.g., the primitive 4-th roots are $\pm i$, which are $\mathbb{Q}$-linearly dependent. For n square free, the primitive n-th roots are a basis. When the primitive roots are not a basis, bases with additional properties can be constructed in a noncanonical way. Most prominently used are the *integral bases* (each element of the ring of integers has its coefficients in $\mathbb{Z}$) and the *power bases* (these have the form $\{1, z, z^2, \ldots, z^{d-1}\}$).

A natural spanning sequence for $\mathbb{Q}(\omega)$ is given by the n-th roots themselves, i.e.,

$$\Phi = (\omega^j)_{j \in \mathbb{Z}_n} = (1, \omega, \omega^2, \ldots, \omega^{n-1}).$$

We now consider the corresponding canonical coordinates $c = c^{\Phi}$. These naturally inherit the geometry of $\mathbb{Q}(\omega)$, e.g., if

$$z = a_0 + a_1\omega + \cdots + a_{n-1}\omega^{n-1}, \qquad a_0, \ldots, a_{n-1} \in \mathbb{Q},$$

then

$$\omega z = a_0\omega + a_1\omega^2 + \cdots + a_{n-1}\omega^n, \qquad \overline{z} = a_0 + a_1\omega^{n-1} + \cdots + a_{n-1}\omega,$$

and so the canonical coordinates satisfy

$$c_j(\omega z) = c_{j+1}(z), \tag{4.14}$$

$$c_j(\overline{z}) = c_{-j}(z), \tag{4.15}$$

i.e., multiplication by ω corresponds to a *forward cyclic shift* of coordinates and complex conjugation to the permutation $j \mapsto -j$ of the indices.

Let μ be the *Möbius function*

$$\mu(n) := \begin{cases} 1, & n = 1; \\ (-1)^n, & \text{n is square free;} \\ 0, & \text{otherwise} \end{cases}$$

which satisfies

$$\sum_{j \in \mathbb{Z}_n^*} \omega^j = \mu(n). \tag{4.16}$$

Here $\mathbb{Z}_n^*$ is the group of units in $\mathbb{Z}_n$ (the primitive n-th roots are ω^j, $j \in \mathbb{Z}_n^*$).

The canonical coordinates can be computed as follows.

Proposition 4.5. *(Calculation). The canonical Gramian for $\Phi = (\omega^j)_{j\in\mathbb{Z}_n}$ is*

$$P_\Phi = \frac{1}{n}\sum_{j\in\mathbb{Z}_n^*} \chi_j\chi_j^*, \qquad \chi_j := (1,\omega^j,\omega^{2j},\ldots,\omega^{(n-1)j})^T, \tag{4.17}$$

which has entries

$$(P_\Phi)_{jk} = \frac{1}{n}\sum_{a\in\mathbb{Z}_n^*}\omega^{a(j-k)} = \frac{1}{n}\varphi(g)\mu\Big(\frac{n}{g}\Big), \qquad g := \gcd(j-k,n).$$

Proof. The vectors χ_j are the characters of $\mathbb{Z}_n$ and hence are orthogonal.
Suppose that $a \in \operatorname{dep}(\Phi)$, i.e.,

$$\chi_{-1}^* a = a_0 + a_1\omega + \cdots a_{n-1}\omega^{n-1} = 0,$$

then applying the Galois action $\omega^{-1}\mapsto\omega^j$, $j\in\mathbb{Z}_n^*$, which fixes $\mathbb{Q}$, gives

$$\chi_j^* a = a_0 + a_1\omega^{-j} + \cdots + a_{n-1}\omega^{-(n-1)} = 0,$$

and so $\chi_j \in \operatorname{dep}(\Phi)^\perp$, $j\in\mathbb{Z}_n$. A dimension count shows that $\{\chi_j : j\in\mathbb{Z}_n^*\}$ is an orthogonal basis for $\operatorname{dep}(\Phi)$, and hence P_Φ is given by (4.17).
Evaluating the entries of P_Φ gives the *Ramanujan sum*

$$(P_\Phi)_{jk} = e_j^*\frac{1}{n}\sum_{a\in\mathbb{Z}_n^*}\chi_a\chi_a^* e_k = \frac{1}{n}\sum_{a\in\mathbb{Z}_n^*}\omega^{-ak}e_j^*\chi_a = \frac{1}{n}\sum_{a\in\mathbb{Z}_n^*}\omega^{a(j-k)}.$$

Using (4.16) and $\varphi(n)=\varphi(g)\varphi(\frac{n}{g})$, this can be simplified to

$$\frac{1}{n}\sum_{a\in\mathbb{Z}_n^*}\omega^{ag\frac{j-k}{g}} = \frac{1}{n}\sum_{a\in\mathbb{Z}_n^*}\omega^{ag} = \frac{1}{n}\varphi(g)\sum_{b\in\mathbb{Z}_{n/g}^*}(\omega^g)^b = \frac{1}{n}\varphi(g)\mu\Big(\frac{n}{g}\Big).$$

□

Example 4.11. The canonical Gramians P_Φ for $n=3,4,5$ are

$$\frac{1}{3}\begin{bmatrix}2&-1&-1\\-1&2&-1\\-1&-1&2\end{bmatrix}, \quad \frac{1}{4}\begin{bmatrix}2&0&-2&0\\0&2&0&-2\\-2&0&2&0\\0&-2&0&2\end{bmatrix}, \quad \frac{1}{5}\begin{bmatrix}4&-1&-1&-1&-1\\-1&4&-1&-1&-1\\-1&-1&4&-1&-1\\-1&-1&-1&4&-1\\-1&-1&-1&-1&4\end{bmatrix}.$$

Example 4.12. For $n=2^k$, the canonical expansion of ω^j is

$$\omega^j = \frac{1}{2}\omega^j - \frac{1}{2}\omega^{j+\frac{n}{2}}.$$

More generally, for $n=p^k$, p a prime, the canonical expansion of ω^j is

$$\omega^j = \frac{1}{p}\{(p-1)\omega^j - \omega^{j+\frac{n}{p}} - \omega^{j+2\frac{n}{p}} - \cdots - \omega^{j+(p-1)\frac{n}{p}}\}.$$

These canonical coordinates have norm

$$\|c(\omega^j)\|_2 = \frac{1}{p}\sqrt{(p-1)^2+(p-1)} = \sqrt{1-\frac{1}{p}} < 1.$$

Multiplication in $\mathbb{Q}(\omega)$ corresponds to convolution of the canonical coordinates:

Proposition 4.6. *The canonical coordinates $c = c^\Phi$ for $\Phi = (\omega^j)_{j\in\mathbb{Z}_n}$ satisfy*

$$c(\alpha x+\beta y) = \alpha c(x)+\beta c(y), \qquad \alpha,\beta \in \mathbb{Q}, \quad x,y \in \mathbb{Q}(\omega), \tag{4.18}$$

$$c(xy) = c(x) * c(y), \qquad x,y \in \mathbb{Q}(\omega), \tag{4.19}$$

where $a*b$ is the **cyclic convolution** *of a and b over $\mathbb{Z}_n$, which is given by*

$$(a*b)_k := \sum_{j=0}^{n-1} a_j b_{k-j}.$$

Proof. The first is just the fact $f \mapsto c(f)$ is linear. For the second, observe that if M is a circulant matrix (such as P_Φ), then

$$M(a*b) = (Ma)*b = a*(Mb).$$

Let $a,b \in \mathbb{Q}^n$ be coordinates for x,y, then

$$xy = \Big(\sum_s a_s\omega^s\Big)\Big(\sum_r b_r\omega^r\Big) = \sum_s\sum_r a_s b_r \omega^{s+r} = \sum_k\sum_j a_j b_{k-j}\omega^k = \sum_k (a*b)_k\omega^k,$$

so that $a*b$ are coordinates for xy, and we have

$$c(xy) = P_\Phi(a*b) = (P_\Phi a)*b = c(x)*b = c(x)*c(y),$$

where for the last equality, we made the particular choice $b = c(y)$. □

The canonical inner product has the following properties.

Proposition 4.7. *(normalised tight frame) The unique $\mathbb{Q}$-inner product on $\mathbb{Q}(\omega)$ for which $\Phi = (\omega^j)_{j\in\mathbb{Z}_n}$ is a normalised tight frame satisfies*

$$\langle xy, z\rangle_\Phi = \langle y, \overline{x}z\rangle_\Phi, \qquad \forall x,y,z \in \mathbb{Q}(\omega) \tag{4.20}$$

$$\langle z, z\rangle_\Phi = \frac{\varphi(n)}{n}|z|^2, \qquad \textit{whenever } |z|^2 \in \mathbb{Q}. \tag{4.21}$$

In particular, multiplication by any $z \in \mathbb{Q}(\omega)$ of unit modulus is a unitary operation.

Proof. In view of (4.15) and (4.19), the first amounts to showing

$$\langle a*b,w\rangle = \langle b,\tilde{a}*w\rangle, \qquad \tilde{a}=(a_{-j}),$$

where $a=c(x)$, $b=c(y)$ and $w=c(z)$. This holds for all a,b,w by direct calculation:

$$\langle a*b,w\rangle = \sum_k\Big(\sum_j a_j b_{k-j}\Big)w_k = \sum_j\sum_k a_j b_{k-j} w_k,$$

$$\langle\!\langle b,\tilde{a}*w\rangle\!\rangle = \sum_k b_k\Big(\sum_j a_{-j}w_{k-j}\Big) = \sum_j\sum_k a_j b_k w_{k+j} = \sum_j\sum_k a_j b_{k-j} w_k.$$

Finally, when $|z|^2\in\mathbb{Q}$, we have

$$\langle z,z\rangle_\Phi = \langle\bar{z}z,1\rangle_\Phi = \langle|z|^2 1,1\rangle_\Phi = |z|^2\langle 1,1\rangle_\Phi.$$

In particular, all the n-roots have the same norm, and so

$$\dim(\mathbb{Q}(\omega)) = \varphi(n) = \operatorname{rank}(P_\Phi) = \operatorname{trace}(P_\Phi) = \sum_j\langle\omega^j,\omega^j\rangle_\Phi = n\langle 1,1\rangle_\Phi.$$

Combining these gives (4.21). □

Example 4.13. This inner product is different from the one induced by viewing the n-th roots of unity as vectors in $\mathbb{R}^2$ (with the Euclidean inner product), which gives a tight frame for $\mathbb{R}^2$. For example, when $n=5$, we have

$$\langle 1,\omega\rangle_\Phi = -\frac{1}{5}, \qquad \langle\begin{pmatrix}1\\0\end{pmatrix},\begin{pmatrix}\cos\frac{2\pi}{5}\\ \sin\frac{2\pi}{5}\end{pmatrix}\rangle = \cos\frac{2\pi}{5}\notin\mathbb{Q}.$$

Here, the coordinates for 1 which minimise $\sum_j|c_j|^2$ over $c\in\mathbb{R}^n$, and over $c\in\mathbb{Q}^n$ (the canonical coordinates), and those given by $1=1\cdot 1+0\cdot\omega+\cdots+0\cdot\omega^{n-1}$ are

$$\begin{pmatrix}\frac{2}{5}\\ \frac{2}{5}\cos\frac{2\pi}{5}\\ \frac{2}{5}\cos\frac{4\pi}{5}\\ \frac{2}{5}\cos\frac{6\pi}{5}\\ \frac{2}{5}\cos\frac{8\pi}{5}\end{pmatrix}, \qquad c^\Phi(1)=\begin{pmatrix}\frac{4}{5}\\ -\frac{1}{5}\\ -\frac{1}{5}\\ -\frac{1}{5}\\ -\frac{1}{5}\end{pmatrix}, \qquad \begin{pmatrix}1\\0\\0\\0\\0\end{pmatrix},$$

respectively, which have norms

$$\frac{2}{5}\sqrt{1+2\cos^2\tfrac{2\pi}{5}+2\cos^2\tfrac{4\pi}{5}}\approx 0.63246, \quad \frac{2}{\sqrt{5}}\approx 0.89442, \frac{2}{\sqrt{5}}\approx 0.89442, \quad 1.$$

Example 4.14. We observe that

$$\omega^j \text{is orthogonal to } \omega^k \iff \mu\Big(\frac{n}{g}\Big)=0, \quad g=\gcd(j-k,n),$$

and so two n-th roots cannot be orthogonal if n is square free.

Example 4.15. (Subfields) Let $n=8$. Then $\omega=\sqrt{i}=e^{\frac{2\pi i}{8}}$ and $\mathbb{Q}(\omega)$ are 4-dimensional, with cyclotomic subfields

$$\mathbb{Q}\subset\mathbb{Q}(i)\subset\mathbb{Q}(\sqrt{i}).$$

The canonical coordinates of $1,\omega,\omega^2,\omega^3$ are

$$c(1)=\begin{pmatrix}\frac{1}{2}\\0\\0\\0\\\frac{1}{2}\\0\\0\\0\end{pmatrix},\quad c(\sqrt{i})=\begin{pmatrix}0\\\frac{1}{2}\\0\\0\\0\\\frac{1}{2}\\0\\0\end{pmatrix},\quad c(i)=\begin{pmatrix}0\\0\\\frac{1}{2}\\0\\0\\0\\\frac{1}{2}\\0\end{pmatrix},\quad c(i\sqrt{i})=\begin{pmatrix}0\\0\\0\\\frac{1}{2}\\0\\0\\0\\\frac{1}{2}\end{pmatrix}.$$

Therefore one can determine what is the smallest cyclotomic subfield of $\mathbb{Q}(\sqrt{i})$ a given element lies in by considering which of its canonical coordinates are zero.

By viewing $\Phi=(\omega^j)_{j\in\mathbb{Z}_n}$ as a normalised tight frame (for the canonical inner product), we can easily show that the natural action of the cyclic group $C_n=\langle a\rangle$ on the $\mathbb{Q}$-vector space $\mathbb{Q}(\omega)$ given by $a\cdot\omega^j=\omega^{j+1}$ is irreducible.

Theorem 4.4. *(Irreducibility) Let* $\langle\!\langle\cdot,\cdot\rangle\!\rangle$ *be the canonical inner product on* $\mathbb{Q}(\omega)$ *given by* $\Phi=(\omega^j)_{j\in\mathbb{Z}_n}$. *For any nonzero* $z\in\mathbb{Q}(\omega)$*, the vectors* $(z,\omega z,\ldots,\omega^{n-1}z)$ *are an equal norm tight frame for* $\mathbb{Q}(\omega)$*, i.e.,*

$$x=\frac{1}{|z|^2}\sum_{j\in\mathbb{Z}_n}\langle\!\langle x,\omega^j z\rangle\!\rangle\omega^j z,\qquad \forall x\in\mathbb{Q}(\omega). \tag{4.22}$$

Proof. Using (4.20) and the fact $(\omega^j)_{j\in\mathbb{Z}_n}$ is a normalised tight frame, we calculate

$$\sum_j\langle\!\langle x,\omega^j z\rangle\!\rangle\omega^j z=\Big(\sum_j\langle\!\langle \bar{z}x,\omega^j\rangle\!\rangle\omega^j\Big)z=(\bar{z}x)z=|z|^2x.$$

Thus every C_n-orbit of $z\neq0$ spans $\mathbb{Q}(\omega)$, i.e., the action is irreducible. □

The *(forward) cyclic shift* operator S on $\mathbb{Q}^n$, which is given by

$$Se_j:=e_{j+1},\qquad j\in\mathbb{Z}_n,$$

and defines a natural action of $C_n=\langle a\rangle$ on $\mathbb{Q}^n$ via

$$a\cdot v=Sv.$$

By (4.14), the canonical coordinates $c=c^\Phi$ for $\Phi=(\omega^j)$ satisfy

$$c(\omega^k z)=S^k c(z). \tag{4.23}$$

Thus it follows that $\operatorname{dep}(\Phi)^\perp$ is an irreducible shift invariant subspace of $\mathbb{Q}^n$.

Corollary 4.1. *(Irreducibility) The shifts of any nonzero* $b \in \operatorname{dep}(\Phi)^\perp$ *are an equal norm tight frame for* $\operatorname{ran}(P_\Phi) = \operatorname{dep}(\Phi)^\perp$ *(with the Euclidean inner product), i.e.,*

$$a = \frac{\varphi(n)}{n} \frac{1}{\langle b,b\rangle} \sum_{j\in\mathbb{Z}_n} \langle a, S^j b\rangle S^j b, \qquad \forall a \in \operatorname{dep}(\Phi)^\perp.$$

Proof. Expanding the canonical inner product in (4.22), using (4.23), gives

$$x = \frac{1}{|z|^2} \sum_{j\in\mathbb{Z}_n} \langle c(x), S^j c(z)\rangle \omega^j z, \qquad \forall x \in \mathbb{Q}(\omega).$$

Applying c to this and letting $a = c(x)$, $b = c(z)$ gives the result. □

The shift invariant subspace $\operatorname{dep}(\Phi)$ of $\mathbb{Q}^n$ is not irreducible when n is not a prime. In this case there is a proper 1-dimensional shift invariant subspace spanned by $(1,1,\ldots,1)$. Nevertheless, we are able to give a single linear dependence a_Φ whose shifts give a tight frame for $\operatorname{dep}(\Phi)$.

Theorem 4.5. *(Shift invariant tight frame) Let* $a_\Phi \in \mathbb{Z}^n$ *be n times the first column of* $I - P_\Phi$*, i.e.,*

$$a_\Phi = \sum_{j\notin\mathbb{Z}_n^*} \chi_j.$$

Then the shifts of a_Φ *are an equal norm tight frame for* $\operatorname{dep}(\Phi)$*, i.e.,*

$$x = \frac{1}{n^2} \sum_{j\in\mathbb{Z}_n} \langle\!\langle x, S^j a_\Phi\rangle\!\rangle S^j a_\Phi, \qquad \forall x \in \operatorname{dep}(\Phi). \tag{4.24}$$

Proof. By (4.17) and the fact the characters χ_j are orthogonal, $I - P_\phi$ is the circulant matrix given by

$$\frac{1}{n} \sum_{j\notin\mathbb{Z}_n^*} \chi_j \chi_j^* = \frac{1}{n}[a_\Phi, S a_\Phi, S^2 a_\Phi, \ldots, S^{n-1} a_\Phi].$$

The columns of the orthogonal projection matrix $I - P_\Phi$ are a normalised tight frame for its range $\operatorname{dep}(\Phi)$ and the corresponding frame expansion is (4.24). □

Example 4.16. For $n = 6$, $\mathbb{Z}_6^* = \{1,5\}$, and

$$I - P_\Phi = \frac{1}{6}\begin{pmatrix} 4 & -1 & 1 & 2 & 1 & -1 \\ -1 & 4 & -1 & 1 & 2 & 1 \\ 1 & -1 & 4 & -1 & 1 & 2 \\ 2 & 1 & -1 & 4 & -1 & 1 \\ 1 & 2 & 1 & -1 & 4 & -1 \\ -1 & 1 & 2 & 1 & -1 & 4 \end{pmatrix}, \qquad a_\Phi = \chi_0 + \chi_2 + \chi_3 + \chi_4 = \begin{pmatrix} 4 \\ -1 \\ 1 \\ 2 \\ 1 \\ -1 \end{pmatrix}.$$

Thus the linear dependencies between the 6-th roots can be expressed as

$$4\omega^j - \omega^{j+1} + \omega^{j+2} + 2\omega^{j+3} + \omega^{j+4} - \omega^{j+5} = 0, \qquad 0 \le j < 6.$$

Here the 4-dimensional subspace dep(Φ) can be decomposed into two 1-dimensional and one 2-dimensional orthogonal shift invariant subspaces, which are generated by

$$(1,1,1,1,1,1)^T, \quad (1,-1,1,-1,1,-1)^T, \quad (0,1,-1,0,1,-1)^T,$$

respectively.

4.7 Generalised barycentric coordinates for affine spaces

We now give the analogue of the canonical coordinates (for a vector space) for affine spaces. An *affine space* X is, in effect, a vector space for which there is no distinguished point that plays the role of the origin in a vector space (or, equivalently, the translation of a vector subspace). As such, we can take *affine combinations* of "points" in X, i.e., linear combinations where the sum of the coefficients is 1 and *differences* of points to obtain "vectors".

Let X be an affine space with *dimension* d, i.e., $d+1$ is the number of points in affinely independent affine spanning set for X. A sequence $v_1,\ldots,v_n$ of $n=d+1$ points in X is affinely independent if and only if each point $x \in X$ can be written uniquely as an affine combination of them, i.e.,

$$x = \sum_j \xi_j(x)\, v_j, \qquad \sum_j \xi_j(x) = 1. \tag{4.25}$$

The affine functions ξ_j, so defined, are the **barycentric coordinates** for $\Theta = (v_j)$.

Definition 4.4. Let X be an affine space over $\mathbb{F}$, with $\overline{\mathbb{F}} = \mathbb{F}$ and $\Theta = (v_1,\ldots,v_n)$ be points with affine span X. Then the **(affine) generalised barycentric coordinates**[1] $\xi^\Theta(x) = (\xi_j(x))_{j=1}^n \in \mathbb{F}^n$ for Θ of a point $x \in X$ are the (unique) coefficients of minimal ℓ_2-norm for which x is an affine combination of Θ, i.e., (4.25) holds.

These are well defined, since the set of vectors $a \in \mathbb{F}^n$ satisfying $x = \sum_j a_j v_j$ and $\sum_j a_j = 1$ form a nonempty affine subspace of $\mathbb{F}^n$.

Proposition 4.8. *(Calculation) The generalised barycentric coordinates $\xi^\Theta = (\xi_j)_{j=1}^n$ for points $\Theta = (v_j)_{j=1}^n$, with barycentre $b = b_\Phi = \frac{1}{n}\sum_{j=1}^n v_j$, are given by*

$$\xi_j^\Theta(x) = c_j^{\Theta - b_\Theta}(x - b_\Theta) + \frac{1}{n}, \tag{4.26}$$

where $c^{\Theta - b_\Theta} = (c_j)$ are the canonical coordinates of $\Theta - b_\Theta = (f_j)$, $f_j := v_j - b_\Theta$. In particular, each $x \mapsto \xi_j^\Theta(x)$ is an affine function.

---

[1] There are many generalisations of barycentric coordinates used in geometric modelling (computer graphics) for nonsimplical polytopes. These are the only ones where each ξ_j is an affine function.

Proof. We seek to minimise $\sum_j |\xi_j(x)|^2$ subject to (4.25). Write $\xi_j(x) = a_j(x) + \frac{1}{n}$. Then $\sum_j \xi_j(x) = 1$ is equivalent to $\sum_j a_j(x) = 0$, and so

$$\sum_j |\xi_j(x)|^2 = \sum_j \Big\{ |a_j(x)|^2 + \frac{1}{n} a_j(x) + \frac{1}{n}\overline{a_j(x)} + \frac{1}{n^2} \Big\} = \sum_j |a_j(x)|^2 + \frac{1}{n}.$$

Since $\sum_j f_j = \sum_j (v_j - b) = \sum_j v_j - nb = 0$, expanding gives

$$x = \sum_j \xi_j(x) v_j = \sum_j \xi_j(x) f_j + \sum_j \xi_j(x) b = \sum_j \{a_j(x) + \frac{1}{n}\} f_j + b = \sum_j a_j(x) f_j + b.$$

Thus we must minimise $\sum_j |a_j(x)|^2$, subject to the constraints

$$x - b = \sum_j a_j(x) f_j, \qquad \sum_j a_j(x) = 0.$$

The minimiser subject to just the first constraint is $a_j(x) = c_j^{\Theta - b_\Phi}(x - b_\Theta)$. But $\sum_j f_j = 0$ implies the dependency $\sum_j c_j = 0$ (by Theorem 4.2), and so the second constraint is also satisfied by this choice for $a_j(x)$. □

The generalised barycentric coordinates have similar properties to those of the canonical coordinates (see Proposition 4.3).

Proposition 4.9. (Properties) *The generalised barycentric coordinates $\xi^\Theta = (\xi_j)$ for points $\Theta = (v_j)_{j=1}^n$ with affine span X satisfy*

1. $\xi_j(v_k) = \overline{\xi_k(v_j)}$.
2. $|\xi_j(v_k)| \le 1$ *with* $|\xi_j(v_k)| = 1$ *if and only if* $k = j$ *and* $v_j \notin \operatorname{aff}(\Theta \setminus v_j)$, *in which case* $\xi_j(v_j) = 1$ *and* $\xi_j = 0$ *on* $\operatorname{aff}(\Theta \setminus v_j)$.
3. $\frac{1}{n} \le \xi_j(v_j) \le 1$, *with* $\xi_j(v_j) = \frac{1}{n}$ *if and only if* $v_j = b_\Theta$ *(the barycentre of* Θ*).*
4. $\sum_j \xi_j(v_j) = d + 1$.
5. $\xi_j = \xi_k$ *if and only if* $v_j = v_k$.

Proof. By Proposition 4.8,

$$\xi_j(v_k) = c_j(f_k) + \frac{1}{n}, \qquad f_k := v_k - b_\Theta, \quad b_\theta := \frac{1}{n}\sum_j v_j,$$

and so, by Proposition 4.3, we immediately obtain 1, the lower bound in 3, 4 and 5.

Since v_k can be written as the affine combination $v_k = 1v_k + \sum_{j \ne k} 0 v_j$, we have

$$|\xi_j(v_k)|^2 \le \sum_j |\xi_j(v_k)|^2 \le 1^2 + \sum_{j \ne k} 0^2 = 1 \quad \Longrightarrow \quad |\xi_j(v_k)| \le 1,$$

with equality if and only if $k = j$ and $\xi_\ell(v_j) = 0$, $\forall \ell \ne j$, i.e., $\xi_j = 0$ on $\operatorname{aff}(\Theta \setminus v_j)$. If $\xi_j = 0$ on $\operatorname{aff}(\Theta \setminus v_j)$, then $v_j \notin \operatorname{aff}(\Theta \setminus v_j)$. Otherwise, $v_j = \sum_{k \ne j} a_k v_k$, $\sum_{k \ne j} a_k = 1$, and v_j can be written as an affine combination

$$v_j = (1-t)v_j + \sum_{k\neq j} t a_k v_k, \qquad t \in \mathbb{R},$$

where the sum of the squares of the coefficients above is

$$(1-t)^2 + \sum_{k\neq j} t^2 |\xi_k|^2 = 1 - 2t + t^2\Big(1 + \sum_{k\neq j} |\xi_k|^2\Big),$$

which is strictly less than 1 for $t > 0$ sufficiently small, and so $\lambda_j(v_j) < 1$. □

From the formula (4.26), we also observe that

- The coordinates of the barycentre $b = b_\Theta$ are $\xi_j(b) = \frac{1}{n}$, $\forall j$.
- ξ_j is constant (equal to $\frac{1}{n}$) if and only if v_j is the barycentre c.

These imply that the set of points where the generalised barycentric coordinates are nonnegative

$$N = N_\Theta := \{x \in X : \xi_j^\Theta(x) \geq 0,\ \forall j\} \tag{4.27}$$

is a convex polytope, with the barycentre as an interior point. For the purpose of illustration, we also define the ellipsoid with centre the barycentre

$$E = E_\Theta := \{x \in X : \sum_j |\xi_j^\Theta(x)|^2 = 1\},$$

inside which the points have generalised barycentric coordinates with ℓ_2-norm < 1.

Example 4.17. (Four points in $\mathbb{R}^2$) (See Figs. 4.1 and 4.2) Suppose, without loss of generality, that

$$\Theta = \Big(\begin{pmatrix}0\\0\end{pmatrix}, \begin{pmatrix}1\\0\end{pmatrix}, \begin{pmatrix}0\\1\end{pmatrix}, \begin{pmatrix}a\\b\end{pmatrix}\Big),$$

where there are no restrictions on (a,b). The generalised barycentric coordinates (indexed by the points) are given by the linear polynomials of (4.10). We observe that these coordinates depend continuously on $(a,b) \in \mathbb{R}^2$.

If the convex hull of the points is a quadrilateral Q, i.e., $a,b > 0$ and $a+b > 1$, then the polytope N_Θ (which depends continuously on Θ) has four vertices, one of which lies on the edge from $(0,0)$ to $(0,1)$, namely

$$\xi_{(0,1)}(x,y) = \xi_{(a,b)}(x,y) = 0 \quad \Longleftrightarrow \quad (x,y) = \big(\tfrac{a+b-1}{2a+b-1}, 0\big).$$

Thus we conclude that N_Θ circumscribes the boundary of Q, with one point on each edge. The barycentre of the vertices of N_Θ is not $b_\Theta = \frac{1}{4}(a+1, b+1)$ in general, since the vertices are

$$\begin{pmatrix}\frac{a+b-1}{2a+b-1}\\ 0\end{pmatrix}, \quad \begin{pmatrix}0\\ \frac{a+b-1}{a+2b-1}\end{pmatrix}, \quad \begin{pmatrix}\frac{ab}{b+1}\\ \frac{b^2+1}{b+1}\end{pmatrix}, \quad \begin{pmatrix}\frac{a^2+1}{a+1}\\ \frac{ab}{a+1}\end{pmatrix}.$$

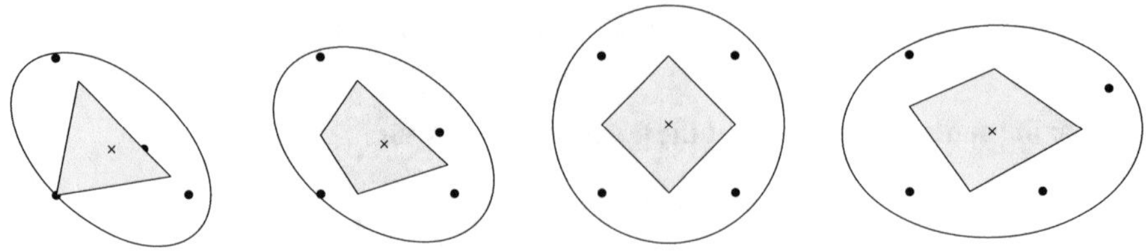

Fig. 4.1: The points Θ, E_Θ, N_Θ, c_Θ for Example 4.17 with $(a,b) = (\frac{2}{3},\frac{1}{3}), (\frac{8}{9},\frac{4}{9}), (1,1), (\frac{3}{2},\frac{3}{4})$.

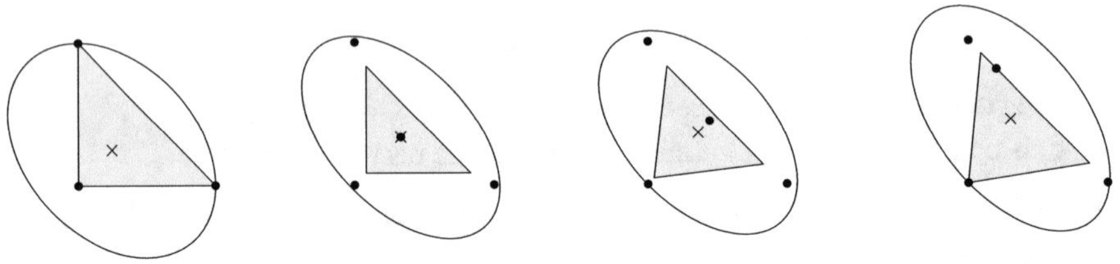

Fig. 4.2: The points Θ, E_Θ, N_Θ, c_Θ for Example 4.17 with $(a,b) = (0,0), (\frac{1}{3},\frac{1}{3}), (\frac{4}{9},\frac{4}{9}), (\frac{1}{5},\frac{4}{5})$.

Example 4.18. (The vertices of a regular polygon) (see Fig. 4.3). Let (v_j) be n equally spaced unit vectors in $\mathbb{R}^2$, say

$$v_j = \begin{pmatrix} \cos \frac{2\pi}{n} j \\ \sin \frac{2\pi}{n} j \end{pmatrix}, \qquad j = 1, \ldots, n.$$

Then

$$\xi_j(x) = \langle x, \tfrac{2}{n} v_j \rangle + \tfrac{1}{n},$$

so that N_Θ is the n-sided regular polygon (inscribing a circle of radius $1/(2\cos\frac{\pi}{n})$) given by

$$N_\Theta = \operatorname{conv}\{w_j : j = 1, \ldots, n\}, \qquad w_j := -\frac{1}{2\cos\frac{\pi}{n}} \begin{pmatrix} \cos \frac{2\pi}{n}(j+\frac{1}{2}) \\ \sin \frac{2\pi}{n}(j+\frac{1}{2}) \end{pmatrix},$$

and E_Θ is the disc of radius $r = \sqrt{\frac{n-1}{2}}$ centred at 0. Here $\xi_j(v_j) = \frac{3}{n}$.

By writing the expansion as

$$x = \frac{1}{n} \sum_{j=1}^{n} (n\xi_j(x))\, p_j, \qquad \frac{1}{n} \sum_{j=1}^{n} (n\xi_j(x)) = 1,$$

we can obtain the limiting case (of points on the unit circle)

$$x = \frac{1}{2\pi} \int_0^{2\pi} \xi_\theta(x)\, v_\theta \, d\theta, \qquad \frac{1}{2\pi} \int_0^{2\pi} \xi_\theta(x)\, d\theta = 1,$$

where

$$\xi_\theta := v_\theta, \quad v_\theta := \begin{pmatrix} \cos\theta \\ \sin\theta \end{pmatrix}, \quad \xi_\theta(x) := 2\langle x, v_\theta \rangle + 1, \qquad 0 \le \theta \le 2\pi.$$

Here the coordinates ξ_θ are nonnegative on the disc with centre 0 and radius $\frac{1}{2}$.

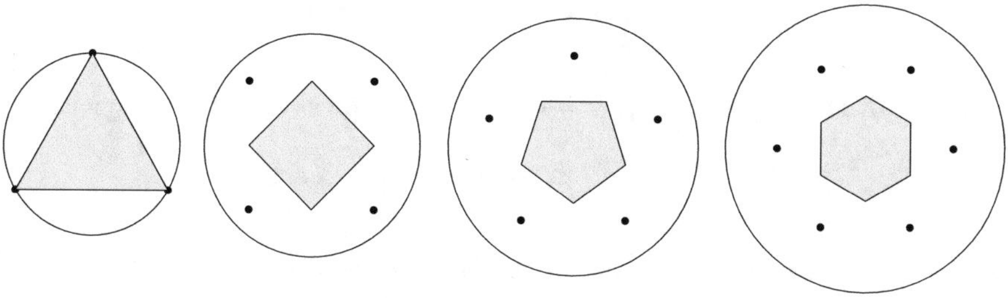

Fig. 4.3: The $n = 3,4,5,6$ equally spaced points Θ of Example 4.18, with N_Θ and E_Θ.

We have the following analogue of Proposition 4.4.

Proposition 4.10. *(Affine maps). Let $A : X \to Y$ be an invertible affine map between affine spaces X and Y, and $\Theta = (v_1, \ldots, v_n)$ be points in X with affine span X. Then the canonical barycentric coordinates for Θ and $A\Theta = (Av_1, \ldots, Av_n)$ satisfy*

$$\xi^{A\Theta}(Ax) = \xi^\Theta(x), \qquad \forall x \in X.$$

Proof. Write $Ax = L(x - b_\Theta) + a$, where L is a linear map (on the vectors in X), and $b_\Theta := \frac{1}{n}\sum_j v_j$ is the barycentre of Θ. Then the barycentre of $A\Theta$ is

$$b_{A\Theta} = \frac{1}{n}\sum_j \bigl(L(v_j - b_\Theta) + a\bigr) = L\Bigl(\frac{1}{n}\sum_j (v_j - b_\Theta)\Bigr) + a = a,$$

and so $Ax - b_{A\Theta} = L(x - b_\Theta)$. Let $\Phi = (v_j - b_\Theta)_{j=1}^n$, then $A\Theta - b_{A\Theta} = L(\Theta - b_\theta)$, and using (4.11), we obtain

$$\begin{aligned} \xi_j^{A\Theta}(Ax) &= c_j^{A\Theta - b_{A\Theta}}(Ax - b_{A\Theta}) + \frac{1}{n} = c_j^{L(\Phi - b_\Theta)}(L(x - b_\Theta)) + \frac{1}{n} \\ &= c_j^{\Phi - b_\Theta}(x - b_\Theta) + \frac{1}{n} = \xi_j^\Theta(x). \end{aligned}$$

□

A similar argument shows that Θ and ξ^Θ have same symmetry group (see §9.2).

4.8 The Bernstein frame

Here we index the generalised barycentric coordinates for $m \geq d+1$ points Θ, with affine span $X = \mathbb{R}^d$, by the points that they correspond to, i.e., $\xi^\Theta = (\xi_v)_{v\in\Theta}$. Let $\Pi_n(\mathbb{R}^d)$ denote the polynomials $\mathbb{R}^d \to \mathbb{R}$ of degree $\leq n$. Since

$$x = \sum_{v\in\Theta} \xi_v(x)v, \qquad \sum_{v\in\Theta} \xi_v(x) = 1, \tag{4.28}$$

it follows that $(\xi_v)_{v\in\Theta}$ spans $\Pi_1(\mathbb{R}^d)$. More generally, for any $n \geq 1$, the polynomials

$$\xi^\alpha = \prod_v \xi_v^{\alpha_v}, \qquad |\alpha| = \sum_v \alpha_v = n \qquad (\alpha \in \mathbb{Z}_+^\Theta)$$

span $\Pi_n(\mathbb{R}^d)$. Here the $\alpha \in \mathbb{Z}_+^\Theta$ are *multiindices*. By the multinomial theorem

$$\Bigl(\sum_v \xi_v\Bigr)^n = \sum_{|\alpha|=n} \binom{n}{\alpha} \xi^\alpha = 1, \qquad \binom{n}{\alpha} := \frac{n!}{\alpha!}.$$

Thus the $\binom{n+m-1}{m-1}$ polynomials (which span $\Pi_n(\mathbb{R}^d)$)

$$B_\alpha^\Theta = B_\alpha := \binom{|\alpha|}{\alpha} \xi^\alpha, \qquad |\alpha| = n, \tag{4.29}$$

form a *partition of unity*, which is nonnegative on the region N_Θ given by (4.27).

Definition 4.5. We call the polynomials $(B_\alpha)_{|\alpha|=n}$ of (4.29) the **Bernstein frame** (of degree n) for $\Pi_n(\mathbb{R}^d)$ given by the points Θ.

By construction (and a dimension count):

The Bernstein frame $(B_\alpha)_{|\alpha|=n}$ is a frame (spanning sequence) for $\Pi_n(\mathbb{R}^d)$. It is a basis if and only if Θ consists of $d+1$ affinely independent points in $\mathbb{R}^d$.

Example 4.19. (Bernstein basis) When the Bernstein frame is a basis, i.e., Θ is $d+1$ affinely independent points in $\mathbb{R}^d$, then it is the usual *Bernstein basis*. For example, when $d=1$, $\Theta = (0,1)$, we have $\xi_0(x)1-x$, $\xi_1 = x$, which gives the univariate Bernstein basis

$$B_{(n-j,j)}(x) = (1-x)^{n-j}x^j, \qquad 0 \leq j \leq n,$$

and when $d=2$, $\Theta = \bigl((0,0),(1,0),(0,1)\bigr)$, we have

$$\xi_{(0,0)}(x,y) = 1-x-y, \quad \xi_{(1,0)}(x,y) = x, \quad \xi_{(0,1)}(x,y) = y,$$

which gives the bivariate Bernstein basis

$$B_{(n-j-k,j,k)}(x) = (1-x-y)^{n-j-k}x^j y^k, \qquad 0 \leq j,k \leq j+k \leq n.$$

Example 4.20. (Vertices of the square) When Θ is the vertices $(0,0),(1,0),(0,1)$, $(1,1)$ of the unit square in $\mathbb{R}^2$, then $a=b=1$ in Example 4.17, gives

$$\xi_{(0,0)}=\frac{1}{4}(-2x-2y+3), \qquad \xi_{(1,0)}=\frac{1}{4}(2x-2y+1),$$

$$\xi_{(0,1)}=\frac{1}{4}(-2x+2y+1), \qquad \xi_{(1,1)}=\frac{1}{4}(2x+2y-1).$$

The corresponding Bernstein frame for the bivariate quadratics has 10 polynomials (a basis has 6). However, the Bernstein frame shares the symmetries of the square, whereas a Bernstein basis cannot.

4.9 Properties of the Bernstein frame

The Bernstein frame shares the following well known and useful properties of the Bernstein basis. Let e_v be the multi-index which is 1 at v and 0 at all other points.

Proposition 4.11. *The Bernstein frame $(B_\alpha)_{|\alpha|=n}$ can be calculated recursively via*

$$B_\alpha=\sum_{v\in\Theta}\xi_v B_{\alpha-e_v}, \qquad B_0=1, \tag{4.30}$$

and expressed in terms of the Bernstein frame for polynomials of degree $n+1$ via

$$B_\alpha=\sum_{v\in V}\frac{\alpha_v+1}{|\alpha|+1}B_{\alpha+e_v}. \tag{4.31}$$

Proof. We calculate

$$\sum_{v\in V}\xi_v B_{\alpha-e_v}=\sum_{v\in V}\binom{|\alpha|-1}{\alpha-e_v}\xi^\alpha=\sum_{v\in V}\frac{\alpha_v}{|\alpha|}\binom{|\alpha|}{\alpha}\xi^\alpha=B_\alpha,$$

and, using $\sum_v \xi_v=1$, that

$$B_\alpha=B_\alpha\sum_{v\in V}\xi_v=\sum_{v\in V}\frac{|\alpha|!}{\alpha!}\xi^\alpha\xi_v=\sum_{v\in V}\frac{\alpha_v+1}{|\alpha|+1}\frac{|\alpha+e_v|!}{(\alpha+e_v)!}\xi^{\alpha+e_v}=\sum_{v\in V}\frac{\alpha_v+1}{|\alpha|+1}B_{\alpha+e_v}.$$

□

Let $D_v f$ denote the directional derivative of f in the direction $v\in\mathbb{R}^d$. We recall that $v\mapsto D_v f$ is linear.

Proposition 4.12. *(Differentiation). For $u,v,w\in V$, we have*

$$D_{v-w}\xi_u=\xi_u(v)-\xi_u(w)=\xi_v(u)-\xi_w(u).$$

Thus the Bernstein frame satisfies

$$D_{v-w}B_\alpha = |\alpha| \sum_{u\in\Theta} (\xi_u(v) - \xi_u(w))B_{\alpha-e_u}.$$

Proof. Since ξ_u is affine

$$\begin{aligned}(D_{v-w}\xi_u)(x) &= \lim_{t\to 0} \frac{\xi_u(x+t(v-w)) - \xi_u(x)}{t} \\ &= \lim_{t\to 0} \frac{\xi_u(x)+t\xi_u(v)-t\xi_u(w)-\xi_u(x)}{t} = \xi_u(v)-\xi_u(w).\end{aligned}$$

By the product and chain rules, we have

$$\begin{aligned}D_{v-w}B_\alpha &= \frac{|\alpha|!}{\alpha!} D_{v-w} \prod_{u\in V} \xi_u^{\alpha_u} = \frac{|\alpha|!}{\alpha!} \sum_{u\in V} \alpha_u \xi_u^{\alpha_u-1} (\xi_u(v)-\xi_u(w))\xi^{\alpha-\alpha_u e_u} \\ &= |\alpha| \sum_{u\in V} (\xi_u(v)-\xi_u(w)) \frac{(|\alpha|-1)!}{(\alpha-e_u)!} \xi^{\alpha-e_u} \\ &= |\alpha| \sum_{u\in V} (\xi_u(v)-\xi_u(w)) B_{\alpha-e_u}.\end{aligned}$$

□

4.10 The generalised Bernstein operator

For a Bernstein frame $(B_\alpha)_{|\alpha|=n}$ given by points Θ in $\mathbb{R}^d$, with convex hull

$$T = \mathrm{conv}(\Theta),$$

we define the (**generalised**) **Bernstein operator** $B_n = B_{n,\Theta} : C(T) \to \Pi_n(\mathbb{R}^d)$ of degree $n \geq 1$ by the usual formula

$$B_n(f) := \sum_{|\alpha|=n} B_\alpha f(v_\alpha), \qquad v_\alpha := \sum_{v\in\Theta} \frac{\alpha_v}{|\alpha|} v, \tag{4.32}$$

which is equivalent to

$$B_n(f) = \sum_{v_1\in\Theta} \cdots \sum_{v_n\in\Theta} f\Big(\frac{v_1+\cdots+v_n}{n}\Big) \xi_{v_1}\cdots\xi_{v_n}.$$

This has the *positivity property*

$$f \geq 0 \quad \text{on } T = \mathrm{conv}(\Theta) \quad \Longrightarrow \quad B_n f \geq 0 \quad \text{on } N_\Theta, \tag{4.33}$$

and reproduces the linear polynomials (see Example 4.21).

We now show that the generalised Bernstein operator is **degree reducing**, i.e.,

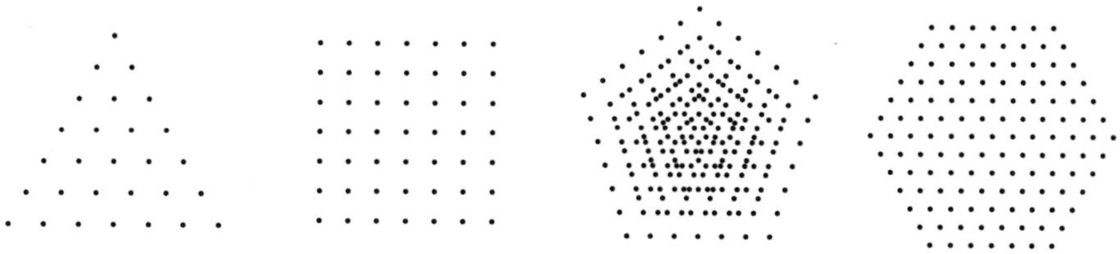

Fig. 4.4: The points $\{v_\alpha\}_{|\alpha|=n}$ used in the definition of $B_{n,V}f$, where $n=7$ and V is the vertices of a triangle, square, pentagon and hexagon (respectively).

$$B_n(f) \in \Pi_k(\mathbb{R}^d), \qquad \forall f \in \Pi_k \quad (k=0,1,\ldots).$$

Define the univariate and multivariate (falling) shifted factorials by

$$[x]^n := x(x-1)\cdots(x-n+1), \qquad [\alpha]^\beta := \prod_{v\in V}[\alpha_v]^{\beta_v},$$

and the multivariate Stirling numbers of the second kind by

$$S(\tau,\beta) := \prod_{v\in V} S(\tau_v,\beta_v),$$

where $S(\tau_v,\beta_v)$ are the Stirling numbers of the second kind. We note that

$$S(\tau,\beta)=0, \qquad \beta \not\leq \tau, \tag{4.34}$$

and define

$$\binom{|\alpha|}{\alpha} := 0, \qquad \alpha \not\geq 0. \tag{4.35}$$

These are related by

$$\alpha^\tau = \sum_{\beta\leq\tau} S(\tau,\beta)[\alpha]^\beta. \tag{4.36}$$

Lemma 4.2. *For any τ and n, we have*

$$\sum_{|\alpha|=n} \alpha^\tau \binom{|\alpha|}{\alpha} \xi^\alpha = \sum_{\beta\leq\tau} S(\tau,\beta)[n]^{|\beta|}\xi^\beta. \tag{4.37}$$

Proof. Since $[|\alpha|]^{|\beta|}\binom{|\alpha-\beta|}{\alpha-\beta} = \binom{|\alpha|}{\alpha}[\alpha]^\beta$ (without restriction on α and β), (4.36) gives

$$\sum_{\beta\leq\tau} S(\tau,\beta)[|\alpha|]^{|\beta|}\binom{|\alpha-\beta|}{\alpha-\beta} = \binom{|\alpha|}{\alpha}\sum_{\beta\leq\tau} S(\tau,\beta)[\alpha]^\beta = \binom{|\alpha|}{\alpha}\alpha^\tau.$$

Thus, we calculate

$$\sum_{|\alpha|=n} \alpha^\tau \binom{|\alpha|}{\alpha} \xi^\alpha = \sum_{|\alpha|=n} \sum_{\beta \le \tau} S(\tau,\beta)[|\alpha|]^{|\beta|} \binom{|\alpha-\beta|}{\alpha-\beta} \xi^\alpha$$
$$= \sum_{\beta\le\tau} S(\tau,\beta)[n]^{|\beta|}\xi^\beta \sum_{\substack{|\alpha|=n\\ \alpha\ge\beta}} \binom{|\alpha-\beta|}{\alpha-\beta}\xi^{\alpha-\beta} = \sum_{\beta\le\tau} S(\tau,\beta)[n]^{|\beta|}\xi^\beta,$$

with the last equality given by the multinomial identity. □

Observe that (4.28) is equivalent to the reproduction formula for affine functions

$$f = \sum_{v\in\Theta} f(v)\xi_v, \qquad \forall f \in \Pi_1(\mathbb{R}^d). \tag{4.38}$$

Theorem 4.6. *(Degree reducing). The generalised Bernstein operator B_n is degree reducing. More precisely,*

$$B_n(\xi^\beta) = \frac{[n]^{|\beta|}}{n^{|\beta|}}\xi^\beta + \sum_{0<|\gamma|<|\beta|} \frac{[n]^{|\gamma|}}{n^{|\beta|}} a(\gamma,\beta)\xi^\gamma, \tag{4.39}$$

where $w_1,\ldots,w_m$ is the sequence of points in V, and

$$a(\gamma,\beta) := \sum_{|\tau_1|=\beta_{w_1}} \cdots \sum_{|\tau_m|=\beta_{w_m}} \binom{\beta_{w_1}}{\tau_1}\xi^{\tau_1}(w_1)\cdots\binom{\beta_{w_m}}{\tau_m}\xi^{\tau_m}(w_m) S(\tau_1+\cdots+\tau_m,\gamma). \tag{4.40}$$

Proof. Since each ξ_w is an affine function, and $\xi_w(v)=\xi_v(w)$, we have

$$\xi_w(v_\alpha) = \xi_w\Big(\sum_{v\in V}\frac{\alpha_v}{|\alpha|}v\Big) = \sum_{v\in V}\frac{\alpha_v}{|\alpha|}\xi_w(v) = \sum_{v\in V}\frac{\alpha_v}{|\alpha|}\xi_v(w),$$

and the multinomial identity gives

$$(\xi^\beta)(v_\alpha) = \prod_{w\in V}\Big(\sum_{v\in V}\frac{\alpha_v}{|\alpha|}\xi_v(w)\Big)^{\beta_w} = \prod_{w\in V}\sum_{|\tau|=\beta_w}\binom{\beta_w}{\tau}\frac{\alpha^\tau}{|\alpha|^{\beta_w}}\xi^\tau(w)$$
$$= \sum_{|\tau_1|=\beta_{w_1}} \cdots \sum_{|\tau_m|=\beta_{w_m}} \binom{\beta_{w_1}}{\tau_1}\cdots\binom{\beta_{w_m}}{\tau_m}\xi^{\tau_1}(w_1)\cdots\xi^{\tau_m}(w_m)\frac{\alpha^{\tau_1+\cdots+\tau_m}}{|\alpha|^{|\beta|}}.$$

Thus, by rearranging (4.32) and Lemma 4.2, we have

$$B_n(\xi^\beta) = \sum_{|\tau_1|=\beta_{w_1}} \cdots \sum_{|\tau_m|=\beta_{w_m}} \binom{\beta_{w_1}}{\tau_1}\xi^{\tau_1}(w_1)\cdots\binom{\beta_{w_m}}{\tau_m}\xi^{\tau_m}(w_m)\sum_{|\alpha|=n}\frac{\alpha^{\tau_1+\cdots+\tau_m}}{n^{|\beta|}}\binom{n}{\alpha}\xi^\alpha$$
$$= \sum_{|\tau_1|=\beta_{w_1}} \cdots \sum_{|\tau_m|=\beta_{w_m}} \binom{\beta_{w_1}}{\tau_1}\xi^{\tau_1}(w_1)\cdots\binom{\beta_{w_m}}{\tau_m}\xi^{\tau_m}(w_m)$$
$$\times \sum_{\gamma\le\tau_1+\cdots+\tau_m}\frac{[n]^{|\gamma|}}{n^{|\beta|}}S(\tau_1+\cdots+\tau_m,\gamma)\xi^\gamma.$$

Here $B_n(\xi^\beta)$ is written as a polynomial in ξ of degree $\le |\beta|$, so that B_n is degree reducing. The terms of degree $|\beta|$ can be simplified using the multinomial identity, $\xi_v(w_j) = \xi_{w_j}(v)$, and (4.38), as follows

$$\sum_{|\tau_1|=\beta_{w_1}} \cdots \sum_{|\tau_m|=\beta_{w_m}} \binom{\beta_{w_1}}{\tau_1} \xi^{\tau_1}(w_1) \cdots \binom{\beta_{w_m}}{\tau_m} \xi^{\tau_m}(w_m) \frac{[n]^{|\beta|}}{n^{|\beta|}} \xi^{\tau_1+\cdots+\tau_m}$$
$$= \frac{[n]^{|\beta|}}{n^{|\beta|}} \prod_{j=1}^m \Big(\sum_{|\tau_j|=\beta_{w_j}} \binom{\beta_{w_j}}{\tau_j} \xi^{\tau_j}(w_j)\xi^{\tau_j} \Big) = \frac{[n]^{|\beta|}}{n^{|\beta|}} \prod_{j=1}^m \Big(\sum_{v\in V} \xi_v(w_j)\xi_v \Big)^{\beta_{w_j}}$$
$$= \frac{[n]^{|\beta|}}{n^{|\beta|}} \prod_{j=1}^m \Big(\sum_{v\in V} \xi_{w_j}(v)\xi_v \Big)^{\beta_{w_j}} = \frac{[n]^{|\beta|}}{n^{|\beta|}} \prod_{j=1}^m \xi_{w_j}^{\beta_{w_j}} = \frac{[n]^{|\beta|}}{n^{|\beta|}} \xi^\beta.$$

By collecting the terms of degree $< |\beta|$, we obtain (4.39). Here, (4.34) allows us to remove the restriction $\gamma \le \tau_1 + \cdots + \tau_m$, and there are no terms of degree 0 since $S(1,0) = 0$. □

Since $[n]^{|\gamma|} = 0$, $|\gamma| > n$, the formula (4.39) implies that B_n is degree reducing.

Example 4.21. (Linear reproduction) For $|\beta| = 1$, we have

$$B_n(\xi_v) = \xi_v, \qquad \forall v \in V, \tag{4.41}$$

i.e., B_n reproduces the linear polynomials $\Pi_1(\mathbb{R}^d) = \operatorname{span}\{\xi_v\}$. This is equivalent to

$$x = \sum_{|\alpha|=n} B_\alpha(x) v_\alpha, \qquad \sum_{|\alpha|=n} B_\alpha(x) = 1, \qquad x \in \mathbb{R}^d. \tag{4.42}$$

Example 4.22. (Quadratics) For $|\beta| = 2$, we recall $S(1,0) = 0$, $S(2,1) = 1$, so that

$$a(e_u, 2e_w) = \xi_u^2(v) = \xi_v^2(u), \quad a(e_u, e_v + e_w) = \xi_u(v)\xi_u(w) = (\xi_v\xi_w)(u),$$

$v \ne w$, and we obtain

$$B_n(\xi^\beta) = \Big(1 - \frac{1}{n}\Big)\xi^\beta + \frac{1}{n} \sum_{u\in V} \xi^\beta(u)\xi_u, \qquad |\beta| = 2. \tag{4.43}$$

Corollary 4.2. (Convergence) *For all polynomials f, $B_n(f) \to f$, as $n \to \infty$.*

Proof. It suffices to consider $f = \xi^\beta$. For $n \ge |\beta|$, (4.39) gives

$$B_n(\xi^\beta) - \xi^\beta = \Big(\frac{[n]^{|\beta|}}{n^{|\beta|}} - 1 \Big) \xi^\beta - \sum_{|\gamma|<|\beta|} \frac{[n]^{|\gamma|}}{n^{|\beta|}} a(\gamma,\beta)\xi^\gamma,$$

where $\frac{[n]^{|\beta|}}{n^{|\beta|}} - 1$, $\frac{[n]^{|\gamma|}}{n^{|\beta|}} = O(\frac{1}{n})$, as $n \to \infty$. □

The spectral structure of B_n can easily be deduced Theorem 4.6 (see [DW15]).

Notes

The *canonical coordinates* were introduced in [Wal11]. They allow the theory of frames to be extended to vector spaces (without an inner product) in a natural way.

Exercises

4.1. *Dual sequences.* Let $(f_j)_{j\in J}$ and $(\lambda_j)_{j\in J}$ be spanning sequences for a vector space X and its algebraic dual X', respectively, with Gramian

$$G := [\lambda_j(f_k)] = \Lambda V, \qquad V := [f_j] : \mathbb{F}^J \to X, \quad \Lambda = (\lambda_j) : X \mapsto \mathbb{F}^J.$$

Show that the following are equivalent
(a) $V\Lambda = I_X$.
(b) $f = \sum_j \lambda_j(f) f_j$, $\forall f \in X$.
(c) $\lambda = \sum_j \lambda(f_j)\lambda_j$, $\forall \lambda \in X'$.
(d) $G^2 = G$.
(e) $V = VG$.
(f) $\Lambda = G\Lambda$.

4.2. Show that if Φ is a finite frame for $\mathscr{H}$, with frame operator $S = VV^*$, then the canonical inner product is given by

$$\langle f, g\rangle_\Phi = \langle S^{-\frac{1}{2}} f, S^{-\frac{1}{2}} g\rangle = \langle f, S^{-1} g\rangle, \qquad \forall f, g \in \mathscr{H}.$$

Hint: Use $c_j^\Phi(f) = \langle f, S^{-1} f_j\rangle$.

4.3. Let Θ be a set of $d+1$ affinely independent points with affine span X, and $\ell = (\ell_v)_{v\in\Theta}$ be the corresponding barycentric coordinates. Suppose that $\Psi = (\psi_j)$ is a sequence of points in Θ, with each $v \in \Theta$ appearing with multiplicity $m_v \geq 1$. Show that the canonical barycentric coordinates (ξ_j) for Ψ are

$$\xi_j = \frac{1}{m_v}\ell_v \quad \text{when} \quad \psi_j = v.$$

Chapter 5
Combining and decomposing frames

We now give a list of ways in which two or more frames can be combined to obtain a new frame, for which the frame and its dual are related in a natural way to those of its constituent parts.

Given a frame, these methods offer a way of decomposing it into simpler parts. Sometimes this can be done in *several* different ways, and at present there is no coherent decomposition theory for frames (or tight frames).

The inner product on the orthogonal direct sum $\mathscr{H}_1 \oplus \mathscr{H}_2$ and the tensor product $\mathscr{H}_1 \otimes \mathscr{H}_2$ are given by

$$\langle (f_1,g_1),(f_2,g_2)\rangle := \langle f_1,f_2\rangle + \langle g_1,g_2\rangle, \qquad \forall (f_1,g_1),(f_2,g_2) \in \mathscr{H}_1 \oplus \mathscr{H}_2,$$

$$\langle f_1 \otimes g_1, f_2 \otimes g_2\rangle := \langle f_1,f_2\rangle\langle g_1,g_2\rangle, \qquad \forall f_1 \otimes g_1, f_2 \otimes g_2 \in \mathscr{H}_1 \otimes \mathscr{H}_2.$$

If no confusion arises, then one can identify $\mathscr{H}_1$ and $\mathscr{H}_2$ as subspaces of $\mathscr{H}_1 \oplus \mathscr{H}_2$, and write f_j or $f_j + 0$ in place of $(f_j, 0)$, etc.

5.1 Unions

The **(disjoint) union** of finite frames $\Phi = (f_j)_{j\in J}$ and $\Psi = (g_k)_{k\in K}$ for $\mathscr{H}_1$ and $\mathscr{H}_2$

$$\Phi \cup \Psi := \Big(\begin{pmatrix} f_j \\ 0 \end{pmatrix}, \begin{pmatrix} 0 \\ g_k \end{pmatrix}\Big)_{j\in J, k\in K} \tag{5.1}$$

is a frame for the orthogonal direct sum $\mathscr{H}_1 \oplus \mathscr{H}_2$ (which is indexed by $J \cup K$), with dual frame $\tilde{\Phi} \cup \tilde{\Psi}$, and

$$\operatorname{Gram}(\Phi \cup \Psi) = \operatorname{Gram}(\Phi) \oplus \operatorname{Gram}(\Psi) := \begin{pmatrix} \operatorname{Gram}(\Phi) & 0 \\ 0 & \operatorname{Gram}(\Psi) \end{pmatrix}. \tag{5.2}$$

Conversely, if a frame for $\mathscr{H}$ has a Gramian matrix which can be block diagonalised as in (5.2), for some partitioning $\Phi \cup \Psi$ of its vectors, then Φ and Ψ are frames for

S.F.D. Waldron, *An Introduction to Finite Tight Frames*, Applied and Numerical Harmonic Analysis, https://doi.org/10.1007/978-0-8176-4815-2_5

their spans $\mathscr{H}_1$ and $\mathscr{H}_2$, which are orthogonal complements, i.e., $\mathscr{H} = \mathscr{H}_1 \oplus \mathscr{H}_2$, and the dual frames of Φ and Ψ are given by the corresponding dual vectors of the original frame (see Exer. 5.2).

Example 5.1. The frame bounds satisfy (see Exer. 5.4)

$$A_{\Phi\cup\Psi} = \min\{A_\Phi, A_\Psi\}, \qquad B_{\Phi\cup\Psi} = \max\{B_\Phi, B_\Psi\},$$

and so a union of normalised tight frames is again a normalised tight frame.

Example 5.2. In the (extreme) case when Gram(Φ) is diagonal, we have

$$\mathscr{H} = \bigoplus_{\phi\in\Phi} \operatorname{span}\{\phi\}, \qquad \tilde{\phi} = \frac{1}{\langle\phi,\phi\rangle}\phi,$$

and Φ is an orthogonal basis.

The union is a natural first candidate for decomposing a frame.

Each frame can be uniquely decomposed into a union of frames for orthogonal subspaces (each corresponds to a component of the frame graph of §8.3).

If $\Phi = (f_j)$ and $\Psi = (g_k)$ are finite tight frames for $\mathscr{H}$, then so is the frame consisting their union as a subset of $\mathscr{H}$. In [LMO14] a finite tight frame for $\mathscr{H}$ is said to be *divisible* if it can be partitioned into two tight frames for $\mathscr{H}$, i.e., it has a proper subset which is a tight frame for $\mathscr{H}$, and otherwise it is *prime*. Clearly, every finite tight frame can be partitioned into prime tight frames (though not uniquely).

5.2 Direct sums

We consider two notions of the sum of frames: the *direct sum* and *sum* (see §5.5). The first of these requires that each summand has the same index set, and the second has no counterpart for infinite dimensional spaces.

Definition 5.1. Let $\Phi = (\phi_j)_{j\in J}$ and $\Psi = (\psi_j)_{j\in J}$ be frames for $\mathscr{H}_1$ and $\mathscr{H}_2$, with the same index set J. Then their **(inner) direct sum** is

$$\Phi \oplus \Psi = (\phi_j + \psi_j)_{j\in J} \subset \mathscr{H}_1 \oplus \mathscr{H}_2.$$

The direct sum may not be a frame for $\mathscr{H}_1 \oplus \mathscr{H}_2$, e.g., when

$$\dim(\mathscr{H}_1 \oplus \mathscr{H}_2) = \dim(\mathscr{H}_1) + \dim(\mathscr{H}_2) > |J|.$$

A necessary and sufficient condition (see Exer. 5.3) is **disjointness**, i.e,

$$\operatorname{ran}(V^*) \cap \operatorname{ran}(W^*) = \{0\}, \qquad V := [\phi_j]_{j\in J},\ W := [\psi_j]_{j\in J}. \tag{5.3}$$

If $\Phi \oplus \Psi$ is a frame, i.e., Φ and Ψ are disjoint, then we say

$\Phi \oplus \Psi$ is an **orthogonal dilation** of Φ and Ψ,
Φ, Ψ are **orthogonal compressions** of $\Phi \oplus \Psi$.

The direct sum of *three* (or more) mutually disjoint frames is not well defined, in general. For example, the frames

$$\Phi_1 = (e_1, 0, e_2, 0,), \quad \Phi_2 = (0, e_1, 0, e_2), \quad \Phi_3 = (e_1, e_1, e_2, e_2)$$

are mutually disjoint frames for $\mathbb{R}^2$, so the direct sum of any two is a basis for $\mathbb{R}^4$, which cannot be disjoint from the remaining frame (by a dimension count).

The stronger condition of **orthogonality** (see Definition 3.5), i.e.,

$$\operatorname{ran}(V^*) \perp \operatorname{ran}(W^*),$$

leads to a direct sum for which the dual frame is the direct sum of the duals.

The direct sum $\oplus_j \Phi_j = \Phi_1 \oplus \cdots \oplus \Phi_n$ of mutually orthogonal frames for $\mathcal{H}_1, \ldots, \mathcal{H}_n$ can be defined in the obvious way, and is associative, since

$$\operatorname{ran}([\phi_j + \psi_j]^*) = \operatorname{ran}([\phi_j]^*) + \operatorname{ran}([\psi_j]^*).$$

We observe by definition (and Proposition 3.2) that:

Disjointness and orthogonality depend only on the frames up to similarity.

Lemma 5.1. *(Orthogonality) Let $\Phi = (\phi_j)_{j\in J}$ and $\Psi = (\psi_j)_{j\in J}$ be finite frames for $\mathcal{H}_1$ and $\mathcal{H}_2$, with $V = [\phi_j]$ and $W = [\psi_j]$. Then the following are equivalent*

1. *Φ and Ψ are orthogonal (strongly disjoint), i.e.,* $\operatorname{ran}(V^*) \perp \operatorname{ran}(W^*)$.
2. *The canonical coordinates of Φ and Ψ are orthogonal, i.e.,* $\operatorname{ran}(P_\Phi) \perp \operatorname{ran}(P_\Psi)$.
3. $\operatorname{Gram}(\Phi^{\mathrm{can}})\operatorname{Gram}(\Psi^{\mathrm{can}}) = 0$.
4. *$WV^* = 0$, i.e., $\sum_j \langle f, \phi_j\rangle \psi_j = 0$, $\forall f \in \mathcal{H}_1$.*
5. *$VW^* = 0$, i.e., $\sum_j \langle g, \psi_j\rangle \phi_j = 0$, $\forall g \in \mathcal{H}_2$.*

Proof. 1.$\Longleftrightarrow$2. We observe that $\operatorname{ran}(V^*) = \ker(V)^\perp = \operatorname{dep}(\Phi)^\perp = \operatorname{ran}(P_\Phi)$, and, similarly, $\operatorname{ran}(W^*) = \operatorname{ran}(P_\Psi)$.

2.$\Longleftrightarrow$3. Since $P_\Phi = \operatorname{Gram}(\Phi^{\mathrm{can}})$ and $P_\Psi = \operatorname{Gram}(\Psi^{\mathrm{can}})$, this follows by the fact that two subspaces are orthogonal if and only if the product of the orthogonal projections onto them is zero.

3.$\Longrightarrow$4. By the factorisation (3.12), we have

$$\sum_j \langle f, \phi_j\rangle \psi_j = WV^*f = WP_\Psi(VP_\Phi)^*f = W(P_\Psi P_\Phi)V^*f = 0.$$

4.$\Longrightarrow$5. We have $VW^* = (WV^*)^* = 0^* = 0$.
5.$\Longrightarrow$1. If $WV^* = 0$, i.e., $W(V^*f) = 0$, $\forall f$, then

$$\operatorname{ran}(V^*) \subset \ker(W) = (\operatorname{ran}(W^*))^\perp \quad \Longrightarrow \quad \operatorname{ran}(V^*) \perp \operatorname{ran}(W^*).$$

□

Theorem 5.1. *The (inner) direct sum $\Phi \oplus \Psi$ is a frame for $\mathscr{H}_1 \oplus \mathscr{H}_2$, with*

$$S_{\Phi\oplus\Psi}(f+g) = S_\Phi(f) + S_\Psi(g), \qquad \forall f \in \mathscr{H}_1,\ \forall g \in \mathscr{H}_2, \tag{5.4}$$

or, equivalently, dual frame given by

$$(\Phi \oplus \Psi)\tilde{} = \tilde{\Phi} \oplus \tilde{\Psi} \tag{5.5}$$

if and only if Φ and Ψ are orthogonal. In this case, the frame bounds are

$$A_{\Phi\oplus\Psi} = \min\{A_\Phi, A_\Psi\}, \qquad B_{\Phi\oplus\Psi} = \max\{B_\Phi, B_\Psi\}, \tag{5.6}$$

and so a direct sum of normalised tight frames is a normalised tight frame.

Proof. Expanding gives

$$S_{\Phi\oplus\Psi}(f+g) = S_\Phi(f) + S_\Psi(g) + \sum_j \langle f, \phi_j \rangle \psi_j + \sum_j \langle g, \psi_j \rangle \phi_j,$$

with the last two sums depending only on f and g, respectively. Thus Φ and Ψ are orthogonal if and only if (5.4) holds. In this case (see Exer. 5.4), $\Phi \oplus \Psi$ is a frame for $\mathscr{H}_1 \oplus \mathscr{H}_2$, with frame bounds (5.6). The condition (5.4) is equivalent to (5.15) for the spanning sequence $\Phi \oplus \Psi$, i.e.,

$$(\phi_j + 0)\tilde{} = S_{\Phi\oplus\Psi}^{-1}(\phi_j + 0) = S_\Phi^{-1}(\phi_j) + S_\Psi^{-1}(0) = \tilde{\phi}_j,$$

and similarly $(0 + \psi_j)\tilde{} = \tilde{\psi}_j$, which gives (5.5). □

Example 5.3. (Gramian) The Gramian and canonical Gramian of a direct sum satisfy

$$\operatorname{Gram}(\Phi \oplus \Psi) = \operatorname{Gram}(\Phi) \oplus \operatorname{Gram}(\Psi),$$

$$\operatorname{ran}(P_{\Phi\oplus\Psi}) = \operatorname{ran}(P_\Phi) + \operatorname{ran}(P_\Psi).$$

Therefore, when Φ and Ψ are orthogonal, we have

$$P_{\Phi\oplus\Psi} = P_\Phi + P_\Psi,$$

and the canonical coordinates satisfy

$$c^{\Phi\oplus\Psi}(f+g) = c^\Phi(f) + c^\Psi(g).$$

Theorem 3.5 can be restated in terms of direct sums as follows.

Theorem 5.2. *Given a finite frame $\Phi = (\phi_j)_{j\in J}$ for $\mathscr{H}$, there exists an orthogonal frame $\Psi = (\psi_j)_{j\in J}$ for some $\mathscr{K}$, such that $\Phi \oplus \Psi$ is a basis for $\mathscr{H} \oplus \mathscr{K}$, with the same frame bounds as Φ. In particular, Φ and its dual are the orthogonal projection of the biorthogonal system given by $\Phi \oplus \Psi$ and its dual.*

The above Ψ is an example of what will be called a *complement* of Φ.

Example 5.4. (Decomposition) Let $\Phi = (f_j)_{j\in J}$ be a finite frame, with $V = [f_j]$, and $P : \mathbb{F}^J \to \mathbb{F}^J$ be any orthogonal projection onto a subspace of $\operatorname{ran}(V^*)$. Then Φ is a direct sum of the orthogonal frames with synthesis operators VP and $V(I-P)$, since

$$\operatorname{ran}((VP)^*) = \operatorname{ran}(PV^*) \perp \operatorname{ran}((V(I-P))^*) = \operatorname{ran}((I-P)V^*).$$

Thus a frame can be decomposed into a direct sum of orthogonal one-dimensional frames in many ways. Therefore the usefulness of direct sums for decompositions seems very limited, unless some additional structure is present (see Example 11.7).

Example 5.5. Consider the frame of four equally spaced unit vectors in $\mathbb{R}^2$

$$\Phi = \left(\begin{bmatrix} 1 \\ 0 \end{bmatrix}, \begin{bmatrix} 0 \\ 1 \end{bmatrix}, \begin{bmatrix} -1 \\ 0 \end{bmatrix}, \begin{bmatrix} 0 \\ -1 \end{bmatrix} \right),$$

and the frames Ψ_1 and Ψ_2 for $\mathbb{R}^1$ given by

$$\Psi_1 = \left(\frac{1}{\sqrt{2}}, \frac{1}{\sqrt{2}}, \frac{1}{\sqrt{2}}, \frac{1}{\sqrt{2}}\right), \qquad \Psi_2 = \left(\frac{1}{\sqrt{2}}, -\frac{1}{\sqrt{2}}, \frac{1}{\sqrt{2}}, -\frac{1}{\sqrt{2}}\right).$$

These are mutually orthogonal tight frames, e.g.,

$$VW^* = \begin{pmatrix} 1 & 0 & -1 & 0 \\ 0 & 1 & 0 & -1 \end{pmatrix} \begin{pmatrix} \frac{1}{\sqrt{2}} & \frac{1}{\sqrt{2}} & \frac{1}{\sqrt{2}} & \frac{1}{\sqrt{2}} \end{pmatrix} = \begin{pmatrix} 0 \\ 0 \end{pmatrix},$$

with the same frame bound. The tight frames $\Phi \oplus \Psi_1$ and $\Phi \oplus \Psi_2$ for $\mathbb{R}^3$ are the lifted four equally spaced vectors (see Example 5.9) and the vertices of the tetrahedron. The direct sum $\Psi_1 \oplus \Psi_2$ consists of two copies of an orthonormal basis for $\mathbb{R}^2$, and $\Phi \oplus \Psi_1 \oplus \Psi_2$ is an orthogonal basis for $\mathbb{R}^4$.

Example 5.6. Let $\Phi = (1,1,1)$ which is a (tight) frame for $\mathbb{F}$. Both of the frames

$$\Psi_1 = (1,-1,0), \qquad \Psi_2 = (2,-1,-1)$$

are orthogonal to Φ, *but* neither Ψ_1 and Ψ_2 are similar, nor are $\Phi \oplus \Psi_1$ and $\Phi \oplus \Psi_2$.

We now consider two special cases of the direct sum: when a summand is for a space of 1-dimension (lifting) and when the direct sum is a basis (complements).

5.3 Lifting

The idea of lifting is to take a frame, and add an additional component to each of its vectors, so as to obtain a frame for a space of dimension one higher, i.e., take the direct sum with a frame for a one-dimensional space.

Definition 5.2. Let $\Phi = (\phi_j)_{j\in J}$ be frame for $\mathscr{H}$, and $\Psi = (\alpha_j \psi)_{j\in J}$ be a frame for $\mathscr{K} = \operatorname{span}\{\psi\}$, $\psi \neq 0$. Then $\Phi \oplus \Psi$ is a **lift** of Φ to $\mathscr{H} \oplus \mathscr{K}$ (by Ψ) if Φ and Ψ are orthogonal, i.e.,

$$\sum_j \bar{\alpha}_j \phi_j = 0, \tag{5.7}$$

and is a **simple lift** when all the α_j are equal.

Since the lift of a frame is an inner direct sum, Theorem 5.1 gives that the dual frame of the lift of Φ by $\Psi = (\alpha_j \psi)_{j\in J}$ is the lift of $\tilde{\Phi}$ by

$$\tilde{\Psi} = (c\alpha_j \psi)_{j\in J}, \qquad c := \frac{1}{\langle \psi, \psi \rangle \sum_j |\alpha_j|^2}.$$

A finite frame (ϕ_j) is said to be **balanced** if $\sum_j \phi_j = 0$. Clearly, a balanced frame cannot be a basis, and a frame is balanced if and only if its dual frame is balanced.

Example 5.7. (Simple lifts) A frame has a simple lift if and only if it is balanced, and (see Exer. 5.5) a frame $\Phi = (f_j)$ is a simple lift if and only if

$$\sum_j f_j \neq 0, \qquad \langle \sum_j f_j, f_k \rangle = C, \quad \forall k.$$

Example 5.8. (Repeated lifts) If $\Phi = (\phi_j)_{j=1}^n$ is in $\mathbb{F}^d$, then condition (5.7) is that $\alpha = (\alpha_j)$ is a nonzero vector orthogonal to the rows of the matrix $[\phi_1, \ldots, \phi_n]$. Thus a finite frame can be lifted if and only if it is not a basis. Moreover, since

$$A_\Psi = B_\Psi = \|\psi\|^2 \sum_j |\alpha_j|^2, \qquad \Psi = (\alpha_j \psi),$$

the lift by Ψ has the same frame bounds as Φ provided

$$A_\Phi \leq \|\psi\|^2 \sum_j |\alpha_j|^2 \leq B_\Phi.$$

In this way, one can successively lift Φ until a basis is obtained, which gives a simple proof of Theorem 3.5 (also see Proposition 5.1). For example, the frame

$$\Phi = \left(\begin{bmatrix} 1 \\ 0 \end{bmatrix}, \begin{bmatrix} 1 \\ -1 \end{bmatrix}, \begin{bmatrix} 1 \\ 1 \end{bmatrix} \right), \qquad \tilde{\Phi} = \left(\begin{bmatrix} 1/3 \\ 0 \end{bmatrix}, \begin{bmatrix} 1/3 \\ -1/2 \end{bmatrix}, \begin{bmatrix} 1/3 \\ 1/2 \end{bmatrix} \right),$$

for $\mathbb{R}^2$ has frame bounds $A_\Phi = 2$, $B_\Phi = 3$. The vector $\alpha = (2c, -c, -c)$ is orthogonal to rows of the matrix with these columns, and so lifts Φ to a basis for $\mathbb{R}^3$, i.e.,

$$\Psi = \left(\begin{bmatrix} 1 \\ 0 \\ 2c \end{bmatrix}, \begin{bmatrix} 1 \\ -1 \\ -c \end{bmatrix}, \begin{bmatrix} 1 \\ 1 \\ -c \end{bmatrix} \right), \qquad \tilde{\Psi} = \left(\begin{bmatrix} 1/3 \\ 0 \\ 1/(3\bar{c}) \end{bmatrix}, \begin{bmatrix} 1/3 \\ -1/2 \\ -1/(6\bar{c}) \end{bmatrix}, \begin{bmatrix} 1/3 \\ 1/2 \\ -1/(6\bar{c}) \end{bmatrix} \right).$$

This has the same frame bounds as Φ if $2 \le \sum_i |\alpha_j|^2 = 6|c|^2 \le 3$, i.e., $\frac{1}{\sqrt{3}} \le |c| \le \frac{1}{\sqrt{2}}$.

Example 5.9. (Lifted roots of unity) Let $\Phi = (u_j)_{j=1}^n$ be $n \ge 3$ equally spaced unit vectors in $\mathbb{R}^2$. Since these vectors sum to zero, they can be lifted by adding a third coordinate, say u_j lifts to $v_j = (u_j, \alpha)$, $\alpha > 0$. The condition on α which ensures that (v_j) is tight is $A_\Phi = \frac{n}{2} = n|\alpha|^2$, i.e., $\alpha = \frac{1}{\sqrt{2}}$. Thus we obtain the equal-norm tight frame

$$\{(\cos\frac{2\pi j}{n}, \sin\frac{2\pi j}{n}, \frac{1}{\sqrt{2}}) : j = 1, \ldots, n\}$$

which we call the **lifted n-th roots of unity** (or the **lifted equally spaced vectors**).

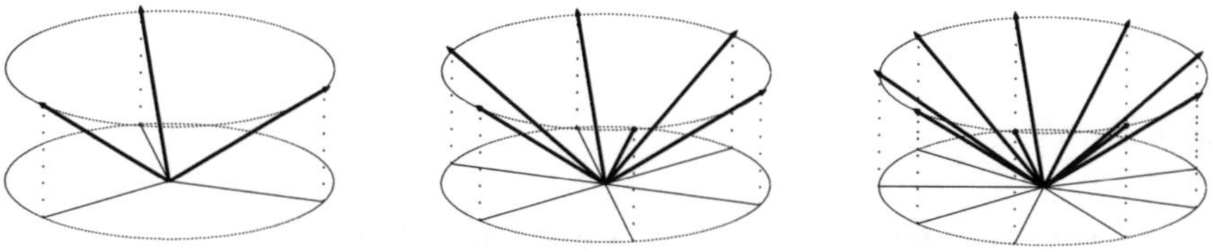

Fig. 5.1: The lifted n equally spaced unit vector in $\mathbb{R}^2$ for $n = 3, 6, 9$.

5.4 Complements

Every normalised tight frame has a unique complement, and each frame is similar to a unique normalised tight frame. Thus we can define a complement (which is unique up to similarity) for an arbitrary frame. This can be thought of as a *maximal* orthogonal direct summand, or as the direct sum of a sequence of lifts which take the frame to a basis.

Definition 5.3. We say that finite frames $\Phi = (\phi_j)_{j \in J}$ and $\Psi = (\psi_j)_{j \in J}$ for $\mathscr{H}_1$ and $\mathscr{H}_2$, with the same index set J, are **complements** of each other if they are disjoint and $\mathscr{H}_1 \oplus \mathscr{H}_2 \approx \ell_2(J)$, i.e., $\dim(\mathscr{H}_1) + \dim(\mathscr{H}_2) = |J|$.

For normalised tight frames, this is equivalent to Definition 2.6. We recall (see §2.9) that the **α-partition frame** is the normalised tight frame of nonzero vectors which is the complement of

$$\Big(\underbrace{\frac{e_1}{\sqrt{\alpha_1}}, \ldots, \frac{e_1}{\sqrt{\alpha_1}}}_{\alpha_1 \text{ times}}, \ldots, \underbrace{\frac{e_k}{\sqrt{\alpha_k}}, \ldots, \frac{e_k}{\sqrt{\alpha_k}}}_{\alpha_k \text{ times}} \Big).$$

Disjoint frames *cannot* be similar, and so, in particular, no normalised tight frame is equal to its complement. By way of contrast, this is not true in infinite dimensions, e.g., [Cas98] shows that every frame for an infinite dimensional Hilbert space can be written as the direct sum of three orthonormal bases.

Here are some equivalences that follow easily from the previous discussion.

Proposition 5.1. *Let $\Phi = (\phi_j)_{j\in J}$, $\Psi = (\psi_j)_{j\in J}$ be a frames for $\mathcal{H}_1$, $\mathcal{H}_2$, with index set J, and synthesis maps $V = [\phi_j]$, $W = [\psi_j]$. Then the following are equivalent*

1. *Φ and Ψ are complements.*
2. *Φ^{can} and Ψ^{can} are complements.*
3. *$\mathrm{Gram}(\Phi^{\mathrm{can}}) + \mathrm{Gram}(\Psi^{\mathrm{can}}) = I$.*
4. *$\ell_2(J) = \mathrm{ran}(P_\Phi) \oplus \mathrm{ran}(P_\Psi)$.*
5. *$\dim(\mathcal{H}_1) + \dim(\mathcal{H}_2) = |J|$ and $VW^* = 0$.*
6. *$\Phi \oplus \Psi$ is a basis and $(\Phi \oplus \Psi)\tilde{} = \tilde{\Phi} \oplus \tilde{\Psi}$.*
7. *$\Psi = Q(I - \mathrm{Gram}(\Phi^{\mathrm{can}}))$, where Q is 1–1 on $\mathrm{ran}(I - \mathrm{Gram}(\Phi^{\mathrm{can}}))$.*
8. *$\Psi = \Psi_1 \oplus \cdots \oplus \Psi_k$, $k = |J| - \dim(\mathcal{H}_1)$, where Ψ_j lifts $\Phi \oplus \Psi_1 \oplus \cdots \oplus \Psi_{j-1}$.*

5.5 Sums

The *sum* of frames requires that at least one of them is *balanced*.

Definition 5.4. Let $\Phi = (\phi_j)_{j=1}^{n_1}$ and $\Psi = (\psi_k)_{k=1}^{n_2}$ be frames for $\mathcal{H}_1$ and $\mathcal{H}_2$, at least one of which is balanced, i.e., has vectors which sum to zero. Then their **sum** is

$$\Phi \hat{+} \Psi := \Bigl(\frac{1}{\sqrt{n_2}}\phi_j + \frac{1}{\sqrt{n_1}}\psi_k\Bigr)_{\substack{1\le j\le n_1\\ 1\le k\le n_2}}. \tag{5.8}$$

This is a frame of $n_1 n_2$ elements for $\mathcal{H}_1 \oplus \mathcal{H}_2$, which is tight if the summands are tight with the same frame bound.

Theorem 5.3. *The sum $\Phi \hat{+} \Psi$ defined by (5.8) is a frame of $n_1 n_2$ vectors for $\mathcal{H}_1 \oplus \mathcal{H}_2$, which is balanced if and only if both Φ and Ψ are balanced, with*

$$S_{\Phi\hat{+}\Psi}(f+g) = S_\Phi f + S_\Psi g, \qquad \forall f \in \mathcal{H}_1,\ \forall g \in \mathcal{H}_2,$$

or, equivalently, dual frame given by

$$(\Phi \hat{+} \Psi)\tilde{} = \tilde{\Phi} \hat{+} \tilde{\Psi}.$$

The frame bounds are

$$A_{\Phi\hat{+}\Psi} = \min\{A_\Phi, A_\Psi\}, \qquad B_{\Phi\hat{+}\Psi} = \max\{B_\Phi, B_\Psi\}. \tag{5.9}$$

In particular, a sum of normalised tight frames is a normalised tight frame.

Proof. Let S be the frame operator for $(\alpha\phi_j+\beta\psi_k)$, with α,β scalars, then

$$\begin{aligned}S(f+g)&=\sum_j\sum_k\langle f+g,\alpha\phi_j+\beta\psi_k\rangle(\alpha\phi_j+\beta\psi_k)\\&=|\alpha|^2n_2S_\Phi f+\bar{\alpha}\beta\langle f,\sum_j\phi_j\rangle\sum_k\psi_k+\alpha\bar{\beta}\langle g,\sum_k\psi_k\rangle\sum_j\phi_j+|\beta|^2n_1S_\Psi g.\end{aligned}$$

If Φ or Ψ is balanced, and we choose $\alpha=\frac{1}{\sqrt{n_2}}$, $\beta=\frac{1}{\sqrt{n_1}}$, then this reduces to

$$S(f+g)=S_\Phi f+S_\Psi g.$$

By Exer. 5.4, $\Phi\hat{+}\Psi$ is a frame for $\mathscr{H}_1\oplus\mathscr{H}_2$, with the asserted properties. □

The sum $\hat{+}$ also satisfies the following rules

$$(\Phi_1\hat{+}\Phi_2)\hat{+}\Phi_3=\Phi_1\hat{+}(\Phi_2\hat{+}\Phi_3),\qquad \alpha(\Phi_1\hat{+}\Phi_2)=(\alpha\Phi_1)\hat{+}(\alpha\Phi_2).$$

Example 5.10. Three equally spaced vectors Φ in $\mathbb{R}^2$ are a balanced tight frame. Taking the sum of Φ with itself gives the following tight frame of 9 vectors for $\mathbb{R}^4$

$$\Phi\hat{+}\Phi:=\left\{\begin{bmatrix}\cos\frac{2\pi}{3}j\\ \sin\frac{2\pi}{3}j\\ \cos\frac{2\pi}{3}k\\ \sin\frac{2\pi}{3}k\end{bmatrix}:1\le j,k\le 3\right\}.$$

Example 5.11. (Roots of unity) The n-th roots of unity sum to zero, and so form a balanced tight frame for $\mathbb{C}$. Let $\omega:=e^{2\pi i/n_1}$, $\mu:=e^{2\pi i/n_2}$ be $n_1,n_2\ge 2$ roots of unity. Then $\Phi:=(\frac{1}{\sqrt{n_1}}\omega^j)_{j=1}^{n_1}$ and $\Psi:=(\frac{1}{\sqrt{n_2}}\mu^k)_{k=1}^{n_2}$ are balanced normalised tight frames, and so their sum

$$\Phi\hat{+}\Psi:=\left(\frac{1}{\sqrt{n_1n_2}}\begin{bmatrix}\omega^j\\ \mu^k\end{bmatrix}\right)_{\substack{1\le j\le n_1\\ 1\le k\le n_2}}$$

is an equal-norm balanced normalised tight frame of n_1n_2 vectors for $\mathbb{C}^2$.

Example 5.12. Taking d summands of $\Phi=(\frac{1}{\sqrt{2}},-\frac{1}{\sqrt{2}})$ gives the **vertices of the cube** in $\mathbb{R}^d$, i.e.,

$$\Phi\hat{+}\cdots\hat{+}\Phi=\left(\frac{1}{\sqrt{2^d}}\begin{bmatrix}(-1)^{j_1}\\ \cdots\\ (-1)^{j_d}\end{bmatrix}\right)_{1\le j_1,\ldots,j_d\le 2}.$$

Example 5.13. (Equal-norm tight frames) Clearly the sum of equal-norm frames is an equal-norm frame, and by (5.9), the sum of equal-norm normalised tight frames (one of them balanced) is an equal-norm normalised tight frame.

5.6 Tensor products

The tensor product of vector spaces V and W is the (abstract) vector space $V \otimes W$ with the property that any bilinear map from $V \times W$ factors uniquely through $V \otimes W$. There is an associated bilinear map

$$\otimes : V \times W \to V \otimes W : (f,g) \mapsto f \otimes g$$

whose images $f \otimes g$ are called *irreducible tensors*, and span $V \otimes W$.

If $\{v_1,\ldots,v_m\}$ and $\{w_1,\ldots,w_n\}$ are bases for V and W, then $V \otimes W$ can be realised as the mn-dimensional vector space with a basis given by (the formal symbols) $v_j \otimes w_k$, $1 \le j \le m$, $1 \le k \le n$, which satisfy the distributive law

$$\Big(\sum_j \alpha_j v_j\Big) \otimes \Big(\sum_k \beta_k w_k\Big) = \sum_j \sum_k \alpha_j \beta_j (v_j \otimes w_k), \qquad \forall \alpha_j, \beta_k \in \mathbb{F}.$$

For Hilbert spaces $\mathscr{H}_1$ and $\mathscr{H}_2$, the *tensor product* Hilbert space $\mathscr{H}_1 \otimes \mathscr{H}_2$ is obtained by taking the inner product on $\mathscr{H}_1 \otimes \mathscr{H}_2$ given by

$$\langle f \otimes g, \phi \otimes \psi \rangle_{\mathscr{H}_1 \otimes \mathscr{H}_2} := \langle f, \phi \rangle_{\mathscr{H}_1} \langle g, \psi \rangle_{\mathscr{H}_2}, \qquad \forall f, \phi \in \mathscr{H}_1, \quad \forall g, \psi \in \mathscr{H}_2.$$

Definition 5.5. The **tensor product** of frames Φ and Ψ for $\mathscr{H}_1$ and $\mathscr{H}_2$ is

$$\Phi \otimes \Psi := (\phi \otimes \psi)_{\phi \in \Phi, \psi \in \Psi} \subset \mathscr{H}_1 \otimes \mathscr{H}_2.$$

The tensor product of frames is a frame.

Theorem 5.4. *If Φ and Ψ are frames for $\mathscr{H}_1$ and $\mathscr{H}_2$, then $\Phi \otimes \Psi$ is a frame for the tensor product Hilbert space $\mathscr{H}_1 \otimes \mathscr{H}_2$, with frame operator*

$$S_{\Phi \otimes \Psi} = S_\Phi \otimes S_\Psi, \tag{5.10}$$

frame bounds

$$A_{\Phi \otimes \Psi} = A_\Phi A_\Psi, \qquad B_{\Phi \otimes \Psi} = B_\Phi B_\Psi, \tag{5.11}$$

and dual frame given by

$$(\phi \otimes \psi)\tilde{} = \tilde{\Phi} \otimes \tilde{\Psi}. \tag{5.12}$$

Proof. For irreducible tensors $f \otimes g \in \mathscr{H}_1 \otimes \mathscr{H}_2$, we have

$$\begin{aligned} S_{\Phi \otimes \Psi}(f \otimes g) &= \sum_{\phi \in \Phi} \sum_{\psi \in \Psi} \langle f \otimes g, \phi \otimes \psi \rangle \, \phi \otimes \psi = \sum_{\phi \in \Phi} \sum_{\psi \in \Psi} (\langle f, \phi \rangle \phi) \otimes (\langle g, \psi \rangle \psi) \\ &= \Big(\sum_{\phi \in \Phi} \langle f, \phi \rangle \phi \Big) \otimes \Big(\sum_{\psi \in \Psi} \langle g, \psi \rangle \psi \Big) = (S_\Phi f) \otimes (S_\Psi g), \end{aligned}$$

and so we obtain $S_{\Phi \otimes \Psi} = S_\Phi \otimes S_\Psi$ by linearity. Since the eigenvalues of a tensor product of operators are the products of eigenvalues of the operators, we obtain the asserted frame bounds.

Finally, the dual frame is $S_{\Phi\otimes\Psi}^{-1}(\phi\otimes\psi)=(S_\Phi^{-1}\phi)\otimes(S_\Psi^{-1}\psi)=\tilde{\Phi}\otimes\tilde{\Psi}$. □

Tensor products of frames with various properties inherit those properties.

Corollary 5.1. *Let Φ and Ψ be frames. Then*

1. *$\Phi\otimes\Psi$ is a tight frame if and only its factors are.*
2. *$\Phi\otimes\Psi$ is an equal-norm tight frame if and only its factors are.*
3. *$\Phi\otimes\Psi$ is an orthonormal basis if and only its factors are.*
4. *$\Phi\otimes\Psi$ is an orthogonal basis if and only its factors are.*
5. *$\Phi\otimes\Psi$ is a real frame if and only if its factors are.*

Proof. Let $\Phi=(\phi_j)$ and $\Psi=(\psi_k)$.
1. Use (5.11) and $A\le B$.
2. Use $\|\phi_j\otimes\psi_k\|=\|\phi_j\|\|\psi_k\|$.
3., 4. Use $\langle\phi_{j_1}\otimes\psi_{k_1},\phi_{j_2}\otimes\psi_{k_2}\rangle=\langle\phi_{j_1},\phi_{j_2}\rangle\langle\psi_{k_1},\psi_{k_2}\rangle$.
5. If $\Phi\otimes\Psi$ is real (and Ψ has a nonzero vector), then

$$\langle\phi_{j_1},\phi_{j_2}\rangle\|\psi_k\|^2=\langle\phi_{j_1}\otimes\psi_k,\phi_{j_2}\otimes\psi_k\rangle\in\mathbb{R}\quad\Longrightarrow\quad\langle\phi_{j_1},\phi_{j_2}\rangle\in\mathbb{R},$$

and similarly $\langle\psi_{k_1},\psi_{k_2}\rangle\in\mathbb{R}$. If Φ and Ψ are real, then clearly $\Phi\otimes\Psi$ is real. □

Example 5.14. (Equally spaced vectors) Let $(v_j)_{j=1}^m$ and $(w_k)_{k=1}^n$ be the tight frames of m and n equally spaced unit vectors in $\mathbb{R}^2$, given by

$$v_j:=\cos\frac{2\pi j}{m}e_1+\sin\frac{2\pi j}{m}e_2,\qquad w_k:=\cos\frac{2\pi k}{n}e_1+\sin\frac{2\pi k}{n}e_2,$$

where $\{e_j\}$ is the standard basis. By taking their tensor product

$$\begin{aligned}v_j\otimes w_k:=\cos\frac{2\pi j}{m}\cos\frac{2\pi k}{n}e_1\otimes e_1+\cos\frac{2\pi j}{m}\sin\frac{2\pi k}{n}e_1\otimes e_2\\+\sin\frac{2\pi j}{m}\cos\frac{2\pi k}{n}e_2\otimes e_1+\sin\frac{2\pi j}{m}\sin\frac{2\pi k}{n}e_2\otimes e_2,\end{aligned}$$

we obtain the tight frame $\Phi=(\phi_{jk})$ of mn unit vectors for $\mathbb{R}^4$, given by

$$\phi_{jk}:=\begin{pmatrix}\cos\frac{2\pi j}{m}\cos\frac{2\pi k}{n}\\ \cos\frac{2\pi j}{m}\sin\frac{2\pi k}{n}\\ \sin\frac{2\pi j}{m}\cos\frac{2\pi k}{n}\\ \sin\frac{2\pi j}{m}\sin\frac{2\pi k}{n}\end{pmatrix},\qquad\begin{matrix}j=1,\ldots,m,\\ k=1,\ldots,n.\end{matrix}$$

Example 5.15. (Orthogonal polynomials) Tensor products of orthogonal polynomials are orthogonal polynomials for the tensor product weight. This can be used to construct frames of orthogonal polynomials in several variables (see [DX01]).

5.7 Decompositions

In general, there is not a unique way to decompose a frame.

Example 5.16. (Nonuniqueness) The tight frame of four equally spaced unit vectors in $\mathbb{R}^2$ (the vertices of the square) can be written as a disjoint union, direct sum (lift), sum and tensor product, e.g.,

$$(1,-1)\cup(1,-1) = \left(\begin{pmatrix}1\\0\end{pmatrix},\begin{pmatrix}-1\\0\end{pmatrix},\begin{pmatrix}0\\1\end{pmatrix},\begin{pmatrix}0\\-1\end{pmatrix}\right),$$

$$(1,-1,0,0)+(0,0,1,-1) = \left(\begin{pmatrix}1\\0\end{pmatrix},\begin{pmatrix}-1\\0\end{pmatrix},\begin{pmatrix}0\\1\end{pmatrix},\begin{pmatrix}0\\-1\end{pmatrix}\right),$$

$$(-1,1)\hat{+}(-1,1) = \left(\begin{pmatrix}1\\1\end{pmatrix},\begin{pmatrix}-1\\1\end{pmatrix},\begin{pmatrix}-1\\-1\end{pmatrix},\begin{pmatrix}1\\-1\end{pmatrix}\right),$$

$$(-1,1)\otimes(e_1,e_2) = (e_1,e_2,-e_1,-e_2).$$

This tight frame is also divisible, e.g., it can be partitioned into two prime tight frames

$$\{e_1,e_2\}\cup\{-e_1,-e_2\}.$$

Notes

The *direct sum*, and the associated notions of a *lift* and *complement* can be found throughout the frame literature. There can be some variation in terminology (when named), e.g., the term *lift* is used for the *simple lift* in [BF03].

Exercises

5.1. Show that a frame (f_j) is a disjoint union of tight frames if and only if each f_j is an eigenvector of the frame operator S (such a frame is said to be **semicritical**).

5.2. *Unions.* Let $\Phi = (\phi_j)$ and $\Psi = (\psi_k)$ be frames for $\mathcal{H}_1$ and $\mathcal{H}_2$.
(a) Show that $\Phi\cup\Psi$, as a sequence in $\mathcal{H}_1\oplus\mathcal{H}_2$, has frame operator S, with

$$S(f+g) = S_\Phi(f)+S_\Psi(g), \qquad \forall f\in\mathcal{H}_1,\ \forall g\in\mathcal{H}_2.$$

(b) Show that the dual frame is $\tilde{\Phi}\cup\tilde{\Psi}$, by using Exer. 5.4.
(c) Show the Gramian is block diagonal, i.e.,

$$\text{Gram}(\Phi \cup \Psi) = \begin{pmatrix} \text{Gram}(\Phi) & 0 \\ 0 & \text{Gram}(\Psi) \end{pmatrix}.$$

5.3. *Direct sums.* Let $\Phi = (\phi_j)_{j\in J}$ and $\Psi = (\psi_j)_{j\in J}$ be finite frames for $\mathscr{H}_1$ and $\mathscr{H}_2$, with $V = [\phi_j]$, $W = [\psi_j]$. Show that $(\phi_j + \psi_j)$ is a frame for $\mathscr{H}_1 \oplus \mathscr{H}_2$ if and only if

$$\text{ran}(V^*) \cap \text{ran}(W^*) = 0,$$

in which case

$$\text{ran}([\phi_j + \psi_j]^*) = \text{ran}(V^*) + \text{ran}(W^*) \qquad \text{(algebraic direct sum).}$$

5.4. Suppose Ξ is a sequence in $\mathscr{H}_1 \oplus \mathscr{H}_2$ for which $S = S_\Xi$ can be decomposed

$$S(f+g) = S_\Phi(f) + S_\Psi(g), \qquad \forall f \in \mathscr{H}_1, \forall g \in \mathscr{H}_2, \tag{5.13}$$

where Φ and Ψ are frames for $\mathscr{H}_1$ and $\mathscr{H}_2$, e.g., the sequences

$$\Phi \cup \Psi, \quad \Phi \oplus \Psi, \quad \Phi \hat{+} \Psi. \tag{5.14}$$

(a) Show that Ξ is a frame for $\mathscr{H}_1 \oplus \mathscr{H}_2$, with frame bounds

$$A = \min\{A_\Phi, A_\Psi\}, \qquad B = \max\{B_\Phi, B_\Psi\}.$$

In particular, this implies that *unions*, *direct sums* and *sums* of normalised tight frames are again normalised tight frames.
(b) Show that (5.13) is equivalent to

$$S^{-1}(f+g) = S_\Phi^{-1}(f) + S_\Psi^{-1}(g), \qquad \forall f+g \in \Xi. \tag{5.15}$$

For the choices (5.14), this gives

$$(\Phi \cup \Psi)\tilde{} = \tilde{\Phi} \cup \tilde{\Psi}, \qquad (\Phi \oplus \Psi)\tilde{} = \tilde{\Phi} \oplus \tilde{\Psi}, \qquad (\Phi \hat{+} \Psi)\tilde{} = \tilde{\Phi} \hat{+} \tilde{\Psi}.$$

5.5. Show that a finite frame $(f_j)_{j\in J}$ is a simple lift if and only if

$$\sum_j f_j \neq 0, \qquad \langle \sum_j f_j, f_k \rangle = C, \quad \forall k.$$

Chapter 6
Variational characterisations of tight frames

If $(f_j)_{j\in J}$ is a finite tight frame for $\mathscr{H}$, then (see Proposition 2.1)

$$\sum_{j\in J}\sum_{k\in J}|\langle f_j,f_k\rangle|^2=\frac{1}{d}\Big(\sum_{j\in J}\langle f_j,f_j\rangle\Big)^2,\qquad d=\dim(\mathscr{H}). \tag{6.1}$$

For a general finite frame, this becomes an inequality, with *equality* if and only if the frame is *tight*. Thus (6.1) characterises finite tight frames.

Important instances (for unit vectors) include Welch bound equality sequences, minimisers of the frame potential, spherical half-designs of order 2, and spherical $(1,1)$-designs.

6.1 Welch bound equality sequences

Let $f_1,\ldots,f_n$ be $n\ge d$ unit vectors in $\mathbb{C}^d$ (signals of unit energy). Then

$$\mathscr{F}(f_1,\ldots,f_n):=\sum_{j=1}^n\sum_{k=1}^n|\langle f_j,f_k\rangle|^2\ge\frac{n^2}{d}, \tag{6.2}$$

which is known as the **Welch bound**, after [Wel74], which used (6.2) to prove

$$\max_{j\neq k}|\langle f_j,f_k\rangle|^2\ge\frac{n^2/d-n}{n^2-n}=\frac{n-d}{d(n-1)}. \tag{6.3}$$

Each gives a lower bound on how small the *cross-correlation* of a set of signals of unit energy can be, i.e., how "spread out" the signals are. Unit vectors $f_1,\ldots,f_n$ which give equality in (6.2) are called **WBE sequences (Welch bound equality sequences)**, see, e.g., [MM93] where they are used for CDMA (code division multiple access) systems. Equality in the Welch bound is the same as the equality (6.1).

S.F.D. Waldron, *An Introduction to Finite Tight Frames*, Applied and Numerical Harmonic Analysis, https://doi.org/10.1007/978-0-8176-4815-2_6

6.2 The variational characterisation

The following extends the Welch bound and frame potential results (see §6.3) to the case where the vectors may have arbitrary lengths (also see Exer. 6.1).

Theorem 6.1. *Let $f_1,\ldots,f_n$ be vectors in $\mathcal{H}$, not all zero, and $d=\dim(\mathcal{H})$. Then*

$$\sum_{j=1}^n\sum_{k=1}^n |\langle f_j,f_k\rangle|^2 \geq \frac{1}{d}\Bigl(\sum_{j=1}^n \|f_j\|^2\Bigr)^2, \tag{6.4}$$

with equality if and only if $(f_j)_{j=1}^n$ is a tight frame for $\mathcal{H}$.

Proof. Let $V=[f_j]$. We recall from (2.7) that the frame operator $S=VV^*$ satisfies

$$\operatorname{trace}(S)=\sum_j \|f_j\|^2, \qquad \operatorname{trace}(S^2)=\sum_j\sum_k |\langle f_j,f_k\rangle|^2.$$

Since $S=VV^*$ is positive definite, it is unitarily diagonalisable with eigenvalues $\lambda_1,\ldots,\lambda_d\geq 0$. By the Cauchy–Schwarz inequality

$$\operatorname{trace}(S)^2=\Bigl(\sum_j \lambda_j\Bigr)^2=\langle(1),(\lambda_j)\rangle^2\leq \|(1)\|^2\|(\lambda_j)\|^2=d\sum_j \lambda_j^2=d\operatorname{trace}(S^2),$$

which is (6.4), with equality if and only if $\lambda_j=A$, $\forall j$, $A>0$, i.e.,

$$S=AI_{\mathcal{H}} \iff (f_j) \text{ is a tight frame for } \mathcal{H}.$$

Note above, since one of the vectors (f_j) is nonzero, $S\neq 0$, and so $A\neq 0$. □

Example 6.1. If all the vectors f_j have unit norm, then (6.1) reduces to the Welch bound (6.2), i.e.,

$$\sum_{j=1}^n\sum_{k=1}^n |\langle f_j,f_k\rangle|^2 \geq \frac{1}{d}\Bigl(\sum_{j=1}^n 1^2\Bigr)^2=\frac{n^2}{d}.$$

Example 6.2. The corresponding generalisation of the Welch bound (6.3) is

$$\max_{j\neq k} |\langle f_j,f_k\rangle|^2 \geq \frac{(\sum_\ell \|f_\ell\|^2)^2/d-\sum_\ell \|f_\ell\|^4}{n^2-n}>0.$$

The tightness of a finite frame can be determined from the absolute values of the entries of its Gramian:

The vectors $(f_j)_{j=1}^n$ are a tight frame for $\mathcal{H}$, $d=\dim(\mathcal{H})$, if and only if

$$\sum_{j=1}^n\sum_{k=1}^n |\langle f_j,f_k\rangle|^2 = \frac{1}{d}\Bigl(\sum_{j=1}^n \|f_j\|^2\Bigr)^2>0. \tag{6.5}$$

6.3 The frame potential

Let $\mathbb{S} = \{f \in \mathscr{H} : \|f\| = 1\}$ be the unit sphere in $\mathscr{H}$, $d = \dim(\mathscr{H})$. The function

$$\mathrm{FP} : \mathbb{S}^n \to [0,\infty) : (f_j)_{j=1}^n \mapsto \mathscr{F}(f_1,\ldots,f_n) = \sum_{j=1}^{n}\sum_{k=1}^{n} |\langle f_j, f_k\rangle|^2 \tag{6.6}$$

of (6.2) was called the **frame potential** by Fickus (see [Fic01]) who derived it from a *frame force* (see §6.14). We recall that unit-norm tight frames exist (Example 2.4). Since a tight frame for $\mathscr{H}$ is a spanning set (which must have at least d elements), Theorem 6.1 implies that for $n \geq d$ the minimum of the frame force is $\frac{n^2}{d}$, which is attained precisely for unit-norm tight frames. For $n < d$ the minimum of n is attained when (f_j) is a (nonspanning) orthonormal sequence (see Exer. 6.5). In summary:

Theorem 6.2. ([BF03]). *Let $\Phi = (f_j)_{j=1}^n$ be n unit vectors in $\mathscr{H}$, $d = \dim(\mathscr{H})$. Then the frame potential is bounded below by*

$$\mathrm{FP}(\Phi) \geq \max\{\frac{n^2}{d}, n\}$$

with equality (a minimum) if and only if either

(a) *Φ is a unit-norm tight frame (and hence $n \geq d$), or*
(b) *Φ is a nonspanning orthonormal sequence (and hence $n < d$).*

A careful analysis of the Lagrange multiplier equations (see §6.15) shows that all local minimisers of the frame potential give the global minimum. This is in contrast with many other well-known potentials, e.g., the Coulomb electrostatic potential, which do not have this property.

In view of (6.4), a **normalised frame potential** $\hat{\mathrm{FP}} : \mathscr{H}^n \setminus \{0\} \to [0,\infty)$ can be defined (see Exer. 6.6) on sequences of vectors which are not all zero by

$$\frac{1}{d} \leq \hat{\mathrm{FP}}(f_1,\ldots,f_n) := \frac{\sum_{j=1}^n \sum_{k=1}^n |\langle f_j, f_k\rangle|^2}{\left(\sum_{j=1}^n \|f_j\|^2\right)^2} \leq 1. \tag{6.7}$$

This leads to the following extension of Theorem 6.2.

Theorem 6.3. ([Wal03]). *For $n \geq d = \dim(\mathscr{H})$ the normalised frame potential $\hat{\mathrm{FP}}(f_1,\ldots,f_n)$ attains its minimum of $\frac{1}{d}$ if and only if $(f_j)_{j=1}^n$ is a tight frame for $\mathscr{H}$, and its maximum of 1 if and only if $\mathrm{span}(f_j)_{j=1}^n$ is 1-dimensional.*

Thus the distance of the normalised frame potential from $\frac{1}{d}$ gives a scaling independent way of measuring how far from being tight it is (also see the discussion of §3.8). In particular, sequences of vectors which do not span $\mathbb{F}^d$ are far from tight, since

$$\hat{\mathrm{FP}}(f_1,\ldots,f_n) \geq \frac{1}{\dim(\mathrm{span}_{1\leq j\leq n}\{f_j\})}.$$

6.4 Spherical t-designs and the Waring formula

We now outline some of the basic results about cubature on the sphere in $\mathbb{R}^d$, and its relationship to tight frames. These are proved, together with the analogous results for the complex sphere, in §6.8. Let $\mathrm{Hom}_j(\mathbb{R}^d)$ be the homogeneous polynomials of total degree j in d real variables.

Definition 6.1. A finite subset Φ of the unit sphere $\mathbb{S}$ in $\mathbb{R}^d$ is a **(real) spherical t-design** if the normalised surface integral satisfies

$$\int_{\mathbb{S}} f\,d\sigma = \frac{1}{|\Phi|}\sum_{\phi\in\Phi} f(\phi), \qquad \forall f\in\mathrm{Hom}_j(\mathbb{R}^d), \quad 0\le j\le t,$$

and it is a **(real) spherical half-design**[1] of order t if

$$\int_{\mathbb{S}} f\,d\sigma = \frac{1}{|\Phi|}\sum_{\phi\in\Phi} f(\phi), \qquad \forall f\in\mathrm{Hom}_t(\mathbb{R}^d).$$

Spherical designs have been studied since the 1970s (see [DGS77] and [BB09]). Their existence for every t (and $d\ge 2$) was proved by [SZ84] (see Theorem 6.5).

Example 6.3. The 12 vertices of the *regular icosahedron* give a spherical 5-design.

Example 6.4. The n equally spaced vectors in $\mathbb{R}^2$ are a spherical $(n-1)$-design, and the $n=t+1$ equally spaced lines in $\mathbb{R}^2$ give a spherical half-design of order $2t$.

The following simple observations are very useful (and illuminating).

- If $f\in\mathrm{Hom}_j(\mathbb{R}^d)$, then $g=\|\cdot\|^2 f\in\mathrm{Hom}_{j+2}$, and the restriction of f and g to the sphere $\mathbb{S}$ is equal. Thus if Φ is a spherical half-design of order t, then it is also a spherical half-design of order $t-2, t-4,\ldots$.
- For j odd, every $f\in\mathrm{Hom}_j(\mathbb{R}^d)$ is an odd function, and so has zero integral. Thus if Φ is **centrally symmetric**, i.e, $\Phi=-\Phi$, then it is a spherical half-design of order $1,3,5,\ldots$.

For t even, a set $\Phi=\{\phi_1,\ldots,\phi_n\}$ of unit vectors in $\mathbb{R}^d$ is a spherical half-design of order t if and only if it satisfies the so-called **Waring formula**

$$\langle x,x\rangle^{t/2} = \frac{d(d+2)\cdots(d+t-2)}{1\cdot 3\cdot 5\cdots(t-1)}\frac{1}{n}\sum_{j=1}^{n}\langle x,\phi_j\rangle^t, \qquad \forall x\in\mathbb{R}^d, \tag{6.8}$$

or, equivalently (see [Sid74], [GS79], [Sei01])

$$\sum_{j=1}^{n}\sum_{k=1}^{n}\langle\phi_j,\phi_k\rangle^t = n^2\frac{1\cdot 3\cdot 5\cdots(t-1)}{d(d+2)\cdots(d+t-2)}. \tag{6.9}$$

---

[1] There is some variation of terminology in the literature, e.g., Seidel [Sei01] refers to a spherical half-design of order t as a "spherical t-design". Spherical t-designs whose number of vectors satisfy the lower bounds of [DGS77] are said to be *tight* (see [BBHS10] for a classification).

For $t=2$, these become

$$\|x\|^2 = \frac{d}{n}\sum_{j=1}^n |\langle x,\phi_j\rangle|^2, \quad \forall x\in\mathbb{R}^d, \qquad \sum_{j=1}^n\sum_{k=1}^n \langle \phi_j,\phi_k\rangle^2 = \frac{n^2}{d}, \tag{6.10}$$

i.e., Φ is a tight frame by definition and by equality in (6.4), respectively.

A spherical half-design of order 2 is precisely a tight frame of distinct unit vectors in $\mathbb{R}^d$ (equivalently, a WBE sequence of distinct vectors).

We now consider the relationship between tight frames and spherical 2-designs.

Proposition 6.1. *A set $\Phi := \{\phi_1,\ldots\phi_n\}$ of unit vectors in $\mathbb{R}^d$ is a spherical 2-design if and only if it is a balanced tight frame, i.e.,*

$$\sum_{j=1}^n \phi_j = 0, \qquad \sum_{j=1}^n\sum_{k=1}^n \langle \phi_j,\phi_k\rangle^2 = \frac{n^2}{d}. \tag{6.11}$$

Moreover, such a Φ is also a 3-design if it is centrally symmetric.

Proof. The second condition in (6.11) ensures that Φ is spherical half-design of order 2 (and hence 0). It therefore remains to add to this a condition which ensures Φ is spherical half-design of order 1. Since the homogeneous polynomials of degree 1 have zero integral and are spanned by $p_y := \langle\cdot,y\rangle$, $y\in\mathbb{R}^d$, the condition is

$$0 = \int p_y\,d\sigma = \frac{1}{n}\sum_{j=1}^n p_y(\phi_j) = \frac{1}{n}\sum_{j=1}^n \langle\phi_j,y\rangle = \frac{1}{n}\langle\sum_{j=1}^n \phi_j, y\rangle, \quad \forall y\in\mathbb{R}^d,$$

i.e., $\sum_j \phi_j = 0$. We already observed that if Φ is centrally symmetric, then it is a spherical half-design of order $t = 1,3,5,\ldots$. □

6.5 Other characterisations of spherical t-designs

The description of homogeneous polynomials on the sphere (and their integrals) is intimately related to the harmonic polynomials (see §16.5 for details). From this connection come various characterisations of spherical t-designs, which we briefly consider. Each homogeneous polynomial $f\in \mathrm{Hom}_t(\mathbb{R}^d) = \Pi_t^\circ(\mathbb{R}^d)$ can be written uniquely

$$f(x) = \sum_{0\le j\le \frac{t}{2}} \|x\|^{2j} f_{t-2j}(x), \qquad f_{t-2j}\in \mathrm{Harm}_{t-2j}(\mathbb{R}^d), \tag{6.12}$$

where $\mathrm{Harm}_k(\mathbb{R}^d) = \mathscr{H}_k(\mathbb{R}^d)$ are the harmonic polynomials of degree k. Since the harmonic polynomials of different degrees are orthogonal with respect to the inner product given by the normalised surface measure σ on $\mathbb{S}$, integrating f gives

$$\int_{\mathbb{S}} f\,d\sigma = \sum_{0\le j\le \frac{t}{2}} \int_{\mathbb{S}} f_{t-2j}\,d\sigma = \begin{cases} 0, & t \text{ is odd;} \\ f_0, & t \text{ is even.} \end{cases}$$

Therefore $\Phi = \{\phi_j\}$ is spherical t-design if and only if

$$\sum_j f(\phi_j) = 0, \qquad \forall f \in \mathrm{Harm}_\ell(\mathbb{R}^d), \quad 1 \le \ell \le t, \tag{6.13}$$

and is a spherical half-design of order t if and only if

$$\sum_j f(\phi_j) = 0, \qquad \forall f \in \mathrm{Harm}_\ell(\mathbb{R}^d), \quad \ell = t, t-2, \ldots \ (\ell \ge 1). \tag{6.14}$$

A set Φ satisfying (6.14) for $\ell = t$, i.e., $\sum_j f(\phi_j) = 0$, $f \in \mathrm{Harm}_t(\mathbb{R}^d)$, is called a *spherical design of harmonic index t* (see [BOT15]).

By using basic properties of reproducing kernels (see §16.3, §16.4 for details), [SW09] have converted the condition (6.13) into a variational characterisation of spherical t-designs similar to that of Theorem 6.2.

Theorem 6.4. ([SW09]) *For each ℓ, let $K_\ell^{(d)}(\langle x, y\rangle)$ be a positive scalar multiple of the reproducing kernel for* $\mathrm{Harm}_\ell(\mathbb{R}^d)$, $\Phi = \{\phi_j\}_{j=1}^n$, *and define $A_t : \mathbb{S}^n \to \mathbb{R}$ by*

$$A_t(\Phi) := \sum_{\ell=1}^{t} \sum_{j=1}^{n} \sum_{k=1}^{n} K_\ell^{(d)}(\langle \phi_j, \phi_k\rangle). \tag{6.15}$$

Then $A_t(\Phi) \ge 0$, and Φ is a spherical t-design if and only if $A_t(\Phi) = 0$.

Proof. Let $(Y_s^{(\ell)})$ be an orthonormal basis for $\mathscr{H}_\ell(\mathbb{R}^d)$. Then Proposition 16.3 gives

$$K_\ell^{(d)}(\langle x, y\rangle) = c_\ell \sum_s Y_s^{(\ell)}(x) Y_s^{(\ell)}(y), \qquad c_\ell > 0,$$

and we compute

$$A_t(\Phi) = \sum_{\ell=1}^{t} \sum_{j=1}^{n} \sum_{k=1}^{n} c_\ell \sum_s Y_s^{(\ell)}(\phi_j) Y_s^{(\ell)}(\phi_k) = \sum_{\ell=1}^{t} c_\ell \sum_s \left(\sum_{j=1}^{n} Y_s^{(\ell)}(\phi_j) \right) \left(\sum_{k=1}^{n} Y_s^{(\ell)}(\phi_k) \right)$$

$$= \sum_{\ell=1}^{t} c_\ell \sum_s \left(\sum_{j=1}^{n} Y_s^{(\ell)}(\phi_j) \right)^2 \ge 0.$$

Clearly, there is equality above if and only if

$$\sum_j Y_s^{(\ell)}(\phi_j) = 0, \quad \forall s, \qquad 1 \le \ell \le t,$$

i.e., by (6.13), Φ is a spherical t-design. □

Example 6.5. By (16.22), $K_\ell^{(d)}$ is a multiple of the *Gegenbauer polynomial* $C_\ell^{(\frac{d-2}{2})}$. By (16.40), we may take $K_1^{(d)}(z) = z$ and $K_2^{(d)}(z) = dz^2 - 1$, which gives

$$A_2(\Phi) = \sum_j \sum_k \langle \phi_j, \phi_k \rangle + \sum_j \sum_k \left(d\langle \phi_j, \phi_k \rangle^2 - 1 \right)$$
$$= \|\sum_j \phi_j\|^2 + d\left(\sum_j \sum_k \langle \phi_j, \phi_k \rangle^2 - \frac{n^2}{d} \right).$$

This is clearly minimised, i.e., gives a spherical 2-design, if and only if (6.11) holds.

Example 6.6. Restricting the sum in (6.15) to $\ell = t, t-2, \ldots$ ($\ell \geq 1$) and to $\ell = t$ gives a nonnegative functional $B_t(\Phi)$ for which $B_t(\Phi) = 0$ if and only if Φ is a spherical half-design of order t and a spherical design of harmonic index t, respectively.

6.6 The existence of cubature formulas

Spherical t-designs $\Phi = \{\phi_j\}_{j=1}^n$ and their variants are examples of *cubature rules*. Their existence (without any estimate of n) is guaranteed by the following result:

Theorem 6.5. ([SZ84]) *Let X be a path-connected topological space, and μ be a finite (positive) measure on X, defined on the open sets, with full support, i.e., $\mu(U) > 0$ for every nonempty open set $U \subset X$. For a continuous integrable function $f : X \to \mathbb{R}^m$, there exists a finite set of samples $A \subset X$ for which*

$$\frac{1}{\mu(X)} \int_X f \, d\mu = \frac{1}{|A|} \sum_{a \in A} f(a).$$

Here $|A|$, the size of A, can be any number with a finite number of exceptions.

This is a generalisation of the integral form of the mean value theorem.

Example 6.7. Let $f = (f_1, \ldots, f_m)$. For $\mu = \sigma$ the normalised surface area on $\mathbb{S}$, choosing $\{f_1, \ldots, f_m\}$ to be a spanning set for $\mathrm{Hom}_t(\mathbb{R}^d)$ gives the existence of a spherical half-design of order t (equal-weight cubature rule for $\Pi_t^\circ(\mathbb{R}^d)$).

Similarly, by choosing a spanning set for the space of functions integrated by the cubature rule, one has

Spherical t-designs, spherical half-designs of order t and spherical designs of harmonic index t exist, for all values of t.

The construction of such designs (cubature rules) with a small number of points (and estimates the smallest number of points) is a subject of ongoing interest.

6.7 Tight frames of symmetric tensors

Here we extend (6.4) to an inequality, for which equality gives a tight frame for the symmetric tensors in $\otimes^t \mathscr{H} := \mathscr{H} \otimes \cdots \otimes \mathscr{H}$ (t times), and a cubature rule for integration of certain homogeneous polynomials on the sphere (see §6.8).

For simplicity, we follow the development of §6.2 as closely as possible. Thus, we define the **symmetric tensors** of rank t to be the subspace of $\otimes^t \mathscr{H}$ given by

$$\operatorname{Sym}^t(\mathscr{H}) := \operatorname{span}\{v^{\otimes t} : v \in \mathscr{H}\}, \qquad v^{\otimes t} := v \otimes \cdots \otimes v \quad (t\,\text{times}).$$

This Hilbert space has dimension

$$\dim(\operatorname{Sym}^t(\mathscr{H})) = \binom{t+d-1}{t}, \qquad d = \dim(\mathscr{H}), \tag{6.16}$$

and we recall (see §5.6) that its inner product satisfies

$$\langle v^{\otimes t}, w^{\otimes t} \rangle = \langle v, w \rangle^t, \qquad \forall v, w \in \mathscr{H}. \tag{6.17}$$

The dual space $(\operatorname{Sym}^t(\mathscr{H}))^* = \operatorname{Sym}^t(\mathscr{H}^*)$ contains $\langle \cdot, v \rangle^{\otimes t}$, $v \in \mathscr{H}$, and its inner product is given by

$$\langle \langle \cdot, v \rangle^{\otimes t}, \langle \cdot, w \rangle^{\otimes t} \rangle = \langle w, v \rangle^t, \qquad \forall v, w \in \mathscr{H}. \tag{6.18}$$

There is a vector space isomorphism between $\operatorname{Sym}^t(\mathscr{H}^*)$ and the space $\mathscr{L}_t(\mathscr{H}, \mathbb{F})$ of **symmetric t-linear maps** from $\mathscr{H}^t \to \mathbb{F}$ which is given by

$$\lambda^{\otimes t} \mapsto L, \quad (\lambda \in \mathscr{H}^*) \qquad L(v_1, \ldots, v_t) := \lambda(v_1) \cdots \lambda(v_t).$$

We define the space of **homogeneous polynomials** on $\mathscr{H}$ of degree t to be

$$\Pi_t^\circ(\mathscr{H}) := \{\hat{L} : L \in \mathscr{L}_t(\mathscr{H}, \mathbb{F})\}, \qquad \hat{L} : \mathscr{H} \to \mathbb{F},\ \hat{L}(v) := L(v, \ldots, v).$$

The map $L \mapsto \hat{L}$ above gives a vector space isomorphism $\mathscr{L}_t(\mathscr{H}, \mathbb{F}) \to \Pi_t^\circ(\mathscr{H})$.

The inner product on $\Pi_t^\circ(\mathscr{H})$ induced from that on $(\operatorname{Sym}^t(\mathscr{H}))^*$ via the above isomorphisms is the **apolar** (or **Bombieri** or **Fisher**) inner product, which is given by

$$\langle \langle \cdot, v \rangle^t, \langle \cdot, w \rangle^t \rangle_\circ := \langle \langle \cdot, v \rangle^{\otimes t}, \langle \cdot, w \rangle^{\otimes t} \rangle = \langle w, v \rangle^t. \tag{6.19}$$

It follows from (6.19) that the apolar inner product satisfies

$$\langle p, \langle \cdot, w \rangle^t \rangle_\circ = p(w), \qquad \forall p \in \Pi_t^\circ(\mathscr{H}), \quad \forall w \in \mathscr{H}, \tag{6.20}$$

i.e., $\langle \cdot, w \rangle^t$ is the Riesz representer of point evaluation at w. With $\tilde{g}(z) := \overline{g(\overline{z})}$, the apolar inner product is given by (see Exer. 6.17)

$$\langle f,g\rangle_\circ = \frac{1}{t!}(f(D)\tilde{g})(0) = \frac{1}{t!}\sum_{|\alpha|=t}\frac{D^\alpha f(0)}{\alpha!}\overline{D^\alpha g(0)}, \qquad \forall f,g\in \Pi_t^\circ(\mathbb{F}^d). \quad (6.21)$$

In particular, the monomials $\{z^\alpha\}_{|\alpha|=t}$ are orthogonal with $\langle z^\alpha, z^\alpha\rangle_\circ = \frac{\alpha!}{|\alpha|!}$.

Theorem 6.6. *Fix $t\in\{1,2,\ldots\}$. Let $f_1,\ldots,f_n$ be vectors in $\mathscr{H}$, not all zero. Then*

$$\sum_{j=1}^n\sum_{k=1}^n |\langle f_j,f_k\rangle|^{2t} \geq \frac{1}{\binom{t+d-1}{t}}\Big(\sum_{j=1}^n \|f_j\|^{2t}\Big)^2, \tag{6.22}$$

with equality precisely when any of the equivalent conditions holds

(a) $(f_j^{\otimes t})_{j=1}^n$ *is a tight frame for the symmetric tensors* $\mathrm{Sym}^t(\mathscr{H})$.
(b) $(\langle\cdot,f_j\rangle^{\otimes t})_{j=1}^n$ *is a tight frame for* $(\mathrm{Sym}^t(\mathscr{H}))^* = \mathrm{Sym}^t(\mathscr{H}^*)$.
(c) $(\langle\cdot,f_j\rangle^t)_{j=1}^n$ *is a tight frame for $\Pi_t^\circ(\mathscr{H})$ with the apolar inner product (6.19).*

Proof. Firstly, we observe that $(f_j^{\otimes t})_{j=1}^n$ is a sequence of vectors in $\mathrm{Sym}^t(\mathscr{H})$ which are not all zero, since $v^{\otimes t}$ is zero if and only if $v=0$.

Thus we may apply Theorem 6.1, using (6.16), to obtain

$$\sum_{j=1}^n\sum_{k=1}^n |\langle f_j^{\otimes t},f_k^{\otimes t}\rangle|^2 \geq \frac{1}{\binom{t+d-1}{t}}\Big(\sum_{j=1}^n \|f_j^{\otimes t}\|^2\Big)^2,$$

with equality if and only if (a) holds. By (6.17), the equation above equals (6.22). A similar argument, using (6.18) and (6.19) in place of (6.17), gives (b) and (c), respectively. □

A simple calculation (Exer. 6.15) shows that equality in (6.22) is also equivalent to the following *generalised Plancherel* and *Bessel identities* (also see Theorem 6.7)

$$\langle x,y\rangle^t = \frac{\binom{d+t-1}{t}}{\sum_{\ell=1}^n\|f_\ell\|^{2t}}\sum_{j=1}^n \langle x,f_j\rangle^t\langle f_j,y\rangle^t, \qquad \forall x,y\in\mathscr{H}, \tag{6.23}$$

$$\|x\|^{2t} = \frac{\binom{d+t-1}{t}}{\sum_{\ell=1}^n\|f_\ell\|^{2t}}\sum_{j=1}^n |\langle x,f_j\rangle|^{2t}, \qquad \forall x\in\mathscr{H}. \tag{6.24}$$

Example 6.8. For $t=1$, Theorem 6.6 says that $(f_j)_{j=1}^n$ is a tight frame for $\mathscr{H}$ if and only if $(\langle\cdot,f_j\rangle)_{j=1}^n$ is a tight frame for its dual $\mathscr{H}^*$, or, equivalently, for the homogeneous linear polynomials $\Pi_1^\circ(\mathscr{H})$ (with the apolar inner product) and that each of these is equivalent to the variational condition (6.5).

For $t>1$, the existence vectors (f_j) giving equality in (6.22) is not immediately obvious (the frames given Theorem 6.6 have a special form). We now investigate this question by showing that (6.22) is equivalent to certain cubature rules.

6.8 Cubature on the real and complex spheres

Here we will show that equality in (6.22) is closely related to cubature formulas for the integrals of certain homogeneous polynomials over the complex sphere. Further, if $\mathscr{H}$ is a *real* Hilbert space and $t > 1$, then it turns out that equality in (6.22) can never be attained. In this case, there is a sharper inequality (Theorem 6.7) for which equality corresponds to cubature formulas for the real sphere.

For simplicity, we suppose that $\mathscr{H} = \mathbb{F}^d$ and use standard multi-index notation, e.g., $(x)_\alpha$ denotes the multivariate Pochhammer symbol. Let σ be the normalised surface area measure on the real or complex unit sphere $\mathbb{S}$. For $\mathscr{H} = \mathbb{C}^d$ the integrals of the monomials in $z=(z_1,\ldots,z_d)$ and $\overline{z}=(\overline{z_1},\ldots,\overline{z_d})$ over the unit sphere are

$$\int_{\mathbb{S}} z^\alpha \overline{z}^\beta \, d\sigma(z) = 0, \quad \alpha \neq \beta, \qquad \int_{\mathbb{S}} |z^\alpha|^2 \, d\sigma(z) = \frac{(d-1)!\alpha!}{(d-1+|\alpha|)!}. \tag{6.25}$$

For $\mathscr{H} = \mathbb{R}^d$ the integrals of the monomials in $x = (x_1,\ldots,x_d)$ are

$$\int_{\mathbb{S}} x^\alpha \, d\sigma(x) = 0, \quad \alpha \notin (2\mathbb{Z})^d, \qquad \int_{\mathbb{S}} x^{2\alpha} \, d\sigma(x) = \frac{(\frac{1}{2})_\alpha}{(\frac{d}{2})_{|\alpha|}}. \tag{6.26}$$

Of interest here is the space of polynomials $\mathbb{F}^d \to \mathbb{F}$

$$\Pi_{t,t}^\circ(\mathbb{F}^d) = \operatorname{Hom}(t,t) := \operatorname{span}\{z \mapsto z^\alpha \overline{z}^\beta : |\alpha| = |\beta| = t\}, \tag{6.27}$$

which are homogeneous of degree t in z and in $\overline{z}$. Equivalently (see Exer. 6.17)

$$\Pi_{t,t}^\circ(\mathbb{F}^d) = \operatorname{span}\{z \mapsto |\langle z, v\rangle|^{2t} : v \in \mathbb{F}^d\}. \tag{6.28}$$

We note $\Pi_{t,t}^\circ(\mathbb{R}^d) = \Pi_{2t}^\circ(\mathbb{R}^d)$. Recall that a homogeneous polynomial f of degree $2t$ is uniquely determined by its values on $\mathbb{S}$ by $f(x) = \|x\|^{2t} f(x/\|x\|)$, $x \neq 0$.

Definition 6.2. A sequence $(f_j)_{j=1}^n$ of vectors in $\mathscr{H}$ is a **cubature rule** for a space P of homogeneous polynomials of degree $2t$, such as $\Pi_{t,t}^\circ(\mathscr{H})$, if

$$\int_{\mathbb{S}} p(x)\, d\sigma(x) = \frac{1}{\sum_k \|f_k\|^{2t}} \sum_{j=1}^n p(f_j) = \sum_{\substack{j=1 \\ f_j \neq 0}}^n \frac{\|f_j\|^{2t}}{\sum_k \|f_k\|^{2t}} p\Big(\frac{f_j}{\|f_j\|}\Big), \qquad \forall p \in P.$$

A cubature rule for which the vectors (f_j) have equal norms gives an *unweighted* cubature rule for the integration of P over $\mathbb{S}$. By Theorem 6.5, these exist (take the coordinates of f to be the real and imaginary parts of a basis for P).

Let $c_t = c_t(d,\mathbb{F})$ denote the integral of the monomial $z \mapsto |z_1|^{2t} = (z_1\overline{z_1})^t$ over $\mathbb{S}$. From (6.25) and (6.26), we have

$$c_t(d,\mathbb{C}) = \frac{1}{\binom{d+t-1}{t}}, \qquad c_t(d,\mathbb{R}) = \frac{1\cdot 3\cdot 5\cdots(2t-1)}{d(d+2)\cdots(d+2(t-1))}.$$

The invariance of surface area measure under unitary maps implies

$$\int_{\mathbb{S}} |\langle x,y\rangle|^{2t}\, d\sigma(y) = \|x\|^{2t} c_t(d,\mathbb{F}), \qquad \forall x \in \mathscr{H}. \tag{6.29}$$

Denote the restriction of a polynomial space P to the unit sphere by $P(\mathbb{S})$.

Theorem 6.7. *Fix $t \in \{1,2,\ldots\}$. Let $f_1,\ldots,f_n$ be vectors in $\mathscr{H} = \mathbb{F}^d$, not all zero. Then*

$$\sum_{j=1}^{n}\sum_{k=1}^{n} |\langle f_j, f_k\rangle|^{2t} \ge c_t(d,\mathbb{F})\Big(\sum_{\ell=1}^{n} \|f_\ell\|^{2t}\Big)^2, \tag{6.30}$$

with equality when any of the following equivalent conditions holds

(a) The generalised Bessel identity

$$c_t(d,\mathbb{F})\|x\|^{2t} = \frac{1}{\sum_{\ell=1}^{n} \|f_\ell\|^{2t}} \sum_{j=1}^{n} |\langle x, f_j\rangle|^{2t}, \qquad \forall x \in \mathscr{H}. \tag{6.31}$$

(b) The generalised Plancherel identity

$$c_t(d,\mathbb{F})\langle x,y\rangle^t = \frac{1}{\sum_{\ell=1}^{n} \|f_\ell\|^{2t}} \sum_{j=1}^{n} \langle x, f_j\rangle^t \langle f_j, y\rangle^t, \qquad \forall x,y \in \mathbb{F}^d. \tag{6.32}$$

(c) The cubature rule for $\Pi_{t,t}(\mathscr{H})$

$$\int_{\mathbb{S}} p(x)\, d\sigma(x) = \frac{1}{\sum_{\ell=1}^{n} \|f_\ell\|^{2t}} \sum_{j=1}^{n} p(f_j), \qquad \forall p \in \Pi_{t,t}^{\circ}(\mathscr{H}), \tag{6.33}$$

or, equivalently, for $\Pi_{t,t}(\mathbb{S})$

$$\int_{\mathbb{S}} p(x)\, d\sigma(x) = \sum_{\substack{j=1 \\ f_j \ne 0}}^{n} \frac{\|f_j\|^{2t}}{\sum_{\ell=1}^{n} \|f_\ell\|^{2t}}\, p\Big(\frac{f_j}{\|f_j\|}\Big), \qquad \forall p \in \Pi_{t,t}^{\circ}(\mathbb{S}). \tag{6.34}$$

(d) The tensor product integration formula

$$\int_{\mathbb{S}} x^{\otimes t} \otimes \overline{x}^{\otimes t}\, d\sigma(x) = \frac{1}{\sum_{\ell=1}^{n} \|f_\ell\|^{2t}} \sum_{j=1}^{n} f_j^{\otimes t} \otimes \overline{f_j}^{\otimes t}. \tag{6.35}$$

(e) The integration formula

$$\int_{\mathbb{S}} \langle \cdot, x^{\otimes t}\rangle x^{\otimes t}\, d\sigma(x) = \frac{1}{\sum_{\ell=1}^{n} \|f_\ell\|^{2t}} \sum_{j=1}^{n} \langle \cdot, f_j^{\otimes t}\rangle f_j^{\otimes t}. \tag{6.36}$$

(f) For all univariate polynomials $g \in \Pi_t(\mathbb{R})$, we have

$$\int_{\mathbb{S}}\int_{\mathbb{S}} g(|\langle x,y\rangle|^2)\,d\sigma(y)\,d\sigma(x) = \sum_{\substack{j=1\\ f_j\neq 0}}^{n}\sum_{\substack{k=1\\ f_k\neq 0}}^{n} \frac{\|f_j\|^{2t}\|f_k\|^{2t}}{(\sum_\ell \|f_\ell\|^{2t})^2} g(|\langle \frac{f_j}{\|f_j\|},\frac{f_k}{\|f_k\|}\rangle|^2). \tag{6.37}$$

Proof. Let $C:=\sum_{\ell=1}^{n}\|f_\ell\|^{2t}$. Define a tensor $\xi\in\operatorname{Sym}^t(\mathscr{H})\otimes\operatorname{Sym}^t(\overline{\mathscr{H}})$ and a self-adjoint operator Q on $\operatorname{Sym}^t(\mathscr{H})$ by

$$\xi:=\int_{\mathbb{S}} x^{\otimes t}\otimes\overline{x}^{\otimes t}\,d\sigma(x)-\frac{1}{C}\sum_{j=1}^{n} f_j^{\otimes t}\otimes\overline{f_j}^{\otimes t},$$

$$Q:=\int_{\mathbb{S}}\langle\cdot,x^{\otimes t}\rangle x^{\otimes t}\,d\sigma(x)-\frac{1}{C}\sum_{j=1}^{n}\langle\cdot,f_j^{\otimes t}\rangle f_j^{\otimes t}.$$

Equip $\operatorname{Sym}^t(\mathscr{H})\otimes\operatorname{Sym}^t(\overline{\mathscr{H}})$ with the apolar inner product, and the space of linear operators on $\operatorname{Sym}^t(\mathscr{H})$ with the Frobenius inner product. Then (see Exer. 6.16) a simple calculation using (6.17) and (6.29) shows that

$$\langle\xi,\xi\rangle_\circ=\langle Q,Q\rangle_F=\frac{1}{C^2}\sum_j\sum_k|\langle f_j,f_k\rangle|^{2t}-c_t(d,\mathbb{F})\geq 0,$$

which is (6.30). Moreover, equality in (6.30) is equivalent to (d) or to (e). By the polarisation identity and (6.18), (a) and (b) are equivalent.

We now complete the proof by showing

$$\text{(d)}\Longrightarrow\text{(c)}\Longrightarrow\text{(a)},\ \text{(f)}\Longrightarrow\text{equality in (6.30)}.$$

(d)$\Longrightarrow$(c): Expand $x^{\otimes t}\otimes\overline{x}^{\otimes t}$ in terms of the coordinates of x. Since

$$x^{\otimes t}=\sum_{k_1=1}^{d}x_{k_1}e_{k_1}\otimes\cdots\otimes\sum_{k_t=1}^{d}x_{k_t}e_{k_t}=\sum_{k\in\{1,\ldots,d\}^t}p_k(x)\eta_k,$$

$$p_k(x):=x_{k_1}x_{k_2}\cdots x_{k_t},\qquad \eta_k:=e_{k_1}\otimes e_{k_2}\otimes\cdots\otimes e_{k_t},$$

we obtain

$$x^{\otimes t}\otimes\overline{x}^{\otimes t}=\sum_{k,\ell}p_k(x)p_\ell(\overline{x})\,\eta_k\otimes\eta_\ell.$$

Thus (d) can be written as

$$\int_{\mathbb{S}}\sum_{k,\ell}p_k(x)p_\ell(\overline{x})\,\eta_k\otimes\eta_\ell\,d\sigma(x)=\frac{1}{C}\sum_{j=1}^{n}\sum_{k,\ell}p_k(f_j)p_\ell(\overline{f_j})\,\eta_k\otimes\eta_\ell.$$

Since the tensors $\eta_k\otimes\eta_\ell$ are linearly independent, equating their coefficients gives the cubature rule for all the polynomials $x\mapsto p_k(x)p_\ell(\overline{x})$, and hence for $\Pi_{t,t}^\circ(\mathscr{H})$.

(c)$\Longrightarrow$(a): Let $p=|\langle x,\cdot\rangle|^{2t}\in\Pi_{t,t}^\circ(\mathbb{F}^d)$ in (6.33) and use (6.29) to obtain

$$c_t(d,\mathbb{F})\|x\|^{2t} = \int_{\mathbb{S}} |\langle x,y\rangle|^{2t}\,d\sigma(y) = \frac{1}{C}\sum_j |\langle x,f_j\rangle|^{2t}.$$

(c) $\Longrightarrow$ (f): Let $p = \|\cdot\|^{2(t-s)}|\langle x,\cdot\rangle|^{2s} \in \Pi^{\circ}_{t,t}$, $0 \le s \le t$ in (6.34) to get

$$\int_{\mathbb{S}} |\langle x,y\rangle|^{2s}\,d\sigma(y) = \sum_k \frac{\|f_k\|^{2t}}{C} |\langle x, \frac{f_k}{\|f_k\|}\rangle|^{2s}.$$

For $x \in \mathbb{S}$, $|\langle x, \frac{f_k}{\|f_k\|}\rangle|^{2s} = \|x\|^{2(t-s)}|\langle x, \frac{f_k}{\|f_k\|}\rangle|^{2s}$, and so using (6.34) again gives

$$\int_{\mathbb{S}}\int_{\mathbb{S}} |\langle x,y\rangle|^{2s}\,d\sigma(y)\,d\sigma(x) = \sum_j \frac{\|f_j\|^{2t}}{C} \sum_k \frac{\|f_k\|^{2t}}{C} |\langle \frac{f_j}{\|f_j\|}, \frac{f_k}{\|f_k\|}\rangle|^{2s}.$$

Thus (6.37) holds for the monomials $(\cdot)^s$, $0 \le s \le t$, and hence for $\Pi_t(\mathbb{R})$.
(a) $\Longrightarrow$ equality in (6.30): Take $x = f_k$ in (a) then sum over k to obtain the required equality

$$c_t(d,\mathbb{F})\|f_k\|^{2t} = \frac{1}{C}\sum_j |\langle f_k,f_j\rangle|^{2t},$$

$$c_t(d,\mathbb{F})C = c_t(d,\mathbb{F})\sum_k \|f_k\|^{2t} = \frac{1}{C}\sum_k\sum_j |\langle f_k,f_j\rangle|^{2t}.$$

(f) $\Longrightarrow$ equality in (6.30): Take $g = (\cdot)^t$ in (f) to obtain the desired equality

$$\begin{aligned} c_t(d,\mathbb{F}) &= \int_{\mathbb{S}}\int_{\mathbb{S}} |\langle x,y\rangle|^{2t}\,d\sigma(y)\,d\sigma(x) = \sum_{\substack{j=1\\ f_j\ne 0}}^{n}\sum_{\substack{k=1\\ f_k\ne 0}}^{n} \frac{\|f_j\|^{2t}\|f_k\|^{2t}}{(\sum_\ell \|f_\ell\|^{2t})^2} |\langle \frac{f_j}{\|f_j\|}, \frac{f_k}{\|f_k\|}\rangle|^{2t} \\ &= \frac{1}{C^2}\sum_j\sum_k |\langle f_j,f_k\rangle|^{2t}. \end{aligned}$$

□

Example 6.9. For *unit vectors* (f_j) in $\mathbb{R}^d$, (c) reduces to the definition for a spherical half-design of order $2t$. The condition (a) is the Waring formula (6.8), and equality in (6.30) is the condition (6.9).

Since $c_t(d,\mathbb{R}) \ge c_t(d,\mathbb{C})$, with strict inequality for $t,d > 1$ (Exer. 6.9), we have:

Let $f_1,\ldots,f_n$ be vectors in $\mathscr{H} = \mathbb{R}^d$, not all zero. Then

$$\sum_{j=1}^{n}\sum_{k=1}^{n} |\langle f_j,f_k\rangle|^{2t} \ge \frac{1\cdot 3\cdot 5\cdots(2t-1)}{d(d+2)\cdots(d+2(t-1))}\Big(\sum_{j=1}^{n}\|f_j\|^{2t}\Big)^2, \tag{6.38}$$

which is a sharper bound than (6.22) when $t,d > 1$. A sequence of unit vectors giving equality in (6.38) is precisely a spherical half-design of order $2t$.

6.9 Spherical (t,t)-designs

We now give examples of sequences giving equality in Theorem 6.7 (also see §6.13 and the numerical study of §6.16).

Definition 6.3. A nonzero sequence (f_j) in $\mathbb{F}^d$ giving equality in Theorem 6.7, i.e.,

$$\sum_{j=1}^{n}\sum_{k=1}^{n}|\langle f_j,f_k\rangle|^{2t}=c_t(d,\mathbb{F})\Bigl(\sum_{\ell=1}^{n}\|f_\ell\|^{2t}\Bigr)^2, \tag{6.39}$$

is called a **(spherical)** (t,t)**-design**[2] for $\mathbb{F}^d$.

Theorem 6.5 implies the (unweighted) cubature rules in Theorem 6.7 exist, i.e.,

For each $t\geq 1$, unit-norm spherical (t,t)-designs for $\mathbb{F}^d$ always exist.
The art is in constructing those with a *small* number of vectors.

Example 6.10. By Theorem 6.6, a unit-norm (t,t)-design (v_j) for $\mathbb{C}^d$ (which always exists) gives a unit-norm tight frame $(v_j^{\otimes t})$ for $\mathrm{Sym}^t(\mathbb{C}^d)$.

Example 6.11. For $d\geq 2$ and $t>1$, there is no tight frame $(v_j^{\otimes t})$ for $\mathrm{Sym}^t(\mathbb{R}^d)$. If there was, then Theorem 6.1 gives

$$\sum_j\sum_k|\langle v_j^{\otimes t},v_k^{\otimes t}\rangle|^2=\sum_j\sum_k|\langle v_j,v_k\rangle|^{2t}=\frac{1}{\binom{t+d-1}{t}}\Bigl(\sum_j\|v_j\|^{2t}\Bigr)^2,$$

which violates the sharpened Welch bound (6.38).

Example 6.12. The unit-norm (t,t)-designs for $\mathbb{R}^d$ are the spherical half-designs of order $2t$ (see §6.4 and Example 6.9).

Example 6.13. In view of (6.31), a $(1,1)$-design is precisely a tight frame for $\mathbb{F}^d$.

Example 6.14. Three equally spaced unit vectors in $\mathbb{R}^2$ are a $(1,1)$-design for $\mathbb{F}^2$. Further, they are a $(2,2)$-design for $\mathbb{R}^2$, but not for $\mathbb{C}^2$, since

$$\sum_j\sum_k|\langle f_j,f_k\rangle|^4=3+6(-\frac{1}{2})^4=\frac{27}{8},\qquad\Bigl(\sum_\ell\|f_\ell\|^4\Bigr)^2=3^2=9,$$

$$c_2(2,\mathbb{R})=\frac{1\cdot3}{2\cdot4}=\frac{3}{8},\qquad c_2(2,\mathbb{C})=\frac{1}{\binom{2+2-1}{2}}=\frac{1}{3}.$$

Example 6.15. There is a $(3,3)$-design of 40 unit vectors for $\mathbb{C}^4$ given by a highly symmetric tight frame (see Exer. 6.12). It has the property that each given vector is orthogonal to 12 others and makes an angle $\frac{1}{\sqrt{3}}$ with 27 others.

---

[2] For $\mathscr{H}=\mathbb{C}^d$, these are also known as **complex t-designs.**

Example 6.16. A SIC (see §2.11, §14.1), i.e., a set of d^2 unit vectors (f_j) in $\mathbb{C}^d$ with

$$|\langle f_j, f_k\rangle|^2 = \frac{1}{d+1}, \qquad j \neq k,$$

is a $(2,2)$-design of d^2 unit vectors for $\mathbb{C}^d$.

Example 6.17. A set of $d+1$ MUBs (mutually unbiased bases) for $\mathbb{C}^d$ (see §2.11), i.e., orthogonal bases with

$$|\langle f, g\rangle| = \frac{1}{\sqrt{d}}, \qquad f \text{ and } g \text{ in different bases},$$

gives a $(2,2)$-design of $d(d+1)$ unit vectors for $\mathbb{C}^d$ [KR04].

Example 6.18. Three MUBs for $\mathbb{C}^2$ give a $(3,3)$-design of six vectors for $\mathbb{C}^2$, e.g., one can take

$$\left\{\begin{bmatrix}1\\0\end{bmatrix}, \begin{bmatrix}0\\1\end{bmatrix}, \frac{1}{\sqrt{2}}\begin{bmatrix}1\\1\end{bmatrix}, \frac{1}{\sqrt{2}}\begin{bmatrix}1\\-1\end{bmatrix}, \frac{1}{\sqrt{2}}\begin{bmatrix}1\\i\end{bmatrix}, \frac{1}{\sqrt{2}}\begin{bmatrix}1\\-i\end{bmatrix}\right\}.$$

For a given (t,t)-design (f_j), the cubature rule (6.34) can be written

$$\int_{\mathbb{S}} p(x)\, d\sigma(x) = \sum_{\substack{j=1\\ f_j\neq 0}}^{n} w_j p(\phi_j), \qquad \forall p \in \Pi^\circ_{t,t}(\mathbb{S}), \tag{6.40}$$

where

$$\phi_j = \frac{f_j}{\|f_j\|}, \qquad w_j = \frac{\|f_j\|^{2t}}{\sum_\ell \|f_\ell\|^{2t}}. \tag{6.41}$$

Since $\|\cdot\|^{2(t-r)} q \in \Pi^\circ_{t,t}(\mathbb{F}^d)$, for $q \in \Pi^\circ_{r,r}(\mathbb{F}^d)$, $1 \leq r \leq t$, we have

$$\Pi^\circ_{r,r}(\mathbb{S}) \subset \Pi^\circ_{t,t}(\mathbb{S}), \qquad 0 \leq r \leq t. \tag{6.42}$$

Combining these observations gives:

Proposition 6.2. *Fix $t \geq 1$. If $(f_j)_{j=1}^n$ is a (t,t)-design for $\mathbb{F}^d$, then $(\|f_j\|^{t/r-1} f_j)$ is an (r,r)-design for $\mathbb{F}^d$, $1 \leq r \leq t$, i.e.,*

$$\sum_{j=1}^{n}\sum_{k=1}^{n} |\langle f_j, f_k\rangle|^{2r} \|f_j\|^{2(t-r)} \|f_k\|^{2(t-r)} = c_r(d,\mathbb{F}) \Big(\sum_{\ell=1}^{n} \|f_\ell\|^{2t}\Big)^2. \tag{6.43}$$

Proof. Let $g_j := \|f_j\|^{t/r-1} f_j$, $q \in \Pi^\circ_{r,r}(\mathbb{S})$. Since $p := \|\cdot\|^{2(t-r)} q \in \Pi^\circ_{t,t}(\mathbb{S})$, we have

$$\sum_{\substack{j=1\\ g_j\neq 0}}^{n} \frac{\|g_j\|^{2r}}{\sum_{\ell=1}^n \|g_\ell\|^{2r}} q\Big(\frac{g_j}{\|g_j\|}\Big) = \sum_{\substack{j=1\\ f_j\neq 0}}^{n} \frac{\|f_j\|^{2t}}{\sum_{\ell=1}^n \|f_\ell\|^{2t}} p\Big(\frac{f_j}{\|f_j\|}\Big) = \int_{\mathbb{S}} p\, d\sigma = \int_{\mathbb{S}} q\, d\sigma,$$

and so, by (6.34), (g_j) is an (r,r)-design. Substituting into (6.39) gives (6.43). □

In particular, we have:

A unit-norm (t,t)-design for $\mathbb{F}^d$ is an (r,r)-design, $1 \le r \le t$.

If (f_j) is a (t,t)-design, $t \ge 1$, then $(\|f_j\|^{t-1} f_j)$ is tight frame.

If the norms of (f_j) are not all equal, then the properties (6.43) for $1 \le r \le t$ of a (t,t)-design and the corresponding equivalent conditions given by Theorem 6.7 are most naturally described in terms of *weighted* (t,t)-designs.

6.10 Weighted (t,t)-designs

Definition 6.4. Suppose that $\Phi = (\phi_j)_{j=1}^n$ are unit vectors in $\mathbb{F}^d$, and $w = (w_j)_{j=1}^n$ satisfy $w_j \ge 0$, $\sum_j w_j = 1$. Then (Φ, w) is a **weighted (spherical)** (t,t)**-design**[3] if

$$\sum_{j=1}^n \sum_{k=1}^n w_j w_k |\langle \phi_j, \phi_k \rangle|^{2t} = c_t(d, \mathbb{F}). \tag{6.44}$$

There is a 1–1 correspondence between the (t,t)-designs (f_j) and the weighted (t,t)-designs (Φ, w) given by (6.41).

In this terminology, Theorem 6.7 becomes:

Corollary 6.1. *(Weighted version) Let $\Phi = (\phi_j)_{j=1}^n$ be a sequence of unit vectors in $\mathbb{F}^d$, and $w = (w_j)_{j=1}^n$ be nonnegative weights, i.e., $w_j \ge 0$, $\sum_j w_j = 1$. Then*

$$\sum_{j=1}^n \sum_{k=1}^n w_j w_k |\langle \phi_j, \phi_k \rangle|^{2t} \ge c_t(d, \mathbb{F}), \tag{6.45}$$

with equality if and only if (Φ, w) is a weighted (t,t)-design, or, equivalently,

$$\int_{\mathbb{S}} p(x)\, d\sigma(x) = \sum_{j=1}^n w_j p(\phi_j), \qquad \forall p \in \Pi_{t,t}^{\circ}(\mathbb{S}). \tag{6.46}$$

If (Φ, w) is a weighted (t,t)-design, then it is a weighted (r,r)-design, $0 \le r \le t$.

Proof. Make the substitution (6.41) in Theorem 6.7 and observe that (c) can be written as (6.46). The last assertion follows from this and (6.42). □

---

[3] These are also known as *weighted spherical half-designs* of order t when $\mathbb{F} = \mathbb{R}$ (see [KP11]).

Example 6.19. A weighted (t,t)-design (Φ, w) satisfies

$$\sum_{j=1}^{n}\sum_{k=1}^{n} w_j w_k |\langle \phi_j, \phi_k \rangle|^{2r} = c_r(d,\mathbb{F}), \qquad 1 \le r \le t,$$

which is the weighted version of (6.43).

Substituting (6.41) into Theorem 6.7 gives a weighted version of each of the equivalent conditions (see Exer. 6.18), e.g., condition (a) becomes

$$c_t(d,F)\|x\|^{2t} = \sum_{j=1}^{n} w_j |\langle x, \phi_j \rangle|^{2t}, \qquad \forall x \in \mathbb{F}^d,$$

or, equivalently

$$\sum_{j=1}^{n} w_j |\langle x, \phi_j \rangle|^{2t} = c_t(d,\mathbb{F}), \qquad \forall x \in \mathbb{S}. \tag{6.47}$$

6.11 Complex projective t-designs

The equality (6.39) defining (t,t)-designs is invariant under multiplying the vectors by unit scalars, and so (t,t)-designs can be extended to a projective setting.

This has been done not only for $\mathbb{R}$ and $\mathbb{C}$, but also the quaternions $\mathbb{H}$ and the octonions $\mathbb{O}$ (see [Hog82]). The **complex projective sphere** $\mathbb{C}P^{d-1}$ can be viewed variously as

- The complex sphere $\mathbb{S}(\mathbb{C}^d)$ with points z and az, $|a| = 1$ identified.
- The 1-dimensional subspaces of $\mathbb{C}^d$ (the complex lines through 0).
- The rank 1 orthogonal projections on $\mathbb{C}^d$.

The polynomials on $\mathbb{S}(\mathbb{C}^d)$ which carry over to this space, i.e., those with

$$p(z) = p(az), \qquad \forall z, \quad \forall a \in \mathbb{F}, |\alpha| = 1$$

are precisely those in $\Pi^\circ_{0,0}(\mathbb{F}^d) \oplus \Pi^\circ_{1,1}(\mathbb{F}^d) \oplus \Pi^\circ_{2,2}(\mathbb{F}^d) \cdots$. There is no notion of odd polynomials on this space (see Exer. 6.25 for details).

We will take the elements of $\mathbb{C}P^{d-1}$ to be rank-one orthogonal projections. There is a unique unitarily invariant probability measure μ on $\mathbb{F}P^{d-1}$ induced from the area measure σ on the sphere $\mathbb{S}(\mathbb{F}^d)$, via

$$\int_{\mathbb{F}P^{d-1}} f(P)\,d\mu(P) = \int_{\mathbb{F}(\mathbb{C}^d)} f(P_x)\,d\sigma(x), \tag{6.48}$$

where $P_x = \langle \cdot, x \rangle x$ denotes the rank-one orthogonal projection onto $\operatorname{span}\{x\}$, $\|x\| = 1$. The Frobenius inner product between rank-one orthogonal projections is

$$\langle P_x, P_y \rangle = \operatorname{trace}(P_x P_y) = |\langle x, y \rangle|^2 \in \mathbb{R}. \tag{6.49}$$

Definition 6.5. Let $\mathscr{P} = (P_j)_{j=1}^n$ be rank-one orthogonal projection on $\mathbb{F}^d$, and $w = (w_j)_{j=1}^n$ satisfy $w_j \geq 0$, $\sum_j w_j = 1$. We say $(\mathscr{P}, w)$ is a **(weighted) projective** (t,t)-**design**[4] if

$$\sum_{j=1}^n \sum_{k=1}^n w_j w_k \langle P_j, P_k \rangle^t = c_t(d, \mathbb{F}).$$

The (t,t)-designs $\Phi = (f_j)$ (up to multiplication by unit scalars) are in 1–1 correspondence with the projective (t,t)-designs $(\mathscr{P}, w)$, via

$$P_j = \frac{1}{\|f_j\|^2} \langle \cdot, f_j \rangle f_j, \qquad w_j = \frac{\|f_j\|^{2t}}{\sum_\ell \|f_\ell\|^{2t}}. \tag{6.50}$$

This gives the projective version of Theorem 6.7 (see [RS07], cf. Corollary 6.1).

Corollary 6.2. *(Projective version) Let $\mathscr{P} = (P_j)_{j=1}^n$ be a sequence of rank-one orthogonal projections in $\mathbb{F}^d$, and $w = (w_j)_{j=1}^n$ satisfy $w_j \geq 0$, $\sum_j w_j = 1$. Then*

$$\sum_{j=1}^n \sum_{k=1}^n w_j w_k \langle P_j, P_k \rangle^t \geq c_t(d, \mathbb{F}), \tag{6.51}$$

with equality if and only if $(\mathscr{P}, w)$ is a projective (t,t)-design.
A projective (t,t)-design is a projective (r,r)-design, $1 \leq r \leq t$.

Proof. With P_j and w_j given by (6.50), (6.49) gives

$$w_j w_k \langle P_j, P_k \rangle^t = \frac{\|f_j\|^{2t} \|f_k\|^{2t}}{(\sum_\ell \|f_\ell\|^{2t})^2} \left(\frac{|\langle f_j, f_k \rangle|^2}{\|f_j\|^2 \|f_k\|^2} \right)^t = \frac{|\langle f_j, f_k \rangle|^{2t}}{(\sum_\ell \|f_\ell\|^{2t})^2}.$$

Thus, making the substitution (6.50) in Theorem 6.7 gives (6.51), with equality for projective (t,t)-designs. The last part follows from Corollary 6.1 and (6.50). □

Other conditions equivalent to being a projective (t,t)-design can be obtained by substituting (6.50) into Theorem 6.7, e.g., by using (6.49), condition (a) becomes

$$c_t(d, \mathbb{F}) = \sum_{j=1}^n w_j \langle Q, P_j \rangle^t, \qquad \forall Q \in \mathbb{F}P^{d-1}. \tag{6.52}$$

The condition (e) gives Levenshtein's definition [Lev98] of a *weighted t-design*

$$\int_{\mathbb{F}P^{d-1}} \int_{\mathbb{F}P^{d-1}} g(\langle P, Q \rangle) \, d\mu(P) \, d\mu(Q) = \sum_{j=1}^n \sum_{k=1}^n w_j w_k \, g(\langle P_j, P_k \rangle), \quad \forall g \in \Pi_t(\mathbb{R}). \tag{6.53}$$

[4] Other terms such as **weighted** or **quantum** t-**design** are also commonly used.

The condition (d) becomes

$$\int_{\mathbb{F}P^{d-1}} P^{\otimes t}\,d\mu(P) = \sum_{j=1}^{n} w_j P_j^{\otimes t}. \tag{6.54}$$

For $\mathbb{F}=\mathbb{C}$, Schur's lemma implies that the left-hand side of (6.54) is

$$\int_{\mathbb{C}P^{d-1}} P^{\otimes t}\,d\mu(P) = \frac{1}{\binom{d+t-1}{t}} \Pi_{\text{sym}}^{(t)},$$

where $\Pi_{\text{sym}}^{(t)}$ is the orthogonal projection of the tensors $\otimes^t \mathbb{C}^d$ onto the symmetric tensors $\text{Sym}^t(\mathbb{C}^d)$.

A naive numerical search (see [Bra11]) for (t,t)-designs in $\mathbb{C}^d$ suggests that in some cases, e.g., $t=4$, $d=4$ and $t=3$, $d=3$, those with the minimal number of vectors do not have constant weights. There are few known constructions of such weighted (t,t)-designs. A very general construction is given in §6.13.

6.12 Isometric embeddings

The generalised Bessel identity (6.31) says that the linear map

$$\mathscr{H} = \ell_2(\mathbb{F}^d) \to \ell_{2t}(\mathbb{F}^n) : x \mapsto (\langle x, f_j\rangle)_{j=1}^n$$

is an **isometric embedding**, i.e., there is a constant $C>0$ with

$$\|(\langle x, f_j\rangle)_{j=1}^n\|_{\ell_{2t}} = C\,\|x\|_{\ell_2}, \qquad \forall x \in \ell_2(\mathbb{F}^d). \tag{6.55}$$

Conversely, any such embedding corresponds (via the Riesz representation) to a sequence of vectors $(f_j)_{j=1}^n$ giving equality in (6.55). Further, isometric embeddings $\ell_2(\mathbb{F}^d) \to \ell_p(\mathbb{F}^n)$ can exist only when $p=2t$ (see [LS04] which considers $\mathbb{F}=\mathbb{R},\mathbb{C}$ and also the quaternions $\mathbb{H}$). Collecting these observations gives:

There is an isometric embedding $\ell_2(\mathbb{F}^d) \to \ell_p(\mathbb{F}^n)$ if and only if $p=2t$ and there is a weighted (t,t)-design for $\mathbb{F}^d$.

We give a little more detail. Suppose that $g=(g_1,\ldots,g_n):\ell_2(\mathbb{F}^d)\to\ell_{2t}(\mathbb{F}^n)$ is an isometric embedding, i.e., a linear map with $\|g(x)\|_{\ell_{2t}} = C\|x\|$, $\forall x\in\ell_2(\mathbb{F}^d)$. By the Riesz representation, the linear maps $x\mapsto g_j(x)$ have the form $g_j(x)=\langle x, f_j\rangle$, where $f_j\in\mathbb{F}^d$. We compute

$$\sum_{j=1}^{n} |\langle x, f_j\rangle|^{2t} = \|g(x)\|_{\ell_{2t}}^{2t} = (C\|x\|)^{2t} = c\|x\|^{2t}.$$

Integrating the above over $x \in \mathbb{S}$, using (6.29), gives $c_t(d,\mathbb{F})\sum_\ell \|f_\ell\|^{2t} = c$, and so (f_j) is a (t,t)-design by virtue of satisfying the generalised Bessel identity (6.31).

Example 6.20. The three equally spaced unit vectors $(1,0), (-\frac{1}{2}, \frac{\sqrt{3}}{2}), (-\frac{1}{2}, \frac{\sqrt{3}}{2})$ are a $(2,2)$-design (4-design) for $\mathbb{R}^2$ and a $(1,1)$-design (tight frame) for $\mathbb{F}^2$, which gives the isometric embeddings

$$g: \ell_2(\mathbb{R}^2) \to \ell_4(\mathbb{R}^3), \qquad g(x,y) = \Bigl(x, -\frac{1}{2}x + \frac{\sqrt{3}}{2}y, -\frac{1}{2}x - \frac{\sqrt{3}}{2}y\Bigr),$$

$$g: \ell_2(\mathbb{F}^2) \to \ell_2(\mathbb{F}^3), \qquad g(z_1,z_2) = \Bigl(z_1, -\frac{1}{2}z_1 + \frac{\sqrt{3}}{2}z_2, -\frac{1}{2}z_1 - \frac{\sqrt{3}}{2}z_2\Bigr).$$

The corresponding generalised Bessel identities are

$$x^4 + \Bigl(-\frac{1}{2}x + \frac{\sqrt{3}}{2}y\Bigr)^4 + \Bigl(-\frac{1}{2}x - \frac{\sqrt{3}}{2}y\Bigr)^4 = \frac{9}{8}(x^2+y^2)^2, \qquad \frac{9}{8} = 3c_2(2,\mathbb{R}),$$

$$|z_1|^2 + \Bigl|-\frac{1}{2}z_1 + \frac{\sqrt{3}}{2}z_2\Bigr|^2 + \Bigl|-\frac{1}{2}z_1 - \frac{\sqrt{3}}{2}z_2\Bigr|^2 = \frac{3}{2}(|z_1|^2+|z_2|^2), \qquad \frac{3}{2} = 3c_1(2,\mathbb{F}).$$

Example 6.21. The E. Lucas identity (1876)

$$6\Bigl(\sum_{j=1}^{4} x_j^2\Bigr)^2 = \sum_{1\le j<k\le 4} (x_j+x_k)^4 + \sum_{1\le j<k\le 4} (x_j-x_k)^4,$$

is the generalised Bessel identity for the $(2,2)$-design of 12 vectors for $\mathbb{R}^4$ given by

$$\{e_j \pm e_k : 1 \le j < k \le 4\}.$$

Equivalently, it provides an isometric embedding $\ell_2(\mathbb{R}^4) \to \ell_4(\mathbb{R}^{12})$.

6.13 Weighted $(2,2)$-designs of orthonormal bases

We now give a very general construction of weighted $(2,2)$-designs (Theorem 6.8). A special case is a construction of $d+1$ MUBs in $\mathbb{C}^d$ for d a prime power.

Let $f: G \to H$ be a function between finite abelian groups with $|G| \le |H|$. The equation

$$f(x+a) - f(x) = b \tag{6.56}$$

has $|G|$ solutions for $(a,b) = (0,0)$. If f is *linear*, i.e., $f(x) = mx$, $m \in \mathbb{Z}$, then it has $|G|$ solutions for $b = f(a)$. In view of this, we have the following notion of a highly nonlinear function.

Definition 6.6. We say a function $f : G \to H$ between finite abelian groups with $|G| \le |H|$ is (**differentially**) 1-**uniform**[5] if

$$f(x+a) - f(x) = b$$

has at most one solution for each $(a,b) \neq (0,0)$.

Example 6.22. The functions $f : \mathbb{Z}_5 \to \mathbb{Z}_6$, $g : \mathbb{Z}_5 \to \mathbb{Z}_5 : x \mapsto x^2$ given by

$$f(0) = 0,\ f(1) = 1,\ f(2) = 0,\ f(3) = 2,\ f(4) = 2,$$

$$g(0) = 0,\ g(1) = 1,\ g(2) = 4,\ g(3) = 4,\ g(4) = 1,$$

are 1-uniform.

We require some basic facts about the character group $\hat{G}$ (see §11). These include:

- $\hat{G}$ is the group homomorphisms $G \to \mathbb{C} \setminus \{0\}$ under pointwise multiplication.
- Different characters are orthogonal (as vectors in $\mathbb{C}^G$).
- $\hat{G}$ is isomorphic to G.
- The *Pontryagin duality map* (11.7) gives a canonical isomorphism $G \to \hat{\hat{G}}$.

Theorem 6.8. ([RS07]) *Let $f : G \to H$ be 1-uniform, $|G| = d$. Then there exists a weighted $(2,2)$-design for $\mathbb{C}^d = \mathbb{C}^G$ given by the $|H|+1$ orthonormal bases*

$$\mathscr{B}_0 = (e_g)_{g \in G}, \quad w_0 = \frac{1}{d(d+1)},$$

$$\mathscr{B}_\psi = (e_g^\psi)_{g \in G}, \quad w_\psi = w_1 = \frac{1}{|H|(d+1)}, \qquad \psi \in \hat{H}$$

where $g \mapsto \chi_g$ is an isomorphism $G \to \hat{G}$, and

$$e_g^\psi := \frac{1}{\sqrt{d}} \sum_{k \in G} \chi_g(k) \psi(f(k)) e_k.$$

Proof. We verify that (6.44) holds for $t = 2$. For $\phi_j, \phi_k \in \mathscr{B}_0$, we have

$$\sum_{a \in G} \sum_{b \in G} w_0^2 |\langle e_a, e_b \rangle|^4 = \frac{d}{d^2(d+1)^2} = \frac{1}{d(d+1)^2},$$

and for $\phi_j \in \mathscr{B}_0$, $\phi_k \in \mathscr{B}_\psi$ and vice versa, we have

$$2 \sum_{\psi \in \hat{H}} w_0 w_1 \sum_{a \in G} \sum_{b \in G} |\langle e_a, e_b^\psi \rangle|^4 = \frac{2|H|}{d(d+1)} \frac{1}{|H|(d+1)} \sum_k \sum_g \frac{1}{d^2} = \frac{2}{d(d+1)^2}.$$

---

[5] The terms *perfect nonlinear* and *maximally nonlinear* are also used when $H = G$.

We now consider the last case $\phi_j \in \mathscr{B}_\psi$, $\phi_k \in \mathscr{B}_\xi$. Expanding $\langle e_a^\psi, e_b^\xi \rangle$ gives

$$\frac{1}{d}\langle \sum_{k\in G} \chi_a(k)\psi(f(k))e_k, \sum_{\ell\in G} \chi_b(\ell)\xi(f(\ell))e_\ell \rangle = \frac{1}{d}\sum_{k\in G} \chi_{a-b}(k)(\psi\xi^{-1})(f(k)),$$

since $\overline{\xi} = \xi^{-1}$ (inverse in the character group). Thus

$$|\langle e_a^\psi, e_b^\xi\rangle|^4 = \frac{1}{d^4} \sum_{w,x,y,z\in G} \chi_{a-b}(w+x-y-z)(\psi\xi^{-1})(f(w)+f(x)-f(y)-f(z)),$$

and so $\sum_{b\in G}\sum_{\xi\in\hat{H}} |\langle e_a^\psi, e_b^\xi\rangle|^4$ equals

$$\frac{1}{d^4} \sum_{w,x,y,z\in G} \sum_{\chi\in\hat{G}} \chi(w+x-y-z) \sum_{\psi\in\hat{H}} \psi(f(w)+f(x)-f(y)-f(z)).$$

By Pontryagin duality and the orthogonality of characters, the two inner sums are nonzero if and only if $w+x-y-z$ and $f(w)+f(x)-f(y)-f(z)=0$, which occurs $d(2d-1)$ times (see Exer. 6.27), giving $d|H|$ each time. Thus

$$\sum_a\sum_\psi\sum_b\sum_\xi w_1^2 |\langle e_a^\psi, e_b^\xi\rangle|^4 = \frac{1}{d^4}\sum_{a\in G}\sum_{\psi\hat{H}} \frac{d(2d-1)d|H|}{|H|^2(d+1)^2} = \frac{2d-1}{d(d+1)^2}.$$

Finally, adding the three contributions gives

$$\sum_j\sum_k w_j w_k |\langle f_j, f_k\rangle|^4 = \frac{1}{d(d+1)^2}\big(1+2+(2d-1)\big) = \frac{2}{d(d+1)} = c_2(d,\mathbb{C}),$$

as claimed. □

Example 6.23. (Maximal MUBs) Let $G = H = F_d$ the finite field of order $d = p^n$, p odd, e.g., $F_p = \mathbb{Z}_p$. The following functions $f : G \to G$ are 1-uniform (see [YCD06])

1. $f(x) = x^2$,
2. $f(x) = x^{p^k+1}$, where $n/\gcd(n,k)$ is odd,
3. $f(x) = x^{(3^k+1)/2}$, where $p = 3$, k is odd, $\gcd(n,k) = 1$,
4. $f(x) = x^{10} - ux^6 - u^2x^2$, where $p = 3$, n is odd, $u \neq 0$.

These give $d+1$ MUBs (mutually unbiased bases) for $\mathbb{C}^d$. The first (which is special case of the second) gives the $d+1$ MUBs for $\mathbb{C}^d$ discussed in §12.19.

Example 6.24. Let $d+1$ be a prime power, and y be a generator for F_{d+1}^* (the multiplicative group for the field of order $d+1$). Then the function $f : \mathbb{Z}_d \to \mathbb{Z}_{d+1}$ defined by

$$f(j) := y^j$$

is 1-uniform (see Exer. 6.26), and we obtain a weighted $(2,2)$-design for $\mathbb{C}^d$ that consists of $d+2$ orthonormal bases, i.e., has $n = d(d+2)$ elements.

6.14 The frame force

Motivated by well distributed points on the sphere, e.g., the n equally spaced vectors in $\mathbb{R}^2$ and the vertices of the Platonic solids in $\mathbb{R}^3$, Fickus [Fic01] gave a central force for which the equilibrium configurations are tight frames. He defined the **frame force** between *unit* vectors in $\mathbb{R}^d$ to be

$$\mathrm{FF}(a,b) := \langle a,b\rangle(a-b), \qquad a,b \in \mathbb{R}^d, \quad \|a\| = \|b\| = 1.$$

Since $\langle a,b\rangle = 1 - \frac{\|a-b\|^2}{2}$ for $\|a\| = \|b\| = 1$,

$$\mathrm{FF}(a,b) = f(\|a-b\|)(a-b), \qquad f(x) := 1 - \frac{x^2}{2}, \tag{6.57}$$

which gives a so-called *central force* on $\mathbb{R}^d$. The frame force between equal vectors is zero, whereas many physical forces are only defined for distinct vectors, e.g., the **Coulomb force** between unit charges at $a \neq b$, given by

$$F_C(a,b) := \frac{(a-b)}{\|a-b\|^3}, \qquad a,b \in \mathbb{R}^d.$$

For a central force on $\mathbb{R}^d$ such as (6.57), the **potential** between a and b is

$$P(a,b) := p(\|a-b\|), \quad \text{where} \quad p'(x) = -xf(x),$$

so that

$$(\nabla P(\cdot,b))(a) = -f(\|a-b\|)(a-b),$$

and the **total potential** for a sequence of vectors $(x_j)_{j=1}^n$ in $\mathbb{R}^d$ is

$$\sum_j \sum_{k\neq j} P(x_j,x_k). \tag{6.58}$$

For the frame force, the choice $p(x) = \frac{x^4}{8} - \frac{x^2}{2}$ gives the potential

$$\begin{aligned} P(a,b) &:= \frac{\|a-b\|^4}{8} - \frac{\|a-b\|^2}{2} \\ &= \frac{1}{2}(\langle a,b\rangle^2 - 1) \qquad (\text{for } \|a\| = \|b\| = 1), \end{aligned}$$

and hence the total potential for unit vectors $x_1,\ldots,x_n$ in $\mathbb{R}^d$

$$\sum_j \sum_{k\neq j} P(x_j,x_k) = \frac{1}{2}\sum_j \sum_k \langle x_j,x_k\rangle^2 - \frac{n}{2}.$$

In this way, Fickus arrived at the **frame potential** (6.6) for unit vectors in $\mathbb{F}^d$.

6.15 Local and global minimisers of the frame potential

The local minimisers of the frame potential (6.6) give a set of unit vectors which are at equilibrium with respect to the frame force (see Exer. 6.28). Moreover, it turns out these local minimisers are in fact global minimisers.

For $f:\mathbb{C}^d\to\mathbb{R}$ with $f(x_1+iy_1,\ldots,x_d+iy_d)$ a differentiable function of the real variables $x_1,y_1,\ldots x_d,y_d\in\mathbb{R}$, define a gradient $\nabla f=2(\overline{\partial}_1 f,\ldots,\overline{\partial}_d f):\mathbb{C}^d\to\mathbb{C}^d$ by

$$\nabla f:=\Bigl(\frac{\partial}{\partial x_j}f(x_1+iy_1,\ldots,x_d+iy_d)+i\frac{\partial}{\partial y_j}f(x_1+iy_1,\ldots,x_d+iy_d)\Bigr)_{j=1}^d. \tag{6.59}$$

Then for both $\mathbb{R}^d$ and $\mathbb{C}^d$, we have

$$\nabla(\|\cdot\|^2)(a)=2a, \qquad \nabla(|\langle\cdot,b\rangle|^2)(a)=2\langle a,b\rangle b. \tag{6.60}$$

The minimisation of the frame potential for unit vectors in $\mathbb{F}^d=\mathbb{R}^d$ or $\mathbb{C}^d$ can be viewed as a constrained optimisation problem:

$$\text{minimise} \quad \sum_j\sum_k|\langle v_j,v_k\rangle|^2 \quad \text{subject to} \quad \|v_1\|=\cdots=\|v_n\|=1.$$

If $(a_1,\ldots,a_n)$ is a local minimiser, then each a_ℓ is a local minimiser of

$$f(v_\ell):=\sum_{k\neq\ell}|\langle v_\ell,a_k\rangle|^2 \quad \text{subject to} \quad g(v_\ell):=\|v_\ell\|^2=1.$$

The critical points of this constrained optimisation of n or $2n$ real variables are given by Lagrange multipliers: $\nabla f(a_\ell)=\lambda\nabla g(a_\ell)$, which by (6.60) becomes

$$\sum_{k\neq\ell}\langle a_\ell,a_k\rangle a_k=\lambda a_\ell, \qquad \lambda=\lambda_\ell\in\mathbb{R}.$$

Thus, each a_ℓ is an eigenvector of the frame operator S for (a_j), i.e.,

$$Sa_\ell=\sum_{k=1}^n\langle a_\ell,a_k\rangle a_k=(\lambda_\ell+\|a_\ell\|^2)a_\ell, \qquad \ell=1,\ldots,n. \tag{6.61}$$

Using this, we now show that local minimisers are global minimisers.

Theorem 6.9. ([BF03]). *Let $\mathbb{S}=\mathbb{S}(\mathcal{H})$ be the unit sphere in $\mathcal{H}$, $d=\dim(\mathcal{H})$. The local minimisers of the frame potential*

$$\mathrm{FP}(v_1,\ldots,v_n):=\sum_{j=1}^n\sum_{k=1}^n|\langle v_j,v_k\rangle|^2, \qquad v_1,\ldots,v_n\in\mathbb{S}$$

are global minimisers, which in turn are tight frames for $\mathcal{H}$ $(n\geq d)$, or nonspanning orthonormal sequences $(n<d)$. In particular, there exist equal-norm tight frames of n vectors in $\mathbb{R}^d$ and $\mathbb{C}^d$, for all values of $n\geq d$.

Proof. Let (a_j) be a local minimiser of the frame potential on $\mathbb{S}$, and S be its frame operator. We have seen, (6.61), that the Lagrange multiplier equations imply each a_j is an eigenvector of S. Moreover, the eigenvalue λ is ≥ 1, since

$$Sa_j = \lambda a_j \quad \Longrightarrow \quad \lambda = \lambda \langle a_j, a_j \rangle = \langle Sa_j, a_j \rangle \geq |\langle a_j, a_j \rangle|^2 = 1.$$

Let $\lambda_{\max}$ be the largest eigenvalue of S. If $\lambda_{\max} = 1$, i.e., it is the only nonzero eigenvalue, then $S = I$ on $\operatorname{span}(a_j)$, and (a_j) is an orthonormal basis for $\operatorname{span}(a_j)$ (see Exer. 2.4). If $\lambda_{\max}$ is the only eigenvalue, i.e., $S = \lambda_{\max} I$, then (a_j) is a tight frame. Thus it reduces to showing that S cannot have a second nonzero eigenvalue $\lambda_2 < \lambda_{\max}$. We suppose that it does and show that (a_j) can be perturbed to obtain (a_j^ε) with a smaller frame potential, thereby giving a contradiction.

Let J be the subset of indices j for which a_j is a $\lambda_{\max}$-eigenvector, and u be a λ_2-eigenvector. For $\varepsilon \in \mathbb{R}$ sufficiently small, define

$$a_j^\varepsilon := \begin{cases} \sqrt{1 - \varepsilon^2 |\beta_j|^2} a_j + \varepsilon \beta_j u, & j \in J; \\ a_j, & j \notin J \end{cases}$$

where the $\beta_j \in \mathbb{C}$ are to be chosen later. Then $\operatorname{FP}\big((a_j^\varepsilon)\big)$ is a C^∞-function of ε for $|\varepsilon| < 1/\max(|\beta_j|)$. Thus, we consider the asymptotic expansion given by the first few terms of the Taylor series. By definition,

$$|\langle a_j^\varepsilon, a_k^\varepsilon \rangle|^2 = |\langle a_j, a_k \rangle|^2, \qquad j \notin J, k \notin J. \tag{6.62}$$

Since S is Hermitian, its eigenspaces are orthogonal to each other, so that

$$\langle a_j, a_k \rangle = 0, \quad j \in J,\ k \notin J, \qquad \langle a_j, u \rangle = 0, \quad j \in J,$$

and we obtain

$$|\langle a_j^\varepsilon, a_k^\varepsilon \rangle|^2 = |\langle a_j, a_k \rangle|^2 + \varepsilon^2 |\beta_j|^2 |\langle u, a_k \rangle|^2, \qquad j \in J,\ k \notin J. \tag{6.63}$$

$$\langle a_j^\varepsilon, a_k^\varepsilon \rangle = \sqrt{1 - \varepsilon^2 |\beta_j|^2} \sqrt{1 - \varepsilon^2 |\beta_k|^2} \langle a_j, a_k \rangle + \varepsilon^2 \beta_j \overline{\beta_k}, \qquad j \in J,\ k \in J.$$

Using the Taylor expansion

$$\sqrt{1 - \varepsilon^2 |\beta_j|^2} = 1 - \frac{1}{2} \varepsilon^2 |\beta_j|^2 + O(\varepsilon^4), \qquad \varepsilon \to 0,$$

we expand the expression for $j \in J$, $k \in J$, to obtain

$$\langle a_j^\varepsilon, a_k^\varepsilon \rangle = \Big(1 - \frac{1}{2} \varepsilon^2 (|\beta_j|^2 + |\beta_k|^2)\Big) \langle a_j, a_k \rangle + \varepsilon^2 \beta_j \overline{\beta_k} + O(\varepsilon^4),$$

which gives

$$|\langle a_j^\varepsilon, a_k^\varepsilon\rangle|^2 = (1-\varepsilon^2(|\beta_j|^2+|\beta_k|^2))|\langle a_j,a_k\rangle|^2 + 2\varepsilon^2\Re(\langle a_j,a_k\rangle\overline{\beta_j}\beta_k) + O(\varepsilon^4), \quad \varepsilon\to 0, \qquad j\in J,\, k\in J. \tag{6.64}$$

Adding (6.62), (6.63) and (6.64) gives

$$\mathrm{FP}\big((a_j^\varepsilon)\big) = \mathrm{FP}\big((a_j)\big) - \varepsilon^2\sum_{j\in J}\sum_{k\in J}(|\beta_j|^2+|\beta_k|^2))|\langle a_j,a_k\rangle|^2 + 2\varepsilon^2\sum_{j\in J}\sum_{k\in J}\Re(\langle a_j,a_k\rangle\overline{\beta_j}\beta_k) + 2\varepsilon^2\sum_{j\in J}\sum_{k\notin J}|\beta_j|^2|\langle u,a_k\rangle|^2 + O(\varepsilon^4).$$

Since the eigenspaces of S are orthogonal, we have

$$\sum_{k\in J}|\langle a_j,a_k\rangle|^2 = \langle Sa_j,a_j\rangle = \lambda_{\max}, \quad j\in J, \qquad \sum_{k\notin J}|\langle u,a_k\rangle|^2 = \langle Su,u\rangle = \lambda_2,$$

and so obtain

$$\mathrm{FP}\big((a_j^\varepsilon)\big) = \mathrm{FP}\big((a_j)\big) - 2\varepsilon^2\Big(\sum_{j\in J}|\beta_j|^2\Big)(\lambda_{\max}-\lambda_2) + 2\varepsilon^2\|\sum_{j\in J}\overline{\beta_j}a_j\|^2 + O(\varepsilon^4).$$

Since $S=\lambda_{\max}I$ on the $\lambda_{\max}$-eigenspace $E=\mathrm{span}\{a_j : j\in J\}$, the vectors $(a_j)_{j\in J}$ are a unit-norm tight frame for E with frame bound $\lambda_{\max}$, and so

$$\lambda_{\max} = \frac{|J|}{\dim(E)} > \lambda_2 \ge 1.$$

Hence the vectors $(a_j)_{j\in J}$ are linearly dependent, and there is $(\beta_j)_{j\in J}\neq 0$ with $\sum_{j\in J}\overline{\beta_j}a_j = 0$, i.e., $(a_j)_{j\in J}$ and $(\beta_j)_{j\in J}$ are mutually orthogonal. Thus

$$\mathrm{FP}\big((a_j^\varepsilon)\big) = \mathrm{FP}\big((a_j)\big) - c\varepsilon^2 + O(\varepsilon^4), \qquad \varepsilon\to 0,$$

where $c = 2(\sum_j|\beta_j|^2)(\lambda_{\max}-\lambda_2) > 0$, which gives the desired contradiction. □

In the motivating case $\mathscr{H}=\mathbb{R}^d$, these minimisers are at equilibrium with respect to the frame force. Indeed, define the **effective frame force** of b on a to be the component of the frame force $\mathrm{FF}(a,b)$ of b on a which is orthogonal to a, i.e.,

$$\mathrm{EFF}(a,b) = \langle a,b\rangle(a\langle a,b\rangle - b),$$

and so the total effective frame force on each point a_j is zero (see Exer. 6.28), i.e.,

$$\sum_k \mathrm{EFF}(a_j,a_k) = 0.$$

Example 6.25. A sequence (a_j) of points on the sphere is said to be **FF critical** (**frame force critical**) if each a_j is an eigenvector of the frame operator of (a_j). The minimisers of the frame force are FF critical. A sequence is FF critical if and only if it is a union of equal-norm tight frames for orthogonal subspaces.

6.16 The numerical construction of spherical (t,t)-designs

Let $V = [v_{\alpha\beta}] = [v_1,\ldots,v_n]$, and $p,g : \mathbb{F}^{d\times n} \to \mathbb{R}$ be the homogeneous polynomials given by

$$p(V) := \sum_j \sum_k |\langle v_j, v_k\rangle|^{2t}, \qquad g(V) := \sum_\ell \|v_\ell\|^{2t}. \tag{6.65}$$

The spherical (t,t)-designs for $\mathbb{F}^d$ are the nontrivial zeros of the nonnegative homogeneous polynomial

$$f(V) := p(V) - c_t(d,\mathbb{F}^d) g(V)^2 \tag{6.66}$$

of degree $4t$ in the real (and imaginary) parts of entries of $V = [v_{\alpha\beta}] \in \mathbb{F}^{d\times n}$.

The minimisers of $p(V) \geq 0$ with $g(V)$ fixed, e.g., $V = [v_j]$ a unit-norm sequence, satisfy the Lagrange equations: $\nabla p(V) = \lambda \nabla g(V)$. Moreover, the ones that give spherical (t,t)-designs are minima of f, and so satisfy $\nabla f(V) = 0$, i.e.,

$$\nabla p(V) = 2c_t(d,\mathbb{F}) g(V) \nabla g(V). \tag{6.67}$$

Thus we obtain the following condition for the existence of spherical (t,t)-designs.

Theorem 6.10. *Let $t \geq 1$ and $f : \mathbb{F}^{d\times n} \to \mathbb{R}$ be the nonnegative function given by*

$$f([v_1,\ldots,v_n]) := \sum_{j=1}^n \sum_{k=1}^n |\langle v_j, v_k\rangle|^{2t} - c_t(d,\mathbb{F}) \Big(\sum_{\ell=1}^n \|v_\ell\|^{2t}\Big)^2.$$

Then the critical points of f satisfy

$$\sum_j |\langle v_j, v_\beta\rangle|^{2(t-1)} \langle v_\beta, v_j\rangle v_j = c_t(d,\mathbb{F}) \Big(\sum_\ell \|v_\ell\|^{2t}\Big) \|v_\beta\|^{2(t-1)} v_\beta, \quad 1 \leq \beta \leq n.$$

In particular, for $t = 1$ the nonzero critical points of f are the tight frames for $\mathbb{F}^d$, which are all global minima.

Proof. The critical point of f is given by (6.67), where ∇f is the gradient of f viewed as a function of real variables, as in (6.59). A calculation (see Exer. 6.29) shows that the β-columns of $\nabla p(V)$ and $\nabla g(V)$ are

$$4t \sum_j |\langle v_j, v_\beta\rangle|^{2(t-1)} \langle v_\beta, v_j\rangle v_j, \qquad 2t\|v_\beta\|^{2(t-1)} v_\beta.$$

Substituting this into (6.67) gives the desired condition.

For $t = 1$, the $V \neq 0$ which are critical points of $f(V)$ satisfies

$$\sum_j \langle v_\beta, v_j\rangle v_j = \frac{1}{d} \Big(\sum_\ell \|v_\ell\|^2\Big) v_\beta, \quad 1 \leq \beta \leq n,$$

and so, by linearity, (v_j) is tight frame for $\mathscr{H} := \operatorname{span}\{v_\beta\}_{1\le\beta\le n} \subset \mathbb{F}^d$, with frame bound $A = \frac{1}{d}\sum_\ell \|v_\ell\|^2$. The trace condition (2.6) gives $\dim(\mathscr{H}) = d$, so that (v_j) is a tight frame for $\mathbb{F}^d$. Thus the nonzero critical points of $f(V)$ are precisely the tight frames for $\mathbb{F}^d$. □

Spherical (t,t)-designs can be found *numerically*, by minimising $f(V)$, with $g(V)$ fixed. This can be done by an iterative algorithm which starts a random V_0, and chooses $V_{k+1} = V_k + W_k$, where W_k is such that $f(V_{k+1}) = f(V_k + W_k) < f(V_k)$. The direction W_k can be random (of an appropriate size) [Bra11], or in the direction of maximal decrease [Hug16] (which is more effective close to a minimum).

The maximal decrease of f at V is in the direction $W = -\nabla f(V)$, where

$$(\nabla f(V))_{\alpha\beta} = 4t\sum_j |\langle v_j, v_\beta\rangle|^{2(t-1)}\langle v_\beta, v_j\rangle v_{\alpha j} - 4tc_t(d,\mathbb{F}^d)\Big(\sum_\ell \|v_\ell\|^{2t}\Big)\|v_\beta\|^{2(t-1)} v_{\alpha\beta}.$$

A summary of these numerical results (which have motivated various analytic constructions) is given in Tables 6.1 and 6.2.

Table 6.1: The minimum numbers n_w and n of vectors in a weighted and in a unit-norm spherical (t,t)-design for $\mathbb{R}^d$ (spherical half-design of order $2t$) as calculated by Daniel Hughes [Hug16].

t	d	n_w	n_e	Comments	
1	d	d	d	orthonormal bases in $\mathbb{R}^d$	
t	2	t+1	t+1	equally spaced lines in $\mathbb{R}^2$	
2	3	6	6	equiangular lines in $\mathbb{R}^3$	
2	4	11	12	no structure	repeated angles
2	5	16	20	group structure	no structure
2	6	22	24	group structure	work in progress
2	7	28	28	equiangular lines in $\mathbb{R}^7$	
2	8	45	>45	no structure	
3	3	11	16	no structure	possible group structure
3	4	23	>23	group structure	
3	5	41	>41	group structure	
4	3	16	25	group structure	no structure
4	4	43	>43	work in progress	
5	3	24	35	no structure	no structure

Table 6.2: The minimum numbers n_w and n of vectors in a weighted and in a unit-norm spherical (t,t)-design for $\mathbb{C}^d$ as calculated by Jennifer Bramwell [Bra11].

t	d	n_w	n_e	Comments
1	d	d	d	orthonormal bases in $\mathbb{R}^d$
2	d	d^2	d^2	SICs
3	2	6	6	three MUBs for $\mathbb{C}^2$
3	3	22	27	some structure
3	4	40	40	highly symmetric tight frame (Example 6.15)
3	5	>100		
4	2	10	12	two orbits, see §9.8
4	3	47	>47	
4	4	>85	>85	
5	2	12	12	group frame, see Exer. 10.12
6	2	18	24	some structure
7	2	22	24	some structure
8	2	37	>37	some structure
9	2	44	>44	some structure

It is also possible to calculate (numerically) the Hessian (second derivative) of f and p at V to investigate the nature of the critical points of f (see Exer. 6.30).

For the real case $V = X \in \mathbb{R}^{d\times n}$, the formulas for the Hessian simplify to

$$\frac{\partial^2 p}{\partial x_{ab}\partial x_{\alpha\beta}}(V) = 4t(2t-1)\langle v_b, v_\beta\rangle^{2(t-1)} v_{\alpha b} v_{a\beta} + 4t\delta_{a\alpha}\langle v_b, v_\beta\rangle^{2t-1}$$
$$+4t(2t-1)\delta_{b\beta}\sum_j \langle v_j, v_\beta\rangle^{2(t-1)} v_{\alpha j} v_{aj},$$

$$\frac{\partial^2 (g^2)}{\partial x_{ab}\partial x_{\alpha\beta}}(V) = 2g(V)\big(2t\delta_{a\alpha}\delta_{b\beta}\|v_\beta\|^{2(t-1)} + 4t(t-1)\delta_{b\beta}\|v_\beta\|^{2(t-2)} v_{\alpha\beta} v_{a\beta}\big)$$
$$+2\big(2t\|v_b\|^{2(t-1)} v_{ab}\big)\big(2t\|v_\beta\|^{2(t-1)} v_{\alpha\beta}\big).$$

Notes

Neil Sloane has a webpage of *putatively optimal* real spherical t-designs

`http://neilsloane.com/sphdesigns/` (see [HS96])

and the author has a similar list for real and complex spherical (t,t)-designs.

There is recent interest in *complex* spherical (t,t)-designs (weighted complex projective t-designs), see, e.g., [KR05], [RS07], [RS14]. Our unified treatment of real spherical half-designs of order $2t$ and complex (t,t)-designs in Theorem 6.7 was adapted from [Kön99], [Wal16] (also see [DHC12], [BH15]). There are other equivalences. Those involving the evaluation of Gegenbauer polynomials can be used to estimate the minimum number of vectors in a (t,t)-design (see [DGS77], [Hog89] for spherical t-designs). Some (t,t)-designs meeting these bounds (for spherical $2t$-designs) are termed *tight* (this is not related to being a tight frame).

Thanks to Aidan Roy, Andreas Klappenecker and Wei–Hsuan Yu for insightful discussions about this chapter.

Exercises

6.1. Let $A \in \mathbb{C}^{n\times n}$ be Hermitian, $A \neq 0$. Show that $\mathrm{rank}(A) \geq \frac{\mathrm{trace}(A)^2}{\mathrm{trace}(A^2)}$ with equality if and only if $A = cUU^*$, $U = [u_1,\ldots,u_r] \in \mathbb{C}^{n\times r}$, with orthonormal columns.
Hint. If $\lambda_1,\ldots,\lambda_r$ are the nonzero eigenvalues of A, then Cauchy–Schwarz gives

$$(\mathrm{trace}(A))^2 = (\sum_{j=1}^{n} \lambda_j)^2 \leq r \sum_{j=1}^{r} \lambda_j^2 = r\,\mathrm{trace}(A^2).$$

6.2. Use the generalised Welch bound (6.4) to prove Example 6.2, i.e., that

$$\max_{j\neq k} |\langle f_j, f_k\rangle|^2 \geq \frac{(\sum_j \|f_j\|)^2/d - \sum_j \|f_j\|^4}{n^2-n} > 0.$$

6.3. ᵐDo a numerical investigation of the variational inequality (6.4)? Does one get close to equality for large numbers of random (unit or otherwise) vectors?

6.4. We may write the variational characterisation (6.4) for tight frames $(v_j)_{j\in J}$ as

$$\sum_{\{j,k\}\subset J} \left(2|\langle v_j, v_k\rangle|^2 + \frac{\|v_j\|^4 + \|v_k\|^4}{n-1} \right) = \frac{1}{d} \sum_{\{j,k\}\subset J} \left(2\|v_j\|^2\|v_k\|^2 + \frac{\|v_j\|^4 + \|v_k\|^4}{n-1} \right).$$

We say that a frame $(v_j)_{j=1}^n$ for $\mathbb{C}^d$ is **perfectly tight** if equality holds for all pairs.
(a) Show that for $d=1$, every tight frame is perfectly tight.
(b) Show that for $d \geq 2$, every perfectly tight frame has nonzero vectors.
(c) Describe the equal-norm perfectly tight frames.
(d) Do there exist perfectly tight frames for $d \geq 2$ which do not have equal norms?

6.5. Let (f_j) be a sequence of $n \leq d$ unit vectors in a $\mathscr{H}$, where $\dim(\mathscr{H}) = d$. Show that

$$\mathrm{FP}(f_1,\ldots,f_n) := \sum_j \sum_k |\langle f_j, f_k \rangle|^2$$

has a minimum value of n, which is attained if and only if (f_j) is orthogonal.

6.6. Show the normalised frame potential satisfies (6.7), i.e.,

$$\frac{1}{d} \leq \hat{\mathrm{FP}}(f_1,\ldots,f_n) \leq 1,$$

and that
(a) $\hat{\mathrm{FP}}(f_1,\ldots,f_n)$ equals $\frac{1}{d}$ if and only if (f_j) is a tight frame.
(b) $\hat{\mathrm{FP}}(f_1,\ldots,f_n)$ equals 1 if and only if $\mathrm{span}(f_j)$ is 1-dimensional.
Remark: For unit-norm vectors $\mathrm{FP} = n^2 \hat{\mathrm{FP}}|_{\mathbb{S}^n}$, and so this extends Theorem 6.2.

6.7. *Real spherical* 2*-designs*. Let $\Phi = \{\phi_1,\ldots,\phi_n\}$ be unit vectors in $\mathbb{R}^d$.
(a) Suppose that Φ is a real spherical 2-design. For $y \in \mathbb{R}^d$, let $p_y \in \Pi_2^\circ(\mathbb{R}^d)$ be given by

$$p_y(x) := |\langle y,x \rangle|^2 = (\langle y,x \rangle)^2.$$

Show that the integral of p_y over the unit sphere $\mathbb{S}$ is $c\|y\|^2$, with $c>0$ independent of y, and hence conclude that Φ is a tight frame.
(b) Now suppose that Φ is a tight frame. By considering the integral of p_y above, or otherwise, show that it is a real spherical 2-design.
Remark: This is a special case of the key arguments of §6.8.

6.8. Let $d \geq 2$. Show that there exists a real spherical 2-design of n points for $\mathbb{R}^d$ unless $n \leq d$ or $n = d+2$ and n is odd (the existence part requires a construction). In particular, there is no real spherical 2-design of five points for $\mathbb{R}^3$.
Hint: For the construction use harmonic frames (see Chapter 11).

6.9. *Real spherical designs and Waring type formulas.*
Here we consider the case of equality in Theorem 6.7 for $\mathscr{H} = \mathbb{R}^d$.
(a) Show that $c_t(d,\mathbb{R}) \geq c_t(d,\mathbb{C})$, with strict inequality when $t>1$, $d>1$.
(b) A sequence (f_j) of unit vectors in $\mathbb{R}^d$ satisfying (6.33) is, by definition, a real spherical half-design of order $2t$. By substituting t for $2t$, write down the equivalent conditions for being a real spherical half-design of order t (for t even) given by equality in (6.30), (6.31) and (6.35).

(c) Show that if (f_j) is centrally symmetric, i.e., of the form $(\pm f_j)_{j=1}^{n/2}$, then the cubature rule (6.33) holds for all odd polynomials, i.e., $p \in \Pi_1^\circ \oplus \Pi_3^\circ \oplus \Pi_5^\circ \oplus \cdots$.

6.10. The n equally spaced (unit) vectors in $\mathbb{R}^2$ are

$$\Phi = (v_j) = \{(\cos\frac{2\pi}{n}j, \sin\frac{2\pi}{n}j) : j = 0,\ldots,n-1\},$$

and the n equally spaced lines in $\mathbb{R}^2$ are

$$\Psi = (w_j) = \{(\cos\frac{\pi}{n}j, \sin\frac{\pi}{n}j) : j = 0,\ldots,n-1\}.$$

(a) Show that the n equally spaced vectors in $\mathbb{R}^2$ are a spherical $(n-1)$-design.
(b) Show that the $n = t+1$ equally spaced lines in $\mathbb{R}^2$ are a spherical half-design of order $2t$, i.e., a (t,t)-design.
Hint: Use the integrals of (6.25) and (6.26).

6.11. A SIC consists of d^2 equiangular unit vectors in $\mathbb{C}^d$, with a common angle $|\langle f_j, f_k\rangle|^2 = \frac{1}{\sqrt{d+1}}$, $j \neq k$, and m MUBs are m orthonormal bases for $\mathbb{C}^d$, with the property that $|\langle f, g\rangle| = \frac{1}{\sqrt{d}}$ for f and g from different bases (see §2.11).
(a) Show that a SIC is a $(2,2)$-design.
(b) Show that m MUBs in $\mathbb{C}^d$ form a $(2,2)$-design if and only if $m = d+1$.
(c) Show that $d+1$ MUBs in $\mathbb{C}^d$ form a $(3,3)$-design if and only if $d = 2$.

6.12. There is a highly symmetric tight frame of 240 vectors for $\mathbb{C}^4$ given as an orbit of the Shephard–Todd group 32 (see §13.8), which gives 40 lines, since the group contains scalar multiplication by the sixth roots of unity. Take a vector from each line. This set Φ of 40 vectors has the property that each is orthogonal to 12 others and makes an angle $\frac{1}{\sqrt{3}}$ with 27 others. Show that Φ is a $(3,3)$-design for $\mathbb{C}^4$.

6.13. Show that an equiangular tight frame $\Phi = (v_j)$ of n unit vectors for $\mathbb{R}^d$ is a spherical $(2,2)$-design for $\mathbb{R}^d$ if and only if $n = \frac{1}{2}d(d+1)$.
Remark: Such equiangular lines are known to exist for $d = 2,3,7,23$ (see §12.1).

6.14. Let Φ be the set of 240 vectors $v \in \mathbb{R}^8$ of with $\|v\|^2 = 2$, and the form

type 1: $v_j \in \{\pm\frac{1}{2}\}$ and v has an even number of positive entries,

type 2: $v_j \in \{0, \pm 1\}$ (such v have two nonzero entries).

Since Φ is centrally symmetric, it can be written $\Phi = \Phi_0 \cup -\Phi_0$.
(a) Show that Φ_0 is a spherical $(3,3)$-design of 120 vectors for $\mathbb{R}^8$.
(b) Show that Φ is a spherical 7-design of 240 vectors $\mathbb{R}^8$.
Remark: This Φ (minimal vectors of the Korkin–Zolotarev lattice) is due to [KP11].

6.15. Show that equality in (6.22) is equivalent to
(a) The *generalised Plancherel identity*

$$\langle x,y\rangle^t = \frac{\binom{d+t-1}{t}}{\sum_{\ell=1}^n \|f_\ell\|^{2t}} \sum_{j=1}^n \langle x,f_j\rangle^t \langle f_j,y\rangle^t, \qquad \forall x,y \in \mathscr{H}.$$

(b) The *generalised Bessel identity*

$$\|x\|^{2t} = \frac{\binom{d+t-1}{t}}{\sum_{\ell=1}^n \|f_\ell\|^{2t}} \sum_{j=1}^n |\langle x,f_j\rangle|^{2t}, \qquad \forall x \in \mathscr{H}.$$

6.16. Define $\xi \in \operatorname{Sym}^t(\mathscr{H}) \otimes \operatorname{Sym}^t(\overline{\mathscr{H}})$ and $Q : \operatorname{Sym}^t(\mathscr{H}) \to \operatorname{Sym}^t(\mathscr{H})$ by

$$\xi := \int_{\mathbb{S}} x^{\otimes t} \otimes \overline{x}^{\otimes t}\, d\sigma(x) - \frac{1}{C}\sum_{j=1}^n f_j^{\otimes t} \otimes \overline{f_j}^{\otimes t},$$

$$Q := \int_{\mathbb{S}} \langle \cdot, x^{\otimes t}\rangle x^{\otimes t}\, d\sigma(x) - \frac{1}{C}\sum_{j=1}^n \langle \cdot, f_j^{\otimes t}\rangle f_j^{\otimes t},$$

where $C := \sum_\ell \|f_\ell\|^{2t}$. Show that

$$\langle \xi,\xi\rangle_\circ = \langle Q,Q\rangle_F = \frac{1}{C^2}\sum_j\sum_k |\langle f_j,f_k\rangle|^{2t} - c_t(d,\mathbb{F}),$$

where the apolar and Frobenius inner products are used, respectively.

6.17. We consider the vector space $\Pi_{t,r}^\circ(\mathbb{C}^d)$ of polynomials $\mathbb{C}^d \to \mathbb{C}$ which are homogeneous of degree t in z and of degree r in $\overline{z}$, i.e.,

$$\Pi_{t,r}^\circ(\mathbb{C}^d) := \operatorname{span}\{z \mapsto z^\alpha \overline{z}^\beta : |\alpha| = t, |\beta| = r\}. \tag{6.68}$$

This absolutely irreducible $\mathscr{U}$-invariant space is denoted by $H(t,r)$ in §16.7. There is a natural identification $\operatorname{Sym}^t(\mathscr{H}^*) \otimes \operatorname{Sym}^r(\overline{\mathscr{H}^*}) \to \Pi_{t,r}^\circ(\mathscr{H})$ given by

$$\langle \cdot, v\rangle^{\otimes t} \otimes \langle \overline{\cdot}, \overline{x}\rangle^{\otimes r} \mapsto \langle \cdot, v\rangle^t \langle \overline{\cdot}, \overline{x}\rangle^r = \langle \cdot, v\rangle^t \langle x, \cdot\rangle^r.$$

For polynomials $p : \mathbb{C}^d \to \mathbb{C}$, we define an associated differential operator $p(\partial)$ by

$$p(\partial) := \sum_{(\alpha,\beta)} c_{\alpha\beta} \partial^\alpha \overline{\partial}^\beta, \qquad \text{where } p(z) = \sum_{(\alpha,\beta)} c_{\alpha\beta} z^\alpha \overline{z}^\beta, \tag{6.69}$$

where ∂ and $\overline{\partial}$ are the Wirtinger complex differential operators given by

$$\partial_j = \frac{\partial}{\partial z_j} = \frac{1}{2}\left(\frac{\partial}{\partial x_j} - i\frac{\partial}{\partial y_j}\right), \qquad \overline{\partial}_j = \frac{\partial}{\partial \overline{z_j}} = \frac{1}{2}\left(\frac{\partial}{\partial x_j} + i\frac{\partial}{\partial y_j}\right).$$

(a) Show that the monomials $z \mapsto z^\alpha \overline{z}^\beta$ in (6.68) are linearly independent, and so

$$\dim(\Pi_{t,r}^{\circ}(\mathbb{F}^d)) = \binom{d+t-1}{t}\binom{d+r-1}{r}. \tag{6.70}$$

(b) By taking the apolar inner product on $\mathrm{Sym}^t(\mathscr{H}^*)\otimes\mathrm{Sym}^r(\overline{\mathscr{H}}^*)$, show that

$$\langle\langle\cdot,v\rangle^t\langle x,\cdot\rangle^r,\langle\cdot,w\rangle^t\langle y,\cdot\rangle^r\rangle_\circ := \langle w,v\rangle^t\langle x,y\rangle^r \tag{6.71}$$

defines an inner product on $\Pi_{t,r}^{\circ}(\mathbb{C}^d)$.
(c) Show the Riesz representer of point evaluation at w is $\langle\cdot,w\rangle^t\langle w,\cdot\rangle^r$, i.e.,

$$\langle p,\langle\cdot,w\rangle^t\langle w,\cdot\rangle^r\rangle_\circ = p(w), \qquad \forall p\in\Pi_{t,r}^{\circ}(\mathbb{C}^d), \quad \forall w\in\mathbb{C}^d.$$

(d) Use this to conclude that $\Pi_{t,r}^{\circ}(\mathbb{C}^d)$ is spanned by ridge functions, i.e.,

$$\Pi_{t,r}^{\circ}(\mathbb{C}^d) = P := \mathrm{span}\{z\mapsto\langle z,v\rangle^t\langle v,z\rangle^r : v\in\mathbb{C}^d\}. \tag{6.72}$$

In particular, $\Pi_{t,t}^{\circ}(\mathbb{C}^d)$ is spanned by ridge functions (plane waves), i.e.,

$$\Pi_{t,t}^{\circ}(\Omega) = \mathrm{span}\{z\mapsto|\langle z,v\rangle|^{2t} : v\in\Omega\}, \quad \text{where } \Omega=\mathbb{C}^d \text{or } \mathbb{S}_{\mathbb{C}}.$$

(e) With $p(\partial)$ given by (6.69) and $\tilde{q}(z) := \overline{q(\overline{z})}$, show that

$$\langle p,q\rangle_\circ = \frac{1}{t!r!}p(\partial)\tilde{q}(0), \qquad \forall p,q\in\Pi_{t,r}^{\circ}(\mathbb{C}^d).$$

In particular, the monomials $z\mapsto z^\alpha\overline{z}^\beta$ in (6.68) form an orthogonal basis.
Remark: It follows from (6.25), that for $\Pi_{t,0}^{\circ}(\mathbb{C}^d)$ and $\Pi_{0,t}^{\circ}(\mathbb{C}^d)$, one has

$$\langle p,q\rangle_\circ = \binom{t+d-1}{t}\int_{\mathbb{S}(\mathbb{C}^d)} p(z)\overline{q(z)}\,d\sigma(z).$$

Cubature rules which integrate $\Pi_{t,r}^{\circ}(\mathbb{C}^d)$, $(t,r)\in\mathscr{T}$ for some set of indices $\mathscr{T}$ are studied in [RS14], where they are called *spherical* $\mathscr{T}$*-designs.*

6.18. Make the substitution (6.41) in Theorem 6.7 to obtain the weighted versions of the conditions (a)–(e).

6.19. Let Δ be the Laplacian for functions $\mathbb{F}^d\to\mathbb{F}$, i.e., for $\mathbb{F}$ equal $\mathbb{R}$ and $\mathbb{C}$

$$\Delta = \sum_{j=1}^d\Big(\frac{\partial}{\partial x_j}\Big)^2, \qquad \Delta = \sum_{j=1}^d\Big(\frac{\partial}{\partial x_j}\Big)^2 + \sum_{j=1}^d\Big(\frac{\partial}{\partial y_j}\Big)^2 = 4\sum_{j=1}^d\partial_j\overline{\partial_j}.$$

(a) Take the Laplacian with respect to $x\in\mathbb{R}^d$ to get

$$\Delta(\|x\|^{2t}) = 2t(d+2t-2)\|x\|^{2t-2}, \qquad \Delta(\langle x,v\rangle^{2t}) = 2t(2t-1)\langle x,v\rangle^{2t-2}\|v\|^2.$$

(b) Take the Laplacian with respect to $z\in\mathbb{R}^d$ to get

$$\Delta(\|z\|^{2t}) = 4t(d+t-1)\|z\|^{2t-2}, \qquad \Delta(|\langle z,v\rangle|^{2t}) = 4t^2|\langle z,v\rangle|^{2(t-1)}\|v\|^2.$$

(c) Using (a) and (b), apply the Laplacian to the Bessel identity (6.31).

6.20. Show that if (v_j) and (w_k) are spherical (t,t)-designs for $\mathbb{F}^d$, with

$$\sum_j \|v_j\|^{2t} = \sum_k \|w_k\|^{2t},$$

then their union $(v_j)\cup(w_k)$ is a spherical (t,t)-design for $\mathbb{F}^d$.

6.21. Use the generalised Bessel identity (6.31) to show that the minimal number n of vectors in a weighted (t,t)-design satisfies

$$n \le \dim(\Pi^\circ_{t,t}(\mathbb{F}^d)) = \begin{cases} \binom{d+t-1}{t}^2, & \mathbb{F}=\mathbb{C} \\ \binom{d+2t-1}{2t}, & \mathbb{F}=\mathbb{R} \end{cases} = O(d^{2t}), \qquad d\to\infty.$$

6.22. Use the generalised Plancherel identity (6.32) to show that the number n of vectors in a weighted (t,t)-design for $\mathbb{F}^d$ satisfies

$$n \ge \dim\big(\Pi^\circ_t(\mathbb{F}^d)\big) = \binom{t+d-1}{d-1} = O(d^t), \qquad t\to\infty.$$

6.23. Suppose that (f_j) is a tight frame for $\mathbb{F}^d$, i.e., is a $(1,1)$-design. Show the condition which ensures it comes from a (t,t)-design, as per Proposition 6.2 is that

$$\sum_{j=1}^n\sum_{k=1}^n \frac{|\langle f_j,f_k\rangle|^{2t}}{\|f_j\|^{2t-2}\|f_k\|^{2t-2}} = c_t(d,\mathbb{F})\Big(\sum_{\ell=1}^n \|f_\ell\|^2\Big)^2.$$

6.24. The lines $\{\mathbb{F}x : x\neq 0\}$ in $\mathscr{H}=\mathbb{F}^d$ are in 1–1 correspondence with the rank-one orthogonal projections, i.e., points in the projective space $\mathbb{F}P^{d-1}$, via

$$\mathbb{F}x \longleftrightarrow P_x := \frac{\langle\cdot,x\rangle}{\langle x,x\rangle}x.$$

(a) Show that the Frobenius inner product between orthogonal projections given by unit vectors is

$$\langle P_x,P_y\rangle = |\langle x,y\rangle|^2 = \langle P_y,P_x\rangle.$$

(b) Show that the metric on $\mathbb{F}P^{d-1}$ given by the Frobenius inner product is

$$\rho(P,Q) = \sqrt{2}\sqrt{1-\langle P,Q\rangle}, \qquad P,Q\in\mathbb{F}P^{d-1}.$$

(c) Show that in terms of lines this metric is

$$\rho(\mathbb{F}x,\mathbb{F}y) = \sqrt{2}\sqrt{1-\Big|\langle\frac{x}{\|x\|},\frac{y}{\|y\|}\rangle\Big|^2}.$$

(d) Show that the set of lines can also be embedded into the real vector space of traceless Hermitian matrices (with the Frobenius norm), via

$$\mathbb{F}x \mapsto P_x - \frac{1}{d}I.$$

6.25. *Polynomials on projective spaces.*
Determine the vector space of polynomials $p : \mathbb{F}^d \to \mathbb{F}$ whose value at each $z \neq 0$ depends only on the 1-dimensional subspace given by z, i.e.,

$$p(z) = p(az), \qquad \forall z \in \mathbb{F}^d, \quad \forall a \in \mathbb{F}, \ |a| = 1.$$

6.26. Let F_{d+1} be the field of order $d+1$, where $d+1$ is a prime power. Suppose that y is a generator for the multiplicative group F_{d+1}^*. Show that $f : \mathbb{Z}_d \to \mathbb{Z}_{d+1} : y \mapsto y^j$ is a 1-uniform function.

6.27. Let $f : G \to H$ be a map between finite abelian groups, with $|G| = d$. Show that if f is 1-uniform, then

$$w + x - y - z = 0, \qquad f(w) + f(x) - f(y) - f(z) = 0$$

has exactly $d(2d+1)$ solutions in $(w,x,y,z) \in G^4$.

6.28. *Equilibrium with respect to the frame force in $\mathbb{R}^d$ and $\mathbb{C}^d$.*
For unit vectors $a, b \in \mathbb{C}^d$, the frame force of b on a can be extended

$$\mathrm{FF}(a,b) := \langle a,b \rangle (a-b),$$

though this is no longer a central force. The frame force between orthogonal vectors and between coincident vectors is zero. The **effective frame force** $\mathrm{EFF}(a,b)$ of b on a is the component of the frame force $v = \mathrm{FF}(a,b)$ which is orthogonal to a.
(a) Calculate $\mathrm{EFF}(a,b)$ for $a, b \in \mathbb{S}$.
(b) Show that if $(a_j)_{j=1}^n$ is a minimiser of the frame potential, then the total effective frame force on each a_j is zero, i.e.,

$$\sum_k \mathrm{EFF}(a_j, a_k) = \sum_{k \neq j} \mathrm{EFF}(a_j, a_k) = 0.$$

6.29. Let $V = [v_{\alpha\beta}] = [v_1, \ldots, v_n]$ and $p, g : \mathbb{F}^{d \times n} \to \mathbb{R}$ be given by

$$p(V) := \sum_j \sum_k |\langle v_j, v_k \rangle|^{2t}, \qquad g(V) := \sum_\ell \|v_\ell\|^{2t}.$$

With ∇f given by (6.59) when $\mathbb{F} = \mathbb{C}$, show the β-columns of $\nabla p(V)$ and $\nabla g(V)$ are

$$4t \sum_j |\langle v_j, v_\beta \rangle|^{2(t-1)} \langle v_\beta, v_j \rangle v_j, \qquad 2t \|v_\beta\|^{2(t-1)} v_\beta.$$

6.30. Here we calculate the Hessian matrix H_f of f of the functions $p, g : \mathbb{F}^{d \times n} \to \mathbb{R}$ given by (6.65). Let X be the real variables with some ordering, i.e.,

$$X = \{x_{\alpha\beta}\} \cup \{y_{\alpha\beta}\} \quad \text{for } \mathbb{F} = \mathbb{C}, \qquad X = \{x_{\alpha\beta}\} \quad \text{for } \mathbb{F} = \mathbb{R}.$$

Then $H_f(V)$ is the $X \times X$ real symmetric matrix with (r,s)-entry given by

$$H_f(V)_{rs} = \frac{\partial^2 f}{\partial r \partial s}(V).$$

Find the Hessian matrix of p, g, and hence the function f given by (6.66).

6.31. A sequence $(f_j)_{j=1}^n$ is a finite tight frame for $\mathbb{F}^d$ if and only if

$$g(x) := \frac{d}{\sum_k \|f_k\|^2} \sum_{j=1}^n |\langle x, f_j \rangle|^2 \frac{1}{\|x\|^2} = 1, \qquad \forall x \neq 0.$$

(a) Show for a general frame that g can take values which are > 1 and < 1. Thus the obvious generalisation of Bessel's inequality does not hold. There are various generalisations in the literature. We now develop a few.
(b) By Cauchy–Schwarz, $|\sum_j \overline{c_j} \langle x, f_j \rangle| = |\langle x, \sum_j c_j f_j \rangle|^2 \leq \|x\|^2 \|\sum_j c_j f_j\|^2$. Use the triangle and Cauchy–Schwarz inequalities to show *Pečarić's inequality*

$$\Big| \sum_{j=1}^n \overline{c_j} \langle x, f_j \rangle \Big|^2 \leq \|x\|^2 \sum_{j=1}^n |c_j|^2 \sum_{k=1}^n |\langle f_j, f_k \rangle|.$$

(c) From this deduce *Selberg's inequality*

$$\sum_{j=1}^n \frac{|\langle x, f_j \rangle|^2}{\sum_{\ell=1}^n |\langle f_\ell, f_j \rangle|} \leq \|x\|^2.$$

(d) From Selberg's inequality, deduce *Bombieri's inequality*

$$\sum_{j=1}^n |\langle x, f_j \rangle|^2 \leq \|x\|^2 \max_{1 \leq \ell \leq n} \sum_{j=1}^n |\langle f_\ell, f_j \rangle|.$$

(e) Vary the argument of (b) to show the inequality.

$$\sum_{j=1}^n |\langle x, f_j \rangle|^2 \leq \|x\|^2 \Big(\sum_{j=1}^n \sum_{k=1}^n |\langle f_j, f_k \rangle|^2 \Big)^{\frac{1}{2}}.$$

Chapter 7
The algebraic variety of tight frames

Let $V = [v_1, \ldots, v_n]$ be the synthesis operator of a normalised tight frame for $\mathbb{F}^d$, i.e., a $d \times n$ matrix with $VV^* = I$ (Proposition 2.1). Since $VV^* = I$, the collection of normalised tight frames of n vectors for a space of dimension d can be viewed as an *algebraic variety* (in $\mathbb{F}^{d \times n}$), as can other classes of frames, such as the equal-norm tight frames. Here we consider some geometry of this algebraic variety, including:

- What norms the vectors v_j can have.
- The dimension of the variety, and how to describe points on it.
- The fact the rational points are dense on the variety, i.e., every normalised tight frame can be arbitrarily well approximated by a normalised tight frame consisting of vectors with rational entries.

Our treatment is based on the following simple observations. If U is unitary, then

$$(VU)(VU)^* = V(UU^*)V^* = VV^* = I,$$

so that $W = VU$ is (the synthesis operator of) a normalised tight frame. Conversely, if V and W are normalised tight frames of n vectors for $\mathbb{F}^d$, then we may choose complementary normalised tight frames V_c and W_c. Since the matrix with rows given by the rows of a normalised tight frame and a complement is unitary, we have

$$\begin{pmatrix} W \\ W_c \end{pmatrix} \begin{pmatrix} W \\ W_c \end{pmatrix}^* = I = \begin{pmatrix} V \\ V_c \end{pmatrix} \begin{pmatrix} V \\ V_c \end{pmatrix}^* \implies \begin{pmatrix} W \\ W_c \end{pmatrix} = \begin{pmatrix} V \\ V_c \end{pmatrix} U, \quad U := \begin{pmatrix} V \\ V_c \end{pmatrix}^* \begin{pmatrix} W \\ W_c \end{pmatrix}.$$

Thus $W = VU$, where $U = V^*W + V_c^* W_c$ is an $n \times n$ unitary matrix. In other words:

Let $V = [v_1, \ldots, v_n]$ be a normalised tight frame for $\mathscr{H}$. Then all normalised tight frames of n vectors for $\mathscr{H}$ have the form $W = VU$, where U is an $n \times n$ unitary matrix.

S.F.D. Waldron, *An Introduction to Finite Tight Frames*, Applied and Numerical Harmonic Analysis, https://doi.org/10.1007/978-0-8176-4815-2_7

7.1 The real algebraic variety of normalised tight frames

Two useful descriptions of the normalised tight frames $V = [v_1,\ldots,v_n] \in \mathbb{F}^{d\times n}$ are the *orthogonality of rows* of V and the *variational characterisation* (see §6), i.e., the systems of equations

$$VV^* = I,$$

$$\sum_{j=1}^{n}\sum_{k=1}^{n} |\langle v_j, v_k\rangle|^2 = \frac{1}{d}\sum_{j=1}^{n}\langle v_j, v_j\rangle^2, \qquad \sum_{j=1}^{n}\|v_j\|^2 = d. \tag{7.1}$$

Since these involve the entries of V and their complex conjugates, it follows that:

The normalised tight frames of n vectors in $\mathbb{F}^d$ are a *real algebraic variety*, denoted by $\mathcal{N}_{n,\mathbb{F}^d}$. For $\mathbb{F} = \mathbb{R}$ it is in dn variables (the entries of V), and for $\mathbb{F} = \mathbb{C}$ it is in $2dn$ variables (the real and imaginary parts of the entries of V).

Since the action of right multiplication by the real Lie group $U(\mathbb{F}^n)$ on $\mathcal{N}_{n,\mathbb{F}^d}$ is transitive, it follows that $\mathcal{N}_{n,\mathbb{F}^d}$ is *irreducible* and *smooth*. Since the stabiliser of $V = [I,0]$ under this action is all $n\times n$ unitary matrices of the form

$$\begin{pmatrix} I & 0 \\ 0 & U \end{pmatrix}, \qquad U \in U(\mathbb{F}^{n-d}),$$

the dimension of this variety is $\dim(U(\mathbb{F}^n)) - \dim(U(\mathbb{F}^{n-d}))$, which gives

$$\dim(\mathcal{N}_{n,\mathbb{R}^d}) = \frac{1}{2}d(2n-d-1), \qquad \dim(\mathcal{N}_{n,\mathbb{C}^d}) = d(2n-d).$$

We have observed that the normalised tight frames of n vectors for $\mathbb{F}^d$ can be indexed by the $n\times n$ unitary matrices (for $n > d$ this is not 1–1). In particular, by choosing V to be the normalised tight frame given by the standard orthonormal basis (and zero vectors), we see (Exer. 7.1) that all normalised tight frames have the form

$$W = [w_1,\ldots,w_n] = [I,0]U = U_1, \quad U = \begin{pmatrix} U_1 \\ U_2 \end{pmatrix} \in U(\mathbb{F}^n), \tag{7.2}$$

i.e., the $d\times n$ submatrices of the $n\times n$ unitary matrices. If $n = d$, i.e., $U_2 = 0$, then the normalised tight frames are given by the unitary matrices (orthonormal bases), i.e., $\mathcal{N}_{d,\mathbb{F}^d} = U(\mathbb{F}^d)$. For $n > d$, a row of the submatrix U_2 can be multiplied by a scalar, so that U has determinant 1. Hence, for $n > d$ the normalised tight frames W can be indexed by elements of the special unitary group $SU(\mathbb{F}^n)$. Since $SU(\mathbb{F}^n)$ and $U(\mathbb{C}^n)$ are path-connected, and $SU(\mathbb{R}^n)$ is not, it follows that:

The set of normalised tight frames $\mathcal{N}_{n,\mathbb{F}^d}$ is *path-connected*, except for when $n = d$ and $\mathbb{F} = \mathbb{R}$.

7.2 The algebraic variety of equal-norm tight frames

We can define the subvariety $\mathscr{F}_{n,\mathbb{F}^d}$ of $\mathscr{N}_{n,\mathbb{F}^d}$ consisting of all equal-norm tight frames of n vectors in $\mathbb{F}^d$ by replacing the norm equation in (7.1), by the n equations

$$\|v_j\|^2 = \frac{d}{n}, \qquad j = 1,\ldots,n.$$

For $n = d$ (equal-norm orthogonal bases) these spaces are the same, and for $n > d$ the dimension reduces by $n-1$ (see [CMS13]), i.e.,

$$\dim(\mathscr{F}_{n,\mathbb{R}^d}) = (n-\frac{d}{2}-1)(d-1), \quad \dim(\mathscr{F}_{n,\mathbb{C}^d}) = d(n-d)+n(d-1)+1, \quad n>d.$$

Example 7.1 The dimensions of the algebraic varieties of equal-norm real tight frames of n vectors for $\mathbb{R}^2$ and $\mathbb{R}^3$ are

$$\dim(\mathscr{F}_{2,\mathbb{R}^2}) = 1, \qquad \dim(\mathscr{F}_{n,\mathbb{R}^2}) = n-2, \quad n>2,$$

$$\dim(\mathscr{F}_{3,\mathbb{R}^3}) = 3, \qquad \dim(\mathscr{F}_{n,\mathbb{R}^3}) = 2n-5, \quad n>3.$$

For $\mathbb{R}^2$ the points on the algebraic variety are determined by the first $n-2$ diagram vectors (see Exer. 2.9), since these are equal-norm complex numbers (w_j) with $\{w_{n-1}, w_n\}$ determined by $w_{n-1}+w_n = -(w_1+\cdots+w_{n-2})$. For $\mathbb{C}^2$ and $\mathbb{C}^3$, we have

$$\dim(\mathscr{F}_{2,\mathbb{R}^2}) = 4, \qquad \dim(\mathscr{F}_{n,\mathbb{R}^2}) = 3n-3, \quad n>2,$$

$$\dim(\mathscr{F}_{3,\mathbb{R}^3}) = 9, \qquad \dim(\mathscr{F}_{n,\mathbb{R}^3}) = 5n-8, \quad n>3.$$

The variety $\mathscr{F}_{n,\mathbb{F}^d}$ has been studied extensively (see [DS06], [Str11], [CMS13]). Classical results of [Whi57] imply that $\mathscr{F}_{n,\mathbb{F}^d}$ is a union of finitely many manifolds, and for n and d relatively prime $\mathscr{F}_{n,\mathbb{F}^d}$ is a manifold [DS06]. We now consider its connectivity (the *frame homotopy problem*).

For a $V \in \mathscr{F}_{n,\mathbb{F}^d}$, its Gramian $P = V^*V$ is an $n\times n$ orthogonal projection matrix with diagonal entries d/n (and hence rank d). Let $\mathscr{G}_{n,\mathbb{F}^d}$ be the set of such orthogonal projections (the *Grassmann manifold* of d-planes in $\mathbb{F}^n$). Since every $P \in \mathscr{G}_{n,\mathbb{F}^d}$ is the Gramian of some equal-norm tight frame for $\mathbb{F}^d$, it follows that:

- The map $\mathscr{F}_{n,\mathbb{F}^d} \to \mathscr{G}_{n,\mathbb{F}^d} : V \mapsto V^*V$ is onto and preserves path-connectivity.
- For $d \geq 2$, the map $\mathscr{G}_{n,\mathbb{F}^d} \to \mathscr{G}_{n,\mathbb{F}^{n-d}} : P \mapsto I-P$ (taking the Gramian to the Gramian of the complementary equal-norm tight frame) is a homeomorphism.

Since tight frames are defined up to unitary equivalence by their Gramians (Corollary 2.1), $\mathscr{G}_{n,\mathbb{F}^d}$ can also be described as the orbit space $\mathscr{F}_{n,\mathbb{F}^d}/U(\mathbb{F}^d)$ for the action of $U(\mathbb{F}^d)$ on $\mathscr{F}_{n,\mathbb{F}^d}$ given by right multiplication. In [DS06] it is shown that $\mathscr{G}_{4,\mathbb{R}^2}$ is homeomorphic to a graph with 12 vertices and 24 edges, and $\mathscr{G}_{5,\mathbb{R}^2}$ is the the orientable surface of genus 25.

Example 7.2 For $n = d$, the space $\mathscr{G}_{d,\mathbb{F}^d}$ has a single point (the identity matrix) and $\mathscr{F}_{d,\mathbb{F}^d} = U(\mathbb{F}^d)$. Thus, $\mathscr{F}_{d,\mathbb{C}^d}$ is path-connected and $\mathscr{F}_{d,\mathbb{R}^d}$ is not.

Example 7.3 For $n = d+1$ (and $d = 1$), we can calculate $\mathscr{G}_{d+1,\mathbb{F}^d} = I - \mathscr{G}_{d+1,\mathbb{F}^1}$. The variety $\mathscr{F}_{n,\mathbb{F}^1}$ consists of all V of the form

$$V = \begin{pmatrix} v_{11} & v_{12} & \cdots & v_{1,n} \end{pmatrix}, \qquad |v_{1j}| = \tfrac{1}{n}, \quad \forall j.$$

Therefore $\mathscr{F}_{n,\mathbb{R}^1}$ consists of 2^n isolated points, and so is not path-connected. The space $\mathscr{G}_{n,\mathbb{R}^1}$ has 2^{n-1} isolated points (and so is not path-connected for $n = d+1$). Hence $\mathscr{F}_{d+1,\mathbb{R}^d}$ and $\mathscr{F}_{n,\mathbb{R}^1}$ are not path-connected.

For $V, W \in \mathscr{F}_{n,\mathbb{C}^1}$, an explicit path from V to W in $\mathscr{F}_{n,\mathbb{C}^1}$ is given by

$$\gamma : [0,1] \to \mathscr{F}_{n,\mathbb{C}^1} : t \mapsto \left(v_{11} \left(\tfrac{w_{11}}{v_{11}}\right)^t \; \ldots \; v_{1,d+1} \left(\tfrac{w_{1n}}{v_{1n}}\right)^t \right).$$

Thus $\mathscr{F}_{n,\mathbb{C}^1}$ and $\mathscr{F}_{d+1,\mathbb{C}^d}$ (by taking complements) are path-connected.

It turns out that the only examples of path-disconnectedness are those above.

Theorem 7.1 *([CMS13]) The variety of equal-norm tight frames satisfies:*

1. *$\mathscr{F}_{n,\mathbb{R}^d}$ is path-connected if and only if $n \geq d+2$ and $d \geq 2$.*
2. *$\mathscr{F}_{n,\mathbb{C}^d}$ is path-connected.*

We will only give an indication of the proof given in [CMS13], which is involved. For a $V = [v_1, \ldots, v_n] \in \mathscr{F}_{n,\mathbb{R}^d}$ let $\lambda_1^{(k)} \leq \cdots \leq \lambda_d^{(k)}$ be the eigenvalues of the k-th partial sum of the frame operator $v_1 v_1^* + \cdots + v_k v_k^*$. These satisfy

(i) $\lambda_j^{(0)} = 0, \forall j$.
(ii) $\lambda_j^{(n)} = 1, \forall j$.
(iii) $\lambda_j^{(k+1)} \leq \lambda_{j+1}^{(k)} \leq \lambda_{j+2}^{(k+1)}, \lambda_j^{(k)} \leq \lambda_{j+1}^{(k+1)} \leq \lambda_{j+2}^{(k)}, 1 \leq j \leq d-2, 1 \leq k \leq n-1$.
(iv) $\sum_{j=1}^n \lambda_j^{(k)} + \frac{d}{n} = \sum_{j=1}^n \lambda_j^{(k+1)}, 1 \leq k \leq n-1$.

The sequences $\lambda = (\lambda_j^{(k)})_{1 \leq j \leq d, 0 \leq k \leq n}$ satisfying the above conditions are called *eigensteps* for $\mathscr{F}_{n,\mathbb{F}^d}$. The set $\Lambda_{n,d}$ of all eigensteps is a convex polytope, and hence it and its interior $\operatorname{int}(\Lambda_{n,d})$ are path-connected. It is shown

- The map $\mathscr{F}_{n,\mathbb{F}^d} \to \Lambda_{n,d} : V \mapsto \lambda = \lambda_V$ of a frame to its eigensteps is onto.
- For any frame $V \in \mathscr{F}_{n,\mathbb{F}^d}$ with eigensteps $\Lambda(V) \in \operatorname{int}(\Lambda_{n,d})$, there is a continuous map $\theta = \theta_V : \operatorname{int}(\Lambda_{n,d}) \to \mathscr{F}_{n,\mathbb{F}^d}$ with $\theta(\Lambda(V)) = V$ and $\Lambda\theta = I$ on $\operatorname{int}(\Lambda_{n,d})$.
- For $n > d$, $\mathscr{F}_{n,\mathbb{F}^d}$ is path-connected if and only if $\mathscr{F}_{n,\mathbb{F}^{n-d}}$ is, and this extends to the subsets of *nonorthodecomposible* frames (those that can't be partitioned into two orthogonal subsets).

These results are used to *lift* paths in $\Lambda_{n,d}$ to paths in $\mathscr{F}_{n,\mathbb{F}^d}$. The case $\mathbb{F} = \mathbb{C}$ exploits the fact that $U(\mathbb{C}^d)$ is connected. The case $\mathbb{F} = \mathbb{R}$ is more technical and is proved using induction on n and d, and some special cases.

7.3 The density of rational tight frames

We have seen that the normalised tight frames $V \in \mathscr{N}_{n,\mathbb{F}^d}$ of n vectors for $\mathbb{C}^d$ can be indexed by the $n \times n$ unitary matrices, e.g., (7.2) gives

$$V = [I_d, 0]U, \qquad U \in U(\mathbb{F}^n), \tag{7.3}$$

where the index U is not unique for $n > d$. Here we use (7.3) to describe points on the variety $\mathscr{N}_{n,\mathbb{F}^d}$. There are various parametrisations of the unitary group $U(\mathbb{F}^n)$, e.g., factorising its elements into Givens rotations or Householder transformations. We now consider the description in terms of the Cayley transform. This allows us to show that the rational points are dense on the $\mathscr{N}_{n,\mathbb{F}^d}$ (but not on $\mathscr{F}_{n,\mathbb{F}^d}$). Let $A \in \mathbb{C}^{n\times n}$ be a *skew Hermitian* matrix, i.e., $A^* = -A$. Then $A+I$ is invertible, and

$$U := \frac{I-A}{I+A} \tag{7.4}$$

is a unitary matrix, called the **Cayley transform** of A. If U is unitary and does not have -1 as an eigenvalue (so that $I+U$ is invertible), then

$$A := \frac{I-U}{I+U} \tag{7.5}$$

is a skew Hermitian matrix. These maps are the inverses of each other, and so

The unitary matrices (without eigenvalue -1) can be parametrised by the skew Hermitian matrices.

Cayley's original presentation (1846) was in the real case, where the Cayley transform (restricted to real matrices) maps into $SO(\mathbb{R}^d)$.

A complex number $x+iy$ is a (Gaussian) rational if $x, y \in \mathbb{Q}$. We now show:

The rational points are dense in the variety $\mathscr{N}_{n,\mathbb{F}^d}$ of normalised tight frames.

Theorem 7.2 *([CFW15]) Every tight frame $V = [v_1, \ldots, v_n]$ for $\mathbb{C}^d$ or $\mathbb{R}^d$ can be approximated arbitrary closely by one with vectors in $(\mathbb{Q}+i\mathbb{Q})^d$ or $\mathbb{Q}^d$, respectively.*

Proof Suppose, without loss of generality, that V is normalised. For $n > d$, we can choose an index $U \in U(\mathbb{F}^n)$ in (7.3) for which U does not have eigenvalue -1 (by multiplying the last row of U by a suitable scalar in $\mathbb{F}$). Thus, for $n > d$, the *truncated Cayley transform*

$$V = [I_d, 0]\frac{I-A}{I+A} \tag{7.6}$$

maps the skew Hermitian matrices onto $\mathscr{N}_{n,\mathbb{F}^d}$. Thus the normalised tight frames $V \in \mathscr{N}_{n,\mathbb{F}^d}$ can be parametrised by the entries which determine the skew symmetric matrices A, i.e., the $\frac{1}{2}n(n-1)$ strictly upper triangular entries and n purely imaginary diagonal entries (for real matrices this reduces to $\frac{1}{2}n(n-1)$ real parameters).

Taking the truncated Cayley transform of such a parametrised matrix A gives a $V \in \mathcal{N}_{n,\mathbb{F}^d}$ with entries in the same field as the parameters. Thus we can approximate the parameters as closely as desired by elements in $\mathbb{Q}+i\mathbb{Q}$ (which is dense in $\mathbb{C}$) or $\mathbb{Q}^d$, and the truncated Cayley transform of the skew Hermitian matrix given by these approximate parameters will approximate V as closely as desired.

For $n=d$, scale the last row of V to obtain a $U \in U(\mathbb{F}^d)$, which does not have -1 as an eigenvalue, approximate as above, and then unscale the row. This can be done since the rational points on the unit circle are dense (this classical result is the special case $d=1$ and $n=1$, incidently). □

Example 7.4 (Three vectors in $\mathbb{R}^2$) The 3×3 skew symmetric matrices A have three real parameters

$$A = \begin{pmatrix} 0 & a & b \\ -a & 0 & c \\ -b & -c & 0 \end{pmatrix}, \qquad a,b,c \in \mathbb{R}.$$

The Cayley transform is the orthogonal matrix

$$U = \frac{I-A}{I+A} = \begin{bmatrix} \frac{1-a^2-b^2+c^2}{1+a^2+b^2+c^2} & \frac{-2(a+bc)}{1+a^2+b^2+c^2} & \frac{2(ac-b)}{1+a^2+b^2+c^2} \\ \frac{2(a-bc)}{1+a^2+b^2+c^2} & \frac{1-a^2+b^2-c^2}{1+a^2+b^2+c^2} & \frac{-2(c+ab)}{1+a^2+b^2+c^2} \\ \frac{2(ac+b)}{1+a^2+b^2+c^2} & \frac{-2(ab-c)}{1+a^2+b^2+c^2} & \frac{1+a^2-b^2-c^2}{1+a^2+b^2+c^2} \end{bmatrix}.$$

The truncated Cayley transform gives the following indexing of $V \in \mathcal{N}_{3,\mathbb{R}^d}$

$$V = \begin{bmatrix} \frac{1-a^2-b^2+c^2}{1+a^2+b^2+c^2} & \frac{-2(a+bc)}{1+a^2+b^2+c^2} & \frac{2(ac-b)}{1+a^2+b^2+c^2} \\ \frac{2(a-bc)}{1+a^2+b^2+c^2} & \frac{1-a^2+b^2-c^2}{1+a^2+b^2+c^2} & \frac{-2(c+ab)}{1+a^2+b^2+c^2} \end{bmatrix}, \qquad a,b,c \in \mathbb{R}.$$

The normalised tight frame of three equally spaced equal-norm vectors is given by

$$V = \sqrt{\frac{2}{3}} \begin{pmatrix} 1 & -\frac{1}{2} & -\frac{1}{2} \\ 0 & \frac{\sqrt{3}}{2} & -\frac{\sqrt{3}}{2} \end{pmatrix}, \qquad \begin{array}{l} a = -2-\sqrt{3}+\sqrt{2}\sqrt{3}+\sqrt{2}, \\ b=\sqrt{3}-\sqrt{2}, \quad c=\sqrt{2}-1. \end{array}$$

We can approximate these parameters to 5 decimal places by rationals

$$\tilde{a} \approx \frac{13165}{100000}, \qquad \tilde{b} \approx \frac{31784}{100000}, \qquad \tilde{c} \approx \frac{41421}{100000}.$$

The corresponding truncated Cayley transform

$$\tilde{V} = \begin{pmatrix} \frac{5266079680}{6449619561} & -\frac{2633025064}{6449619561} & -\frac{2633092535}{6449619561} \\ -\frac{25064}{6449619561} & \frac{4560603095}{6449619561} & -\frac{4560536360}{6449619561} \end{pmatrix},$$

gives a *normalised tight frame* which approximates the equally spaced vectors of V to 5 decimal places. It is *not* an equal-norm frame. Theorem 7.2 does not imply that there is a dense set of rational points in $\mathcal{F}_{n,\mathbb{F}^d}$ (there are none on $\mathcal{F}_{3,\mathbb{R}^2}$).

7.4 The existence of tight frames with given norms

If $[v_1,\ldots,v_n]$ is a normalised tight frame for $\mathcal{H} = \mathbb{F}^d$, then its norms must satisfy

$$\|v_1\|, \|v_2\|, \ldots, \|v_n\| \leq 1, \qquad \|v_1\|^2 + \|v_2\|^2 + \cdots + \|v_n\|^2 = d, \tag{7.7}$$

since it is the projection of an orthonormal basis (Theorem 2.2) and by (2.9).

It is natural to ask whether there exists a normalised tight frame with some given norms which satisfy this condition, e.g., an equal-norm frame. We will show there is always such a frame. First we show the existence of such a tight frame follows from the Schur–Horn majorisation theorem, and then give a simple constructive proof.

A vector $\beta \in \mathbb{R}^n$ is said to **majorise** a vector $\alpha \in \mathbb{R}^n$ if after reordering so that their entries are increasing, one has

$$\alpha_1 + \alpha_2 + \cdots + \alpha_k \leq \beta_1 + \beta_2 + \cdots + \beta_k, \quad 1 \leq k \leq n, \qquad \alpha_1 + \cdots + \alpha_n = \beta_1 + \cdots + \beta_n.$$

Theorem 7.3 *(Schur–Horn) Let $a \in \mathbb{R}^n$ be a vector which majorises $\lambda \in \mathbb{R}^n$, then there is positive semidefinite real $n \times n$ matrix A with diagonal a and eigenvalues λ.*

Example 7.5 Let $a \in \mathbb{R}^n$ be a vector with

$$0 \leq a_1, a_2, \ldots, a_n \leq 1, \qquad a_1 + a_2 + \cdots + a_n = d,$$

and $\lambda = e_{n-d+1} + \cdots + e_n = (0,\ldots,0,1,\ldots,1)$ ($n-d$ zeros and d ones). Then a majorises λ, by the calculation

$$\sum_{j=1}^{k} a_j \geq \sum_{j=1}^{k} \lambda_j = 0, \quad 1 \leq k \leq n-d,$$

$$\sum_{j=1}^{k} a_j = d - \sum_{j=k+1}^{n} a_j \geq d - (n-k) = d - n + k = \sum_{j=1}^{k} \lambda_j, \quad n-d+1 \leq k \leq n.$$

Thus there exists a semidefinite matrix A with eigenvalues λ, i.e., an orthogonal projection of rank d and diagonal a. This matrix is the Gramian of a normalised tight frame $V = [v_1,\ldots,v_n]$ for $\mathbb{R}^d$ with $\|v_j\|^2 = a_j$, $1 \leq j \leq n$.

Thus, we have:

Corollary 7.1 *(Existence) There is a normalised tight frame $V = [v_1,\ldots,v_n]$ for $\mathcal{H} = \mathbb{F}^d$ with norms $\|v_j\|^2 = a_j$, $1 \leq j \leq n$, if and only if*

$$0 \leq a_1, a_2, \ldots, a_n \leq 1, \qquad a_1 + a_2 + \cdots + a_n = d. \tag{7.8}$$

Example 7.6 (Equal-norm tight frames). By taking $a_j = \frac{d}{n}$, $1 \leq j \leq n$, we conclude that an equal-norm tight frame of $n \geq d$ vectors for $\mathbb{F}^d$ exists (for every n and d), i.e., $\mathscr{F}_{n,\mathbb{F}^d}$ is a nontrivial variety. Explicit constructions are given by Example 2.4, or, more generally, group frames (see §10).

7.5 The construction of tight frames with given norms

All normalised tight frames are the orbit of any given one $V = [v_1, \ldots, v_n] \in \mathscr{N}_{n,\mathbb{F}^d}$ under the action right multiplication by the $n \times n$ unitary matrices. It is therefore natural to try and move along the variety $\mathscr{N}_{n,\mathbb{F}^d}$ by using unitary matrices U that only make small (hence controlled) changes. Here we consider U of the form $U_{\alpha\beta} = \delta_{\alpha\beta}$, $\alpha, \beta \notin \{j,k\}$, i.e.,

$$U = \begin{pmatrix} 1 & & & & & \\ & \ddots & & & & \\ & & u_{jj} & & u_{jk} & \\ & & & \ddots & & \\ & & u_{kj} & & u_{kk} & \\ & & & & & \ddots & \\ & & & & & & 1 \end{pmatrix} \begin{matrix} \\ \\ \leftarrow j \\ \\ \leftarrow k \\ \\ \\ \end{matrix} \tag{7.9}$$

These have the effect of fixing all vectors of $[v_1, \ldots, v_n]$ except for v_j and v_k, which transform to

$$v_j' = av_j + bv_k, \qquad v_k' = -e^{i\theta}(\overline{b}v_j - \overline{a}v_k), \tag{7.10}$$

where $VU = [v_1', \ldots, v_n']$ and

$$U|_{\{j,k\}^2} = \begin{pmatrix} u_{jj} & u_{jk} \\ u_{kj} & u_{kk} \end{pmatrix} = \begin{pmatrix} a & -e^{i\theta}\overline{b} \\ b & e^{i\theta}\overline{a} \end{pmatrix}, \qquad |a|^2 + |b|^2 + 1, \quad \theta \in \mathbb{R}. \tag{7.11}$$

Since $(VU)^*(VU) = V^*(U^*U)V = V^*V$, we have the U *conservation of norms*

$$\|v_1'\|^2 + \|v_2'\|^2 + \cdots + \|v_n'\|^2 = \|v_1\|^2 + \|v_2\|^2 + \cdots + \|v_n\|^2,$$

which for matrices of the form (7.9) gives

$$\|v_j'\|^2 + \|v_k'\|^2 = \|v_j\|^2 + \|v_k\|^2. \tag{7.12}$$

We now investigate precisely what norms the v_j' (and hence v_k') of (7.10) can take.

Lemma 7.1 *Let $v, w \in \mathbb{F}^d$. Then for $a, b \in \mathbb{F}$ with $|a|^2 + |b|^2 = 1$, we have*

$$\Big| \|av + bw\|^2 - \frac{1}{2}(\|v\|^2 + \|w\|^2) \Big| \le \frac{1}{2}\sqrt{(\|v\|^2 - \|w\|^2)^2 + 4|\langle v, w\rangle|^2}, \tag{7.13}$$

where the maximum and minimum of $\|av + bw\|$ over $|a|^2 + |b|^2 = 1$ give equality.

Proof Without loss of generality, we suppose that

$$a = t, \quad b = \sigma\sqrt{1 - t^2}, \qquad 0 \le t \le 1, \quad \sigma \in \mathbb{F}, \quad |\sigma| = 1.$$

It therefore suffices to find the maxima and minima of $f = f_\sigma : [0,1] \to \mathbb{R}$ given by

$$f(t) := \|tv + \sigma\sqrt{1-t^2}w\|^2 = t^2\|v\|^2 + (1-t^2)\|w\|^2 + 2t\sqrt{1-t^2}\alpha,$$

where $\alpha = \alpha_\sigma := \Re(\overline{\sigma}\langle v, w\rangle)$, and then to optimise these values over $|\sigma| = 1$. These calculations (see Exer. 7.2) give the result. □

Example 7.7 The interval (7.13) for the norms $\|av + bw\|$, $|a|^2 + |b|^2 = 1$ is smallest when v and w are orthogonal, i.e., $\langle v, w\rangle = 0$, which gives

$$\min\{\|v\|, \|w\|\} \le \|av + bw\| \le \max\{\|v\|, \|w\|\}, \tag{7.14}$$

and largest when v and w are linearly dependent, i.e., $|\langle v, w\rangle| = \|v\|\|w\|$, which gives

$$0 \le \|av + bw\|^2 \le \|v\|^2 + \|w\|^2. \tag{7.15}$$

Without any knowledge of $\langle v, w\rangle$, it is only possible to construct vectors in the interval (7.14). This is sufficient for our purposes.

Lemma 7.2 *If $V = [v_1, \ldots, v_n]$ is a normalised tight frame for $\mathbb{F}^d$, and*

$$\min\{\|v_j\|, \|v_k\|\} \le r \le \max\{\|v_j\|, \|v_k\|\}, \tag{7.16}$$

then there exists a real $n \times n$ unitary matrix U of the form (7.9) with

$$\|v_j'\| = r, \qquad \|v_k'\| = \sqrt{\|v_j\|^2 + \|v_k\|^2 - r^2},$$

where $[v_1', \ldots, v_n'] := VU$.

Proof By Lemma 7.1 there are $a, b \in \mathbb{F}$ with $\|av + bw\| = r$, $|a|^2 + |b|^2 = 1$. Since

$$g(t) := \|tv_j \pm \sqrt{1-t^2}v_k\|$$

is a continuous function of t with $g(0) = \|v_k\|$, $g(1) = \|v_j\|$, we can choose such $a, b \in \mathbb{R}$ and determine them by solving $g(t) = r$. A unitary matrix U with the desired properties is then given by (7.11). □

Example 7.8 One can take the U above to be a *Givens rotation*, by the choice

$$a = \cos\psi, \quad b = -\sin\psi, \quad e^{i\theta} = -1,$$

or to be a *Householder transformation* $I - 2ww^T$, by the choice

$$w_j = \sqrt{\frac{1-a}{2}}, \qquad w_k = \begin{cases} -\frac{b}{\sqrt{2(1-a)}}, & a \ne 1; \\ 1, & a = 1 \end{cases} \qquad w_\ell = 0, \quad \ell \ne j, k.$$

We now use Lemma 7.2 to give a constructive proof of Corollary 7.1, i.e., to construct a normalised tight frame $V = [v_1, \ldots, v_n]$ with given norms $\|v_j\|^2 = a_j$, $1 \le j \le n$, where $a \in \mathbb{R}^n$ satisfies (7.8). Since $\|v_j'\|$ is an average of $\|v_j\|$ and $\|v_k\|$, one must take some care.

We now describe the algorithm of [FWW06], which starts with a normalised tight frame with large intervals (7.16) and moves it closer to one with the desired norms. Another algorithm based on the factorisation of elements of $U(\mathbb{R}^d)$ into Givens rotations was given earlier by [Cas04], [CL06].

Algorithm (for constructing a tight frame with given norms): Suppose that

$$1 \ge a_1 \ge a_2 \ge \cdots \ge a_n \ge 0, \qquad a_1 + a_2 + \cdots + a_n = d.$$

Starting with $V^{(0)}$, we construct iterates $V^{(k)} = [v_1^{(k)}, \ldots, v_n^{(k)}] \in \mathscr{N}_{n,\mathbb{F}^d}$ with

(i) $\|v_j^{(k)}\|^2 = a_j$, $1 \le j \le k$.
(ii) $\|v_j^{(k)}\|^2 \ge a_{k+1}$ or $\|v_j^{(k)}\| = 0$, $k+2 \le j \le n$.

In view of (7.12), $V^{(n-1)}$ has norms $\|v_j\|^2 = a_j$, $\forall j$, and so gives the desired frame.

- Let $V^{(0)} = [B, 0]$ with B unitary, e.g., $V^{(0)} = [e_1, \ldots, e_d, 0, \ldots, 0]$.
- Suppose that $V^{(k)}$, $0 \le k \le n-2$, has been constructed. There are three cases:

1. If $\|v_{k+1}^{(k)}\|^2 = a_{k+1}$, then we can take $V^{(k+1)} = V^{(k)}$.
2. If $\|v_{k+1}^{(k)}\|^2 < a_{k+1}$, then $a_{k+1} \le \|v_{j_0}^{(k)}\|^2$, for some $k+2 \le j_0 \le n$.
3. If $\|v_{k+1}^{(k)}\|^2 > a_{k+1}$, then $a_{k+1} \ge \|v_{j_0}^{(k)}\|^2$ or $\|v_{j_0}^{(k)}\|^2 = 0$, for some $k+2 \le j_0 \le n$.

(see Exer. 7.3) Interchange the vectors $v_{k+2}^{(k)}$ and $v_{j_0}^{(k)}$ in the cases 2 and 3.

- Since a_{k+1} is in the interval (7.16) given by $v_{k+1}^{(k)}$ and $v_{k+2}^{(k)}$, there is a unitary matrix U of the form (7.9) for which $V^{(k+1)} := V^{(k)}U$ satisfies

$$v_j^{(k+1)} = v_j^{(k)}, \quad j \ne k+1, k+2, \qquad \|v_{k+1}^{(k+1)}\|^2 = a_{k+1}.$$

  Since $a_{k+1} \ge a_{k+2}$, it follows that $V^{(k+1)} \in \mathscr{N}_{n,\mathbb{F}^d}$ satisfies properties (i) and (ii).

Example 7.9 Here we construct an equal-norm tight frame of three vectors for $\mathbb{R}^2$. We start with $V^{(0)} = [e_1, e_2, 0]$. Since the first two columns are orthonormal, every vector $av_1^{(0)} + bv_2^{(0)}$, $a^2 + b^2 = 1$, has norm 1 (the interval of Lemma 7.1 is a single point). Thus we take a linear combination of the first and third columns given by a Givens rotation

$$V^{(0)}U = \begin{pmatrix} 1 & 0 & 0 \\ 0 & 1 & 0 \end{pmatrix} \begin{pmatrix} \cos\theta & 0 & \sin\theta \\ 0 & 1 & 0 \end{pmatrix}, \qquad U := \begin{pmatrix} \cos\theta & 0 & \sin\theta \\ 0 & 1 & 0 \\ -\sin\theta & 0 & \cos\theta \end{pmatrix}.$$

Clearly the first column can take any norm between 0 and 1. We want it to be $\sqrt{2/3}$, and so take $\cos\theta = \sqrt{2/3}$, $\sin\theta = \sqrt{1/3}$, to obtain

$$V^{(1)} = \begin{pmatrix} \sqrt{\frac{2}{3}} & 0 & \sqrt{\frac{1}{3}} \\ 0 & 1 & 0 \end{pmatrix}.$$

Applying a rotation with fixes the first column (which has the desired norm) gives

$$V^{(1)}U = \begin{pmatrix} \sqrt{\frac{2}{3}} & -\sqrt{\frac{1}{3}}\sin\psi & \sqrt{\frac{1}{3}}\cos\psi \\ 0 & \cos\psi & \sin\psi \end{pmatrix}, \qquad U := \begin{pmatrix} 1 & 0 & 0 \\ 0 & \cos\psi & \sin\psi \\ 0 & -\sin\psi & \cos\psi \end{pmatrix}.$$

The squared norm of the second column above can take any value between $\frac{1}{3}$ and 1. Choosing $\cos\psi = -1/\sqrt{2}$, $\sin\psi = 1/\sqrt{2}$ gives the equal-norm tight frame

$$V^{(2)} = \sqrt{\frac{2}{3}} \begin{pmatrix} 1 & -\frac{1}{2} & -\frac{1}{2} \\ 0 & -\frac{\sqrt{3}}{2} & \frac{\sqrt{3}}{2} \end{pmatrix}.$$

The algorithm presented moves between certain elements of the variety $\mathscr{N}_{n,\mathbb{F}^d}$, i.e., those satisfying (i) and (ii), by using Lemma 7.2. The next example indicates how one can move on the variety by using the more general Lemma 7.1.

Example 7.10 Consider the tight frame of three equally spaced unit vectors

$$V = [v_1, v_2, v_3] = \begin{pmatrix} 1 & -\frac{1}{2} & -\frac{1}{2} \\ 0 & \frac{\sqrt{3}}{2} & -\frac{\sqrt{3}}{2} \end{pmatrix}.$$

We will transform this to a tight frame with a two orthogonal vectors and the zero vector, by right multiplication by unitary matrices. Since the vectors v_j have unit norm, Lemma 7.2 cannot be applied. Since $|\langle v_j, v_k\rangle| = \frac{1}{2}$, $j \neq k$, by Lemma 7.1 we can choose $a^2+b^2 = 1$ so that $\|av_j + bv_k\|^2$ is as small as $\frac{1}{2}$ and as large as $\frac{3}{2}$. The choice

$$V^{(1)} := VU = \begin{pmatrix} 1 & 0 & -\frac{1}{\sqrt{2}} \\ 0 & -\frac{\sqrt{3}}{\sqrt{2}} & 0 \end{pmatrix}, \qquad U = \begin{pmatrix} 1 & 0 & 0 \\ 0 & \frac{1}{\sqrt{2}} & \frac{1}{\sqrt{2}} \\ 0 & -\frac{1}{\sqrt{2}} & \frac{1}{\sqrt{2}} \end{pmatrix},$$

gives squared norms $1, \frac{3}{2}, \frac{1}{2}$ and $|\langle v_1^{(1)}, v_3^{(1)}\rangle| = \frac{1}{\sqrt{2}}$. Thus, $0 \le \|av_1^{(1)} + bv_3^{(1)}\|^2 \le \frac{3}{2}$, we obtain

$$V^{(2)} = V^{(1)}U = \begin{pmatrix} \frac{\sqrt{3}}{\sqrt{2}} & 0 & 0 \\ 0 & -\frac{\sqrt{3}}{\sqrt{2}} & 0 \end{pmatrix}, \qquad U = \begin{pmatrix} \frac{\sqrt{2}}{\sqrt{3}} & 0 & \frac{1}{\sqrt{3}} \\ 0 & 1 & 0 \\ -\frac{1}{\sqrt{3}} & 0 & \frac{\sqrt{2}}{\sqrt{3}} \end{pmatrix}.$$

Since every unitary matrix is a product of matrices of the form (7.9), it is not difficult to imagine that one could move from any element of $\mathcal{N}_{n,\mathbb{F}^d}$ to any other by right multiplying by a finite sequence of such matrices.

Notes

The existence of equal-norm tight frames was not widely known until recently. This question was raised at Bommerholz in September 2000 (see [BF03]) and was "settled" in various ways: retrospectively [GVT98], by explicit constructions [Zim01], [RW02] and by minimisation of the frame potential [BF03]. The connection with results such as the Schur–Horn majorisation theorem are now well known, and there are sophisticated algorithms [Str12], [CFM12], [CFM+13], [FMP16] for moving over the varieties $\mathcal{N}_{n,\mathbb{F}^d}$ and $\mathcal{F}_{n,\mathbb{F}^d}$. The corresponding algebraic varieties of spherical (t,t)-designs for $t \neq 1$ (see §6.9) are far less studied.

Exercises

7.1 Let $U(\mathbb{F}^n)$, the real Lie group of $n \times n$ unitary matrices over $\mathbb{F}$, act on $\mathcal{N}_{n,\mathbb{F}^d}$ (the normalised tight frames of n vectors for $\mathbb{F}^d$) via right multiplication.
(a) Show that the stabiliser of $V = [I, 0] \in \mathcal{N}_{n,\mathbb{F}^d}$ is

$$\mathrm{Stab}(V) = \{ \begin{pmatrix} I & 0 \\ 0 & U \end{pmatrix} : U \in U(\mathbb{F}^{n-d} \}.$$

(b) Since the action is irreducible, it follows that $\mathcal{N}_{n,\mathbb{F}^d}$ is isomorphic to $U(\mathbb{F}^n)/\mathrm{Stab}(V)$. Use this to calculate its dimension.
Hint: $\dim(U(\mathbb{C}^n)) = n^2$, $\dim(U(\mathbb{R}^n)) = \frac{1}{2}n(n-1)$.

7.2 For $v, w \in \mathbb{F}^d$ and $\sigma \in \mathbb{F}$, $|\sigma| = 1$, define $f = f_\sigma : [0,1] \to \mathbb{R}$ by

$$f(t) := \|tv + \sigma\sqrt{1-t^2}w\|^2 = t^2\|v\|^2 + (1-t^2)\|w\|^2 + 2t\sqrt{1-t^2}\alpha,$$

where $\alpha = \alpha_\sigma := \Re(\overline{\sigma}\langle v, w\rangle)$.

(a) For a fixed σ, find a possible local maximum and minimum of $f = f_\sigma$ over $[0,1]$.
(b) Optimise the possible local maximum and minimum from (a) over all $|\sigma| = 1$.

(c) By considering the end points $t = 0, 1$ find the maximum and minimum of $f_\sigma(t)$ over t and σ.

7.3 Suppose that $1 \geq a_1 \geq a_2 \geq \cdots \geq a_n \geq 0$, $a_1 + a_2 + \cdots + a_n = d$, and there is a normalised tight frame $V^{(k)} = [v_1^{(k)}, \ldots, v_n^{(k)}] \in \mathcal{N}_{n,\mathbb{F}^d}$ with

(i) $\|v_j^{(k)}\|^2 = a_j$, $1 \leq j \leq k$.
(ii) $\|v_j^{(k)}\|^2 \geq a_{k+1}$ or $\|v_j^{(k)}\| = 0$, $k+2 \leq j \leq n$.

Show that

(a) If $\|v_{k+1}^{(k)}\|^2 < a_{k+1}$, then $a_{k+1} \leq \|v_{j_0}^{(k)}\|^2$, for some $k+2 \leq j_0 \leq n$.
(b) If $\|v_{k+1}^{(k)}\|^2 > a_{k+1}$, then $a_{k+1} \geq \|v_{j_0}^{(k)}\|^2$ or $\|v_{j_0}^{(k)}\|^2 = 0$, for some $k+2 \leq j_0 \leq n$.

Chapter 8
Projective unitary equivalence and fusion frames

Two finite sequences of vectors $\Phi = (v_j)$ and $\Psi = (w_j)$ in inner product spaces are unitarily equivalent if and only if their respective inner products (Gramian matrices) are equal (Corollary 2.1, §3.4). For *projective unitary equivalence*, i.e.,

$$v_j = \alpha_j U w_j, \qquad \forall j,$$

where $|\alpha_j| = 1$, $\forall j$, and U is unitary, the inner products are not projective unitary invariants, since

$$\langle v_j, v_k \rangle = \langle \alpha_j U w_j, \alpha_k U w_k \rangle = \alpha_j \overline{\alpha_k} \langle w_j, w_k \rangle.$$

Obvious projective invariants are

$$\langle v_j, v_j \rangle = \|v_j\|^2,$$
$$\langle v_j, v_k \rangle \langle v_j, v_k \rangle = |\langle v_j, v_k \rangle|^2,$$

but these don't characterise projective unitary equivalence, unless (v_j) is orthogonal. A projective unitary invariant is given by

$$\langle v_j, v_k \rangle \langle v_k, v_\ell \rangle \langle v_\ell, v_j \rangle.$$

These "triple products" do characterise projective unitary equivalence when none of the inner products $\langle v_j, v_k \rangle$ are zero, e.g., for equiangular frames, but not in general.

Here we show that finite sequences of vectors (lines) in inner product spaces are projectively unitarily equivalent if and only if certain projective unitary invariants (called m-products) are equal (Theorem 8.1). This is proved by giving an algorithm to recover a sequence of vectors (up to projective unitary equivalence) from a small subset of these projective invariants, which are determined by a spanning tree for the "frame graph". We also extend our results to the projective similarity of vectors in $\mathbb{F}$-vector spaces (where $\overline{\mathbb{F}} = \mathbb{F}$).

S.F.D. Waldron, *An Introduction to Finite Tight Frames*, Applied and Numerical Harmonic Analysis, https://doi.org/10.1007/978-0-8176-4815-2_8

8.1 Projective unitary equivalence

As in §2.3 and §3.4, finite sequences of vectors $\Phi = (v_j)$ and $\Psi = (w_j)$ in (real or complex) inner product spaces $\mathscr{H}_1$ and $\mathscr{H}_2$ are **projectively unitarily equivalent** if there is a unitary map $U : \mathscr{H}_1 \to \mathscr{H}_2$ and unit scalars α_j, such that

$$w_j = \alpha_j U v_j, \qquad \forall j, \tag{8.1}$$

or, equivalently,

$$w_j w_j^* = U(v_j v_j^*)U^*, \qquad \forall j.$$

The study of lines in $\mathbb{R}^d$ and $\mathbb{C}^d$, in particular, equiangular lines (Chapter 12), is effectively the study of configurations of unit vectors up to projective unitary equivalence. Many applications, such as signal analysis, depend only on frames up to projective unitary equivalence. For example, if the vectors are multiplied by unit modulus scalars, say $w_j = \alpha_j v_j$, then the frame operator is unchanged, and so the frame expansion is essentially unchanged, i.e.,

$$f = \sum_j \langle f, \tilde{w}_j \rangle w_j = \sum_j \langle f, \alpha_j \tilde{v}_j \rangle \alpha_j v_j = \sum_j \langle f, \tilde{v}_j \rangle v_j, \qquad \forall f.$$

The condition (8.1) can be written as $w_j = U(\alpha_j v_j)$, i.e., (w_j) and $(\alpha_j v_j)$ are unitarily equivalent, which by the Gramian condition (see §3.4) is equivalent to

$$\langle w_k, w_j \rangle = \alpha_k \overline{\alpha_j} \langle v_k, v_j \rangle, \qquad \forall j,k. \tag{8.2}$$

Equivalently, in terms of the Gramian:

Frames Φ and Ψ are projectively unitarily equivalent if and only if

$$\mathrm{Gram}(\Psi) = C^* \mathrm{Gram}(\Phi) C, \tag{8.3}$$

where C is the diagonal matrix with diagonal entries α_j.

This is a practicable method for determining projective unitary equivalence only when the inner product space is real. In this case $\alpha_j = \pm 1$, and so there are only finitely many possible matrices C.

In view of (8.2), the inner products between vectors are *not* projective unitary invariants (in general). However, some products of them are, e.g.,

$$\begin{aligned}\langle w_j, w_k \rangle \langle w_k, w_\ell \rangle \langle w_\ell, w_j \rangle &= \langle \alpha_j U v_j, \alpha_k U v_k \rangle \langle \alpha_k U v_k, \alpha_\ell U v_\ell \rangle \langle \alpha_\ell U v_\ell, \alpha_j U v_j \rangle \\ &= \alpha_j \overline{\alpha_k} \langle U v_j, U v_k \rangle \alpha_k \overline{\alpha_\ell} \langle U v_k, U v_\ell \rangle \alpha_\ell \overline{\alpha_j} \langle U v_\ell, U v_j \rangle \\ &= \langle v_j, v_k \rangle \langle v_k, v_\ell \rangle \langle v_\ell, v_j \rangle. \end{aligned} \tag{8.4}$$

This projective unitary invariant generalises in the obvious way.

8.2 The m-products

Definition 8.1. Let $\Phi = (v_j)$ be a sequence of n vectors (in a Hilbert space). Then, the m**-products** (or the m**-vertex Bargmann invariants**) of Φ are

$$\Delta(v_{j_1}, v_{j_2}, \ldots, v_{j_m}) := \langle v_{j_1}, v_{j_2} \rangle \langle v_{j_2}, v_{j_3} \rangle \cdots \langle v_{j_m}, v_{j_1} \rangle, \quad 1 \leq j_1, \ldots, j_m \leq n. \tag{8.5}$$

The 3-products will also be called **triple products**.

We observe that there are only *finitely* many m-products, and by the argument of (8.4), we have

The m-products of Φ are projective unitary invariants, i.e., if Φ and Ψ are projectively unitarily equivalent, then they have the same m-products.

The main result of this chapter (Theorem 8.1) is the converse of this, i.e., that if Φ and Ψ have the same m-products then they are projectively unitarily equivalent.

Example 8.1. The 1-products, 2-products and 3-products of $\Phi = (v_j)$ are

$$\begin{aligned} \Delta(v_j) &= \langle v_j, v_j \rangle = \|v_j\|^2, \\ \Delta(v_j, v_k) &= \langle v_j, v_k \rangle \langle v_j, v_k \rangle = |\langle v_j, v_k \rangle|^2, \\ \Delta(v_j, v_k, v_\ell) &= \langle v_j, v_k \rangle \langle v_k, v_\ell \rangle \langle v_\ell, v_j \rangle. \end{aligned}$$

The 1-products and 2-products can be deduced from the 3-products, since

$$\Delta(v_j, v_j, v_j) = \langle v_j, v_j \rangle^3, \qquad \Delta(v_j, v_j, v_k) = \langle v_j, v_j \rangle |\langle v_j, v_k \rangle|^2. \tag{8.6}$$

Example 8.2. (Conjugation) The m-products are closed under complex conjugation, i.e.,

$$\overline{\Delta(v_{j_1}, v_{j_2}, \ldots, v_{j_m})} = \Delta(v_{j_m}, \ldots, v_{j_2}, v_{j_1}). \tag{8.7}$$

Example 8.3. (Decompositions) We observe that in some cases m-products can be decomposed into products of smaller ones, e.g.,

$$\Delta(v_1, v_2, \ldots, v_n) = \frac{\Delta(v_1, v_2, \ldots, v_{n-1}) \Delta(v_1, v_{n-1}, v_n)}{\Delta(v_1, v_{n-1})}, \tag{8.8}$$

provided $\Delta(v_1, v_{n-1}) \neq 0$ and $n \geq 2$.

We now define the notion of a "generating set" for the m-products.

Definition 8.2. A subset of the m-products of Φ (or the corresponding indices/cycles) is a **determining set** if all the m-products can be determined from them.

Example 8.4. If all the inner products between the vectors in Φ are nonzero, then (8.8) implies that the triple products are determining set for the m-products of Φ.

8.3 The frame graph

Here we give examples which show that orthogonality (zero inner products) between vectors in a frame (v_j) affect which m-products determine it up to projective unitary equivalence. This motivates the following definition (see [Str11], [AN13]).

Definition 8.3. The **frame graph** (or **correlation network**) of a sequence of vectors (v_j) is the graph with vertices $\{v_j\}$ (or the indices j themselves) and

$$\text{an edge between}\, v_j \,\text{and}\, v_k,\, j \neq k \quad \iff \quad \langle v_j, v_k \rangle \neq 0.$$

Clearly, the frame graph is determined by the 2-products, and so projectively unitarily equivalent frames have the same frame graph.

- Edges in the frame graph correspond to inner products which are nonzero.
- m-cycles in the frame graph correspond to m-products which are nonzero.

Example 8.5. (Empty graph) The frame graph of (v_j) is empty (edgeless) if and only if all the inner products between different vectors are zero, i.e., the nonzero vectors are orthogonal. In this case, the Gramian is diagonal, and so by (8.3) all projectively unitarily equivalent frames have the same Gramian, i.e., the frame (v_j) is determined up to projective unitary equivalence by its 2-products.

Example 8.6. (Complete graph) We will see (Example 8.11) that if a frame has a complete frame graph, then it is determined up to projective unitary equivalence by its triple products.

Example 8.7. (n-cycle) Let (e_j) be the standard basis vectors in $\mathbb{C}^n$. Fix $|z| = 1$, and let

$$v_j := \begin{cases} e_j + e_{j+1}, & 1 \leq j < n, \\ e_n + z e_1, & j = n. \end{cases}$$

Then, the frame graph of (v_j) is the n-cycle $(v_1, \ldots, v_n)$, and so the only nonzero m-products for distinct vectors are

$$\Delta(v_j) = \|v_j\|^2 = 2, \qquad 1 \leq j \leq n, \tag{8.9}$$

$$\Delta(v_j, v_{j+1}) = |\langle v_j, v_{j+1} \rangle|^2 = 1, \quad 1 \leq j < n, \tag{8.10}$$

$$\Delta(v_1, v_2, \ldots, v_n) = z, \tag{8.11}$$

and their complex conjugates. Therefore, different choices for z give projectively unitarily inequivalent frames. Thus, for $n > 3$, the vectors (v_j) are not defined up to projective unitary equivalence by their triple products.

Example 8.8. (Connected components) The vectors in a connected component of the frame graph are orthogonal to all the other vectors of the frame, and hence a frame is union (see §5.1) of the frames given by the vertices of each connected component.

8.4 Characterisation of projective unitary equivalence

We now show that a sequence of n vectors is determined up to projective unitary equivalence by its m-products for $1 \leq m \leq n$ (or a determining set).

This is done by using the m-products of $\Phi = (v_j)$ to construct all the possible Gramians $G = [\langle w_k, w_j \rangle]$ given by sequences of vectors (w_j) which are projectively unitarily equivalent to Φ (and so have the same m-products as Φ).

We motivate the proof with an example. Let $\Phi = (v_j)$ be the frame for $\mathbb{R}^3$ given

$$\Phi = \left(\begin{pmatrix} 1 \\ 0 \\ 0 \end{pmatrix}, \begin{pmatrix} -\frac{1}{2} \\ \frac{\sqrt{3}}{2} \\ 0 \end{pmatrix}, \begin{pmatrix} -\frac{1}{2} \\ -\frac{\sqrt{3}}{2} \\ 0 \end{pmatrix}, \begin{pmatrix} 0 \\ 0 \\ 1 \end{pmatrix} \right),$$

i.e., three equally spaced unit vectors in a 2-dimensional subspace together with a unit vector orthogonal to them all. The frame graph of Φ has an edge between each pair of the points v_1, v_2, v_3 and v_4 as an isolated point. By (8.3), all the possible Gramians of frames $\Psi = (w_j)$ which are projectively equivalent to Φ are given by

$$G = [\langle w_k, w_j \rangle] = C\,\mathrm{Gram}(\Phi)C^* = \begin{pmatrix} 1 & -\frac{1}{2}\alpha_1\overline{\alpha_2} & -\frac{1}{2}\alpha_1\overline{\alpha_3} & 0 \\ -\frac{1}{2}\alpha_2\overline{\alpha_1} & 1 & -\frac{1}{2}\alpha_2\overline{\alpha_3} & 0 \\ -\frac{1}{2}\alpha_3\overline{\alpha_1} & -\frac{1}{2}\alpha_3\overline{\alpha_2} & 1 & 0 \\ 0 & 0 & 0 & 1 \end{pmatrix}, \qquad |\alpha_j| = 1.$$

Here we assume that only the m-products of Φ are known, so above we only know the modulus of the inner products between vectors in Φ (this happens to be $\frac{1}{2}$ for vectors which are not orthogonal). Clearly, any of the nonzero inner products (which correspond to edges) is a free variable of the form $\langle w_j, w_k \rangle = |\langle v_j, v_k \rangle| a$, $|a| = 1$. We suppose that $\langle w_1, w_2 \rangle = \frac{1}{2}a$, $|a| = 1$ (we could choose any edge). Effectively, we have fixed α_1 and α_2 (without knowing $\langle v_1, v_2 \rangle$). We now consider an edge from one of the points v_1, v_2 (which have already been scaled) to an unscaled point, say v_2 to v_3. Since v_3 has not been scaled, we can choose α_3, so that $\langle w_2, w_3 \rangle = \frac{1}{2}b$, $|b| = 1$, is a second free variable. We now have a spanning tree for the connected component of the frame graph which involves the vertices v_1, v_2, v_3. The remaining edge (from v_3 to v_1) no longer corresponds to a free variable ($\alpha_1, \alpha_2, \alpha_3$ have been fixed). This edge is in a cycle (v_1, v_2, v_3), where the inner products corresponding to the other edges of the cycle are free variables (i.e., belong to the spanning tree), and so the inner product $\langle w_3, w_1 \rangle$ given by this edge is determined by the m-product given by the cycle

$$\begin{aligned} \Delta(w_1, w_2, w_3) = \Delta(v_1, v_2, v_3) \quad &\Longrightarrow \quad \left(\frac{1}{2}a\right)\left(\frac{1}{2}b\right)\left(\langle w_3, w_1 \rangle\right) = \left(-\frac{1}{2}\right)^3 \\ &\Longrightarrow \quad \langle w_1, w_3 \rangle = -\frac{1}{2}ab. \end{aligned}$$

This example illustrates the main points of the proof below:

The edges in a spanning tree for a connected component of the frame graph of Φ correspond to inner products which can be taken as free variables. Once these are chosen (with the appropriate moduli), then there is a unique choice for the others (determined by the m-products of Φ) which gives the Gramian of a frame which is projectively unitarily equivalent to Φ.

We recall the following facts:

- Every finite graph Γ has a spanning tree (forest) $\mathscr{T}$.
- For each edge $e \in \Gamma \setminus \mathscr{T}$, there is a unique cycle in $e \cup \mathscr{T}$ called the **fundamental cycle** (corresponding to e).

Theorem 8.1. *(Characterisation) Sequences $\Phi = (v_j)$ and $\Psi = (w_j)$ of n vectors are projectively unitarily equivalent if and only if their m-products are equal, i.e.,*

$$\Delta(v_{j_1}, v_{j_2}, \ldots, v_{j_m}) = \Delta(w_{j_1}, w_{j_2}, \ldots, w_{j_m}), \qquad 1 \leq j_1, \ldots, j_m \leq n, \quad 1 \leq m \leq n.$$

Proof. We have already observed that projectively unitarily equivalent sequences have the same m-products. We therefore suppose that Φ and Ψ have the same m-products, and will show that we can choose $\alpha_1, \ldots, \alpha_n$ so that (8.2) holds. The Gramians of Φ and Ψ are block diagonal (with entries having the same moduli), with blocks given by the vertices of the connected components of the common frame graph. We therefore assume without loss of generality that there is a single block, i.e., the frame graph Γ is connected.

Spanning tree argument. Find a spanning tree $\mathscr{T}$ of Γ with root vertex r. By working outwards from the root r, we can multiply the vertices $v \in \Gamma \setminus \{r\}$ by unit scalars α_v so that for an edge $\{v_j, v_k\} \in \mathscr{T}$, (8.2) holds, i.e.,

$$\langle w_k, w_j \rangle = \alpha_k \overline{\alpha_j} \langle v_k, v_j \rangle.$$

In this way, we can choose $\alpha_1, \ldots, \alpha_n$ so that (8.2) holds for all edges $\{v_j, v_k\} \in \mathscr{T}$.

Completing cycles. It remains only to show that (8.2) also holds for all edges $e = \{v_j, v_k\} \in \Gamma \setminus \mathscr{T}$. Let $(v_j, v_k, v_{\ell_1}, \ldots, v_{\ell_r})$ be the fundamental cycle given by the edge $e = \{v_j, v_k\}$. Since the m-products are equal, and the other edges in this cycle belong to $\mathscr{T}$, we obtain

$$\begin{aligned}
\Delta(w_j, w_k, w_{\ell_1}, \ldots, w_{\ell_r}) &= \langle w_j, w_k \rangle \langle w_k, w_{\ell_1} \rangle \langle w_{\ell_1}, w_{\ell_2} \rangle \cdots \langle w_{\ell_r}, w_j \rangle \\
&= \langle w_j, w_k \rangle \alpha_k \overline{\alpha_{\ell_1}} \langle v_k, v_{\ell_1} \rangle \alpha_{\ell_1} \overline{\alpha_{\ell_2}} \langle v_{\ell_1}, v_{\ell_2} \rangle \cdots \alpha_{\ell_r} \overline{\alpha_j} \langle v_{\ell_r}, v_j \rangle \\
&= (\alpha_k \overline{\alpha_j} \langle w_j, w_k \rangle) \langle v_k, v_{\ell_1} \rangle \langle v_{\ell_1}, v_{\ell_2} \rangle \cdots \langle v_{\ell_r}, v_j \rangle \\
&= \langle v_j, v_k \rangle \langle v_k, v_{\ell_1} \rangle \langle v_{\ell_1}, v_{\ell_2} \rangle \cdots \langle v_{\ell_r}, v_j \rangle \\
&= \Delta(v_j, v_k, v_{\ell_1}, \ldots, v_{\ell_r}),
\end{aligned}$$

and cancellation gives (8.2) for the edge $\{v_j, v_k\} \in \Gamma \setminus \mathscr{T}$. □

Above we associated the (directed) m-cycle $(v_{j_1},\ldots,v_{j_m})$ in the frame graph with the nonzero m-product $\Delta(v_{j_1},\ldots,v_{j_m})$. For $m \geq 3$, all m-products can be calculated from those corresponding to simple cycles, since if a cycle crosses at a, we have

$$\Delta(v_1,\ldots,v_s,a,w_1,\ldots,w_t,a) = \Delta(v_1,\ldots,v_s,a)\Delta(w_1,\ldots,w_t,a). \tag{8.12}$$

The **cycle space** of a finite graph Γ is the set of its **Eulerian subgraphs** (those with vertices of even degree). This can be viewed as a $\mathbb{Z}_2$-vector space, where the addition is the symmetric difference of sets. From this, it follows that the cycle space is spanned by the simple cycles (and its elements are disjoint unions of cycles). If the sum of two simple cycles is a simple cycle (in the frame graph), then corresponding m-product can be determined from those of the summands

$$\begin{aligned}&\Delta(v_1,\ldots,v_s,e_1,\ldots,e_r)\Delta(w_1,\ldots,w_t,e_r,\ldots,e_1)\\&\quad=\Delta(e_1,e_2)\cdots\Delta(e_{r-1},e_r)\Delta(v_1,\ldots,v_s,e_1,w_1,\ldots,w_t,e_r).\end{aligned}$$

Combining these observations, we have:

A determining set for the m-products of Φ is given by the 2-products and the m-products corresponding to a basis for the cycle space of the frame graph.

The fundamental cycles corresponding to a spanning tree (forest) of a finite graph form a basis for the cycle space called a **fundamental cycle basis**. We therefore have the following strengthening of Theorem 8.1.

Corollary 8.1. *A finite frame Φ, with frame graph Γ, is determined up to projective unitary equivalence by a determining set for the m-products, e.g.,*

1. *The 2-products.*
2. *The m-products, $3 \leq m \leq n$, corresponding to a fundamental cycle basis (for the cycle space of Γ) formed from a spanning tree (forest) $\mathscr{T}$ for Γ.*

In particular, if M is the number of edges of $\Gamma \setminus \mathscr{T}$, then it is sufficient to know all of the 2-products, and M of the m-products, $3 \leq m \leq n$.

Proof. It suffices to verify the condition of Theorem 8.1 for a determining set. □

Example 8.9. Let $\Phi = (v_j)$ be four equiangular vectors with $C > 0$. The frame graph of Φ is complete, and $M = 6 - 3 = 3$. Spanning trees (see Figure 8.1) include

$$\begin{aligned}&\mathscr{T}_p := \text{the path } v_1,v_2,v_3,v_4,\\&\mathscr{T}_s := \text{the star graph with internal vertex } v_1 \text{ and leaves } v_2,v_3,v_4.\end{aligned}$$

For $\mathscr{T}_p$, the fundamental cycles given by the edges $\{v_1,v_4\}$, $\{v_1,v_3\}$, $\{v_2,v_4\}$ are

$$(v_1,v_2,v_3,v_4), \quad (v_1,v_2,v_3), \quad (v_2,v_3,v_4).$$

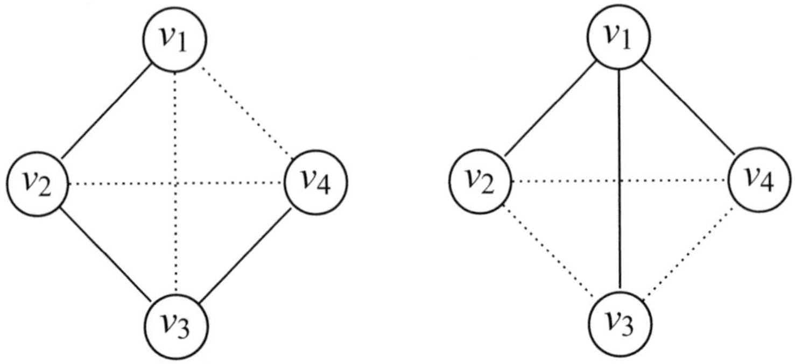

Fig. 8.1: The spanning trees $\mathscr{T}_p$ and $\mathscr{T}_s$ (and cycle completions) of Example 8.9.

For $\mathscr{T}_s$, the fundamental cycles given by the edges $\{v_2,v_3\}$, $\{v_2,v_4\}$, $\{v_3,v_4\}$ are

$$(v_1,v_2,v_3), \quad (v_1,v_2,v_4), \quad (v_1,v_3,v_4).$$

Thus Φ is determined up to projective unitary equivalence by its 2-products, and the either of the following sets of m-products

$$\Delta(v_1,v_2,v_3,v_4), \quad \Delta(v_1,v_2,v_3), \quad \Delta(v_2,v_3,v_4),$$

$$\Delta(v_1,v_2,v_3), \quad \Delta(v_1,v_2,v_4), \quad \Delta(v_1,v_3,v_4).$$

8.5 Reconstruction from the m-products

We now state the characterisation in way which summarises how all projectively unitarily frames can be constructed from a small determining set of m-products.

Theorem 8.2. *(Reconstruction) Suppose $\Phi = (v_j)$ is a frame of n vectors. Let Γ be the frame graph of Φ, $\mathscr{T}$ be a spanning tree (forest) for Γ and*

$$N = \text{the number of edges in } \mathscr{T},$$
$$M = \text{the number of edges in } \Gamma \setminus \mathscr{T}.$$

Then the collection of all Gramians $G = [\langle w_k, w_j\rangle]$ of frames $\Psi = (w_j)$ which are unitarily projectively equivalent to Φ can be parameterised by N free variables. More precisely, for each of the N edges $\{v_j, v_k\} \in \mathscr{T}$ (choose an order) we have a free variable

$$\langle w_j, w_k\rangle = |\langle v_j, v_k\rangle| a_{(j,k)}, \qquad |a_{(j,k)}| = 1,$$

and for the remaining M edges $e = \{v_j, v_k\} \in \Gamma \setminus \mathscr{T}$, $\langle w_j, w_k\rangle$ is uniquely determined by equality of the m-products of Φ and Ψ for the fundamental cycle given by e.

Example 8.10. Let $\Phi = (v_j)$ be the "two mutually unbiased bases" (see §8.6) for $\mathbb{C}^2$ given by

$$\Phi = \left(\begin{pmatrix} 1 \\ 0 \end{pmatrix}, \begin{pmatrix} 0 \\ 1 \end{pmatrix}, \begin{pmatrix} \frac{1}{\sqrt{2}} \\ \frac{1}{\sqrt{2}} \end{pmatrix}, \begin{pmatrix} \frac{1}{\sqrt{2}} \\ -\frac{1}{\sqrt{2}} \end{pmatrix} \right), \qquad \mathrm{Gram}(\Phi) = \begin{pmatrix} 1 & 0 & \frac{1}{\sqrt{2}} & \frac{1}{\sqrt{2}} \\ 0 & 1 & \frac{1}{\sqrt{2}} & -\frac{1}{\sqrt{2}} \\ \frac{1}{\sqrt{2}} & \frac{1}{\sqrt{2}} & 1 & 0 \\ \frac{1}{\sqrt{2}} & -\frac{1}{\sqrt{2}} & 0 & 1 \end{pmatrix}.$$

The frame graph Γ of Φ is the 4-cycle (v_1, v_3, v_2, v_4). A spanning tree $\mathcal{T}$ is given by the path v_1, v_3, v_2, v_4. Corresponding to the three edges of $\mathcal{T}$, we have three free variables

$$\langle w_1, w_3 \rangle = \frac{a}{\sqrt{2}}, \qquad \langle w_3, w_2 \rangle = \frac{\overline{b}}{\sqrt{2}}, \qquad \langle w_2, w_4 \rangle = \frac{c}{\sqrt{2}}.$$

The remaining inner product $\langle w_4, w_1 \rangle = \frac{1}{\sqrt{2}}\overline{z}$ is determined the fundamental cycle given by $\{v_1, v_4\}$, i.e., by completing the 4-cycle

$$\langle w_1, w_3 \rangle \langle w_3, w_2 \rangle \langle w_2, w_4 \rangle \langle w_4, w_1 \rangle = \langle v_1, v_3 \rangle \langle v_3, v_2 \rangle \langle v_2, v_4 \rangle \langle v_4, v_1 \rangle,$$

which gives

$$a\overline{b}c\overline{z} = -1.$$

Thus all the Gramians of Ψ which are projectively unitarily equivalent to Φ are given by

$$G = \begin{pmatrix} 1 & 0 & \frac{\overline{a}}{\sqrt{2}} & -\frac{\overline{a}b\overline{c}}{\sqrt{2}} \\ 0 & 1 & \frac{\overline{b}}{\sqrt{2}} & \frac{\overline{c}}{\sqrt{2}} \\ \frac{a}{\sqrt{2}} & \frac{b}{\sqrt{2}} & 1 & 0 \\ -\frac{a\overline{b}c}{\sqrt{2}} & \frac{c}{\sqrt{2}} & 0 & 1 \end{pmatrix}, \qquad |a| = |b| = |c| = 1.$$

This particular Φ is in fact determined up to projective unitary equivalence by just its 2-products. This is because Sylvester's criterion for G (as function of a, b, c, z) to be positive semidefinite gives

$$\det(G) = -\frac{1}{4}\frac{(bz+ac)^2}{abcz} = -\frac{1}{4}\left|\frac{bz}{ac}+1\right|^2 \geq 0 \implies \frac{bz}{ac}+1 = 0 \implies z = -\frac{ac}{b}.$$

In contrast, the (v_j) of Example 8.7 for $n = 4$ also has frame graph a 4-cycle, but it is not determined up to projective unitary equivalence by its 2-products (and triple products).

We now consider those frames which are determined up to projective unitary equivalence by their 2-products and triple products.

8.6 Triple products, equiangular lines, SICs and MUBs

The following special case of Corollary 8.1 is often useful.

Corollary 8.2. *(Triple products) A finite frame Φ is determined up to projective unitary equivalence by its triple products (3-products) if the cycle space of its frame graph is spanned by 3-cycles (and so the cycle space has a basis of 3-cycles).*

Proof. The 2-products can be deduced from the triple products by (8.6). □

Example 8.11. (Chordal graphs) A graph is said to be **chordal** (or **triangulated**) if each of its cycles of four or more vertices has a chord, and so the cycle space is spanned by the 3-cycles. Hence a frame is determined by its triple products if its frame graph is chordal. The extreme cases are the empty graph (orthogonal bases) where there are no cycles, and the complete graph where all subsets of three vectors lie on a 3-cycle (equiangular lines).

A set of equiangular lines given by a frame Φ is determined up to projective unitary equivalence by the triple products of Φ.

We now give an example (Corollary 8.3) where the cycle space of the frame graph has a basis of 3-cycles, but the frame graph is not chordal.

Definition 8.4. A family of orthonormal bases $\mathscr{B}_1, \mathscr{B}_2, \ldots, \mathscr{B}_k$ for $\mathbb{C}^d$ is said to be **mutually unbiased** if for $r \neq j$

$$|\langle v, w\rangle|^2 = \frac{1}{d}, \qquad v \in B_r, \quad w \in B_j.$$

We call $\mathscr{B}_1, \ldots, \mathscr{B}_k$ a sequence of k **MUBs** (**mutually unbiased bases**).

The maximal number of MUBs is a question of considerable interest (see §2.11). The frame graph of two or more MUBs ($d > 1$) is not chordal, because there is a 4-cycle (v_1, w_1, v_2, w_2), $v_1, v_2 \in \mathscr{B}_r$, $w_1, w_2 \in \mathscr{B}_s$ not containing a chord.

We now show for three or more MUBs the cycle space of the frame graph is spanned by the 3-cycles. This is not case for two MUBs (see Example 8.10).

Corollary 8.3. *(MUBs) A frame Φ consisting of three or more MUBs in $\mathbb{C}^d$, $d \geq 2$, is determined up to projective unitary equivalence by its triple products.*

Proof. It suffices to show that the cycle space of the frame graph Γ of Φ has a basis of 3-cycles. To this end, let $\mathscr{B}_j$, $j = 1, \ldots, k$, be the MUBs for $\mathbb{C}^d$, so that Γ is a complete k-partite graph (with partite sets $\mathscr{B}_j$). Fix $v_1 \in \mathscr{B}_1$ and $v_2 \in \mathscr{B}_2$. A spanning tree $\mathscr{T}$ for Γ is given by taking an edge from v_1 to each vertex of $\mathscr{B}_j$, $j \neq 1$, and an edge from v_2 to each vertex of $\mathscr{B}_1 \setminus v_1$. Each of the remaining edges of $\Gamma \setminus \mathscr{T}$ gives a fundamental cycle. These have two types (see Figure 8.2):

1. $\frac{1}{2}d^2(k-1)(k-2)$ edges between vertices in $\mathscr{B}_r$ and $\mathscr{B}_s$, $r,s \neq 1$, which give fundamental 3-cycles (involving v_1).
2. $(d-1)((k-1)d-1)$ edges between vertices $u \in \mathscr{B}_1 \setminus v_1$ and $w \in \cup_{j\neq 1}\mathscr{B}_j \setminus v_2$, which give fundamental 4-cycles (u,w,v_1,v_2). These can be written as a sum (symmetric difference) of the 3-cycles (u,w,v_2) and (v_1,v_2,w).

Thus the cycle space is spanned by 3-cycles. □

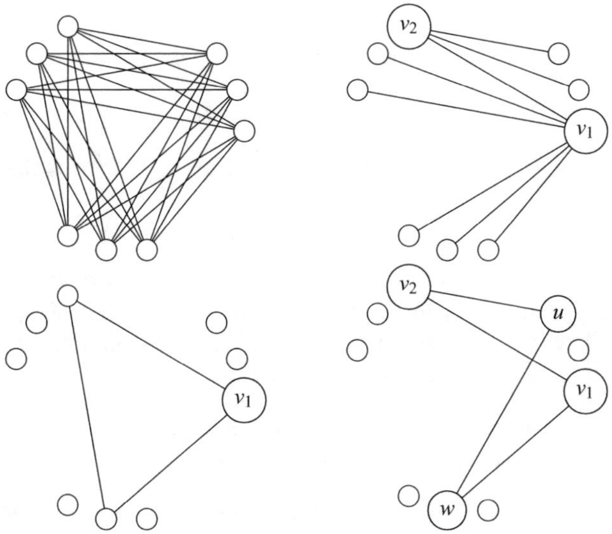

Fig. 8.2: Details from the proof of Corollary 8.3 for MUBs $\mathscr{B}_1, \mathscr{B}_2, \mathscr{B}_3$ in $\mathbb{C}^3$. The frame graph Γ, the spanning tree $\mathscr{T}$, and fundamental cycles of type 1 and 2.

There exist graphs which are not chordal, with every edge on a 3-cycle (as is the case for the frame graph of three or more MUBs), but for which the cycle space is not spanned by 3-cycles (see Figure 8.3).

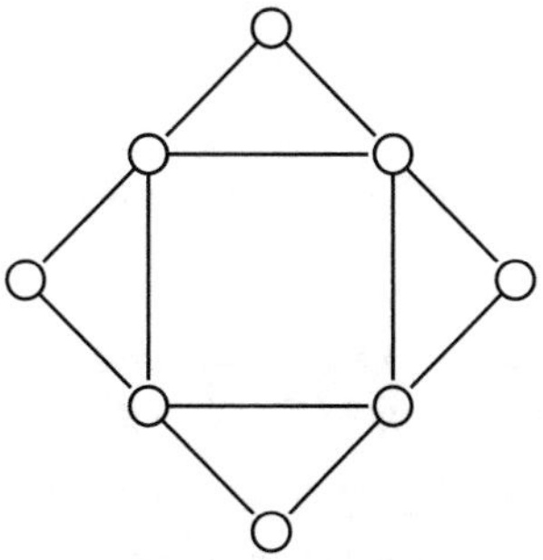

Fig. 8.3: A nonchordal graph for which each edge is on a 3-cycle.

8.7 Projective similarity and canonical m-products for vector spaces

We now use the previous results to characterise *projective similarity.*

Let $\Phi = (v_j)_{j\in J}$ and $\Psi = (w_j)_{j\in J}$ be finite sequences of vectors which span vector spaces X and Y over a subfield $\mathbb{F}$ of $\mathbb{C}$. We say that Φ and Ψ are **similar** if there is an invertible linear map $Q : X \to Y$ with

$$w_j = Qv_j, \qquad \forall j,$$

and are **projectively similar** if there unit scalars α_j and Q invertible, with

$$w_j = \alpha_j Q v_j, \qquad \forall j.$$

Assume that $\overline{\mathbb{F}} = \mathbb{F}$. Then the canonical Gramian $P_\Phi \in \mathbb{F}^{J\times J}$ is defined (see §), and

$$L_\Phi : X \to \operatorname{ran}(P_\Phi) : v_j \mapsto Pe_j$$

is an invertible linear map. Now $(P_\Phi e_j)$ and $(P_\Psi e_j)$ are projectively similar if and only if $(\alpha_j P_\Phi e_j)$ and $(P_\Psi e_j)$ are similar (for α_j as above). But $(\alpha_j P_\Phi e_j)$ and $(P_\Psi e_j)$ are normalised tight frames, and so are similar if and only if they are unitarily equivalent (see Exer. 2.5). Combining these observations gives:

$$\begin{aligned} &\Phi = (v_j) \text{ and } \Psi = (w_j) \text{ are projectively similiar} \\ \iff\quad &(P_\Phi e_j) \text{ and } (P_\Psi e_j) \text{ are projectively similiar} \\ \iff\quad &(P_\Phi e_j) \text{ and } (P_\Psi e_j) \text{ are projectively unitarily equivalent.} \end{aligned} \tag{8.13}$$

This motivates the definition:

Definition 8.5. Let $\Phi = (v_j)$ be a finite sequence of vectors in an $\mathbb{F}$-vector space, with $\overline{\mathbb{F}} = \mathbb{F}$. Then the **canonical m-products** of Φ are the m-products of $(P_\Phi e_j)$, which we denote by

$$\Delta_C(v_{j_1}, \ldots, v_{j_m}) := \Delta(P_\Phi e_{j_1}, \ldots, P_\Phi e_{j_m}) = p_{j_1 j_2} p_{j_2 j_3} \cdots p_{j_m j_1}, \tag{8.14}$$

where $P_\Phi = [p_{kj}]$.

These depend on Φ as well as $v_{j_1}, \ldots, v_{j_m}$, (unlike the usual m-products), and one could use notation such as $\Delta_C^\Phi = \Delta_C$ the emphasise this. In this way, we may apply Theorem 8.1.

Theorem 8.3. *(Characterisation) Let $\Phi = (v_j)$ and $\Psi = (w_j)$ be finite sequences of vectors in vector spaces over a subfield $\mathbb{F}$ of $\mathbb{C}$ with $\overline{\mathbb{F}} = \mathbb{F}$. Then*

1. *Φ and Ψ are similar if and only if $P_\Phi = P_\Psi$ (the canonical Gramians are equal).*
2. *Φ and Ψ are projectively similar if and only if their canonical m-products (for a determining set) are equal.*

Proof. The first follows from Proposition 4.1, and second from the observation (8.13) and Theorem 8.1. □

For projective similarity, one can calculate the α_j and Q in $w_j = \alpha_j Q v_j$ explicitly:

Corollary 8.4. *(Construction) Suppose that $\Phi = (v_j)$ and $\Psi = (w_j)$ are projectively similar, i.e., $w_j = \alpha_j Q v_j$, $\forall j$, and Γ is the frame graph of $(P_\Phi e_j)$. Then the unit scalars α_j are unique up to multiplication of those corresponding to a component of Γ by a unit scalar. All possible choices for (α_j) can be constructed as follows*

1. *Fix the α_j corresponding to the root(s) of a spanning tree (forest) for Γ.*
2. *Determine the remaining α_j by the applying the spanning tree argument to*

$$P_\Psi = C^* P_\Phi C, \qquad C = \mathrm{diag}(\alpha_j).$$

The invertible linear map Q is then defined by $Q(\alpha_j v_j) = w_j$, $\forall j$.

We illustrate this with a simple example (also see §9.7).

Example 8.12. Suppose that $\Phi = (v_j)$ and $\Psi = (w_j)$ span 2-dimensional spaces, say, for simplicity,

$$\begin{aligned} a_1 v_1 + a_2 v_2 + a_3 v_3 &= 0, \qquad |a_1|^2 + |a_2|^2 + |a_3|^2 = 1, \\ b_1 w_1 + b_2 w_2 + b_3 w_3 &= 0, \qquad |b_1|^2 + |b_2|^2 + |b_3|^2 = 1. \end{aligned} \tag{8.15}$$

For a general field $\mathbb{F}$, it may not be possible to normalise the vector $a = (a_1, a_2, a_3)^T$, which spans $\mathrm{dep}(\Phi)$, in which case one can modify the argument below. We have

$$P_\Phi = I - aa^* = \begin{pmatrix} 1-|a_1|^2 & -a_1\overline{a_2} & -a_1\overline{a_3} \\ -\overline{a_1}a_2 & 1-|a_2|^2 & -a_2\overline{a_3} \\ -\overline{a_1}a_3 & -\overline{a_2}a_3 & 1-|a_3|^2 \end{pmatrix}. \tag{8.16}$$

The canonical 2-products are uniquely determined by the $|a_j|$, since

$$\Delta_C(v_j, v_j) = (1-|a_j|^2)^2, \qquad \Delta_C(v_j, v_k) = |-a_j\overline{a_k}|^2 = |a_j|^2|a_k|^2, \ j \neq k,$$

as are the canonical 3-products corresponding to a 3-cycle

$$\Delta_C(v_j, v_k, v_\ell) = (-a_j\overline{a_k})(-a_k\overline{a_\ell})(-a_\ell\overline{a_j}) = -|a_j|^2|a_k|^2|a_\ell|^2.$$

Thus

1. Φ and Ψ are similar if and only if $a_j\overline{a_k} = b_j\overline{b_k}$, $\forall j,k$.
2. Φ and Ψ are projectively similar if and only if $|a_j| = |b_j|$, $\forall j$.

We now suppose that Φ and Ψ are projectively similar, i.e., $w_j = \alpha_j Q v_j$, $\forall j$, and calculate (α_j) and Q from

$$P_\Psi = C^* P_\Phi C = \begin{pmatrix} 1-|a_1|^2 & \overline{\alpha_1}\alpha_2(-a_1\overline{a_2}) & \overline{\alpha_1}\alpha_3(-a_1\overline{a_3}) \\ \overline{\alpha_2}\alpha_1(-\overline{a_1}a_2) & 1-|a_2|^2 & \overline{\alpha_2}\alpha_3(-a_2\overline{a_3}) \\ \overline{\alpha_3}\alpha_1(-\overline{a_1}a_3) & \overline{\alpha_3}\alpha_2(-\overline{a_2}a_3) & 1-|a_3|^2 \end{pmatrix}, \tag{8.17}$$

and the frame graph Γ of $(P_\Phi e_j)$. In view of (8.15) and (8.16), there only are three possibilities for Γ (up to graph isomorphism).

Γ is complete, i.e., $a_1, a_2, a_3 \neq 0$. A spanning tree for Γ is given by the path $P_\Phi e_1, P_\Phi e_2, P_\Phi e_3$. Fix α_1 (corresponding to the root) then α_2, α_3 are determined by the entries of (8.17) corresponding to the edges, i.e.,

$$\overline{\alpha_1}\alpha_2(-a_1\overline{a_2}) = -b_1\overline{b_2}, \qquad \overline{\alpha_2}\alpha_3(-a_2\overline{a_3}) = -b_2\overline{b_3}.$$

Solving these gives

$$\alpha_2 = \frac{b_1\overline{b_2}}{a_1\overline{a_2}}\alpha_1, \qquad \alpha_3 = \frac{b_2\overline{b_3}}{a_2\overline{a_3}}\alpha_2 = \frac{b_2\overline{b_3}}{a_2\overline{a_3}}\frac{b_1\overline{b_2}}{a_1\overline{a_2}}\alpha_1 = \frac{b_1\overline{b_3}}{a_1\overline{a_3}}\alpha_1.$$

Thus, with α_1 fixed, say $\alpha_1 = 1$, Q is given by

$$Q(v_1) = w_1, \qquad Q\Big(\frac{b_1\overline{b_2}}{a_1\overline{a_2}}v_2\Big) = w_2.$$

Suppose Γ is not complete, say $-a_2\overline{a_3} = 0$ with $a_3 = 0$, then (8.17) reduces to

$$P_\Psi = C^* P_\Phi C = \begin{pmatrix} 1-|a_1|^2 & \overline{\alpha_1}\alpha_2(-a_1\overline{a_2}) & 0 \\ \overline{\alpha_2}\alpha_1(-\overline{a_1}a_2) & 1-|a_2|^2 & 0 \\ 0 & 0 & 1-|a_3|^2 \end{pmatrix},$$

where one of a_1, a_2 must be nonzero, say $a_1 \neq 0$. This gives the remaining two cases.

Γ has one edge, i.e., $a_1, a_2 \neq 0$, $a_3 = 0$. The edge $P_\Phi e_1, P_\Phi e_2$ together with $P\Phi e_3$ is a spanning forest. Fix α_1, α_3 (corresponding to the roots). Then, α_2 is given by

$$\overline{\alpha_1}\alpha_2(-a_1\overline{a_2}) = -b_1\overline{b_2} \quad \Longrightarrow \quad \alpha_2 = \frac{b_1\overline{b_2}}{a_1\overline{a_2}}\alpha_1.$$

Thus, with α_1, α_3 fixed, say $\alpha_1, \alpha_3 = 1$, Q is given by

$$Q(v_1) = w_1, \qquad Q(v_3) = w_3.$$

Γ has no edges, i.e., $a_1 \neq 0$, $a_2, a_3 = 0$. The vertices $P_\Phi e_1, P_\Phi e_2, P_\Phi e_3$ are a spanning forest. We can make any choice for $\alpha_1, \alpha_2, \alpha_3$, and then Q is determined by $Q(\alpha_j v_j) = w_j$. Here $a_1 v_1 = 0$, so $v_1 = 0$, and so the $j = 1$ equation is vacuous. Thus, for $\alpha_2 = \alpha_3 = 1$, Q is given by

$$Q(v_2) = w_2, \qquad Q(v_3) = w_3.$$

8.8 Fusion frames

As observed in (2.6), the frame expansion for a finite tight frame (f_j) for $\mathscr{H}$ can be written

$$f = \frac{1}{A}\sum_j \langle f, f_j\rangle f_j = \sum_j c_j P_{W_j} f, \qquad \forall f \in \mathscr{H},$$

where $c_j = \frac{1}{A}\|f_j\|^2 \geq 0$, and P_{W_j} is the orthogonal projection onto $W_j := \mathrm{span}\{f_j\}$. This expansion is *projectively invariant*, i.e., it only depends on the vectors up to multiplication by unit modulus scalars. It can be generalised by letting the W_j be subspaces with any possible dimension, to obtain a *tight fusion frame*.

Let $(W_j)_{j\in J}$ be subspaces of $\mathscr{H}$ and $(c_j)_{j\in J}$ be nonnegative weights. Then the collection of pairs $\{(W_j, c_j)\}$ is a **tight fusion frame** for $\mathscr{H}$ if for some $A > 0$

$$f = \frac{1}{A}\sum_{j\in J} c_j P_{W_j} f, \qquad \forall f \in \mathscr{H}. \tag{8.18}$$

Many (projectively invariant) results for tight frames extend to tight fusion frames, e.g., taking the trace of the linear operators in (8.18) generalises (2.9) to

$$\sum_j c_j \dim(W_j) = dA, \qquad d := \dim(\mathscr{H}). \tag{8.19}$$

A tight fusion frame $(W_j), (c_j)$ for $\mathscr{H}$ is **normalised** if $A = 1$, i.e., $\sum_j c_j = \dim(\mathscr{H})$.

The finite normalised tight frames (f_j) for $\mathscr{H}$ (up to projective equivalence) are equivalent to the tight fusion frames $(W_j), (c_j)$ for $\mathscr{H}$ with

$$\dim(W_j) = 1, \quad c_j \neq 0, \qquad \sum_j c_j = \dim(\mathscr{H}),$$

via $c_j = \|f_j\|^2$, $W_j = \mathrm{span}\{f_j\}$.

Example 8.13. A trivial tight fusion frame is given by taking a single subspace, i.e., $W_1 = \mathscr{H}$, $c_1 = 1$.

Example 8.14. If $\{(W_j, c_j)\}$ is a normalised tight fusion frame for $\mathscr{H}$, then (8.18) and (8.19) give

$$\sum_j c_j P_{W_j} = I = I_{\mathscr{H}} \quad \Longrightarrow \quad \sum_j c_j (I - P_{W_j}) = (d-1)I \quad \Longrightarrow \quad \sum_j \frac{c_j}{d-1} P_{W_j^\perp} = I.$$

Thus $\{(W_j^\perp, \frac{c_j}{d-1})\}$ is a normalised tight fusion frame for $\mathscr{H}$, $d = \dim(\mathscr{H}) > 1$. When $\{(W_j, c_j)\}$ is a tight frame, this fusion frame expresses the identity as a weighted sum of orthogonal projections onto hyperplanes.

Example 8.15. There is no tight fusion frame for $\mathbb{C}^3$ consisting two 2-dimensional subspaces. If there were, then the construction of Example 8.14 would give a tight frame of two vectors for $\mathbb{C}^3$.

The study of nontight fusion frames (frames of subspaces) was initiated by [AK05], [CKL08]. It parallels theory of frames. We now give a few of the salient definitions and details. The **fusion frame operator** of the subspace and nonnegative weight pairs $\{(W_j,c_j)\}_{j\in J}$ is the positive semidefinite operator $S:\mathscr{H}\to\mathscr{H}$ given by

$$Sf := \sum_j c_j P_{W_j} f, \qquad \forall f\in\mathscr{H}.$$

The collection $\{(W_j,c_j)\}$ is a **fusion frame** for $\mathscr{H}$ if S is boundedly invertible, i.e.,

$$AI_{\mathscr{H}} \le S \le BI_{\mathscr{H}} \qquad \text{(Loewner order)},$$

which is equivalent to

$$A\|f\|^2 \le \langle Sf,f\rangle = \sum_j c_j \|P_{W_j}f\|^2 \le B\|f\|^2, \qquad \forall f\in\mathscr{H}.$$

To calculate the orthogonal projections P_{W_j} in the fusion frame operator S one can introduce a *local* frame $\mathscr{F}_j=(f_{jk})_{k\in K_j}$ for each W_j, so that (see Exer. 3.12)

$$P_{W_j}f = \sum_{k\in K_j}\langle f,\tilde f_{jk}\rangle f_{jk} = \sum_{k\in K_j}\langle f,f_{jk}\rangle \tilde f_{jk}, \qquad \forall f\in\mathscr{H},$$

where $\tilde f_{jk}=S_{\mathscr{F}_j}^{-1}f_{jk}$. The collection of triples $\{(W_j,c_j,\mathscr{F}_j)\}$ is known as a **fusion frame system**, and each $\mathscr{F}_j$ as a **local frame** (for W_j). The calculation of Sf can be *distributed* by using a fusion frame system, which is natural for sensor networks (see [CKL08]). Let $(f_{jk}^{\text{can}})_{k\in K_j}$, $f_{jk}^{\text{can}} := S_{\mathscr{F}_j}^{-1/2} f_{jk}$ be the canonical tight frame for a local frame $\mathscr{F}_j$ for W_j. Then the fusion frame operator can be written as

$$Sf = \sum_j c_j \sum_{k\in K_j}\langle f, f_{jk}^{\text{can}}\rangle f_{jk}^{\text{can}} = \sum_j\sum_{k\in K_j}\langle f,\sqrt{c_j}f_{jk}^{\text{can}}\rangle\sqrt{c_j}f_{jk}^{\text{can}}, \qquad \forall f\in\mathscr{H},$$

so that the fusion frame operator can be viewed as the frame operator of the frame $(f_{jk}^{\text{can}})_{j\in J,k\in K_j}$ for $\mathscr{H}$, which has frame bounds A and B as above. Thus $S^{-1}f$ can be calculated using the *frame algorithm* (see §3.9) and the distributed calculation (parallel processing) of Sf outlined above. The *fusion frame expansion* is

$$f = S^{-1}Sf = \sum_j c_j S^{-1}P_{W_j}f = \sum_j c_j \sum_{k\in K_j}\langle f,\tilde f_{jk}\rangle S^{-1}f_{jk}, \qquad \forall f\in\mathscr{H}.$$

The frame operator of a frame can similarly be calculated by a parallel algorithm.

8.9 Signed frames

The tight frame expansion (8.8) can be generalised by letting the scalars $c_j \geq 0$ take (possibly negative) real values. We call such a configuration:

$$f = \sum_j \sigma_j \langle f, f_j \rangle f_j = \sum_j c_j P_{W_j} f, \qquad \forall f \in \mathscr{H}, \tag{8.20}$$

$$\sigma_j \in \{\pm 1\}, \quad f_j \in \mathscr{H}, \qquad c_j := \sigma_j \|f_j\|^2, \quad W_j := \mathrm{span}\{f_j\}, \tag{8.21}$$

a **tight signed frame** for $\mathscr{H}$ with **signature** $\sigma = (\sigma_j)$. By the polarisation identity, the condition (8.20) is equivalent to

$$\|f\|^2 = \sum_j \sigma_j |\langle f, f_j \rangle|^2, \qquad \forall f \in \mathscr{H}. \tag{8.22}$$

Using this presentation, one can develop a theory of *signed frames* along the lines of that for frames (see [PW02], Exercises 8.1, 8.2).

Example 8.16. The tight signed frames with positive signature $\sigma = (1)$ are precisely the normalised tight frames.

Example 8.17. Take any three unit vectors in $\mathbb{R}^2$ none of which are multiples of each other. Then there are unique c_j that give a tight signed frame, which are given by

$$c_j = \frac{\cos(\beta - \alpha)}{\sin \alpha \sin \beta},$$

where $-\pi/2 \leq \alpha < \beta \leq \pi/2$ are the (acute) angles from the subspace spanned by this vector to those spanned by the other two. This is negative if $\alpha < 0$, $\beta > 0$, $\beta - \alpha < \pi/2$, i.e., the subspace generated by the vector lies in the region between the acute angle made by the other two.

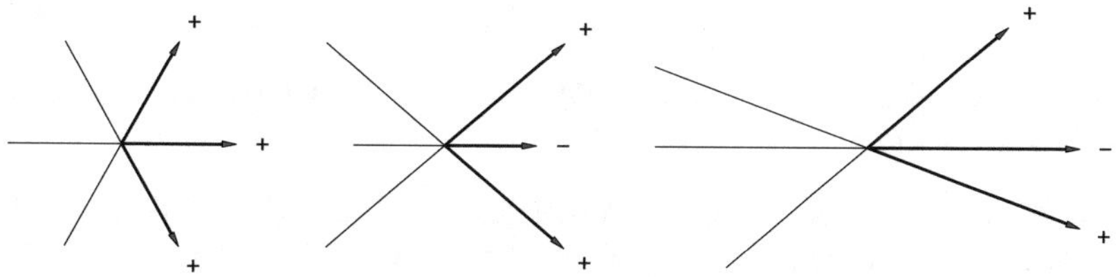

Fig. 8.4: Tight signed frames of three vectors in $\mathbb{R}^2$ with the signature indicated.

Example 8.18. There exist tight signed frames of n vectors for $\mathbb{F}^d$ for any signature which takes the value $+1$ at least d times (see Exer. 8.3).

8.10 Scaling the vectors of a frame to obtain a tight frame

We say a finite frame (f_j) of (nonzero) vectors for $\mathcal{H}$ can be *scaled to a tight frame*[1] if there are scalars $c_j \geq 0$, such that

$$I = I_{\mathcal{H}} = \sum_j c_j P_j, \qquad \text{where} \quad P_j f := \langle f, u_j \rangle u_j, \quad u_j := \frac{f_j}{\|f_j\|}, \tag{8.23}$$

i.e., $(\sqrt{c_j} f_j)$ is a normalised tight frame for $\mathcal{H}$. Similarly, we say that (f_j) can be *scaled to a tight signed frame* if (8.23) holds for some choice of $c_j \in \mathbb{R}$.

We now use tight signed frames to consider the question:

When can a finite frame of n vectors for $\mathcal{H} = \mathbb{F}^d$ be scaled to a tight frame?

It turns out (Corollary 8.5), that for the *particular value* of n

$$n = \begin{cases} \frac{1}{2} d(d+1), & \mathcal{H} \text{ real;} \\ d^2, & \mathcal{H} \text{ complex} \end{cases} \tag{8.24}$$

almost all sequences of n vectors have a unique scaling to a tight signed frame for $\mathcal{H} = \mathbb{F}^d$. In the generic situation, there is no scaling for less than n vectors, and infinitely many for more than n vectors.

We first consider the geometry of the set of best possible scalings.

Proposition 8.1. (Best approximation) *Let $u_1, \ldots, u_n$ be unit vectors in $\mathcal{H}$. Then the coefficients $c = (c_j)_{j=1}^n \in \mathbb{F}^n$ which minimise the Frobenius (matrix) norm*

$$\|I - \sum_{j=1}^n c_j P_j\|_F, \qquad P_j f := \langle f, u_j \rangle u_j, \tag{8.25}$$

are the solutions of the $n \times n$ linear system

$$Ac = [1], \qquad A := [|\langle u_k, u_j \rangle|^2]_{j,k=1}^n. \tag{8.26}$$

In particular, (u_j) can be scaled to a tight frame if and only if this minimum is zero and there is a solution with $c_j \geq 0$.

Proof. We first recall (see Exer. 3.1), that $\langle P_j, P_k \rangle_F = |\langle u_j, u_k \rangle|^2$, $\langle I, P_k \rangle_F = 1$. The minimum (least squares solution) of (8.25) occurs when (the error) $I - \sum_j c_j P_j$ is orthogonal to all the P_k, i.e., $\forall k$

$$I - \sum_j c_j P_j \perp P_k \iff \sum_j c_j \langle P_j, P_k \rangle_F = \langle I, P_k \rangle_F \iff \sum_j c_j |\langle u_j, u_k \rangle|^2 = 1.$$

□

[1] The term *scalable* is used in [KOPT13].

We now seek a condition on the vectors (u_j) which ensures the *normal equations* (8.26) have a unique solution, i.e., there is a unique scaling to a tight signed frame. Let $\circ$ be the Hadamard (pointwise) product of matrices: $(S \circ T)_{jk} := s_{jk} t_{jk}$. Then the matrix A in the normal equations is the Hadamard product

$$A := [|\langle u_k, u_j \rangle|^2] = B \circ \overline{B}, \qquad B := [\langle u_k, u_j \rangle], \tag{8.27}$$

where B is the Gramian of (u_j). We will use the Schur product theorem (cf [HJ91]).

Theorem 8.4. *(Schur product) If A and B are positive semidefinite, then so is $A \circ B$. If, in addition, B is positive definite and A has no diagonal entry equal to zero, then $A \circ B$ is positive definite.*

The following Lemma gives a condition which ensures that $[|\langle u_k, u_j \rangle|^2]$ (and other matrices) is invertible. It uses Lebesgue measure on $\mathbb{F}^d \times \cdots \times \mathbb{F}^d$, and the fact that the zero set of a nonzero polynomial has measure zero.

Lemma 8.1. *For almost every $v_1, \ldots, v_n \in \mathbb{F}^d$*

$$\operatorname{rank}([\langle v_k, v_j \rangle^r]) = \min\{n, \binom{d+r-1}{r}\}, \qquad r \geq 0.$$

For almost every $v_1, \ldots, v_n \in \mathbb{C}^d$

$$\operatorname{rank}([\langle v_k, v_j \rangle^r \overline{\langle v_k, v_j \rangle}^s]) = \min\{n, \binom{d+r-1}{r}\binom{d+s-1}{s}\}, \qquad r, s \geq 0.$$

Proof. Let $B = V^*V$ be the Gramian of $V = [v_1, \ldots, v_n]$. The matrices above are Hadamard products of B and $\overline{B}$, i.e., respectively

$$A := [\langle v_k, v_j \rangle^r] = \underbrace{B \circ B \circ \cdots \circ B}_{r \text{ times}},$$

$$A := [\langle v_k, v_j \rangle^r \overline{\langle v_k, v_j \rangle}^s] = \underbrace{B \circ B \circ \cdots \circ B}_{r \text{ times}} \circ \underbrace{\overline{B} \circ \overline{B} \circ \cdots \circ \overline{B}}_{s \text{ times}}.$$

Since B (and hence $\overline{B}$) is positive semidefinite, it follows from the Schur product theorem that A is also. Almost every choice of $(v_j)_{j=1}^n$ is in general position, and so we may assume without loss of generality that they are chosen to be so.

First suppose that $n \leq d$. Then the (v_j) are linearly independent, so that B is positive definite, and by the Schur product theorem A is positive definite, giving $\operatorname{rank}(A) = n$, as asserted.

Hence it suffices to suppose that $n > d$. Clearly, $\operatorname{rank}(A) \leq n$. Since B and V have the same kernel, and $\operatorname{rank}(V) = d$, the positive semidefinite matrix $B = V^*V$ has rank d. Thus B can be written

$$B = \sum_{j=1}^{d} u_j u_j^*,$$

where $\{u_j,\ldots,u_j\}$ is an orthogonal basis for the range of B.

We now consider only the second case (the first is similar, following from the algebra for $s=0$). Expanding (see Exer. 8.4) gives

$$A=[\langle v_k,v_j\rangle^r\overline{\langle v_k,v_j\rangle}^s]=\underbrace{B\circ B\circ\cdots\circ B}_{r \text{ times}}\circ\underbrace{\overline{B}\circ\overline{B}\circ\cdots\circ\overline{B}}_{s \text{ times}}$$

$$=\sum_{k_1=1}^{d}\cdots\sum_{k_r=1}^{d}\sum_{j_1=1}^{d}\cdots\sum_{j_s=1}^{d}(u_{k_1}\circ\cdots\circ u_{k_r}\circ\overline{u_{j_1}}\circ\cdots\circ\overline{u_{j_s}})(u_{k_1}\circ\cdots\circ u_{k_r}\circ\overline{u_{j_1}}\circ\cdots\circ\overline{u_{j_s}})^*,$$

a sum of at most $\binom{d+r-1}{r}\binom{d+s-1}{s}$ rank-one matrices ($\circ$ is commutative), giving

$$\operatorname{rank}(A)\le\binom{d+r-1}{r}\binom{d+s-1}{s}.$$

Thus, it suffices to show that $\operatorname{rank}(A)=n$, where

$$n=\binom{d+r-1}{r}\binom{d+s-1}{s},$$

for some choice of nonzero vectors (v_j). Since $\det(A)$ is a polynomial in $v_1,\ldots,v_n$ and $\overline{v_1},\ldots,\overline{v_n}$, this then implies that $\det(A)$ will be nonzero for almost every choice of (v_j), which gives the result.

The existence of vectors (v_j) for which A is invertible follows from the fact that $\Pi_{r,s}^\circ(\mathbb{C}^d)$ has a basis of ridge polynomials $z\mapsto\langle z,v\rangle^r\overline{\langle z,v\rangle}^s$ (see Exer. 8.5). □

Theorem 8.5. *(Equivalence) Let $u_1,\ldots,u_n$ be unit vectors in a Hilbert space $\mathcal{H}$ of dimension d, where*

$$n=\begin{cases}\frac{1}{2}d(d+1), & \mathcal{H}\text{ real};\\ d^2, & \mathcal{H}\text{ complex}.\end{cases}$$

Let $A:=[|\langle u_k,u_j\rangle|^2]$. Then the following are equivalent

(a) The $n\times n$ positive semidefinite matrix A is invertible.
(b) The vectors $u_1,\ldots,u_n$ have a unique scaling which gives a tight signed frame, with the c of (8.23) given by $c=A^{-1}[1]$.
(c) The Hermitian forms on $\mathcal{H}$ have a basis given by

$$(f,g)\mapsto\langle f,u_j\rangle\langle u_j,g\rangle,\qquad j=1,\ldots,n.$$

(d) The Hermitian operators on $\mathcal{H}$ have a basis given by the rank 1 orthogonal projections

$$P_j:f\mapsto\langle f,u_j\rangle u_j,\qquad j=1,\ldots,n.$$

Proof. (a)$\Longleftrightarrow$(b) As discussed in Proposition 8.1, A is the matrix giving the normal equations for finding a best scaling. If A is invertible, then the system $\sum_j c_jP_j=I$ has a unique solution given by $c=(c_j)=A^{-1}[1]$.

We now consider the case $\mathcal{H} = \mathbb{C}^d$. The case $\mathcal{H} = \mathbb{R}^d$ is similar, and easier. It uses the indentification of $\Pi_2^\circ(\mathbb{R}^d)$ with the symmetric bilinear forms on $\mathbb{R}^d$ (real Hermitian forms).

(a)$\Longleftrightarrow$(c) By taking $r = s = 1$ in Exer. 8.5, the invertibility of A is equivalent to the polynomials $z \mapsto \langle z, u_j\rangle\langle u_j, z\rangle$, $1 \le j \le n$, being a basis for $\Pi_{1,1}^\circ(\mathbb{C}^d)$. Using the indentification of $\Pi_{1,1}^\circ(\mathbb{C}^d)$ with the Hermitian forms (see §6.7), we conclude that the Hermitian forms $(f,g) \mapsto \langle f, u_j\rangle\langle u_j, g\rangle$, $1 \le j \le n$ are a basis for the (real vector space) of Hermitian forms on $\mathbb{C}^d$.

(c)$\Longleftrightarrow$(d) We observe the Hermitian operator corresponding to the Hermitian form $(f,g) \mapsto \langle f, u_j\rangle\langle u_j, g\rangle$ is $P = u_j u_j^*$. □

Corollary 8.5. *(Scaling to a tight signed frame). Let $\mathcal{H}$ be a Hilbert space of dimension d, and*

$$n = \begin{cases} \frac{1}{2}d(d+1), & \mathcal{H} \text{ real;} \\ d^2, & \mathcal{H} \text{ complex.} \end{cases}$$

Then for almost every choice of unit vectors $(u_1,\ldots,u_n)$ in $\mathcal{H}$ there is a unique scaling that gives a tight signed frame, with the constants c_j in (8.23) given by

$$c = A^{-1}[1], \qquad A := [|\langle u_k, u_j\rangle|^2]. \tag{8.28}$$

Proof. We observe that $\det(A)$ is a polynomial in $u_1,\ldots,u_n$ (and $\overline{u_1},\ldots,\overline{u_n}$ for $\mathcal{H}$ complex). By Lemma 8.1, this polynomial is nonzero for almost every choice for $u_1,\ldots,u_n$. When it is nonzero, i.e., A is invertible, Theorem 8.5 implies that there is unique scaling of (u_j) to a tight signed frame given by (8.28). □

Since the set of scalings c of (u_j) which give a tight signed frame is an affine subspace (by Proposition 8.1), the number n above is a *cut off*. Almost all sequences of less than n vectors don't have a scaling to a tight signed frame, and almost all sequences of more than n vectors have infinitely many such scalings.

Example 8.19. (2 dimensions). Almost every set of three vectors in $\mathbb{R}^2$ can be uniquely scaled to a tight signed frame. See Example 8.17 (and Figure 8.4) for a description on when this scaling is a tight frame. Two vectors in $\mathbb{R}^2$ can be scaled to a tight signed frame if and only if they are orthogonal. This shows directly that almost every set of two vectors in $\mathbb{R}^2$ cannot be scaled to a tight signed frame.

Almost every set of four vectors in $\mathbb{C}^2$ can be uniquely scaled to a tight signed frame for $\mathbb{C}^2$. The possible signatures are $++++$ (a tight frame), $+++-$, and $++--$ (see Exer. 8.3).

Notes

The characterisation of SICs up to projective unitary equivalence by their triple products was given by [AFF11]. This work was adapted to the general case (which includes MUBs) by [CW16] (see Theorems 8.1 and 8.2). The results of this chapter allow projective objects such as spherical (t,t)-designs and frames viewed as fusion frames to be classified (up to projective unitary equivalence) and their projective symmetries to be determined (see §9.3).

There is ongoing interest in fusion frames, e.g., see [BE15], the *Fusion frame* page of the *Frame Research Centre*, and www.fusionframe.org.

Tight signed frames were introduced in [PW02], where their relationship to the question of scaling to a tight frame (as presented here) was studied. The scaling question was also addressed in [KOPT13], who gave geometric descriptions of when a frame can be scaled to a tight frame.

Exercises

8.1. Let $(\phi_j)_{j=1}^n$ be a sequence of vectors in $\mathscr{H}$, and $c_j \in \mathbb{F}$ be scalars.
(a) Show that there exists a representation of the form

$$f = \sum_j c_j \langle f, \phi_j \rangle \phi_j, \qquad \forall f \in \mathscr{H}, \tag{8.29}$$

if and only if

$$\|f\|^2 = \sum_j c_j |\langle f, \phi_j \rangle|^2, \qquad \forall f \in \mathscr{H}. \tag{8.30}$$

(b) Suppose that (8.29) holds. Show that there is a unique choice for (c_j) which minimises $\sum_j |c_j|^2$, and that this satisfies $c_j \in \mathbb{R}$, $\forall j$. Prove the analogue of (2.9), i.e.,

$$\sum_j c_j \|\phi_j\|^2 = \dim(\mathscr{H}).$$

8.2. Let $(f_j)_{j=1}^n$ be vectors in $\mathscr{H}$, and $\sigma = (\sigma_j)$, $\sigma_j \in \{\pm 1\}$. We say that (f_j) is a **signed frame** with **signature** σ for $\mathscr{H}$ if there exist (signed frame bounds) $A, B > 0$ with

$$A\|f\|^2 \le \sum_j \sigma_j |\langle f, f_j \rangle|^2 \le B\|f\|^2, \forall f \in \mathscr{H}. \tag{8.31}$$

The **signed frame operator** $S; \mathscr{H} \to \mathscr{H}$ of a vector, signature pair (f_j), (σ_j) is given by

$$Sf := \sum_j \sigma_j \langle f, f_j \rangle f_j, \qquad \forall f \in \mathscr{H}.$$

(a) Show that the frame operator S of a signed frame with bounds A, B is invertible, with $(1/B) I_{\mathscr{H}} \le S^{-1} \le (1/A) I_{\mathscr{H}}$.

(b) For a signed frame (f_j) with signature σ and frame operator S, define the **dual signed frame** to be $(\tilde{f}_j)$ with signature σ, where $\tilde{f}_j := S^{-1}f_j$. Show that the dual signed frame is a signed frame with frame operator S^{-1}, and one has the expansion

$$f = \sum_j \sigma_j \langle f, \tilde{f}_j \rangle f_j = \sum_j \sigma_j \langle f, f_j \rangle \tilde{f}_j, \qquad \forall f \in \mathcal{H}.$$

Define the **canonical tight signed frame** to be (f_j^{can}) with signature σ, where $f_j^{\text{can}} := S^{-1/2}f_j$, and show that this is a tight signed frame.

8.3. Show that there is a tight signed frame of n vectors for $\mathbb{F}^d$ with signature σ if and only if σ takes the value $+1$ at least d times.

8.4. Show that the Hadamard product satisfies

$$(aa^*) \circ (bb^*) = (a \circ b)(a \circ b)^*, \qquad \forall a, b \in \mathbb{F}^d.$$

8.5. Here we consider the space $\Pi^\circ_{r,s}(\mathbb{C}^d)$ of Exer. 6.17, which has dimension

$$n = \binom{d+r-1}{r}\binom{d+s-1}{s}.$$

Let $v_1, \ldots, v_n \in \mathbb{C}^d$. Show that the following are equivalent
(a) The polynomials $p_j : z \mapsto \langle z, v_j \rangle^r \langle v_j, z \rangle^s$ are a basis for $\Pi^\circ_{r,s}(\mathbb{C}^d)$.
(b) The point evaluations $\delta_j : f \mapsto f(v_j)$ are a basis for dual space $\Pi^\circ_{r,s}(\mathbb{C}^d)'$.
(c) The $n \times n$ positive semidefinite matrix $A = [\langle v_j, v_k \rangle^r \overline{\langle v_j, v_k \rangle}^s]$ is invertible.
Remark: Since $\Pi^\circ_{r,s}(\mathbb{C}^d)$ has a basis of ridge polynomials $z \mapsto \langle z, v \rangle^r \langle v, z \rangle^s$, it follows that A is invertible for some choices of (v_j).

Chapter 9
Symmetries of tight frames

The angle preserving transformations of $\mathbb{R}^2$ form the *real orthogonal group*

$$O(2) := \{A \in \mathbb{R}^{2\times 2} : A^T A = I\},$$

which can be thought of as the symmetries of the inner product space $\mathscr{H} = \mathbb{R}^2$.

One might reasonably hope that an expansion for this space would reflect this structure as much as possible. To understand the issues involved here, consider an orthonormal basis and the tight frame of three equally spaced vectors.

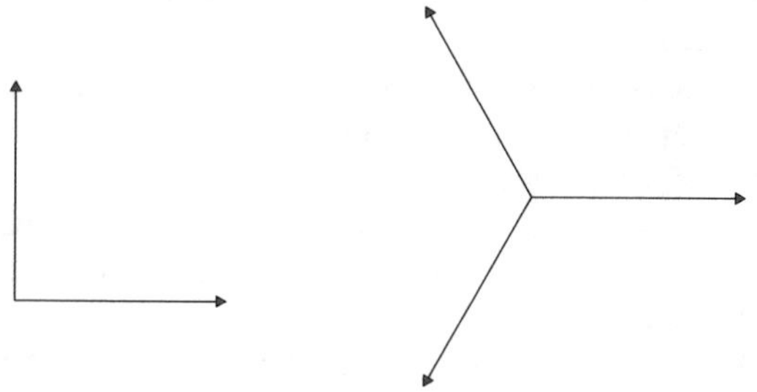

The first is invariant (mapped to itself) under a reflection (which generates a group of order 2), and the second is invariant under the dihedral group (of order 6) generated by a rotation (through $\frac{2\pi}{3}$) and a reflection. Thus the three equally spaced vectors have more of the symmetry of the space $\mathbb{R}^2$ than an orthonormal basis does. Further, this "large" symmetry group is closely related to the vectors of the frame being *equiangular*, which is desirable. It is even possible to have an expansion with *all* of the symmetry of the space, as in (1.3), but necessarily this requires one to use an (uncountably) infinite set of vectors.

We now make these (intuitively obvious) ideas precise by defining the *symmetry group* and *projective symmetry group* of a frame (which can be calculated from a small set of invariants). For simplicity, we suppose that the frame is finite, though the theory extends to infinite frames without any complications (see Chapter 16).

S.F.D. Waldron, *An Introduction to Finite Tight Frames*, Applied and Numerical Harmonic Analysis, https://doi.org/10.1007/978-0-8176-4815-2_9

9.1 The symmetries of a sequence of vectors

We define the *symmetry group* and the *projective symmetry group* of a finite frame in a very general setting (which includes versions allowing antilinear symmetries). Each of these "symmetry groups" has the following key features:

- It is defined for *all* finite frames as a group of permutations on the index set.
- It is simple to calculate from a small set of invariants.
- The symmetry group of a frame and all similar frames are equal. In particular, a frame, its dual frame and canonical tight frame have the same symmetry group.
- The symmetry group of various combinations of frames, such as tensor products and direct sums, are related to those of the constituent frames in a natural way.
- The symmetry group of a frame and its complementary frame are equal.

Frames with a large symmetry group have a simple structure (which embodies underlying symmetries of the space). Often the symmetry group can be very useful in constructing the frame itself. A special case is when the action of the symmetry group is transitive, which gives a *group frame* (see Chapter 10). If the symmetry group is abelian, then the group frames are the *harmonic frames* (see Chapter 11), all of which can be constructed from the *abstract* abelian groups. All the known maximal sets of complex equiangular lines come as the orbit of the projective action of an abelian group (see Chapter 14). Another example is multivariate orthogonal polynomials, where tight frames sharing the symmetries of the weight function can be constructed (see Chapter 15).

Let S_J denote the **symmetric group** on the set J, i.e., the group of all bijections $J \to J$ (called permutations) under composition.

Throughout, let $\mathscr{H}$ be a finite dimensional vector space over $\mathbb{F}$, where $\overline{\mathbb{F}} = \mathbb{F}$. Any finite sequence $\Phi = (f_j)_{j\in J}$ in $\mathscr{H}$ can be thought of as normalised tight frame for an appropriate inner product on $\mathscr{H}$ (see §4.5). Thus, one can suppose that Φ is a tight frame (and thereby bypass Chapter 4).

Let $C : v \to \overline{v}$ be the *complex conjugation map* (see §2.3). A product of C and a linear (unitary) map is called an **antilinear** (**antiunitary map**). In this way, we can extend the linear and unitary maps:

$$EGL(\mathscr{H}) := \{L, LC : L \in GL(\mathscr{H})\} \quad \text{(Extended general linear group)},$$
$$EU(\mathscr{H}) := \{U, UC : L \in U(\mathscr{H})\} \quad \text{(Extended unitary group)}.$$

Of course these are groups, with $GL(\mathscr{H}) \subset EGL(\mathscr{H})$, $U(\mathscr{H}) \subset EU(\mathscr{H})$, where there is strict inclusion if and only if $\mathbb{F} \not\subset \mathbb{R}$ (i.e., $C \neq I$).

It is very convenient to have the possibility of a symmetry of $\Phi = (f_j)_{j\in J}$ which takes f_j to f_k where $j \neq k$, *but* $f_j = f_k$, i.e., nonidentity symmetries which map repeated vectors to themselves. For this reason, we define the symmetry groups to be permutations of the indices (given by a possibly unfaithful action).

9.2 The symmetry group of a sequence of vectors

Definition 9.1. Let $\Phi = (f_j)_{j\in J}$ be a finite sequence of vectors which spans $\mathcal{H}$. Then the **symmetry group** and the **extended symmetry group** of Φ are the groups

$$\mathrm{Sym}(\Phi) := \{\sigma \in S_J : \exists L_\sigma \in GL(\mathcal{H}) \,\mathrm{with}\, L_\sigma f_j = f_{\sigma j}, \forall j \in J\},$$

$$\mathrm{Sym}_{\mathrm{E}}(\Phi) := \{\sigma \in S_J : \exists L_\sigma \in EGL(\mathcal{H}) \,\mathrm{with}\, L_\sigma f_j = f_{\sigma j}, \forall j \in J\}.$$

In other words, $\sigma \in \mathrm{Sym}(\Phi)$ if (f_j) and $(f_{\sigma j})$ are similar. If (f_j) is a tight frame, i.e., $S_\Phi = [f_j][f_j]^* = AI$, $A > 0$, then this similarity becomes unitary equivalence, since

$$L_\sigma L_\sigma^* = \frac{1}{A} L_\sigma [f_j][f_j]^* L_\sigma^* = \frac{1}{A}[L_\sigma f_j][L_\sigma f_j]^* = \frac{1}{A}[f_{\sigma j}][f_{\sigma j}]^* = \frac{1}{A}(AI) = I.$$

It is easy to check that these symmetry groups are indeed groups, with

$$\mathrm{Sym}(\Phi) \subset \mathrm{Sym}_{\mathrm{E}}(\Phi) \subset S_J.$$

Since linear and antilinear maps are determined by their action on a spanning set, it follows that if σ is a symmetry, then there is a unique linear or antilinear map $L_\sigma : \mathcal{H} \to \mathcal{H}$ with

$$L_\sigma f_j = f_{\sigma j}, \qquad \forall j \in J.$$

and the linear map

$$\pi_\Phi : \mathrm{Sym}(\Phi) \to GL(\mathcal{H}) : \sigma \mapsto L_\sigma \tag{9.1}$$

is a group homomorphism, i.e., a linear representation of $G = \mathrm{Sym}(\Phi)$ on $\mathcal{H}$. We will refer to both σ and L_σ (the action of σ) as a **symmetry** of Φ when L_σ is linear, and as an **antisymmetry** when L_σ is antilinear. With this understanding, we have:

If Φ is tight frame, then its symmetry group consists of unitary maps, i.e., the action of $\mathrm{Sym}(\Phi)$ is unitary (and its extended symmetry group consists of unitary and antiunitary maps).

If the vectors in Φ are distinct, then π_Φ is injective, i.e., the representation is faithful (see Exer. 9.2). In this case $\mathrm{Sym}(\Phi)$ can be identified with its image.[1]

Proposition 9.1. *(Similarity) If finite sequences Φ and Ψ are similar, then*

$$\mathrm{Sym}(\Psi) = \mathrm{Sym}(\Phi), \qquad \mathrm{Sym}_{\mathrm{E}}(\Psi) = \mathrm{Sym}_{\mathrm{E}}(\Phi).$$

In particular, if Φ is a frame, then its dual and canonical tight frame have the same symmetry group, i.e., $\mathrm{Sym}(\Phi) = \mathrm{Sym}(\tilde{\Phi}) = \mathrm{Sym}(\Phi^{\mathrm{can}})$.

---

[1]The group of linear maps $\pi_\Phi(\mathrm{Sym}(\Phi))$ is sometimes defined to be the "symmetry group" of Φ.

Proof. Suppose that $\Phi = (f_j)$ is similar to Ψ, say $\Psi = (Qf_j)$ for some invertible linear map Q. If $\sigma \in \mathrm{Sym}(\Phi)$, then

$$L_\sigma f_j = f_{\sigma j}, \forall j \quad \Longrightarrow \quad (QL_\sigma Q^{-1})Qf_j = Qf_{\sigma j}, \forall j,$$

so that $\sigma \in \mathrm{Sym}(\Psi)$, and $\mathrm{Sym}(\Phi) \subset \mathrm{Sym}(\Psi)$. The reverse inclusion follows since Ψ is similar to Φ. A similar argument shows that $\mathrm{Sym}_\mathrm{E}(\Psi) = \mathrm{Sym}_\mathrm{E}(\Phi)$. □

In other words:

The (extended) symmetry group of Φ depends only on its similarity class.

Example 9.1. (Bases) If $\Phi = (f_j)_{j \in J}$ is a basis, then for each $\sigma \in S_J$, $L_\sigma f_j := f_{\sigma j}$ defines a linear map, and hence $\mathrm{Sym}(\Phi) = S_J$.

Example 9.2. (Vertices of a simplex) Suppose that $\Phi = (f_j)_{j \in J}$ is the vertices of a simplex, i.e., the vectors have a single linear dependence $\sum_j f_j = 0$. Fix an index k, then for each $\sigma \in S_J$, $L_\sigma f_j := f_{\sigma j}$, $j \neq k$ defines a linear map, with

$$L_\sigma(f_k) = L_\sigma\big(-\sum_{j \neq k} f_j\big) = -\sum_{j \neq k} f_{\sigma j} = f_{\sigma k},$$

and hence $\mathrm{Sym}(\Phi) = S_J$.

Example 9.3. (see Figure 9.1) Let $\Phi = (v_1, v_2, v_3)$ be the tight frame of three equally spaced unit vectors for $\mathbb{R}^2$, and $\Psi = (v_1, v_2, -v_3)$. Then

$$\mathrm{Sym}(\Phi) = S_3 = S_{\{1,2,3\}} \quad \text{(order 6)}, \qquad \mathrm{Sym}(\Psi) = \{1, (12)\} \quad \text{(order 2)}.$$

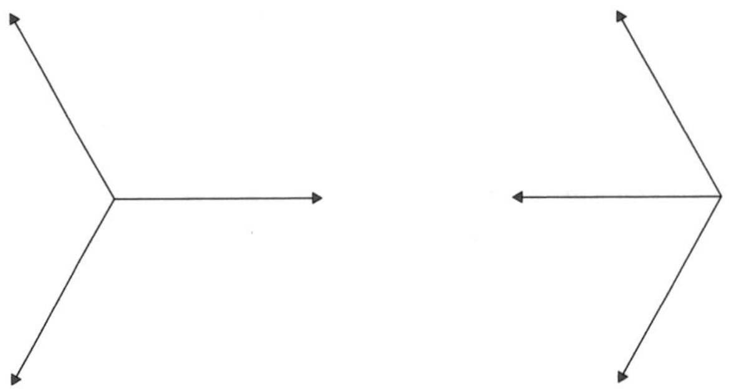

Fig. 9.1: The frames Φ and Ψ of Example 9.3, which have $|\mathrm{Sym}(\Phi)| = 6$ and $|\mathrm{Sym}(\Psi)| = 2$.

Example 9.4. The symmetry group of the tight frame $\Phi = (v_j)_{j=1}^n$ consisting of n equally spaced unit vectors, say

$$v_j = \begin{pmatrix} \sin \frac{2\pi j}{n} \\ \cos \frac{2\pi j}{n} \end{pmatrix} \in \mathbb{R}^2,$$

is the *dihedral group* of order $2n$, i.e., $D_n := \langle a, b : a^n = 1, b^2 = 1, b^{-1}ab = a^{-1} \rangle$, where $a = (12\cdots n)$ acts as rotation through $\frac{2\pi}{n}$, and b as a reflection.

We now explain how the symmetry group of Φ can be calculated directly from its canonical Gramian P_Φ (see §4.1). There is a bijection between the permutations $\sigma \in S_J$ and the so-called $J \times J$ *permutation matrices*, given by $\sigma \mapsto P_\sigma$, where

$$P_\sigma e_j := e_{\sigma j}.$$

Let $V = [f_j]$, and $f_{\sigma j} = L_\sigma f_j$, then $VP_\sigma = [Ve_{\sigma j}] = [f_{\sigma j}] = [L_\sigma f_j] = L_\sigma V$, so that

$$\sigma \in \mathrm{Sym}(\Phi) \quad \iff \quad VP_\sigma = L_\sigma V, \quad \text{for some } L_\sigma \in GL(\mathscr{H}). \tag{9.2}$$

Lemma 9.1. *(Calculation) Suppose that $\Phi = (f_j)$ is a finite sequence of vectors, with canonical Gramian P_Φ. For Φ a frame, $P_\Phi = \mathrm{Gram}(\Phi^{\mathrm{can}})$. Then*

$$\sigma \in \mathrm{Sym}(\Phi) \quad \iff \quad P_\sigma^* P_\Phi P_\sigma = P_\Phi. \tag{9.3}$$

Proof. By Propositions 4.1 and 9.1, we can suppose $\Phi = (f_j)$ is the normalised tight frame given by the columns of P_Φ (an orthogonal projection), so that $V = [f_j] = P_\Phi$.

($\Longrightarrow$) If $\sigma \in \mathrm{Sym}(\Phi)$, then L_σ is unitary, and we have

$$P_\sigma^* P_\Phi P_\sigma = (P_\Phi P_\sigma)^* P_\Phi P_\sigma = (L_\sigma P_\Phi)^* L_\sigma P_\Phi = P_\Phi^* (L_\sigma^* L_\sigma) P_\Phi = P_\Phi.$$

($\Longleftarrow$) If $P_\sigma^* P_\Phi P_\sigma = P_\Phi$, then by choosing $L_\sigma := P_\sigma$, and writing $P_\Phi = V$, we have

$$VP_\sigma = P_\Phi P_\sigma = P_\sigma P_\Phi = L_\sigma V,$$

and so, by (9.2), we have that $\sigma \in \mathrm{Sym}(\Phi)$. □

If Φ is tight frame, then P_Φ is a scalar multiple of $\mathrm{Gram}(\Phi)$, and hence we have:

$$\begin{aligned} \sigma \in \mathrm{Sym}(\Phi) \quad &\iff \quad P_\sigma^* \mathrm{Gram}(\Phi) P_\sigma = \mathrm{Gram}(\Phi) \\ &\iff \quad \langle f_{\sigma j}, f_{\sigma k} \rangle = \langle f_j, f_k \rangle, \quad \forall j,k. \end{aligned}$$

The condition for σ to be an extended symmetry (for $\mathbb{R} \not\subset \mathbb{F}$) is that

$$P_\sigma^* P_\Phi P_\sigma = P_\Phi^T, \tag{9.4}$$

which for a tight frame becomes

$$\langle f_{\sigma j}, f_{\sigma k} \rangle = \overline{\langle f_j, f_k \rangle} = \langle f_k, f_j \rangle, \quad \forall j,k.$$

Example 9.5. (Canonical coordinates) A sequence of vectors Φ and its canonical coordinates $\Psi = c^\Phi$ (see §4.2) have the same symmetries, i.e., $\mathrm{Sym}(\Phi) = \mathrm{Sym}(c^\Phi)$. This follows since $P_{c^\Phi} = P_\Phi^T$ (Theorem 4.2) and $P_\sigma^T = P_\sigma^*$ give

$$\begin{aligned}\sigma \in \mathrm{Sym}(\Phi) \quad &\Longleftrightarrow \quad P_\sigma^* P_\Phi P_\sigma = P_\Phi \quad \Longleftrightarrow \quad P_\sigma^* P_\Psi^T P_\sigma = P_\Psi^T \\ &\Longleftrightarrow \quad P_\sigma^* P_\Psi P_\sigma = P_\Psi \quad \Longleftrightarrow \quad \sigma \in \mathrm{Sym}(\Psi).\end{aligned}$$

We now show that a frame and its complement have the same symmetry group. Such a result is not possible if the symmetry group is defined to be a group of linear transformations (as in [VW05]).

Theorem 9.1. *(Complements) Suppose that Φ is a finite sequence of vectors, and Ψ is a complementary sequence, i.e., $P_\Phi + P_\Psi = I$. Then*

$$\mathrm{Sym}(\Phi) = \mathrm{Sym}(\Psi), \qquad \mathrm{Sym}_{\mathrm{E}}(\Phi) = \mathrm{Sym}_{\mathrm{E}}(\Psi).$$

Proof. By Lemma 9.1 and $P_\sigma^* P_\sigma = I$, we have

$$\begin{aligned}\sigma \in \mathrm{Sym}(\Phi) \quad &\Longleftrightarrow \quad P_\sigma^* P_\Phi P_\sigma = P_\Phi \quad \Longleftrightarrow \quad P_\sigma^*(I - P_\Psi) P_\sigma = I - P_\Psi \\ &\Longleftrightarrow \quad P_\sigma^* P_\Psi P_\sigma = P_\Psi \quad \Longleftrightarrow \quad \sigma \in \mathrm{Sym}(\Psi).\end{aligned}$$

For $\mathrm{Sym}_{\mathrm{E}}(\Phi)$, a similar argument using (9.4) gives the result. □

Example 9.6. We consider the equal-norm tight frames Φ of four vectors for $\mathbb{C}^3$ with nontrivial symmetries, i.e., $|\mathrm{Sym}(\Phi)| > 1$. A complement Ψ consists of four equal-norm vectors for $\mathbb{C}$. The symmetries of Ψ (and hence of Φ) are given by those permutations of Ψ which can be realised by multiplication by a unit modulus complex number, e.g., the permutation $\sigma = (12)(34)$ is a symmetry of $\Psi = ([1],[-1],[1],[-1])$ corresponding to multiplication by -1. Therefore, the only possibilities for these complementary frames (up to similarity) are

$$([1],[1],[1],[z]),\ z \neq 1, \quad ([1],[1],[z],[z]),\ z \neq \pm 1, \quad ([1],[1],[z],[w]),\ z \neq w,\ z,w \neq 1,$$

$$([1],[1],[1],[1]), \quad ([1],[-1],[1],[-1]), \quad ([1],[i],[-1],[-i]).$$

The corresponding symmetry groups are (up to group isomorphism)

$$S_3, \quad S_2 \times S_2, \quad S_2, \quad S_4, \quad D_4 \text{ (dihedral group of order 8)}, \quad C_4.$$

The Gramian matrices for the last three (which are harmonic frames) are

$$\frac{1}{4}\begin{pmatrix} 3 & -1 & -1 & -1 \\ -1 & 3 & -1 & -1 \\ -1 & -1 & 3 & -1 \\ -1 & -1 & -1 & 3 \end{pmatrix}, \quad \frac{1}{4}\begin{pmatrix} 3 & 1 & -1 & 1 \\ 1 & 3 & 1 & -1 \\ -1 & 1 & 3 & 1 \\ 1 & -1 & 1 & 3 \end{pmatrix}, \quad \frac{1}{4}\begin{pmatrix} 3 & -i & 1 & i \\ i & 3 & -i & 1 \\ 1 & i & 3 & -i \\ -i & 1 & i & 3 \end{pmatrix}.$$

This example can be generalised to give all possible symmetry groups for a tight frame of n vectors in $\mathbb{C}^{n-1}$ (see [VW10]).

9.3 The projective symmetry group of a sequence of vectors

We now define *projective* symmetries of a sequence of vectors $\Phi = (f_j)$, and give a parallel theory to that for (nonprojective) symmetries (here representations become projective representations, inner products are replaced by m-products, etc.).

Definition 9.2. Let $\Phi = (f_j)_{j \in J}$ be a finite sequence of vectors which spans $\mathscr{H}$. The **projective symmetry group** and **extended projective symmetry group** of Φ are

$$\mathrm{Sym}_\mathrm{P}(\Phi) := \{\sigma \in S_J : \exists L_\sigma \in GL(\mathscr{H}), |\alpha_j| = 1 \,\text{with}\, L_\sigma f_j = \alpha_j f_{\sigma j}, \forall j \in J\},$$

$$\mathrm{Sym}_\mathrm{EP}(\Phi) := \{\sigma \in S_J : \exists L_\sigma \in EGL(\mathscr{H}), |\alpha_j| = 1 \,\text{with}\, L_\sigma f_j = \alpha_j f_{\sigma j}, \forall j \in J\}.$$

In other words, a permutation $\sigma \in \mathrm{Sym}_\mathrm{P}(\Phi)$ if (f_j) and $(f_{\sigma j})$ are projectively similar (equivalently, projectively unitarily equivalent, when Φ is a tight frame). These projective symmetry groups are groups, which contain the corresponding symmetry groups, i.e.,

$$\mathrm{Sym}(\Phi) \subset \mathrm{Sym}_\mathrm{P}(\Phi), \qquad \mathrm{Sym}_\mathrm{E}(\Phi) \subset \mathrm{Sym}_\mathrm{EP}(\Phi).$$

Example 9.7. Let $\Phi = (v_1, v_2, v_3)$ and $\Psi = (v_1, v_2, -v_3)$ be the tight frames for $\mathbb{R}^2$ of Example 9.3. These clearly have the same projective symmetries, so that

$$S_3 = \mathrm{Sym}(\Phi) \subset \mathrm{Sym}_\mathrm{P}(\Phi) = \mathrm{Sym}_\mathrm{P}(\Psi) \subset S_3.$$

Hence $\mathrm{Sym}_\mathrm{P}(\Psi) = S_3$ (of order 6) properly contains $\mathrm{Sym}(\Psi)$ (of order 2).

It is possible to associate with $\sigma \in \mathrm{Sym}_\mathrm{P}(\Phi)$ the projective linear map induced by L_σ, thereby obtaining a projective linear representation. We won't labour this point, but do observe that the representation gives a projective unitary action of $\mathrm{Sym}_\mathrm{P}(\Phi)$ (on the lines of $\mathscr{H}$) when Φ is a tight frame.

We have the projective analogue of Proposition 9.1.

Proposition 9.2. *(Projective similarity) If finite sequences Φ and Ψ are projectively similar, then*

$$\mathrm{Sym}_\mathrm{P}(\Psi) = \mathrm{Sym}_\mathrm{P}(\Phi), \qquad \mathrm{Sym}_\mathrm{EP}(\Psi) = \mathrm{Sym}_\mathrm{EP}(\Phi).$$

In particular, if Φ is a frame, then its dual and canonical tight frame have the same (extended) projective symmetry group.

In other words:

$\mathrm{Sym}_\mathrm{P}(\Phi)$ and $\mathrm{Sym}_\mathrm{EP}(\Phi)$ depend only on Φ up to projective similarity.

We now show how the projective symmetry group of Φ can be calculated directly from its canonical m-products (see §8.7).

Lemma 9.2. *(Calculation) Suppose that $\Phi = (f_j)$ is a finite sequence of vectors, with canonical m-products $\Delta_C(f_{j_1},\ldots,f_{j_m})$. Then $\sigma \in \mathrm{Sym}_\mathrm{P}(\Phi)$ if and only if*

$$\Delta_C(f_{j_1},\ldots,f_{j_m}) = \Delta_C(f_{\sigma j_1},\ldots,f_{\sigma j_m}), \tag{9.5}$$

for a determining set of canonical m-products.

Proof. The condition $\sigma \in \mathrm{Sym}_\mathrm{P}(\Phi)$ is that (f_j) and $(f_{\sigma j})$ are projectively similar, and so we can apply Theorem 8.3. □

For $\sigma \in \mathrm{Sym}_\mathrm{EP}(\Phi)$ (an antisymmetry), the condition (9.5) is replaced by

$$\Delta_C(f_{j_1},\ldots,f_{j_m}) = \overline{\Delta_C(f_{\sigma j_1},\ldots,f_{\sigma j_m})} = \Delta_C(f_{\sigma j_m},\ldots,f_{\sigma j_1}). \tag{9.6}$$

For Φ a tight frame, the canonical m-products are a nonzero scalar multiple of the m-products, so that:

If $\Phi = (f_j)$ is tight frame, then $\sigma \in \mathrm{Sym}_\mathrm{P}(\Phi)$ if and only if

$$\Delta(f_{j_1},\ldots,f_{j_m}) = \Delta(f_{\sigma j_1},\ldots,f_{\sigma j_m}), \tag{9.7}$$

for a determining set of m-products for Φ.

A frame and its complement have the same projective symmetry group.

Theorem 9.2. *(Complements) If $\Phi = (v_j)_{j\in J}$ is a finite sequence of vectors, and $\Psi = (w_j)_{j\in J}$ is a complement up to projective similarity, i.e., $P_\Phi + CP_\Psi C^* = I$, where $C = \mathrm{diag}(\alpha_j)$ is a unitary diagonal matrix, then*

$$\mathrm{Sym}_\mathrm{P}(\Psi) = \mathrm{Sym}_\mathrm{P}(\Phi), \qquad \mathrm{Sym}_\mathrm{EP}(\Psi) = \mathrm{Sym}_\mathrm{EP}(\Phi).$$

Proof. Let $P_\Phi = [p_{kj}]$, $P_\Psi = [q_{kj}]$. Then the canonical inner products, which are given by (8.14), satisfy

$$\begin{aligned}
\Delta_C^\Phi(v_{j_1},\ldots,v_{j_m}) &= p_{j_1 j_2} p_{j_2 j_3} \cdots p_{j_m j_1} \\
&= (-\alpha_{j_1}\overline{\alpha_{j_2}} q_{j_1 j_2})(-\alpha_{j_2}\overline{\alpha_{j_3}} q_{j_2 j_3}) \cdots (-\alpha_{j_m}\overline{\alpha_{j_1}} q_{j_m j_1}) \\
&= (-1)^m q_{j_1 j_2} q_{j_2 j_3} \cdots q_{j_m j_1} \\
&= (-1)^m \Delta_C^\Psi(w_{j_1},\ldots,w_{j_m}).
\end{aligned}$$

The result then follows from Lemma 9.2 and (9.6). □

Example 9.8. Let $\Phi = (v_j)$ be an equal-norm tight frame of $d+1$ vectors for $\mathbb{C}^d$, e.g., the vertices of the regular simplex, and $\Psi = (w_j)$ be the complementary tight frame for $\mathbb{C}^1$. Since P_Φ has a constant diagonal, the vectors of Ψ are equal-norm, say $w_j = (a_j)$, $|a_j| = r > 0$. Therefore, the m-products of Ψ are

$$\Delta(w_{j_1},\ldots,w_{j_m}) = (a_{j_1}\overline{a_{j_2}})(a_{j_2}\overline{a_{j_3}})\cdots(a_{j_m}\overline{a_{j_1}}) = r^{2m}.$$

Thus all equal-norm tight frames $\Phi = (v_j)$ of $d+1$ vectors in $\mathbb{C}^d$ are projectively similar (to the vertices of the simplex), with $\mathrm{Sym}_P(\Phi) = S_J$.

9.4 Symmetries of combinations of frames

We now consider how the symmetry groups of a combination of frames Φ and Ψ (see Chapter 5) is related to their symmetry groups. In view of Theorem 4.1, these results also hold for spanning sequences of vectors.

Let $\Phi = (\phi_j)_{j\in J}$ and $\Psi = (\psi_k)_{k\in K}$ be finite frames for $\mathscr{H}_1$ and $\mathscr{H}_2$. The inner products on the orthogonal direct sum $\mathscr{H}_1 \oplus \mathscr{H}_2$ and tensor product $\mathscr{H}_1 \otimes \mathscr{H}_2$ are given by

$$\langle (f_1,g_1),(f_2,g_2)\rangle := \langle f_1,f_2\rangle + \langle g_1,g_2\rangle, \qquad \forall (f_1,g_1),(f_2,g_2) \in \mathscr{H}_1 \oplus \mathscr{H}_2,$$

$$\langle f_1\otimes g_1, f_2\otimes g_2\rangle := \langle f_1,f_2\rangle\langle g_1,g_2\rangle, \qquad \forall f_1\otimes g_1, f_2\otimes g_2 \in \mathscr{H}_1 \otimes \mathscr{H}_2.$$

For $\sigma \in \mathrm{Sym}(\Phi)$, $\tau \in \mathrm{Sym}(\Psi)$, with corresponding $L_\sigma \in GL(\mathscr{H}_1)$, $L_\tau \in GL(\mathscr{H}_2)$, let $L_{(\sigma,\tau)} = L_\sigma \oplus L_\tau \in GL(\mathscr{H}_1 \oplus \mathscr{H}_2)$, i.e.,

$$L_{(\sigma,\tau)}\begin{pmatrix} f \\ g \end{pmatrix} := \begin{pmatrix} L_\sigma f \\ L_\tau g \end{pmatrix}, \qquad \forall f \in \mathscr{H}_1,\ \forall g \in \mathscr{H}_2. \tag{9.8}$$

We interpret (σ,τ) as a permutation on $J \cup K$ in the obvious way. This induces symmetry on the *union* and *sum* of Φ and Ψ, via

$$L_{(\sigma,\tau)}\begin{pmatrix} \phi_j \\ 0 \end{pmatrix} = \begin{pmatrix} L_\sigma \phi_j \\ 0 \end{pmatrix} = \begin{pmatrix} \phi_{\sigma j} \\ 0 \end{pmatrix}, \qquad L_{(\sigma,\tau)}\begin{pmatrix} 0 \\ \psi_k \end{pmatrix} = \begin{pmatrix} 0 \\ L_\tau \psi_k \end{pmatrix} = \begin{pmatrix} 0 \\ \psi_{\tau k} \end{pmatrix},$$

$$L_{(\sigma,\tau)}\begin{pmatrix} \frac{1}{\sqrt{n_2}}\phi_j \\ \frac{1}{\sqrt{n_1}}\psi_k \end{pmatrix} = \begin{pmatrix} \frac{1}{\sqrt{n_2}}L_\sigma \phi_j \\ \frac{1}{\sqrt{n_1}}L_\tau \psi_k \end{pmatrix} = \begin{pmatrix} \frac{1}{\sqrt{n_2}}\phi_{\sigma j} \\ \frac{1}{\sqrt{n_1}}\psi_{\tau k} \end{pmatrix}.$$

In this way, we have

$$\mathrm{Sym}(\Phi)\times\mathrm{Sym}(\Psi) \subset \mathrm{Sym}(\Phi\cup\Psi), \qquad \mathrm{Sym}(\Phi)\times\mathrm{Sym}(\Psi) \subset \mathrm{Sym}(\Phi\,\hat{+}\,\Psi).$$

For the *direct sum*, where $J = K$, we have

$$L_{(\sigma,\tau)}\begin{pmatrix} \phi_j \\ \psi_k \end{pmatrix} = \begin{pmatrix} L_\sigma \phi_j \\ L_\tau \psi_k \end{pmatrix} = \begin{pmatrix} \phi_{\sigma j} \\ \psi_{\tau k} \end{pmatrix},$$

which is a permutation of the direct sum provided $\sigma = \tau$. In this way, we have

$$\mathrm{Sym}(\Phi) \cap \mathrm{Sym}(\Psi) \subset \mathrm{Sym}(\Phi \oplus \Psi).$$

For the *tensor product*, define $L_{(\sigma,\tau)} \in GL(\mathcal{H}_1 \otimes \mathcal{H}_2)$ by $L_{(\sigma,\tau)} = L_\sigma \otimes L_\tau$. Then

$$L_{(\sigma,\tau)}(\phi_j \otimes \psi_k) = (L_\sigma \phi_j) \otimes (L_\tau \psi_k) = \phi_{\sigma j} \otimes \psi_{\tau k}$$

and so we obtain

$$\mathrm{Sym}(\Phi) \times \mathrm{Sym}(\Psi) \subset \mathrm{Sym}(\Phi \otimes \Psi).$$

In summary, we have:

Proposition 9.3. *The symmetry group of a finite frame satisfies*

$$\begin{aligned}
\mathrm{Sym}(\Phi) \times \mathrm{Sym}(\Psi) &\subset \mathrm{Sym}(\Phi \cup \Psi), \\
\mathrm{Sym}(\Phi) \times \mathrm{Sym}(\Psi) &\subset \mathrm{Sym}(\Phi \hat{+} \Psi), \\
\mathrm{Sym}(\Phi) \times \mathrm{Sym}(\Psi) &\subset \mathrm{Sym}(\Phi \otimes \Psi), \\
\mathrm{Sym}(\Phi) \cap \mathrm{Sym}(\Psi) &\subset \mathrm{Sym}(\Phi \oplus \Psi).
\end{aligned}$$

Moreover, these inclusions also hold for the other symmetry groups.

Each of these inclusions can be strict (see Exer. 9.1).

9.5 Maximally symmetric tight frames

If Φ is a frame of n vectors, then

$$\mathrm{Sym}(\Phi) \subset S_n \quad \Longrightarrow \quad |\mathrm{Sym}(\Phi)| \,\big|\, n! \quad \Longrightarrow \quad |\mathrm{Sym}(\Phi)| \le n!.$$

Thus, there are *maximally symmetric* frames in any class of such frames.

Definition 9.3. Let $\mathcal{C}$ be a class of frames of n vectors, e.g., the tight frames or equal-norm frames in $\mathbb{F}^d$. We say that $\Phi \in \mathcal{C}$ is **maximally symmetric** if

$$|\mathrm{Sym}(\Phi)| = \max_{\Psi \in \mathcal{C}} |\mathrm{Sym}(\Psi)|.$$

This definition should be treated with a little caution for frames with repeated vectors. For example, the frame of n vectors for $\mathbb{R}^2$ consisting e_1 repeated $n-1$ times and e_2 has symmetry group of order $(n-1)!$, while that of the n equally spaced unit vectors has order $2n$.

Example 9.9. The only cases when a frame Φ of n vectors for $\mathbb{F}^d$ $(d > 1)$ can have maximal symmetry by virtue of $\mathrm{Sym}(\Phi) = S_n$ is when $n = d$, i.e., Φ is a basis, or when $n = d+1$, i.e., Φ is the vertices of the simplex (see Example 9.8). This follows since the canonical 3-products (for distinct vectors) are all equal (see Exer. 9.4).

Example 9.10. The tight frame of n equally spaced unit vectors in $\mathbb{R}^2$ is a maximally symmetric tight frame of n distinct vectors for $\mathbb{R}^2$. This is because the unitary transformations on $\mathbb{R}^2$ are products of rotations and reflections.

Example 9.11. The n equally spaced unit vectors in $\mathbb{R}^2$ are not always maximally symmetric tight frames of n distinct vectors in $\mathbb{C}^2$. For n even, the harmonic frame

$$\left\{\begin{pmatrix}1\\1\end{pmatrix},\begin{pmatrix}\omega\\-\omega\end{pmatrix},\begin{pmatrix}\omega^2\\\omega^2\end{pmatrix},\begin{pmatrix}\omega^3\\-\omega^3\end{pmatrix},\begin{pmatrix}\omega^4\\\omega^4\end{pmatrix},\ldots\begin{pmatrix}\omega^{n-2}\\\omega^{n-2}\end{pmatrix},\begin{pmatrix}\omega^{n-1}\\-\omega^{n-1}\end{pmatrix}\right\},\qquad \omega:=e^{\frac{2\pi i}{n}}$$

has a symmetry group of order $\frac{1}{2}n^2$.

Example 9.12. (Five vectors in $\mathbb{C}^3$). We consider the maximally symmetric tight frames Φ of five vectors in $\mathbb{C}^3$, by considering the complementary tight frames Ψ (which have the same symmetry group). We note $|\mathrm{Sym}(\Phi)|=|\mathrm{Sym}(\Psi)|$ divides 5!

Since zero vectors are fixed by a symmetry, the most symmetric Φ with a zero vector is given by the $(1,4)$-partition frame corresponding to

$$\Psi=\left\{\begin{pmatrix}1\\0\end{pmatrix},\begin{pmatrix}0\\\frac{1}{2}\end{pmatrix},\begin{pmatrix}0\\\frac{1}{2}\end{pmatrix},\begin{pmatrix}0\\\frac{1}{2}\end{pmatrix},\begin{pmatrix}0\\\frac{1}{2}\end{pmatrix}\right\},\qquad |\mathrm{Sym}(\Phi)|=4!=24,$$

which is the vertices of the *tetrahedron* and a zero vector. If $\mathrm{Sym}(\Phi)$ does not have an element of order 5, then the next most symmetric is the $(2,3)$-partition frame given by

$$\Psi=\left\{\begin{pmatrix}\frac{1}{\sqrt{2}}\\0\end{pmatrix},\begin{pmatrix}\frac{1}{\sqrt{2}}\\0\end{pmatrix},\begin{pmatrix}0\\\frac{1}{\sqrt{3}}\end{pmatrix},\begin{pmatrix}0\\\frac{1}{\sqrt{3}}\end{pmatrix},\begin{pmatrix}0\\\frac{1}{\sqrt{3}}\end{pmatrix}\right\},\qquad |\mathrm{Sym}(\Phi)|=2!3!=12,$$

which is the vertices of the *trigonal bipyramid*, followed by

$$\Psi=\left\{\begin{pmatrix}\frac{1}{\sqrt{2}}\\0\end{pmatrix},\begin{pmatrix}\frac{1}{\sqrt{2}}\\0\end{pmatrix},\begin{pmatrix}0\\\frac{1}{\sqrt{2}}\end{pmatrix},\begin{pmatrix}0\\\frac{1}{\sqrt{2}}\end{pmatrix},\begin{pmatrix}0\\0\end{pmatrix}\right\},\qquad |\mathrm{Sym}(\Phi)|=8,$$

which is *four equally spaced vectors and one orthogonal*.

If $\mathrm{Sym}(\Phi)$ has an element of order 5, then Ψ must be a harmonic frame with the largest possible symmetry group (see Chapter 11), i.e., five equally spaced vectors

$$\Psi=\left\{\sqrt{\frac{2}{5}}\begin{pmatrix}\cos\frac{2\pi j}{5}\\\sin\frac{2\pi j}{5}\end{pmatrix}:j=1,\ldots,5\right\}\qquad |\mathrm{Sym}(\Phi)|=|D_5|=10,$$

and Φ is the *lifted five equally spaced vectors*. We therefore conclude:

The most symmetric tight frame of five (nonzero) vectors in $\mathbb{C}^3$ is the vertices of the trigonal bipyramid (the solution of Tammes' problem).

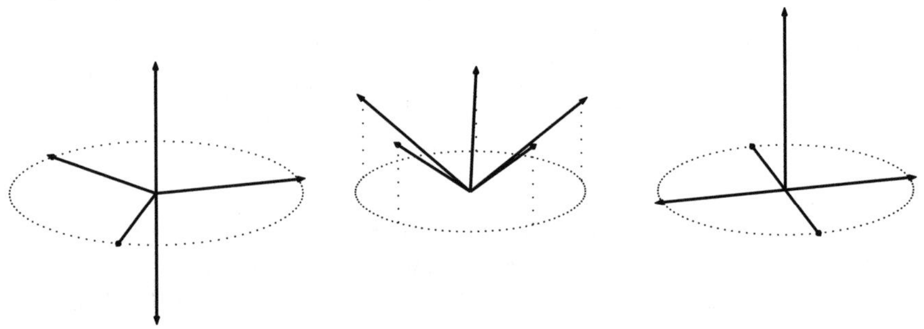

Fig. 9.2: The most symmetric tight frames of five distinct nonzero vectors in $\mathbb{R}^3$. The trigonal bipyramid (12 symmetries), five equally spaced vectors lifted (10 symmetries), and four equally spaced vectors and one orthogonal (8 symmetries).

The known examples of maximally symmetric tight frames (such as those above) suggest a close relationship with group frames (see Chapter 10):

Conjecture 9.1. A maximally symmetric tight frame is a union of group frames.

9.6 Algorithms and examples

To calculate the symmetry groups of a finite frame $\Phi = (v_j)$ of n vectors, one must determine which of the $n!$ permutations σ give a frame $(v_{\sigma j})$ which is (extended) (projectively) unitarily equivalent to Φ. This can be checked (in theory) by applying Theorem 8.3. To make this feasible (for large n) requires an algorithm which checks the inner product or m-product condition efficiently, i.e., for many permutations at a time. We mention two such algorithms.

The projective symmetry group of a frame can be viewed as the *stabiliser* (under the action of the symmetric group) of the frame graph with its m-cycles *labelled* by the corresponding m-products. It is not necessary to fully label the frame graph, e.g., the triple products suffice when the cycle space of the frame graph is spanned by the 3-cycles (Corollary 8.2). Algorithms are being developed (with Markus Grassl) which are efficient, and will be made publicly available.

We now give a simple algorithm, which is suited to hand calculations, and cases where the symmetry group is small. For frames $\Phi = (v_j)_{j\in J}$ and $\Psi = (w_j)_{j\in J}$ of n vectors, the algorithm determines the set of $\sigma \in S_J$ for which Φ and $(w_{\sigma j})$ are projectively similar, i.e.,

$$\Delta_C(v_{j_1},\ldots,v_{j_m}) = \Delta_C(w_{\sigma j_1},\ldots,w_{\sigma j_m}), \tag{9.9}$$

for all cycles $(j_1,\ldots,j_m)$ from a determining set for Φ. In particular, for $\Psi = \Phi$ it calculates $\mathrm{Sym}_\mathrm{P}(\Phi)$, and if there is some σ then Φ and Ψ are projectively similar. There are two cases:

1. $\mathrm{Sym}_\mathrm{P}(\Phi)$ is large, i.e., the m-products take few different values.
2. $\mathrm{Sym}_\mathrm{P}(\Phi)$ is small, i.e., the m-products take many different values.

An extreme example of the first is the vertices of a d-simplex (see Example 9.8), where

$$(P_\Phi)_{jk} = \begin{cases} \frac{d}{d+1}, & j=k; \\ -\frac{1}{d+1}, & j\neq k, \end{cases} \qquad \mathrm{Sym}_\mathrm{P}(\Phi) = S_J.$$

Here the m-products are all equal (for fixed m), and so it is easy to check that each $\sigma \in S_J$ is a projective symmetry.

Our algorithm is best suited to the second case: when $\mathrm{Sym}_\mathrm{P}(\Phi)$ is small, and the m-products take many different values. This is the generic situation. Indeed, if the diagonal entries of P_Φ (the 1-products) are distinct, then $|\mathrm{Sym}_\mathrm{P}(\Phi)| = 1$.

For an index set J of size n, we define a ***k*-flag** f to be an ordering of k distinct elements of J

$$f = (j_1, j_2, \ldots, j_k).$$

For a given fixed n-flag

$$f_b = (j_1,\ldots,j_n),$$

we can represent the permutation $\sigma : j_\ell \mapsto \sigma j_\ell$ (giving a projective similarity or symmetry) by the n-flag

$$f_\sigma = (\sigma j_1,\ldots,\sigma j_n).$$

Determining whether $\Phi = (v_j)$ and $\Psi = (w_{\sigma j})$ are projectively similar is equivalent to determining which of the $n!$ permutations σ, i.e., n-flags f_σ, satisfy (9.9).

We think of each possible n-flag $f_\sigma = (\sigma j_1,\ldots,\sigma j_n)$ as being built up from the 0-flag () by successively adding entries

$$f_\sigma^0 = (), \quad f_\sigma^1 = (\sigma j_1), \quad f_\sigma^2 = (\sigma j_1, \sigma j_2), \quad \ldots \quad f_\sigma^n = (\sigma j_1, \sigma j_2, \ldots, \sigma j_n).$$

We will call the operation of going from a set $\mathscr{F}_{k-1}$ of $(k-1)$-flags to a set $\mathscr{F}_k$ of k-flags as **growing**. At the k-th stage there are $n-k+1$ choices for the next entry, so that

$$|\mathscr{F}_k| = (n-k+1)|\mathscr{F}_{k-1}|.$$

If $|\mathrm{Sym}(\Phi)| < n!$, then, at some stage, not all $f_\sigma^k \in \mathscr{F}_k$ will be extendable to an n-flag satisfying (9.9). A necessary condition for such an extension to exist is that (9.9) hold for all cycles (of length $\leq k$) on the first k indices of the fixed flag $f_b = (j_1,\ldots,j_n)$ from a determining set for $(v_{j_1},\ldots,v_{j_k})$. Removing elements from $\mathscr{F}_k$ because they fail this condition (either in full or in part) will be called **pruning**. When the full condition is imposed we have a **full pruning**, otherwise a **partial**

pruning. In these terms, our algorithm for finding the set $\mathscr{F}_n$ of n-flags f_σ giving the permutations σ that Φ and $(w_{\sigma j})$ are projectively similar is:

Algorithm (to determine the n-flags $\mathscr{F}_n$ giving a projective similarity).

```
Let 𝓕₀ := {()} consist of the empty flag
for k from 1 to n do
  Grow 𝓕_{k−1} to 𝓕_k
  Prune 𝓕_k
end for
Fully prune 𝓕_n, if necessary.
```

The art is in balancing the cost of pruning, with that of growing the set of possible k-flags overly large. One can do this on a case by case basis, or by using an adaptive algorithm. The algorithm can easily be parallelised: simply partition $\mathscr{F}_k$ in any way, at any stage k, and apply the algorithm to each subset.

We now illustrate our algorithm with a couple of examples, where $\Psi = \Phi$. As a pruning rule we ask that a k-flag $(j_1,\ldots,j_k)$ match

$$\Delta_C(v_{j_1},\ldots,v_{j_k}) = \Delta_C(w_{\sigma j_1},\ldots,w_{\sigma j_k}).$$

Thus at each stage we check only one new m-product, which is easily calculated.

Example 9.13. (SICs) Consider the equiangular tight frame $\Phi := (v, Sv, \Omega v, S\Omega v)$ of four vectors for $\mathbb{C}^2$ (the second prototypical example), where

$$v := \frac{1}{\sqrt{6}}\begin{pmatrix}\sqrt{3+\sqrt{3}} \\ e^{\frac{\pi}{4}i}\sqrt{3-\sqrt{3}}\end{pmatrix}, \qquad S := \begin{pmatrix}0 & 1 \\ 1 & 0\end{pmatrix}, \qquad \Omega := \begin{pmatrix}1 & \\ & -1\end{pmatrix}.$$

We have

$$P_\Phi = \frac{1}{2}\begin{pmatrix} 1 & \frac{1}{\sqrt{3}} & \frac{1}{\sqrt{3}} & -\frac{i}{\sqrt{3}} \\ \frac{1}{\sqrt{3}} & 1 & \frac{-i}{\sqrt{3}} & \frac{1}{\sqrt{3}} \\ \frac{1}{\sqrt{3}} & \frac{i}{\sqrt{3}} & 1 & -\frac{1}{\sqrt{3}} \\ \frac{i}{\sqrt{3}} & \frac{1}{\sqrt{3}} & -\frac{1}{\sqrt{3}} & 1 \end{pmatrix}.$$

Take the base flag to be $(1,2,3,4)$. The empty flag (0-flag) $\mathscr{F}_0 = \{()\}$ grows to the set of 1-flags

$$\mathscr{F}_1 = \{(1),(2),(3),(4)\}.$$

The pruning rule is that $\langle v_1, v_1\rangle = \langle w_{\sigma 1}, w_{\sigma_1}\rangle$, i.e., the norm is preserved, and so there is no pruning. Growing gives

$$\mathscr{F}_2 = \{(1,2),(1,3),(1,4),(2,1),(2,3),(2,4),(3,1),(3,2),(3,4),(4,1),(4,2),(4,3)\},$$

and pruning gives no reduction since Φ is equiangular. We now consider growing the 2-flag $(3,2)$, the others being similar. This grows to the 3-flags $(3,2,1)$, $(3,2,4)$. Since

$$\Delta_C(v_1,v_2,v_3) = \frac{i}{24\sqrt{3}}, \quad \Delta_C(w_3,w_2,w_1) = -\frac{i}{24\sqrt{3}}, \quad \Delta_C(w_3,w_2,w_4) = \frac{i}{24\sqrt{3}},$$

the 3-flag $(3,2,1)$ is pruned. Continuing in this way gives

$$\mathscr{F}_3 = \{(1,2,3),(1,3,4),(1,4,2),(2,1,4),(2,3,1),(2,4,3),\\ (3,1,2),(3,2,4),(3,4,1),(4,1,3),(4,2,1),(4,3,2)\}.$$

The final stage $k=n$, growing does not increase the size of $\mathscr{F}_{n-1}$ and in this case, nothing gets pruned, by the rule used, or a full prune. Thus we have

$$\mathrm{Sym}_{\mathrm{P}}(\Phi) = \mathscr{F}_4 = \{(1,2,3,4),(1,3,4,2),(1,4,2,3),(2,1,4,3),(2,3,1,4),(2,4,3,1),\\ (3,1,2,4),(3,2,4,1),(3,4,1,2),(4,1,3,2),(4,2,1,3),(4,3,2,1)\}.$$

This is the alternating group A_4.

Example 9.14. (Anti projective symmetries) Applying the full pruning algorithm to the previous example, with the m-products $\Delta_C(v_{j_1},\ldots,v_{j_m})$ replaced by their conjugates, and base flag $(1,2,3,4)$ gives the following antiprojective symmetries

$$\mathscr{F}_4 = \{(1,2,4,3),(1,3,2,4),(1,4,3,2),(2,1,3,4),(2,3,4,1),(2,4,1,3),\\ (3,1,4,2),(3,2,1,4),(3,4,2,1),(4,1,2,3),(4,2,3,1),(4,3,1,2)\}.$$

Hence, we have

$$\mathrm{Sym}_{\mathrm{P}}(\Phi) = A_4 \subset \mathrm{Sym}_{\mathrm{EP}}(\Phi) = S_4.$$

In general, since the product of two antiprojective symmetries is a projective symmetry, $\mathrm{Sym}_{\mathrm{EP}}(\Phi)$ is generated by $\mathrm{Sym}_{\mathrm{P}}(\Phi)$ together with any antiprojective symmetry (if they exist).

Example 9.15. (MUBS) Let $\Phi = (v_j)$ be the following two MUBs in $\mathbb{C}^2$ (see §8.6)

$$v_1 = e_1, \quad v_2 = e_2, \quad v_3 = \frac{1}{\sqrt{2}}(e_1+e_2), \quad v_4 = \frac{1}{\sqrt{2}}(e_1-e_2).$$

Here

$$P_\Phi = \frac{1}{2}\begin{pmatrix} 1 & 0 & \frac{1}{\sqrt{2}} & -\frac{1}{\sqrt{2}} \\ 0 & 1 & \frac{1}{\sqrt{2}} & \frac{1}{\sqrt{2}} \\ \frac{1}{\sqrt{2}} & \frac{1}{\sqrt{2}} & 1 & 0 \\ -\frac{1}{\sqrt{2}} & \frac{1}{\sqrt{2}} & 0 & 1 \end{pmatrix}.$$

We arrive at the same $\mathscr{F}_2$ as in Example 9.13, without pruning. The pruning rule says that modulus of the inner product between v_1 and v_2 must be preserved. Since this is zero, the index pairs in $\mathscr{F}_2$ must correspond to pairs of orthogonal vectors, which leads to the pruning

$$\mathscr{F}_2 = \{(1,2),(2,1),(3,4),(4,3)\}.$$

Growing this gives

$$\mathscr{F}_3 = \{(1,2,3),(1,2,4),(2,1,3),(2,1,4),(3,4,1),(3,4,2),(4,3,1),(4,3,2)\}.$$

All 3-products for distinct vectors are zero, and so there is no pruning at this stage. Growing, then full pruning leads to

$$\begin{aligned}\mathrm{Sym}_{\mathrm{P}}(\Phi) = \mathscr{F}_4 = \{&(1,2,3,4),(1,2,4,3),(2,1,3,4),(2,1,4,3),\\ &(3,4,1,2),(3,4,2,1),(4,3,1,2),(4,3,2,1)\}.\end{aligned}$$

This group is the dihedral group of order 8 (the only subgroup of S_4 of order 8), which is generated by the following permutations

$$(1324) \quad \text{(rotation through 90 degrees)}, \qquad (34) \quad \text{(reflection in the } x\text{-axis)}.$$

For d a prime power, $d+1$ MUBs for $\mathbb{C}^d$ can be constructed from the columns of elements $R^\ell F$ from the Clifford group (see Theorem 12.22, §14.7). The projective symmetry groups, as calculated by our algorithm, for the first few d are given in Table 9.1.

Table 9.1: The projective symmetry groups $\mathrm{Sym}_{\mathrm{P}}(\Phi)$ and $\mathrm{Sym}_{\mathrm{EP}}(\Phi)$ for Φ the tight frame of n vectors given by $d+1$ MUBs in $\mathbb{C}^d$, including the transitive subgroups of $\mathrm{Sym}_{\mathrm{P}}(\Phi)$.

d	n	$\mathrm{Sym}_{\mathrm{P}}(\Phi)$	$\mathrm{Sym}_{\mathrm{EP}}(\Phi)$	transitive subgroups of $\mathrm{Sym}_{\mathrm{P}}(\Phi)$
2	6	$<24,12>$	$<48,48>$	$<6,1>, <12,3>$
3	12	$<216,153>$	$<432,734>$	$<72,41>$
4	20	1920	3840	$<20,3>, <60,5>, <80,49>, <120,34>,$ $<160,234>, <320,1635>, <960,11357>$
5	30	3000	6000	$<600,150>$

The projective symmetry group of harmonic frames is considered in §11.12.

9.7 Case study: 16 equiangular lines in $\mathbb{R}^6$

We consider the 16 equiangular lines in $\mathbb{R}^6$ as presented by Janet Tremain [Tre08]

$$V = [v_j] = \begin{bmatrix} 1 & -1 & -1 & -1 & -1 & -1 & 1 & 1 & 1 & 1 & 1 & 1 & 1 & 1 & 1 & 1 \\ 1 & -1 & 1 & 1 & 1 & 1 & -1 & -1 & -1 & -1 & 1 & 1 & 1 & 1 & 1 & 1 \\ 1 & 1 & -1 & 1 & 1 & 1 & -1 & 1 & 1 & 1 & -1 & -1 & -1 & 1 & 1 & 1 \\ 1 & 1 & 1 & -1 & 1 & 1 & 1 & -1 & 1 & 1 & -1 & 1 & 1 & -1 & -1 & 1 \\ 1 & 1 & 1 & 1 & -1 & 1 & 1 & 1 & -1 & 1 & 1 & -1 & 1 & -1 & 1 & -1 \\ 1 & 1 & 1 & 1 & 1 & -1 & 1 & 1 & 1 & -1 & 1 & 1 & -1 & 1 & -1 & -1 \end{bmatrix}$$

The projective symmetry group has order $11520 = 2^8 \cdot 3 \cdot 5$, and is generated by the two permutations

$$a = (1,2,6,11,16,9)(3,12,8)(4,14,7)(5,10),$$
$$b = (1,12,15,16,5,10)(2,11,4)(3,6,14,8,13,9).$$

We now seek a unitary matrix L_σ, $\sigma \in \{a,b\}$, with

$$L_\sigma v_j = \alpha_j v_{\sigma j}, \qquad \forall j.$$

By Corollary 8.4, $Q = L_\sigma$ is unique up to a scalar ± 1, and can be calculated from its action on a basis of the v_j, once a suitable choice of the scalars $\alpha_j = \pm 1$ is known. We suppose that $\alpha_1 = 1$, so that

$$\langle v_1, v_j \rangle = \langle L_\sigma v_1, L_\sigma v_j \rangle = \langle \alpha_1 v_{\sigma 1}, \alpha_j v_{\sigma j} \rangle \quad \Longrightarrow \quad \alpha_j = \frac{\langle v_{\sigma 1}, v_{\sigma j} \rangle}{\langle v_1, v_j \rangle}, \quad \forall j.$$

In this way, we obtain

$$L_a = \begin{bmatrix} 0 & 0 & 0 & 0 & -1 & 0 \\ 0 & -1 & 0 & 0 & 0 & 0 \\ 0 & 0 & 0 & 1 & 0 & 0 \\ 0 & 0 & 1 & 0 & 0 & 0 \\ 0 & 0 & 0 & 0 & 0 & 1 \\ 1 & 0 & 0 & 0 & 0 & 0 \end{bmatrix}, \qquad L_b = \begin{bmatrix} 0 & 0 & 0 & 0 & 1 & 0 \\ 0 & 0 & 1 & 0 & 0 & 0 \\ 0 & 0 & 0 & 0 & 0 & -1 \\ 1 & 0 & 0 & 0 & 0 & 0 \\ 0 & -1 & 0 & 0 & 0 & 0 \\ 0 & 0 & 0 & 1 & 0 & 0 \end{bmatrix}.$$

The group $G = \langle L_a, L_b \rangle$ is irreducible and has order 23040 (it contains the scalar matrix $-I$). Since the projective symmetry group is transitive, it follows (see §10.7) that the 16 equiangular lines $\{\pm v_j\}$ are an irreducible G-frame.

9.8 Case study: A spherical $(4,4)$-design of 12 lines in $\mathbb{C}^2$

Several unit-norm spherical $(4,4)$-designs of 12 vectors (lines) in $\mathbb{C}^2$ were computed numerically using the techniques of §6.16. The projective symmetry group for each was calculated (as described in §9.6) to be the dihedral group of order 10, with the projective action giving two orbits: one of size 2 (with the vectors orthogonal), and one of size 10. This suggests there is a $(4,4)$-design of the form

$$\Phi = (v, av, a^2v, a^3v, a^4v, bv, abv, a^2bv, a^3bv, a^4bv) \cup (u_1, u_2), \qquad (9.10)$$

where $v \in \mathbb{C}^2$ is a unit vector, a (a rotation) and b (a reflection) are generators of the dihedral group and $\{u_1, u_2\}$ is an orthonormal basis. Taking

$$a = \begin{pmatrix} \omega & 0 \\ 0 & \overline{\omega} \end{pmatrix}, \quad \omega := e^{\frac{2\pi i}{5}}, \qquad b = \begin{pmatrix} 0 & 1 \\ 1 & 0 \end{pmatrix}, \quad u_1 = \begin{pmatrix} 1 \\ 0 \end{pmatrix}, \; u_2 = \begin{pmatrix} 0 \\ 1 \end{pmatrix}, \qquad (9.11)$$

and optimising over v to obtain a $(4,4)$-design numerically suggested that the ratio of the components of a suitable v was the *golden ratio* $\frac{\sqrt{5}+1}{2}$, i.e.,

$$v := \frac{1}{\sqrt{10+2\sqrt{5}}} \begin{pmatrix} (1+\sqrt{5})\zeta \\ 2 \end{pmatrix}, \qquad |\zeta| = 1. \qquad (9.12)$$

An elementary calculation (see Exer. 9.7) shows that (9.10), (9.11), (9.12) define a spherical $(4,4)$-design Φ of 12 unit vectors for $\mathbb{C}^2$.

The general method used here is known as *precision bumping* (see §14.20, §14.24).

Notes

The symmetry group of a finite tight frame (which is a natural notion) was first studied in [VW05], where it was considered as a group of unitary matrices. The definition given here (a group of permutations acting on the frame) was introduced in [VW10]. The projective symmetry group has been studied for SICs [AFF11], [Zhu12], and by [WC14] for a general finite frame. The calculation of the projective symmetry group as the stabiliser of a suitably labelled frame graph was initiated by Markus Grassl.

Exercises

9.1. Show that following inclusions can be strict
(a) $\mathrm{Sym}(\Phi)\times\mathrm{Sym}(\Psi)\subset\mathrm{Sym}(\Phi\cup\Psi)$.
(b) $\mathrm{Sym}(\Phi)\times\mathrm{Sym}(\Psi)\subset\mathrm{Sym}(\Phi\hat{+}\Psi)$.
(c) $\mathrm{Sym}(\Phi)\times\mathrm{Sym}(\Psi)\subset\mathrm{Sym}(\Phi\otimes\Psi)$.
(d) $\mathrm{Sym}(\Phi)\cap\mathrm{Sym}(\Psi)\subset\mathrm{Sym}(\Phi\oplus\Psi)$.

9.2. Let Φ be a frame of n vectors for $\mathcal{H}$, $\dim(\mathcal{H})=d$.
(a) Show that $|\mathrm{Sym}(\Phi)|$ divides $n!$.
(b) Let π_Φ be the representation $\sigma\mapsto L_\sigma$ of (9.1). Show that if Φ has distinct vectors, then π_Φ is faithful, and

$$|\pi_\Phi(\mathrm{Sym}(\Phi))|\le m(m-1)\cdots(m-d+1),$$

where m is the number of distinct vectors in Φ.

9.3. Here we consider the map $\pi_\Phi:\mathrm{Sym}(\Phi)\to GL(\mathcal{H}):\sigma\mapsto L_\sigma$ given by (9.1).
(a) Show that π_Φ is a group homomorphism, i.e., is a linear representation (action) of the group $G=\mathrm{Sym}(\Phi)$ on $\mathcal{H}$.
(b) Let S_Φ be the frame operator of Φ. Show that if $g\in GL(\mathcal{H})$, then

$$S_\Phi(gf)=(g^*)^{-1}S_{g^*\Phi}(f),\qquad \forall f\in\mathcal{H}.$$

(c) Show that if Φ is tight, then the action of $\mathrm{Sym}(\Phi)$ on $\mathcal{H}$ commutes with S_Φ, i.e.,

$$S_\Phi(\sigma f)=\sigma S_\Phi(f),\qquad \forall\sigma\in\mathrm{Sym}(\Phi),\quad \forall f\in\mathcal{H}.$$

9.4. Let Φ be a sequence of n vectors. Show that if $\mathrm{Sym}(\Phi)=S_n$, then either Φ is a basis or the vertices of the simplex (up to projective unitary equivalence).
Remark: Therefore, the symmetry group of θ-isogonal configuration of n vectors (see Example 3.9) is S_n.

9.5. Describe the symmetry group of an α-partition frame for $\mathbb{R}^d$ (see §2.9).

9.6. Find the symmetry groups and their action on $\mathbb{C}^2$ of the tight frames

$$\Phi:=\left(\begin{bmatrix}1\\1\end{bmatrix},\begin{bmatrix}\omega\\\omega^2\end{bmatrix},\begin{bmatrix}\omega^2\\\omega\end{bmatrix}\right),\qquad \Psi:=\left(\begin{bmatrix}1\\1\end{bmatrix},\begin{bmatrix}1\\\omega\end{bmatrix},\begin{bmatrix}1\\\omega^2\end{bmatrix}\right),\qquad \omega:=e^{\frac{2\pi i}{3}},$$

of Example 2.8.

9.7. Show that

$$\Phi=(v,av,a^2v,a^3v,a^4v,bv,abv,a^2bv,a^3bv,a^4bv)\cup(u_1,u_2),$$

as given by (9.11) and (9.12) is a spherical $(4,4)$-design of 12 unit vectors for $\mathbb{C}^2$.

Chapter 10
Group frames

Here we introduce an important example of a *structured frame*, i.e., one in which the frame vectors can be obtained from the index set in a simple way. Prototypical examples of such frames include Gabor and wavelet systems.

We consider the analogue of a Gabor system, where the index set is a finite group G, and the vectors are an orbit under a unitary action of G. This is equivalent to a frame whose symmetry group acts transitively on its vectors (Theorem 10.4). A useful example to keep in mind is the n equally spaced unit vectors in $\mathbb{R}^2$

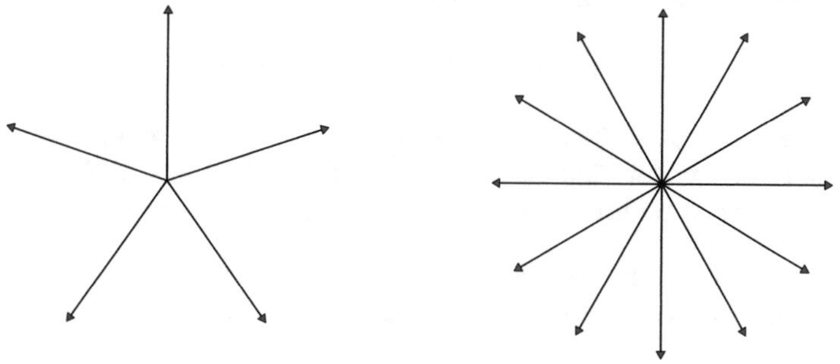

or the vertices of a Platonic solid in $\mathbb{R}^3$.

A *group frame* Φ for $\mathscr{H}$ will be, effectively, a frame of the form

$$\Phi = (gv)_{g \in G}, \qquad v \in \mathscr{H}, \tag{10.1}$$

where $G \subset GL(\mathscr{H})$ is a finite group of linear transformations, i.e., one which is the orbit of a *single* vector v under the linear action of G. Many important frames come in this way, e.g., the vertices of the Platonic solids, the harmonic frames, many equiangular tight frames (including all the known SICs), and all the known MUBs.

By using representation theory, we will consider increasingly more general cases. We start with the case when the G-orbit of all vectors $v \neq 0$ gives a frame for $\mathscr{H}$ and finish with a complete characterisation of the G-invariant frames (Theorem 10.9).

S.F.D. Waldron, *An Introduction to Finite Tight Frames*, Applied and Numerical Harmonic Analysis, https://doi.org/10.1007/978-0-8176-4815-2_10

10.1 Representations and G-frames

Suppose that G is a finite abstract group. A **representation** or **linear action** of G on a (finite dimensional) vector space $\mathcal{H}$ (over $\mathbb{F}$) is a group homomorphism

$$\rho_{\mathcal{H}} : G \to GL(\mathcal{H}).$$

The *linear action* of G on $\mathcal{H}$ given by $\rho = \rho_{\mathcal{H}}$ will often be written

$$gv := \rho(g)(v), \qquad g \in G,\ v \in \mathcal{H}.$$

Equivalent terminology is that $\mathcal{H}$ is an $\mathbb{F}G$-**module**, or just a G-**module** (if the field $\mathbb{F}$ and action of G is clear from the context). This is because $\mathcal{H}$ is a module over the **group algebra** $\mathbb{F}G$ (the $\mathbb{F}$-vector space with a basis given by the elements of G and multiplication given by extending the multiplication of G linearly). A subspace (or set) $V \subset \mathcal{H}$ is said to be G-**invariant** if $gv \in V$, $\forall v \in V$, and so

The G-invariant subspaces of $\mathcal{H}$ are precisely the $\mathbb{F}G$-submodules of $\mathcal{H}$.

Two representations $(\mathcal{H}, \rho_{\mathcal{H}})$ and $(\mathcal{K}, \rho_{\mathcal{K}})$ are said to be *equivalent* if there is an invertible linear map $T : \mathcal{H} \to \mathcal{K}$ such that

$$\rho_{\mathcal{K}}(g) = T\rho_{\mathcal{H}}(g)T^{-1}, \qquad \forall g \in G.$$

Any representation of G is equivalent to one in which $\mathcal{H}$ is a Hilbert space, and all the $\rho_{\mathcal{H}}(g)$ are unitary transformations (see Exer. 10.4). It therefore suffices to consider frames of the form (10.1) where the action is unitary (see Corollary 10.1).

Definition 10.1. Let G be a finite group. We say that a frame $(\phi_g)_{g\in G}$ for $\mathcal{H}$ is a **group frame** or G-**frame** if there exists a unitary representation $\rho : G \to \mathcal{U}(\mathcal{H})$ such that

$$g\phi_h := \rho(g)\phi_h = \phi_{gh}, \qquad \forall g,h \in G.$$

Example 10.1. Let $C_n = \langle a \rangle$ be the cyclic group of order n. Then the n equally spaced unit vectors are a C_n-frame $(gv)_{g\in C_n}$ for $\mathbb{R}^2$ given by the unitary group action

$$a^j v := \rho_{\mathbb{R}^2}(a^j)v = \rho_{\mathbb{R}^2}(a)^j v, \qquad \rho_{\mathbb{R}^2}(a) := \begin{pmatrix} \cos\frac{2\pi}{n} & -\sin\frac{2\pi}{n} \\ \sin\frac{2\pi}{n} & \cos\frac{2\pi}{n} \end{pmatrix}. \tag{10.2}$$

Since the representation (10.2) is *faithful*, i.e., injective, we can take G to be the abstract group C_n, or the isomorphic copy $\rho_{\mathbb{R}^2}(G) = \langle \rho_{\mathbb{R}^2}(a) \rangle$. In contrast, $([1])_{g\in C_n}$ is a C_n-frame for $\mathbb{R}$, via the trivial representation ($a^j v := v$, $\forall v \in \mathbb{R}$), but here G cannot be taken to be $\rho_{\mathbb{R}}(G) = \{[1]\} \subset \mathcal{U}(\mathbb{R})$ (the trivial group). This "repetition" has certain technical advantages, e.g., for taking combinations (see Theorem 10.2), or decomposing a G-frame into its constitute parts (see Theorem 10.7).

10.2 The frame operator of a G-frame

Since unitary maps preserve lengths, the definition implies that

A G-frame is an equal-norm frame.

The *unitary* action of G in a G-frame gives:

Lemma 10.1. *The frame operator S of a G-frame commutes with (the unitary action of) G, i.e.,*

$$S(hf) = hS(f), \qquad \forall h \in G,\ \forall f \in \mathscr{H}.$$

Proof. Let $\Phi = (\phi_g)_{g\in G}$ be a G-frame for $\mathscr{H}$, with frame operator $S = S_\Phi$. Then $\rho(h)^* = \rho(h)^{-1} = \rho(h^{-1})$ gives

$$\begin{aligned} S(hf) &= \sum_{g\in G} \langle hf, \phi_g\rangle \phi_g = h \sum_{g\in G} \langle f, h^{-1}\phi_g\rangle h^{-1}\phi_g \\ &= h \sum_{g\in G} \langle f, \phi_{h^{-1}g}\rangle \phi_{h^{-1}g} = hS(f), \end{aligned}$$

as supposed. □

The G-frame structure carries over to the dual and canonical tight frames.

Theorem 10.1. *If Φ is G-frame for $\mathscr{H}$, then so is*

- *The dual frame $\tilde{\Phi}$.*
- *The canonical tight frame Φ^{can}.*
- *Any unitarily equivalent frame Ψ.*

In particular, Φ^{can} is an equal-norm tight frame.

Proof. Let $\Phi = (\phi_g)_{g\in G}$ be a G-frame for $\mathscr{H}$. By Lemma 10.1, the frame operator $S = S_\Phi$ commutes with G, and hence with S^{-1} and $S^{-\frac{1}{2}}$ (as these can be written as power series in S). Thus the dual frame $\tilde{\Phi} = (\tilde{\phi}_g)_{g\in G}$ satisfies

$$g\tilde{\phi}_h = gS^{-1}\phi_h = S^{-1}g\phi_h = S^{-1}\phi_{gh} = \tilde{\phi}_{gh}, \qquad \forall g,h \in G,$$

and so is a G-frame. A similar argument (with $S^{-\frac{1}{2}}$) shows that Φ^{can} is a G-frame.

Suppose that Ψ is unitarily equivalent to Φ, say $\Psi = cU\Phi$, $c > 0$, $U \in \mathscr{U}(\mathscr{H})$. Let ρ_Φ be the representation with $\rho_\Phi(g)\phi_h = \phi_{gh}$, $\forall g,h$, and define a representation $\rho_\Psi : G \to \mathscr{U}(\mathscr{H})$ by $\rho_\Psi(g) := U\rho_\Phi U^{-1}$. Then ρ_Ψ is unitary, and

$$\rho_\Psi(g)\psi_h = U\rho_\Phi(g)U^{-1}cU\phi_h = cU\rho_\Phi(g)\phi_h = cU\phi_{gh} = \psi_{gh}, \qquad \forall g,h \in G,$$

so that Ψ is a G-frame. □

Example 10.2. A θ-isogonal configuration $(x_j)_{j=1}^d$ of d unit vectors for $\mathbb{R}^d$, i.e., one with

$$\langle x_j, x_k \rangle = \cos\theta \neq -\frac{1}{d-1}, \qquad j \neq k$$

is a group frame (Exer. 10.13), and hence so is its dual. Its dual is also isogonal (see Exer. 3.19). The canonical tight frame, an orthonormal basis, is also a group frame.

A frame given by a linear action $\rho : G \to GL(\mathscr{H})$ is similar to one given by a unitary action:

Corollary 10.1. *Let G be a finite group, and $\Phi = (\phi_g)_{g\in G}$ be a frame for $\mathscr{H}$. Then the following are equivalent*

1. $\Phi = (\rho(g)v)_{g\in G}$ *for* $\rho : G \to GL(\mathscr{H})$ *a group homomorphism and* $v \in \mathscr{H}$.
2. Φ^{can} *is a (tight) G-frame.*

Proof. 1$\Longrightarrow$2. If $\Phi = (\rho(g)v)_{g\in G}$ is a frame, then it is similar to a tight G-frame Ψ^{can} (see Exer. 10.4). Since similar frames have unitarily equivalent canonical tight frames (Theorem 3.4), Φ^{can} is unitarily equivalent to Ψ^{can}, and hence is a G-frame.

2$\Longrightarrow$1. Suppose that $\Phi^{\mathrm{can}} = (\tau(g)v)_{g\in G}$ is a G-frame, and let S be the frame operator for Φ. Then

$$\Phi = S^{\frac{1}{2}}\Phi^{\mathrm{can}} = (S^{\frac{1}{2}}\tau(g)S^{-\frac{1}{2}}w)_{g\in G}, \qquad w := S^{\frac{1}{2}}v,$$

where $G \to GL(\mathscr{H}) : g \mapsto S^{\frac{1}{2}}\tau(g)S^{-\frac{1}{2}}$ is a group homomorphism. □

For many of the methods of combining frames (Section 5), a combination of group frames is again a group frame.

Theorem 10.2. *Group frames can be combined as follows.*

1. *The direct sum of disjoint G-frames is a G-frame.*
2. *The sum of a G_1-frame and a G_2-frame is a $G_1 \times G_2$-frame.*
3. *The tensor product of G_1-frame with a G_2-frame is a $G_1 \times G_2$-frame.*

Proof. Use the notation $x \oplus y$ and $x \otimes y$ for the elements of $\mathscr{H}_1 \oplus \mathscr{H}_2$ and $\mathscr{H}_1 \otimes \mathscr{H}_2$.

1. If $\Phi = (gv)_{g\in G}$ and $\Psi = (gw)_{g\in G}$ are disjoint G-frames, then their direct sum is $\Phi \oplus \Psi = (g(v\oplus w))_{g\in G}$, where $g(x\oplus y) := gx \oplus gy$ is a unitary action.

2 and 3. Let $\Phi_j = (g_j v_j)_{g_j\in G_j}$ be G_j-frames. Then

$$(g_1,g_2)(x\oplus y) := g_1x \oplus g_2y, \qquad (g_1,g_2)(x\otimes y) := g_1x \otimes g_2y$$

define a unitary action of $G_1 \times G_2$ on $\mathscr{H}_1 \oplus \mathscr{H}_2$ and $\mathscr{H}_1 \otimes \mathscr{H}_2$, respectively. Thus

$$\Phi_1 \hat{+} \Phi_2 = \Big((g_1,g_2)(\frac{1}{\sqrt{|G_2|}}v_1 \oplus \frac{1}{\sqrt{|G_1|}}v_2)\Big)_{(g_1,g_2)\in G_1\times G_2},$$
$$\Phi_1 \otimes \Phi_2 = \big((g_1,g_2)(v_1\otimes v_2)\big)_{(g_1,g_2)\in G_1\times G_2},$$

and so the sum and tensor product of group frames is again a group frame. □

10.3 Group matrices and the Gramian of a G-frame

If $\Phi = (\phi_g)_{g\in G}$ is a G-frame for $\mathcal{H}$, then the unitary action of G gives

$$g^*v := \rho(g)^*v = \rho(g)^{-1}v = \rho(g^{-1})v = g^{-1}v, \qquad \forall f \in \mathcal{H},$$

and so the entries of its Gramian $[\langle \phi_h, \phi_g \rangle]_{g,h\in G}$ have the special form

$$\langle \phi_h, \phi_g \rangle = \langle h\phi_1, g\phi_1 \rangle = \langle g^{-1}h\phi_1, \phi_1 \rangle. \tag{10.3}$$

This is an example of what is called a group matrix or G-matrix.

Definition 10.2. Let G be a finite group. A matrix $A = [a_{gh}]_{g,h\in G}$ is a G-**matrix** (or **group matrix**[1]) if there exists a function $\nu : G \to \mathbb{C}$ such that

$$a_{gh} = \nu(g^{-1}h), \qquad \forall g,h \in G.$$

It is natural to think of an $n \times n$ matrix A as a G-matrix for some G of order n if there is an indexing of its entries by G which yields a G-matrix. However, we do not take this as our definition, as it complicates the description of the algebraic properties of group matrices.

Example 10.3. Let $G = D_3 \cong S_3$ be the dihedral group of order 6, i.e.,

$$G = D_3 = \langle a,b : a^3 = 1, b^2 = 1, b^{-1}ab = a^{-1} \rangle, \tag{10.4}$$

and order its elements $1, a, a^2, b, ab, a^2b$. Then each G-matrix has the form

$$\begin{array}{c|cccccc} & 1 & a & a^2 & b & ab & a^2b \\ \hline 1 & \nu(1) & \nu(a) & \nu(a^2) & \nu(b) & \nu(ab) & \nu(a^2b) \\ a & \nu(a^2) & \nu(1) & \nu(a) & \nu(a^2b) & \nu(b) & \nu(ab) \\ a^2 & \nu(a) & \nu(a^2) & \nu(1) & \nu(ab) & \nu(a^2b) & \nu(b) \\ b & \nu(b) & \nu(a^2b) & \nu(ab) & \nu(1) & \nu(a^2) & \nu(a) \\ ab & \nu(ab) & \nu(b) & \nu(a^2b) & \nu(a) & \nu(1) & \nu(a^2) \\ a^2b & \nu(a^2b) & \nu(ab) & \nu(b) & \nu(a^2) & \nu(a) & \nu(1) \end{array}. \tag{10.5}$$

Example 10.4. If G is a cyclic group, say $\mathbb{Z}_n$ with its elements ordered $0, 1, \ldots, n-1$, then a G-matrix is a **circulant** matrix (see [Dav79])

[1] Recently there has been a revival of interest in group matrices, see, e.g., [BR04] and [Joh07]. Some authors write the (g,h)-entry as $\nu(gh^{-1})$, and variations thereof.

$$A = \begin{bmatrix} c_0 & c_1 & \cdots & c_{n-2} & c_{n-1} \\ c_{n-1} & c_0 & c_1 & & c_{n-2} \\ \vdots & c_{n-1} & c_0 & \ddots & \vdots \\ c_2 & & \ddots & \ddots & c_1 \\ c_1 & c_2 & \cdots & c_{n-1} & c_0 \end{bmatrix}, \qquad a_{jk} = c_{k-j} := \nu(k-j).$$

We now show:

A frame is a G-frame if and only if its Gramian is a G-matrix.

Theorem 10.3. *(Characterisation) Let G be a finite group. Then $\Phi = (\phi_g)_{g\in G}$ is a G-frame (for its span $\mathscr{H}$) if and only if its Gramian is a G-matrix.*

Proof. If Φ is a G-frame, then (10.3) implies that its Gramian is a G-matrix.

Conversely, suppose that the Gramian of a frame Φ for $\mathscr{H}$ is a G-matrix. Let $\tilde{\Phi} = (\tilde{\phi}_g)_{g\in G}$ be the dual frame, so that

$$f = \sum_{g\in G} \langle f, \tilde{\phi}_g \rangle \phi_g, \qquad \forall f \in \mathscr{H}. \tag{10.6}$$

For each $g \in G$, define a linear operator $U_g : \mathscr{H} \to \mathscr{H}$ by

$$U_g(f) := \sum_{h_1\in G} \langle f, \tilde{\phi}_{h_1} \rangle \phi_{gh_1}, \qquad \forall f \in \mathscr{H}.$$

Since $\operatorname{Gram}(\Phi) = [\langle \phi_h, \phi_g \rangle]_{g,h\in G}$ is a G-matrix, we have

$$\langle \phi_{gh_1}, \phi_{gh_2} \rangle = \nu((gh_2)^{-1}gh_1) = \nu(h_2^{-1}h_1) = \langle \phi_{h_1}, \phi_{h_2} \rangle. \tag{10.7}$$

It follows from (10.6) and (10.7) that U_g is unitary, by the calculation

$$\begin{aligned} \langle U_g(f_1), U_g(f_2) \rangle &= \langle \sum_{h_1\in G} \langle f_1, \tilde{\phi}_{h_1} \rangle \phi_{gh_1}, \sum_{h_2\in G} \langle f_2, \tilde{\phi}_{h_2} \rangle \phi_{gh_2} \rangle \\ &= \sum_{h_1\in G} \sum_{h_2\in G} \langle f_1, \tilde{\phi}_{h_1} \rangle \overline{\langle f_2, \tilde{\phi}_{h_2} \rangle} \langle \phi_{gh_1}, \phi_{gh_2} \rangle \\ &= \sum_{h_1\in G} \sum_{h_2\in G} \langle f_1, \tilde{\phi}_{h_1} \rangle \overline{\langle f_2, \tilde{\phi}_{h_2} \rangle} \langle \phi_{h_1}, \phi_{h_2} \rangle \\ &= \langle \sum_{h_1\in G} \langle f_1, \tilde{\phi}_{h_1} \rangle \phi_{h_1}, \sum_{h_2\in G} \langle f_2, \tilde{\phi}_{h_2} \rangle \phi_{h_2} \rangle = \langle f_1, f_2 \rangle. \end{aligned}$$

Similarly, we have

$$U_g \phi_h = \sum_{h_1\in G} \langle \phi_h, \tilde{\phi}_{h_1} \rangle \phi_{gh_1} = \sum_{h_1\in G} \langle \phi_{gh}, \tilde{\phi}_{gh_1} \rangle \phi_{gh_1} = \phi_{gh}.$$

This implies that $\rho : G \to \mathscr{U}(\mathscr{H}) : g \mapsto U_g$ is a group homomorphism, since

$$U_{g_1 g_2}\phi_h = \phi_{g_1 g_2 h} = U_{g_1}\phi_{g_2 h} = U_{g_1}U_{g_2}\phi_h, \qquad \mathscr{H} = \mathrm{span}\{\phi_h\}_{h\in G}.$$

Thus ρ is a unitary representation of G on $\mathscr{H}$, with

$$\rho(g)\phi_h = \phi_{gh}, \qquad \forall g,h \in G,$$

i.e., Φ is a G-frame for $\mathscr{H}$. □

In particular, we can characterise normalised tight G-frames:

Corollary 10.2. *Let G be a finite group. Then $\Phi = (\phi_g)_{g\in G}$ is a normalised tight G-frame (for its span $\mathscr{H}$) if and only if its Gramian P is a G-matrix which is a projection, i.e., $P^2 = P$.*

Proof. A finite frame is a normalised tight frame if and only if its Gramian P is an orthogonal projection matrix (Theorem 2.1), i.e., $P^2 = P$ and $P = P^*$. The result then follows, since every Gramian satisfies $P = P^*$. □

This extends to spanning sequences via the canonical coordinates (Chapter 4).

Corollary 10.3. *A spanning sequence $\Phi = (\phi_g)_{g\in G}$ for an $\mathbb{F}$-vector space (with $\overline{\mathbb{F}} = \mathbb{F}$) is similar to a G-frame if and only if the canonical Gramian $P = P_\Phi$ is a G-matrix. In this case, we can take $\Psi = (P_\Phi e_g)_{g\in G}$ with the unitary action given by*

$$g(Pe_h) = Pe_{gh}.$$

We will see in §13.1 that the G-matrices form an algebra, which is isomorphic to the *group algebra* $\mathbb{C}G$, via the correspondence

$$A = [\nu(g^{-1}h)]_{g,h\in G} \quad \Longleftrightarrow \quad \sum_{g\in G} \nu(g)g \in \mathbb{C}G. \tag{10.8}$$

Thus G-frames (up to similarity) can also be identified with elements of $\mathbb{C}G$.

We now show that if Φ is similar to a G-frame, then its complement is similar to a G-frame.

Corollary 10.4. *The complement of a tight G-frame is a tight G-frame.*

Proof. Let Ψ be the complement of a normalised tight G-frame $\Phi = (\phi_g)_{g\in G}$. Since $\mathrm{Gram}(\Phi) + \mathrm{Gram}(\Psi)$ is the identity matrix, we have

$$\langle \psi_h, \psi_g \rangle = \begin{cases} -\langle \phi_h, \phi_g \rangle = -\langle g^{-1}h\phi_1, \phi_1 \rangle, & h \neq g; \\ 1 - \langle \phi_h, \phi_g \rangle = 1 - \langle g^{-1}h\phi_1, \phi_1 \rangle, & h = g \end{cases}$$

i.e., Ψ is a G-frame. □

10.4 Identifying G-frames from their Gramian

The previous results imply that we can determine whether a frame Φ is similar to a G-frame by considering its (canonical) Gramian P_Φ.

This gives simple *necessary conditions* for Φ to be similar to a G-frame, e.g.,

- The diagonal entries of P_Φ must be constant.
- The entries of every row/column of P_Φ must be the same (up to reordering).

By investigating the symmetry group $\mathrm{Sym}(\Phi)$ (which can be calculated), one can determine whether a frame Φ is similar to a G-frame (possibly with repetitions).

Theorem 10.4. *A frame Φ with n vectors is similar to a G-frame if and only if its symmetry group* $\mathrm{Sym}(\Phi)$ *has a transitive subgroup G. In this case, Φ with each of its vectors repeated $|G|/n$ times is similar to the tight G-frame*

$$(\sigma P_\Phi e_j)_{\sigma\in G} = (P_\Phi e_{\sigma j})_{\sigma\in G} \qquad (j \text{ fixed}).$$

Proof. Let $\Phi = (\phi_j)_{j=1}^n$. The action of $\sigma \in \mathrm{Sym}(\Phi)$ on $\mathrm{ran}(P_\Phi)$ given by

$$\sigma(P_\Phi e_j) := P_\Phi e_{\sigma j}$$

is unitary (since $(P_\Phi e_j)_{j=1}^n$ is a tight frame). Hence, for j fixed, $(\sigma P_\Phi e_j)_{\sigma\in G}$ is a tight G-frame, which is similar to $\Psi = (\phi_{\sigma j})_{\sigma\in G}$. If G is transitive, i.e., for any $j,k \in \{1,\ldots,n\}$ there is a permutation $\sigma \in G$ mapping j to k, then Ψ consists of $|G|/n$ copies of Φ.

Conversely, if $\mathrm{Sym}(\Phi)$ is not transitive, then Φ is not a group frame. □

Example 10.5. Let $\Phi = (u_j)_{j=1}^3$ be the three equally spaced unit vectors in $\mathbb{R}^2$ (Mercedes-Benz frame), which has symmetry group $\mathrm{Sym}(\Phi) = S_3$ (Exer. 9.2).

The transitive group S_3 has one transitive proper subgroup, i.e., the cyclic group

$$C_3 = \langle a \rangle, \qquad a = (123).$$

The canonical Gramian of Φ, which determines $\mathrm{Sym}(\Phi)$, is

$$P_\Phi = \frac{2}{3}\begin{pmatrix} 1 & -\frac{1}{2} & -\frac{1}{2} \\ -\frac{1}{2} & 1 & -\frac{1}{2} \\ -\frac{1}{2} & -\frac{1}{2} & 1 \end{pmatrix} = \frac{2}{3}[v_1, v_2, v_3].$$

Thus (by fixing $j = 1$) Φ can be thought of as a S_3-frame $(v_{\sigma 1})_{\sigma\in S_3}$ of six vectors, or as a C_3-frame $(v_{\sigma 1})_{\sigma\in C_3}$ of three vectors.

Suppose the action of a on Φ is rotation through $\frac{2\pi}{3}$, and $b = (23)$ acts as the reflection which fixes u_1, i.e., $bu_1 = u_1$, $bu_2 = u_3$, $bu_3 = u_2$. Then the S_3-frame $(v_{\sigma 1})_{\sigma\in S_3}$ is similar to the S_3-frame $(\sigma u_1)_{\sigma\in S_3}$ (with the above unitary action). As these two frames are tight, they have the same Gramian (up to a scalar), which is an S_3-matrix. This matrix, with the same indexing as in (10.5), is given by

$$\begin{array}{c} \\ 1 \\ a \\ a^2 \\ b \\ ab \\ a^2b \end{array}\begin{array}{c} \begin{array}{cccccc} 1 & a & a^2 & b & ab & a^2b \end{array} \\ \begin{bmatrix} 1 & -\frac{1}{2} & -\frac{1}{2} & 1 & -\frac{1}{2} & -\frac{1}{2} \\ -\frac{1}{2} & 1 & -\frac{1}{2} & -\frac{1}{2} & 1 & -\frac{1}{2} \\ -\frac{1}{2} & -\frac{1}{2} & 1 & -\frac{1}{2} & -\frac{1}{2} & 1 \\ 1 & -\frac{1}{2} & -\frac{1}{2} & 1 & -\frac{1}{2} & -\frac{1}{2} \\ -\frac{1}{2} & 1 & -\frac{1}{2} & -\frac{1}{2} & 1 & -\frac{1}{2} \\ -\frac{1}{2} & -\frac{1}{2} & 1 & -\frac{1}{2} & -\frac{1}{2} & 1 \end{bmatrix} \end{array}. \tag{10.9}$$

This can be calculated via

$$\nu(g^{-1}h) := \langle hu_1, gu_1 \rangle = \langle g^{-1}hu_1, u_1 \rangle = \frac{2}{3}\langle v_{g^{-1}h1}, v_1 \rangle,$$

e.g.,

$$\nu(ab) = \frac{2}{3}\langle v_{(123)(23)1}, v_1 \rangle, = \frac{2}{3}\langle v_2, v_1 \rangle = \frac{2}{3}\left(-\frac{3}{4}\right) = -\frac{1}{2}.$$

It is easy to check that this matrix is a rank 2 orthogonal projection (up to a scalar). The element of the group algebra $\mathbb{C}S_3$ that it corresponds to via (10.8) is

$$\sum_{g \in S_3} \nu(g)g = 1 - \frac{1}{2}a - \frac{1}{2}a^2 + b - \frac{1}{2}ab - \frac{1}{2}a^2b = (1+b)(1 - \frac{1}{2}a - \frac{1}{2}a^2).$$

Example 10.6. (Repeated vectors) If Φ is a frame of n vectors, then it may be that $\mathrm{Sym}(\Phi)$ is transitive, but doesn't contain a transitive subgroup G of order n. Thus it may be necessary to the repeat the vectors in Φ in order to view it as a group frame. For example, let Φ be the tight frame given by the 20 vertices of dodecahedron (see §10.6). With the labelling of Figure 10.2,

$$\mathrm{Sym}(\Phi) = \langle a, b, r \rangle \cong A_5 \times \mathbb{Z}_2, \qquad |\mathrm{Sym}(\Phi)| = 120,$$

where a and b act as rotations through 72°, c as a reflection, and are given by

$$\begin{aligned} a &= (1,2,3,4,5)(6,8,10,12,14)(7,9,11,13,15)(16,17,18,19,20), \\ b &= (1,6,7,8,2)(3,5,15,16,9)(4,14,20,17,10)(12,13,19,18,11), \\ c &= (2,5)(3,4)(7,15)(8,14)(9,13)(10,12)(16,20)(17,19). \end{aligned}$$

This group has one subgroup of order 20, which is not transitive, no subgroup of order 40, and unique and transitive subgroups of orders 60 and 120. Thus the vertices of the dodecahedron can be viewed as an A_5-frame (vectors repeated three times) and as an $A_5 \times \mathbb{Z}_2$-frame (vectors repeated six times).

10.5 Irreducible G-frames

Given a unitary action of a finite group G on $\mathscr{H}$, and some $v \in \mathscr{H}$, one has

$$\Phi = (gv)_{g\in G} \text{ is a } G\text{-frame for its span.}$$

In §10.9, we will answer the questions:

- When is Φ a frame for $\mathscr{H}$?
- When is Φ tight?

The answer is particularly simple, and instructive, in the following situation.

Definition 10.3. A linear action (or representation) of a group G on $\mathscr{H} \neq 0$ is said to be **irreducible** if the only G-invariant subspaces of $\mathscr{H}$ are $\{0\}$ and $\mathscr{H}$, i.e.,

$$\operatorname{span}\{gv\}_{g\in G} = \mathscr{H}, \qquad \forall v \neq 0,\ v \in \mathscr{H}.$$

A G-frame given by such a unitary action is called an **irreducible G-frame**.

Theorem 10.5. *(Irreducible G-frames). Suppose a unitary action of a group G on $\mathscr{H}$ is irreducible. Then $(gv)_{g\in G}$ is a tight G-frame for $\mathscr{H}$ for any $v \neq 0$, i.e.,*

$$f = \frac{\dim(\mathscr{H})}{|G|}\frac{1}{\|v\|^2}\sum_{g\in G}\langle f, gv\rangle gv, \qquad \forall f \in \mathscr{H}.$$

Proof. Fix a nonzero vector v, and let S be the frame operator of $(gv)_{g\in G}$. Since S is positive, it has an eigenvalue $\lambda > 0$ with eigenvector w. By Lemma 10.1, S commutes with (the action of) G, and so gw is also an eigenvector for λ, for any $g \in G$, by the calculation

$$S(gw) = gS(w) = g(\lambda w) = \lambda(gw).$$

But $\{gw\}_{g\in G}$ spans $\mathscr{H}$, and so $S = \lambda I_{\mathscr{H}}$, i.e., $(gv)_{g\in G}$ is a tight frame. Taking traces gives

$$\operatorname{trace}(S) = \sum_g \|gv\|^2 = |G|\,\|v\|^2 = \lambda \dim(\mathscr{H}) = \operatorname{trace}(\lambda I_{\mathscr{H}}),$$

which determines λ. □

Corollary 10.5. *All irreducible G-frames are tight.*

Example 10.7. (Equally spaced vectors) The n equally spaced unit vectors in $\mathbb{R}^2$ (vertices of an n sided regular polygon) are an irreducible (tight) C_n-frame. They are an orbit of the cyclic group of order n acting via rotations as in (10.2), which is clearly irreducible. By way of comparison, to show this frame is tight by direct calculation would require the identities

$$\sum_{j=1}^{n}\left(\cos\frac{2\pi j}{n}\right)^2=\sum_{j=1}^{n}\left(\sin\frac{2\pi j}{n}\right)^2=\frac{n}{2}, \qquad \sum_{j=1}^{n}\cos\frac{2\pi j}{n}\sin\frac{2\pi j}{n}=0,$$

which can now be viewed as a consequence of Theorem 10.5.

We note that all irreducible representations of abelian groups (such as C_n) over complex vector spaces are one-dimensional. From this it follows (see Chapter 11) that there are only finitely many G-frames for G abelian, i.e., the harmonic frames. We now show:

If G is nonabelian, then there are uncountably many unitarily inequivalent irreducible G-frames.

Proposition 10.1. *Let G be a finite nonabelian group with an irreducible unitary action on $\mathbb{C}^d$, $d \geq 2$ (such an action always exists). Then there are uncountably many irreducible G-frames for $\mathbb{C}^d$ (up to projective unitary equivalence).*

Proof. Since all irreducible G-frames are tight, they are unitarily equivalent if and only if their Gramians are equal (when the generating vectors have equal norms). Let ρ be the representation. Then some $\rho(g)$ has two distinct eigenvalues $\lambda_1 \neq \lambda_2$, otherwise each $\rho(g)$ would be a scalar matrix, and

$$\operatorname{span}\{gv\}_{g\in G}=\operatorname{span}\{v\}_{g\in G}\neq\mathbb{C}^d,$$

contradicting the irreducibility of ρ. Let $u_1 \perp u_2$ be corresponding unit eigenvectors, and v_α be the unit vector

$$v_\alpha := \alpha u_1+\sqrt{1-|\alpha|^2}u_2, \qquad |\alpha|\leq 1.$$

The $(1,g)$-entry of the Gramian of the irreducible G-frame $\Phi_\alpha=(gv_\alpha)_{g\in G}$ is

$$\langle gv_\alpha, v_\alpha\rangle=\langle \alpha\lambda_1 u_1+\sqrt{1-|\alpha|^2}\lambda_2 u_2, \alpha u_1+\sqrt{1-|\alpha|^2}u_2\rangle=\alpha\lambda_1+\sqrt{1-|\alpha|^2}\lambda_2,$$

□

and so for different α, the frames Φ_α are unitarily inequivalent.

Example 10.8. The above argument extends to some irreducible actions ρ on $\mathbb{R}^d$. For example, if there is an element $\rho(g)\neq -I$ of order 2, then it has eigenvalues ± 1 (with orthogonal eigenvectors in $\mathbb{R}^d$). An example of this is the action of the dihedral group D_n on $\mathbb{R}^2$ (as reflections and rotations) given in Example 9.4, which gives uncountably many D_n-frames for $\mathbb{R}^2$ (see Figure 2.2 for the case $n=3$).

Example 10.9. Let G be the nonabelian group of orthogonal matrices generated by a the rotation through $2\pi/3$ and b the reflection in the y-axis, i.e.,

$$a = \frac{1}{2}\begin{pmatrix} -1 & -\sqrt{3} \\ \sqrt{3} & -1 \end{pmatrix}, \qquad b = \begin{pmatrix} -1 & \\ & 1 \end{pmatrix},$$

which is the dihedral group D_3. This action of D_3 on $\mathbb{R}^2$ is irreducible. Thus

$$\Phi := (gv)_{g\in D_3} = (v, av, a^2v, bv, abv, a^2bv)$$

is an irreducible D_3-frame for every nonzero vector $v \in \mathbb{R}^2$. For $v = (x,y)$ a unit vector, the Gramian of Φ is the D_3-matrix

$$\mathrm{Gram}(\Phi) = \begin{pmatrix} 1 & -1/2 & -1/2 & \alpha+\beta & -\alpha & -\beta \\ -1/2 & 1 & -1/2 & -\beta & \alpha+\beta & -\alpha \\ -1/2 & -1/2 & 1 & -\alpha & -\beta & \alpha+\beta \\ \alpha+\beta & -\beta & -\alpha & 1 & -1/2 & -1/2 \\ -\alpha & \alpha+\beta & -\beta & -1/2 & 1 & -1/2 \\ -\beta & -\alpha & \alpha+\beta & -1/2 & -1/2 & 1 \end{pmatrix},$$

where

$$-\alpha := \frac{1}{2}x^2 - \sqrt{3}xy - \frac{1}{2}y^2, \quad -\beta := \frac{1}{2}x^2 + \sqrt{3}xy - \frac{1}{2}y^2, \quad \alpha+\beta = y^2 - x^2.$$

We observe that this gives uncountably many unitarily inequivalent D_3-frames (see Figure 2.2). By considering the leading principal minor of order 3, we can see that the irreducible unitary action of the abelian subgroup $C_3 = \langle a \rangle$ gives just one irreducible C_3-frame up to unitary equivalence (the Mercedes-Benz frame).

For irreducible actions on $\mathscr{H} = \mathbb{C}^d$ there is an associated tight frame for $\mathbb{C}^{d\times d}$ (with the Frobenius inner product).

Theorem 10.6. *Suppose that there is a unitary action ρ of a finite group G on $\mathbb{C}^d$. Then the following are equivalent*

1. *The action of ρ is irreducible.*
2. *If $\langle v,w\rangle \neq 0$, then $(gv)_{g\in G}$ and $(gw)_{g\in G}$ are dual tight G-frames, i.e.,*

$$f = \frac{d}{|G|}\frac{1}{\langle v,w\rangle}\sum_{g\in G}\langle f, gw\rangle gv, \qquad \forall f \in \mathbb{C}^d. \tag{10.10}$$

3. *The unitary maps $(\rho(g))_{g\in G}$ are a tight G-frame for the $d\times d$ matrices, i.e.,*

$$A = \frac{d}{|G|}\sum_{g\in G}\langle A, \rho(g)\rangle \rho(g), \qquad \forall A \in \mathbb{C}^{d\times d}. \tag{10.11}$$

Proof. Let S be the operator $\mathbb{C}^d \to \mathbb{C}^d$ given by

$$Sf := \sum_{g\in G} \langle f, gw\rangle gv, \qquad f \in \mathbb{C}^d.$$

This commutes with the action of G on $\mathbb{C}^d$ (see Exer. 10.1).

1$\Longrightarrow$2. If ρ is irreducible, then it is absolutely irreducible, and Schur's lemma (see Lemma 10.4) implies that $S = \lambda I$. Take the trace of this, using Exer. 3.1 and the fact the action of G is unitary, to get

$$\lambda d = \text{trace}(\lambda I) = \text{trace}(S) = \sum_{g\in G} \langle gv, gw\rangle = \sum_{g\in G} \langle v, w\rangle = |G|\langle v, w\rangle.$$

2$\Longrightarrow$3. We have $\langle f, gw\rangle = \text{trace}(f(\rho(g)w)^*) = \text{trace}(fw^*\rho(g)^*) = \langle fw^*, \rho(g)\rangle$, $\langle v, w\rangle = w^*v$, and so (10.10) can be written as

$$fw^*v = \frac{d}{|G|}\sum_g \langle fw^*, \rho(g)\rangle\rho(g)v, \quad \forall f, w, v \in \mathbb{C}^d$$

$$\iff \quad fw^* = \frac{d}{|G|}\sum_g \langle fw^*, \rho(g)\rangle\rho(g), \quad \forall f, w \in \mathbb{C}^d$$

$$\iff \quad A = \frac{d}{|G|}\sum_g \langle A, \rho(g)\rangle\rho(g), \quad \forall A \in \mathbb{C}^{d\times d} = \text{span}\{e_j e_k^*\}_{1\le j,k\le d}.$$

3$\Longrightarrow$1. Take $A = fv^*$, and use $\langle fv^*, \rho(g)\rangle = \langle f, \rho(g)v\rangle$, $v^*v = \|v\|^2$ to get

$$fv^* = \frac{d}{|G|}\sum_g \langle fv^*, \rho(g)\rangle\rho(g) \quad \Longrightarrow \quad f\|v\|^2 = \frac{d}{|G|}\sum_g \langle f, \rho(g)v\rangle\rho(g)v,$$

i.e., $f \in \text{span}\{gv\}_{g\in G}$, $\forall v \neq 0$, and so the action is irreducible. □

Example 10.10. The real unitary matrices

$$\begin{pmatrix} 0 & 1 & 0 \\ 0 & 0 & 1 \\ 1 & 0 & 0 \end{pmatrix}, \quad \begin{pmatrix} 1 & 0 & 0 \\ 0 & -1 & 0 \\ 0 & 0 & -1 \end{pmatrix}$$

generate a group G of order 12 (isomorphic to A_4) whose action on $\mathbb{C}^3$ is irreducible. Thus the 12 matrices in G form a tight frame for the 9-dimensional space $\mathbb{R}^{3\times 3}$.

Example 10.11. (Nice error bases) If ρ is an irreducible unitary action of G on $\mathbb{C}^d$, then, by Schur's lemma, the centre $Z(\rho(G))$ of the matrix group $\rho(G)$ consists of scalar matrices, and ρ gives a projective representation of $H = \rho(G)/Z(\rho(G))$. When H has order d^2, i.e., the d^2 matrices of $\rho(G)$ (up to scalar multiples) are an orthogonal basis for $\mathbb{C}^{d\times d}$, the matrices are known as a *nice error basis* with index group H (see §13.13).

10.6 The vertices of the Platonic solids

The unitary action of the symmetry groups of the five *Platonic solids* (with centre of gravity at the origin) on $\mathbb{R}^3$ is irreducible (see Exer. 10.8), and so the vertices of the Platonic solids are irreducible G-frames. Similarly, the vertices of the truncated icosahedron (aka the "soccer ball" or "Buckyball") form a tight frame for $\mathbb{R}^3$.

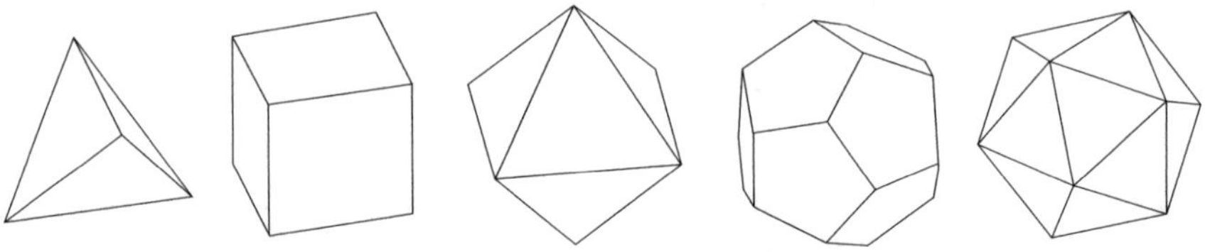

Fig. 10.1: The five Platonic solids: *tetrahedron*, *cube*, *octahedron*, *dodecahedron* and *icosahedron*.

We may apply Theorem 10.4 to determine for which groups G the vertices of a given Platonic solid are a (possibly not irreducible) G-frame (see Table 10.1). These calculations are outlined in Figure 10.2, which were adapted from [Mor04] (who kindly allowed us to reproduce his figures here).

Table 10.1: The groups G for which the vertices Φ of a Platonic solid are a G-frame.

Platonic solid	vertices	faces	transitive subgroups G of $\mathrm{Sym}(\Phi)$
Tetrahedron	4	4	S_4 (order 24), A_4 (order 12),
			D_4 (order 8), $\mathbb{Z}_2 \times \mathbb{Z}_2$, $\mathbb{Z}_4$ (order 4).
Cube	8	6	$S_4 \times \mathbb{Z}_2$ (order 48),
			S_4, $A_4 \times \mathbb{Z}_2$ (order 24), $D_4 \times \mathbb{Z}_2$ (order 16),
			$\mathbb{Z}_2 \times \mathbb{Z}_2 \times \mathbb{Z}_2$, $\mathbb{Z}_2 \times \mathbb{Z}_4$, D_4 (order 8).
Octahedron	6	8	$S_4 \times \mathbb{Z}_2$ (order 48), S_4, $A_4 \times \mathbb{Z}_2$ (order 24),
			A_4, D_6 (order 12), S_3, $\mathbb{Z}_6$ (order 6).
Dodecahedron	20	12	$A_5 \times \mathbb{Z}_2$ (order 120), A_5 (order 60).
Icosahedron	12	20	$A_5 \times \mathbb{Z}_2$ (order 120), A_5 (order 60).
			$A_4 \times \mathbb{Z}_2$ (order 24), A_4 (order 12).

Let G be the symmetry group of a Platonic solid acting on $\mathbb{R}^3$ (the order of G can be $|G| = 24, 48, 120$). For a generic vector $v \in \mathbb{R}^3$, the irreducible G-frame $(gv)_{g\in G}$ has $|G|$ distinct vectors (none are repeated), and there are uncountably many such unitarily inequivalent frames. The vertices of the Platonic solids have the additional property that they are stabilised by a *proper* subgroup of G. Generalising this leads to the finite class of *highly symmetric tight frames* (see §13.8).

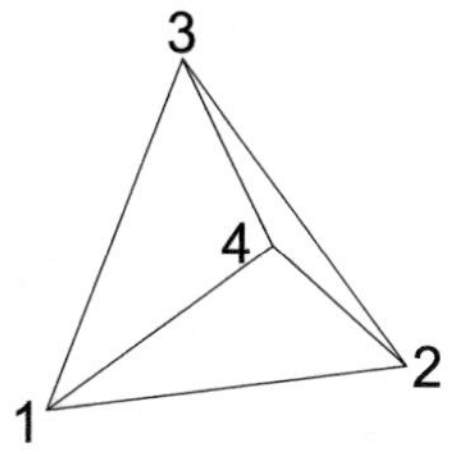

Tetrahedron

$a := (1\ 2\ 3)$,
$b := (4\ 2\ 3)$,
$c := (1\ 2)$,
$\langle a,b,c\rangle \cong S_4$ (tetrahedral group).

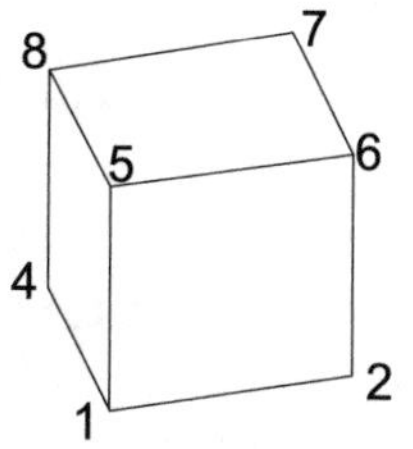

Cube

$a := (1\ 2\ 3\ 4)(5\ 6\ 7\ 8)$,
$b := (1\ 4\ 8\ 5)(2\ 3\ 7\ 6)$,
$c := (2\ 4)(6\ 8)$,
$\langle a,b,c\rangle \cong S_4 \times \mathbb{Z}_2$ (octohedral group).

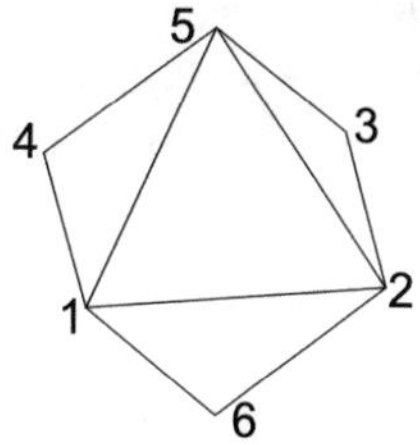

Octahedron

$a := (1\ 2\ 3\ 4)$,
$b := (5\ 1\ 2)(6\ 3\ 4)$,
$c := (1\ 2)(3\ 4)$,
$\langle a,b,c\rangle \cong S_4 \times \mathbb{Z}_2$ (octohedral group).

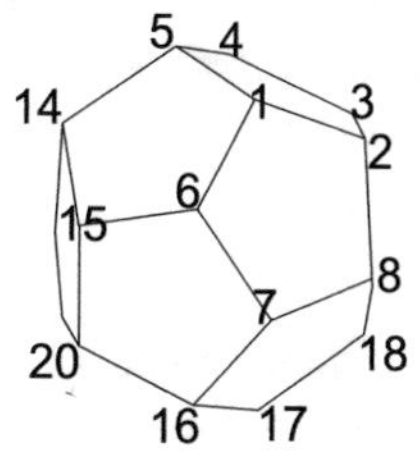

Dodecahedron

$a := (1\ 2\ 3\ 4\ 5)(6\ 8\ 10\ 12\ 14)(7\ 9\ 11\ 13\ 15)(16\ 17\ 18\ 19\ 20)$,
$b := (1\ 6\ 7\ 8\ 2)(3\ 5\ 15\ 16\ 9)(4\ 14\ 20\ 17\ 10)(12\ 13\ 19\ 18\ 11)$,
$c := (2\ 5)(3\ 4)(7\ 15)(8\ 14)(9\ 13)(10\ 12)(16\ 20)(17\ 19)$,
$\langle a,b,c\rangle \cong A_5 \times \mathbb{Z}_2$ (icosohedral group).

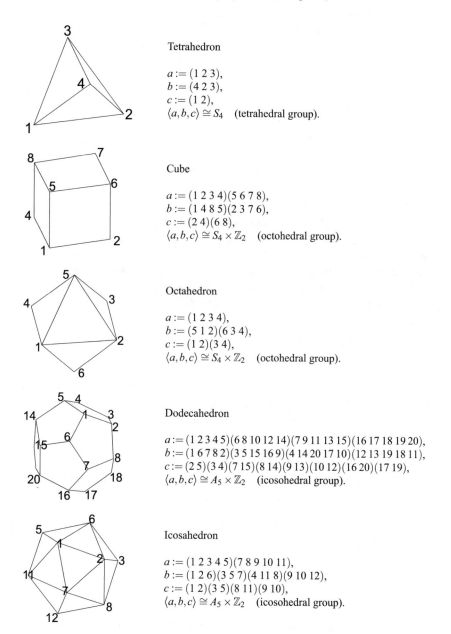

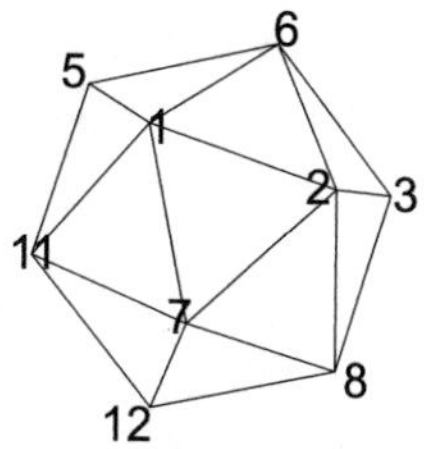

Icosahedron

$a := (1\ 2\ 3\ 4\ 5)(7\ 8\ 9\ 10\ 11)$,
$b := (1\ 2\ 6)(3\ 5\ 7)(4\ 11\ 8)(9\ 10\ 12)$,
$c := (1\ 2)(3\ 5)(8\ 11)(9\ 10)$,
$\langle a,b,c\rangle \cong A_5 \times \mathbb{Z}_2$ (icosohedral group).

Fig. 10.2: Generators for the symmetry groups of the vertices of the Platonic solids.

10.7 Irreducible G-frames from vertex-transitive graphs

Let Γ be a graph on n vertices. There is a unitary action of the automorphism group $\mathrm{Aut}(\Gamma)$ of Γ on $\mathbb{R}^n$ (where the vertices are labelled $1,2,\ldots,n$) given by

$$\sigma e_j := e_{\sigma j}, \qquad \forall j.$$

Let $A = [a_{jk}]$ be the $(0,1)$-adjacency matrix of the graph Γ ($a_{jk} = 1$ if and only if there is an edge from j to k). For $\sigma \in \mathrm{Aut}(\Gamma)$, $a_{\sigma j,\sigma k} = a_{jk}$, and so we have

$$\begin{aligned}\sigma(Ae_j) = \sigma\Bigl(\sum_r a_{rj}e_r\Bigr) &= \sum_r a_{rj}\sigma e_r = \sum_r a_{rj}e_{\sigma r} = \sum_s a_{\sigma^{-1}s,j}e_s \\ &= \sum_s a_{s,\sigma j}e_s = Ae_{\sigma j} = A(\sigma e_j),\end{aligned} \tag{10.12}$$

i.e., the action of $\mathrm{Aut}(\Gamma)$ commutes with the action A.

Now suppose that Γ is a **vertex-transitive graph**, i.e., a graph for which there is an automorphism taking any vertex to any other. Let G be any subgroup of $\mathrm{Aut}(\Gamma)$ whose action on the vertices of Γ is transitive. Then by (10.12), the columns of A are a G-frame $(\sigma Ae_j)_{\sigma\in G} = (Ae_{\sigma j})_{\sigma\in G}$ (where each column is repeated $|G|/n$ times). Since A is symmetric, it is unitarily diagonisable. Let E_λ be the orthogonal projection onto the λ-eigenspace. The E_λ are called the primitive idempotents. By (10.12),

$$\sigma\Bigl(\sum_\lambda \lambda E_\lambda e_j\Bigr) = \sum_\lambda E_\lambda e_{\sigma j} \quad\Longrightarrow\quad \sigma E_\lambda e_j = E_\lambda e_{\sigma j}.$$

The unitary action of G on the λ-eigenspace is irreducible (this is true for any distance-transitive graph), and so the G-orbit of any λ-eigenvector v is an irreducible G-frame for the λ-eigenspace (Theorem 10.5). Taking the choice $v = E_\lambda e_j$ and letting G be any transitive subgroup of $\mathrm{Aut}(\Gamma)$ gives a tight G-frame

$$(\sigma E_\lambda e_j)_{\sigma\in G} = (E_\lambda e_{\sigma j})_{\sigma\in G}$$

for the λ-eigenspace, which has at most n distinct vectors. The Gramian of the normalised tight frame $(E_\lambda e_j)_{j=1}^n$ is E_λ. From these observations, we obtain:

Proposition 10.2. *Let Γ be a vertex-transitive graph, with adjacency matrix A, and primitive idempotents E_λ. For each choice $\alpha_\lambda \in \{0,1\}$ (not all $\alpha_\lambda = 0$), define an orthogonal projection matrix*

$$P_\alpha := \sum_\lambda \alpha_\lambda E_\lambda, \qquad \mathrm{rank}(P_\alpha) = \sum_\lambda \alpha_\lambda \dim(E_\lambda).$$

Then the n columns of P_α can be viewed as a normalised tight G-frame $(P_\alpha e_{\sigma j})_{\sigma\in G}$, where G is any transitive subgroup of $\mathrm{Aut}(\Gamma)$.

By construction, the tight frame $\Phi = (P_\alpha e_j)_{j=1}^n$ has $\mathrm{Aut}(\Gamma) \subset \mathrm{Sym}(\Phi)$.

10.8 Maschke's theorem and homogeneous G-frames

To understand the G-frames for unitary actions which are not irreducible, we need to decompose a space into G-invariant subspaces ($\mathbb{F}G$-submodules). This is often presented in the language of $\mathbb{F}G$-modules (see Exer. 10.9, [JL01]). We will use the following terminology:

Definition 10.4. Suppose that there is a linear action of a group G on $\mathbb{F}$-vector spaces V and W. Then a linear map $\sigma : V \to W$ is an $\mathbb{F}G$-**homomorphism** if it commutes with the action of G, i.e.,

$$\sigma(gf) = g(\sigma f), \qquad \forall g \in G, \forall f \in \mathscr{H}.$$

A bijective $\mathbb{F}G$-homomorphism is an $\mathbb{F}G$-**isomorphism**, written $V \cong W$.

An $\mathbb{F}G$-homomorphism is also called a G-**morphism**, a G-**equivariant map**, or a G-**map**, where the field $\mathbb{F}$ is understood from the context.

Example 10.12. The frame operator S of a G-frame is an $\mathbb{F}G$-isomorphism (Lemma 10.1), and if there is a unitary action of G on V and W, then

$$Sf := \sum_{g \in G} \langle f, gv \rangle gw, \qquad \forall f \in V,$$

defines an $\mathbb{F}G$-homomorphism $S : V \to W$ for any $v \in V$, $w \in W$ (see Exer. 10.1).

We now give a version of Maschke's theorem, where the unitary action ensures that the direct sum is *orthogonal.*

Lemma 10.2. *(Maschke). Suppose that there is unitary action of finite group G on $V = \mathscr{H}$. Then V can be written as an orthogonal (internal) direct sum*

$$V = V_1 \oplus V_2 \oplus \cdots \oplus V_m \qquad (10.13)$$

of irreducible G-invariant subspaces V_j, where the V_j are unique up to ordering and $\mathbb{F}G$-isomorphism, and the homogeneous components

$$H_V(W) := \sum_{\substack{X \subset V \\ X \cong W}} X = \bigoplus_{V_j \cong W} V_j \qquad (10.14)$$

corresponding to an irreducible W are unique.

Proof. Use strong induction on $d = \dim(\mathscr{H})$. The case $d = 0$ is trivial. For $d > 0$, let $V_1 \neq 0$ be an irreducible G-invariant subspace. Then $V = V_1 \oplus X$, with X the orthogonal complement of V_1. We now show that X is G-invariant, which gives (10.13). Since the action is unitary, for $g \in G$ and $x \in X$, we have

$$\langle gx, v \rangle = \langle x, g^{-1}v \rangle = 0, \qquad \forall v \in V_1 \quad \Longrightarrow \quad gx \in V_1^{\perp} = X.$$

The usual arguments to show that the V_j are unique up to $\mathbb{F}G$-isomorphism (Jordan-Hölder), and $H_V(W) = \bigoplus_{V_j \cong W} V_j$, apply without modification. □

Thus there is a unique orthogonal decomposition of the space $\mathscr{H} = V$ into its homogeneous components

$$V = \bigoplus_{W \in \mathscr{W}} H_V(W),$$

where $\mathscr{W}$ is the collection of different irreducible G-invariant subspaces of $\mathscr{H}$ (up to $\mathbb{F}G$-isomorphism). The $H_V(W)$ can be calculated by Theorem 13.2.

It is easy to check that a G-invariant subspace of some $H_W(V)$ is again a sum of irreducible G-invariant subspaces which are $\mathbb{F}G$-isomorphic to W.

Definition 10.5. A G-frame for a space $\mathscr{H}$ with just one nonzero homogeneous component is called a **homogeneous G-frame**.

Example 10.13. An irreducible G-frame is a homogeneous G-frame. For G non-abelian, there are uncountably many homogeneous G-frames (see Proposition 10.1).

To decompose a G-frame into homogeneous G-frames, we need:

Lemma 10.3. *(Schur). Suppose that $A : V \to W$ is a linear map between irreducible G-invariant subspaces of $\mathscr{H}$ which commutes with the linear action of G, i.e., an $\mathbb{F}G$-homomorphism. Then either*

1. *$A = 0$.*
2. *A is invertible, i.e., V and W are $\mathbb{F}G$-isomorphic.*

Proof. Suppose that $A \neq 0$. Then $\ker(A) \neq V$ and $\operatorname{ran}(A) \neq 0$. Since the map A is an $\mathbb{F}G$-homomorphism, its kernel and range are G-invariant subspaces of V and W. Hence, by irreducibility, $\ker(A) = 0$, $\operatorname{ran}(A) = W$, i.e., A is a bijection. □

Theorem 10.7. *(Homogeneous decomposition). Suppose there is a unitary action of a finite group G on*

$$V = \mathscr{H} = \bigoplus_{W \in \mathscr{W}} H_V(W).$$

If $v = \sum_W v_W$, $v_W \in H_V(W)$, then the G-frame $\Phi = (gv)_{g \in G}$ (for its span) can be decomposed as an orthogonal direct sum of homogeneous G-frames, i.e.,

$$\Phi = \bigoplus_{W \in \mathscr{W}} \Phi_W, \qquad \Phi_W := (gv_W)_{g \in G}.$$

In particular, G-frames in different homogeneous components are orthogonal.

Proof. Since $g(\sum_W v_W) = \sum_W g v_W$, it follows immediately that Φ is the direct sum of the homogeneous G-frames Φ_W for $\mathrm{span}(\Phi_W) \subset H_W(V)$. It remains only to show the orthogonality of this direct sum, i.e., that the maps

$$B : \mathrm{span}(\Phi_{W_1}) \to \mathrm{span}(\Phi_{W_2}) : f \mapsto \sum_{g\in G} \langle f, g v_{W_1}\rangle g v_{W_2}, \qquad W_1 \not\cong W_2$$

are zero. By Maschke's theorem, it suffices to show that $A := P_{V_2} B|_{V_1} = 0$, where $V_j \subset \mathrm{span}(\Phi_{W_j}) \subset H_V(W_j)$, $V_j \cong W_j$, are irreducible G-invariant subspaces, and P_{V_j} is the orthogonal projection onto V_j. The map $A : V_1 \to V_2$ commutes with the action of G, since P_{V_2} and $A|_{V_1}$ do (the latter by the argument of Lemma 10.1). By Schur's lemma (Lemma 10.3), we have $A = 0$, as otherwise V_1 and V_2 would be $\mathbb{F}G$-isomorphic, contradicting $V_j \cong W_j$. □

In other words:

A G-frame is the *orthogonal* direct sum of homogeneous G-frames.

10.9 The characterisation of all tight G-frames

If there is a G-frame for $\mathscr{H}$, then there is one which is tight:

Proposition 10.3. *Suppose that there is a unitary action of a finite group G on $\mathscr{H}$. Then the following are equivalent*

1. *For some $v \in \mathscr{H}$, $\{gv\}_{g\in G}$ spans $\mathscr{H}$.*
2. *There is a G-frame for $\mathscr{H}$.*
3. *There is a tight G-frame for $\mathscr{H}$.*

Proof. Since finite frames are simply spanning sequences, Theorem 10.1 gives

$$\begin{aligned}\mathrm{span}\{gv\}_{g\in G} = \mathscr{H} \quad &\iff \quad \Phi = (gv)_{g\in G} \text{ is a } G\text{-frame for } \mathscr{H} \\ &\iff \quad \Phi^{\mathrm{can}} \text{ is a tight } G\text{-frame for } \mathscr{H}.\end{aligned}$$

□

This gives a simple necessary condition for the existence of a tight G-frame for $\mathscr{H}$. We now give a *constructive* characterisation of which G-orbits under a unitary action are tight G-frames.

We recall that a real vector space $\mathscr{H}$ can be complexified $\mathscr{H}^{\mathbb{C}} \cong \mathscr{H} \oplus i\mathscr{H}$, and a linear/unitary map U on $\mathscr{H}$ extends to $\mathscr{H}^{\mathbb{C}}$ via $U(v_1 \oplus iv_2) := (Uv_1) \oplus (iUv_2)$. The **complexification** of $\mathscr{H}$ is $\mathscr{H}^{\mathbb{C}}$ when $\mathbb{F} = \mathbb{R}$, and $\mathscr{H}$ when $\mathbb{F} = \mathbb{C}$.

In this way, an $\mathbb{F}G$-module V can always be thought of as a $\mathbb{C}G$-module.

Definition 10.6. A linear action of a finite group G on $\mathscr{H}$ (or representation, or $\mathbb{F}G$-module) is **absolutely irreducible** if the action on $\mathscr{H}^{\mathbb{C}}$ is irreducible.

Example 10.14. For $\mathbb{F}=\mathbb{C}$ irreducibility and absolute irreducibility coincide. The unitary action (10.2) of the cyclic group C_n on $\mathbb{R}^2$ is irreducible, but is not absolutely irreducible, since the eigenspaces of $\rho_{\mathbb{R}^2}(a)$ as a $\mathbb{C}$-linear map are G-invariant.

Schur's lemma (Lemma 10.3) implies that the $\mathbb{F}$-vector space $\mathrm{Hom}_{\mathbb{F}G}(V,W)$ of the $\mathbb{F}G$-homomorphisms between the $\mathbb{F}G$-isomorphic irreducibles V and W is a *division ring*. When $\mathbb{F}=\mathbb{C}$ (or V and W are absolutely irreducible), then this division ring is one-dimensional. This leads to the following.

Lemma 10.4. *(Schur) Suppose that $S:V_j\to V_k$ is an $\mathbb{F}G$-homomorphism between absolutely irreducible $\mathbb{F}G$-modules, which are $\mathbb{F}G$-isomorphic via $\sigma:V_j\to V_k$. Then*

$$S=c\sigma, \qquad \text{for some } c\in\mathbb{F}.$$

Proof. Since V_j and V_k are absolutely irreducible, $\mathrm{Hom}_{\mathbb{F}G}(V_j,V_k)$ is one-dimensional, and so spanned by σ. This gives the result (with c possibly zero). □

We now show that every tight G-frame is a direct sum of irreducible G-frames:

Theorem 10.8. *(Characterisation). Let there be a unitary action of a finite group G on $\mathcal{H}=V_1\oplus V_2\oplus\cdots\oplus V_m$, an orthogonal direct sum of irreducible G-invariant subspaces. Then*

$$(gv)_{g\in G}, \qquad v=v_1+\cdots+v_m, \quad v_j\in V_j$$

is a tight G-frame for $\mathcal{H}$ if and only if

$$v_j\neq 0, \quad \forall j, \qquad \frac{\|v_j\|^2}{\|v_k\|^2}=\frac{\dim(V_j)}{\dim(V_k)}, \quad j\neq k, \tag{10.15}$$

and when $V_j\neq V_k$ are $\mathbb{F}G$-isomorphic, $(gv_j)_{g\in G}$ and $(gv_k)_{g\in G}$ are orthogonal, i.e.,

$$\sum_{g\in G}\langle v_j,gv_j\rangle gv_k=0. \tag{10.16}$$

Moreover, if V_j is absolutely irreducible, then (10.16) *can be replaced by*

$$\langle\sigma v_j,v_k\rangle=0, \tag{10.17}$$

where $\sigma:V_j\to V_k$ is any $\mathbb{F}G$-isomorphism.

Proof. $\Phi=(gv)_{g\in G}$ is a tight frame for $\mathcal{H}$ if and only if there exists a $\lambda>0$ with

$$S_\Phi(f)=\sum_{g\in G}\langle f,gv\rangle gv=\lambda f, \qquad \forall f\in\mathcal{H}.$$

By linearity, it suffices to show this for $f_j\in V_j$, $1\le j\le m$, i.e.,

$$\sum_{g\in G}\langle f_j,gv\rangle gv=\sum_{g\in G}\langle f_j,gv_j\rangle gv_j+\sum_{g\in G}\sum_{k\neq j}\langle f_j,gv_j\rangle gv_k=\lambda f_j, \tag{10.18}$$

since $gv = \sum_j gv_j$. By equating the V_k components, (10.18) holds if and only if

$$\sum_{g\in G} \langle f_j, gv_j\rangle gv_j = \lambda f_j, \qquad \sum_{g\in G} \langle f_j, gv_j\rangle gv_k = 0, \quad k \neq j. \tag{10.19}$$

By Theorem 10.5, the first condition will hold for all $f_j \in V_j$ provided $v_j \neq 0$, with a $\lambda = \lambda_j > 0$, which depends on j, given by

$$\lambda_j = \frac{|G|\,\|v_j\|^2}{\dim(V_j)}.$$

Therefore λ_j is independent of j if and only if the second part of (10.15) is satisfied. By Theorem 10.7, the second condition in 10.19 automatically holds if $V_j \not\cong V_k$. Since $\{hv_j\}_{h\in G}$ spans V_j, the second condition in 10.19 can be rewritten

$$\sum_{g\in G} \langle hv_j, gv_j\rangle gv_k = h \sum_{g\in G} \langle v_j, h^{-1}gv_j\rangle h^{-1}gv_k = h \sum_{g\in G} \langle v_j, gv_j\rangle gv_k = 0, \quad k \neq j,$$

which gives (10.16), since h maps $V_k \to V_k$ bijectively.

Finally, if V_j is *absolutely* irreducible, then

$$S : V_j \to V_k : f_j \mapsto \sum_{g\in G} \langle f_j, gv_j\rangle gv_k$$

is a nonzero $\mathbb{F}G$-homomorphism (Exer. 10.1), so Lemma 10.4 gives $S = c\sigma$, $c \in \mathbb{F}$. We now determine c. Since the action is unitary, using Theorem 10.5, we calculate

$$\begin{aligned}
\langle Sv_j, \sigma v_j\rangle &= \langle \sum_{g\in G} \langle v_j, gv_j\rangle gv_k, \sigma v_j\rangle = \sum_{g\in G} \langle v_j, gv_j\rangle \langle gv_k, \sigma v_j\rangle \\
&= \sum_{g\in G} \langle g^{-1}v_j, v_j\rangle \langle v_k, \sigma g^{-1}v_j\rangle = \langle v_k, \sigma \sum_{g\in G} \langle v_j, g^{-1}v_j\rangle g^{-1}v_j\rangle \\
&= \langle v_k, \sigma \frac{|G|\,\|v_j\|^2}{\dim(V_j)} v_j\rangle = \frac{|G|\,\|v_j\|^2}{\dim(V_j)} \langle v_k, \sigma v_j\rangle.
\end{aligned}$$

Since $\langle Sv_j, \sigma v_j\rangle = \langle c\sigma v_j, \sigma v_j\rangle = c\|\sigma v_j\|^2$, we have

$$Sf = \sum_{g\in G} \langle f, gv_j\rangle gv_k = \frac{|G|\,\|v_j\|^2}{\dim(V_j)\|\sigma v_j\|^2} \langle v_k, \sigma v_j\rangle \sigma f,$$

and the condition (10.16), i.e., $Sv_j = 0$, is equivalent to (10.17). □

Example 10.15. (One summand). For $m = 1$ this reduces to Theorem 10.5.

Example 10.16. (Harmonic frames) If G is an abelian group, then all the absolutely irreducible $\mathbb{F}G$-modules are *one*-dimensional, with the action of G given by

$$gv_j = \xi_j(g)v_j, \qquad v_j \in V_j,$$

where ξ_j is a character of G (see §11.2). The condition (10.16) becomes

$$\sum_{g\in G}\langle v_j,\xi_j(g)v_j\rangle\xi_k(g)v_k=\sum_{g\in G}\xi_k(g)\overline{\xi_j(g)}\|v_j\|^2v_k=0 \quad\Longrightarrow\quad \sum_{g\in G}\xi_k(g)\overline{\xi_j(g)}=0,$$

i.e., the characters are orthogonal, and so each character can be taken at most once. These *harmonic frames* are studied in Chapter 11.

Example 10.17. (Dimension of a G-frame) For an absolutely irreducible $W\in\mathscr{W}$ (for which the action of G is unitary) and $\sigma_j:V_j\to W$ any $\mathbb{F}G$-isomorphism, the condition (10.17) can be written (see Exer. 10.2) as

$$\langle\sigma_jv_j,\sigma_kv_k\rangle=0,\qquad j\neq k.$$

Therefore the maximal number of summands which can be $\mathbb{F}G$-isomorphic to W is $\dim(W)$. When the number of summands in $\mathscr{H}=\oplus_jV_j$ that are $\mathbb{F}G$-isomorphic to a given W is zero or $\dim(W)$, then one obtains a *central* G-frame (see §13.4).

Taking each absolutely irreducible $W\in\mathscr{W}$ as a summand $\dim(W)$ times gives rise to tight G-frame, so that

$$\sum_{W\in\mathscr{W}}\dim(W)^2\le|G|. \tag{10.20}$$

On the other hand, taking $P=I$ in Corollary 10.3 gives the tight G-frame $(e_g)_{g\in G}$ for $\mathbb{C}^G$, so there is equality. More generally, all of the possible dimensions for a G-frame are

$$\sum_{W\in\mathscr{W}}a_W\dim(W),\qquad 0\le a_W\le\dim(W),$$

where a_W is the multiplicity of the absolutely irreducible W as a summand.

Example 10.18. Suppose that there is an irreducible unitary action of G on $W=\mathbb{C}^d$. This induces a unitary action ρ on $d\times d$ matrices via matrix multiplication

$$\rho(g)[u_1,\ldots,u_d]=[gu_1,\ldots,gu_d],\qquad u_j\in\mathbb{C}^d.$$

Let V_j be the G-invariant subspace of $\mathbb{C}^{d\times d}$ consisting of the matrices which are zero in all but the j-th column (and the zero matrix),

$$v_j=[0,\ldots,u_j,\ldots,0],\qquad \sigma_j:V_j\to\mathbb{C}^d:v_j\mapsto u_j.$$

Then σ_j is a $\mathbb{C}G$-isomorphism, and so the G-orbit of $U=[u_1,\ldots,u_d]=v_1+\cdots+v_d$ is a tight G-frame for $\mathbb{C}^{d\times d}$ if and only if $\|v_j\|=\|u_j\|$ is constant (for all j) and

$$\langle\sigma_jv_j,\sigma_kv_k\rangle=\langle u_j,u_k\rangle=0,\qquad j\neq k,$$

i.e., U is a unitary matrix (up to a scalar). Thus, with U unitary, (10.11) generalises to

$$A=\frac{d}{|G|}\sum_{g\in G}\langle A,\rho(g)U\rangle\rho(g)U,\qquad \forall A\in\mathbb{C}^{d\times d}.$$

10.10 *G*-frames of multivariate orthogonal polynomials

Here we apply Theorem 10.8 to construct tight frames of orthogonal polynomials of several variables which share the symmetries of the weight (also see §15 and §16).

Let $\Pi := \Pi(\mathbb{R}^d)$ be the polynomials in d real variables, and $\Pi_k := \Pi_k(\mathbb{R}^d)$ those of degree $\leq k$. Let μ be a measure on $\mathbb{R}^d$, for which

$$\langle f_1, f_2 \rangle_\mu := \int f_1 f_2 \, d\mu$$

defines an inner product on Π. The space of **orthogonal polynomials** of degree k with respect to the measure μ is

$$\mathscr{V}_k(\mu) := \{ f \in \Pi_k : \langle f, p \rangle_\mu = 0, \ \forall p \in \Pi_{k-1} \}.$$

This has dimension

$$\dim(\mathscr{V}_k(\mu)) = \binom{k+d-1}{d-1}.$$

The **symmetry group** of the measure μ is

$$G = \mathrm{Sym}(\mu) := \{ g \in \mathrm{Aff}(\mathbb{R}^d) : \int f \circ g \, d\mu = \int f \, d\mu, \forall f \in \Pi \},$$

where $\mathrm{Aff}(\mathbb{R}^d)$ denotes the group of affine transformations on $\mathbb{R}^d$. This acts on Π via $g \cdot f := f \circ g^{-1}$, with each $g \in \mathrm{Sym}(\mu)$ inducing a unitary transformation

$$\langle g \cdot f_1, g \cdot f_2 \rangle_\mu = \int (f_1 \circ g^{-1})(f_2 \circ g^{-1}) \, d\mu = \int (f_1 f_2) \circ g^{-1} \, d\mu = \int f_1 f_2 \, d\mu = \langle f_1, f_2 \rangle_\mu,$$

which maps $\mathscr{V}_k(\mu)$ onto itself since $g \cdot \Pi_{k-1} = \Pi_{k-1}$, i.e., $\mathscr{V}_k(\mu)$ is a G-invariant subspace. It is therefore natural to seek a G-invariant tight frame for $\mathscr{V}_k(\mu)$, which has a small number of vectors.

Example 10.19. (Legendre polynomials on a square) Let $\mathscr{P}_2$ be the 3-dimensional space of quadratic Legendre polynomials on the square $[-1,1]^2 \subset \mathbb{R}^2$. Here

$$\langle f_1, f_2 \rangle = \frac{1}{4} \int_{[-1,1]^2} f_1 f_2,$$

so that $\|1\| = 1$, and $G = D_4$ acts as the symmetries of the square.

The orthogonal decomposition of $\mathscr{P}_2$ into its homogeneous components consists of three 1-dimensional G-invariant subspaces, which are given by

$$p_1(x,y) = x^2 + y^2 - \frac{2}{3}, \qquad p_2(x,y) = xy, \qquad p_3(x,y) = x^2 - y^2.$$

None of these spaces are $\mathbb{R}G$-isomorphic. To see this, observe that D_4 acts as the identity on p_1, and the symmetry $(x,y) \mapsto (y,x)$ fixes p_2 but not p_3.

By Theorem 10.8, any f which is a sum of vectors of equal length from these subspaces gives a G-frame $(gf)_{g\in G}$ for $\mathscr{P}_2$. Since $\|p_1\| = \|p_3\| = \frac{2\sqrt{2}}{3\sqrt{5}}$, $\|p_2\| = \frac{1}{3}$, we can take

$$f(x,y) = \frac{\sqrt{5}}{2}(x^2+y^2-\frac{2}{3}) + \sqrt{2}xy + \frac{\sqrt{5}}{2}(x^2-y^2) = \sqrt{5}(x^2-\frac{1}{3}) + \sqrt{2}xy.$$

This gives a D_4-invariant tight frame of four vectors for $\mathscr{P}_2$. Since the size of an orbit divides the order of the group, there is no D_4-invariant basis for $\mathscr{P}_2$.

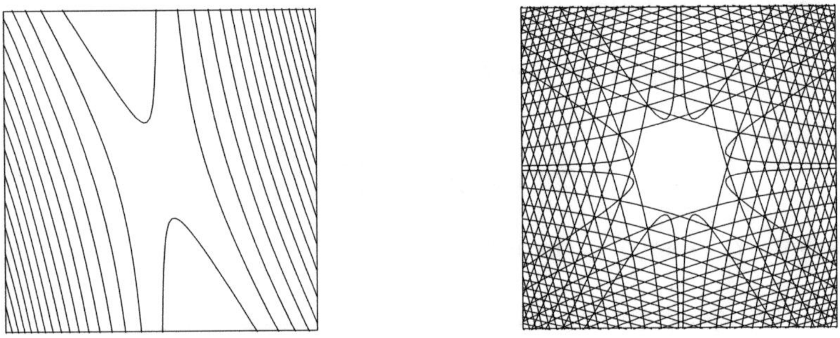

Fig. 10.3: Contour plots of the f of Example 10.19, and its orbit (showing the square symmetry).

Example 10.20. (Legendre polynomials on a triangle). Let $\mathscr{P}_2$ be the 3-dimensional space of quadratic Legendre polynomials on a triangle T with vertices V. Here

$$\langle f_1, f_2\rangle = \frac{1}{\text{Area}(T)}\int_T f_1 f_2,$$

and $G = S_V = D_3$ acts as the symmetries of the triangle, i.e., for $\sigma \in S_V$

$$\sigma\Big(\sum_{v\in V} a_v v\Big) = \sum_{v\in V} a_v \sigma v, \qquad (\text{where } \textstyle\sum_v a_v = 1).$$

The action of G on the barycentric coordinates $\xi = (\xi_v)_{v\in V}$ for V (see §4.7) is

$$\sigma \cdot \xi_v = \xi_{\sigma v}, \qquad \forall \sigma \in S_V. \tag{10.21}$$

The inner product between powers of the barycentric coordinates (which are linear polynomials) is given by the special case $d = 2$ of the formula

$$\langle \xi^\alpha, \xi^\beta\rangle = \frac{(\alpha+\beta)!}{(d+1)_{|\alpha|+|\beta|}},$$

for (normalised) integration over a simplex in $\mathbb{R}^d$ with vertices V. Let $V = \{u, v, w\}$.

In view of (10.21), the 1-dimensional subspace V_1 of $\mathscr{P}_2$ spanned by

$$f_1 := \xi_u^2 + \xi_v^2 + \xi_w^2 - \frac{1}{2}, \qquad \|f_1\|^2 = \frac{1}{60}$$

is G-invariant. Further, its G-invariant orthogonal complement $V_2 := V_1^\perp = \mathscr{P}_2 \ominus V_1$ is irreducible. This follows since the action of G on $\mathscr{P}_2$ is faithful, and so if V_2 was not irreducible, then linear maps $\mathscr{P}_2 \to \mathscr{P}_2$ induced by G would be simultaneously diagonalisable, and so $G \cong S_3$ would be abelian. The orthogonal projection of ξ_v^2 onto V_2 is given by

$$f_2 := \Big(\xi_v^2 - \frac{4}{5}\xi_v + \frac{1}{10}\Big) - \frac{1}{3}f_1, \qquad \|f_2\|^2 = \frac{1}{675}.$$

By Theorem 10.8, the G-orbit of the unit vector

$$f := \frac{f_1}{\|f_1\|}\frac{1}{\sqrt{3}} \pm \frac{f_2}{\|f_2\|}\sqrt{\frac{2}{3}} = (2\sqrt{5} \mp 5\sqrt{2})f_1 \pm 15\sqrt{2}\Big(\xi_v^2 - \frac{4}{5}\xi_v + \frac{1}{10}\Big),$$

is a tight frame for the 3-dimensional space $\mathscr{P}_2$. The polynomial f is fixed by any permutation which fixes v, and so this orbit has three distinct vectors. These give an orthonormal basis for $\mathscr{P}_2$ (the quadratic Legendre polynomials on the triangle), which is invariant under the symmetries of the triangle (see Figure 10.4).

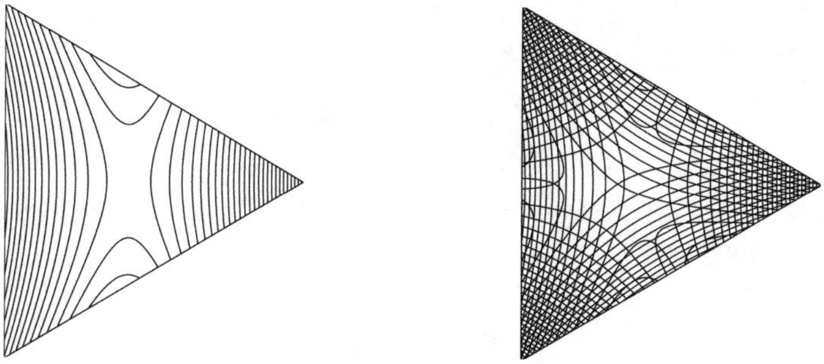

Fig. 10.4: Contour plots of the f of Example 10.20, and its orbit (showing the triangular symmetry).

Since the orders of the symmetry groups G of the square and triangle are 8 and 6, and the size of a G-frame must divide $|G|$, it is not possible to find a tight G-frame for the space of orthogonal polynomials of degree k on the square and triangle (or for any weight with a finite symmetry group) as soon as its dimension is greater than $|G|$. In these cases, it is natural to seek tight frame which is the G-orbit of a small number of vectors (see §10.11 and Chapter 15).

10.11 G-invariant frames

Suppose there is unitary action of a finite group G on $\mathscr{H}$. If there is no G-frame $(gv)_{g\in G}$ for $\mathscr{H}$, then it is natural to seek a G-invariant (tight) frame for $\mathscr{H}$, which is the orbit of a *small number* of vectors. Here we give a complete characterisation of such G-invariant tight frames (Theorem 10.9). This allows one to calculate the minimal number of generators for a G-invariant frame (§10.13).

Definition 10.7. The **number of generators** of a G-invariant frame is the number of orbits under the action of G on the set of its vectors.

We now prove the main result: a characterisation of which G-orbits of vectors $w_1,\ldots,w_r$ give tight frames.

Theorem 10.9. *(Characterisation). Let $\mathscr{H}$ be a Hilbert space over $\mathbb{F}=\mathbb{C}$ or $\mathbb{R}$. Suppose there is a unitary action of a finite group G on $\mathscr{H}=V_1\oplus V_2\oplus\cdots\oplus V_m$, an orthogonal direct sum of irreducible G-invariant subspaces. Let $P_j=P_{V_j}$ be the orthogonal projection of $\mathscr{H}$ onto V_j. Then*

$$\Phi=(gw_s)_{g\in G,1\le s\le r},\qquad w_1,\ldots,w_r\in\mathscr{H},$$

is a tight G-invariant frame for $\mathscr{H}$ if and only if

$$\sum_{s=1}^{r}\|P_jw_s\|^2\ne 0,\quad \forall j,\qquad \frac{\sum_{s=1}^{r}\|P_jw_s\|^2}{\sum_{s=1}^{r}\|P_kw_s\|^2}=\frac{\dim(V_j)}{\dim(V_k)},\quad j\ne k, \tag{10.22}$$

and when $V_j\ne V_k$ are $\mathbb{F}G$-isomorphic

$$\sum_{s}\sum_{g\in G}\langle v_j,gP_jw_s\rangle gP_kw_s=0, \tag{10.23}$$

for any (and hence all) nonzero $v_j\in V_j$. Moreover, if V_j is absolutely irreducible, then (10.23) can be replaced by

$$\sum_{s=1}^{r}\langle\sigma P_jw_s,P_kw_s\rangle=0, \tag{10.24}$$

where $\sigma:V_j\to V_k$ is any choice of $\mathbb{F}G$-isomorphism.

Proof. Φ is a tight frame for $\mathscr{H}$ if and only if there exists a $\lambda>0$ with

$$S_\Phi(f)=\sum_{s}\sum_{g\in G}\langle f,gw_s\rangle gw_s=\lambda f,\qquad \forall f\in\mathscr{H}.$$

By linearity, it suffices to show this for $f_j\in V_j$, $1\le j\le m$, i.e. to show that there exists λ (independent of j) such that,

$$\sum_{s}\sum_{g\in G}\langle f_j,gw_s\rangle gw_s=\sum_{s}\sum_{g\in G}\sum_{k}\langle f_j,gP_jw_s\rangle gP_kw_s=\lambda f_j, \tag{10.25}$$

since $w_s = \sum_k P_k w_s$. By equating the V_k components, (10.25) holds if and only if

$$\sum_s \sum_{g\in G} \langle f_j, gP_j w_s\rangle gP_j w_s = \lambda f_j, \qquad \sum_s \sum_{g\in G} \langle f_j, gP_j w_s\rangle gP_k w_s = 0, \quad k \neq j. \tag{10.26}$$

By Theorem 10.5, the first part of (10.26) will hold for all $f_j \in V_j$ provided that some w_s has a nonzero V_j-component, i.e., $\sum_s \|P_j w_s\|^2 \neq 0$, with a $\lambda = \lambda_j > 0$, which depends on j, given by

$$\lambda_j = \frac{|G|}{\dim(V_j)} \sum_s \|P_j w_s\|^2.$$

This λ_j is independent of j if and only if (10.22) holds. By Theorem 10.7, the second part of (10.26) automatically holds if $V_j \not\cong V_k$, and so reduces to

$$\sum_s \sum_{g\in G} \langle f_j, gP_j w_s\rangle gP_k w_s = 0, \quad \forall f_j \in V_j, \qquad k \neq j, \quad V_j \cong V_k. \tag{10.27}$$

By Lemma 10.3, this holds if and only if it holds for f_j some nonzero $v_j \in V_j$. We now seek to simplify (10.27) in the case that V_j is absolutely irreducible. Let $\tau : V_j \to V_k$ be the $\mathbb{F}G$-homomorphism (see Exer. 10.1)

$$\tau f := \sum_s \sum_g \langle f, gP_j w_s\rangle gP_k w_s.$$

Then for $\sigma : V_j \to V_k$ an $\mathbb{F}G$-isomorphism, we calculate

$$\begin{aligned}
\langle \tau v_j, \sigma v_j\rangle &= \langle \sum_s \sum_g \langle v_j, gP_j w_s\rangle gP_k w_s, \sigma v_j\rangle \\
&= \sum_s \sum_g \langle v_j, gP_j w_s\rangle \langle gP_k w_s, \sigma v_j\rangle \\
&= \sum_s \sum_g \langle g^{-1} v_j, P_j w_s\rangle \langle P_k w_s, \sigma g^{-1} v_j\rangle \\
&= \sum_s \langle P_k w_s, \sigma \sum_g \langle P_j w_s, g^{-1} v_j\rangle g^{-1} v_j\rangle \\
&= \sum_s \langle P_k w_s, \sigma \frac{\|v_j\|^2 |G|}{\dim(V_j)} P_j w_s\rangle,
\end{aligned}$$

with the last equality given by Theorem 10.5. Since V_j is absolutely irreducible, Lemma 10.4 implies that $\tau = c\sigma$, for some $c \in \mathbb{F}$ (possibly zero). Substituting $\tau = c\sigma$ into the above gives

$$\tau = \frac{|G|\,\|v_j\|^2}{\dim(V_j)\|\sigma v_j\|^2} \sum_s \langle P_k w_s, \sigma P_j w_s\rangle \sigma.$$

Thus (10.27), which is equivalent to $\tau = 0$, holds if and only if (10.24) does. □

For a single generator ($r = 1$), Theorem 10.9 reduces to Theorem 10.8.

10.12 Frames invariant under the action of an abelian group

We now consider Theorem 10.9 when G is *abelian*. In this case, all the absolutely irreducible G-invariant subspaces V_j are 1-dimensional, with the action of G given by

$$gv = \xi(g)v, \qquad v \in V_j,$$

where ξ is a **character** of G, i.e., a homomorphism $G \to \mathbb{C}$ (see §11.2).

For one generator, there are a *finite* number of tight G-frames for $\mathbb{C}^d$, the so called *harmonic frames* (see Example 10.16 and Chapter 11).

We now use Theorem 10.9 to describe the situation for two or more generators.

Corollary 10.6. *(G abelian) Suppose that there is a unitary action of a finite abelian group G on $\mathbb{C}^d$, and, without loss of generality, that the irreducible G-invariant subspaces are $V_j = \operatorname{span}\{e_j\}$, with the action on V_j given by $ge_j = \xi_j(g)e_j$, where $\xi_j : G \to \mathbb{C}$ is a character of G. Let $w_1, \ldots, w_r \in \mathbb{C}^d$. Then*

$$\Phi = (gw_s)_{1 \le s \le r, g \in G}$$

is a tight frame for $\mathbb{C}^d$ if and only if

1. *The matrix $W = [w_1, \ldots, w_r]$ has rows of equal norm.*
2. *The rows of W corresponding to the same character are orthogonal.*

In particular, there is a G-invariant tight frame for $\mathbb{C}^d$ with r generators if and only if $d \le r|G|$.

Proof. Since $P_j w_s = (w_s)_j e_j$, we have

$$\sum_s \|P_j w_s\|^2 = \sum_s |(w_s)_j|^2 = (\text{norm of the } j\text{-th row of } W)^2,$$

and so (10.22) reduces to 1. If V_j and V_k are $\mathbb{C}G$-isomorphic, i.e., correspond to the same character, then $\sigma : V_j \to V_k : e_j \mapsto e_k$ is a $\mathbb{C}G$-isomorphism, since

$$\sigma(ge_j) = \sigma(\xi_j(g)e_j) = \xi_j(g)\sigma(e_j) = \xi_k(g)e_k = g(\sigma e_j).$$

In this case, $\sigma P_j w_s = \sigma(w_s)_j e_j = e(w_s)_j e_k$, and so (10.24) becomes

$$\sum_s \langle \sigma P_j w_s, P_k w_s \rangle = \sum_s \langle (w_s)_j e_k, (w_s)_k e_k \rangle = \sum_s (w_s)_j \overline{(w_s)_k} = 0,$$

i.e., the j and k rows of W are orthogonal, which is condition 2. □

Example 10.21. For $G = \langle a \rangle$ the cyclic group of order 2, define a unitary action of G on $\mathbb{C}^3$ by

$$av := \begin{pmatrix} 1 & 0 & 0 \\ 0 & 1 & 0 \\ 0 & 0 & -1 \end{pmatrix} v = \begin{pmatrix} v_1 \\ v_2 \\ -v_3 \end{pmatrix},$$

i.e., take the trivial representation on $V_1 = \mathbb{C}e_1$, $V_2 = \mathbb{C}e_2$, and the sign representation on $V_3 = \mathbb{C}e_3$. There is no G-frame $(gv)_{g\in G}$ for $\mathbb{C}^3$. However, there are many choices for w_1, w_2 so that $(gw_j)_{1\le j\le 2, g\in G}$ is a tight G-invariant frame for $\mathbb{C}^3$, e.g.,

$$W = [w_1, w_2] = \begin{bmatrix} 1 & 0 \\ 0 & 1 \\ u & \sqrt{1-|u|^2} \end{bmatrix}, \qquad |u| \le 1.$$

Here $\langle w_1, w_2 \rangle = u\sqrt{1-|u|^2}$, and so *infinitely* many unitarily inequivalent G-invariant tight frames for $\mathbb{C}^3$ can be constructed in this way.

Given some $w_1 = (x, y, z)$, a suitable w_2 can be chosen provided $|z|^2 \le |x|^2 + |y|^2$, since we can take $w_2 = (\overline{y}, -\overline{x}, u)$ where $|u|^2 = |x|^2 + |y|^2 - |z|^2$. For example, if $w_1 = (1,2,2)$, then choosing $w_2 = (2,-1,1)$ gives the G-invariant tight frame

$$(w_1, w_2, aw_1, aw_2) = \left(\begin{bmatrix} 1 \\ 2 \\ 2 \end{bmatrix}, \begin{bmatrix} 2 \\ -1 \\ 1 \end{bmatrix}, \begin{bmatrix} 1 \\ 2 \\ -2 \end{bmatrix}, \begin{bmatrix} 2 \\ -1 \\ -1 \end{bmatrix} \right).$$

Example 10.22. Let $G = \langle a \rangle$ be the cyclic group of order n. An irreducible unitary action of G on $\mathbb{R}^2$ is given by (10.2), i.e.,

$$av = Av, \qquad A := \begin{pmatrix} \cos\frac{2\pi}{n} & -\sin\frac{2\pi}{n} \\ \sin\frac{2\pi}{n} & \cos\frac{2\pi}{n} \end{pmatrix}.$$

Suppose that G acts on $\mathbb{R}^4 = \mathbb{R}^2 \times \mathbb{R}^2$ componentwise, i.e., $a(v,w) = (Av, Aw)$. Then the conditions for the G-orbit of (v,w) to be a tight frame for $\mathbb{R}^4$ are

$$\|v\| = \|w\| \ne 0, \qquad \sum_{j=1}^{n} \langle f_1, A^j v \rangle A^j w = \Big(\sum_{j=1}^{n} A^j w v^* A^{-j} \Big) f_1 = 0, \quad \forall f_1 \in \mathbb{R}^2.$$

A calculation shows that this is not possible, and so no G-orbit is a tight frame.

This can also be seen by appealing to Theorem 10.9, as follows. The action of G on $\mathbb{R}^2$ given by $av = Av$ is not absolutely irreducible. On the complexification $\mathbb{C}^2$ there are two orthogonal G-invariant subspaces: the eigenspaces of A corresponding to the eigenvalues $\lambda = \omega, \overline{\omega}$, $\omega := e^{\frac{2\pi i}{n}}$. For $n > 2$, $\omega \ne \overline{\omega}$, and so these subspaces are not $\mathbb{C}G$-isomorphic. Thus the complexification $\mathbb{C}^4$ of $\mathbb{R}^2 \times \mathbb{R}^2$ decomposes as the sum of four 1-dimensional G-invariant subspaces, with two pairs $\mathbb{C}G$-isomorphic to each other. By Theorem 10.9, it is therefore not possible to find a G-frame for $\mathbb{C}^4$, and hence neither for $\mathbb{R}^4$.

10.13 The minimal number of generators for a G-invariant frame

Suppose that there is a unitary action of a finite group G on a Hilbert space $\mathscr{H}$. Let $\mathscr{W}$ be a **complete set of nonisomorphic irreducible $\mathbb{C}G$-modules**, i.e., each irreducible $\mathbb{C}G$-module occurs once in $\mathscr{W}$ up to $\mathbb{C}G$-isomorphism. We denote the direct sum of k copies of an $\mathbb{F}G$-module V by V^k.

Example 10.17 shows that there is a tight G-frame for $\mathscr{H}$ if and only if $\mathscr{H}$ is $\mathbb{C}G$-isomorphic to an orthogonal direct sum

$$\bigoplus_{W\in\mathscr{W}} W^{a_W}, \qquad 0\le a_W\le \dim(W). \tag{10.28}$$

On the other hand, Wedderburn's Theorem states that $\mathbb{C}G$ is $\mathbb{C}G$-isomorphic to

$$\bigoplus_{W\in\mathscr{W}} W^{\dim(W)}.$$

Combining these observations gives the following.

Proposition 10.4. *The following are equivalent:*

1. *There is a tight frame $(gv)_{g\in G}$ for $\mathscr{H}$.*
2. *There is a frame (spanning set) $(gv)_{g\in G}$ for $\mathscr{H}$.*
3. *$\mathscr{H}$ is $\mathbb{C}G$-isomorphic to a submodule of $\mathbb{C}G$.*

In other words:

The existence of a tight G-frame for $\mathscr{H}$ depends only on the $\mathbb{C}G$-module structure of $\mathscr{H}$. It does not depend on the inner product on $\mathscr{H}$.

We now generalise this. Denote the number of times an irreducible $\mathbb{C}G$-module appears in a direct sum decomposition of $\mathscr{H}$ into irreducible $\mathbb{C}G$-modules by

$$\mathrm{mult}(W,\mathscr{H}) = \dim(\mathrm{Hom}_{\mathbb{C}G}(W,\mathscr{H})) = \dim(\mathrm{Hom}_{\mathbb{C}G}(\mathscr{H},W)).$$

Theorem 10.10. *(Minimal number of generators) The following are equivalent.*

1. *There is a G-invariant tight frame $(gv_j)_{g\in G,1\le j\le r}$ for $\mathscr{H}$ with r generators.*
2. *There is a G-invariant frame $(gv_j)_{g\in G,1\le j\le r}$ for $\mathscr{H}$ with r generators.*
3. *$\mathscr{H}$ is $\mathbb{C}G$-isomorphic to a submodule of $(\mathbb{C}G)^r$, i.e.,*

$$\mathrm{mult}(W,\mathscr{H}) \le r\dim(W), \tag{10.29}$$

for every irreducible $\mathbb{C}G$-module W.

Proof. ($1\Longleftrightarrow 2$) This follows from a variation of the argument for Theorem 10.1, i.e., if $\Phi = (gw_j)_{g\in G,1\le j\le r}$ spans $\mathscr{H}$ and S is the frame operator of Φ, then $(gS^{-\frac{1}{2}}w_j)_{g\in G,1\le j\le r}$ is a tight frame for $\mathscr{H}$.

($2\Longleftrightarrow 3$) First, suppose that $(gw_j)_{g\in G,1\le j\le r}$ spans $\mathscr{H}$. Then $\mathscr{H}$ is a quotient of $\bigoplus_j W_j$, where $W_j := \operatorname{span}\{gv_j\}_{g\in G}$. By Proposition 10.4, each W_j is $\mathbb{C}G$-isomorphic to a submodule of $\mathbb{C}G$, and hence $\mathscr{H}$ is $\mathbb{C}G$-isomorphic to a submodule of $(\mathbb{C}G)^r$.

Conversely, now suppose that $\operatorname{mult}(W,\mathscr{H}) \le r\dim(W)$ for every irreducible $\mathbb{C}G$-module W. Then we can write $\mathscr{H} = Z_1\oplus\cdots\oplus Z_r$, where each submodule Z_j has $\operatorname{mult}(W,Z_j)\le\dim(W)$, and so is $\mathbb{C}G$-isomorphic to a submodule of $\mathbb{C}G$. We can choose vectors $w_j\in Z_j$ for which $(gw_j)_{g\in G}$ spans Z_j (by Proposition 10.4). Thus $(gw_j)_{g\in G,1\le j\le r}$ spans $\mathscr{H}$. □

We now briefly indicate how Theorems 10.9 and 10.10 can be used to construct tight frames of multivariate orthogonal polynomials (see §10.10) which are invariant under the symmetries of the weight and have a minimal number of generators.

Recall from §10.10, that the symmetry group G of a measure μ has a unitary action on the d-variate polynomials $\Pi = \Pi_k(\mathbb{R}^d)$ given by $g\cdot f := f\circ g^{-1}$. We are interested in finding G-invariant tight frames for space $\mathscr{V}_k(\mu)$ of orthogonal polynomials of degree k for this measure.

We denote the space of homogeneous polynomials of degree k by $\Pi_k^\circ = \Pi_k^\circ(\mathbb{R}^d)$. This is a G-invariant subspace of $\Pi(\mathbb{R}^d)$, indeed it is $\mathbb{F}G$-isomorphic to $\mathscr{V}_k(\mu)$.

Proposition 10.5. *Let G be a finite subgroup of the symmetry group of a measure μ. Then $\mathscr{V}_k(\mu)$ is $\mathbb{F}G$-isomorphic to Π_k°.*

Proof. Let P_V be the orthogonal projection onto a subspace V. If V is G-invariant, then P_V commutes with the action of G. Thus

$$\Pi_k^\circ \to \mathscr{V}_k(\mu) : f\mapsto f - P_{\Pi_{k-1}}f = P_{\mathscr{V}_k(\mu)}(f)$$

is an $\mathbb{F}G$-homomorphism onto $\mathscr{V}_k(\mu)$. Moreover, this is an $\mathbb{F}G$-isomorphism since $\dim(\Pi_k^\circ) = \dim(\mathscr{V}_k(\mu))$. □

Thus the problem of finding G-invariant frames for $\mathscr{V}_k(\mu)$ therefore reduces to understanding the $\mathbb{C}G$-module structure of Π_k°. To this end, denote the G-invariant polynomials by

$$\Pi^G := \{f\in\Pi : g\cdot f = f, \forall g\in G\}.$$

Example 10.23. (Trivial representation). Let W be the trivial irreducible $\mathbb{C}G$-module. This is one-dimensional, with the G fixing every vector in W. Thus (10.29) becomes

$$\operatorname{mult}(W,\Pi_k^0) = \dim(\Pi^G\cap\Pi_k^0) \le r\dim(W) = r,$$

and so the minimal number of generators r for $\mathscr{V}_k(\mu)$ satisfies $r\ge\dim(\Pi^G\cap\Pi_k^0)$.

When G is a finite reflection group, there are techniques to decompose Π_k° into irreducible $\mathbb{C}G$-submodules. We now briefly outline these.

10.14 The coinvariants of a finite reflection group

A linear transformation on a complex vector space is a **complex reflection** if it has finite order and fixes a hyperplane. A group G generated by complex reflections is called a complex reflection group. A complete classification of the finite irreducible complex reflection groups was given by Shephard and Todd.

Let Π_+^G be the G-invariant polynomials with zero constant term, and $\mathscr{I}_G$ be the ideal generated by Π_+^G in Π. The **ring of coinvariants** (or **coinvariant space**) is the quotient ring

$$\Pi_G := \frac{\Pi}{\mathscr{I}_G}.$$

This is naturally graded with respect to degree and inherits the action of G (since $\mathscr{I}_G$ is a homogeneous ideal which is G-invariant).

We now suppose that G is a finite reflection group. This implies

- Π^G is a ring (this characterises reflection groups).
- $\Pi_G \cong \mathbb{C}G$ ($\mathbb{C}G$-isomorphism).
- There is a decomposition of Π as a tensor product of graded $\mathbb{C}G$-modules:

$$\Pi \cong \Pi^G \otimes_{\mathbb{C}} \Pi_G. \tag{10.30}$$

We identify Π_G as a $\mathbb{C}G$-submodule of Π. Then (10.30) gives the $\mathbb{C}G$-module decomposition

$$\Pi_k^\circ = \bigoplus_{j=0}^{k} (\Pi^G \cap \Pi_j^\circ) \otimes_{\mathbb{C}} (\Pi_G \cap \Pi_{k-j}^\circ).$$

We observe that $\Pi^G \cap \Pi_j^\circ$ consists of $\dim(\Pi^G \cap \Pi_j^\circ)$ copies of the trivial module. Hence, for W an irreducible $\mathbb{C}G$-module,

$$\operatorname{mult}(W, \Pi_k^\circ) = \sum_{j=0}^{k} \dim(\Pi^G \cap \Pi_j^\circ) \operatorname{mult}(W, \Pi_G \cap \Pi_{k-j}^\circ).$$

where $\sum_j \operatorname{mult}(W, \Pi_G \cap \Pi_j^\circ) \le \operatorname{mult}(W, \Pi_G) = \operatorname{mult}(W, \mathbb{C}G) = \dim(W)$. This gives:

For $G \subset \operatorname{Sym}(\mu)$ a finite reflection group, determining the minimal number of generators r for a G-invariant tight frame for $\mathscr{V}_k(\mu)$ (via Theorem 10.10) amounts to calculating $\operatorname{mult}(W, \Pi_G \cap \Pi_j^\circ)$, $0 \le j \le k$, for each irreducible W.

These calculations are done for the orthogonal polynomials on a regular polygon in $\mathbb{R}^2$ and the cube in $\mathbb{R}^3$ (see [VW16]).

The discussion above also gives the following estimate for r

$$r \le \max_{0 \le j \le k} \dim(\Pi^G \cap \Pi_j^\circ).$$

Notes

The basic theory of groups frame was given in [VW05], [Han07] (one generator), and [VW16] (multiple generators). Thanks to Patrick Morandi for the graphics used in Figures 10.1 and 10.2.

Exercises

10.1. Suppose that there are unitary actions of a finite group G on V_j and V_j. For any $v_j \in V_j$, $v_k \in V_k$, show that $S : V_j \to V_k$ given by

$$Sf := \sum_{g\in G} \langle f, gv_j\rangle gv_k, \qquad \forall f \in V_j,$$

is an $\mathbb{F}G$-homomorphism, i.e., commutes with the action of G.

10.2. Show that if W is absolutely irreducible (with the action of G unitary) and $\sigma_j : V_j \to W$ is an $\mathbb{F}G$-isomorphism, then the condition (10.17) can be written as

$$\langle \sigma_j v_j, \sigma_k v_k\rangle = 0.$$

10.3. Let $\mathscr{H}$ be the subspace of $\mathbb{R}^8$ consisting of vectors x with $\sum_j x_j = 0$.
(a) Show there is unitary action of $G = S_8$ on $\mathscr{H}$ given by $\rho(\sigma)x := (x_{\sigma j})_{j=1}^8$.
(b) Show that there are 28 distinct vectors in the G-frame $(gv)_{g\in G}$ for $\mathscr{H}$ given by the vector $v = (3,3,-1,-1,-1,-1,-1,-1)$.
Remark: These 28 vectors are an equiangular tight frame for $\mathbb{R}^7$ (see Example 12.3). They can be thought of as a G-frame (apply the permutation matrices to the vectors in $\mathbb{R}^8$ and calculate the size of the stabiliser).

10.4. Let $\Phi = (gv)_{g\in G} = (\rho(g)v)_{g\in G}$ be a finite frame for the vector space $\mathscr{H}$, where $\rho : G \to GL(\mathscr{H})$ is a representation (group homomorphism). Choose some inner product $\langle\cdot,\cdot\rangle$ on $\mathscr{H}$, and let A^* be the Hermitian transpose with respect to it.
(a) Show that a second inner product on $\mathscr{H}$ can be defined by

$$\langle x,y\rangle_\rho := \frac{1}{|G|}\sum_{g\in G}\langle \rho(g)x, \rho(g)y\rangle = \langle Ax, y\rangle, \quad A = A_\rho = \frac{1}{|G|}\sum_{g\in G}\rho(g)^*\rho(g).$$

(b) Show that each $\rho(g)$ is unitary with respect to $\langle\cdot,\cdot\rangle_\rho$, i.e., it is G-invariant

$$\langle \rho(h)x, \rho(h)y\rangle_\rho = \langle x,y\rangle_\rho, \qquad \forall h \in G.$$

(c) Let $B = A^{\frac{1}{2}}$ be the positive square root of the positive definite A above. Show that $\tilde\rho(g) := B\rho(g)B^{-1}$ defines a representation $\tilde\rho : G \to \mathscr{U}(\mathscr{H})$.
(d) Show that Φ is similar to the G-frame $\Psi = (\tilde\rho(g)Bv)_{g\in G}$, and hence is similar to the tight G-frame Ψ^{can}.

10.5. Let $\langle\cdot,\cdot\rangle_\rho$ be the G-invariant inner product of Exer. 10.4. Show that

$$\langle\!\langle x,y\rangle\!\rangle := \langle Mx,y\rangle_\rho, \qquad M \text{ positive definite with respect to } \langle\cdot,\cdot\rangle_\rho$$

gives a G-invariant inner product on $\mathscr{H}$ if and only if M commutes with the action of G, i.e., $\rho(g)M = M\rho(g)$, $\forall g \in G$. In particular, if ρ is absolutely irreducible, then there is a unique G-invariant inner product.

10.6. Let $(gv)_{g\in G}$ be a G-frame for $\mathscr{H}$.
(a) Show that the variational condition (6.4) for tightness becomes

$$\frac{\dim(\mathscr{H})}{|G|}\sum_{g\in G}|\langle v,gv\rangle|^2 \geq \|v\|^4.$$

(b) Show that $(gv)_{g\in G}$ is a (t,t)-design if and only if

$$\frac{1}{|G|}\sum_{g\in G}|\langle v,gv\rangle|^{2t} = c_t(d,\mathbb{F})\|v\|^{4t}.$$

10.7. Prove the assertions of Theorem 10.2, i.e.,
(a) The direct sum of disjoint G-frames is a G-frame.
(b) The sum of a G_1-frame and a G_2-frame is a $G_1 \times G_2$-frame.
(c) The tensor product of G_1-frame with a G_2-frame is a $G_1 \times G_2$-frame.
(d) The complement of a tight G-frame is a tight G-frame.

10.8. Let G be the symmetry group of a Platonic solid $T \subset \mathbb{R}^3$ (with centre of gravity the origin) acting as unitary transformations. Use the fact that finite subgroups of $\mathscr{U}(\mathbb{R}^2)$ are cyclic or dihedral to prove this action is irreducible.

10.9. $\mathbb{F}G$*-modules*. Let G be a group. An **action** of the group G on a set X is a map $G\times X \to X : (g,x) \mapsto gx$ satisfying $g_1(g_2x) = (g_1g_2)x$, $1x = x$. A vector space V over $\mathbb{F}$ is an $\mathbb{F}G$-**module** if there is a multiplication gv, $g \in G$, $v \in V$ for which $(g,v) \mapsto gv$ is an *action* and $v \mapsto gv$ is a *linear map*, $\forall g \in G$.
(a) Show that V is an $\mathbb{F}G$-module if and only if $\rho : G \to GL(V)$, $\rho(g)v := gv$ is a (linear) representation/action.
(b) Show that $\mathbb{F}G$-submodules are the same as G-invariant subspaces.
Remark: Other terminology carries over in the obvious way, e.g., the action of G on $V \neq 0$ given by some representation is irreducible if and only if the only the $\mathbb{F}G$-module V is *simple*, i.e., has no $\mathbb{F}G$-submodules other than 0 and V.

10.10. Suppose that V_j and V_k are absolutely irreducible G-invariant subspaces of $\mathscr{H}$, and σ, τ are $\mathbb{F}G$-isomorphisms $V_j \to V_k$. Prove that

$$\sigma = \lambda\tau, \qquad \text{for some } \lambda \in \mathbb{C}.$$

10.11. Let $\Phi = (gv)_{g\in G}$ be a G-frame for $\mathscr{H}$ given by ρ, Ψ be the set of vectors in Φ, and $H := \{g \in G : gv = v\}$ be the stabiliser of v.
(a) Show the vectors in Ψ can be indexed by the set of left cosets of H in G

$$\Psi = (gv)_{gH \in \mathscr{C}}, \qquad \mathscr{C} := \{gH : g \in G\}. \tag{10.31}$$

(b) Suppose H is normal, so that $\mathscr{C}$ becomes the group G/H. Show that (10.31) gives a G/H-frame if and only $H = N$, where N is the kernel of ρ.
(c) Suppose G is abelian. Show that $H = N$, which is independent of v, and hence Ψ is a G/N-frame of distinct vectors.

10.12. Let $G = \langle a, b \rangle$ be the *binary icosahedral group* of order 120 generated by

$$a = \frac{1}{2}\begin{pmatrix} t^{-1} - ti & 1 \\ -1 & t^{-1} + ti \end{pmatrix}, \qquad b = \begin{pmatrix} -i & 0 \\ 0 & i \end{pmatrix},$$

where $t = \frac{1}{2}(1 + \sqrt{5})$ is the golden ratio.
(a) Show that for every unit vector $v \in \mathbb{C}^2$, the G-frame $(gv)_{g \in G}$ is a $(5,5)$-design.
(b) Show that there is $(5,5)$-design of 12 vectors for $\mathbb{C}^2$.

10.13. Show that a θ-isogonal configuration (see Example 10.2) is a group frame by determining the group G and its unitary action.

Chapter 11
Harmonic frames

11.1 Introduction

Here we consider the tight G-frames for G *abelian*. We will see that:

- There are *finitely* many such frames (they will be called the *harmonic frames*).
- Each is given by a subset of the characters of G.
- Each is given by a subset of G.

We first motivate the definitions to come, by considering the Example 2.4 for $n=3$. The character table (Fourier matrix) of the cyclic group $G=C_3$ of order 3 is

$$\begin{bmatrix} 1 & 1 & 1 \\ 1 & \omega & \omega^2 \\ 1 & \omega^2 & \omega \end{bmatrix}, \qquad \omega \neq 1, \quad \omega^3 = 1,$$

which has orthogonal columns of equal length. Since the orthogonal projection of an orthonormal basis is a tight frame, removing rows (i.e., characters) from this character table gives a submatrix whose columns are an equal-norm tight frame, e.g., removing rows 1 or 2 gives the equal-norm tight frames for $\mathbb{C}^2$

$$\left(\begin{bmatrix} 1 \\ 1 \end{bmatrix}, \begin{bmatrix} \omega \\ \omega^2 \end{bmatrix}, \begin{bmatrix} \omega^2 \\ \omega \end{bmatrix} \right), \qquad \left(\begin{bmatrix} 1 \\ 1 \end{bmatrix}, \begin{bmatrix} 1 \\ \omega \end{bmatrix}, \begin{bmatrix} 1 \\ \omega^2 \end{bmatrix} \right). \tag{11.1}$$

Equivalently, one could remove columns (i.e., elements of G). This amounts to restricting the characters to a subset J of G, which is the most convenient way to describe harmonic frames (we take this as the definition). The tight frames of (11.1) are clearly G-frames under the respective unitary actions of $G=\langle a \rangle$ given by

$$\rho(a) = \begin{pmatrix} \omega & \\ & \omega^2 \end{pmatrix}, \qquad \rho(a) = \begin{pmatrix} 1 & \\ & \omega \end{pmatrix}.$$

S.F.D. Waldron, *An Introduction to Finite Tight Frames*, Applied and Numerical Harmonic Analysis, https://doi.org/10.1007/978-0-8176-4815-2_11

11.2 Character tables

We first outline the basics of *character theory* for finite abelian groups (cf. [Rud90]). Let G be a finite abelian group. The **(irreducible) characters** of G are the group homomorphisms $\xi : G \to \mathbb{C} \setminus \{0\}$, where $\mathbb{C} \setminus \{0\}$ is a group under multiplication. Here we think of them as vectors $\xi \in \mathbb{C}^G$ (with the Euclidean inner product), which satisfy

$$\xi(gh) = \xi(g)\xi(h), \qquad \forall g, h \in G. \tag{11.2}$$

The set of irreducible characters of the abelian group G is denoted by $\hat{G}$.

The characters $\hat{G}$ form group under the multiplication $(\xi\eta)(g) := \xi(g)\eta(g)$, which is called the **character group**. The character group $\hat{G}$ is isomorphic to G. For $\chi \in \hat{G}$, (11.2) implies that $\chi(g)$ is a $|G|$-th root of unity, and so the inverse of χ satisfies

$$\chi^{-1}(g) = \frac{1}{\chi(g)} = \overline{\chi(g)}. \tag{11.3}$$

The square matrix with the irreducible characters of G as rows is referred to as the **character table** of G. For example, if $G = \langle a \rangle$ is the cyclic group of order n, with its elements ordered $1, a, \ldots, a^{n-1}$, then its character table is

$$\begin{bmatrix} 1 & 1 & 1 & \cdots & 1 \\ 1 & \omega & \omega^2 & \cdots & \omega^{n-1} \\ 1 & \omega^2 & \omega^4 & \cdots & \omega^{2(n-1)} \\ \vdots & \vdots & \vdots & & \vdots \\ 1 & \omega^{n-1} & \omega^{2(n-1)} & \cdots & \omega^{(n-1)(n-1)} \end{bmatrix}, \qquad \omega := e^{\frac{2\pi i}{n}}. \tag{11.4}$$

This (and the scalar multiple by $\frac{1}{\sqrt{n}}$) is also known as the **Fourier matrix**.

Example 11.1. If all elements of G have order 2, i.e., G is an elementary abelian 2-group $\mathbb{Z}_2 \times \cdots \times \mathbb{Z}_2$, then the entries of the character table of G are ± 1.

The rows and columns of the character table are orthogonal, i.e.,

$$\langle \xi, \eta \rangle := \sum_{g \in G} \xi(g)\overline{\eta(g)} = 0, \qquad \xi \neq \eta, \tag{11.5}$$

$$\sum_{\chi \in \hat{G}} \chi(g)\overline{\chi(h)} = 0, \qquad g \neq h. \tag{11.6}$$

These are referred to as the **orthogonality** and **column orthogonality** relations.

The **Pontryagin duality map** (canonical group isomorphism) is given by

$$G \to \hat{\hat{G}} : g \mapsto \hat{\hat{g}}, \qquad \hat{\hat{g}}(\chi) := \chi(g), \quad \forall \chi \in \hat{G},\ g \in G. \tag{11.7}$$

11.3 Harmonic frames

For G abelian, the absolutely irreducible $\mathbb{F}G$-modules are *one*-dimensional, with the action of G given by

$$gv_j = \xi_j(g)v_j, \qquad \xi_j \in \hat{G}, \tag{11.8}$$

where $\xi_j \in \hat{G}$. As outlined in Example 10.16, the condition (10.16) in Theorem 10.8 (the characterisation of tight G-frames) is that the characters ξ_j in (11.8) be orthogonal, and hence can be taken at most once. By choosing $v_j = e_j$, we therefore conclude that all tight G-frames for $\mathbb{C}^d$ (up to unitary equivalence) are given by

$$\Psi = (\psi_g)_{g\in G}, \qquad \psi_g := (\xi_j(g))_{j=1}^d \in \mathbb{C}^d, \tag{11.9}$$

where $\xi_1,\ldots,\xi_d \in \hat{G}$ are *distinct*. In view of (11.2), the action of G on $\Psi = (\psi_g)$ is

$$g\psi_h := \begin{pmatrix} \xi_1(g) & & \\ & \ddots & \\ & & \xi_d(g) \end{pmatrix} \begin{pmatrix} \xi_1(h) \\ \vdots \\ \xi_d(h) \end{pmatrix} = \begin{pmatrix} \xi_1(gh) \\ \vdots \\ \xi_d(gh) \end{pmatrix} = \psi_{gh}.$$

The construction (11.9) amounts to taking d *rows* of the character table of G, i.e., a subset of d elements of $\hat{G}$. Equivalently (since G and $\hat{G}$ are isomorphic), one can select *columns* of the character table, i.e., a subset $J \subset G$. The latter is the most convenient (cf. Theorem 11.2), and so we take it as our definition.

Definition 11.1. Let G be a finite abelian group of order n. A tight frame for $\mathbb{C}^d$ which is unitarily equivalent to

$$\Phi_J = (\xi|_J)_{\xi\in\hat{G}} \subset \mathbb{C}^J \approx \mathbb{C}^d, \tag{11.10}$$

where $J \subset G$, $|J| = d$, is called a **harmonic frame** (given by $J \subset G$). If G can be a cyclic group, then we say that the harmonic frame is **cyclic**[1].

Our observations lead to the following:

Theorem 11.1. *(Characterisation of harmonic frames) Let Φ be an equal-norm tight frame for $\mathscr{H} \approx \mathbb{C}^d$. Then the following are equivalent:*

1. *Φ is a G-frame, where G is abelian.*
2. *Φ is given by a submatrix of d rows of the character table of an abelian G.*
3. *Φ is harmonic, given by some $J \subset G$, $|J| = d$.*
4. *The symmetry group* $\mathrm{Sym}(\Phi)$ *has a transitive abelian subgroup G.*

For each Φ, G can be taken to be the same in 1,2,3 and 4, but it need not be unique.

---

[1]Harmonic frames are also called *geometrically uniform frames* [EB03] (orbits of abelian matrix groups), and the term harmonic frame is also used for what we call a *cyclic* harmonic frame.

Proof. The equivalence of 1 and 2 follows from Theorem 10.8, as already discussed. The equivalence of 1 and 4 follows from Theorem 10.4.

($3 \Longrightarrow 1$) The Φ_J given by (11.10) is a $\hat{G}$-frame, via the unitary action

$$\chi \cdot (\xi|_J) := (\chi|_J)(\xi|_J) = (\chi\xi)|_J.$$

Since $\hat{G}$ and G are isomorphic, it therefore follows that Φ_J is a G-frame.

($2 \Longrightarrow 3$) We use the *Pontryagin duality map* (11.7). Suppose that Φ is given by a submatrix of the character table of G. Since $\hat{G}$ and G are isomorphic, Φ is given by a submatrix of the character table of $\hat{G}$, say

$$[\hat{\hat{g}}(\xi)]_{g\in J,\xi\in\hat{G}} = [\xi(g)]_{g\in J,\xi\in\hat{G}} = [\xi|_J]_{\xi\in\hat{G}}, \qquad J \subset G,$$

and so Φ is given by the harmonic frame $\Phi_J = (\xi|_J)_{\xi\in\hat{G}}$.

For the possible nonuniqueness of G, see Examples 11.2, 11.3 and 11.4. □

Early applications of cyclic harmonic frames include robust signal transmission with quantisation and erasures [GVT98], [GKK01], [CK03] (which introduces the term harmonic tight frame), and multiple-antenna code design [HMR+00].

Example 11.2. (Nonuniqueness of G) Let $\Phi = (e_1, e_2, -e_1, -e_2)$ be the equal-norm tight frame of four equally spaced unit vectors for $\mathbb{R}^2$. This has symmetry group the dihedral group $D_8 = \langle a, b\rangle$ (see Exercise 9.4), where the action of a and b on $\mathbb{R}^2$ is given by

$$a = \begin{pmatrix} 0 & -1 \\ 1 & 0 \end{pmatrix}, \qquad b = \begin{pmatrix} 0 & 1 \\ 1 & 0 \end{pmatrix}.$$

The frame Φ is the G-orbit of the nonisomorphic abelian subgroups

$$\langle a^2, b\rangle \approx C_2 \times C_2, \qquad \langle a\rangle \approx C_4,$$

and hence the group G in Theorem 11.1 need not be unique.

Example 11.3. (Orthonormal bases) Since the columns of a character table have equal norm and are orthogonal, any orthonormal basis for $\mathbb{R}^n$ (or $\mathbb{C}^n$) is a G-frame for any abelian group G of order n.

Example 11.4. (Simplex) By removing the first row from (11.4), i.e., the trivial character $\chi = 1$, one obtains the $n = d+1$ vertices of the regular simplex in $\mathbb{R}^d$, which is therefore a tight C_{d+1}-frame. Similarly, the vertices of the simplex are a G-frame for any abelian group G of order n.

Example 11.5. (Noncyclic harmononic frames) Consider the harmonic frame given by the eight vertices of the *cube* $[-1,1]^3$. This is a both a $\mathbb{Z}_2 \times \mathbb{Z}_2 \times \mathbb{Z}_2$-frame and a $\mathbb{Z}_2 \times \mathbb{Z}_4$-frame, but not a cyclic harmonic frame since its symmetry group (the octahedral group $S_4 \times \mathbb{Z}_2$) contains no elements of order 8.

11.4 Harmonic frames with distinct vectors and with real vectors

The condition for a harmonic frame to have distinct vectors, be real, or lifted are:

Theorem 11.2. *Let G be an abelian group of order n, and $\Phi = \Phi_J = (\xi|_J)_{\xi\in\hat{G}}$ be the harmonic frame of n vectors for $\mathbb{C}^d$ given by a choice $J \subset G$, $|J| = d$. Then*

1. *Φ has distinct vectors if and only if J generates G.*
2. *Φ is a real frame if and only J is closed under taking inverses.*
3. *Φ is a lifted frame if and only if the identity is an element of J.*

Proof. We write G additively and observe that $\overline{\xi(j)} = \xi(-j)$, $\forall \xi \in \hat{G}$, $\forall j \in G$.

1. Let H be the subgroup of G generated by J. Then Φ has distinct vectors if and only if the composition of maps $\hat{G} \mapsto \hat{H} \mapsto \mathbb{C}^J : \xi \mapsto \xi|_H \mapsto \xi|_J$ is 1–1. Since each $h \in H$ can be written as a sum of elements in J, and ξ is a character, $\xi(h)$ is determined by $\xi|_J$, and so $\xi|_H \mapsto \xi|_J$ is 1–1. Hence $\xi \mapsto \xi|_J$ is 1–1 if and only if the group homomorphism given by $\hat{G} \mapsto \hat{H} : \xi \mapsto \xi|_H$ is 1–1, i.e., $\hat{G} = \hat{H}$, and so $G = H = \langle J \rangle$.

2. The frame Φ is real if and only if $\langle \chi|_J, \eta|_J \rangle = \langle (\chi\eta^{-1})|_J, 1|_J \rangle \in \mathbb{R}$, $\forall \chi\eta^{-1} \in \hat{G}$ i.e.,

$$\psi := \sum_{j\in J} \hat{\hat{j}} \in \mathbb{R}^{\hat{G}}. \tag{11.11}$$

First, suppose that J is closed under taking inverses, and $j \in J$. Then either j is its own inverse, so $\xi(j) = \xi(-j) = \overline{\xi(j)} \in \mathbb{R}$, or the pair $\{j, -j\} \subset J$ contributes $\xi(j) + \xi(-j) = \xi(j) + \overline{\xi(j)} \in \mathbb{R}$ to the sum above. Thus we conclude each inner product is real.

Conversely, suppose the inner products are real, i.e., (11.11) holds, and $\overline{\psi} = \psi$. Let $\langle \zeta, \tau \rangle$ be the Euclidean inner product on $\mathbb{C}^{\hat{G}}$ normalised so that the characters of $\hat{G}$ are orthonormal. Then

$$j \in J \iff \langle \psi, \hat{\hat{j}} \rangle = 1 \iff \langle \overline{\psi}, \overline{\hat{\hat{j}}} \rangle = \langle \psi, (-j)\hat{\hat{}} \rangle = 1 \iff -j \in J.$$

3. By the column orthogonality relation (11.6) for characters, Φ is unlifted if and only if

$$\sum_{\xi\in\hat{G}} \xi|_J = 0 \iff \sum_{\xi\in\hat{G}} \xi(j) = \sum_{\xi\in\hat{G}} \xi(j)\overline{\xi(0)} = 0, \quad \forall j \in J \iff 0 \notin J.$$

□

It suffices to consider harmonic frames with distinct vectors:

Corollary 11.1. *Let $\Phi_J = (\xi|_J)_{\xi\in\hat{G}}$ be a harmonic frame, and H be the subgroup of G generated by H. Then the $|H|$ distinct vectors of Φ_J are an H-frame $(\chi|_J)_{\chi\in\hat{H}}$.*

Proof. For $\xi \in \hat{G}$, $\chi = \xi|_H$ is a character of H with the property that $\chi|_J$ appears exactly $|G|/|H|$ times in Φ_J. □

11.5 Combining and decomposing harmonic frames

We first observe that Theorem 11.2 guarantees the existence of harmonic frames of any number of distinct vectors for $\mathbb{C}^d$.

Corollary 11.2. *(Existence) Let G be a finite abelian group, with minimal number of generators d^*. Then there is a G-frame of distinct vectors for $\mathbb{C}^d$ if and only if*

$$d^* \leq d \leq |G|.$$

Example 11.6. Let G be an *elementary abelian p-group* (p a prime), i.e.,

$$G = \mathbb{Z}_p \times \cdots \times \mathbb{Z}_p \qquad (k \text{ times}),$$

Then G gives a harmonic frame of *distinct* vectors for $\mathbb{C}^d$ only for $k \leq d \leq p^k$.

Since harmonic frames are G-frames, we have a special case of Theorem 10.2.

Theorem 11.3. *Harmonic frames can be combined as follows.*

- *The direct sum of disjoint harmonic frames is a harmonic frame.*
- *The sum of harmonic frames a harmonic frame.*
- *The tensor product of harmonic frames is a harmonic frame.*
- *The complement of a harmonic frame is a harmonic frame.*

Proof. For the first three, use Theorem 10.2 and observe that products of abelian groups are abelian. For the last, use Corollary 10.4. □

Example 11.7. (Direct sums) The orthogonality of the irreducible characters $\xi_j \in \hat{G}$ of an abelian group G is equivalent to the harmonic frames $(\xi_j(g))_{g\in G}$ for $\mathbb{C}^1$ being orthogonal (see Example 3.11). Thus the construction of harmonic frames by taking rows ξ_j of the character table can be interpreted as taking a direct sum of these harmonic frames for $\mathbb{C}^1$.

The decomposition of a harmonic frame as a direct sum of irreducible characters given above is unique (for G given). Its decomposition into sums and tensor products may not be unique:

Example 11.8. The four equally spaced vectors in $\mathbb{R}^2$ (the vertices of the square) can be written as a sum $\{-1,1\}\dot{+}\{-1,1\}$, and as a tensor product $\{-1,1\} \otimes \{e_1, e_2\}$, where $\{e_1, e_2\}$ is an orthonormal basis.

Example 11.9. (Symmetries) The symmetry group of a harmonic frame which is combination of harmonic frames can be larger than that guaranteed by Proposition 9.3. For example, the harmonic frame of 9 vectors for $\mathbb{R}^4$ given by the tensor product (see Example 5.14) and by the sum of the three equally spaced unit vectors in $\mathbb{R}^2$ with themselves has symmetry group of order $3!3^3 = 162$. These two harmonic frames are unitarily equivalent (up to a reordering).

The projective symmetry group of a harmonic frame is considered in §11.12.

11.6 Real harmonic frames

Most harmonic frames are complex (see [WH06], §11.10), but *real* harmonic frames can always be constructed by using Theorem 11.2.

Corollary 11.3. *(Existence of real cyclic harmonic frames). For all $d \geq 2$ and $n \geq 2$, there exists a cyclic harmonic frame of n distinct vectors for $\mathbb{R}^d$.*

Proof. Let $G = \mathbb{Z}_n$. By choosing $J \subset \mathbb{Z}_n$, $|J| = d$, to be a union of the disjoint sets

$$\{0\}, \qquad \{j,-j\}, \quad 1 \leq j \leq \tfrac{n}{2} - 1, \qquad \{\tfrac{n}{2}\} \quad (\text{for } n \text{ even}),$$

we obtain a real cyclic harmonic frame for $\mathbb{R}^d$ (they all come in this way). We can ensure that this frame has distinct vectors by making J a generating set for $\mathbb{Z}_n$, e.g., by choosing $\{1,-1\} \subset J$. □

The construction of Corollary 11.3 gives *all* real cyclic harmonic frames. It is equivalent to selecting real rows and complex conjugate pairs of rows from the Fourier matrix (11.4). The real rows corresponding to $\{0\}$ and $\{\frac{n}{2}\}$ (n even) are

$$\begin{bmatrix} 1 \; 1 \; 1 \; 1 \; \cdots \; 1 \; 1 \end{bmatrix}, \qquad \begin{bmatrix} 1 \; -1 \; 1 \; -1 \; \cdots \; 1 \; -1 \end{bmatrix}.$$

To obtain a copy of the frame explicitly in $\mathbb{R}^d$, one can apply the unitary map

$$U := \frac{1}{\sqrt{2}} \begin{bmatrix} 1 & 1 \\ -i & i \end{bmatrix}, \qquad U \begin{bmatrix} z \\ \overline{z} \end{bmatrix} = \sqrt{2} \begin{bmatrix} \Re z \\ \Im z \end{bmatrix},$$

to the complex conjugate pairs of rows corresponding to $\{j,-j\}$ to obtain real rows

$$U \begin{bmatrix} 1 & \omega^j & \omega^{2j} & \cdots & \omega^{(n-1)j} \\ 1 & \omega^{-j} & \omega^{-2j} & \cdots & \omega^{-(n-1)j} \end{bmatrix} = \sqrt{2} \begin{bmatrix} 1 & \cos(2\pi\frac{j}{n}) & \cos(2\pi\frac{2j}{n}) & \cdots & \cos(2\pi\frac{(n-1)j}{n}) \\ 1 & \sin(2\pi\frac{j}{n}) & \sin(2\pi\frac{2j}{n}) & \cdots & \sin(2\pi\frac{(n-1)j}{n}) \end{bmatrix}.$$

The orthogonality of the rows of V above can be viewed as orthogonality of the corresponding trigonometric polynomials with respect to the discrete inner product

$$\langle f, g \rangle := \sum_{k=0}^{n-1} f\Big(\frac{2\pi k}{n}\Big) g\Big(\frac{2\pi k}{n}\Big).$$

Example 11.10. (Equally spaced vectors) The n equally spaced unit vectors in $\mathbb{R}^2$ a cyclic harmonic frame, since they are a $\mathbb{Z}_n$-frame (Example 10.1). By the above calculation, they are given by $J = \{1,-1\} \subset \mathbb{Z}_n$ (or $\{j,-j\}$ for j a generator of $\mathbb{Z}_n$).

Example 11.11. For the special case $p = 2$ in Example 11.6, all nonzero elements of $G = \mathbb{Z}_2^k$ have order 2, and so are equal to their inverse. Thus all harmonic frames given by $G = \mathbb{Z}_2^k$ are real. These are noncyclic harmonic frames when $k \geq 3$, and the vectors are distinct (cf. Example 11.5). The character table of the elementary abelian 2-groups $\mathbb{Z}_2 \times \cdots \times \mathbb{Z}_2$ can be calculated by taking *Kronecker product* of that for $\mathbb{Z}_2$, e.g.,

$$\begin{bmatrix} 1 & 1 \\ 1 & -1 \end{bmatrix} \otimes \begin{bmatrix} 1 & 1 \\ 1 & -1 \end{bmatrix} = \begin{bmatrix} 1 & 1 & 1 & 1 \\ 1 & -1 & 1 & -1 \\ 1 & 1 & -1 & -1 \\ 1 & -1 & -1 & 1 \end{bmatrix}. \tag{11.12}$$

Example 11.12. The vertices of the *Platonic solids* (see §10.6) give tight frames Φ for $\mathbb{R}^3$ (with a high degree of symmetry). By Theorem 11.1, these are harmonic if and only if $\mathrm{Sym}(\Phi)$ has a transitive abelian subgroup G. By Table 10.1, we have that the *tetrahedron*, the *cube* and the *octahedron* are harmonic frames, and the *icosahedron* (12 vertices) and *dodecahedron* (20 vertices) are not harmonic frames.

The eight vertices of the cube give a *noncyclic* harmonic frame for $\mathbb{R}^3$ (since $\mathbb{Z}_8$ is not a transitive subgroup of their symmetry group). See Table 11.1.

We observe that the complex unlifted cyclic harmonic frame of 12 vectors given by $\{1,5,9\} \subset \mathbb{Z}_{12}$ has a symmetry group of order 384, and the complex lifted and unlifted cyclic harmonic frames of 20 vectors given by $\{0,1,11\}, \{1,10,11\} \subset \mathbb{Z}_{20}$ have a symmetry group of order 200 (see Exer. 11.2).

Example 11.13. There exist real harmonic frames with "large" symmetry groups, e.g, the real cyclic harmonic frames of 14 vectors for $\mathbb{C}^5$ given by $J = \{0, \pm 1, \pm 6\}$ and $J = \{\pm 1, \pm 6, 7\}$ have symmetry groups of order 392. The remaining 334 cyclic harmonic frames of 14 vectors for $\mathbb{C}^5$ have symmetry group orders $98, 42, 28, 14$.

Table 11.1: The numbers of real harmonic frames of $n \leq 12$ distinct vectors for $\mathbb{R}^d$, $1 \leq d \leq 7$ (up to unitary equivalence and reordering), with the number which are not cyclic in brackets.

$n \backslash d$	$d=1$	$d=2$	$d=3$	$d=4$	$d=5$	$d=6$	$d=7$
$n=4$	0(0)	1(0)	2(0)	1(0)			
$n=5$	0	1	1	1	1		
$n=6$	0	1	3	3	2	1	
$n=7$	0	1	1	1	1	1	1
$n=8$	0(0)	1(0)	3(1)	5(2)	5(1)	3(0)	2(0)
$n=9$	0(0)	1(0)	1(0)	3(1)	3(1)	2(0)	2(0)
$n=10$	0	1	3	5	7	6	4
$n=11$	0	1	1	2	2	2	2
$n=12$	0	1(0)	3(1)	9(2)	15(3)	17(4)	17(3)

11.7 Unitary equivalence preserving the group structure

Here we outline techniques for efficiently determining whether harmonic frames are unitarily equivalent to each other.

Since a harmonic frame $\Phi = (\phi_g)_{g\in G}$ is a G-frame, its Gramian is a G-matrix (Theorem 10.3). Thus each row and column of the Gramian has the same entries. We call this *multiset* minus the diagonal entry the **angle multiset** of Φ

$$\mathrm{Ang}(\Phi) := \text{multiset of off diagonal entries of any row/column of } \mathrm{Gram}(\Phi).$$

The angle multiset of the cyclic harmonic frame given by $J = \{j_1, \ldots, j_d\} \subset \mathbb{Z}_n$ is

$$\mathrm{Ang}(\Phi_J) = \{\omega^{aj_1} + \omega^{aj_2} + \cdots + \omega^{aj_d} : 1 \le a \le n-1\}, \qquad \omega = e^{\frac{2\pi}{n}}. \tag{11.13}$$

Since the Gramian determines a frame up to unitary equivalence, we have:

Harmonic frames which are unitarily equivalent up to a reordering *must* have the same angle multiset.

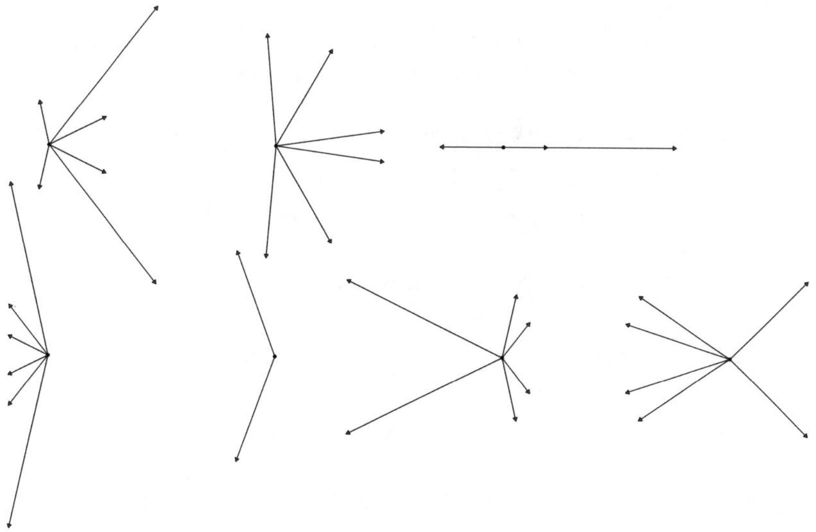

Fig. 11.1: The angle multisets of the unitarily inequivalent harmonic frames of 7 vectors for $\mathbb{C}^3$. Note that one is real, and three are equiangular.

Example 11.14. (7 vectors in $\mathbb{C}^3$) The angle multiset for the 7 unitarily inequivalent (cyclic) harmonic frames of 7 vectors for $\mathbb{C}^3$ are depicted in Figure 11.1. Observe that one is real (Corollary 11.3) and three are equiangular.

Most, but not all, reindexings (reorderings) of a given harmonic frame which make it unitarily equivalent to another are *automorphisms*. Let $\mathrm{Aut}(G)$ denote the group of automorphisms of G, i.e., isomorphisms $\sigma : G \to G$.

Definition 11.2. We say G-frames $(v_g)_{g\in G}$ and $(w_g)_{g\in G}$ are **unitarily equivalent via an automorphism** $\sigma \in \mathrm{Aut}(G)$ if $(v_g)_{g\in G}$ and $(w_{\sigma g})_{g\in G}$ are unitarily equivalent.

Subsets J and K of G are **multiplicatively equivalent** if there is a $\sigma \in \mathrm{Aut}(G)$ for which $K = \sigma J$. In this case, $\hat{\sigma} : \hat{G} \to \hat{G} : \chi \mapsto \chi \circ \sigma^{-1}$ is an automorphism of $\hat{G}$, and

$$\langle \xi|_J, \eta|_J \rangle = \langle \hat{\sigma}\xi|_K, \hat{\sigma}\eta|_K \rangle,$$

i.e., Φ_J and Φ_K are unitarily equivalent after reindexing by the automorphism $\hat{\sigma}$.

For $G=\mathbb{Z}_n$, each $\sigma \in \mathrm{Aut}(G)$ has the form $g \mapsto ag$, with $a \in \mathbb{Z}_n^*$ a unit, and hence J and K are multiplicatively equivalent if and only if $K = aJ$ for some $a \in \mathbb{Z}_n^*$.

We now give a simple condition which ensures harmonic frames are unitarily equivalent via an automorphism.

Theorem 11.4. *(Multiplicative equivalence) Let G be a finite abelian group, and Φ_J and Φ_K be the harmonic frames for $J, K \subset G$. Then the following are equivalent:*

1. *The harmonic frames Φ_J and Φ_K are unitarily equivalent via an automorphism.*
2. *The subsets J and K are multiplicatively equivalent.*

Proof. Let $\sigma \in \mathrm{Aut}(G)$. Using the Pontryagin duality map (11.7), and the fact that the characters are a basis (they are orthogonal), we calculate:

$$\begin{aligned}
&\Phi_{\sigma J} \text{ and } \Phi_K \text{ are unitarily equivalent} \\
\iff\quad & \langle \xi|_{\sigma J}, \eta|_{\sigma J} \rangle = \langle \xi|_K, \eta|_K \rangle, \quad \forall \xi, \eta \in \hat{G} \\
\iff\quad & \sum_{j\in J} \xi(\sigma j)\overline{\eta(\sigma j)} = \sum_{k\in K} \xi(k)\overline{\eta(k)}, \quad \forall \xi, \eta \in \hat{G} \\
\iff\quad & \sum_{j\in J} (\xi\eta^{-1})(\sigma j) = \sum_{k\in K} (\xi\eta^{-1})(k), \quad \forall \xi, \eta \in \hat{G} \qquad \text{[by (11.3)]} \\
\iff\quad & \sum_{j\in J} \chi(\sigma j) = \sum_{k\in K} \chi(k), \quad \forall \chi \in \hat{G} \qquad (\text{take } \chi = \xi\eta^{-1}) \\
\iff\quad & \sum_{j\in J} \widehat{(\sigma j)} = \sum_{k\in K} \hat{\hat{k}} \qquad (\text{by Pontryagin duality}) \\
\iff\quad & \sigma J = \{\sigma j : j \in J\} = K \qquad (\text{the characters of } \hat{G} \text{ are a basis})
\end{aligned}$$

i.e., J and K are multiplicatively equivalent. □

Multiplicative equivalence is an equivalence relation, with the equivalence classes being the orbits of the natural action of $\mathrm{Aut}(G)$ on the d-element subsets of G. The number of multiplicative equivalence classes of d-element subsets of a group G which generate G is essentially Hall's *Eulerian function* $\Phi_d(G)$, which counts the ordered d-element generating subsets of G.

In view of Theorem 11.4, the question of determining all the harmonic frames (for a given group) up to unitary equivalence (and reordering) reduces to determining whether harmonic frames corresponding to different multipicative equivalence classes are unitarily equivalent. This can most often be done by a simple calculation, e.g., comparing their angle multisets. Examples where there is unitary equivalence via a permutation which is not an automorphism are considered in §11.2.

Example 11.15. (Four vectors in $\mathbb{C}^2$) First consider $G = \mathbb{Z}_4$. The automorphism group $\mathrm{Aut}(G)$ has order 2 and is generated by $\sigma : g \mapsto 3g$ ($\mathbb{Z}_4^* = \{1,3\}$). Thus the multiplicative equivalence classes of 2-element subsets of G are

$$\{\{0,1\},\{0,3\}\}, \qquad \{\{1,2\},\{2,3\}\}, \qquad \{\{1,3\}\}, \qquad \{\{0,2\}\}.$$

The first three give cyclic harmonic frames with distinct vectors (1 generates G), while the last does not. None are unitarily equivalent, since their angle multisets are

$$\{-i+1,0,i+1\}, \qquad \{0,-i-1,i-1\}, \qquad \{0,0,-2\}, \qquad \{0,0,2\}.$$

Now consider $G = \mathbb{Z}_2 \times \mathbb{Z}_2$, which is generated by any two of its three elements $\{a,b,a+b\}$ of order 2. The automorphism group of G has order 6, with an automorphism corresponding to each permutation of $\{a,b,a+b\}$. Thus the multiplicative equivalence classes are

$$\{\{a,b\},\{a,a+b\},\{b,a+b\}\}, \qquad \{\{0,a\},\{0,b\},\{0,a+b\}\}.$$

Only the first gives a harmonic frame with distinct vectors. This frame is unitarily equivalent to the cyclic harmonic frame of four equally spaced unit vectors for $\mathbb{R}^2$.

Example 11.16. (Seven vectors in $\mathbb{C}^3$) For $G = \mathbb{Z}_7$, there are seven multiplicative equivalence classes of 3-element subsets, with representatives

$$\{1,2,6\},\ \{1,2,3\},\ \{0,1,2\},\ \{0,1,3\},\ \{1,2,5\} \quad (\text{size}\,6)$$

$$\{0,1,6\} \quad (\text{size}\,3) \qquad \{1,2,4\} \quad (\text{size}\,2).$$

Each gives a cyclic harmonic frame with distinct vectors (nonzero elements generate G). None are unitarily equivalent since their angle multisets differ (see Fig. 11.1).

There is just one harmonic frame of n distinct vectors for $\mathbb{C}^1$.

Example 11.17. (Cyclic harmonic frames for $\mathbb{C}^1$) There is a unique harmonic frame of n distinct vectors for $\mathbb{C}^1$, namely the cyclic harmonic frame given by the n-th roots of unity. This follows since such a frame must be given by $J = \{g\}$, where g generates G (which is therefore cyclic). There is a automorphism of G taking any generator to any other.

Example 11.18. (Cyclic harmonic frames for $\mathbb{C}^2$) The cyclic harmonic frames of n distinct vectors for $\mathbb{C}^2$ are unitarily equivalent (up to reordering) if and only if the subsets of $\mathbb{Z}_n$ that give them are multiplicatively equivalent. This follows by considering the angle multisets (see [CW11] for details).

11.8 Noncyclic harmonic frames

The eight vertices of cube are a noncyclic harmonic frame for $\mathbb{R}^3$ (Example 11.12).

Example 11.19. (Noncyclic harmonic frame for $\mathbb{C}^2$) There is a noncyclic complex harmonic frame of eight vectors for $\mathbb{C}^2$ given by $J=\{(0,1),(1,0)\} \subset G=\mathbb{Z}_4 \times \mathbb{Z}_2$, i.e.,

$$\left(\begin{bmatrix} 1 \\ 1 \end{bmatrix}, \begin{bmatrix} 1 \\ -1 \end{bmatrix}, \begin{bmatrix} i \\ 1 \end{bmatrix}, \begin{bmatrix} i \\ -1 \end{bmatrix}, \begin{bmatrix} -1 \\ 1 \end{bmatrix}, \begin{bmatrix} -1 \\ -1 \end{bmatrix}, \begin{bmatrix} -i \\ 1 \end{bmatrix}, \begin{bmatrix} -i \\ -1 \end{bmatrix} \right).$$

A finite abelian group G can be written as a direct sum of p-groups

$$G_p = \mathbb{Z}_{p^{e_1}} \oplus \mathbb{Z}_{p^{e_2}} \oplus \cdots \oplus \mathbb{Z}_{p^{e_m}}$$

where p are the prime divisors of $|G|$. The automorphism group of G_p has order

$$|\mathrm{Aut}(G_p)| = \prod_{k=1}^{m} (p^{d_k} - p^{k-1}) \prod_{j=1}^{m} (p^{e_j})^{m-d_j} \prod_{i=1}^{m} (p^{e_i-1})^{m-c_i+1}, \tag{11.14}$$

where $c_k := \min\{r : e_r = e_k\} \leq k$, $d_k := \max\{r : e_r = e_k\} \geq k$, and so the order of $\mathrm{Aut}(G)$ is the product of these orders (see [HR07]). In effect, the less cyclic an abelian group is, the larger its automorphism group becomes.

Table 11.2: The numbers of *non*cyclic, *cyc*lic harmonic frames of $n \leq 35$ distinct vectors for $\mathbb{C}^d$, $d = 2,3,4$ (up to unitary equivalence and reordering) when there is a noncyclic abelian group.

$d = 2$

n	non	cyc	harm
4	0	3	3
8	1	7	8
9	1	6	7
12	2	13	15
16	4	13	17
18	2	18	20
20	3	19	22
24	6	27	33
25	1	15	16
27	3	18	21
28	4	25	29
32	9	25	34

$d = 3$

n	non	cyc	harm
4	0	3	3
8	5	16	21
9	3	15	18
12	11	57	68
16	28	74	102
18	19	121	140
20	29	137	166
24	89	241	330
25	8	115	123
27	33	159	192
28	57	255	312
32	158	278	436

$d = 4$

n	non	cyc	harm
4	0	1	1
8	8	21	29
9	5	23	28
12	30	141	171
16	139	228	367
18	80	494	574
20	154	622	776
24	604	1349	1953
25	37	636	673
27	202	973	1175
28	443	1697	2140
32	1379	2152	3531

It was observed in [WH06] that most harmonic frames seem to be cyclic, with increasingly fewer as G becomes less cyclic. A heuristic explanation for this is that: as the group G becomes less cyclic, its automorphism group becomes larger (so the number of multiplicative equivalence classes becomes smaller), and the orders of its elements become smaller (so that $J \subset G$ is less likely to generate G, and hence give a harmonic frame with distinct vectors).

11.9 Unitary equivalence not preserving the group structure

By considering the angle multisets (see [CW11]), it can be shown that:

Theorem 11.5. *Cyclic harmonic frames of n distinct vectors for $\mathbb{C}^2$ are unitarily equivalent (up to a reordering) if and only if the subsets of $\mathbb{Z}_n$ that give them are multiplicatively equivalent.*

Theorem 11.5 implies that unitary equivalence (up to reordering) and multiplicative equivalence are the same for cyclic harmonic frames for $\mathbb{C}^3$, except if both frames are unlifted. There do exist examples of unlifted cyclic harmonic frames which are unitarily equivalent after a reordering which is not an automorphism.

Example 11.20. For $\mathbb{Z}_8$ there are 17 multiplicative equivalence classes of 3-element subsets which generate it. Only two of these give frames with the same angles, namely

$$\{\{1,2,5\},\{3,6,7\}\}, \qquad \{\{1,5,6\},\{2,3,7\}\}.$$

The common angle multiset is $\{-1,i,i,-i,-i,-2i-1,2i-1\}$. These two frames are unitarily equivalent up to a reordering. Since they are given by multiplicatively inequivalent subsets, this reordering *cannot* be an automorphism (see Table 11.3).

Three infinite families of such "exceptional cases" where unitary equivalence (up to reordering) does not imply multiplicative equivalence are given in [Chi10]. It is not known whether or not these are all of them.

If n is *square free*, i.e., is a product of distinct primes, then the primitive n-th roots of unity are a basis for the cyclotomic field $\mathbb{Q}(\omega)$, $\omega = e^{\frac{2\pi i}{n}}$. This leads to:

Theorem 11.6. *Let n be square free. Then the cyclic harmonic frames of n vectors for $\mathbb{C}^d$ given by $J,K \subset \mathbb{Z}_n^*$ (the units) are unitarily equivalent (up to a reordering) if and only if J and K are multiplicatively equivalent.*

Proof. Suppose that Φ_J and Φ_K are unitarily equivalent (after reordering), but are not multiplicatively equivalent. Then the angle $\sum_{j\in J}\omega^j$ of Φ_J is an angle of Φ_K, so that $\sum_{j\in J}\omega^j = \sum_{k\in K}\omega^{ak}$, where $a \notin \mathbb{Z}_n^*$. Since $\gcd(ak,n) = \gcd(a,n)$, $\forall k \in K$, it follows that $\sum_{k\in K}\omega^{ak} = \sum_{b\in\mathbb{Z}_n^*} c_b\omega^b$, where each $c_b \in \mathbb{Z}$ has the same sign (see [Chi10] for details). Thus it follows that $\sum_{j\in J}\omega^j \neq \sum_{k\in K}\omega^{ak}$ (since n is not even), which is a contradiction. □

Computations of [Chi10] suggest that Theorem 11.6 also holds when n is not square free, i.e., all examples of cyclic harmonic frames for which multiplicative equivalence and unitary equivalence are not equivalent (such as Example 11.20) involve cyclic harmonic frames Φ_J for which J contains a nonunit.

11.10 The number of cyclic harmonic frames

Computations of [WH06] suggest that the number of harmonic frames of n distinct vectors for $\mathbb{C}^d$ grows like n^{d-1} (for d fixed), and *most harmonic frames are cyclic*. In [MW16], it was established that the number $h_{n,d}$ of *cyclic* harmonic frames of n distinct vectors for $\mathbb{C}^d$ up to unitary equivalence (and reordering) grows like

$$h_{n,d} \approx \frac{n^d}{\varphi(n)} \geq n^{d-1}, \qquad n \to \infty,$$

where $\varphi(n)$ is Euler's totient function. The proof uses a correspondence between unitarily equivalent cyclic harmonic frames and points on the torus $\mathbb{T}^{2d}$, in which multiplicative equivalence gives a torsion point. Here we consider the particularly simple case of when n is a prime, for which all harmonic frames are cyclic and there is an explicit formula for $h_{n,d}$ (see [MW16], [Hir10]).

Let $\mathbb{Z}_n^*$ (as the automorphisms of $\mathbb{Z}_n$) act on the d-element subsets of $\mathbb{Z}_n$, and S be invariant under this action. Then, by Burnside's counting lemma, the number of multiplicative equivalence classes in S is

$$|S/\mathbb{Z}_n^*| = \frac{1}{\varphi(n)} \sum_{a\in\mathbb{Z}_n^*} |\mathrm{Fix}(a)|, \tag{11.15}$$

where $\mathrm{Fix}(a)$ is the set of elements of S fixed by $a \in \mathbb{Z}_n^*$.

Theorem 11.7. *Let p be a prime, and $h^{\mathrm{u}}_{p,d}$ and $h^{\mathrm{l}}_{p,d}$ be the numbers of unlifted and lifted (cyclic) harmonic frames of $p \geq d$ distinct vectors for $\mathbb{C}^d$, $d > 1$. Then*

$$h^{\mathrm{u}}_{p,d} = \frac{1}{p-1} \sum_{j|\gcd(p-1,d)} \binom{\frac{p-1}{j}}{\frac{d}{j}} \varphi(j) \approx p^{d-1}, \qquad p \to \infty,$$

$$h^{\mathrm{l}}_{p,d} = \frac{1}{p-1} \sum_{j|\gcd(p-1,d-1)} \binom{\frac{p-1}{j}}{\frac{d-1}{j}} \varphi(j) \approx p^{d-2}, \qquad p \to \infty.$$

In particular, the number of harmonic frames $h_{p,d} = h^{\mathrm{u}}_{p,d} + h^{\mathrm{l}}_{p,d} \approx p^{d-1}$, $p \to \infty$.

Proof. Let $a \in \mathbb{Z}_n^*$ be one of the $\varphi(j)$ elements of order j. The orbit of a nonzero element of $\mathbb{Z}_n$ under the action of a has size j, and the orbit of zero is a singleton. Thus if a d-element subset $J \subset \mathbb{Z}_n$ is fixed by a, then either

- J is unlifted and is the union of $\frac{d}{j}$ of the $\langle a\rangle$-orbits of nonzero singletons.
- J is lifted and is the union of $\{0\}$ and $\frac{d-1}{j}$ of the $\langle a\rangle$-orbits of nonzero singletons.

Since $d>1$ and p is prime, all the harmonic frames have distinct vectors. Now $\mathbb{Z}_p^*$ is cyclic of order $\varphi(p)=p-1$ (so j divides $p-1$), and (11.15) gives the result. □

Example 11.21. For $d=2$, we have $h_{p,2}=\frac{1}{2}(p+1)$, since

$$h^{\mathrm{u}}_{p,1}=h^{\mathrm{l}}_{p,2}=1,\qquad h^{\mathrm{u}}_{p,2}=h^{\mathrm{l}}_{p,3}=\frac{1}{p-1}\left\{\binom{p-1}{2}+\binom{\frac{p-1}{2}}{1}\right\}=\frac{1}{2}(p-1).$$

For $d=3$, we have

$$h^{\mathrm{u}}_{p,3}=h^{\mathrm{l}}_{p,4}=\frac{1}{p-1}\begin{cases}\binom{p-1}{3}, & p\not\equiv 1 \pmod 3;\\ \binom{p-1}{3}+2\binom{\frac{p-1}{3}}{1}, & p\equiv 1 \pmod 3.\end{cases}$$

Hence

$$h_{p,3}=\begin{cases}\frac{1}{6}(p^2-2p+3), & p\not\equiv 1 \pmod 3;\\ \frac{1}{6}(p^2-2p+7), & p\equiv 1 \pmod 3.\end{cases}$$

As indicated above, formulas for $h_{p,d}$ depending on p modulo d and $d-1$ can be always be constructed. It is also possible to modify the proof of Theorem 11.7 to count the real harmonic frames:

Proposition 11.1. *Let p be an odd prime. For d even, the number of real harmonic (unlifted) frames of p vectors for $\mathbb{R}^d$ (up to unitary equivalence) is*

$$h^{\mathbb{R}}_{p,d}=\frac{1}{p-1}\Bigg\{\sum_{\substack{j\mid\gcd(p-1,d)\\ j\text{ even}}}\binom{\frac{p-1}{j}}{\frac{d}{j}}\varphi(j)+\sum_{\substack{j\mid\gcd(p-1,\frac{d}{2})\\ j\text{ odd}}}\binom{\frac{p-1}{2j}}{\frac{d}{2j}}\varphi(j)\Bigg\}.$$

For d odd, the number of real harmonic (lifted) frames of p vectors for $\mathbb{R}^d$ is

$$h^{\mathbb{R}}_{p,d}=\frac{1}{p-1}\Bigg\{\sum_{\substack{j\mid\gcd(p-1,d-1)\\ j\text{ even}}}\binom{\frac{p-1}{j}}{\frac{d-1}{j}}\varphi(j)+\sum_{\substack{j\mid\gcd(p-1,\frac{d-1}{2})\\ j\text{ odd}}}\binom{\frac{p-1}{2j}}{\frac{d-1}{2j}}\varphi(j)\Bigg\}.$$

Proof. See Exer. 11.11 for detail. □

Example 11.22. For $d\le 3$, there is a single real harmonic frame of p vectors, i.e.,

$$h^{\mathbb{R}}_{p,1}=h^{\mathbb{R}}_{p,2}=h^{\mathbb{R}}_{p,3}=1.$$

For d even, $d\ge 4$, we have the estimate

$$h^{\mathbb{R}}_{p,d}=h^{\mathbb{R}}_{p,d+1}\approx p^{\frac{d}{2}-1},\qquad p\to\infty.$$

11.11 Projective unitary equivalence of harmonic frames

We have seen that multiplicatively equivalent subsets $J \subset G$ give harmonic frames which are unitarilly equivalent up to a reindexing (Theorem 11.4). We now show that the *translates* of J gives projectively unitarily equivalent harmonic frames.

The **translation** of a finite abelian group G by b is the bijection

$$\tau_b : G \to G : j \mapsto j+b, \qquad b \in G.$$

We say that K is a **translate** of J if $K = J+b$, i.e., $K = \tau_b J$.

Theorem 11.8. *(Translates) Let G be a finite abelian group. If K is a translate of J, then the harmonic frames Φ_J and Φ_K are projectively unitarily equivalent.*

Proof. Suppose that $K = J+b$. Since $\Phi_J = (\xi|_J)_{\xi\in\hat{G}}$, we need to show

$$\xi|_K = c_\xi U(\xi|_J), \qquad \xi \in \hat{G},$$

where $U : \mathbb{C}^J \to \mathbb{C}^K$ is unitary. Let $U_b = \mathbb{C}^J \to \mathbb{C}^K$ be the unitary map given by $(U_b v)(k) := v(k-b)$, $k \in K$. Since ξ is a character, we have

$$(U_b \xi|_J)(k) = \xi|_J(k-b) = \xi(k-b) = \xi(k)\xi(-b) = \overline{\xi(b)}\,\xi|_K(k),$$

and so we can take $U = U_b$ and $c_\xi = \xi(b)$. □

The converse: that projective unitary equivalence (without reordering) implies that J and K are translates of each other, appears to be true.

Example 11.23. (Seven vectors in $\mathbb{C}^3$) We now revisit Example 11.16. The seven multiplicative equivalence classes of 3-element subsets of $G = \mathbb{Z}_7$ give rise to seven (cyclic) harmonic frames up to unitary equivalence and reordering. By translating a given subset, we obtain a projectively unitarily equivalent harmonic frame. Since the representatives of the multiplicative equivalence classes are related by

$$\{0,1,3\} = 6\{1,2,6\}+2, \quad \{1,2,4\} = 6\{1,2,6\}+3,$$

$$\{0,1,2\} = \{1,2,3\}+6, \quad \{1,2,5\} = 3\{1,2,3\}+6, \quad \{0,1,6\} = \{1,2,3\}+5,$$

we conclude that there are exactly two harmonic tight frames of seven vectors for $\mathbb{C}^3$ up to projective unitary equivalence and reordering. By considering the angle multisets of Figure 11.1, we have that one is real, and the other is equiangular.

Example 11.24. Let $p > 2$ be a prime. Then all harmonic frames of p vectors for $\mathbb{C}^2$ are projectively unitarily equivalent up to reindexing to the p equally spaced unit vectors in $\mathbb{R}^2$. This follows since there is a unique affine map, taking a sequence of two distinct elements of $\mathbb{Z}_p$ to any other. In particular the two harmonic frames of three vectors in $\mathbb{C}^2$ which are unitarily inequivalent (one is real, one is complex) are projectively unitarily equivalent up to a reordering.

We define the **affine group** of G to be the group of bijections $\theta: G \to G$ generated by the translations and automorphisms of G, i.e., the $|G|\,|\mathrm{Aut}(G)|$ maps of the form

$$\theta(g) = \sigma(g) + b, \qquad \sigma \in \mathrm{Aut}(G), \quad b \in G.$$

If $K = \theta J$, for some θ in the affine group, we say J and K are **affinely equivalent**, i.e., they are in the same orbit under the natural action of the affine group of G on the subsets of G. Combining Theorems 11.4 and 11.8 gives the following:

Theorem 11.9. *(Affine equivalence) Let G be a finite abelian group. If J and K are affinely equivalent subsets of G, then the harmonic frames Φ_J and Φ_K that they give are projectively unitarily equivalent up to reindexing by an automorphism.*

In view of Theorem 11.9, the question of determining the harmonic frames up to projective unitary equivalence (and reordering) reduces to determining whether frames given by affinely inequivalent subsets are projectively unitarily equivalent. In all cases considered (see, e.g., Table 11.3) this can be done by using a simple test, such as comparing the multiset of m-products.

Table 11.3: The number of unitary and projective unitary equivalence classes (up to reindexing) of cyclic harmonic frames of n vectors for $\mathbb{C}^d$ ($d \leq 6$). When the group theoretic estimate given by Theorem 11.9 is larger (i.e., there are reindexings which are not automorphisms), then it is given in the row below.

$d = 2$

n	uni	proj
2	1	1
3	2	1
4	3	2
5	3	1
6	6	3
7	4	1
8	7	3
9	6	2
10	9	3
11	6	1
12	13	5
13	7	1
14	12	3
15	13	3

$d = 3$

n	uni	proj
3	1	1
4	3	1
5	3	1
6	11	3
7	7	2
8	16	4
	17	
9	15	3
10	29	4
11	17	2
12	56	9
	57	
13	25	3

$d = 4$

n	uni	proj
5	2	1
6	9	3
7	7	2
8	21	6
	23	5
9	23	4
	24	
10	53	9
	54	
11	34	4
12	138	21
	141	

$d = 5$

n	uni	proj
5	1	1
6	4	1
7	4	1
8	19	4
	20	
9	23	4
	24	
10	67	9
11	48	6

$d = 6$

n	uni	proj
6	1	1
7	2	1
8	11	3
9	16	3
10	55	9
	56	
11	48	6

11.12 The projective symmetry group of a harmonic frame

The projective symmetry group of various harmonic frames was calculated in [WC14] (see Table 11.4). We say that a frame has **projectively repeated** vectors if some vectors are scalar multiples of each other (in this case the projective symmetry group is not given). When a harmonic frame is complex, the extended projective symmetry group is given also. The number of **erasures** of a frame for $\mathbb{C}^d$ is the maximum number of vectors that can be removed from it so that those remaining still span $\mathbb{C}^d$.

Table 11.4: The cyclic harmonic frames of n vectors for $\mathbb{C}^3$. If the frame doesn't have projectively repeated vectors, then its projective symmetry group is given, and when it is complex, then its extended projective symmetry group is given. We also indicate if the frame has orthogonal vectors, and its number of erasures.

d	n	real	orth	reps	$\mathrm{Sym}_P(\Phi)$	$\mathrm{Sym}_{EP}(\Phi)$	J	erasures
3	3	$\mathbb{R}$	y		$<6,1>$		$\{0,1,2\}$	0
3	4	$\mathbb{R}$			$<24,12>$		$\{1,2,3\}$	1
3	5	$\mathbb{R}$			$<10,1>$		$\{0,1,3\}$	2
3	6				$<18,3>$	$<36,10>$	$\{0,1,4\}$	2
		$\mathbb{R}$	y		$<12,4>$		$\{1,2,3\}$	3
		$\mathbb{R}$	y	y			$\{1,3,5\}$	1
3	7				$<21,1>$	$<42,1>$	$\{1,2,6\}$	4
		$\mathbb{R}$			$<14,1>$		$\{1,3,5\}$	4
3	8				$<16,8>$	$<32,43>$	$\{1,3,4\}$	5
					$<32,11>$	$<64,134>$	$\{0,1,4\}$	3
		$\mathbb{R}$			$<16,7>$		$\{0,1,2\}$	5
		$\mathbb{R}$		y			$\{1,3,5\}$	3
3	9				$<9,1>$	$<18,1>$	$\{1,4,6\}$	5
		$\mathbb{R}$	y		$<18,1>$		$\{0,1,2\}$	6
		$\mathbb{R}$	y	y			$\{1,4,7\}$	2
3	10				$<50,3>$	$<100,13>$	$\{0,1,5\}$	4
		$\mathbb{R}$			$<20,4>$		$\{0,1,9\}$	7
					$<10,2>$	$<20,4>$	$\{0,1,8\}$	7
		$\mathbb{R}$		y			$\{1,5,7\}$	5
3	11				$<11,1>$	$<22,1>$	$\{0,1,3\}$	8
		$\mathbb{R}$			$<22,1>$		$\{1,2,3\}$	8
3	12				$<12,2>$	$<24,6>$	$\{1,2,11\}$	8
		$\mathbb{R}$	y		$<24,6>$		$\{1,2,3\}$	9
		$\mathbb{R}$		y			$\{1,4,10\}$	5
					$<12,2>$	$<24,6>$	$\{0,3,4\}$	7
				y			$\{1,5,7\}$	5
			y		$<24,5>$	$<48,38>$	$\{0,1,8\}$	7
		$\mathbb{R}$	y	y			$\{1,3,5\}$	7
					$<72,30>$	$<144,154>$	$\{2,3,8\}$	5
		$\mathbb{R}$	y	y			$\{1,5,9\}$	3
3	13	$\mathbb{R}$			$<26,1>$		$\{0,1,12\}$	10
					$<13,1>$	$<26,1>$	$\{0,1,3\}$	10
					$<39,1>$	$<78,1>$	$\{1,2,11\}$	10

Notes

It has been known (at least) since [Zim01] that there exists at least one real harmonic frame of $n \geq d$ vectors for $\mathbb{C}^d$.

$$p_{n,d} \approx \frac{n^{d-1}}{\varphi(n)}$$

Exercises

11.1. An equal-norm frame $\Phi = (v_j)$ for $\mathbb{F}^d$ is said to be **equispaced** if the distances $\|v_j - v_k\|$ are constant for $j \neq k$ (this is equivalent to $\Re\langle v_j, v_k\rangle$ being constant).
(a) Show that the real and complex harmonic frames of three vectors for $\mathbb{C}^2$ are equispaced (and also equiangular), but that the spacings are different in each case.
(b) Let $\Phi = (v_j)$ be the cyclic harmonic frame for $\mathbb{C}^2$ given by $\{1,2\} \subset \mathbb{Z}_5$, i.e.,

$$\Phi = \left(\begin{bmatrix} 1 \\ 1 \end{bmatrix}, \begin{bmatrix} \omega \\ \omega^2 \end{bmatrix}, \begin{bmatrix} \omega^2 \\ \omega^4 \end{bmatrix}, \begin{bmatrix} \omega^3 \\ \omega \end{bmatrix}, \begin{bmatrix} \omega^4 \\ \omega^3 \end{bmatrix} \right), \qquad \omega := e^{2\pi i/5}.$$

Show that Φ is equispaced.
(c) The equispaced harmonic frames of (a) and (b) can be generalised as follows. Let $\Phi = (v_j)$ be the cyclic harmonic frame of $2d-1$ vectors for $\mathbb{C}^d$ given by $\{0,1,\ldots,d-1\} \subset \mathbb{Z}_{2d-1}$, and $\Psi = (w_j)$ be the cyclic harmonic frame of $2d+1$ vectors for $\mathbb{C}^d$ given by $\{1,2,\ldots,d\} \subset \mathbb{Z}_{2d+1}$. Show that Φ and Ψ are equispaced.

11.2.^{m} Write a function to compute the symmetry group of a frame from its Gramian. Apply it in the following cases:
(a) The unlifted cyclic harmonic frame of 12 vectors for $\mathbb{C}^3$ given by $\{1,5,9\} \subset \mathbb{Z}_{12}$.
(b) The cyclic frames of 20 vectors for $\mathbb{C}^3$ given by $\{0,1,11\}, \{1,10,11\} \subset \mathbb{Z}_{20}$.

11.3. Let U be a unitary matrix and $v \in \mathbb{F}^d$. Show that $(U^j v)_{j=0}^{n-1}$ is an equal-norm tight frame for $\mathbb{F}^d$ if and only if it is projectively unitarily equivalent to a cyclic harmonic frame, i.e., $U^n = cI$.

11.4. Show that the set of n-th roots of unity is the unique harmonic frame of n distinct vectors for $\mathbb{C}^1$.

11.5. Show that the only real harmonic frame of n distinct vectors for $\mathbb{C}^2$ is the n equally spaced unit vectors in $\mathbb{R}^2$.

11.6. Find all the (unitarily inequivalent) harmonic frames of six vectors for $\mathbb{C}^2$. Show that none are complex conjugates of each other, by computing the distances between their vectors, or otherwise.

11.7. Suppose that $\Phi = (gv)_{g\in G}$ is a frame for $\mathbb{F}^d$ generated by a finite abelian group of matrices $G \subset GL(\mathscr{H})$. Show that Φ has *distinct* vectors. Does this hold for nonabelian groups?

11.8. Show that if Φ is a harmonic frame given by $J \subset G$, then the complementary tight frame is the harmonic frame given by $G \setminus J$ (the complement of J).

11.9. Suppose that Φ is an unlifted harmonic frame with $|\mathrm{Sym}(\Phi)| = m$. Show that $\Psi := \Phi \hat{+} \cdots \hat{+} \Phi$ (d summands) has a symmetry group of order at least $d!m^d$. *Remark:* The *vertices of the cube* $[-1,1]^d$ in $\mathbb{C}^d$ (see Example 5.12) are a balanced tight frame of 2^d vectors for $\mathbb{R}^d$ with symmetry group of order $d!2^d$.

11.10. Let G be a finite abelian group.
(a) Show that a harmonic frame $(v_g)_{g\in G}$ satisfies

$$\langle v_j, v_k \rangle = \langle v_{j+a}, v_{k+a} \rangle, \qquad \forall j,k,a \in G.$$

(b) Suppose that harmonic frames $(v_g)_{g\in G}$ and $(w_g)_{g\in G}$ are unitarily equivalent up to reindexing. Show that for all $j,k,b \in G$ there exists some $a \in G$ with

$$\langle v_j, v_k \rangle = \langle w_a, w_b \rangle.$$

(c) Suppose that the harmonic frames given by $\{\xi_1,\ldots,\xi_d\},\{\eta_1,\ldots,\eta_d\} \subset \hat{G}$ are unitarily equivalent up to reindexing. Show that

$$\sum_{j=1}^{d} \xi_j(1) = \sum_{j=1}^{d} \eta_j(a),$$

for some $a \in G$.

11.11. Let p be an odd prime.
(a) Show that if $J \subset \mathbb{Z}_p$, $|J| = d$ gives a real cyclic harmonic frame, then $J = K \cup -K$ when d is even, and $J = \{0\} \cup K \cup -K$ when d is odd (disjoint unions), where $K \subset \mathbb{Z}_p \setminus \{0\}$ generates $\mathbb{Z}_p$ (for $d > 1$).
(b) Use Burnside orbit counting to show that the number of real (cyclic) harmonic frames of p vectors for $\mathbb{R}^d$ is given by

$$h_{p,d}^{\mathbb{R}} = \frac{1}{p-1}\Big\{ \sum_{\substack{j \mid \gcd(p-1,d) \\ j \text{ even}}} \binom{\frac{p-1}{j}}{\frac{d}{j}} \varphi(j) + \sum_{\substack{j \mid \gcd(p-1,\frac{d}{2}) \\ j \text{ odd}}} \binom{\frac{p-1}{2j}}{\frac{d}{2j}} \varphi(j) \Big\} \qquad (d \text{ even}),$$

$$h_{p,d}^{\mathbb{R}} = \frac{1}{p-1}\Big\{ \sum_{\substack{j \mid \gcd(p-1,d-1) \\ j \text{ even}}} \binom{\frac{p-1}{j}}{\frac{d-1}{j}} \varphi(j) + \sum_{\substack{j \mid \gcd(p-1,\frac{d-1}{2}) \\ j \text{ odd}}} \binom{\frac{p-1}{2j}}{\frac{d-1}{2j}} \varphi(j) \Big\}, \qquad (d \text{ odd}).$$

Chapter 12
Equiangular and Grassmannian frames

A set of n unit vectors $\{f_j\}$ in $\mathbb{F}^d$ is said to define a set of **equiangular lines** if the angles between the subspaces (lines) they determine are equal, i.e., $\exists C \geq 0$ with

$$|\langle f_j, f_k \rangle| = C, \qquad \forall j \neq k. \tag{12.1}$$

An illuminating example is three isogonal vectors in $\mathbb{R}^3$ (see Example 3.9),

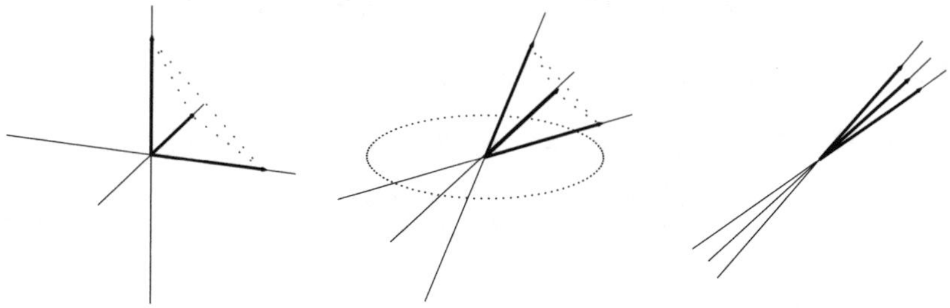

where one can choose $0 \leq C < 1$, or even $C = 1$ (in which case the vectors are equal). By the variational characterisation of tight frames (Theorem 6.1), it follows that

$$C \geq \sqrt{\frac{n-d}{d(n-1)}}, \tag{12.2}$$

with equality in (12.2) if and only if $\{f_j\}$ is a tight frame for $\mathbb{F}^d$. Moreover, the maximum C can be is 1, when the vectors lie in a common line.

Here we consider those sets of unit vectors that define a set of equiangular lines, which, in addition, give equality in (12.2), i.e., are a finite tight frame for $\mathbb{F}^d$. The corresponding sets of *tight* equiangular lines can be thought as being spread out in $\mathbb{F}^d$ as much as possible. This extra structure gives sets of equiangular lines with special properties, e.g., the Gramian of the vectors defining the lines is an orthogonal projection matrix, and so a complementary set of equiangular lines exists.

S.F.D. Waldron, *An Introduction to Finite Tight Frames*, Applied and Numerical Harmonic Analysis, https://doi.org/10.1007/978-0-8176-4815-2_12

12.1 Equiangular lines and frames

Sets $\{f_j\}$ of n unit vectors in $\mathbb{F}^d$ for which the **maximum cross-correlation**

$$M := \max_{j\neq k} |\langle f_j, f_k\rangle| \in [0,1]$$

is *small* have applications to wireless communication and coding theory [SH03]. These correspond to sets of lines $L_j = \{cf_j : c \in \mathbb{F}\}$ (through 0 and f_j) for which the **minimal angle** between them

$$\theta := \min_{j\neq k} \cos^{-1}(|\langle f_j, f_k\rangle|) \in [0, \frac{1}{2}\pi]$$

is *large*. The lower bound (6.3) for M is attained as follows.

Definition 12.1. A set $\{f_j\}$ of equal-norm vectors in $\mathscr{H}$ is **equiangular** if $\exists C \geq 0$ with

$$|\langle f_j, f_k\rangle| = C, \qquad \forall j \neq k.$$

For a set of equiangular vectors of unit length, the constant is often written

$$C = \alpha = \cos\theta, \qquad \alpha \in [0,1], \quad \theta \in [0, \pi/2],$$

with both $C = \alpha$ and θ referred to as the *angle* between the vectors (or lines).

Theorem 12.1. *Let $(f_j)_{j=1}^n$ be a sequence of n unit vectors in $\mathscr{H}$, $d = \dim(\mathscr{H})$. Then*

$$M = \max_{j\neq k} |\langle f_j, f_k\rangle| \geq \sqrt{\frac{n-d}{d(n-1)}}, \tag{12.3}$$

or, equivalently,

$$\theta = \min_{j\neq k} \cos^{-1} |\langle f_j, f_k\rangle| \leq \cos^{-1} \sqrt{\frac{n-d}{d(n-1)}}, \tag{12.4}$$

with equality if and only if $(f_j)_{j=1}^n$ is an equiangular tight frame.

Proof. We consider the case of equality in Exer. 6.2. Firstly, observe that

$$M^2 = \max_{j\neq k} |\langle f_j, f_k\rangle|^2 \geq \frac{1}{n^2-n} \sum_{j\neq k} |\langle f_j, f_k\rangle|^2 = \frac{1}{n^2-n} \Big\{\sum_j \sum_k |\langle f_j, f_k\rangle|^2 - n\Big\}$$

with equality if and only if (f_j) is equiangular. By Theorem 6.1, we have

$$\frac{1}{n^2-n} \Big\{\sum_j \sum_k |\langle f_j, f_k\rangle|^2 - n\Big\} \geq \frac{1}{n^2-n} \Big(\frac{n^2}{d} - n\Big) = \frac{n-d}{d(n-1)},$$

with equality if and only if (f_j) is a tight frame. The two inequalities above give (12.3), and the fact that $\cos^{-1}$ is strictly decreasing on $[0, \frac{1}{2}\pi]$ gives (12.4). □

Example 12.1. The d vectors of an *orthonormal basis* in $\mathbb{F}^d$ and the $d+1$ vertices of the regular *simplex* in $\mathbb{R}^d$ give real equiangular tight frames with $C=0$ and $C=\frac{1}{d}$. These exist in every dimension d and are the unique equiangular tight frames of d and $d+1$ vectors in $\mathbb{C}^d$ up to projective unitary equivalence (see Example 9.8).

The equiangularity condition on a set of lines ensures that the corresponding set of one-dimensional orthogonal projections is linearly independent, which in turn gives a bound on the possible number of equiangular lines in $\mathbb{F}^d$.

Theorem 12.2. *Suppose $d>1$. Let (f_j) be a sequence of n unit vectors in $\mathbb{F}^d$ giving a set of n equiangular lines, then the orthogonal projections*

$$P_j : f \mapsto \langle f, f_j \rangle f_j, \qquad j = 1, \ldots, n$$

are linearly independent, and hence

$$n \le \begin{cases} \frac{1}{2}d(d+1), & \mathbb{F}=\mathbb{R}; \\ d^2, & \mathbb{F}=\mathbb{C} \end{cases} \tag{12.5}$$

with equality if and only if $\{P_j\}_{j=1}^n$ is a basis for the Hermitian matrices.

Proof. Since $d>1$, the equiangularity constant C is less than 1. By Exer. 3.1

$$\operatorname{trace}(P_j P_k^*) = |\langle f_j, f_k \rangle|^2 = C^2, \qquad j \ne k,$$

and so the Frobenius norm of the linear combination $\sum_j c_j P_j$ is

$$\begin{aligned} \|\sum_j c_j P_j\|_F^2 &= \operatorname{trace}(\sum_j c_j P_j \sum_k \overline{c_k} P_k^*) = \sum_j \sum_k c_j \overline{c_k} \operatorname{trace}(P_j P_k^*) \\ &= \sum_j \sum_k c_j \overline{c_k} C^2 + \sum_j c_j \overline{c_j}(1-C^2) \\ &= C^2 |\sum_j c_j|^2 + (1-C^2) \sum_j |c_j|^2, \end{aligned}$$

which is zero only for the trivial linear combination.

The projections $\{P_j\}$ belong to the real vector space of Hermitian matrices, which has dimension (see Exer. 12.1) given by the right-hand side of (12.5). □

The right-hand side of (12.5) is the number n of (8.24) which ensures that a generic sequence of n vectors in $\mathbb{F}^d$ has a unique scaling to a tight signed frame.

If the upper bound (12.5) is attained, then (f_j) is a unit-norm tight frame for $\mathbb{F}^d$ (see Exer. 12.2), i.e., the identity matrix can be written

$$I = \frac{d}{n} \sum_{j=1}^n P_j.$$

The upper bound (12.5) for the maximum number of equiangular lines in $\mathbb{R}^d$ is called the **absolute bound** (or **Gerzon bound**). It is rarely attained: only the cases $d = 2,3,7,23$ are known (see Example 12.35). Many maximal set of equiangular lines turn out to be tight frames (see Table 12.3). By way of contrast, the upper bound (12.5) for complex equiangular lines is conjectured to always be attained (Zauner's conjecture), by a tight frame known as a SIC (see Chapter 14).

Example 12.2. There are sets of two and three equiangular lines in $\mathbb{R}^2$. Theorem 12.2 confirms the familiar geometric fact that there are *not* four (or more) equiangular lines in $\mathbb{R}^2$ and hence no equiangular tight frame of four or more vectors for $\mathbb{R}^2$. There is a tight frame giving four equiangular lines in $\mathbb{C}^2$ (see Example 2.16).

Example 12.3. An equiangular tight frame of 28 vectors for $\mathbb{R}^7$ can be constructed as follows. Let (f_j) be the 28 unit vectors of the form

$$\frac{1}{\sqrt{24}}(x_1,x_2,\ldots,x_8), \qquad x_j \in \{-1,3\}, \quad \sum_j x_j = 0. \tag{12.6}$$

These vectors span a 7-dimensional subspace of $\mathbb{R}^8$ and give an equiangular tight frame (see Exer. 12.4, Example 12.56), which gives equality $n = \frac{1}{2}d(d+1)$ in (12.5).

12.2 Grassmannian frames

If an equiangular tight frame does not exist for a given $n \geq d$, then a good substitute is a *Grassmannian frame* (see [SH03], [BK06]) or a variation thereof.

Definition 12.2. A frame $\Phi = (f_j)_{j=1}^n$ of n unit vectors for $\mathscr{H}$ is a **Grassmannian frame** if it minimises

$$\mathscr{M}_\infty(\Phi) := \max_{j \neq k} |\langle f_j, f_k \rangle|, \tag{12.7}$$

and is **optimal** if it gives equality in (12.3), i.e., is an equiangular tight frame.

Grassmannian frames *exist* for all $n \geq d$, for $\mathscr{H}$ real or complex, since the set of n element unit-norm frames is a compact subset of $\mathscr{H}^n$. By the same reasoning, there exist equal-norm normalised tight frames which minimise (12.7). These were frames were called 2-*optimal* frames by [HP04], who proved that they are optimal for the 2-erasure problem.

The **Grassmannian space** $\mathscr{G}(\mathscr{H},\ell) = \mathscr{G}(d,\ell)$ is the set of all ℓ-dimensional subspaces of the d-dimensional space $\mathscr{H} = \mathbb{F}^d$ (usually real). The **Grassmannian packing problem** is to find the best packing of n subspaces of dimension ℓ in $\mathscr{H}$, so that the angle between any two is as large as possible (see [CHS96], [DHST08]). Clearly, for $\ell = 1$ this equivalent to finding a Grassmannian frame.

Example 12.4. The first n vertices of the regular $2n$-gon (multiplied by ± 1), i.e., the n equally spaced lines in $\mathbb{R}^2$ (see Exer. 6.10) are the *unique* Grassmannian frame of n vectors for $\mathbb{R}^2$ (see Exer. 12.5). This is tight, and it is equiangular only for $n = 3$.

Example 12.5. By Theorem 12.1, an equiangular tight frame is Grassmannian.

Example 12.6. The only known example of a *nontight* Grassmannian frame is that of 5 vectors in $\mathbb{R}^3$ which lie on the diagonals of the regular icosahedron, which is the optimal Grassmannian line packing (see [CHS96], Exer. 12.7). This is the nontight equiangular frame given by the graph of the 5-cycle (see Example 12.44).

There is an extensive literature on the closely related problem of finding the distributions of points on a sphere that minimise a given potential (see [SK97]). A classical example is **Tammes' problem** of determining how to place n points on the unit sphere in $\mathbb{R}^3$ so as to maximise the minimum distance between them. For five vectors the solution to Tammes' problem is the north and south poles together with three equally spaced points on the equator. This set of five vectors is *not* a tight frame or a Grassmannian frame for $\mathbb{R}^3$.

12.3 Equiangular harmonic frames and difference sets

Here we show that a harmonic frame $\Phi_J = (\xi|_J)_{\xi\in\hat{G}}$ is equiangular if and only if $J \subset G$ is a *difference set* for the abelian group G. This leads to some infinite families of equiangular tight frames.

Definition 12.3. A d element subset J of a finite group G of order n is said to be a (n,d,λ)**-difference set** if every nonidentity element of G can be written as a difference $j_1 j_2^{-1}$ of two elements $j_1, j_2 \in J$ in exactly λ ways. The difference set is said to be **abelian**, **cyclic**, etc, when the group G is.

The complement $G \setminus J$ of a (n,d,λ)-difference set J is a

$$(n, n-d, n-2d+\lambda)\text{-difference set}$$

and so lists of difference sets (Tables 12.1 and 12.2) usually assume $2d \le n$, $\lambda > 0$.

Since there are exactly $d^2 - d$ pairs of elements from J whose difference is not the identity, a (n,d,λ)-difference set must satisfy

$$d(d-1) = (n-1)\lambda. \tag{12.8}$$

The set of all translates of a difference set J gives a *symmetric block design* (not all symmetric block designs appear in this way). In particular, the parameters of a difference set must satisfy the *Bruck–Ryser–Chowla theorem*, which gives

- If n is even, then its *order* $d-\lambda$ is a square.
- If n is odd, the *Diophantine equation*

$$x^2 + (d-\lambda)y^2 - (-1)^{(n-1)/2}\lambda z^2 = 0$$

has a nontrivial solution (x,y,z).

We assume the basic theory of harmonic frames (and characters) as detailed in Chapter 11. The orthogonality between a character $\chi \neq 1$ and the trivial character 1 gives

$$\sum_{g\in G}\chi(g)=0 \quad\Longrightarrow\quad \sum_{g\neq 0}\chi(g)=-1, \quad \chi\neq 1. \tag{12.9}$$

The characters of an abelian group G sum to the so-called **regular character**

$$\sum_{\chi\in\hat{G}}\chi=\chi_{\mathrm{reg}}, \qquad \chi_{\mathrm{reg}}(g):=\begin{cases}|G|, & g=1;\\ 0, & g\neq 1.\end{cases} \tag{12.10}$$

Theorem 12.3. *Let G be an abelian group of order n, and $\Phi_J=(\xi|_J)_{\xi\in\hat{G}}$ be the harmonic frame given by $J\subset G$, $|J|=d$. Then the following are equivalent*

1. *Φ_J is an equiangular tight frame of n vectors for $\mathbb{C}^d$.*
2. *J is a (n,d,λ)-difference set for G.*
3. *$|\sum_{j\in J}\chi(j)|^2=\frac{d(n-d)}{n-1}$, for all $\chi\neq 1$, $\chi\in\hat{G}$.*

Proof. ($1\iff 3$) Suppose that $\xi\neq\eta$, i.e., $\chi:=\xi\eta^{-1}\neq 1$. We calculate

$$\langle\xi|_J,\eta|_J\rangle=\sum_{j\in J}\xi(j)\overline{\eta(j)}=\sum_{j\in J}(\xi\eta^{-1})(j)=\sum_{j\in J}\chi(j),$$

and so Φ_J is equiangular if and only if

$$\Big|\langle\frac{\xi|_J}{\sqrt{d}},\frac{\eta|_J}{\sqrt{d}}\rangle\Big|=\frac{1}{d}\Big|\sum_{j\in J}\chi(j)\Big|=\sqrt{\frac{n-d}{d(n-1)}},$$

which gives the equivalence of 1 and 3.

We now show the equivalence of 1 and 2. Since $\overline{\chi(k)}=\chi(-k)$, we have that

$$|\langle\xi|_J,\eta|_J\rangle|^2=\sum_{j\in J}\sum_{k\in J}\chi(j-k)=\sum_{g\in G}\alpha_g\chi(g), \qquad \chi\neq 1, \tag{12.11}$$

where $\alpha_g:=\#\{(j,k)\in J\times J: j-k=g\}$. Observe that $\alpha_0=d$, $\sum_{g\neq 0}\alpha_g=d^2-d$.

($2\Longleftarrow 1$) Suppose that J is a difference set, i.e., $\alpha_g=\lambda$, $g\neq 0$. Then (12.11), together with (12.9), gives

$$|\langle\xi|_J,\eta|_J\rangle|^2=\lambda\sum_{g\neq 0}\chi(g)+d=-\lambda+d,$$

i.e., Φ_J is equiangular.

($1\Longrightarrow 2$) Suppose that Φ_J is equiangular. Then (12.11) gives

$$|\langle\xi|_J,\eta|_J\rangle|^2=\sum_{g\neq 0}\alpha_g\chi(g)+d=-\lambda+d, \qquad \chi\neq 1,$$

for some fixed λ. By Pontryagin duality, this can be written as

$$\sum_{g\neq 0} \alpha_g \hat{\hat{g}}(\chi) = -\lambda, \qquad \chi \neq 1.$$

Moreover, by counting the number of differences

$$\sum_{g\neq 0} \alpha_g \hat{\hat{0}}(\chi) = \sum_{g\neq 0} \alpha_g = d^2 - d.$$

These combine to give the regular character of $\hat{G}$ as a sum of characters

$$\sum_{g\neq 0} \alpha_g \hat{\hat{g}} + \lambda \hat{\hat{0}} = \frac{d^2 - d + \lambda}{n} \chi_{\text{reg}}^{\hat{G}}.$$

Since the (irreducible) characters of $\hat{G}$ are linearly independent, (12.10) implies that $\alpha_g = \lambda$, $\forall g \neq 0$, i.e., J is a (n,d,λ)-difference set for G. □

The condition 3 for difference sets is well known [Tur65], [LM90]. However, its interpretation in terms of equiangular lines (harmonic frames) seems to be more recent [Kön99], [XZG05], [Kal06]. In [XZG05], equiangular tight frames are called *MWBE (maximum–Welch–bound–equality) codebooks.*

Example 12.7. The cyclic harmonic frame of n vectors for $\mathbb{C}^d$ given by

$$\Phi_J = \left\{ \begin{pmatrix} 1 \\ \vdots \\ 1 \end{pmatrix}, \begin{pmatrix} \omega^{j_1} \\ \vdots \\ \omega^{j_d} \end{pmatrix}, \begin{pmatrix} \omega^{2j_1} \\ \vdots \\ \omega^{2j_d} \end{pmatrix}, \ldots, \begin{pmatrix} \omega^{(n-1)j_1} \\ \vdots \\ \omega^{(n-1)j_d} \end{pmatrix} \right\}, \qquad \omega := e^{\frac{2\pi i}{n}}$$

is equiangular if and only if $J = \{j_1, j_2, \ldots, j_d\}$ is a (n,d,λ)-difference set for $\mathbb{Z}_n$. The cyclic difference sets for $d \leq 10$ are given in Table 12.1.

Example 12.8. (Orthonormal bases) The choice $J = G$ is an (n,n,n)-difference set which gives an orthonormal basis.

Example 12.9. (Simplex) There is a unique difference set with $\lambda = 0$, $d \geq 1$, i.e., the $(n,1,0)$-difference set consisting of the identity element of G. The complementary $(n,n-1,n-2)$-difference set (of the nonidentity elements) gives the vertices of the simplex. Both of these difference sets (and those with $d = 0,n$) are said to be *trivial.*

Difference sets J_1 and J_2 for a group G are said to be **equivalent** if there is an automorphism σ of G and some $g \in G$ with

$$\sigma J_1 = g J_2, \qquad \text{for some } g \in G.$$

By Theorem 11.8, it follows that

Equivalent difference sets give projectively unitarily equivalent harmonic frames (up to reordering by an automorphism).

Table 12.1: List of the cyclic (n,d,λ)-difference sets J for $d \le 10$, $\lambda > 0$ (including complements). These give the cyclic equiangular harmonic frames of $n > d+1$ vectors for $\mathbb{C}^d$.

n	d	λ	$J \subset \mathbb{Z}_n$
7	3	1	{1,2,4}
7	4	2	{0,3,5,6}
13	4	1	{0,1,3,9}
11	5	2	{1,3,4,5,9}
21	5	1	{3,6,7,12,14}
11	6	3	{0,2,6,7,8,10}
31	6	1	{1,5,11,24,25,27}
15	7	3	{0,1,2,4,5,8,10}

n	d	λ	$J \subset \mathbb{Z}_n$
15	8	4	{3,6,7,9,11,12,13,14}
57	8	1	{0,1,6,15,22,26,45,55}
13	9	6	{2,4,5,6,7,8,10,11,12}
19	9	4	{1,4,5,6,7,9,11,16,17}
37	9	2	{1,7,9,10,12,16,26,33,34}
73	9	1	{ 0,1,12,20,26,30,33,35,57}
19	10	5	{0,2,3,8,10,12,13,14,15,18}
91	10	1	{0,2,6,7,18,21,31,54,63,71}

Table 12.2: The abelian (n,d,λ)-difference sets for $d \le 50$, $n \ge 2d$, $\lambda > 0$, where G has invariant factors $[a_1,\ldots,a_m]$. These give the equiangular harmonic frames of $n > d+1$ vectors for $\mathbb{C}^d$.

n	d	λ	G
7	3	1	[7]
13	4	1	[13]
21	5	1	[21]
11	5	2	[11]
16	6	2	[4,4]
16	6	2	[2,8]
16	6	2	[2,2,4]
16	6	2	[2,2,2,2]
31	6	1	[31]
15	7	3	[15]
57	8	1	[57]
73	9	1	[73]
37	9	2	[37]
19	9	4	[19]
91	10	1	[91]
23	11	5	[23]
45	12	3	[3,15]
133	12	1	[133]
27	13	6	[3,3,3]
40	13	4	[40]
183	14	1	[183]
36	15	6	[3,12]
36	15	6	[6,6]

n	d	λ	G
31	15	7	[31]
273	17	1	[273]
35	17	8	[35]
307	18	1	[307]
96	20	4	[2,4,12]
96	20	4	[2,2,2,12]
96	20	4	[2,2,2,2,6]
381	20	1	[381]
85	21	5	[85]
43	21	10	[43]
47	23	11	[47]
553	24	1	[553]
101	25	6	[101]
651	26	1	[651]
64	28	12	[2,2,16]
64	28	12	[2,4,8]
64	28	12	[4,4,4]
64	28	12	[2,2,4,4]
64	28	12	[8,8]
64	28	12	[4,16]
64	28	12	[2,2,2,2,2,2]
64	28	12	[2,2,2,2,4]
64	28	12	[2,2,2,8]

n	d	λ	G
757	28	1	[757]
109	28	7	[109]
59	29	14	[59]
175	30	5	[5,35]
871	30	1	[871]
156	31	6	[156]
63	31	15	[63]
63	31	15	[3,21]
993	32	1	[993]
133	33	8	[133]
67	33	16	[67]
71	35	17	[71]
1407	38	1	[1407]
79	39	19	[79]
121	40	13	[121]
121	40	13	[11,11]
83	41	20	[83]
1723	42	1	[1723]
1893	44	1	[1893]
2257	48	1	[2257]
99	49	24	[3,33]
197	49	12	[197]
2451	50	1	[2451]

By (12.8), the number n of vectors in an equiangular harmonic frame for $\mathbb{C}^d$ is

$$n = \frac{1}{\lambda}(d^2 - d) + 1 \leq d^2 - d + 1. \tag{12.12}$$

This is largest when $\lambda = 1$, i.e., J is a **planar** difference set. Here the corresponding symmetric block design gives a *projective plane* (the blocks are the points on a line, and so each pair of distinct lines intersects in a single point). For example, the $(7,3,1)$-difference set $J = \{1,2,4\} \subset \mathbb{Z}_7$ gives the *Fano plane* (7 points, 7 lines, 3 points on each line), whose lines are the translates of J, i.e.,

$$\{1,2,4\},\ \{2,3,5\},\ \{3,4,6\},\ \{4,5,0\},\ \{5,6,1\},\ \{6,0,2\},\ \{0,1,3\}.$$

There is considerable work on difference sets (for both abelian and nonabelian groups). The online *La Jolla Difference Set Repository* maintained by Dan Gordon is an excellent resource. From this, we obtained the Table 12.2 of all abelian difference sets for $n \leq 50$, i.e., all equiangular harmonic frames of $n \leq 50$ vectors (for brevity, the complementary difference sets are not included).

There are several infinite families of difference sets. These are often classified via their parameters, e.g., difference sets with parameters $(4m-1, 2m-1, m-1)$ are of *Paley-type*[1], and those with parameters $(4m^2, 2m^2 - m, m^2 - m)$ are *Hadamard difference sets* (see [DJ96]). We now give a few examples.

Example 12.10. (Singer) Let $PG_m(q)$ be the m-dimensional projective geometry over $\mathbb{F}_q$. The intersection of two hyperplanes is an $(m-2)$-dimensional subspace containing $\frac{q^{m-2}-1}{q-1}$ projective points, and so one obtains a symmetric block design where the blocks are the projective points on a hyperplane. A classical result of Singer shows that there is a cyclic subgroup of the projective linear transformations which acts regularly on the points and hyperplanes of $PG_m(q)$. From this, it follows that there exists a cyclic $(\frac{q^m-1}{q-1}, \frac{q^{m-1}-1}{q-1}, \frac{q^{m-2}-1}{q-1})$-difference set. In particular, by taking $m = 3$, and letting $d = q+1$, one obtains a cyclic $(d^2 - d + 1, d, 1)$-difference set, i.e., an equiangular cyclic harmonic frame of $n = d^2 - d + 1$ vectors for $\mathbb{C}^d$. The *Prime Power Conjecture* is that every abelian planar difference set (such as these) has $d - 1$ a prime power. This has been verified for $d \leq 2{,}000{,}000$ [BG04].

Example 12.11. (Paley) Let $q = 4m - 1$ be a prime power ($q \equiv 3 \mod 4$). Then the set of all nonzero squares in $(\mathrm{GF}(q), +)$ is a $(4m-1, 2m-1, m-1)$-difference set.

Example 12.12. (Twin prime powers) Let q and $q+2$ be *twin prime powers*, i.e., a pair of odd integers, each of which is a prime power, e.g, $q = 7, q+2 = 9 = 3^2$. Then there exists a $(q(q+2), \frac{q(q+2)-1}{2}, \frac{q(q+2)-3}{4})$-difference set.

The last two examples and the $q = 2$ Singer difference sets are of Paley-type.

Example 12.13. ($n = 2d+1$, d odd) For a difference set of Paley-type, by letting $d = 2m - 1$ (d is odd), the parameters become $(2d+1, d, \frac{d-1}{2})$. Hence, Paley-type

[1] The term *Hadamard type* difference set is also used.

difference sets give equiangular harmonic frames of $n = 2d+1$ vectors for $\mathbb{C}^d$. These exist when $2d+2$ is a power of 2, or $2d+1$ is a prime power congruent to 3 mod 4, or $2d+1$ is a product of twin prime powers. The first odd d where there is not an equiangular harmonic frame of $2d+1$ vectors for $\mathbb{C}^d$ is $d = 19$ $(2d+1 = 39)$.

Example 12.14. ($n = 2d-1$, d even) The complement of a Paley-type difference set is a $(4m-1, 2m, m)$-difference set. By letting $d = 2m$ (d is even), these parameters become $(2d-1, d, \frac{d}{2})$, and so Paley-type difference sets give rise to equiangular harmonic frames of $n = 2d-1$ vectors for $\mathbb{C}^d$.

Example 12.15. (Reversible difference sets) A difference set J is **reversible** if it is closed under taking inverses, i.e., the corresponding equiangular harmonic frame is real (Theorem 11.2). A reversible difference set for $G = \mathbb{Z}_4 \times \mathbb{Z}_4$ is given by

$$J = \{(1,0),(2,0),(3,0),(0,1),(0,2),(0,3)\},$$

and so there exists a real equiangular harmonic frame of 16 vectors for $\mathbb{C}^6$. There is also a $(16,6,2)$-difference set for $G = \mathbb{Z}_2^4$, which must be reversible (as are all difference sets for groups of exponent 2), since each element of $\mathbb{Z}_2^4$ is its own inverse. The set

$$J = \{(x,0),(0,x),(x,x) : x \neq 0\} \subset \mathbb{Z}_6 \times \mathbb{Z}_6$$

is a reversible $(36,15,6)$-difference set for $G = \mathbb{Z}_6 \times \mathbb{Z}_6$, and so there exists a real equiangular harmonic frame of 36 vectors for $\mathbb{R}^{15}$.

The bound $n \leq d^2 - d + 1$ of (12.12) for equiangular We say that a set of unit vectors in $\mathbb{C}^d$ (or the lines they determine) is **flat** if each vector entry has constant modulus (of $\frac{1}{\sqrt{d}}$). Harmonic frames are flat.

Proposition 12.1. *([GR09]) There can be at most $d^2 - d + 1$ flat equiangular lines in $\mathbb{C}^d$, and any such set must be tight.*

Proof. Suppose that the unit vectors $\Phi = (v_j)$ give n flat equiangular lines in $\mathbb{C}^d$, i.e.,$|\langle v_j, v_k \rangle| = \alpha$, $j \neq k$. The Gram matrix of the rank one orthogonal projections $v_1 v_1^*, \ldots, v_n v_n^*$ and $e_1 e_1^*, \ldots, d_d e_d^*$ (with the Frobenius inner product) is

$$A = \begin{pmatrix} \alpha^2 J + (1-\alpha^2) I_n & \frac{1}{d} J \\ \frac{1}{d} J & I_d \end{pmatrix}, \qquad J = [1].$$

Since these vectors lie in the d^2-dimensional space $\mathbb{C}^{d \times d}$, we have $\operatorname{rank}(A) \leq d^2$.

A calculation (see Exer. 12.9) shows that $\operatorname{rank}(A) = n+d+1$ and Φ is tight, or $\operatorname{rank}(A) = n+d$. Thus $n+d-1 \leq \operatorname{rank}(A) \leq d^2$, which gives the result. □

Example 12.16. For $2 \leq d \leq 6$, there are $d^2 - d + 1$ flat equiangular lines for $\mathbb{C}^d$, given by an equiangular harmonic frame corresponding to a cyclic difference set (see Table 12.1).

12.4 Equiangular tight frames from block designs

We call an equiangular tight frame (v_j) of n unit vectors for $\mathbb{C}^d$ **subsimplicial** (of dimension r) if

$$|\langle v_j, v_k\rangle| = \frac{1}{r}, \quad j \neq k, \qquad \text{for some integer } r,$$

i.e., if every $r+1$ element subset of (v_j) is an r-simplex, up to projective unitary equivalence. The possible d,n for a subsimplicial frame are restricted by the fact that

$$r = \sqrt{\frac{d(n-1)}{n-d}} \in \mathbb{Z}. \tag{12.13}$$

A single d-simplex is a trivial example of a subsimplicial frame.

Example 12.17. (Complementary frames) If d,n satisfy the necessary condition (12.13) for being a subsimplicial frame, then the complementary frame may not. For example, for $d=8$, $n=15$ there is a harmonic equiangular tight frame given by the difference set $\{3,6,7,9,11,12,13,14\} \subset \mathbb{Z}_{15}$ which is subsimplicial with $r=4$, but the complementary tight frame is not subsimplicial ($r=\frac{7}{2}$).

Example 12.18. (Real equiangular lines) A set of real equiangular lines meeting the absolute bound $n=\frac{1}{2}d(d+1)$ is subsimplicial (with $r=\sqrt{d+2}$) if and only if $d+2$ is a square. Many other maximal sets of tight real equiangular lines are also subsimplicial, e.g., all those for $d \neq 3$ listed in Table 12.3 (including the conjectured sets of lines in grey).

Example 12.19. (SICs) A set of complex equiangular lines meeting the absolute bound $n=d^2$ is subsimplicial (with $r=\sqrt{d+1}$) if and only if $d+1$ is a square. The first two examples $d=3,8$ are exceptional cases for SICs (see Chapter 14).

We now outline a construction of [FMT12] which uses Steiner systems (a type of block design) to construct subsimplicial equiangular tight frames directly in $\mathbb{C}^d$.

A **Steiner system** $S(t,k,v)$ is a collection $\mathscr{B}$ of k-element subsets (called **blocks**) of an v element set $\mathscr{V}$, with the property that each element of $\mathscr{V}$ is in exactly r blocks and each t-element subset of $\mathscr{V}$ is contained in exactly one block. A count shows

$$r = \frac{\binom{v-1}{t-1}}{\binom{k-1}{t-1}}.$$

We consider Steiner systems $\mathscr{B}$ on $\mathscr{V}$ with $t=2$ (so $r=\frac{v-1}{k-1}$), i.e., the property that every pair of points lies in exactly one block. We illustrate the construction using an $S(2,2,4)$ system (all two element subsets of a four element set). Each $\mathscr{B}$ can be represented by a $\mathscr{B}\times\mathscr{V}$ matrix A^T with a 1 in the (β,a) entry if the point a is in the block β, and a 0 otherwise (A is the incidence matrix of the system). For each point $a \in \mathscr{V}$, choose a Hadamard matrix of size $r+1$, i.e.,

$$H^{(a)} = \begin{pmatrix} h_0^{(a)} \\ h_{\beta_1}^{(a)} \\ \vdots \\ h_{\beta_r}^{(a)} \end{pmatrix}, \qquad h_\beta^{(a)} = \begin{bmatrix} h_{\beta,1}^{(a)} \; h_{\beta,2}^{(a)} \cdots h_{\beta,r+1}^{(a)} \end{bmatrix},$$

where $\beta_1,\ldots,\beta_r \in \mathscr{B}$ are the r blocks containing a. The $r+1$ rows of $H^{(a)}$ are orthogonal and of length $\sqrt{r+1}$. Let V be the $\mathscr{B} \times \nu(r+1)$ block matrix obtained by replacing the nonzero (β,a)-entries of A^T by the $1\times(r+1)$ blocks $h_\beta^{(a)}$ and replacing the zero entries by the $1\times(r+1)$ zero matrix.

$$A^T = \begin{pmatrix} 1\,1\,0\,0 \\ 1\,0\,1\,0 \\ 1\,0\,0\,1 \\ 0\,1\,1\,0 \\ 0\,1\,0\,1 \\ 0\,0\,1\,1 \end{pmatrix} \begin{matrix} \textit{blocks} \\ \{1,2\} \\ \{1,3\} \\ \{1,4\} \\ \{2,3\} \\ \{2,4\} \\ \{3,4\} \end{matrix} \quad \longrightarrow \quad V = \begin{pmatrix} h_{\{1,2\}}^{(1)} & h_{\{1,2\}}^{(2)} & 0 & 0 \\ h_{\{1,3\}}^{(1)} & 0 & h_{\{1,3\}}^{(3)} & 0 \\ h_{\{1,4\}}^{(1)} & 0 & 0 & h_{\{1,4\}}^{(4)} \\ 0 & h_{\{2,3\}}^{(2)} & h_{\{2,3\}}^{(3)} & 0 \\ 0 & h_{\{2,4\}}^{(2)} & 0 & h_{\{2,4\}}^{(4)} \\ 0 & 0 & h_{\{3,4\}}^{(3)} & h_{\{3,4\}}^{(4)} \end{pmatrix}.$$

The columns of $V = [v_{a,j}]_{a\in\mathscr{V},1\le j\le r+1}$ are the vectors in $\mathbb{C}^{\mathscr{B}}$, $|\mathscr{B}| = \frac{\nu(\nu-1)}{k(k-1)}$, given by

$$v_{a,j}(\beta) = \begin{cases} h_{\beta,j}^{(a)}, & a \in \beta; \\ 0, & a \notin \beta \end{cases} \tag{12.14}$$

Recall that the rows and columns of a Hadamard matrix are orthogonal. Thus, the rows of V are orthogonal, and so $(v_{a,j})$ is an equal-norm tight frame. The inner product between different vectors indexed by the same point a is the inner product between the corresponding columns of $H^{(a)}$ with the first entries removed, giving

$$\langle v_{a,j}, v_{a,k}\rangle = -h_{0,j}^{(a)}\overline{h_{0,k}^{(a)}}, \qquad j \neq k. \tag{12.15}$$

Vectors indexed by different points a and b are both nonzero only in the β-entry, where β is the unique block containing both points, which gives

$$\langle v_{a,j}, v_{b,k}\rangle = h_{\beta,j}^{(a)}\overline{h_{\beta,k}^{(b)}}, \qquad a \neq b. \tag{12.16}$$

We have just proved the following result.

Theorem 12.4. *(Steiner systems). For a $(2,k,\nu)$-Steiner system $\mathscr{B}$, $r = \frac{\nu-1}{k-1}$, and one can construct an equiangular tight frame $(v_{a,j})$ of $n = \nu(r+1)$ vectors for a space of dimension $d = \frac{\nu(\nu-1)}{k(k-1)}$, via (12.14). Moreover, the inner products (scaled to have unit modulus) are given by (12.15) and (12.16).*

The frame $(v_{a,j})$ above is called a **Steiner equiangular tight frame** (constructed from r-simplices). It is subsimplicial of dimension r.

A natural choice for the Hadamard matrices $H^{(a)}$ in the construction above is the Fourier matrix for an abelian group of order $r+1$. In particular, if $p_1,\ldots,p_j$ are the prime factors of $r+1$ and $m := p_1 p_2 \cdots p_j$, then one can choose $H^{(a)}$ to have m-th roots of unity entries. The scaled inner products of the resulting Steiner equiangular tight frame are then m-th roots of unity and their negatives.

Example 12.20. (Real case) There exists a real Steiner equiangular tight frame if and only if there is a real Hadamard matrix of size $r+1$, e.g., when $r+1=2^j$.

Example 12.21. For the Steiner system $S(2,2,4)$ of our worked example $r=3$, and $H^{(a)}$ can be chosen to be one of

$$\begin{pmatrix} 1 & 1 \\ 1 & -1 \end{pmatrix} \otimes \begin{pmatrix} 1 & 1 \\ 1 & -1 \end{pmatrix} = \begin{pmatrix} 1 & 1 & 1 & 1 \\ 1 & -1 & 1 & -1 \\ 1 & 1 & -1 & -1 \\ 1 & -1 & -1 & 1 \end{pmatrix}, \qquad \begin{pmatrix} 1 & 1 & 1 & 1 \\ 1 & i & -1 & -i \\ 1 & -1 & 1 & -1 \\ 1 & -i & -1 & i \end{pmatrix}.$$

The first choice gives a real Steiner equiangular tight frame of 16 vectors in $\mathbb{R}^6$, i.e.,

$$V = \begin{pmatrix} 1 & -1 & 1 & -1 & 1 & -1 & 1 & -1 & 0 & 0 & 0 & 0 & 0 & 0 & 0 & 0 \\ 1 & 1 & -1 & -1 & 0 & 0 & 0 & 0 & 1 & -1 & 1 & -1 & 0 & 0 & 0 & 0 \\ 1 & -1 & -1 & 1 & 0 & 0 & 0 & 0 & 0 & 0 & 0 & 0 & 1 & -1 & 1 & -1 \\ 0 & 0 & 0 & 0 & 1 & 1 & -1 & -1 & 1 & 1 & -1 & -1 & 0 & 0 & 0 & 0 \\ 0 & 0 & 0 & 0 & 1 & -1 & -1 & 1 & 0 & 0 & 0 & 0 & 1 & 1 & -1 & -1 \\ 0 & 0 & 0 & 0 & 0 & 0 & 0 & 0 & 1 & -1 & -1 & 1 & 1 & -1 & -1 & 1. \end{pmatrix}$$

Example 12.22. The collection $\mathscr{B}$ of all 2-element subsets of $\{1,\ldots,v\}$, $v \geq 2$ is an $S(2,2v)$ Steiner system, with $r+1=v$. The corresponding Steiner equiangular tight frames have $n=v^2$ vectors in a space of dimension $d=\frac{1}{2}v(v-1)$. For $v=3$, taking the Fourier matrix gives

$$A^T = \begin{pmatrix} 1 & 1 & 0 \\ 1 & 0 & 1 \\ 0 & 1 & 1 \end{pmatrix}, \; H^{(a)} = \begin{pmatrix} 1 & 1 & 1 \\ 1 & \omega & \omega^2 \\ 1 & \omega^2 & \omega \end{pmatrix}, \quad \longrightarrow \quad V = \begin{pmatrix} 1 & \omega & \omega^2 & 1 & \omega & \omega^2 & 0 & 0 & 0 \\ 1 & \omega^2 & \omega & 0 & 0 & 0 & 1 & \omega & \omega^2 \\ 0 & 0 & 0 & 1 & \omega^2 & \omega & 1 & \omega^2 & \omega \end{pmatrix},$$

where $\omega = e^{\frac{2\pi i}{3}}$.

Example 12.23. There are eight infinite families of Steiner equiangular tight frames arising from known infinite families of Steiner systems $S(2,k,v)$ (see [FMT12]).

We now use certain Steiner equiangular tight frames to obtain equiangular tight frames with more vectors in a higher dimension.

12.5 Tremain equiangular tight frames

Let $(v_{a,j})$ be the Steiner equiangular tight frame given by a *Steiner triple system* $\mathscr{B}$ on v points $\mathscr{V}$ (blocks of size three) and Hadamard matrices $H^{(a)}$, $a \in \mathscr{V}$. We now outline a construction of [Tre08] (see [FJMP16]) which gives an equiangular tight frame of $\frac{1}{2}(v+1)(v+2)$ vectors for a space of dimension $\frac{1}{6}(v+2)(v+3)$.

For the vectors $v_{a,j} \in \mathbb{C}^{\beta}$, $a \in \mathscr{V}$, $1 \leq j \leq r+1$, given by (12.14), let

$$\tilde{v}_{a,j} := (v_{a,j}, \sqrt{2}h_{0,j}^{(a)}e_a, 0) \in \mathbb{C}^{\mathscr{B}} \oplus \mathbb{C}^{\mathscr{V}} \oplus \mathbb{C}. \tag{12.17}$$

For a (complex) Hadamard matrix $\begin{pmatrix} \alpha_1 & \alpha_2 & \cdots \alpha_{v+1} \\ w_1 & w_2 & \cdots w_{v+1} \end{pmatrix}$, $\alpha_j \in \mathbb{C}$, $w_j \in \mathbb{C}^v$, let

$$\tilde{w}_\ell := (0, \frac{1}{\sqrt{2}}w_\ell, \sqrt{\frac{3}{2}}\alpha_\ell) \in \mathbb{C}^{\mathscr{B}} \oplus \mathbb{C}^{\mathscr{V}} \oplus \mathbb{C}. \tag{12.18}$$

Since $\langle w_\ell, w_m \rangle = -\alpha_\ell \overline{\alpha_m}$, $\ell \neq m$, $\|w_\ell\|^2 = v$, the vectors $(w_\ell)_{1 \leq \ell \leq v+1}$ are said to be the vertices of a *unimodular simplex* in $\mathbb{C}^v$ with *complement* (α_j).

Theorem 12.5. *(Tremain frames) Let $(v_{a,j})$ be the Steiner equiangular tight frame given by a Steiner triple system $\mathscr{B}$ on v points $\mathscr{V}$ and Hadamard matrices $H^{(a)}$, and (w_ℓ) be the vertices of a unimodular simplex for $\mathbb{C}^v$, with complement (α_ℓ). Define $\tilde{v}_{a,j}$ and $\tilde{w}_\ell$ by (12.17) and (12.18). Then*

$$(\tilde{v}_{a,j})_{a \in \mathscr{V}, 1 \leq j \leq r+1} \cup (\tilde{w}_\ell)_{1 \leq \ell \leq v+1} \tag{12.19}$$

is an equiangular tight frame of $\frac{1}{2}(v+1)(v+2)$ vectors for $\mathbb{C}^{\frac{1}{6}(v+2)(v+3)}$, which is subsimplicial of dimension $r+2 = \frac{1}{2}(v+3)$.

Proof. A calculation shows that the vectors are equiangular and tight, with angle $\alpha = \frac{1}{r+2}$ (see Exer. 12.10). □

Subsimplicial equiangular tight frames of the form (12.19) are known as **Tremain (equiangular tight) frames**. Steiner triple systems on v points exist if and only if $v \equiv 1, 3 \pmod 6$, $v \geq 3$. Thus (since Hadamard matrices of all sizes exist), we have

There exists a Tremain equiangular tight frame of $\frac{1}{2}(v+1)(v+2)$ vectors for $\mathbb{C}^{\frac{1}{6}(v+2)(v+3)}$ if and only if $v \equiv 1, 3 \pmod 6$, $v \geq 3$.

Example 12.24. The construction gives a real equiangular tight frame, provided there exists a real Hadamard matrix of size $r+1 = \frac{1}{2}(v+1)$ (and hence one of size $v+1$). Thus there is a Tremain equiangular tight frame of $\frac{1}{2}(v+1)(v+2)$ vectors for $\mathbb{R}^{\frac{1}{6}(v+2)(v+3)}$ if and only if there is a real Hadamard matrix of size $\frac{1}{2}(v+1)$, where $\frac{1}{2}(v+1) \equiv 1, 2 \pmod 3$, e.g., there are 820 equiangular lines in $\mathbb{R}^{287}$ $(v = 39)$.

12.6 Equiangular frames and their signature matrices

Since frames are determined up to unitary equivalence by their Gramian matrices, the Gramian of an equiangular frame Φ with $C > 0$ has the form

$$Q = \begin{pmatrix} 1 & Cz_{12} & Cz_{13} & \cdots & Cz_{1n} \\ C\overline{z_{12}} & 1 & Cz_{23} & \cdots & Cz_{2n} \\ C\overline{z_{13}} & C\overline{z_{23}} & 1 & & \\ \vdots & \vdots & & \ddots & \\ C\overline{z_{1n}} & C\overline{z_{2n}} & & & 1 \end{pmatrix} = I + C\Sigma, \qquad |z_{jk}| = 1.$$

We call any $n \times n$ Hermitian matrix Σ of the above form, i.e., with zero diagonal and off diagonal entries of modulus 1 a **signature matrix**. Since signature matrices are nonzero Hermitian matrices with zero trace, they have at least two eigenvalues: one negative and one positive. Let $\mathbb{F}$ stand for $\mathbb{R}$ or $\mathbb{C}$. The $n \times n$ signature matrices with entries from $\mathbb{F}$ are in 1–1 correspondence with the equiangular frames of n vectors (which are not orthogonal bases).

Theorem 12.6. *(Signature matrices) Let Σ be an $n \times n$ signature matrix (over $\mathbb{F}$), with smallest eigenvalue $-\lambda$ of multiplicity $n-d$, $d \geq 1$, then*

$$Q := I + \frac{1}{\lambda}\Sigma = \frac{1}{\lambda}(\Sigma - (-\lambda)I), \qquad r > 0$$

is the Gramian matrix of an equiangular frame of n vectors for $\mathbb{F}^d$, and every Gramian of an equiangular frame of $n > d$ vectors for $\mathbb{F}^d$ can be constructed in this way. Further, the frame is tight if and only if Σ has (exactly) two eigenvalues λ_1, λ_2, in which case

$$\lambda_1 = -\lambda = -\sqrt{\frac{d(n-1)}{n-d}}, \qquad \lambda_2 = \sqrt{\frac{(n-d)(n-1)}{d}}. \tag{12.20}$$

Proof. By construction, the matrix Q is positive semidefinite of rank $d > 0$ and so has a positive square root $B = Q^{\frac{1}{2}}$. Since $Q = B^2 = B^*B$, Q is the Gramian matrix of the frame given by the columns of B (which span a d-dimensional subspace).

Conversely, a frame of $n > d$ vectors is tight if only if its Gramian has a zero eigenvalue of multiplicity $n-d$ and exactly one nonzero eigenvalue (see Exer. 2.17). Hence an equiangular frame (f_j) of $n > d$ vectors for a d-dimensional space is tight if and only if its signature matrix has exactly two eigenvalues. Moreover, by Exercise 2.16, we have

$$1 = \langle f_j, f_j \rangle = \frac{dA}{n}, \quad \forall j, \qquad \frac{1}{\lambda} = |\langle f_j, f_k \rangle| = \frac{A}{n}\sqrt{\frac{d(n-d)}{n-1}}, \quad j \neq k,$$

which gives the formula for $\lambda = -\lambda_1$. Since $\operatorname{trace}(\Sigma) = (n-d)\lambda_1 + d\lambda_2 = 0$, we have

$$\lambda_2 = -\frac{n-d}{d}\lambda_1 = \sqrt{\frac{(n-d)(n-1)}{d}}.$$

Thus we have (12.20). □

Theorem 12.6 gives rise to a system of $\frac{1}{2}n(n-1)$ equations in the $\frac{1}{2}n(n-1)$ entries of Σ (see [BP05]).

Corollary 12.1. *Let $(z_{jk})_{1\le j<k\le n}$ be scalars of modulus 1, then the signature matrix*

$$\Sigma = \begin{pmatrix} 0 & z_{12} & z_{13} & \cdots & z_{1n} \\ \overline{z_{12}} & 0 & z_{23} & \cdots & z_{2n} \\ \overline{z_{13}} & \overline{z_{23}} & 0 & & \\ \vdots & \vdots & & \ddots & \\ \overline{z_{1n}} & \overline{z_{2n}} & & & 0 \end{pmatrix}$$

gives an equiangular tight frame of $n>d$ vectors for $\mathbb{F}^d$ if and only if

$$\Sigma^2-(\lambda_1+\lambda_2)\Sigma-(n-1)I=0, \qquad \lambda_1+\lambda_2 := (n-2d)\sqrt{\frac{n-1}{d(n-d)}}, \tag{12.21}$$

which is equivalent to

$$(n-2d)\sqrt{\frac{n-1}{d(n-d)}}z_{jk} = \sum_{\ell=1}^{j-1}\overline{z_{\ell j}}z_{\ell k} + \sum_{\ell=j+1}^{k-1} z_{j\ell}z_{\ell k} + \sum_{\ell=k+1}^{n} z_{j\ell}\overline{z_{k\ell}}, \qquad 1\le j<k\le n. \tag{12.22}$$

Proof. The signature matrix has two eigenvalues $\lambda_1=-\lambda$ (with multiplicity $n-d$) and λ_2 if and only if it satisfies the minimal polynomial

$$\Sigma^2-(\lambda_1+\lambda_2)\Sigma+\lambda_1\lambda_2 I=0, \tag{12.23}$$

where $\lambda_1\lambda_2=n-1$. From the entries of the matrix equation (12.23), we therefore obtain n^2 equations in the z_{jk}, with real coefficients depending only on n and d. Those from the diagonal entries hold automatically, and since the (j,k) and (k,j) entries are complex conjugates, we obtain the equivalent system

$$(\lambda_1+\lambda_2)z_{jk} = (\Sigma^2)_{jk}, \qquad 1\le j<k\le n,$$

which can be written as (12.22). □

Example 12.25. For $n=4$, $d=2$, (12.22) gives 6 equations. Let $z_{12}=a$, $z_{13}=b$, $z_{14}=c$. Then the $(j,k)=(1,2)$ and $(1,3)$ equations are

$$z_{13}\overline{z_{23}}+z_{14}\overline{z_{24}}=0, \quad z_{13}z_{23}+z_{14}\overline{z_{34}}=0 \quad \Longrightarrow \quad z_{24}=-\overline{b}cz_{23}, \quad z_{34}=-\overline{a}c\overline{z_{23}}.$$

Making the above substitutions for z_{24} and z_{34} reduces the other 4 equations to one

$$(a\bar{b}z_{23})^2 = -1 \quad \Longrightarrow \quad z_{23} = \pm i\bar{a}b.$$

Hence there is a three-parameter family of unitarily inequivalent equiangular tight frames of four vectors for $\mathbb{C}^2$ given by the signature matrices

$$\Sigma = \begin{pmatrix} 0 & a & b & c \\ \bar{a} & 0 & \pm i\bar{a}b & \mp i\bar{a}c \\ \bar{b} & \mp ia\bar{b} & 0 & \pm i\bar{b}c \\ \bar{c} & \pm ia\bar{c} & \mp ib\bar{c} & 0 \end{pmatrix}, \qquad |a| = |b| = |c| = 1. \tag{12.24}$$

If an $n \times n$ signature matrix Σ satisfies $\Sigma^2 = \mu\Sigma + \tau I$ for some μ, τ, then μ is real and $\tau = n-1$ (by considering entries), so that Σ has two eigenvalues with sum μ. Therefore, Corollary 12.1 can be stated in the following convenient form.

Theorem 12.7. *(Characterisation) Let Σ be an $n \times n$ signature matrix. Then the following are equivalent*

1. *Σ is the signature matrix of an equiangular tight frame for $\mathbb{C}^d$.*
2. *$\Sigma^2 = (n-1)I + \mu\Sigma$ for some necessarily real μ.*
3. *Σ has exactly two eigenvalues (with sum μ).*

Further, when these hold, we have

$$d = \frac{n}{2} - \frac{n\mu}{2\sqrt{4(n-1)+\mu^2}}. \tag{12.25}$$

Proof. Solving

$$\mu^2 = (n-2d)^2 \frac{n-1}{d(n-d)}$$

for d gives

$$d = \frac{n}{2} \pm \frac{n\mu}{2\sqrt{4(n-1)+\mu^2}}.$$

Substituting this formula for d back into

$$\mu = (n-2d)\sqrt{\frac{n-1}{d(n-d)}},$$

shows that the $-$ choice must be made. □

Example 12.26. (Hermitian complex Hadamard matrices) If $\mu = \lambda_1 + \lambda_2 = -2$, so that $\Sigma^2 + 2\Sigma = (n-1)I$, then $\Sigma + I$ is a Hermitian complex Hadamard matrix, via the calculation

$$(\Sigma+I)(\Sigma+I)^* = (\Sigma+I)^2 = \Sigma^2 + 2\Sigma + I = (n-1)I + I = nI.$$

This example generalises to a correspondence between the signature matrices with $|\mu| \leq 2$ and the complex Hadamard matrices with constant diagonal [Szö13].

Corollary 12.2. *(Complex Hadamard matrices) Let Σ be an $n \times n$ signature matrix. Then the following are equivalent*

1. *Σ gives an equiangular tight frame for $\mathbb{C}^d$, where $\frac{n}{2} - \frac{\sqrt{n}}{2} \leq d \leq \frac{n}{2} + \frac{\sqrt{n}}{2}$.*
2. *$\Sigma^2 = (n-1)I + \mu\Sigma$, where $-2 \leq \mu \leq 2$.*
3. *$\Sigma + \zeta I$ is a complex Hadamard matrix for $\zeta = \frac{1}{2}(-\mu \pm \sqrt{4 - |\mu|^2 i})$.*

Proof. The equivalence of the first two follows from the observation that

$$\mu^2 = (n-2d)^2 \frac{n-1}{d(n-d)} \leq 4 \iff \left(d - \frac{n}{2}\right)^2 \leq \frac{n}{4}.$$

($2 \implies 3$) Suppose that 2 holds, then $\zeta = \frac{1}{2}(-\mu \pm \sqrt{4 - |\mu|^2 i})$ is a unit modulus complex number, with $\zeta + \overline{\zeta} = -\mu$, and

$$\begin{aligned}(\Sigma + \zeta I)^*(\Sigma + \zeta I) &= (\Sigma + \overline{\zeta} I)(\Sigma + \zeta I) = \Sigma^2 + (\zeta + \overline{\zeta})\Sigma + I \\ &= ((n-1)I + \mu\Sigma) - \mu\Sigma + I = nI,\end{aligned}$$

i.e., $\Sigma + \zeta I$ is a complex Hadamard matrix.

($3 \implies 2$) Suppose that 3 holds, then $\zeta = \frac{1}{2}(-\mu \pm \sqrt{4 - |\mu|^2 i})$ has unit modulus, and $-2 \leq \mu = -(\zeta + \overline{\zeta}) = 2\Re(-\zeta) \leq 2$, which gives

$$\Sigma^2 = (\Sigma + \zeta I)^*(\Sigma + \zeta I) - (\zeta + \overline{\zeta})\Sigma - I = nI + \mu\Sigma - I = (n-1)I + \mu\Sigma.$$

□

Example 12.27. For a SIC, i.e., d^2 equiangular lines in $\mathbb{C}^d$, the signature matrix satisfies $\Sigma^2 = I$ when $d = 2$ ($\mu = 0$), and $\Sigma^2 = 2I + 2\Sigma$ when $d = 3$ ($\mu = 2$).

Example 12.28. Let Σ be the signature matrix for the equiangular harmonic frame of 7 vectors for $\mathbb{C}^3$ given by the cyclic difference set $J = \{1,2,4\}$. Here $\mu = \frac{1}{\sqrt{2}}$, and we obtain the constant diagonal Hadamard matrix

$$\Sigma + \zeta I = \begin{pmatrix} \zeta & \zeta & \zeta & \overline{\zeta} & \zeta & \overline{\zeta} & \overline{\zeta} \\ \overline{\zeta} & \zeta & \zeta & \zeta & \overline{\zeta} & \zeta & \overline{\zeta} \\ \overline{\zeta} & \overline{\zeta} & \zeta & \zeta & \zeta & \overline{\zeta} & \zeta \\ \zeta & \overline{\zeta} & \overline{\zeta} & \zeta & \zeta & \zeta & \overline{\zeta} \\ \overline{\zeta} & \zeta & \overline{\zeta} & \overline{\zeta} & \zeta & \zeta & \zeta \\ \zeta & \overline{\zeta} & \zeta & \overline{\zeta} & \overline{\zeta} & \zeta & \zeta \\ \zeta & \zeta & \overline{\zeta} & \zeta & \overline{\zeta} & \overline{\zeta} & \zeta \end{pmatrix}, \qquad \zeta := -\frac{1}{2\sqrt{2}} + \frac{\sqrt{7}}{2\sqrt{2}} i,$$

where $\zeta = \frac{1}{\sqrt{2}}(\omega + \omega^2 + \omega^4)$, $\omega = e^{\frac{2\pi i}{7}}$.

For V a complex Hadamard matrix, we now define a variant of the Kronecker product of V and V^* which yields Hadamard matrices with constant diagonal.

Theorem 12.8. *Let $V = [v_1, \ldots, v_m]$ be a complex Hadamard matrix of order m. Then*

$$K = \begin{bmatrix} v_1 v_1^* & \cdots & v_m v_1^* \\ \vdots & & \vdots \\ v_1 v_m^* & \cdots & v_m v_m^* \end{bmatrix} \tag{12.26}$$

is a Hermitian complex Hadamard matrix of order m^2, with 1's on its diagonal. Thus $\Sigma = \pm(K - I)$ is the signature matrix of equiangular tight frame of $n = m^2$ vectors for $\mathbb{C}^d$, where $d = \frac{1}{2}m(m \pm 1)$.

Proof. By construction K is Hermitian with entries of unit modulus, and 1's on its diagonal. The (j,k)-block of K^2 is

$$\sum_\ell v_\ell v_j^* v_k v_\ell^* = (v_j^* v_k) \sum_\ell v_\ell v_\ell^* = (v_j^* v_k) V V^* = (v_j^* v_k) m I = \begin{cases} m^2 I, & j = k; \\ 0, & j \neq k \end{cases}$$

i.e., $K = \Sigma + I$ is a complex Hadamard matrix. Corollary 12.2 implies that Σ is the signature matrix of an equiangular tight frame of $n = m^2$ vectors for $\mathbb{C}^d$, where $\mu = -2$, and, by (12.25),

$$d = \frac{n}{2} - \frac{n\mu}{2\sqrt{4(n-1)+\mu^2}} = \frac{n}{2} + \frac{\sqrt{n}}{2} = \frac{1}{2}m(m+1).$$

The complementary equiangular tight frame for $d = m^2 - \frac{1}{2}m(m+1) = \frac{1}{2}m(m-1)$ dimensions has signature matrix $-(K - I)$. □

Example 12.29. (Cube root signature matrices) Taking V to be the 3×3 Hadamard matrix given by the Fourier matrix gives

$$V = \begin{pmatrix} 1 & 1 & 1 \\ 1 & \omega & \omega^2 \\ 1 & \omega^2 & \omega \end{pmatrix}, \quad \omega = e^{\frac{2\pi i}{3}} \qquad \longrightarrow \qquad K = \begin{pmatrix} 1 & 1 & 1 & 1 & 1 & 1 & 1 & 1 & 1 \\ 1 & 1 & 1 & \omega & \omega & \omega & \omega^2 & \omega^2 & \omega^2 \\ 1 & 1 & 1 & \omega^2 & \omega^2 & \omega^2 & \omega & \omega & \omega \\ 1 & \omega^2 & \omega & 1 & \omega & \omega^2 & 1 & \omega & \omega^2 \\ 1 & \omega^2 & \omega & \omega^2 & 1 & \omega & \omega & \omega^2 & 1 \\ 1 & \omega^2 & \omega & \omega & \omega^2 & 1 & \omega^2 & 1 & \omega \\ 1 & \omega & \omega^2 & 1 & \omega^2 & \omega & 1 & \omega^2 & \omega \\ 1 & \omega & \omega^2 & \omega^2 & \omega & 1 & \omega & 1 & \omega^2 \\ 1 & \omega & \omega^2 & \omega & 1 & \omega^2 & \omega^2 & \omega & 1 \end{pmatrix}$$

with $K - I$ the signature matrix for an equiangular tight frame of 9 vectors in $\mathbb{C}^6$ and $-K + I$ the signature matrix for an equiangular tight frame of 9 vectors in $\mathbb{C}^3$.

12.7 The reduced signature matrix and projective unitary equivalence

Recall from §2.3 (and §8.1) that frames Ψ are Φ are projectively unitarily equivalent if and only if their Gramians satisfy

$$\mathrm{Gram}(\Psi) = \Lambda^* \mathrm{Gram}(\Phi)\Lambda, \qquad \Lambda = \mathrm{diag}(\alpha_j),$$

where $|\alpha_j| = 1, \forall j$. For equiangular frames, this condition can be expressed in terms of the signature matrices

$$\Sigma_\Psi = \Lambda^* \Sigma_\Phi \Lambda.$$

Hence every equiangular frame is projectively unitarily equivalent to one with a signature matrix of the form

$$\Lambda\Sigma\Lambda^* = \begin{pmatrix} 0 & \mathbf{1}^* \\ \mathbf{1} & \hat{\Sigma} \end{pmatrix}, \qquad \Lambda := \mathrm{diag}(1, z_{12}, z_{13}, \ldots, z_{1n}), \quad \mathbf{1} := (1, 1, \ldots, 1)^*. \tag{12.27}$$

We will call the above matrix $\hat{\Sigma}$ the **reduced signature matrix** of the equiangular frame (and its projective unitary equivalence class).

An equiangular frame is uniquely determined by its reduced signature matrix.

Theorem 12.9. *(Triple products) An equiangular frame is uniquely determined up to projective unitary equivalence by its reduced signature matrix.*

Proof. Each equiangular frame Φ has a reduced signature matrix $\hat{\Sigma}$. The matrix

$$\Sigma := \begin{pmatrix} 0 & \mathbf{1}^* \\ \mathbf{1} & \hat{\Sigma} \end{pmatrix}$$

is the signature matrix of an equiangular frame Ψ which is projectively unitarily equivalent to Φ. It therefore suffices to show that the entries of the above Σ are projectively unitarily invariant. Without loss of generality, assume that $\Phi = (v_j)$ has be scaled so that $\langle v_j, v_k \rangle = \Sigma_{kj}$. Then

$$\langle v_1, v_j \rangle = 1, \qquad \forall j.$$

Thus the (k, j)-entry of Σ can be written as

$$\langle v_j, v_k \rangle = \langle v_1, v_j \rangle \langle v_j, v_k \rangle \langle v_k, v_1 \rangle.$$

Clearly, the "triple products" on the right-hand side depend only on (v_j) up to projective unitary equivalence (see §for the general theory). □

Example 12.30. The reduced signature matrix for the Σ of (12.24) is

$$\hat{\Sigma} = \begin{pmatrix} 0 & \pm i & \mp i \\ \mp i & 0 & \pm i \\ \pm i & \mp i & 0 \end{pmatrix}. \tag{12.28}$$

Thus there are just *two* equiangular tight frames of four vectors in $\mathbb{C}^2$ up to projective unitary equivalence (and reordering). Moreover, these can be obtained from each other by taking the entrywise complex conjugate of the reduced signature matrix.

Example 12.31. The complement of a *tight* equiangular frame is a tight equiangular frame. If its signature and reduced signature matrices are Σ and $\hat{\Sigma}$, then those of the complementary equiangular tight frame are $-\Sigma$ and $-\hat{\Sigma}$.

By Theorem 12.6, the construction of equiangular frames with large numbers of vectors (compared to the dimension) is equivalent to finding signature matrices Σ whose smallest eigenvalue has a large multiplicity. We now seek to translate this to the reduced signature matrix $\hat{\Sigma}$.

Proposition 12.2. *(Spectral structure of $\hat{\Sigma}$) Let $\hat{\Sigma}$ be the reduced signature matrix of a Σ satisfying (12.27), and v be a λ-eigenvector of $\hat{\Sigma}$. Then*

- $(0,v)$ *is a λ-eigenvector of $\Lambda\Sigma\Lambda^*$ if and only if $v \perp \mathbf{1}$.*
- $(a,\mathbf{1})$ *is an eigenvector of $\Lambda\Sigma\Lambda^*$ if and only if $\mathbf{1}$ is a β-eigenvector of $\hat{\Sigma}$ and*

$$a = \frac{1}{2}\left(-\beta \pm \sqrt{\beta^2 + 4(n-1)}\right). \tag{12.29}$$

Proof. The first follows since

$$\begin{pmatrix} 0 & \mathbf{1}^* \\ \mathbf{1} & \hat{\Sigma} \end{pmatrix} \begin{pmatrix} 0 \\ v \end{pmatrix} = \begin{pmatrix} \mathbf{1}^* v \\ \hat{\Sigma} v \end{pmatrix} = \begin{pmatrix} \mathbf{1}^* v \\ \lambda v \end{pmatrix}.$$

For the second, observe that

$$\begin{pmatrix} 0 & \mathbf{1}^* \\ \mathbf{1} & \hat{\Sigma} \end{pmatrix} \begin{pmatrix} a \\ \mathbf{1} \end{pmatrix} = \begin{pmatrix} \mathbf{1}^*\mathbf{1} \\ a\mathbf{1} + \hat{\Sigma}\mathbf{1} \end{pmatrix} = \begin{pmatrix} n-1 \\ a\mathbf{1} + \hat{\Sigma}\mathbf{1} \end{pmatrix}.$$

Thus $(a,\mathbf{1})$ can be an eigenvector only if $a\mathbf{1} + \hat{\Sigma}\mathbf{1}$ is a multiple of $\mathbf{1}$, i.e., $\mathbf{1}$ is a β-eigenvector of $\hat{\Sigma}$. In this case, $(a,\mathbf{1})$ is a α-eigenvector if and only if

$$\alpha a = n-1, \qquad (a+\beta) = \alpha,$$

which gives (12.29), where $\alpha = a + \beta$. □

Since the eigenspaces of the Hermitian matrix $\hat{\Sigma}$ are orthogonal, we obtain a correspondence between the eigenstructures of Σ and $\hat{\Sigma}$ when $\mathbf{1}$ is a μ-eigenvector of $\hat{\Sigma}$.

Example 12.32. The reduced signature matrix $\hat{\Sigma}$ of (12.28) has eigenvector $\mathbf{1}$ for eigenvalue $\beta = 0$, and the other eigenvalues are $\pm\sqrt{3}$. Thus the eigenvalues of any Σ (of size $n = 4$) with this reduced signature matrix are

$$\alpha = a + \beta = \frac{1}{2}\left(\beta \pm \sqrt{\beta^2 + 4(n-1)}\right) = \pm\sqrt{3},$$

together with $\pm\sqrt{3}$ (which happen to be the same).

The factorisation (12.27) allows the number of variables in (12.22) to be reduced by $n-1$ to $\frac{1}{2}(n-1)(n-2)$. We now express these equations in terms of the reduced signature matrix $\hat{\Sigma}$ and show that $\mathbf{1}$ must be an eigenvector of $\hat{\Sigma}$.

Proposition 12.3. *There exists an equiangular tight frame of $n > d$ vectors for $\mathbb{F}^d$ with reduced signature matrix $\hat{\Sigma}$ if and only if*

$$(\lambda_1 + \lambda_2)\hat{\Sigma} = \hat{\Sigma}^2 + J - (n-1)I, \qquad \lambda_1 + \lambda_2 := (n-2d)\sqrt{\frac{n-1}{d(n-d)}}, \qquad J := \mathbf{1}\mathbf{1}^*,$$

and $\mathbf{1}$ is an eigenvector of $\hat{\Sigma}$ for the eigenvalue $\lambda_1 + \lambda_2$.

Proof. Substitute $\Sigma = \begin{pmatrix} 0 & \mathbf{1}^* \\ \mathbf{1} & \hat{\Sigma} \end{pmatrix}$ into (12.23)

$$\begin{pmatrix} n-1 & \mathbf{1}^*\hat{\Sigma} \\ \hat{\Sigma}\mathbf{1} & J + \hat{\Sigma}^2 \end{pmatrix} - (\lambda_1 + \lambda_2)\begin{pmatrix} 0 & \mathbf{1}^* \\ \mathbf{1} & \hat{\Sigma} \end{pmatrix} - (n-1)\begin{pmatrix} 1 & 0 \\ 0 & I \end{pmatrix} = 0,$$

and equate the blocks. □

Example 12.33. (The simplex) Up to projective unitary equivalence and reordering, there is a unique equiangular tight frame of $n = d+1$ vectors for $\mathbb{R}^d$ or $\mathbb{C}^d$, which is given by the vertices of the regular simplex. For $n = d+1$, the $d \times d$ reduced signature matrix $\hat{\Sigma}$ has $\mathbf{1}$ as an eigenvector for $\lambda_1 + \lambda_2 = -(d-1)$, and so all its off diagonal entries must be -1. In particular, there is a unique equiangular tight frame of three vectors in $\mathbb{C}^2$, which is a real frame (the Mercedes–Benz frame).

In summary:

Each equiangular frame is uniquely determined up to unitary equivalence by its signature matrix, and it is uniquely determined up to projective unitary equivalence by its reduced signature matrix.

12.8 The relative bound on the number of real equiangular lines

Equiangular lines in $\mathbb{R}^d$ are given by a set of unit vectors $\{f_j\}$ with

$$|\langle f_j, f_j \rangle| = \alpha = \cos\theta, \qquad j \neq k.$$

The variational characterisation of tight frames (see Theorem 6.1) immediately gives the following estimate on the maximum number of equiangular lines.

Theorem 12.10. *(Relative bound) Suppose that $\{f_j\}$ gives n equiangular lines in $\mathbb{R}^d$ with angle $\alpha = \cos\theta$. If $\alpha < 1/\sqrt{d}$, then*

$$n \leq \frac{d - d\alpha^2}{1 - d\alpha^2}, \tag{12.30}$$

with equality if and only if $\{f_j\}$ is a tight frame for $\mathbb{R}^d$.

Proof. By the variational characterisation, we have

$$\sum_j \sum_k |\langle f_j, f_k \rangle|^2 = n + (n^2 - n)\alpha^2 \geq \frac{1}{d} n^2 = \frac{1}{d} \Big(\sum_j \|f_j\|^2 \Big)^2,$$

with equality if and only if $\{f_j\}$ is a tight frame. The above inequality is equivalent to

$$n\Big(\frac{1}{d} - \alpha^2\Big) \leq 1 - \alpha^2.$$

Hence if $\alpha < 1/\sqrt{d}$, then we can divide (without changing the sign) to obtain the result. □

Example 12.34. The isogonal vectors of Example 3.9 give d equiangular lines in $\mathbb{R}^d$ with $\frac{1}{\sqrt{d}} \leq \alpha < 1$.

We observe that the relative bound is not a generalisation of the absolute bound (Theorem 12.2). Indeed, it also holds (by the proof) for complex equiangular lines.

In the algebraic graph theory literature (12.30) is called the **relative bound**, and the tight frame condition is usually stated as the orthogonal projections $P_j := f_j f_j^*$ satisfy

$$\sum_{j=1}^n P_j = \frac{n}{d} I.$$

Thus tightness (equality) in (12.30) occurs if and only the frame $\{f_j\}$ is tight.

We say that a set of equiangular lines in $\mathbb{R}^d$ is **tight** if the vectors defining them are a tight frame, equivalently, there is equality in the relative bound.

The relative bound is very useful when used in conjunction with the following.

Theorem 12.11. *Suppose there are n equiangular lines in $\mathbb{R}^d$ with angle $\alpha = \cos\theta$. If $n > 2d$, then $1/\alpha$ is an odd integer.*

Proof. Let $Q = I + \alpha\Sigma$ be the Gramian of n unit vectors giving the equiangular lines in $\mathbb{R}^d$. Since these vectors span a subspace of dimension $d' \le d$, by Theorem 12.6, the signature matrix

$$\Sigma = \frac{1}{\alpha}(Q - I)$$

has smallest eigenvalue $-1/\alpha$, with multiplicity $m = n - d' \ge n - d$. Since Σ is an integer matrix, its eigenvalues are algebraic integers. The eigenvalue $-1/\alpha$ must be rational and hence is an integer. Otherwise, the conjugate of $-1/\alpha$ would be a second eigenvalue with multiplicity m, so $n \ge 2m \ge 2(n-d)$, which contradicts $n > 2d$.

To see $1/\alpha$ is odd, consider the integer matrix

$$B = \frac{1}{2}\big(\Sigma + \mathbf{1}\mathbf{1}^* + I\big).$$

Since Σ has a $-1/\alpha$ eigenspace of dimension $m \ge n - d > 2d - d \ge 1$ and $\mathbf{1}\mathbf{1}^*$ has a 0 eigenspace of dimension $n-1$, the matrix B has a $\beta = \frac{1}{2}\big(-\frac{1}{\alpha} + 0 + 1\big)$ eigenvalue. Since β is a rational algebraic integer, it is an integer, and so $1/\alpha = -2\beta + 1$ is odd. □

Unlike the relative bound, Theorem 12.11 does not hold for complex equiangular lines in $\mathbb{C}^d$, e.g., for $n = d^2$ such lines $1/\alpha = \sqrt{d+1}$.

Example 12.35. If the absolute and relative bounds hold for $n > 2d$, then

$$d = \Big(\frac{1}{\alpha}\Big)^2 - 2, \qquad \text{where } \frac{1}{\alpha} \text{ is an odd integer,}$$

i.e., $d = 7, 23, 47, 79, \ldots$. Thus the absolute bound (Theorem 12.2) for the maximal number of equiangular lines in $\mathbb{R}^d$ can only hold when $d = 2, 3$ or $d+2$ is the square of an odd integer.

The maximum size $M(d)$ of a set of n real equiangular lines in $\mathbb{R}^d$ has been investigated since [Haa48] showed that $M(2) = 3$ (the Mercedes–Benz frame) and $M(3) = 6$ (lines through the opposite vertices of an icosahedron). As of [BY14], [GKMS16], the best estimates of $M(d)$, $d \le 47$ are given in Table 12.3.

We observe (see Example 12.41) that the first instance where $M(d)$ is known to be not attained by a tight frame (in a possibly lower dimension) is $d = 16$.

A nontight set of $\frac{2}{9}(d+1)^2$ equiangular lines for $\mathbb{C}^d$ was constructed by de Caen [dC00] for $d = \frac{3}{2}4^t - 1$, $t \ge 1$. By using the existence of a set of $\frac{d}{2} + 1$ mutually unbiased bases in $\mathbb{R}^d$ for $d = 4^t$, $t \ge 1$, [GKMS16] adapted this example to construct nontight equiangular lines in $\mathbb{C}^d$ giving the following lower bound

$$M(d) \ge \frac{32d^2 + 328d + 296}{1089} > \frac{(d+2)^2}{72}.$$

Table 12.3: The maximum number $M(d)$ of equiangular lines in $\mathbb{R}^d$, $d \leq 47$. The last column gives the number of vectors in tight frame in $\mathbb{R}^d$ given by a strongly regular graph (see §12.10). When the existence of the graph (and hence the equiangular lines) is still unknown, then it is in grey.

d	$M(d)$	$\frac{1}{\alpha}$	tight	d	$M(d)$	$\frac{1}{\alpha}$	tight
2	3	2	3	19	72–75	5	
3	6	$\sqrt{5}$	6	20	90–95	5	
4	6	$\sqrt{5},3$		21	126	5	126
5	10	3	10	22	176	5	176
6	16	3	16	23–41	276	5	276
7–13	28	3	28	42	276–288	5,7	288
14	28–29	3,5		43	344	7	344
15	36	5	36	44	344–422	7	
16	40–41	5		45	344–540	7	540
17	48–50	5		46	344–736	7	736
18	54–60	5		47	344–1127	7	

12.9 The connection with algebraic graph theory

The **Seidel matrix** $\Sigma = \text{Seid}(\Gamma)$ of a graph Γ with n vertices is the $n \times n$ matrix with a -1 in the (j,k)-entry if the j and k vertices are adjacent (connected by an edge), a 1 if they are nonadjacent, and 0 diagonal entries. Clearly, Seidel matrices are signature matrices over $\mathbb{R}$ and vice versa.

For $\mathbb{F} = \mathbb{R}$, there are *finitely* many possible $n \times n$ signature matrices and hence finitely many real equiangular frames of n vectors, each in 1–1 correspondence with a graph on n vertices, namely the graph whose Seidel matrix is its signature matrix. This correspondence between vectors defining a set of real equiangular lines and a graph goes back to the foundation of algebraic graph theory (see [GR01]).

The set of n vectors $\{f_j\}$ defining a set of equiangular lines in $\mathbb{R}^d$ can each be multiplied by $\sigma_j = \pm 1$ to obtain a set defining the *same* equiangular lines, but possibly with *different* Gramian matrices

$$\text{Gram}(\{\sigma_j f_j\}) = \Lambda \text{Gram}(\{f_j\}) \Lambda, \qquad \Lambda := \text{diag}(\sigma_1, \ldots, \sigma_n), \tag{12.31}$$

and hence possibly *different* corresponding graphs. Any two graphs related in this way (for some ordering of their points) are said to be **switching equivalent** (by **switching** on the vertices $\{j : \sigma_j = -1\}$). This is an equivalence relation. Switching on a vertex j of a graph entails changing all edges from j to nonedges and all nonedges from j to edges. The set of all graphs switching equivalent to Γ is called the **switching class**[2] of Γ In view of (12.31), $\mathrm{Seid}(\{\sigma_j f_j\}) = \Lambda^{-1}\mathrm{Seid}(\{f_j\})\Lambda$, and so the eigenvalues of the Seidel matrices of any two switching equivalent graphs are the same (reindexing a set of vectors does not change the eigenvalues of their Gramian). The eigenvalues of the Seidel matrix are called the **Seidel spectrum** of the graph (or its by switching class).

Each real equiangular frame which is not a basis is uniquely determined up to projective unitary equivalence and reordering by a switching class of graphs (the one containing a graph whose Seidel matrix is its signature matrix). It is tight if and only if its Seidel matrix has exactly two eigenvalues.

Example 12.36. (The simplex) Let Γ be the complete graph K_n on $n \geq 2$ vertices. Then its Seidel matrix Σ has *two* eigenvalues: $-(n-1)$ of multiplicity 1 and 1. The corresponding tight frame of n vectors in $\mathbb{R}^{n-1}$ (cf. Example 12.33) is the vertices of a regular simplex. Similarly, its complement, the empty graph on n vertices, gives the equiangular tight frame for $\mathbb{R}^1$ consisting of a nonzero vector repeated n times. By switching on a vertex of K_n one obtains the graph consisting of a point together with K_{n-1}. For $n = 3$, Γ is the 3-cycle (see Figure 12.1). Its Seidel matrix S and those obtained by switching on the first, second and third vertices are

$$S = \begin{pmatrix} 0 & -1 & -1 \\ -1 & 0 & -1 \\ -1 & -1 & 0 \end{pmatrix}, \quad \begin{pmatrix} 0 & 1 & 1 \\ 1 & 0 & -1 \\ 1 & -1 & 0 \end{pmatrix}, \begin{pmatrix} 0 & 1 & -1 \\ 1 & 0 & 1 \\ -1 & 1 & 0 \end{pmatrix}, \begin{pmatrix} 0 & -1 & 1 \\ -1 & 0 & 1 \\ 1 & 1 & 0 \end{pmatrix}.$$

Fig. 12.1: The graphs in the switching class of the complete graph K_3 and the corresponding equiangular (tight) frames of three vectors in $\mathbb{R}^2$.

Example 12.37. For $n = 4$, the eleven graphs lie in three switching classes. The one containing the complete graph gives the vertices of the regular tetrahedron, and the one containing the empty graph gives equiangular lines in $\mathbb{R}^1$. The third switching class, which contains the *path graph*, gives four lines at an angle of $\cos^{-1}(\frac{1}{\sqrt{5}}) \approx 63.4°$, which we recognise as diagonals of the regular icosahedron (see Example 12.44).

[2] The *switching class* is also known as a *two-graph*.

12.10 Real equiangular tight frames and strongly regular graphs

Let Σ be the Seidel matrix of a graph Γ, and $\hat{\Sigma}$ given by (12.27) be its *reduced Seidel matrix*. The condition that Σ have two eigenvalues, and hence give a real equiangular tight frame, is most easily expressed in terms of the graph with Seidel matrix $\hat{\Sigma}$ (which can depend on the ordering of the points of Γ). This graph must be *strongly regular*, with particular parameters (Theorem 12.12).

A regular graph of degree k with v vertices is said to be **strongly regular**, or a $\mathrm{srg}(v,k,\lambda,\mu)$, if there are integers λ and μ such that

- Every two adjacent vertices have λ common neighbours.
- Every two nonadjacent vertices have μ common neighbours.

The adjacency matrix $A = \mathrm{Adj}(\Gamma)$ (which has a 1 for adjacency, and a 0 otherwise) of a strongly regular graph Γ which is not complete or empty is characterised by

$$AJ = kJ, \qquad A^2 + (\mu-\lambda)A + (\mu-k)I = \mu J, \tag{12.32}$$

where $J = J_v$ is the $v \times v$ matrix of all 1's and $I = I_v$ is the identity.

The eigenvalues of the adjacency matrix A are

$$\theta = \frac{\lambda-\mu+\sqrt{\Delta}}{2}, \quad \tau = \frac{\lambda-\mu-\sqrt{\Delta}}{2}, \qquad \Delta := (\lambda-\mu)^2 + 4(k-\mu), \tag{12.33}$$

with multiplicities

$$m_\theta = \frac{1}{2}\Big(v-1-\frac{2k+(v-1)(\lambda-\mu)}{\sqrt{\Delta}}\Big), \qquad m_\tau = \frac{1}{2}\Big(v-1+\frac{2k+(v-1)(\lambda-\mu)}{\sqrt{\Delta}}\Big), \tag{12.34}$$

and k with eigenvector $\mathbf{1}$ (see [GR01]). Since the trace of A is zero, its smallest eigenvalue is $\tau < 0$. We note that the adjacency and Seidel matrices $A = \mathrm{Adj}(\Gamma)$ and $\Sigma = \mathrm{Seid}(\Gamma)$ of a graph Γ are related by $A = \frac{1}{2}(J-I-\Sigma)$, $\Sigma = J-I-2A$. The spectral structure of one can be deduced from the other when Γ is k-regular, i.e., $\mathbf{1} = (1,\ldots,1)$ is an eigenvector of A for eigenvalue k (cf. Proposition 12.2).

Theorem 12.12. *Let Σ be the Seidel matrix of a graph Γ on n vertices which is not switching equivalent to the complete or empty graph, and $\hat{\Sigma}$ be given by (12.27). Then Σ has two eigenvalues (and so corresponds with an equiangular tight frame of $n > d+1$ vectors for $\mathbb{R}^d$) if and only if $\hat{\Sigma}$ is the Seidel matrix of a strongly regular graph $\hat{\Gamma}$ of the type*

$$\mathrm{srg}(n-1,k,\lambda,\mu), \qquad \lambda = \frac{3k-n}{2}, \qquad \mu = \frac{k}{2}. \tag{12.35}$$

The n, k, d above are related as follows

$$d = \frac{1}{2}n - \frac{1}{2}\frac{n(n-2k-2)}{\sqrt{(n-2k)^2+8k}} > 1, \qquad k = \frac{1}{2}n - 1 + \Big(1-\frac{n}{2d}\Big)\sqrt{\frac{d(n-1)}{n-d}}. \tag{12.36}$$

Proof. By Proposition 12.3, Σ has two eigenvalues (i.e., gives rise to an equiangular tight frame for $\mathbb{R}^d$), if and only if $\mathbf{1}$ is an eigenvector of $\hat{\Sigma}$ for the eigenvalue $\lambda_1+\lambda_2$, and

$$(\lambda_1+\lambda_2)\hat{\Sigma}=\hat{\Sigma}^2+J-(n-1)I, \qquad J:=\mathbf{1}\mathbf{1}^*. \tag{12.37}$$

Thus $\mathbf{1}$ is an eigenvector of the adjacency matrix $A=\frac{1}{2}(J-I-\hat{\Sigma})$ of $\hat{\Gamma}$ for the eigenvalue

$$k=\frac{(n-1)-1-(\lambda_1+\lambda_2)}{2}=\frac{1}{2}n-1+\Big(1-\frac{n}{2d}\Big)\sqrt{\frac{d(n-1)}{n-d}}, \tag{12.38}$$

with k a positive integer (since the nonzero entries of A are 1). Hence $\hat{\Gamma}$ is a regular graph of degree k, which is not the complete or empty graph (by our assumption). Using $\hat{\Sigma}=J-I-2A$, $AJ=JA=kJ$ and (12.38), we can rewrite (12.37) as

$$A^2+\Big(\frac{n}{2}-k\Big)A-\frac{k}{2}I=A^2+\Big(\frac{k}{2}-\frac{3k-n}{2}\Big)A+\Big(\frac{k}{2}-k\Big)I=\frac{k}{2}J,$$

which is equivalent to $\hat{\Gamma}$ being a $\operatorname{srg}(n-1,k,\lambda,\mu)$, $\lambda=\frac{1}{2}(3k-n)$, $\mu=\frac{k}{2}$.

Finally, solving (12.38) for d gives

$$d=\frac{1}{2}n\pm\frac{1}{2}\frac{n(n-2k-2)}{\sqrt{(n-2k)^2+8k}},$$

with the choice of sign determined by the multiplicities of the eigenvalues of Σ. More precisely, d is the multiplicity of the largest eigenvalue of Σ, which by Proposition 12.2 and $A=\frac{1}{2}(J-I-\hat{\Sigma})$ is 1 plus the multiplicity of the smallest eigenvalue of A, i.e.,

$$1+\frac{1}{2}\left(n-2+\frac{2k+(n-2)(\frac{1}{2}(2k-n))}{\sqrt{\frac{1}{4}(2k-n)^2+4\frac{1}{2}k}}\right),$$

which simplifies to the formula for d in (12.36). □

The linear equations of (12.35) can be solved for n and k, giving

$$n=-2\lambda+6\mu, \qquad k=2\mu.$$

In this way, the equiangular tight frames be indexed by the pair of integers (λ,μ).

Example 12.38. (Complements) Let $\hat{\Sigma}$ be the reduced signature matrix of an equiangular tight frame of $n>d+1$ vectors for $\mathbb{R}^d$, which corresponds to a strongly regular graph with the parameters (12.35). The complementary equiangular tight frame of n vectors for $\mathbb{R}^{n-d}$ has reduced signature matrix $-\hat{\Sigma}$ (see Exer. 12.14) and so corresponds to the complementary strongly regular graph, which has the parameters

$$\operatorname{srg}\Big(n-1,n-k-2,\frac{2n-3k-6}{2},\frac{n-k-2}{2}\Big)$$

where k is given by (12.36).

By Theorem 12.12, the existence and construction of equiangular tight frames in $\mathbb{R}^d$ can be expressed in terms of strongly regular graphs with certain parameters.

Strongly regular graphs which are the union m complete graphs on w vertices, and their complements, are considered *boring*. The strongly regular graphs which are not boring are said to be **primitive**. These graphs have parameters

$$\mathrm{srg}(mw, w-1, w-2, 0), \qquad \mathrm{srg}(mw, (m-1)w, (m-2)w, (m-1)w),$$

and so give rise to equiangular tight frames (as above) if and only if $m = w = 1$.

Corollary 12.3. *There exists an equiangular tight frame of $n > d+1$ vectors for $\mathbb{R}^d$ if and only if there exists a strongly regular graph $\hat{\Gamma}$, with Seidel matrix $\hat{\Sigma}$, of the type*

$$\mathrm{srg}(n-1, k, \frac{3k-n}{2}, \frac{k}{2}), \qquad k := \frac{1}{2}n - 1 + \left(1 - \frac{n}{2d}\right)\sqrt{\frac{d(n-1)}{n-d}}.$$

Moreover, all graphs Γ giving an equiangular tight frame of $n > d+1$ vectors for $\mathbb{R}^d$ have Seidel matrices of the form

$$P^{-1}\Lambda^{-1}\begin{pmatrix} 0 & \mathbf{1}^* \\ \mathbf{1} & \hat{\Sigma} \end{pmatrix}\Lambda P, \qquad \Lambda = \mathrm{diag}(\sigma_1, \ldots, \sigma_n), \quad \sigma_j = \pm 1, \tag{12.39}$$

where P is a permutation matrix and $\hat{\Sigma}$ is as above. In particular, we can take Γ to be $\hat{\Gamma}$ together with an isolated point.

Example 12.39. The switching class of the graph Γ obtained by adding an isolated point to the unique $\mathrm{srg}(5,2,0,1)$, i.e., the 5-cycle, is given in the Figure 12.2. The corresponding tight frame of $n = 6$ vectors for $\mathbb{R}^3$ consists of vectors which are on the six diagonals of the regular icosahedron.

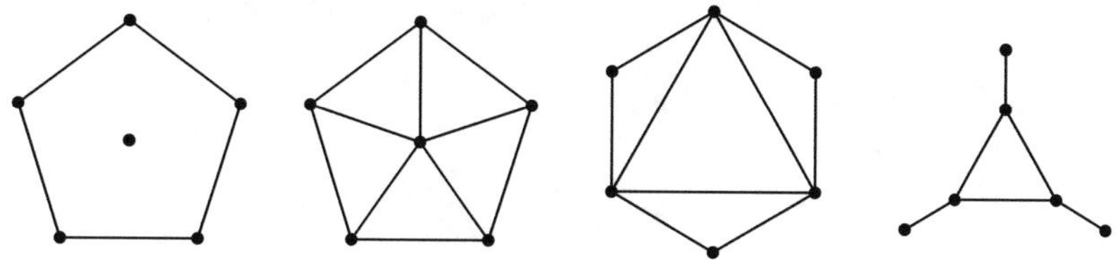

Fig. 12.2: The switching class of the graph Γ consisting of a 5-cycle and an isolated point.

We observe that graph Γ consisting of a strongly regular graph $\hat{\Gamma}$ together with an isolated point is switching equivalent to $\hat{\Gamma}$ together with a point which is adjacent to all points of $\hat{\Gamma}$.

The switching class of a graph Γ giving an equiangular tight frame of n vectors is called a **regular two-graph**. It may contain graphs consisting of an *isolated point* (one with no neighbours) together various nonisomorphic strongly regular graphs on $n-1$ points (called *neighbourhoods*). This is sometimes used as a method to construct strongly regular graphs. Thus reordering the vectors of an equiangular tight frame may yield a reduced signature matrix for a different strongly regular graph (with the same parameters).

Since the unique equiangular tight frame of $n = d+1$ vectors in $\mathbb{R}^d$ (the vertices of the simplex) is given by the boring strongly regular graph

$$\hat{\Gamma} = K_{n-1} = \mathrm{srg}(n-1, n-2, n-3, 0) \qquad \text{(the complete graph)},$$

we can summarise our results as follows.

Up to projective unitary equivalence and reordering, each equiangular tight frame (which isn't an orthonormal basis) corresponds to a collection of strongly regular graphs. These strongly regular graphs have the property that the graphs obtained by adding an isolated point are switching equivalent.

A list of the known real equiangular tight frames (for $d \le 50$) obtained by this correspondence is given in Table 12.5.

The construction of Theorem 12.12 also gives *nontight* equiangular frames.

Corollary 12.4. *Let $\hat{\Sigma}$ be the Seidel matrix of a strongly regular graph $\hat{\Gamma}$ of the type*

$$\mathrm{srg}(n-1, k, \lambda, \mu), \qquad \lambda \ne \frac{3k-n}{2}. \tag{12.40}$$

Then there exists a nontight equiangular tight frame of $n > d+1$ vectors for $\mathbb{R}^d$ with reduced signature matrix $\hat{\Sigma}$ if and only if

$$-1 - \lambda + \mu - \sqrt{\Delta} < \frac{1}{2}\left(n - 2 - 2k - \sqrt{(n-2-2k)^2 + 4(n-1)}\right),$$

where $\Delta := (\lambda - \mu)^2 + 4(k - \mu)$, and

$$d = \frac{1}{2}\left(n + 2 + \frac{2k + (n-2)(\lambda - \mu)}{\sqrt{\Delta}}\right).$$

Proof. Let $\nu = n-1$, then by (12.33), (12.34) and Proposition 12.2, the eigenvalues of Σ are $-1 - 2\theta = -1 - \lambda + \mu - \sqrt{\Delta}$, $-1 - 2\tau = -1 - \lambda + \mu + \sqrt{\Delta}$, and

$$\beta + \frac{1}{2}\left(-\beta \pm \sqrt{\beta^2 + 4(n-1)}\right), \qquad \beta := n - 2 - 2k = \nu - 1 - 2k,$$

with multiplicities $m_\theta, m_\tau, 1, 1$.

The condition for $-1 - 2\theta$ to be the smallest eigenvalue is

$$-1-2\theta=-1-\lambda+\mu-\sqrt{\Delta}\leq\frac{1}{2}\big(\beta-\sqrt{\beta^2+4(n-1)}\big).$$

If there is equality, then the frame is tight, and if the inequality doesn't hold, then Σ has rank $n-1$ and so corresponds to the $n=d+1$ vertices of the simplex. For strict inequality $n-d=m_\theta$, which completes the proof. □

Table 12.4: Selected examples of nontight equiangular frames of $n>d+1$ vectors for $\mathbb{R}^d$ given by Corollary 12.4, including the Seidel spectrum of Σ.

n	d	srg parameters	eigenvalues	multiplicities
11	6	(10, 3, 0, 1)	$-3,3,-2,5$	$5,4,1,1$
17	7	(16, 5, 0, 2)	$-3,5,\frac{1}{2}(1\pm\sqrt{89})$	$10,5,1,1$
17	11	(16, 6, 2, 2)	$-5,3\ \frac{1}{2}(3\pm\sqrt{73})$	$6,9,1,1$
27	14	(26, 10, 3, 4)	$-5,5,\frac{1}{2}(5\pm\sqrt{129})$	$13,12,1,1$
37	14	(36, 14, 4, 6)	$-5,7,\frac{1}{2}(7\pm\sqrt{193})$	$21,14,1,1$
41	17	(40, 12, 2, 4)	$-5,7,\frac{1}{2}(15\pm\sqrt{385})$	$24,15,1,1$
65	20	(64, 18, 2, 6)	$-5,11,\frac{1}{2}(27\pm\sqrt{985})$	$45,18,1,1$

12.11 Conditions for the existence of real equiangular tight frames

For *tight* real equiangular lines, we refine the absolute bound and Theorem 12.11.

Theorem 12.13. *Suppose $d>1$. Then a necessary condition for an equiangular tight frame of $n>d+1$ vectors for $\mathbb{R}^d$ to exist is that n be even, and*

$$n\leq\min\{\frac{1}{2}d(d+1),\frac{1}{2}(n-d)(n-d+1)\},\tag{12.41}$$

or, equivalently,

$$d+2<\frac{2d+1+\sqrt{8d+1}}{2}\leq n\leq\frac{1}{2}d(d+1).\tag{12.42}$$

Moreover, for $n\neq 2d$, one must have

- *The eigenvalues $\lambda_1=-\sqrt{\frac{d(n-1)}{n-d}}$, $\lambda_2=\sqrt{\frac{(n-d)(n-1)}{d}}$ of the signature matrix Σ are odd integers.*
- *$n-1$ is odd, but not a prime.*
- *$\frac{1}{4}\frac{n^2(n-1)}{d(n-d)}$ and $\frac{1}{4}\frac{(n-2d)^2(n-1)}{d(n-d)}$ are perfect squares.*

Proof. Since the complement of an equiangular tight frame of n vectors for $\mathbb{R}^d$ is one for $\mathbb{R}^{n-d}$, from the bound (12.5) of Theorem 12.2, we get (12.41). Now

$$n \le \frac{1}{2}(n-d)(n-d+1) \quad \iff \quad f(n) := n^2-(2d+1)n+d^2-d \ge 0.$$

The quadratic f is nonnegative if and only if n is less than its smallest root or greater than its largest, i.e.,

$$n \le d+\frac{1}{2}-\frac{1}{2}\sqrt{8d+1} < d, \qquad n \ge d+\frac{1}{2}+\frac{1}{2}\sqrt{8d+1},$$

which gives (12.42).

The eigenvalues λ_1,λ_2 of Σ are given by (12.20) and are real algebraic integers with multiplicities $n-d, d \ge 2$. By Theorem 12.12, Proposition 12.2 and the spectral structure of strongly regular graphs λ_1,λ_2 are eigenvalues of $\hat{\Sigma}$ with multiplicities $n-d-1, d-1 \ge 1$, and the corresponding eigenvectors are orthogonal to $\mathbf{1}$. Hence, the adjacency matrix of the graph given by $\hat{\Sigma}$, i.e., $A := \frac{1}{2}(J-I-\hat{\Sigma})$ has eigenvalues $\frac{1}{2}(-1-\lambda_1)$, $\frac{1}{2}(-1-\lambda_2)$ with multiplicities $n-d-1, d-1 \ge 1$.

For $n \ne 2d$, the pairs of multiplicities above are not equal, and so the pairs of eigenvalues are algebraic integers which are not algebraic conjugates. A simple calculation shows $\lambda_1 \ne -1$, $\lambda_2 \ne 1$. The three conditions then follow from the calculations

$$\lambda_j = -2\frac{-1-\lambda_j}{2}-1, \qquad \lambda_1\lambda_2 = -(n-1),$$

$$\Bigl(\frac{-1-\lambda_1}{2}-\frac{-1-\lambda_2}{2}\Bigr)^2 = \frac{1}{4}(\lambda_2-\lambda_1)^2 = \frac{1}{4}\frac{n^2(n-1)}{d(n-d)},$$

$$\Bigl(\frac{-1-\lambda_1}{2}+\frac{-1-\lambda_2}{2}\Bigr)^2 = \frac{1}{4}(\lambda_1+\lambda_2)^2 = \frac{1}{4}\frac{(n-2d)^2(n-1)}{d(n-d)}.$$

□

For $n=2d$, the eigenvalues of Σ are $\pm\sqrt{2d-1}$. In this case, we obtain an infinite family of equiangular lines, corresponding to the *conference graphs*.

An $n \times n$ matrix C is a **conference matrix** if its diagonal entries are 0, its off diagonal entries are ± 1, and $C^*C=(n-1)I$. The reduced signature matrix $\hat{\Sigma}$ of a symmetric conference matrix $C=\Sigma$ (for which n must be even) is a strongly regular graph

$$\operatorname{srg}\Bigl(n-1, \frac{n-2}{2}, \frac{n-6}{4}, \frac{n-2}{4}\Bigr),$$

called a **conference graph**. Equivalently, conference graphs are the strongly regular graphs $\operatorname{srg}(v,k,\lambda,\mu)$ with

$$2k+(v-1)(\lambda-\mu)=0,$$

i.e., the multiplicities (12.34) of the eigenvalues θ,τ of the adjacency matrix equal.

We now characterise equiangular tight frames of $n=2d$ vectors for $\mathbb{R}^d$.

Proposition 12.4. *For $d \geq 3$, the following are equivalent*

1. *There is an equiangular tight frame of $2d$ vectors for $\mathbb{R}^d$.*
2. *There exists a $2d \times 2d$ symmetric conference matrix.*
3. *There exists a conference graph with $2d-1$ vertices, i.e., a*

$$\mathrm{srg}(2d-1, d-1, \frac{d-3}{2}, \frac{d-1}{2}).$$

For these to hold it is necessary that d be odd, and $2d-1$ be a sum of two squares.

Proof. Suppose that $n = 2d$, $d \geq 3$. Let $\hat{\Sigma}$ be the reduced signature matrix of an $n \times n$ signature matrix Σ. By Corollary 12.1 and Proposition 12.3, the following conditions are equivalent to Σ giving an equiangular tight frame of $n = 2d$ vectors for $\mathbb{R}^d$

$$\Sigma^2 - (n-1)I = 0,$$
$$\hat{\Sigma}\mathbf{1} = 0, \qquad \hat{\Sigma}^2 + J - (n-1)I = 0.$$

The first says that Σ is a symmetric conference matrix. Writing the second in terms of the adjacency matrix $A = \frac{1}{2}(J - I - \hat{\Sigma})$ gives

$$AJ = \frac{1}{2}(n-2)J, \qquad A^2 + A + \frac{1}{4}(2-n)I = \frac{1}{4}(n-2)J.$$

By the characterisation (12.32), this says that $\hat{\Sigma}$ gives a $\mathrm{srg}(2d-1, k, \lambda, \mu)$ with $k = \frac{1}{2}(n-2) = d-1$, $\lambda = \frac{1}{4}(n-6) = \frac{1}{2}(d-3)$, $\mu = \frac{1}{4}(n-2) = \frac{d-1}{2}$.

A well-known necessary condition for an $n \times n$ conference matrix to exist is that $n \equiv 2 \pmod 4$ and $n-1$ be a sum of squares. This completes the proof. □

An infinite family of conference graphs are given by the *Paley graphs*.

Example 12.40. (Paley graphs) Let $q = p^m$ be a prime power with $q \equiv 1 \pmod 4$. This implies $\mathbb{F}_q$, the unique finite field of order q, has a square root of -1 and so $b-a$ is a square if and only if $b-a$ is a square. The **Paley graph** of order $n-1 = q$ (or q-**Paley**) is the graph with vertices $\mathbb{F}_q$, and an edge between a and b if $a-b$ is a square (aka quadratic residue mod p if $q = p$). When q is prime its Paley graph is a *Hamiltonian circulant graph*. The Paley graph on $n-1 = q$ vertices is a self-complementary conference graph.

12.12 A list of real equiangular tight frames

There may or may not be a real equiangular tight frame of $n \geq d > 1$ vectors for $\mathbb{R}^d$. For example, for

- $n = d$ one has orthonormal bases.
- $n = d+1$ one has the vertices of the regular simplex.
- $n = d+2$ there are *no* equiangular tight frames, by Theorem 12.13.

If there did exist an equiangular tight frame of $n = d+2$ vectors for $\mathbb{R}^d$, then its complement would be a real equiangular tight frame of $n > 3$ vectors for $\mathbb{R}^2$.

In Table 12.5, we list all equiangular tight frames of $n > d+1$ vectors known to exist for $d \leq 50$. This was constructed using the literature on strongly regular graphs (see Brouwer [Bro07] and the associated internet page). We also include (in grey) those which are conjectured to exist, e.g., the long-standing open question in algebraic graph theory:

Does there exist a conference graph srg$(65,32,15,16)$, i.e., an equiangular tight frame of 66 vectors for $\mathbb{R}^{33}$?

Table 12.5: The equiangular tight frames of $n > d+1$ vectors for $\mathbb{R}^d$ ($d \leq 50$), for $n \neq 2d$ and $n = 2d$ (conference graphs). Here # gives the number of associated strongly regular graphs $\hat{G}$ (with + indicating at least one exists). The existence of those in grey is open problems.

n	d	#	λ_1	λ_2	$\hat{G}$
16	6	1	−3	5	srg(15,6,1,3)
16	10	1	−5	3	srg(15,8,4,4)
28	7	1	−3	9	srg(27,10,1,5)
28	21	1	−9	3	srg(27,16,10,8)
36	15	3854	−5	7	srg(35,16,6,8)
36	21	+	−7	5	srg(35,18,9,9)
126	21	+	−5	25	srg(125,52,15,26)
176	22	+	−5	35	srg(175,72,20,36)
276	23	1	−5	55	srg(275,112,30,56)
64	28	+	−7	9	srg(63,30,13,15)
64	36	+	−9	7	srg(63,32,16,16)
120	35	+	−7	17	srg(119,54,21,27)
148	37	?	−7	21	srg(147,66,25,33)
246	41	?	−7	35	srg(245,108,39,54)
288	42	?	−7	41	srg(287,126,45,63)
344	43	+	−7	49	srg(343,150,53,75)
100	45	+	−9	11	srg(99,48,22,24)
540	45	?	−7	77	srg(539,234,81,117)
736	46	?	−7	105	srg(735,318,109,159)

n	d	#	$\hat{\Gamma}$	$2d-1$
6	3	1	Paley(5)	1^2+2^2
10	5	1	Paley(9)	0^2+3^2
14	7	1	Paley(13)	2^2+3^2
18	9	1	Paley(17)	1^2+4^2
26	13	15	Paley(25)	3^2+4^2
30	15	41	Paley(29)	2^2+5^2
38	19	6760	Paley(37)	1^2+6^2
42	21	+	Paley(41)	4^2+5^2
46	23	+	Conference	3^2+6^2
50	25	+	Paley(49)	0^2+7^2
54	27	+	Paley(53)	2^2+7^2
62	31	+	Paley(61)	5^2+6^2
66	33	?	Conference	1^2+8^2
74	37	+	Paley(73)	3^2+8^2
82	41	+	Paley(81)	0^2+9^2
86	43	?	Conference	2^2+9^2
90	45	+	Paley(89)	5^2+8^2
98	49	+	Paley(97)	4^2+9^2

Example 12.41. There is no equiangular tight frame of 40 or 41 vectors for $\mathbb{R}^{16}$. Therefore (by Table 12.3), the maximal number of equiangular lines in $\mathbb{R}^{16}$ is given by a nontight equiangular frame. This is currently the only known case where the maximum number of equiangular lines is not given by a tight frame.

12.13 Nontight real equiangular frames from graphs

We now consider some nontight real equiangular frames and the graphs that they come from. We recall Theorem 12.6 as it applies:

Proposition 12.5. *Let Γ be a graph (with two or more vertices) whose $n \times n$ Seidel matrix Σ has smallest eigenvalue $-\lambda$ of multiplicity $n-d$. Then Σ is the signature matrix of an equiangular frame of $n > d$ vectors for $\mathbb{R}^d$, which is tight if and only if Σ has only two eigenvalues.*

Example 12.42. The construction of Corollary 12.4 (see Table 12.4) gives examples of nontight frames whose associated graphs have four Seidel eigenvalues.

Example 12.43. The lower bounds on the maximum number of real equiangular lines in $\mathbb{R}^d$ given by Table 12.3 are attained for $d = 16, 17, 19, 20$ by graphs with three (but not two) eigenvalues. See Table 12.6 (and [GKMS16] for details).

Table 12.6: The graphs corresponding to the nontight equiangular frames giving the maximum number n of equiangular lines in $\mathbb{R}^d$ currently known (see [GKMS16]).

n	d	Origin of graph	eigenvalues	multiplicities
40	16	$\mathrm{srg}(40,12,2,4)$	$-5,7,12$	$24,15,1$
48	17	Netto triples	$-5,7,11$	$31,8,9$
72	19	Witt design	$-5,13,19$	$53,16,3$
90	20	Taylor [Tay72]	$-5,15,19,25$	$70,9,10,1$

Example 12.44. Let Φ be the nontight equiangular frame of five vectors for $\mathbb{R}^3$ given by Γ the 5-cycle, whose Seidel matrix Σ has eigenvalues $-\sqrt{5}, -\sqrt{5}, 0, \sqrt{5}, \sqrt{5}$. This a *nontight Grassmannian frame* (see Example 12.6). The dual $\tilde{\Phi}$ and canonical tight frame Φ^{can} are equal-norm frames. The minimal angles for Φ, $\tilde{\Phi}$ and Φ^{can} are

$$\cos^{-1}\frac{1}{\sqrt{5}} \approx 63.4^\circ, \qquad \cos^{-1}\frac{3+\sqrt{5}}{8} \approx 49.1^\circ, \qquad \cos^{-1}\frac{1+\sqrt{5}}{6} \approx 57.4^\circ.$$

It is easy to verify that Φ consists of vectors that lie in five of the six diagonals of the regular icosahedron, and that the tight frame Φ^{can} is the harmonic frame given by the lifted fifth roots of unity (see Exer. 12.7).

We now generalise Example 12.44 to when Γ is a strongly regular graph (these have adjacency matrices with three eigenvalues).

We say that an equal-norm frame $\Phi = (f_j)$ has **two distances** (cf §12.14) if $\langle f_j, f_k \rangle$, $j \neq k$ takes two values.

Lemma 12.1. *Let A be the adjacency matrix of a* $\operatorname{srg}(v,k,\lambda,\mu)$, *and P_θ, P_τ, P_k be the orthogonal projections onto its θ, τ, k eigenspaces, where θ, τ are given by (12.33). Then*

$$P_k = \frac{J}{v}, \qquad P_\theta = \frac{1}{\theta-\tau}\Big(A - \tau I - (k-\tau)\frac{J}{v}\Big), \qquad P_\tau = I - P_\theta - \frac{J}{v}. \tag{12.43}$$

In particular, the diagonal entries of P_θ and P_τ are constant, and their off diagonal entries take only two values.

Proof. We have, $P_k = (\mathbf{1}/\sqrt{v})(\mathbf{1}/\sqrt{v})^* = J/v$, and $I = P_k + P_\theta + P_\tau$, so it suffices to prove the formula for P_θ. From the spectral decomposition of A, we have

$$A = \theta P_\theta + \tau P_\tau + k\frac{J}{v}, \qquad \tau I = \tau P_\theta + \tau P_\tau + \tau \frac{J}{v}.$$

Eliminating τP_τ and solving for P_θ gives the desired formula for P_θ. □

Proposition 12.6. *Let Γ be a strongly regular graph* $\operatorname{srg}(v,k,\lambda,\mu)$, *and Φ be the equiangular frame of v vectors for $\mathbb{R}^d$ that it determines. Then Φ is tight if and only if*

$$\lambda - \mu - 2k + v = \pm\sqrt{(\lambda-\mu)^2 + 4(k-\mu)}. \tag{12.44}$$

Otherwise, either

$$\sqrt{(\lambda-\mu)^2 + 4(k-\mu)} > \mu - \lambda + 2k - v, \tag{12.45}$$

and the frame, its dual and canonical tight frame are equal-norm frames with two distances, where

$$d = \frac{1}{2}\Big(v + 1 + \frac{2k + (v-1)(\lambda-\mu)}{\sqrt{(\lambda-\mu)^2 + 4(k-\mu)}}\Big), \tag{12.46}$$

or $d = v - 1$.

Proof. By (12.33), (12.34), the eigenvalues of the Seidel matrix Σ of Γ are

$$-1-2\theta, \quad -1-2\tau, \quad v-1-2k,$$

with multiplicities $m_\theta, m_\tau, 1$. These are distinct, unless $-1-2\theta$ or $-1-2\tau$ equals $v-1-2k$, which is equivalent to (12.44), and Φ is tight, with $d = v-(m_\theta+1)$ or $d = v - m_\theta$, respectively (by Theorem 12.6).

Otherwise, the minimal eigenvalue of Σ is $-(1+2\theta)$ (with multiplicity m_θ) when $-1-2\theta < v-1-k$, i.e., (12.45) holds. Here, the spectral decomposition of the symmetric matrix Σ is

$$\Sigma = -(1+2\theta)P_\theta + (-1-2\tau)P_\tau + (v-1-2k)\frac{J}{v},$$

where P_θ and P_τ are the orthogonal projections onto the θ and τ eigenspaces of A. Since $P_\theta = I - P_\tau - J/v$, the Gramian of the associated equiangular frame for $\mathbb{R}^d$, $d = v - m_\theta$, has the form

$$I + \frac{1}{1+2\theta}\Sigma = \alpha P_\tau + \beta\frac{J}{v}, \qquad \alpha = \frac{2\theta - 2\tau}{1+2\theta}, \quad \beta = \frac{2\theta + v - 2k}{1+2\theta}.$$

The dual and canonical tight frames have Gramians

$$(I + \frac{1}{\lambda}\Sigma)^\dagger = \frac{1}{\alpha}P_\tau + \frac{1}{\beta}\frac{J}{v}, \qquad (I + \frac{1}{\lambda}\Sigma)^\dagger(I + \frac{1}{\lambda}\Sigma) = P_\tau + \frac{J}{v}.$$

By Lemma 12.1, P_τ has a constant diagonal and off diagonal entries taking two possible values. Hence Φ, its dual and canonical tight frame are equal-norm frames with two distances (all the entries of J are 1).

The only remaining case is when $v - 1 - k < -1 - 2\theta$, in which case the minimal eigenvalue $v - 1 - k$ has multiplicity one, and so $d = v - 1$. □

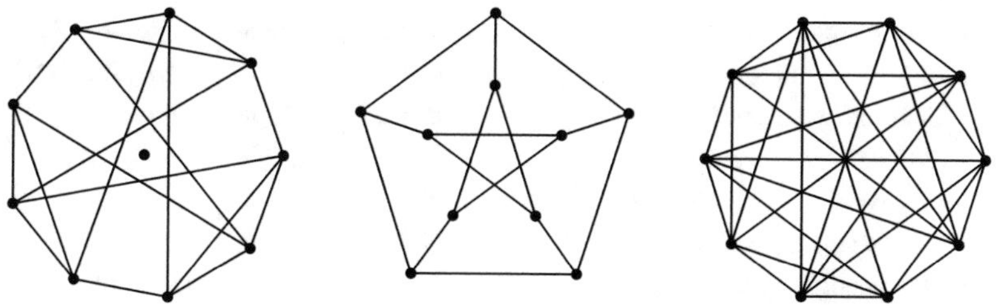

Fig. 12.3: Three switching equivalent graphs that give 10 lines in $\mathbb{R}^5$: The Paley graph on 9 vertices and a point, the Petersen graph and the triangular graph T_5.

Example 12.45. For $n \le 50$, there are 28 equal-norm tight frames with two angles that can be constructed by Theorem 12.6 (see Table 12.7). There are also many equiangular tight frames that can be constructed in this way, e.g., for $n = 10$ the unique graphs srg$(10,3,0,1)$ (the *Petersen graph*) and srg$(10,6,3,4)$ (the *triangular graph* T_5) give a set of 10 equiangular vectors in $\mathbb{R}^5$. Since there is a unique such frame up to switching equivalence of the graphs, it follows these two graphs are switching equivalent to that obtained by taking the Paley graph on nine vertices and adding an isolated vertex (see Figure 12.3).

Example 12.46. The strongly regular graph srg$(40,12,2,4)$ gives 40 equiangular lines in $\mathbb{R}^{16}$, which are not tight. There are no strongly regular graphs giving tight frames of 40 or 41 vectors in $\mathbb{R}^{16}$ (see Table 12.5). Therefore the maximal set of equiangular lines in 16 dimensions is not tight (see Table 12.3).

Table 12.7: The list of all equiangular frames of $n \le 50$ vectors for $\mathbb{R}^d$ constructed from strongly regular graphs Γ. Here the type refers to the three cases in Theorem 12.6.

n	d	type	Γ
5	3	nontight	srg(5,2,0,1)
9	5	nontight	srg(9,4,1,2)
10	5	tight	srg(10,3,0,1)
10	5	tight	srg(10,6,3,4)
13	7	nontight	srg(13,6,2,3)
15	6	nontight	srg(15,6,1,3)
15	10	nontight	srg(15,8,4,4)
16	6	tight	srg(16,5,0,2)
16	10	tight	srg(16,10,6,6)
16	10	tight	srg(16,6,2,2)
16	6	tight	srg(16,9,4,6)
17	9	nontight	srg(17,8,3,4)
21	7	nontight	srg(21,10,3,6)
21	15	nontight	srg(21,10,5,4)
25	17	nontight	srg(25,8,3,2)
25	24	–	srg(25,16,9,12)
25	13	nontight	srg(25,12,5,6)
26	13	tight	srg(26,10,3,4)
26	13	tight	srg(26,15,8,9)
27	7	nontight	srg(27,10,1,5)
27	21	nontight	srg(27,16,10,8)
28	21	tight	srg(28,12,6,4)
28	7	tight	srg(28,15,6,10)
29	15	nontight	srg(29,14,6,7)
35	15	nontight	srg(35,16,6,8)
35	21	nontight	srg(35,18,9,9)
36	26	nontight	srg(36,10,4,2)
36	35	–	srg(36,25,16,20)
36	15	tight	srg(36,14,4,6)
36	21	tight	srg(36,21,12,12)
36	28	nontight	srg(36,14,7,4)
36	35	–	srg(36,21,10,15)
36	21	tight	srg(36,15,6,6)
36	15	tight	srg(36,20,10,12)
37	19	nontight	srg(37,18,8,9)
40	16	nontight	srg(40,12,2,4)
40	39	–	srg(40,27,18,18)
41	21	nontight	srg(41,20,9,10)
45	25	nontight	srg(45,12,3,3)
45	44	–	srg(45,32,22,24)
45	36	nontight	srg(45,16,8,4)
45	44	–	srg(45,28,15,21)
45	23	nontight	srg(45,22,10,11)
49	37	nontight	srg(49,12,5,2)
49	48	–	srg(49,36,25,30)
49	31	nontight	srg(49,18,7,6)
49	48	–	srg(49,30,17,20)
49	25	nontight	srg(49,24,11,12)
50	22	nontight	srg(50,7,0,1)
50	49	–	srg(50,42,35,36)
50	25	nontight	srg(50,21,8,9)
50	25	tight	srg(50,28,15,16)

Example 12.47. The strongly regular graphs srg(76,30,8,14) and srg(76,40,18,24) give equiangular tight frames of 76 vectors for $\mathbb{C}^{19}$, i.e., a srg(75,32,10,16). Since there are no other strongly regular graphs on 75 or 76 vertices (other than the complements) [Hae93], it follows that there is a strongly regular graph on 75 points if and only if there is a strongly regular graph on 76 points. In [BPR14] it was shown that there is no srg(76,30,8,14)) and in [AM15] that there is no srg(75,32,10,16). It therefore follows that there is no srg(76,40,18,24) (and hence no strongly regular graphs on 75 or 76 points) and no equiangular tight frames of 76 vectors for $\mathbb{C}^{19}$.

Example 12.48. In [AM16], it was shown that no srg$(95,40,12,20)$ exists, and so there are not 96 equiangular lines in $\mathbb{R}^{20}$. From Proposition 12.6, it follows that no srg$(96,38,10,18)$ exists (see [Deg07]) and no srg$(96,50,22,30)$ exists.

The next open problem of the above type is the following.

Does there exist 148 equiangular lines in $\mathbb{R}^{37}$, i.e., a srg$(147,66,2533)$? The existence of such an equiangular tight frame implies the existence of a srg$(148,63,22,30)$ or a srg$(148,77,36,44)$, and the nonexistence implies that there is no strongly regular graph with 148 vertices.

The 6-cycle is a regular, but *not* strongly regular graph. Its Seidel matrix has three eigenvalues and gives an equiangular frame of six vectors for $\mathbb{R}^4$. The dual and canonical tight frames have equal norms (and more than two angles). This is a consequence of the 6-cycle being a *circulant graph*.

Let C be a subset of $\mathbb{Z}_n$ which is closed under taking additive inverses, i.e., $-c\in C$, $\forall c \in C$. Then the **circulant graph** G with **connection set** C is the graph with vertices $\mathbb{Z}_n$ and an edge from j to k if $j-k \in C$. The choice $C = \{-1,1\}$ gives the n-cycle.

Proposition 12.7. *(Circulant graphs). Let Γ be a circulant graph, and Φ be the real equiangular frame that it determines. Then the dual frame $\tilde{\Phi}$ and canonical tight frame Φ^{can} are equal-norm frames.*

Proof. Since Γ is a circulant graph, the Gramian of Φ is an $n\times n$ circulant matrix and hence is diagonalised by the Fourier matrix F, i.e.,

$$F^{-1}(I+\frac{1}{\lambda}\Sigma)F = \mathrm{diag}(\lambda_1,\ldots,\lambda_n), \qquad F := \frac{1}{\sqrt{n}}[\omega^{jk}]_{j,k\in\mathbb{Z}_n}, \ \omega := e^{\frac{2\pi i}{n}}.$$

Since F is unitary, we can write this spectral decomposition as

$$I+\frac{1}{\lambda}\Sigma = \sum_j \lambda_j P_j, \qquad P_j := f_j f_j^*, \ \ f_j := \frac{1}{\sqrt{n}}(\omega^{jk})_{k\in\mathbb{Z}_n}.$$

The rank one projection matrices P_j have constant diagonal entries (equal to $\frac{1}{n}$), and so the dual and canonical tight frames have equal norms (their Gramians are $\sum_j \frac{1}{\lambda_j}P_j$ and $\sum_j P_j$). □

Example 12.49. (n-cycle graph) The Seidel matrix of the n-cycle has a minimal eigenvalue $-1-4\cos\frac{2\pi}{n}$ of multiplicity 2 with corresponding eigenvectors f_{-1}, f_1. The corresponding equiangular frame is of n vectors for $\mathbb{R}^{n-2}$, and its complement is the tight frame of n equally spaced vectors for $\mathbb{R}^2$ (up to similarity).

Example 12.50. (Paley graphs) For n a prime congruent to 1 (mod 4), the *Paley graph* is circulant, with connection set C given by the quadratic residues modulo n.

12.14 Spherical two-distance tight frames

A set $\Phi = (x_j)$ of n unit vectors in $\mathbb{R}^d$ is called a (**spherical**) **two-distance set** if the inner products between distinct vectors take two values a and b, i.e.,

$$\{\langle x_j, x_k\rangle : j \neq k\} = \{a, b\}, \qquad a \neq b.$$

The spherical distance between points x_j and x_k is $\cos^{-1}\langle x_j, x_k\rangle$, and so these sets have precisely two spherical distances between their points. The maximum size $g(d)$ of a spherical two-distance set has been studied for several decades [BY13].

When $b = -a$, a spherical two-distance set is equivalent to a real equiangular frame (see §12.9). We therefore primarily consider the case when $a+b \neq 0$.

Let $e_1, \ldots, e_{d+1}$ be the standard basis for $\mathbb{R}^{d+1}$. Then the $n = \frac{1}{2}d(d+1)$ points $e_j + e_k$, $j \neq k$ give a spherical two-distance set (after scaling), as does the orthogonal projection of these points onto the d-dimensional subspace given by the orthogonal complement of $e_1 + \cdots + e_{d+1}$ (see §12.16). We therefore have the lower bound

$$g(d) \geq \frac{1}{2}d(d+1).$$

Delsarte, Goethals and Seidel [DGS77] proved the "harmonic" upper bound

$$g(d) \leq \frac{1}{2}d(d+3).$$

Various improvements and refinements of this have been made, e.g., Musin [Mus09] proved the upper bound

$$g(d) \leq \frac{1}{2}d(d+1), \qquad \text{when } a+b \geq 0.$$

This generalises the estimate (12.2) for the maximal number of real equiangular lines (the case $a+b=0$). As of the improvements of [BY13], the best estimates of $g(d)$ for $d \leq 93$ are as follows.

Theorem 12.14. *The maximum size $g(d)$ of a spherical two-distance set in $\mathbb{R}^d$ satisfies*

$$g(2) = 5, \quad g(3) = 6, \quad g(4) = 10, \quad g(5) = 16, \quad g(6) = 27,$$

$$g(d) = \frac{1}{2}d(d+1), \qquad 7 \leq d \leq 93,\ d \neq 22, 46, 78,$$

$$g(22) = 275,$$

$$\frac{1}{2}d(d+1) \leq g(d) \leq \frac{1}{2}d(d+3) - 1, \qquad d = 46, 78.$$

All the known maximal configurations are tight.

Given a spherical two-distance set $\Phi = (x_j)$, one can associate with it the graph Γ_a with points $\{x_j\}$ and an edge from x_j to x_k if $\langle x_j, x_k\rangle = a$. The following result of

[BGOY14] shows that if Φ is tight and $b \neq -a$ (Φ is not equiangular), then Γ_a is a strongly regular graph, and that all tight nonequiangular two-distance frames can be constructed in this way.

Theorem 12.15. *Let $\Phi = (x_j)$ be a two-distance set, with $a \neq -b$ and associated graph Γ_a. Then Φ is tight if and only if Γ_a is a strongly regular graph* $\mathrm{srg}(v,k,\lambda,\mu)$*, in which case (after normalisation) its Gramian is one of*

$$P_\theta, \quad P_\theta + \frac{J}{v}, \quad P_\tau, \quad P_\tau + \frac{J}{v}, \tag{12.47}$$

where $P_\theta, P_\tau, \frac{J}{v}$ are the orthogonal projections onto the eigenspaces of the adjacency matrix A of Γ_a. Further, all the projections of (12.47) give tight two distance sets.

Proof. Suppose that $\Phi = (x_j)_{j=1}^v$ is a *tight* two-distance frame for $\mathbb{R}^d$.

Let N_a be the number of points at a distance a from a given point x_j. Since Φ is a unit-norm tight frame (with $A = \frac{v}{d}$), the norm of the j-th column of its Gramian is

$$1 + N_a a^2 + (v - 1 - N_a) b^2 = \frac{v}{d}.$$

Since $a \neq -b$, $a^2 - b^2 \neq 0$, and so the above can be solved for N_a, which is therefore independent of j, i.e., Γ_a is a regular graph.

Fix a pair of indices k, ℓ with $\langle x_k, x_\ell \rangle = a$. Let C_a be the number of indices $j \neq k, \ell$ for which the distances between x_j and x_k, x_l are both a. Since Φ is tight, Parseval gives

$$\frac{v}{d}\langle x_k, x_\ell \rangle = \sum_{j=1}^{v} \langle x_k, x_j \rangle \langle x_j, x_\ell \rangle,$$

from which we obtain

$$\begin{aligned} \frac{v}{d} a &= 2a + 2(N_a - C_a - 1)ab + C_a a^2 + (v - 2N_a + C_a) b^2 \\ &= 2a + 2(N_a - 1)ab + (v - 2N_a) b^2 + (a - b)^2 C_a. \end{aligned}$$

Since $a \neq b$, this gives a unique solution for C_a, i.e., every two adjacent vertices have $\lambda = C_a$ common neighbours. Similarly, every two nonadjacent vertices has the same number of adjacent vertices, and so Γ_a is strongly regular.

The (normalised) Gram matrix of Φ is

$$P = \frac{d}{v}\big((1-b)I + (a-b)A + bJ\big).$$

The spectral decomposition of A (see Lemma 12.1) gives

$$I = P_\theta + P_\tau + \frac{J}{v}, \qquad A = \theta P_\theta + \tau P_\tau + k\frac{J}{v},$$

and so P can be written in the form

$$P = \alpha P_\theta + \beta P_\tau + \gamma \frac{J}{v}.$$

The condition $P^2 = P$ implies that $\alpha^2 = \alpha$, $\beta^2 = \beta$, $\gamma^2 = \gamma$ and so α, β, γ must be 0 or 1. The choices $P = P_\theta + P_\tau + \frac{J}{v} = I$ and $P = P_\theta + P_\tau = I - \frac{J}{v}$ give an orthonormal basis and the vertices of the simplex, which are not two distance sets. The other choices give two distance sets. For example, when $P = P_\theta + \frac{J}{v}$ (which has constant diagonal), we have $d = \operatorname{rank}(P_\theta) + \operatorname{rank}(J) = m_\theta + 1$, and, by Lemma 12.1,

$$P = \frac{m_\theta + 1}{v}\big((1-b)I + (a-b)A + bJ\big) = \frac{1}{\theta - \tau}\Big(A - \tau I - (k-\tau)\frac{J}{v}\Big) + \frac{J}{v}.$$

By taking an entry with $A_{jk} = 1$, we obtain

$$\frac{m_\theta + 1}{v} a = \frac{1}{\theta - \tau}\Big(1 - (k-\tau)\frac{1}{v}\Big) + \frac{1}{v} \quad \Longrightarrow \quad a = \frac{v - k + \theta}{(m_\theta + 1)(\theta - \tau)}.$$

Similarly, taking an off diagonal entry with $A_{jk} = 0$ gives a formula for b. The results of these calculations for the four possible choices are given in Table 12.8. We observe that, since $v \neq 0$, each of these choices gives a two-distance set. □

Table 12.8: The parameters of the tight two distance sets for $\mathbb{R}^d$ given by Γ_a a $\operatorname{srg}(v,k,\lambda,\mu)$. Here the Gramian matrix (after normalisation) is the orthogonal projection matrix P, and N_a is the number of neighbours of a. The condition for these to be equiangular is given in the last column.

Gramian matrix P	d	N_a	a	b	$a = -b$ holds
P_θ	m_θ	k	$\frac{v-k+\tau}{m_\theta(\theta-\tau)}$	$\frac{-k+\tau}{m_\theta(\theta-\tau)}$	$v = 2(k-\tau)$
$P_\theta + \frac{J}{v}$	$m_\theta + 1$	k	$\frac{v-k+\theta}{(m_\theta+1)(\theta-\tau)}$	$\frac{-k+\theta}{(m_\theta+1)(\theta-\tau)}$	$v = 2(k-\theta)$
$P_\tau = I - P_\theta - \frac{J}{v}$	$v - m_\theta - 1$	k	$\frac{-v+k-\theta}{(v-m_\theta-1)(\theta-\tau)}$	$\frac{k-\theta}{(v-m_\theta-1)(\theta-\tau)}$	$v = 2(k-\theta)$
$P_\tau + \frac{J}{v} = I - P_\theta$	$v - m_\theta$	k	$\frac{-v+k-\tau}{(v-m_\theta)(\theta-\tau)}$	$\frac{k-\tau}{(v-m_\theta)(\theta-\tau)}$	$v = 2(k-\tau)$

By combining Corollary 12.3 and Theorem 12.15, we obtain a characterisation of all spherical two-distance tight frames in terms of the strongly regular graphs.

Corollary 12.5. *All spherical two-distance tight frames of $n > d+1$ vectors for $\mathbb{R}^d$ can be constructed from the strongly regular graphs as follows*

1. *(Equiangular) From the* $\operatorname{srg}(n-1, k, \frac{3k-n}{2}, \frac{k}{2})$, $k := \frac{1}{2}n - 1 + \big(1 - \frac{n}{2d}\big)\sqrt{\frac{d(n-1)}{n-d}}$.
2. *(Nonequiangular) From the* $\operatorname{srg}(v,k,\lambda,\mu)$, $v = n$.

Each complementary pair of strongly regular graphs gives four nonequiangular spherical two-distance tight frames, unless $v = 2(k-\tau)$ or $v = 2(k-\theta)$, in which case there are only two (the other two being equiangular).

Example 12.51. For a srg$(10,3,0,1)$, we have $\theta = 1$, $\tau = -2$, so that

$$v = 10 = 2(3-(-2)) = 2(k-\tau).$$

Hence the spherical two-distance tight frames with Gramians P_θ and $P_\tau + \frac{J}{v}$ are equiangular, and those with Gramians $P_\theta + \frac{J}{v}$ and P_τ are nonequiangular. Here, P_θ is the tight frame constructed by Proposition 12.6, and $P_\tau + \frac{J}{v}$ is the one constructed from the complementary graph.

Example 12.52. Here we consider the nonprimitive strongly regular graph given by the union of m complete graphs on w vertices, which is a srg$(mw, w-1, w-2, 0)$. We have $\theta = w-1$, $m_\theta = m-1$, $\tau = -1$, $m_\tau = m(w-1)$ and

$$P_\theta = \frac{1}{w}(A+I) - \frac{J}{v}.$$

In this case, there are only two eigenvalues, since $k = w-1 = \theta$. The corresponding spherical two-distance tight frames are described in Table 12.9.

Table 12.9: The parameters of the two-distance tight frames for $\mathbb{R}^d$ given by the nonprimitive $\Gamma_a = \text{srg}(mw, w-1, w-2, 0)$, which consists of m copies of a complete graph on w points.

Gramian matrix P	d	N_a	a	b	description
$Q = \frac{1}{w}(A+I) - \frac{J}{v}$	$m-1$	$w-1$	1	$-\frac{1}{m-1}$	w copies of m-simplex
$Q + \frac{J}{v}$	m	$w-1$	1	0	w copies of $\{e_1, \ldots, e_m\}$
$R = I - (Q + \frac{J}{v})$	$n-m$	$w-1$	$-\frac{1}{w-1}$	0	complement of second
$R + \frac{J}{v} = I - Q$	$n-m+1$	$w-1$	$-\frac{m-1}{n-m+1}$	$\frac{1}{n-m+1}$	complement of first

12.15 Two-distance tight frames and partial difference sets

We now show that nonequiangular two-distance tight frames which are harmonic can be constructed from *partial difference sets* (cf. Theorem 12.3).

Definition 12.4. A d element subset J of a finite group G of order n is said to be a (n, d, λ, μ)**-partial difference set** if the multiset of (nonzero) differences

$$\{j_2 j_2^{-1} : j_1, j_2 \in J, j_1 \neq j_2\}$$

contains each nonidentity element of J exactly λ times and each nonidentity element of $G \setminus J$ exactly μ times. If $\lambda = \mu$, then one has a difference set.

Partial difference sets with $\lambda \neq \mu$ are reversible (i.e., closed under taking inverses). A partial difference set J for which $1 \notin J$ is said to be **regular**. It suffices to study the regular partial difference sets, since if J is a partial difference set with $1 \in J$, then $J \setminus \{1\}$ is a (regular) partial difference set.

Proposition 12.8. *Let G be an abelian group of order n, and $\Phi_J = (\xi|_J)_{\xi \in \hat{G}}$ be the harmonic frame given by $J \subset G$, $|J| = d$. Then the following are equivalent*

1. *J is a (n,d,λ,μ)-partial difference set for G.*
2. *For all $\chi \neq 1$, $\chi \in \hat{G}$,*

$$\sum_{j\in J} \chi(j) = \begin{cases} \frac{\lambda-\mu\pm\sqrt{(\lambda-\mu)^2+4(d-\lambda)}}{2}, & 0 \in J; \\ \frac{\lambda-\mu\pm\sqrt{(\lambda-\mu)^2+4(d-\mu)}}{2}, & 0 \notin J. \end{cases} \tag{12.48}$$

In particular, if these hold with $\lambda \neq \mu$, then Φ_J is a nonequiangular two-distance tight frame of n vectors for $\mathbb{R}^d$.

Proof. The equivalence of 1 and 2 is a standard characterisation of partial difference sets [LM90], [Ma94]. By (12.48), the harmonic frame given by a partial difference set is real, with two distances, and is equiangular if and only if $\lambda = \mu$. □

Example 12.53. The set $J = \{1,4\} \subset \mathbb{Z}_5$ (of nonzero quadratic residues) is a regular cyclic $(5,2,0,1)$-partial difference set. The regular $(9,4,1,2)$-partial difference set

$$J = \{(0,1),(0,2),(1,0),(2,0)\} \subset \mathbb{Z}_3 \times \mathbb{Z}_3.$$

is a noncyclic example.

Table 12.10 lists the regular abelian partial difference sets for $d \leq 21$.

Table 12.10: The regular (n,d,λ,μ)-partial difference sets for $d \leq 21$, where $0 \leq \lambda \leq k-1$, $1 \leq \mu \leq k-1$, $k \leq \frac{n-1}{2}$. This was adapted from [Ma94] (which lists those with $d \leq 100$).

n	d	λ	μ
5	2	0	1
9	4	1	2
16	5	0	2
16	6	2	2
13	6	2	3
25	8	3	2

n	d	λ	μ
17	8	3	4
36	10	4	2
49	12	5	2
25	12	5	6
36	14	4	6
64	14	6	2

n	d	λ	μ
29	14	6	7
36	15	6	6
81	16	7	2
64	18	2	6
49	18	7	6
100	18	8	2

n	d	λ	μ
37	18	8	9
81	20	1	6
121	20	9	2
41	20	9	10
64	21	8	6
243	22	1	2

Example 12.54. The set $J = \{0,1,4\} \subset \mathbb{Z}_5$ is a cyclic $(5,2,0,1)$-partial difference set. It is not regular, since it contains the identity 0.

In view of Theorem 12.15, the nonequiangular two-distance harmonic frames must be associated with strongly regular graphs. These turn out to be Cayley graphs.

For J a generating subset of a group G with $1 \notin J$, the (uncoloured) **Cayley graph** has vertices G and a directed edge from g_1 to g_2 if $g_2 = g_1 j$ for some $j \in J$ (J is called the *connection set*). For J with $J^{-1} = J$ (e.g., a partial difference set with $\lambda \neq \mu$), the Cayley graph given by J is undirected.

Proposition 12.9. *If J is a regular (n,d,λ,μ)-partial difference set for a group G with $\lambda \neq \mu$, then the (undirected) Cayley graph given by J is a* $\mathrm{srg}(n,d,\lambda,\mu)$.

There is a converse of Proposition 12.9 (the condition $\lambda \neq \mu$ is replaced by the condition that the partial difference set be reversible) [Ma94].

12.16 The standard m-distance tight frame

We say that a set $\Phi = (x_j)$ of n unit vectors in $\mathbb{R}^d$ is a **(spherical) m-distance set** if the inner products between distinct vectors take m values. This generalisation of the isogonal configurations of Example 3.9 ($m = 1$) and spherical two-distance sets ($m = 2$) are not well studied. However, we can provide one nice example.

Let $e_1, \ldots, e_{d+1}$ be the standard basis for $\mathbb{R}^{d+1}$. Then the $n = \binom{d+1}{m}$ points

$$e_{j_1} + \cdots + e_{j_m}, \qquad 1 \leq j_1 < \cdots < j_m \leq d+1$$

give an m-distance set (after scaling). The orthogonal projection of these points onto $(e_1 + \cdots + e_{d+1})^\perp$ gives an m-distance tight frame for $\mathbb{R}^d$.

Proposition 12.10. *Suppose $1 \leq m \leq d$. For $J = \{j_1, \ldots, j_m\} \subset \{1,2,\ldots,d+1\}$, let*

$$w_J := e_{j_1} + e_{j_2} + \cdots + e_{j_m} - \frac{m}{d+1}(e_1 + \cdots e_{d+1}).$$

Then $(w_J)_{|J|=m}$ is a tight frame for $(e_1 + \cdots + e_{d+1})^\perp$, with Gramian given by

$$\langle w_J, w_K \rangle = r - \frac{m^2}{d+1}, \qquad |J \cap K| = r.$$

Proof. This frame is the highly symmetric tight frame of Example xx. In particular, it is tight, since it is the orbit of an irreducible unitary action. Writing

$$W_J = \sum_{j \in J \cap K} e_j + \sum_{j \in J \setminus K} e_j - \frac{m}{d+1}(e_1 + \cdots e_{d+1}),$$

and expanding the inner product gives

$$\langle w_J, w_K \rangle = \left(r - \frac{m}{d+1} m \right) - \frac{m}{d+1} \left(m - \frac{m}{d+1}(d+1) \right) = r - \frac{m^2}{d+1}, \quad |J \cap K| = r. \quad \square$$

The number of r entries in a given column of the Gramian is $\binom{m}{r}\binom{d+1-m}{m-r}$, and so for there to be at least one r entry, one must have

$$d+1-m \ge m-r \quad \Longleftrightarrow \quad d+1 \ge 2m-r.$$

In particular, for $d+1 \ge 2m$,

$$\Phi = (cw_J)_{|J|=m}, \qquad c := \sqrt{\frac{d+1}{m(d+1-m)}}$$

is an m-distance tight frame of $\binom{d+1}{m}$ vectors for $\mathbb{R}^d$, with distances

$$\alpha_r := \frac{d+1}{m(d+1-m)}\Big(r - \frac{m^2}{d+1}\Big), \qquad r = 0,1,\ldots,m-1,$$

which we call the **standard m-distance tight frame.**

Example 12.55. For $d \ge 3$, the standard 2-distance tight frame has

$$a = \alpha_1 = \frac{d-3}{2(d-1)}, \qquad b = \alpha_0 = -\frac{2}{d-1}, \tag{12.49}$$

and corresponds (via P_Θ in Theorem 12.15) to $\Gamma_a = T_{d+1}$, the *triangular graph*, which is a $\operatorname{srg}\big(\frac{1}{2}d(d+1), 2(d-1), d-1, 4\big)$. The triangular graph is the unique graph with these parameters, except for when $d+1=8$, in which case there are three other $\operatorname{srg}(28,12,6,4)$ (the Chang graphs).

Example 12.56. (28 equiangular lines in $\mathbb{R}^7$) The standard 2-distance tight frame gives a set of equiangular lines if and only if $a=b$ in (12.49), i.e., $d=7$, which gives the 28 lines of (12.6).

12.17 Complex equiangular tight frames

We first observe that the complex analogue of Theorem 12.13 holds.

Proposition 12.11. *Suppose $d>1$. A necessary condition for an equiangular tight frame of $n>d+1$ vectors for $\mathbb{C}^d$ to exist is that*

$$n \le \min\{d^2, (n-d)^2\}, \tag{12.50}$$

or, equivalently,

$$d + \frac{1+\sqrt{4d+1}}{2} = \frac{2d+1+\sqrt{4d+1}}{2} \le n \le d^2. \tag{12.51}$$

Example 12.57. There is no equiangular tight frame of $n = d+2$ vectors for $\mathbb{C}^d$ when $d \geq 3$.

The existence and construction of equiangular tight frames of $n = d^2$ vectors for $\mathbb{C}^d$ is a compelling subject (see Chapter 14).

There have been attempts to adapt the methods of constructing real equiangular tight frames to complex ones. Most notably, by restricting the entries of the Gramian (or the vectors) to be m-th roots of unity, thereby giving a finite set of possible n vector equiangular tight frames for $\mathbb{C}^d$. When the entries of the vectors are roots of unity (as for harmonic frames), then $n \leq d^2 - d + 1$ (Theorem 12.16). When the entries of the signature matrix are roots of unity (generalising real equiangular tight frames), then there exist some intriguing examples, including a maximal set of $n = d^2$ equiangular lines for $d = 3, 8$.

We now give some details on the known constructions.

Theorem 12.16. *Suppose that $\Phi = (f_j)$ is an equiangular tight frame of n (unit) vectors for $\mathbb{C}^d$, for which the entries of $d\,\mathrm{Gram}(\Phi)$ are contained in a subring $\mathscr{A}$ of the algebraic integers. Then*

1. *$\mathscr{A}$ has elements of modulus $\sqrt{\frac{d(n-d)}{n-1}}$.*
2. *$\frac{d(n-d)}{n-1} \in \mathscr{A} \cap \mathbb{Z}$.*
3. *$n = \frac{1}{\lambda} d(d-1) + 1$, for some positive integer λ. In particular, $n \leq d^2 - d + 1$.*

Proof. By our assumption $a_{jk} := d\langle f_j, f_k\rangle \in \mathscr{A}$, $j \neq k$. Since Φ is equiangular,

$$|a_{jk}| = d|\langle f_j, f_k\rangle| = d\sqrt{\frac{n-d}{d(n-1)}} = \sqrt{\frac{d(n-d)}{n-1}}.$$

For any fixed $j \neq k$,

$$\frac{d(n-d)}{n-1} = |a_{jk}|^2 = a_{jk}a_{kj} \in \mathscr{A},$$

so that $\frac{d(n-d)}{n-1}$ is a rational algebraic integer and hence is an integer (in $\mathscr{A}$). Further,

$$\lambda = d - \frac{d(n-d)}{n-1} = \frac{d(d-1)}{n-1}$$

is a rational algebraic integer and hence is a positive integer. □

Example 12.58. (Unital frames) If the n vectors of Φ have the form

$$v = \frac{1}{\sqrt{d}} \begin{pmatrix} v_1 \\ \vdots \\ v_d \end{pmatrix}, \qquad v_1, \ldots, v_d \in \mathscr{A},$$

where $\mathscr{A}$ is a subring of the algebraic integers which is closed under conjugation, then $d\,\mathrm{Gram}(\Phi)$ has entries in $\mathscr{A}$, and so $n \le d^2 - d + 1$.

A special case is when the entries v_j are m-th roots of unity, which is known as a *unital* equiangular tight frame (of degree m). These are a generalisation of the equiangular harmonic frames (see §12.3). The only known example of a unital equiangular tight frame which is not harmonic is one of 576 vectors for $\mathbb{R}^{276}$, given by a *Kirkman frame* [JMF14]. For example, [STDH07] found a unital equiangular tight frame of 27 vectors for $\mathbb{C}^{13}$ with $m=3$ via a computer search. This corresponds to a $(27,13,6)$-difference set for $\mathbb{Z}_3^4$.

Theorem 12.17. *(m-th root signature matrices) Let Σ be the signature matrix of an equiangular tight frame of n vectors for $\mathbb{C}^d$. Then*

$$\lambda_1 + \lambda_2 = (n-2d)\sqrt{\frac{n-1}{d(n-d)}} \in \mathbb{Z}[\Sigma] \quad (\textit{the ring generated by the entries of } \Sigma).$$

If $n \ne 2d$, and the entries of Σ are algebraic integers, then eigenvalues of Σ

$$\lambda_1 = -\sqrt{\frac{d(n-1)}{n-d}}, \qquad \lambda_2 = \sqrt{\frac{(n-d)(n-1)}{d}}$$

are algebraic integers in the subfield of $\mathbb{C}$ generated by the entries of Σ. Thus, if the entries of Σ are powers of the primitive m-th root of unity $\omega = e^{\frac{2\pi i}{m}}$, then

$$\lambda_1, \lambda_2 \in \mathbb{Z}[\omega + \overline{\omega}] = \mathbb{Z}[\cos\frac{2\pi}{m}].$$

Proof. Multiply (12.22) by $z_{kj} = \overline{z_{jk}}$ to obtain $\lambda_1 + \lambda_2 \in \mathbb{Z}[\Sigma]$. If the entries of Σ are algebraic integers, then so are the coefficients of the characteristic polynomial of Σ, whose roots λ_1, λ_2 are therefore algebraic integers. For $n \ne 2d$, solving

$$\lambda_1 + \lambda_2 = (n-2d)\sqrt{\frac{n-1}{d(n-d)}} \in \mathbb{Z}[\Sigma], \qquad (n-d)\lambda_1 + d\lambda_2 = 0,$$

shows that λ_1, λ_2 are in the field generated by the entries of Σ. Finally, when Σ consists of m-th roots of unity, we have that λ_1, λ_2 are in the cyclotomic field $\mathbb{Q}(\omega)$. Since λ_1, λ_2 are also real algebraic integers, they are in $\mathbb{Z}(\omega + \overline{\omega})$ (the intersection of $\mathbb{Q}(\omega)$ and the real algebraic integers). □

Example 12.59. (SICs) For $n = d^2$, we have

$$\lambda_1 + \lambda_2 = (d-2)\sqrt{d+1}.$$

Thus, for $d \ne 2$, $\sqrt{d+1}$ must be in the field generated by the entries of Σ.

Example 12.60. (Cube roots) If the entries of a Σ giving an equiangular tight frame are third roots of unity, then λ_1, λ_2 must be integers. It can be shown [BPT09] that

$$\lambda_1, \lambda_2 \equiv 2 \pmod 3.$$

In particular, since $\lambda_1 \lambda_2 = n-1$, one must have $n \equiv 0 \pmod 3$. Let $\omega = e^{\frac{2\pi i}{3}}$. For $n = 9$, $d = 6$ there is a cube root signature matrix Σ given by the Kronecker product-type construction of Example 12.29 (taking the product again gives $n = 81$, $d = 45$). This example can also be obtained by associating Σ with a directed graph (1 corresponds to a nonedge and ω, ω^2 to directed edges) [BPT09]. The choice $n = 33$, $d = 11$ gives $\lambda_1 = -4$, $\lambda_2 = 8$. It is unknown whether or not there is a cube root signature matrix with these parameters.

The above condition on λ_1, λ_2 has been generalised by [BE10] as follows.

Theorem 12.18. *Let $p > 2$ be prime. If the entries of Σ are p-th roots of unity and Σ is the signature matrix of an equiangular tight frame of n vectors for $\mathbb{C}^d$, then*

1. *λ_1, λ_2 are integers, with $\lambda_1, \lambda_2 \equiv p-1 \pmod p$.*
2. *$4(n-1) + (\lambda_1 - \lambda_2)^2$ is a perfect square, which is 0 modulo p^2.*

Many of the known p-th root signature matrices for equiangular tight frames are given by Theorem 12.8 applied to *Butson type* complex Hadamard matrices V, i.e., those with p-th root entries.

Corollary 12.6. *Let $V \in \mathrm{H}(p,m)$, i.e., be a complex Hadamard matrix of order m with p-th root entries, and $K \in \mathrm{H}(p,m^2)$ be given by (12.26). Then*

1. *$K - I$ is a p-th root signature matrix for an equiangular tight frame of $n = m^2$ vectors for a $\frac{1}{2}m(m+1)$-dimensional space.*
2. *$I - K$ is a $2p$-th root (or a p-th root, when p is even) signature matrix for an equiangular tight frame of $n = m^2$ vectors for a $\frac{1}{2}m(m-1)$-dimensional space.*

Proof. By construction, the entries of $K - I$ are p-th roots of unity, and those of $I - K$ are the negatives of p-th roots of unity (which are $2p$-th roots when p is odd and are p-th roots when p is even). □

The equiangular tight frames of Corollary 12.6 are subsimplicial with $r = m \pm 1$.

Example 12.61. (Butson) Let p be a prime. There are complex Hadamard matrices V of order $m = 2^j p^k$, $0 \le j \le k$, whose entries are p-th roots of unity [But62]. Thus, with

$$n = (2^j p^k)^2, \qquad 0 \le j \le k.$$

there is a p-th root signature matrix of size n giving an equiangular tight frame of n vectors in $d = \frac{1}{2} 2^j p^k (2^j p^k + 1)$ dimensions. The complementary equiangular tight frame in $\frac{1}{2} 2^j p^k (2^j p^k - 1)$ dimensions has a $2p$-th (or p-th) root signature matrix.

Example 12.62. ($n = 36$) There exist 6×6 complex Hadamard matrices with cube root entries and with fourth root entries. Thus there is a p-th root signature matrix for an equiangular tight frame of 36 vectors in $\mathbb{C}^{21}$ for $p = 2, 3, 4$ and one in $\mathbb{C}^{15}$ for $p = 2, 6, 4$ (the complement). The real equiangular tight frames ($p = 2$) can be obtained from a reversible difference set (see Example 12.15).

Example 12.63. (Conference matrices) Let $C = -C^T$ be an $n \times n$ skew-symmetric conference matrix (n must be a multiple of 4 when $n \neq 2$). Then

$$\Sigma = iC$$

is the signature matrix (with entries $\pm i$) of an equiangular tight frame of $n = 2d$ vectors for $\mathbb{C}^d$. This follows by Theorem 12.7, via the calculation (see Exer. 12.11)

$$\Sigma^2 = -(-C^T)C = (n-1)I.$$

These frames are not subsimplicial, since $r = \sqrt{2d-1}$ implies that d is odd, and so $n = 2d$ is not a multiple of 4. Given an $n \times n$ skew-symmetric conference matrix C, one of size $2n$ is given by

$$\begin{pmatrix} C & C-I \\ C+I & -C \end{pmatrix}.$$

For example,

$$C = \begin{pmatrix} 0 & -1 \\ 1 & 0 \end{pmatrix} \quad \longrightarrow \quad \Sigma = i \begin{pmatrix} C & C-I \\ C+I & -C \end{pmatrix} = \begin{pmatrix} 0 & -i & -i & -i \\ i & 0 & i & -i \\ i & -i & 0 & i \\ i & i & -i & 0 \end{pmatrix},$$

which is the signature matrix of an equiangular tight frame of four vectors for $\mathbb{C}^2$. Fourth root Seidel matrices are studied in [DHS10].

A suitable choice for $H^{(a)}$ in Theorem 12.4 gives p-th root signature matrices.

Corollary 12.7. *For a $(2,k,v)$-Steiner system $\mathscr{B}$, let m be the product of the prime factors of $r+1 = \frac{v-1}{k-1}+1$. Then there exists a Steiner equiangular tight frame of $n = v(r+1)$ vectors for a space of dimension $d = \frac{v(v-1)}{k(k-1)}$ whose signature matrix has nonzero entries the m-th roots of unity and their negatives.*

Proof. In the construction of Theorem 12.4, take each $H^{(a)}$ to be a Fourier matrix with m-th root entries. □

Example 12.64. There exist Steiner equiangular tight frames of n vectors for $\mathbb{C}^d$ with m-th root signature matrices in cases where such frames cannot be constructed by the methods already considered, e.g., when (n,d) is $(45,12)$ $(r=4,\ m=10)$, $(65,13)$ $(r=4,\ m=10)$, $(65,13)$ $(r=5,\ m=6)$, $(65,13)$ $(r=6,\ m=14)$.

Example 12.65. The inner products between the vectors in a Tremain equiangular tight frame are products of the entries of the Hadamard matrices that are used in their construction (see Exer. 12.10). The Hadamard matrices $H^{(a)}$ can be taken as in Corollary 12.7, and the Hadamard matrix giving the unimodular simplex in $\mathbb{C}^v$ can be taken to be one of these tensored by a 2×2 real Hadamard matrix. In this way, Tremain equiangular tight frames can be constructed with signature matrix having nonzero entries given by the m-th roots of unity and their negatives.

Motivated by Theorem 12.18, in Table 12.11 we list all n,d ($d < 30$) for which the eigenvalues λ_1, λ_2 of the signature matrix Σ of an equiangular tight frame of $n > d+1$ vectors for $\mathbb{C}^d$ are integers. In view of (12.13), this condition is equivalent to both the frame and its complementary frame being subsimplicial. We indicate when an m-th root a signature matrix of such a frame is known to exist, by Corollaries 12.3, 12.6 and 12.7, or the Hoggar lines ($n = 64$, $d = 8$ see §14.6).

Table 12.11: The n,d ($d < 30$) for which the eigenvalues λ_1, λ_2 of the signature matrix Σ of an equiangular tight frame of $n > d+1$ vectors for $\mathbb{C}^d$ are integers. We indicate when such a signature matrix is known to exist with entries which are m-th roots, and use grey when it is unknown.

n	d	λ_1	λ_2	exists
9	3	-2	4	$m=6$
10	5	-3	3	$m=2$
9	6	-4	2	$m=3$
16	6	-3	5	$m=2$
28	7	-3	9	$m=2$
64	8	-3	21	$m=4$
16	10	-5	3	$m=2$
25	10	-4	6	$m=10$
33	11	-4	8	$m \geq 3$
45	12	-4	11	$m=10$
26	13	-5	5	$m=2$
65	13	-4	16	$m-10$
105	14	-4	26	$m \geq 3$
25	15	-6	4	$m=5$
36	15	-5	7	$m=2$
225	15	-4	56	$m \geq 2$
51	17	-5	10	$m \geq 4$
76	19	-5	15	$m \geq 2$
96	20	-5	19	$m=6$
28	21	-9	3	$m=2$
36	21	-7	5	$m=2$
49	21	-6	8	$m=14$
126	21	-5	25	$m=2$
33	22	-8	4	$m \geq 4$
55	22	-6	9	$m \geq 4$
176	22	-5	35	$m=2$
276	23	-5	55	$m=2$
576	24	-5	115	$m \geq 4$
50	25	-7	7	$m=2$
91	26	-6	15	$m=14$
49	28	-8	6	$m=7$
64	28	-7	9	$m=2$
145	29	-6	24	$m \geq 4$

Example 12.66. (SICs) For $n = d^2$ and $d = j^2 - 1$, $j = 2, 3, \ldots$, we have

$$\lambda_1 = -j, \qquad \lambda_2 = j(j^2-1), \qquad \sqrt{d+1} = j.$$

The first two cases are:

- $d = 3$, where there is a cube root signature matrix,
- $d = 8$, where there is a fourth root signature matrix (the Hoggar lines).

Both are exceptional cases for SICs. In view of this, one can speculate about the existence of a SIC for

$$d = 15, 24, 35, \ldots$$

with a simple form, which is not a Heisenberg SIC (see §14.5).

12.18 Algebraic equations for tight complex equiangular lines

The equations (12.22) determining the signature matrix Σ of an equiangular tight frame of n vectors for $\mathbb{C}^d$ each involve variables z_{jk} from each row/column of Σ. The number of these variables can be reduced by considering a reduced signature matrix $\hat{\Sigma}$ (Proposition 12.3). Here we consider necessary equations that involve only variables from principal submatrices of $\Sigma = [z_{jk}]$. These have been used to good effect by Szöllősi [Szö14] to investigate the existence of complex equiangular tight frames with small numbers of vectors. We now consider these ideas.

The Gramian of an equiangular tight frame of n unit vectors for $\mathbb{C}^d$, $n > d$ has the form

$$Q = I + \alpha\Sigma = \begin{pmatrix} 1 & \alpha z_{12} & \alpha z_{13} & \cdots & \alpha z_{1n} \\ \alpha\overline{z_{12}} & 1 & \alpha z_{23} & \cdots & \alpha z_{2n} \\ \alpha\overline{z_{13}} & \alpha\overline{z_{23}} & 1 & & \\ \vdots & \vdots & & \ddots & \\ \alpha\overline{z_{1n}} & \alpha\overline{z_{2n}} & & & 1 \end{pmatrix}, \qquad \alpha = \sqrt{\frac{n-d}{d(n-1)}}, \quad |z_{jk}| = 1, \tag{12.52}$$

where $P := \frac{d}{n}Q$ is a rank d orthogonal projection matrix (see §12.6). In particular,

(i) The $(d+1)\times(d+1)$ minors of $I+\alpha\Sigma$ are zero, giving $\binom{n}{d+1}^2$ equations.

From $P^2 = P$, i.e., $dQ^2 - nQ = 0$, a block matrix calculation (see Exer. 12.15) gives

(ii) If Q_{n-r} is a principal submatrix of $Q = I+\alpha\Sigma$ of size $n-r$, $0 \le r \le \frac{n}{2}$, then

$$\operatorname{rank}(dQ_{n-r}^2 - nQ_{n-r}) \le r.$$

For a given r, this gives $\binom{n}{r+1}^2$ equations.

Less obvious equations involving principal submatrices of Σ of size $n-2$ can be obtained from the equations (12.22) which characterise equiangular tight frames, by using the following identity for the triple product of three flat vectors in $\mathbb{C}^2$.

Lemma 12.2. *Let $x, y, z \in \mathbb{C}^2$, with $|x_1| = |x_2| = |y_1| = |y_2| = |z_1| = |z_2| = 1$, then*

$$\langle x,y\rangle\langle y,z\rangle\langle z,x\rangle = |\langle x,y\rangle|^2 + |\langle y,z\rangle|^2 + |\langle z,x\rangle|^2 - 4.$$

Proof. By direct calculation, using the fact that $\overline{\xi} = 1/\xi$, when $|\xi| = 1$. □

With $z_{jj} := 0$, the equations (12.21), which characterise equiangular tight frames, can be written

$$(n-2d)\sqrt{\frac{n-1}{d(n-d)}}z_{ab} = \sum_{\substack{s=1 \\ s\neq a,b}}^{n} z_{as}z_{sb} - (n-1)\delta_{ab}, \qquad 1 \le a,b \le n.$$

Thus, for $1 \le a,b \le n-2$, we have

$$c_{ab} := (n-2d)\sqrt{\frac{n-1}{d(n-d)}}z_{ab} - \sum_{\substack{s=1 \\ s\neq a,b}}^{n-2} z_{as}z_{sb} + (n-1)\delta_{ab} = \langle v_a, v_b \rangle, \qquad (12.53)$$

where $v_a := (z_{a,n-1}, z_{a,n}) \in \mathbb{C}^2$. Since each v_a is flat, we can apply Lemma 12.2 to obtain equations which depend only on the principal submatrix of the signature matrix $\Sigma = [z_{jk}]$ of size $n-2$.

Theorem 12.19. *([Szö14]) Let $n \ge 5$ and $\Sigma = [z_{jk}]$ be the signature matrix of an equiangular tight frame of n vectors for $\mathbb{C}^d$, $n > d$ (here $z_{jj} = 0$, $z_{jk} = \overline{z_{kj}}$, $j \neq k$). For $1 \le a,b \le n-2$, $a \neq b$, define*

$$c_{ab} := (n-2d)\sqrt{\frac{n-1}{d(n-d)}}z_{ab} - \sum_{\substack{s=1 \\ s\neq a,b}}^{n-2} z_{as}z_{sb}, \quad a \neq b, \qquad c_{aa} = 2. \qquad (12.54)$$

Then Σ satisfies

$$c_{jk}c_{k\ell}c_{\ell j} = |c_{jk}|^2 + |c_{k\ell}|^2 + |c_{\ell j}|^2 - 4, \qquad 1 \le j,k,\ell \le n-2. \qquad (12.55)$$

Proof. By Lemma 12.2, with $x = v_j$, $y = v_k$, $z = v_\ell$, we have (12.55), where the formula (12.53) defining c_{ab} can be written as (12.54). □

In [Szö14], the equations (i), (ii) and (12.55) for an $(n-2) \times (n-2)$ principal submatrix of the Gramian were used to give a description of all possible sets of n tight complex equiangular lines in $\mathbb{C}^3$. We now outline what they are.

Example 12.67. There exists a unique set of n tight equiangular lines in $\mathbb{C}^3$ (up to projective unitary equivalence) for $n = 3$ (orthonormal basis) and $n = 4$ (vertices of the tetrahedron). For $n = 5$, there are no tight equiangular lines (Example 12.57). For $n = 6$, there are tight real equiangular lines given by the diagonals of the regular icosahedron (Example 12.39). Further, a calculation (using Groebner bases) of the possible 4×4 principal submatrices shows that all the Gramian matrices for equiangular tight frames six vectors in $\mathbb{C}^3$ have the (reduced signature) form

$$G_6^{(1)}(a)=\frac{1}{\sqrt{5}}\left[\begin{array}{cccc|cc}\sqrt{5} & 1 & 1 & 1 & 1 & 1\\ 1 & \sqrt{5} & a & -a & -1 & 1\\ 1 & \bar{a} & \sqrt{5} & 1 & -\bar{a} & -1\\ 1 & -\bar{a} & 1 & \sqrt{5} & \bar{a} & -1\\ \hline 1 & -1 & -a & a & \sqrt{5} & 1\\ 1 & 1 & -1 & -1 & 1 & \sqrt{5}\end{array}\right], \qquad a\in\mathbb{C}, \quad |a|=1.$$

For $n=7$, there is a unique set of 7 equiangular lines (given by the harmonic frame for the $(7,3,1)$-difference set). For $n=8$, there are no tight equiangular lines in $\mathbb{C}^3$ (the equations have no solution). For $n=9$, there is a one-parameter family of SICs (see [BW07], [Zhu12], [Szö14]).

Example 12.68. For $\mathbb{C}^4$, sets of n tight equiangular lines are known for $n=4,5$ (orthonormal basis, simplex), $n=7,13$ (harmonic frames given by difference sets), $n=8$ (Example 12.63) and $n=16$ (a SIC). There are no tight equiangular lines for $n=6$ (Example 12.57), and so we can pose the following elementary open problem:

Is there an equiangular tight frame of 9 vectors for $\mathbb{C}^4$?

12.19 Mutually unbiased bases and s-angular tight frames

For an equal-norm frame Φ (or set of lines), its set of **angles** is

$$\mathrm{Ang}(\Phi):=\{|\langle v,w\rangle| : v,w\in\Phi \text{ are not scalar multiples}\}.$$

It is said to be s-**angular** (or has s angles) if $\mathrm{Ang}(\Phi)$ has s elements.

The 1-angular frames/lines are precisely the equiangular frames/lines, and MUBs are 2-angular tight frames. We now generalise Theorem 12.2.

Theorem 12.20. *Let Φ be a set of n unit vectors in $\mathbb{C}^d$ $(d>1)$ giving a system of n (distinct) lines with s angles $A=\mathrm{Ang}(\Phi)$, and*

$$p_v(x):=\langle x,v\rangle^k\prod_{\alpha\in A\setminus 0}\frac{|\langle x,v\rangle|^2-\alpha\langle x,x\rangle}{1-\alpha}, \qquad k:=\begin{cases}1, & 0\in A;\\ 0, & 0\notin A.\end{cases}$$

Then the polynomials $\{p_v\}_{v\in\Phi}$ are linearly independent, and hence

$$n\le\begin{cases}\binom{d+s-1}{s}^2, & 0\notin A;\\ \binom{d+s-1}{s}\binom{d+s-2}{s-1}, & 0\in A.\end{cases} \qquad (12.56)$$

Proof. Each p_v is a polynomial which is homogeneous of degree s in $x_1,\ldots,x_d$ and is homogeneous of degree $s-k$ in $\overline{x_1},\ldots,\overline{x_d}$. The space $\Pi^\circ_{s,s-k}(\mathbb{C}^d)$ of such polynomials has complex dimension $\binom{s+d-1}{d-1}\binom{s-k+d-1}{d-1}$ (see Exer. 6.17). By construction,

$$\delta_w(p_v) = p_v(w) = \delta_{v,w}, \qquad v,w \in \Phi,$$

so the $\{p_v\}_{v\in\Phi}$ are linearly independent, and we obtain (12.56). □

Remark 12.1. The same argument gives the following bound for vectors giving systems of lines in $\mathbb{R}^d$

$$n \le \begin{cases} \binom{2s+d-1}{d-1}, & 0 \notin A; \\ \binom{2s+d-2}{d-1}, & 0 \in A. \end{cases} \tag{12.57}$$

Example 12.69. For nonorthogonal equiangular lines ($s=1$, $0 \notin A$), (12.56) and (12.57) give Theorem 12.2.

Example 12.70. Spherical two-distance sets (and tight frames) are 2-angular frames for $\mathbb{R}^d$ (see §12.14). The estimate of the maximum size of a spherical two-distance set in $\mathbb{R}^d$ given by (12.57) has a higher order of growth in d than that given by Theorem 12.14 (though it is exact for $d=2$).

Example 12.71. The highly symmetric tight frames given by finite reflection groups (see §13.8, §13.10) are s-angular tight frames with s *small* (see Tables 13.1, 13.2).

For the 2-angular frame give by m mutually unbiased bases for $\mathbb{C}^d$ (see §2.11), the bound (12.56) gives

$$md \le \binom{d+1}{2}\binom{d}{1} = \frac{1}{2}d^2(d+1) \quad \Longrightarrow \quad m \le \frac{1}{2}d(d+1).$$

The following argument of [WF89] gives the sharper bound $m \le d+1$.

Proposition 12.12. *Let $\mathscr{B}$ be m mutually unbiased bases for $\mathbb{C}^d$. Then $m \le d+1$.*

Proof. Let P_v be the orthogonal projection onto $v \in \mathbb{C}^d$, so that $P_v - \frac{I}{d}$ is a traceless Hermitian operator. Since $B \in \mathscr{B}$ is an orthonormal basis, $\sum_{v\in\mathscr{B}} P_v = I$, so that the matrices $P_v - \frac{I}{d}$, $v \in B$ are linearly dependent and span a space V_B of dimension $\le d-1$. The spaces V_B, $B \in \mathscr{B}$ are orthogonal in the Frobenius norm, since for v and w in mutually unbiased bases, we have

$$\begin{aligned}\langle P_v - \frac{I}{d}, P_w - \frac{I}{d}\rangle &= \operatorname{trace}(P_vP_w) - \frac{1}{d}\operatorname{trace}(P_v) - \frac{1}{d}\operatorname{trace}(P_w) - \frac{1}{d^2}\operatorname{trace}(I) \\ &= \frac{1}{d} - \frac{2}{d} + \frac{1}{d} = 0.\end{aligned}$$

The dimension of the real vector space of traceless of Hermitian operators is d^2-1. Thus a dimension count gives $\dim(\oplus_{B\in\mathscr{B}} V_B \le m(d-1) \le d^2-1 = (d+1)(d-1)$, and cancelling $d-1$ gives the result. □

12.20 Mutually unbiased bases and Hadamard matrices

We now investigate MUBs in more detail. First consider the case of two MUBs (v_j) and (w_j) for $\mathbb{C}^d$. The synthesis operator of this tight frame $(v_j)\cup(w_j)$ is $[V,W]$, where $V=[v_j]$, $W=[w_j]$, and so its Gramian has the form

$$[V,W]^*[V,W]=\begin{pmatrix} V^*V & V^*W \\ W^*V & W^*W \end{pmatrix}=\begin{pmatrix} I & V^*W \\ W^*V & I \end{pmatrix}.$$

Thus, to understand two MUBs up to unitary equivalence, it suffices to consider the $d\times d$ matrix V^*W. This leads to the following connection with Hadamard matrices.

Lemma 12.3. *Two orthonormal bases (v_j) and (w_j) for $\mathbb{C}^d$ are mutually unbiased if and only if*

$$H:=\sqrt{d}V^*W=\sqrt{d}\begin{pmatrix} \langle w_1,v_1\rangle & \cdots & \langle w_d,v_1\rangle \\ \vdots & \ddots & \vdots \\ \langle w_1,v_d\rangle & \cdots & \langle w_d,v_d\rangle \end{pmatrix} \tag{12.58}$$

is a (complex) Hadamard matrix (of size d).

Proof. If $V=[v_j]$ and $W=[w_j]$ are the synthesis operators of orthonormal bases for $\mathbb{C}^d$, i.e., $V^*V=I$ and $WW^*=I$, then $H:=\sqrt{d}V^*W$ satisfies

$$H^*H=d(W^*V)(V^*W)=dW^*(VV^*)W=dWW^*=dI.$$

Thus (v_j) and (w_j) are mutually unbiased if and only if H has entries of modulus 1, i.e., H is a Hadamard matrix. □

We will refer to the H of (12.58) as the **Hadamard matrix of** (v_j) **and** (w_j). Conversely, if H is a (complex) Hadamard matrix of size d, then

$$Q=\begin{pmatrix} I & \frac{1}{\sqrt{d}}H \\ \frac{1}{\sqrt{d}}H^* & I \end{pmatrix} \tag{12.59}$$

is the Gramian of two mutually unbiased bases for $\mathbb{C}^d$. This follows since Q has the correct form, and is a tight frame for $\mathbb{C}^d$, since $P=\frac{1}{2}Q$ is an orthogonal projection of rank d, by the calculation

$$P^2=\frac{1}{4}\begin{pmatrix} I+\frac{1}{d}HH^* & \frac{1}{\sqrt{d}}H+\frac{1}{\sqrt{d}}H \\ \frac{1}{\sqrt{d}}H^*+\frac{1}{\sqrt{d}}H^* & \frac{1}{d}HH^*+I \end{pmatrix}=\frac{1}{2}\begin{pmatrix} I & \frac{1}{\sqrt{d}}H \\ \frac{1}{\sqrt{d}}H^* & I \end{pmatrix}=P.$$

Two MUBs for $\mathbb{C}^d$ are determined up to unitary equivalence by their Hadamard matrix, and each Hadamard matrix corresponds to two MUBs.

The vectors (v_j) and (w_j) can be multiplied by unit scalars (α_j) and (β_j) so that the Hadamard matrix of (12.58) has each entry of it first row and column 1. This is called a **dephased Hadamard matrix**. In this way, each pair of MUBs corresponds to a *unique* dephased Hadamard matrix: once α_1 is chosen the remaining scalars are determined by

$$\langle \beta_k w_k, \alpha_1 v_1 \rangle = 1, \quad 1 \leq k \leq d, \qquad \langle \beta_1 w_1, \alpha_j v_j \rangle = 1, \quad 2 \leq j \leq d, \tag{12.60}$$

with each choice of α_1 giving the same dephased Hadamard matrix, which we call the **dephased Hadamard matrix of** (v_j) **and** (w_k).

Proposition 12.13. *Two mutually unbiased bases for $\mathbb{C}^d$ are uniquely determined up to projective unitary equivalence by their dephased Hadamard matrix.*

Proof. Let (v_j) and (w_j) be mutually unbiased bases for $\mathbb{C}^d$. We already observed that their dephased Hadamard matrix is projectively unitarily invariant. Indeed, by solving (12.60), we see that the (j,k)-entry of the dephased Hadamard matrix is the 4-product

$$\langle \beta_k w_k, \alpha_j v_j \rangle = \langle w_k, v_j \rangle \langle v_j, w_1 \rangle \langle w_1, v_1 \rangle \langle v_1, w_k \rangle.$$

It therefore suffices to show that (v_j) and (w_j) are determined (up to projective unitary equivalence) by their dephased Hadamard matrix H. This follows from the formula (12.59), which gives the Gramian Q of a frame that is projectively unitarily equivalent to $(v_j) \cup (w_j)$. $\square$

Hadamard matrices H_1 and H_2 are said to be **equivalent** if there are unitary diagonal matrices $\Lambda_1 = \operatorname{diag}(\alpha_j)$, $\Lambda_2 = \operatorname{diag}(\beta_j)$ and permutation matrices P_σ, P_τ, with

$$H_1 = \Lambda_1^{-1} P_\sigma^{-1} H_2 P_\tau \Lambda_2$$

In particular, for the Hadamard matrix $H = \sqrt{d}[\langle w_k, v_j \rangle]$ of (12.58) is equivalent to

$$\Lambda_1^{-1} P_\sigma^{-1} H P_\tau \Lambda_2 = \Lambda_1^{-1} [\langle w_{\tau k}, v_{\sigma j} \rangle] \tau \Lambda_2 = [\langle \beta_k w_{\tau k}, \alpha_j v_{\sigma j} \rangle].$$

Thus we have:

Two MUBs for $\mathbb{C}^d$ are determined up to projective unitary equivalence by their dephased Hadamard matrix H, and are determined to projective unitary equivalence after a reordering by H up to Hadamard matrix equivalence.

A *catalogue of complex Hadamard matrices* was given in [TZ06], and an online version (for $2 \leq d \leq 16$) is maintained by Bruzda, Tadej and Życzkowski.

Example 12.72. For $d = 2,3,5$, there is a unique complex Hadamard matrix of size d up to equivalence, which is given by the Fourier matrix (see [Haa97], which uses the language of *maximal abelian $*$-algebras*). For $d = 4$, there is one-parameter family of inequivalent Hadamard matrices given by

$$F_4^{(1)}(a) = \begin{pmatrix} 1 & 1 & 1 & 1 \\ 1 & ie^{ia} & -1 & -ie^{ia} \\ 1 & -1 & 1 & -1 \\ 1 & -ie^{ia} & -1 & ie^{ia} \end{pmatrix}, \qquad 0 \le a < \pi.$$

Motivated by this example, it was conjectured that there is a unique Hadamard matrix of size d (up to equivalence), given by the Fourier matrix, when d is a prime. However, this is not the case for primes $p \ge 7$ (see the discussion of [Szö10]).

Example 12.73. The Hadamard matrices of size $d = 6$ have not been fully classified. There is evidence that their variety has dimension 4 (see [SNS09], [Szö12]).

Let H be $d \times d$ matrix with complex entries of unit modulus. Then equations

$$H^*H = dI$$

characterise when H is a Hadamard matrix. These are the analogue of the equations $\Sigma^2 - (\lambda_1 + \lambda_2)\Sigma - (n-1)I = 0$ for the signature matrix of a equiangular tight frame (Corollary 12.1). Define the dephased form of H as for when it is a Hadamard matrix

$$\begin{pmatrix} 1 & \mathbf{1}^* \\ \mathbf{1} & \hat{H} \end{pmatrix}, \qquad \mathbf{1} = (1,\ldots,1)^t \in \mathbb{R}^{d-1}, \quad \hat{H} \in \mathbb{C}^{(d-1)\times(d-1)}.$$

We call $\hat{H}$ (which has unit modulus entries) the **reduced Hadamard matrix** of H. The analogue of the equations of Proposition 12.3 is as follows.

Proposition 12.14. *Let $\hat{H}$ be the reduced Hadamard matrix of a $d \times d$ matrix with entries of unit modulus. Then $\hat{H}$ gives a $d \times d$ Hadamard matrix if and only if*

1. $\hat{H}^*\hat{H} = dI - J$, $J := \mathbf{1}^*\mathbf{1}$.
2. $\mathbf{1}$ *is an eigenvector of $\hat{H}$ and of $\hat{H}^*$ for eigenvalue* -1.

Proof. Block multiplication of the condition for being a Hadamard matrix gives

$$\begin{pmatrix} 1 & \mathbf{1}^* \\ \mathbf{1} & \hat{H} \end{pmatrix}^* \begin{pmatrix} 1 & \mathbf{1}^* \\ \mathbf{1} & \hat{H} \end{pmatrix} = \begin{pmatrix} d & \mathbf{1}^* + \mathbf{1}^*\hat{H} \\ \mathbf{1} + \hat{H}^*\mathbf{1} & \mathbf{1}\mathbf{1}^* + \hat{H}^*H \end{pmatrix} = dI = \begin{pmatrix} d & 0 \\ 0 & dI_{d-1} \end{pmatrix}.$$

Equating the blocks gives 1, and that $\mathbf{1}$ is an eigenvector of $\hat{H}^*$ for eigenvalue -1. By considering the condition $HH^* = dI$ instead, we conclude that $\mathbf{1}$ must also be an eigenvector of $\hat{H}$ (for eigenvalue -1). □

By using Lemma 12.2 (as in Theorem 12.19), one can obtain necessary equations for being a Hadamard matrix that depend only on the submatrices of H of size $d-2$ (see [Haa97]). We now briefly consider how to go from *two* mutually unbiased bases (a Hadamard matrix) to *three* or more.

Theorem 12.21. *There are m mutually unbiased bases for $\mathbb{C}^d$ if and only if there are $m-1$ Hadamard matrices $H_2,\ldots,H_m$ of size d, for which the $\frac{1}{2}(m-1)(m-2)$ matrices*

$$\frac{1}{\sqrt{d}}H_j^*H_k, \qquad 2\le j<k\le m,$$

have entries of modulus of 1.

Proof. Let $V_1,\ldots,V_m$ be the synthesis maps for m mutually unbiased bases $\mathscr{B}_1,\ldots,\mathscr{B}_m$ for $\mathbb{C}^d$. Let H_j be the Hadamard matrix for $\mathscr{B}_1$ and $\mathscr{B}_m$, i.e., $H_m = \sqrt{d}V_1^*V_m$. Then the Hadamard matrix for $\mathscr{B}_j$ and $\mathscr{B}_k$ is

$$H_{jk} := \sqrt{d}V_j^*V_k = \sqrt{d}V_j^*(V_1V_1^*)V_k = \frac{1}{\sqrt{d}}(\sqrt{d}V_1^*V_j)^*(\sqrt{d}V_1^*V_k) = \frac{1}{\sqrt{d}}H_j^*H_k.$$

Thus all the Hadamard matrices for the $\mathscr{B}_1,\ldots,\mathscr{B}_m$ are determined by $H_2,\ldots,H_m$ by the above formula for $2\le j<k\le m$. These matrices satisfy

$$H_{jk}^*H_{jk} = \frac{1}{d}(H_j^*H_k)(H_k^*H_j) = \frac{1}{d}H_j^*(dI)H_j = \frac{1}{d}H_j^*(dI)H_j = dI,$$

and so are Hadamard (giving mutually unbiased bases) if and only if they have entries of modulus 1. □

The condition that $H_j^*H_k$ have entries of constant modulus gives the following concrete construction for the mutually unbiased bases.

If $H_2,\ldots,H_m$ are the Hadamard matrices of Theorem 12.21, then the columns of the $d\times d$ matrices

$$I,\frac{1}{\sqrt{d}}H_2,\ldots,\frac{1}{\sqrt{d}}H_m$$

give m mutually unbiased bases for $\mathbb{C}^d$.

Example 12.74. For the three MUBs for $\mathbb{C}^2$ of Example 2.18, i.e.,

$$\mathscr{B}_1 = \left\{\begin{bmatrix}1\\0\end{bmatrix},\begin{bmatrix}0\\1\end{bmatrix}\right\},\quad \mathscr{B}_2 = \left\{\frac{1}{\sqrt{2}}\begin{bmatrix}1\\1\end{bmatrix},\frac{1}{\sqrt{2}}\begin{bmatrix}1\\-1\end{bmatrix}\right\},\quad \mathscr{B}_3 = \left\{\frac{1}{\sqrt{2}}\begin{bmatrix}1\\i\end{bmatrix},\frac{1}{\sqrt{2}}\begin{bmatrix}1\\-i\end{bmatrix}\right\},$$

the Hadamard matrices of Theorem 12.21 (which are all equivalent) are

$$H_1 = \begin{pmatrix}1 & 1\\ 1 & -1\end{pmatrix},\quad H_2 = \begin{pmatrix}1 & 1\\ i & -i\end{pmatrix},\quad \text{with}\quad H_{23} = \frac{1}{\sqrt{2}}H_2^*H_3 = \frac{1}{\sqrt{2}}\begin{pmatrix}1+i & 1-i\\ 1-i & 1+i\end{pmatrix}.$$

12.21 Examples of MUBs

Let F be the $d \times d$ Fourier matrix as defined by (2.5), i.e.,

$$F_{jk} := \omega^{jk}, \qquad j,k \in \mathbb{Z}_d, \quad \omega := e^{\frac{2\pi i}{d}}.$$

The columns of F are an orthonormal basis for $\mathbb{C}^d$. Since F has entries of constant modulus, this basis is mutually unbiased to the standard basis (e_j). This still holds if the rows/columns of F are multiplied by unit scalars. In this way, we seek further mutually unbiased bases. Suitable scalars can be described in terms of the diagonal matrix R of §14.7, which is given by

$$R_{jk} := \mu^{j(j+d)} \delta_{jk}, \qquad j,k \in \mathbb{Z}_d, \quad \mu := e^{\frac{2\pi i}{2d}}. \tag{12.61}$$

This matrix plays a key role in the description of the known SICs (see Chapter 14).

Theorem 12.22. *Let F be the $d \times d$ Fourier matrix, R be the $d \times d$ diagonal matrix given by (12.61) and define orthonormal bases by*

$$\mathscr{E} := \{e_j\}, \qquad \mathscr{B}_\ell := \{R^\ell F e_j\}.$$

Then for $\ell, m \in \mathbb{Z}_d$, the following pairs are mutually unbiased bases

1. $\{\mathscr{B}_\ell, \mathscr{E}\}$.
2. $\{\mathscr{B}_\ell, \mathscr{B}_m\}$, $\ell - m \in \mathbb{Z}_d^*$.

In particular, for d a prime, $\mathscr{E}, \mathscr{B}_0, \ldots, \mathscr{B}_{d-1}$ are $d+1$ mutually unbiased bases.

Proof. We have already observed that $\mathscr{B}_\ell$ and $\mathscr{E}$ are mutually unbiased, since

$$|\langle R^\ell F e_k, e_j \rangle| = |(R^\ell F)_{jk}| = \frac{1}{\sqrt{d}}.$$

To show that $\mathscr{B}_\ell$ and $\mathscr{B}_k$ are mutually unbiased, we require

$$|\langle R^\ell F e_k, R^m F e_j \rangle| = |(F^{-1} R^{\ell - m} F)_{jk}| = \frac{1}{\sqrt{d}}, \qquad a := \ell - m \in \mathbb{Z}_d,$$

i.e., the matrix $F^{-1} R^a F$, $a \in \mathbb{Z}_d$, has entries of constant modulus (for $a \neq 0$). In other words, $H_a := \sqrt{d} F^{-1} R^a F$ is a Hadamard matrix. This can be done by a direct calculation using Gauss sums (see Exercise 12.19), or as follows.

A calculation (see Exer. 12.19) with Gauss sum shows that

$$(F^{-1} R^a F)_{jk} = \mu^{a(d-1)} \omega^{a+j-k} (F^{-1} R^a F)_{j,k-a}. \tag{12.62}$$

Since $F^{-1} R^a F$ is circulant (it is diagonalised by the Fourier matrix), this implies that it has entries of constant modulus when a is a unit (and so generates $\mathbb{Z}_d$). □

Example 12.75. For $d = 3$, the diagonal matrix is $R = \mathrm{diag}(1, \omega^2, \omega^2)$, $\omega = e^{\frac{2\pi i}{3}}$. Thus four mutually unbiased bases for $\mathbb{C}^3$ are given by the standard basis, together with the columns of the matrices

$$F = \frac{1}{\sqrt{3}} \begin{bmatrix} 1 & 1 & 1 \\ 1 & \omega & \omega^2 \\ 1 & \omega^2 & \omega \end{bmatrix}, \quad RF = \frac{1}{\sqrt{3}} \begin{bmatrix} 1 & 1 & 1 \\ \omega^2 & 1 & \omega \\ \omega^2 & \omega & 1 \end{bmatrix}, \quad R^2F = \frac{1}{\sqrt{3}} \begin{bmatrix} 1 & 1 & 1 \\ \omega & \omega^2 & 1 \\ \omega & 1 & \omega^2 \end{bmatrix}.$$

Example 12.76. For $d = 6$ (which is not a prime power), any pair of (cyclically) consecutive bases $\mathscr{B}_0, \ldots, \mathscr{B}_5$ are mutually unbiased. Therefore $\mathscr{E}$ together with any consecutive pair $\mathscr{B}_j, \mathscr{B}_{j+1}$ gives *three* mutually unbiased bases for $\mathbb{C}^6$.

A long-standing open question is the *MUB problem*:

Do there exist more than three mutually unbiased bases in $\mathbb{C}^6$?

There are various constructions of $d+1$ MUBs for $\mathbb{C}^d$ when d is a prime, or a prime power (see [GR09]). All of these constructions are cases of a construction of [CCKS97] based on *symplectic spreads* and $\mathbb{Z}_4$-*Kerdock codes*, which gives many projectively unitarily inequivalent sets of $d+1$ MUBs for $\mathbb{C}^d$, when d is an *odd power of two*. We now give the construction of [WF89] (also see [KR04]).

Example 12.77. Let $d = p^n$ be an odd prime power, $\mathbb{F}_d$ be the Galois field of order d (viewed as an extension of $\mathbb{Z}_p$), with *field trace* $\mathrm{tr} : \mathbb{F}_d \to \mathbb{Z}_p$. Define $d \times d$ matrices

$$V_a = \frac{1}{\sqrt{d}} \left[\omega^{\mathrm{tr}(aj^2+jk)}\right]_{j,k\in\mathbb{F}_d}, \qquad a \in \mathbb{F}_d.$$

Then I, V_a, $a \in \mathbb{F}_d$ are the synthesis maps of $d+1$ mutually unbiased bases for $\mathbb{C}^d$. For $d = p > 2$, the field trace is the identity, and so the matrices V_a are given by

$$V_a = R^{2a}F, \qquad a \in \mathbb{Z}_d,$$

which are the bases of Theorem 12.22 (since 2 is a unit).

Example 12.78. Let $d = 2^n$ be an even prime power, and $\mathscr{R}$ be the Galois ring $\mathrm{GR}(2^2, n)$ with Teichmüller set $\mathscr{T}_n$ and trace $\mathrm{tr} : \mathscr{R} \to \mathbb{Z}_4$. Define $d \times d$ matrices by

$$W_a = \frac{1}{\sqrt{d}} \left[i^{\mathrm{tr}(aj^2+2jk)}\right]_{j,k\in\mathbb{T}_n},$$

Then I, W_a, $a \in \mathscr{T}_n$ are the synthesis maps of $d+1$ mutually unbiased bases for $\mathbb{C}^d$.

The examples of this section can be viewed as the eigenvectors of orthogonal (in the Frobenius inner product) commuting (up to a scalar) unitary matrices, e.g., see [BBRV02], [KR04], [GR09], Exer. 14.7 and Theorem 12.22).

Notes

There is currently considerable activity on determining estimates of the maximal number $M(d)$ of real equiangular lines in $\mathbb{R}^d$ (see §12.8 and Table 12.3), e.g., [Gre16] reduced the bound for equiangular lines in $\mathbb{R}^{18}$ from 48–61 to 48–60, then [Szö17] improved it to 54–60, [BY14] give bounds on $M(d)$ obtained by semidefinite programming, and [Buk16], [BDKS16] give asymptotic estimates in d for when the angle α is fixed. Maximal relative projection constants and equiangular tight frames are studied in [FS17].

There are various internet sites with lists of equiangular tight frames (also see [FM15]) and the associated geometric structures such as *difference sets*, *Steiner systems*, *strongly regular graphs* and *Hadamard matrices*. An excellent resource is the talks and follow-up preprints on the webpage of the workshop *Systems of Lines: Applications of Algebraic Combinatorics* organised by Bill Martin.

Thanks to Alexander Barg, Peter Cameron, Simon Foucart, Gary Greaves, John Jasper, Ferenc Szöllősi and Wei-Hsuan Yu for many insightful discussions.

Exercises

12.1. Show that the dimension of the real vector space of $d \times d$ Hermitian matrices is $\frac{1}{2}d(d+1)$ when $\mathbb{F} = \mathbb{R}$, and is d^2 when $\mathbb{F} = \mathbb{C}$.

12.2. Suppose that $(P_j)_{j=1}^n$ are the projections onto a set of equiangular lines in $\mathbb{F}^d$, $d > 1$, with constant $C = \alpha$, and the identity can be written as $I = \sum_{j=1}^n c_j P_j$.
(a) Show that $c_j = \frac{d}{n}$, $\forall j$.
(b) Show that

$$n(1 - d\alpha^2) = d(1 - \alpha^2),$$

and hence $\alpha^2 < \frac{1}{d}$,

$$n = \frac{d(1-\alpha^2)}{1-\alpha^2}, \qquad \alpha^2 = \frac{n-d}{d(n-1)}.$$

12.3. Suppose that (f_j) is a finite normalised tight frame of *nonzero* vectors for $\mathbb{F}^d$ $(d \geq 2)$ which satisfy the equiangularity condition

$$\Big|\langle \frac{f_j}{\|f_j\|}, \frac{f_k}{\|f_k\|} \rangle\Big| = C, \quad j \neq k.$$

Show that (f_j) is an equal-norm frame, i.e., it is an equiangular tight frame.

12.4. Show that the 28 unit vectors of (12.6) are equiangular.

12.5. Let $\Phi = (f_j)$ be n unit vectors in $\mathbb{R}^2$. Show that (12.3) can be sharpened to

$$\mathscr{M}_\infty(\Phi) := \max_{j\neq k} |\langle f_j, f_k\rangle| \geq \cos\Big(\frac{\pi}{n}\Big) > \sqrt{\frac{n-2}{2(n-1)}}, \qquad n > 3$$

with equality if and only if the f_j are the first n vertices of the regular $2n$-gon (up to multiplication by ± 1), i.e., they give n equally spaced lines in $\mathbb{R}^2$.

12.6. Show that if $\Phi = (f_j)$ is a unit-norm tight frame of n vectors for $\mathbb{F}^d$ which minimises

$$\mathscr{M}_\infty(\Phi) := \max_{j\neq k} |\langle f_j, f_k\rangle|,$$

then so is the complementary tight frame $\Psi = (g_j)$ for $\mathbb{F}^{n-d}$ (scaled appropriately).

12.7. Let Φ be the nontight equiangular frame of five vectors for $\mathbb{R}^3$ given by Γ the 5-cycle (see §12.13).
(a) Calculate the minimal angle between the five lines given by Φ, $\tilde{\Phi}$, and Φ^{can}.
(b) Show the vectors in Φ lie on five of the six diagonals of the regular icosahedron.
Remark: Since the solution to the Grassmannian packing problem for five lines in $\mathbb{R}^3$ is given by five diagonals of the regular icosahedron (see [CHS96]), we conclude that Φ is a *nontight* Grassmannian frame.
(c) Show that Φ^{can} is the harmonic frame given by the fifth roots of unity.
Hint: The vertices of the regular icosahedron are given by the cyclic permutations of the vector $(0, \pm 1, \pm c)$, where $c = \frac{1+\sqrt{5}}{2}$ is the *golden ratio*.

12.8. If a (nontight) sequence of unit vectors defines a set of equiangular lines, then does the dual frame define a set of equiangular lines?

12.9. Suppose that (v_j) is a set of n flat equiangular lines (see Proposition 12.1). Let A be the Gram matrix of the rank one orthogonal projections $v_1 v_1^*, \ldots, v_n v_n^*$ and $e_1 e_1^*, \ldots, d_d e_d^*$ (with the Frobenius inner product), i.e.,

$$A = \begin{pmatrix} \alpha^2 J + (1-\alpha^2) I_n & \frac{1}{d} J \\ \frac{1}{d} J & I_d \end{pmatrix}, \qquad J = [1].$$

(a) Row reduce A to an upper triangular matrix B (to determine its rank).
(b) Show that $n+d-1 \leq \mathrm{rank}(A) \leq n+d$, and Φ is tight when $\mathrm{rank}(A) = n+d-1$.

12.10. Let $(v_{a,j})$ be the Steiner equiangular tight frame of Theorem 12.4 given by a $(2,3,v)$-Steiner system $\mathscr{B}$ on points $\mathscr{V}$ (a Steiner triple system) and Hadamard matrices $H^{(a)}$, $a \in \mathscr{V}$. Show that the vectors

$$(\tilde{v}_{a,j})_{a\in\mathscr{V}, 1\leq j\leq r+1} \cup (\tilde{w}_\ell)_{1\leq \ell\leq v+1}$$

given by (12.17) and (12.18) form an equiangular tight frame for $\mathbb{C}^{\mathscr{B}} \oplus \mathbb{C}^{\mathscr{V}} \oplus \mathbb{C}$, with angle $\alpha = \frac{1}{r+2}$.

12.11. (Conference matrices) Let $\Sigma = iC$ be an $n \times n$ complex signature matrix with off diagonal entries $\pm i$. Here C has zero diagonal and off diagonal entries ± 1.
(a) Show that $C = -C^T$, i.e., C must be skew-symmetric.
(b) Show that Σ gives an equiangular tight frame of n vectors for $\mathbb{C}^d$ if and only if $C^T C = (n-1)I$, i.e., C is skew-symmetric conference matrix, where $n = 2d$.

12.12. Let C be a skew-symmetric conference matrix of size $n+1$ in the standard form

$$C = \begin{pmatrix} 0 & \mathbf{1}^* \\ -\mathbf{1} & A \end{pmatrix}, \qquad \mathbf{1} := (1,1,\ldots,1).$$

These exist for $n = 2^k - 1$ (Example 12.63) and $n = p^m$ an odd prime power with $n \equiv 3 \pmod 4$ (with A the Seidel adjacency matrix (12.64) of the Paley digraph).
(a) Show that A satisfies $A^2 = J - nI$ and $AJ = JA = 0$, where $J = \mathbf{1}\mathbf{1}^*$.
(b) Show that the $n \times n$ matrix

$$\Sigma := \frac{1}{\sqrt{n+1}}\big(\pm(J-I) + i\sqrt{n}A\big) \tag{12.63}$$

is the signature matrix of an equiangular tight frame of n vectors for $\mathbb{C}^d$, where $d = \frac{n\pm 1}{2}$ $(n = 2d \mp 1)$.
(c) Show that $\Sigma + \zeta I$, $\zeta = \frac{1}{\sqrt{n+1}}(\sqrt{n}i \pm 1)$, is an $n \times n$ complex Hadamard matrix.

12.13. Let $q = p^m = 4m-1$ be a (necessarily odd) prime power. Let S be the set of all nonzero squares in the Galois field $\mathrm{GF}(q)$, i.e.,

$$S := \{x^2 : x \in \mathrm{GF}(q), x \neq 0\}.$$

This is a $(4m-1, 2m-1, m-1)$-difference set for $G := (\mathrm{GF}(q),+) \cong \mathbb{Z}_p^m$, i.e., gives an equiangular harmonic frame of $4m-1$ vectors for $\mathbb{C}^{2m-1}$ (by Theorem 12.3). Denote the nonsquares in G by $N := G \setminus (S \cup \{0\})$. The $q \times q$ matrix given by

$$A_{jk} := \begin{cases} 0, & j = k; \\ 1, & j-k \in S; \\ -1, & j-k \in N, \end{cases} \tag{12.64}$$

is the Seidel adjacency matrix of the Paley digraph.
(a) Show that $S \cup \{0\}$ is a $(4m-1, 2m, m)$-difference set for G.
(b) Show that N and $N \cup \{0\}$ are difference sets for G.
(c) Show that there are equiangular tight frames of $4m-1$ vectors for $\mathbb{C}^{2m-1}$ and $\mathbb{C}^{2m}$ for which the inner products between their vectors have a constant real part.
(d) Show that there is an equiangular tight frame of $4m$ vectors for $\mathbb{C}^{2m}$.

12.14. Let $\hat{\Sigma}$ be the reduced signature matrix of an equiangular tight frame of $n > d+1$ vectors for $\mathbb{R}^d$, which corresponds (by Theorem 12.12) to a strongly regular graph

$$\mathrm{srg}\big(n-1,k,\frac{3k-n}{2},\frac{k}{2}\big), \qquad k:=\frac{1}{2}n-1+\big(1-\frac{n}{2d}\big)\sqrt{\frac{d(n-1)}{n-d}}.$$

Show that the complementary equiangular tight frame of n vectors for $\mathbb{R}^{n-d}$ has reduced signature matrix $-\hat{\Sigma}$ and so is given by the complementary strongly regular graph, which has the parameters

$$\mathrm{srg}\big(n-1,n-k-2,\frac{2n-3k-6}{2},\frac{n-k-2}{2}\big).$$

12.15. The Gramian matrix Q of an equiangular tight frame of n *unit* vectors for $\mathbb{C}^d$ satisfies $dQ^2-nQ=0$. Show that this implies

$$\mathrm{rank}(dQ_{n-r}^2-nQ_{n-r})\le r,$$

where Q_{n-r} is any principal submatrix of Q of size $n-r$, $0\le r<n$.

12.16. Let G be the Gramian of an equiangular tight frame of n unit vectors for $\mathbb{C}^d$, and U be the $n\times n$ matrix

$$U:=I-2\frac{d}{n}G.$$

(a) Show that U is Hermitian, unitary, with constant diagonal entries $1-2\frac{d}{n}$ and constant modulus off diagonal entries.
(b) Suppose that U is an $n\times n$ Hermitian unitary matrix with constant diagonal entries $\lambda=1-2\frac{d}{n}\in[-1,1]$ and constant modulus off diagonal entries. Show that

$$G:=\frac{n}{2d}(I-U)$$

is the Gramian of an equiangular tight frame of n unit vectors for $\mathbb{C}^d$.
Remark: A matrix B is *unistochastic* if it has the form $B_{jk}=|U_{jk}|^2$, for U unitary. Such a B is *bistochastic* (its rows and columns sum to 1). The above correspondence between equiangular tight frames and unistochastic matrices with constant diagonal and constant modulus off diagonal entries is considered by [GT16].

12.17. Let (v_j) be a unit-norm tight frame of n vectors for $\mathbb{F}^d$.
(a) For $0<p<2$, show that

$$\sum_j\sum_k|\langle v_j,v_k\rangle|^p\le\frac{(\frac{n^2}{d}-n)^{\frac{p}{2}}}{(n^2-n)^{\frac{p}{2}-1}}+n,$$

with equality if and only if (v_j) is an equiangular tight frame.
(b) For $2<p<\infty$, show that

$$\sum_j\sum_k|\langle v_j,v_k\rangle|^p\ge\frac{(\frac{n^2}{d}-n)^{\frac{p}{2}}}{(n^2-n)^{\frac{p}{2}-1}}+n,$$

with equality if and only if (v_j) is an equiangular tight frame.
(c) Show that for $p = 2$ equality holds in the inequalities of parts (a) and (b).

12.18. Let H_1 and H_2 be $d \times d$ (complex) Hadamard matrices. By (12.58), these are given by pairs of MUBs for $\mathbb{C}^d$ with synthesis operators $[V_1, W_1]$ and $[V_2, W_2]$. Suppose that these two pairs of MUBs are projectively unitarily equivalent up to a reordering which maps the respective bases to each other, i.e., there is a unitary map U, unitary diagonal matrices $\Lambda_1 = \operatorname{diag}(\alpha_j)$, $\Lambda_2 = \operatorname{diag}(\beta_j)$ and permutation matrices P_σ, P_τ, with

$$U[V_1, W_1] = [V_2, W_2] \begin{pmatrix} P_\sigma \Lambda_1 & \\ & P_\tau \Lambda_2 \end{pmatrix}. \tag{12.65}$$

(a) Show that the condition (12.65) is equivalent to H_1 and H_2 being Hadamard matrix equivalent.
(b) Use the fact that a pair of MUBs for $\mathbb{C}^d$ is determined up to projective unitary equivalence by its frame graph and the nonzero 4-products (the 4-cycles span the frame graph) to show that H_1 and H_2 are Hadamard matrix equivalent if and only if

$$(H_1)_{jk}(H_1)_{k\ell}(H_1)_{\ell m}(H_1)_{mj} = (H_2)_{\sigma j,\tau k}(H_2)_{\sigma k,\tau \ell}(H_2)_{\sigma \ell,\tau m}(H_2)_{\sigma m,\tau j},$$

for some permutations $\sigma, \tau \in S_d$. In particular, the multiset

$$\{H_{jk}H_{k\ell}H_{\ell m}H_{mj} : 1 \le j,k,\ell,m \le d\}$$

must be the same for all equivalent Hadamard matrices H.

12.19. *Gauss sums.* Let F and R be given by (2.5) and (12.61), and $a \in \mathbb{Z}_d^*$ be a unit. Since $F^{-1}R^aF$ is diagonalised by the Fourier matrix F, it is a circulant matrix.
(a) Show that the entries of $F^{-1}R^aF$ are *generalised Gauss sums*, i.e.,

$$(F^{-1}R^aF)_{jk} = \frac{1}{2d} G(a, ad + 2(k-j), 2d), \quad G(a,b,c) := \sum_{n=0}^{c-1} e^{\frac{2\pi i}{c}(an^2+bn)}.$$

(b) Let $\overline{b}$ be the multiplicative inverse of a unit $b \in \mathbb{Z}_d^*$. Show that

$$G(a, ad + 2(k-j), 2d) = \begin{cases} 2\omega^{\frac{d-1}{2}\overline{a}(k-j)^2} G(2a, 0, d), & d \text{ odd;} \\ \mu^{-a(\frac{d}{2}+\overline{a}(k-j))^2} G(a, 0, 2d), & d \text{ even.} \end{cases}$$

(c) Since $2a$ is a unit for d odd, and a is odd for d even, the formulas for classical Gauss sums give

$$|G(2a,0,d)| = \sqrt{d}, \quad d \text{ odd}, \qquad |G(a,0,2d)| = \sqrt{2}\sqrt{2d}, \quad d \text{ even}.$$

Use this to prove that the entries of $F^{-1}R^aF$ have constant modulus.

Chapter 13
Tight frames generated by nonabelian groups

If G is a finite *abelian* group, then there are a *finite* number of tight G-frames, i.e., the harmonic frames (see §11). If G is *nonabelian*, then there is an *uncountable* number of unitarily inequivalent G-frames (see Proposition 10.1). To illustrate this, consider the smallest nonabelian group $G = D_3 = \langle a, b \rangle \cong S_3$, the dihedral group of order 6, acting on $\mathbb{R}^2$ as unitary transformations via:

$$a = \text{rotation through } \tfrac{2\pi}{3}, \qquad b = \text{reflection in the } x\text{-axis}. \tag{13.1}$$

Then (see Figure 13.1), for each of the unit vectors $v_\theta := (\cos\theta, \sin\theta), 0 \le \theta \le \frac{\pi}{6}$, the tight D_3-frames $(gv_\theta)_{g \in D_3}$ are unitarily inequivalent (since their angles differ).

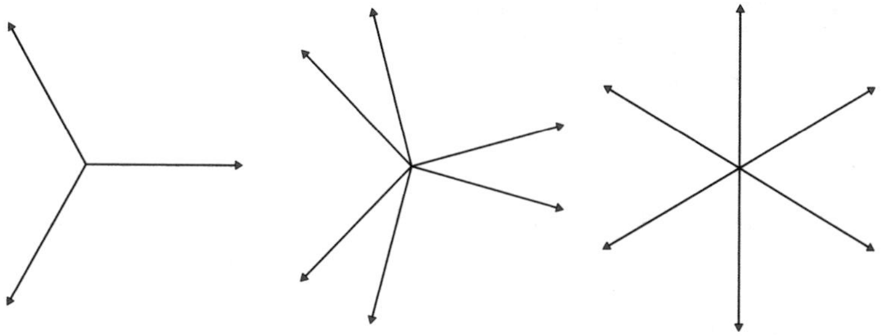

Fig. 13.1: The unitarily inequivalent tight D_3-frames given by v_θ, for $\theta = 0, \frac{\pi}{12}, \frac{\pi}{6}$.

Here we study the tight G-frames $(gv)_{g \in G}$ for G nonabelian by:

- Showing that $(gv)_{g \in G}$ corresponds to an element of the group algebra $\mathbb{C}G$.
- Putting additional restrictions on $(gv)_{g \in G}$ to obtain a *finite* set of G-frames, e.g., the *central* G-frames and the *highly symmetric* G-frames.
- Investigating nonabelian groups G which come as projective representations of nice groups. These give interesting tight frames, e.g., all the known SICs are G-frames for a projective representation of an abelian group (see §14.5).

S.F.D. Waldron, *An Introduction to Finite Tight Frames*, Applied and Numerical Harmonic Analysis, https://doi.org/10.1007/978-0-8176-4815-2_13

13.1 The identification of the G-matrices with the group algebra

Let G be a finite group. By Corollary 10.2, $(\phi_g)_{g\in G}$ is a normalised tight G-frame if and only if its Gramian is a G-matrix which is a projection. To understand such projections, we now consider the structure of the algebra of G-matrices.

Definition 13.1. Given a function $\nu : G \to \mathbb{C}$, let $M(\nu)$ be the G-matrix

$$M(\nu) := [\nu(g^{-1}h)]_{g,h\in G}. \tag{13.2}$$

The **group algebra** $\mathbb{C}G$ of G is the algebra obtained from the complex vector space with basis the elements of G, and the multiplication given by extending the multiplication in G linearly.

Proposition 13.1. *(G-matrix algebra) Let G be a finite group. Then the G-matrices form an algebra, i.e., the sum and product of G-matrices is again a G-matrix. This algebra is isomorphic to the group algebra $\mathbb{C}G$, via the map*

$$\pi : M(\nu) \mapsto \sum_{g\in G} \nu(g)g. \tag{13.3}$$

Proof. Clearly, M is an injective linear map from $\mathbb{C}^G$ onto the G-matrices, which therefore form a vector space. The product of G-matrices is a G-matrix since

$$M(\nu)M(\mu) = M(\nu * \mu), \qquad (\nu * \mu)(g) := \sum_{h\in G} \nu(gh)\mu(h^{-1}), \tag{13.4}$$

and so the G-matrices form an algebra.

Let $e_g : G \to \mathbb{C}$, $g \in G$ be the standard basis vectors for $\mathbb{C}^G$. In view of the natural vector space isomorphism between $\mathbb{C}^G$ and $\mathbb{C}G$, it follows that

$$g \mapsto M(e_g), \qquad g \in G \tag{13.5}$$

gives a vector space isomorphism between $\mathbb{C}G$ and the G-matrices. Further, this is an isomorphism of algebras since $M(e_{g_1})M(e_{g_2}) = M(e_{g_1} * e_{g_2})$ where

$$(e_{g_1} * e_{g_2})(g) = \sum_{h\in G} e_{g_1}(gh)e_{g_2}(h^{-1}) = e_{g_1}(gg_2^{-1}) = e_{g_1g_2}(g) \quad \Longrightarrow \quad e_{g_1} * e_{g_2} = e_{g_1g_2}.$$

Finally, we observe that the inverse of this isomorphism (13.5) is (13.3). □

The corresponding element of the group algebra for the three equally spaced unit vectors viewed as an S_3-frame was calculated in Example 10.5.

We observe that the Hermitian transpose of a G-matrix is given by the formula

$$M(\nu)^* = M(\tilde{\nu}), \qquad \tilde{\nu}(g) := \overline{\nu(g^{-1})}. \tag{13.6}$$

Next we characterise the Gramians of the normalised tight G-frames when viewed as elements of the group algebra via (13.3).

13.2 Tight G-frames as idempotents of the group algebra

We now show that:

> The normalised tight G-frames, i.e., the G-matrices which are orthogonal projections, correspond to idempotents of the group algebra $\mathbb{C}G$.

Proposition 13.2. *(Characterisation) The normalised tight G-frames $\Phi = (\phi_g)_{g\in G}$ are in a 1–1 correspondence with the elements $p = \sum_g c_g g$ of the group algebra $\mathbb{C}G$ satisfying $c_{g^{-1}} = \overline{c_g}$, $\forall g \in G$, and $p^2 = p$, i.e.,*

$$\sum_{h\in G} c_h c_{h^{-1}g} = c_g, \qquad \forall g \in G, \tag{13.7}$$

given by

$$P = \mathrm{Gram}(\Phi) = M(v) \qquad \longleftrightarrow \qquad p = \sum_g v(g)g, \tag{13.8}$$

where $v(g) = \langle \phi_g, \phi_1 \rangle = \langle g\phi_1, \phi_1 \rangle$.

Proof. Here (13.8) is the correspondence $p = \pi P$ of Proposition 13.1. Since Φ is a normalised tight frame if and only if its Gramian is an orthogonal projection matrix (Corollary 10.2), it suffices to determine the conditions on $p = \sum_g c_g g \in \mathbb{C}G$, which ensure that P is an orthogonal projection, i.e., $P^* = P$, $P^2 = P$. By (13.6), the first condition is that $c_g = \overline{c_{g^{-1}}}$. Since the π of (13.3) is an isomorphism of algebras, the second condition is that

$$p^2 = \sum_{h_1\in G}\sum_{h_2\in G} c_{h_1}c_{h_2}h_1h_2 = \sum_{g\in G} c_g g = p.$$

This can be rewritten as (13.7). □

Example 13.1. Let $G = C_3 = \langle a \rangle$ be the cyclic group of order 3. The first condition gives $c_1 \in \mathbb{R}$ (this always the case) and $c_{a^2} = \overline{c_a}$. The second condition (13.7) is that

$$c_1^2 + c_a c_{a^2} + c_{a^2} c_a = c_1, \quad c_1 c_a + c_a c_1 + c_{a^2}^2 = c_a, \quad c_1 c_{a^2} + c_a^2 + c_{a^2} c_1 = c_{a^2}.$$

Solving these equations gives the following six choices for p

$$0, \quad \frac{1}{3}(1+a+a^2), \quad \frac{1}{3}(1+\omega a+\omega^2 a^2), \quad \frac{1}{3}(2-\omega^2 a-\omega a^2), \quad \frac{1}{3}(2-a-a^2), \quad 1.$$

The ranks of the corresponding orthogonal projections P are $0,1,1,2,2,3$.

Example 13.2. The SIC of (1.7) viewed as a G-frame for $G = \langle S, \Omega \rangle$ corresponds to

$$\sum_g v(g)g = \frac{1}{4\sqrt{3}}\big(\sqrt{3}I + S + \Omega - iS\Omega - \sqrt{3}(-I) - (-S) - (-\Omega) + i(-S\omega)\big).$$

13.3 Characters of nonabelian groups

The (linear) characters of an abelian group generalise as follows.

The **character** of a representation $\rho : G \to GL(\mathscr{H})$ of a finite group G (or the $\mathbb{F}G$-module $\mathscr{H}$) is the map $\chi = \chi_\rho : G \to \mathbb{C}$ defined by

$$\chi(g) := \operatorname{trace}(\rho(g)).$$

The **degree** of χ is $\deg(\chi) := \dim(\mathscr{H})$. A character is said to be **irreducible** if the corresponding representation is irreducible. The character χ of G satisfies

- χ is constant on the conjugacy classes of G.
- $\chi(g^{-1}) = \overline{\chi(g)}$.
- $\chi(1) = \deg(\chi) = \dim(\mathscr{H})$.

Characters are important in study of $\mathbb{C}G$-modules, in particular

- $\mathbb{C}G$-modules are $\mathbb{C}G$-isomorphic if and only if they have the same character.
- If χ is a character of G, then $\overline{\chi}$ is a character of G.

The following example motivates the class of *central G-frames* (see §13.4).

Proposition 13.3. *Let W be an irreducible $\mathbb{C}G$-module of dimension d. Suppose that there is a unitary action of G on $\mathscr{H} \cong W^d = W \oplus \cdots \oplus W$ ($\mathbb{C}G$-isomorphism). Then all normalised tight G-frames Φ for $\mathscr{H}$ are unitarily equivalent, with*

$$P = P_\chi := \operatorname{Gram}(\Phi) = [\nu(g^{-1}h)]_{g,h \in G}, \qquad \nu(g) := \frac{\chi(1)}{|G|}\chi(g), \tag{13.9}$$

where χ is the character of W.

Proof. Without loss of generality, assume the action ρ on W is unitary, and that the unitary action on $V = W^d$ is given by

$$g \cdot (w_1, \ldots, w_d) := (\rho(g)w_1, \ldots, \rho(g)w_d).$$

Let V_j be the absolutely irreducible subspace of vectors

$$v_j = (0, \ldots, w_j, \ldots, 0), \qquad w_j \in W,$$

which are zero in all but the j-th coordinate (and the zero vector). Clearly, these are orthogonal, and $\sigma_j : V_j \to W : v_j \mapsto w_j$ is a $\mathbb{C}G$-isomorphism.

Let $v = \sum_j v_j \in \oplus_j V_j$. By Theorem 10.8, $\Phi = (gv)_{g \in G}$ is a normalised tight frame for $\mathscr{H}$ if and only if

$$\|v_j\|^2 = \|w_j\|^2 = \frac{d}{|G|}, \qquad \langle \sigma_j v_j, \sigma_k v_k \rangle = \langle w_j, w_k \rangle = 0, \quad j \neq k.$$

Thus the $\nu : G \to \mathbb{C}$ defining the Gramian P of the G-frame Φ is given by

$$v(g) := \langle gv, v\rangle = \langle g\sum_j v_j, \sum_k v_k\rangle = \sum_j \langle gv_j, v_j\rangle = \frac{d}{|G|}\sum_j \frac{\langle \rho(g)w_j, w_j\rangle}{\|w_j\|^2}$$
$$= \frac{\dim(W)}{|G|}\mathrm{trace}(\rho(g)) = \frac{\chi(1)}{|G|}\chi(g).$$

□

By Theorem 10.6 (or from the proof of Proposition 13.3), it follows that the unique normalised tight G-frame Φ for $\mathscr{H} \cong W^d$, W an irreducible $\mathbb{C}G$-module of dimension d and character χ, can be realised by

$$\Phi = \Phi_\chi := \sqrt{\frac{\chi(1)}{|G|}}\big(\rho(g)\big)_{g\in G}, \qquad \langle A, B\rangle = \mathrm{trace}(AB^*), \tag{13.10}$$

where $\rho : G \to \mathscr{U}(\mathbb{C}^d)$ is a unitary representation equivalent to W.

Example 13.3. Let $G = D_3 \cong S_3$ be the dihedral group of order 6 (see Example 10.3) and order its elements $1, a, a^2, b, ab, a^2b$. An irreducible representation of D_3

$$\rho : D_3 \to \mathscr{U}(\mathbb{C}^2) \subset \mathbb{C}^{2\times 2} \approx \mathbb{C}^4$$

with character $\chi = (2, -1, -1, 0, 0, 0)$ is given by

$$\rho(1) = \begin{pmatrix} 1 & 0 \\ 0 & 1 \end{pmatrix} \approx \begin{bmatrix} 1 \\ 0 \\ 0 \\ 1 \end{bmatrix}, \quad \rho(a) = \begin{pmatrix} \omega & 0 \\ 0 & \omega^2 \end{pmatrix} \approx \begin{bmatrix} \omega \\ 0 \\ 0 \\ \omega^2 \end{bmatrix}, \quad \rho(a^2) = \begin{pmatrix} \omega^2 & 0 \\ 0 & \omega \end{pmatrix} \approx \begin{bmatrix} \omega^2 \\ 0 \\ 0 \\ \omega \end{bmatrix},$$

$$\rho(b) = \begin{pmatrix} 0 & 1 \\ 1 & 0 \end{pmatrix} \approx \begin{bmatrix} 0 \\ 1 \\ 1 \\ 0 \end{bmatrix}, \quad \rho(ab) = \begin{pmatrix} 0 & \omega \\ \omega^2 & 0 \end{pmatrix} \approx \begin{bmatrix} 0 \\ \omega \\ \omega^2 \\ 0 \end{bmatrix}, \quad \rho(a^2b) = \begin{pmatrix} 0 & \omega^2 \\ \omega & 0 \end{pmatrix} \approx \begin{bmatrix} 0 \\ \omega^2 \\ \omega \\ 0 \end{bmatrix},$$

where $\omega := e^{\frac{2\pi i}{3}}$. Thus from (13.10) we obtain the normalised tight D_3-frame

$$\Phi = \frac{1}{\sqrt{3}}\left(\begin{bmatrix} 1 \\ 0 \\ 0 \\ 1 \end{bmatrix}, \begin{bmatrix} \omega \\ 0 \\ 0 \\ \omega^2 \end{bmatrix}, \begin{bmatrix} \omega^2 \\ 0 \\ 0 \\ \omega \end{bmatrix}, \begin{bmatrix} 0 \\ 1 \\ 1 \\ 0 \end{bmatrix}, \begin{bmatrix} 0 \\ \omega \\ \omega^2 \\ 0 \end{bmatrix}, \begin{bmatrix} 0 \\ \omega^2 \\ \omega \\ 0 \end{bmatrix}\right)$$

for $\mathbb{C}^4$, which has $P = \mathrm{Gram}(\Phi) = [\frac{1}{3}\chi(g^{-1}h)]_{g,h\in D_3}$.

13.4 Central G-frames

The G-frame of Proposition 13.3 with Gramian $P = [\nu(g^{-1}h)]_{g,h\in G}$ has the property that $\nu : G \to \mathbb{C}$ is a **class function**, i.e., is constant on the conjugacy classes of G. The irreducible characters form a basis for the vector space of class functions on G, and ν is a class function if and only if

$$\sum_{g\in G} \nu(g)g \in Z(\mathbb{C}G), \qquad (13.11)$$

where $Z(\mathbb{C}G)$ denotes the centre of the group algebra $\mathbb{C}G$.

Definition 13.2. A G-frame $\Phi = (\phi_g)_{g\in G}$ is said to be **central** if $\nu : G \to \mathbb{C}$ defined by

$$\nu(g) := \langle \phi_g, \phi_1 \rangle = \langle g\phi_1, \phi_1 \rangle$$

is a class function.

In view of Proposition 13.3 and (13.11), each of the following conditions on a G-frame Φ with $\mathrm{Gram}(\Phi) = [\nu(g^{-1}h)]_{g,h\in H}$ are equivalent to it being central

- ν is a class function.
- $\sum_{g\in G} \nu(g)g \in Z(\mathbb{C}G)$.
- $\mathrm{Gram}(\Phi)$ is in the centre of the algebra of group matrices.
- The symmetry condition $\langle gv, hv \rangle = \langle gw, hw \rangle$, $\forall g, h \in G$, $\forall v, w \in \Phi$.

For G abelian, all G-frames are central (the conjugacy classes are singletons), and so the central G-frames are a generalisation of the harmonic frames.

We will show (Theorem 13.1) that there are a finite number of tight central G-frames (for a given G). To this end, consider the homogeneous normalised tight central G-frame of Proposition 13.3, which has Gramian

$$P_\chi := [\frac{\chi(1)}{|G|}\chi(g^{-1}h)]_{g,h\in G} = \frac{\chi(1)}{|G|}M(\chi), \qquad (13.12)$$

where χ is an irreducible character of G. Since χ is a class function, the idempotent $p_\chi \in \mathbb{C}G$ that it corresponds to (via Proposition 13.2) is in the centre of the group algebra, i.e.,

$$p_\chi := \frac{\chi(1)}{|G|} \sum_{g\in G} \chi(g)g \in Z(\mathbb{C}G). \qquad (13.13)$$

Moreover, for different characters, these homogeneous G-frames are orthogonal (Theorem 10.7), and so the product of their Gramians is zero (Lemma 5.1), which gives

$$p_{\chi_j} p_{\chi_k} = 0, \qquad \chi_j \neq \chi_k. \qquad (13.14)$$

Thus, if $\chi_1, \ldots, \chi_r$ are the irreducible characters of G, then $\{p_{\chi_j}\}_{1\leq j\leq r}$ is a basis of (orthogonal) idempotents for $Z(\mathbb{C}G)$.

13.5 The classification of central tight G-frames

The central tight G-frames can be characterised in terms of the Gramian.

Theorem 13.1. *(Classification) Let G be a finite group with irreducible characters $\chi_1,\ldots,\chi_r$. Then Φ is a central normalised tight G-frame if and only if its Gramian is given by*

$$\operatorname{Gram}(\Phi)=\sum_{j\in J}P_{\chi_j}=M\Big(\sum_{j\in J}\frac{\chi_j(1)}{|G|}\chi_j\Big), \tag{13.15}$$

for some $J\subset\{1,\ldots,r\}$, where P_χ is defined by (13.12), and M by (13.2).

Proof. The two formulas given for $\operatorname{Gram}(\Phi)$ are equal. Since $P_{\chi_j}P_{\chi_k}=0$, $j\neq k$, the first gives the Gramian of a normalised tight G-frame, and this is central (by the second formula). Thus, it suffices to assume that Φ is a central normalised tight G-frame and to show that $P:=\operatorname{Gram}(\Phi)=M(v)$ is given by (13.15).

Since Φ is central, v is a class function, and so the idempotent $p=\sum_g v(g)g$ corresponding to P is in $Z(\mathbb{C}G)$. Write p in terms of the basis $\{p_j\}_{1\le j\le r}$, $p_j:=p_{\chi_j}$ for $\mathbb{C}G$

$$p=\sum_j\alpha_jp_j,\qquad \alpha_j\in\mathbb{C}.$$

Since p is an idempotent, (13.14) gives

$$p^2=\sum_j\sum_k\alpha_j\alpha_kp_jp_k=\sum_j\alpha_j^2p_j=p=\sum_j\alpha_jp_j\quad\Longrightarrow\quad\alpha_j^2=\alpha_j.$$

Hence $\alpha_j\in\{0,1\}$, and $p=\sum_{j\in J}p_j$, where $J:=\{j:\alpha_j=1\}$. We therefore have

$$P=\pi^{-1}(p)=\pi^{-1}\Big(\sum_{j\in J}p_{\chi_j}\Big)=\sum_{j\in J}P_{\chi_j}.$$

□

The G-frame Φ of Example 13.3 is a central normalised tight D_3-frame for $\mathbb{C}^4$ (by construction). Excluding the frame $(0)_{g\in G}$, there are six others. More generally:

For a given finite group G with r distinct irreducible characters there are 2^r-1 nontrivial central normalised tight G-frames (up to unitary equivalence).

Theorem 13.1 leads to the following count.

Corollary 13.1. *Suppose there is a unitary action of G on the complex space $\mathscr{H}$. Then either*

1. *There is no G-frame for $\mathscr{H}$ ($\mathscr{H}$ is not $\mathbb{C}G$-isomorphic to a submodule of $\mathbb{C}G$).*
2. *There is one tight G-frame for $\mathscr{H}$ (which is central)*
3. *There are uncountably many tight G-frames for $\mathscr{H}$ (none of which are central).*

Proof. In terms of the the homogeneous decomposition $\mathscr{H} = \oplus_{W \in \mathscr{W}} W^{a_W}$ of (10.28), the three cases are

1. $a_W > \dim(W)$, for some W (apply Proposition 10.4).
2. $a_W \in \{0, \dim(W)\}$, for all W (apply Proposition 13.3).
3. $1 < a_W < \dim(W)$, for some W (apply Proposition 10.1).

In the second case, $J = \{j : \chi_j = \mathrm{char}(W) \text{ and } a_W = \dim(W),\ W \in \mathscr{W}\}$ in (13.15), where $\mathrm{char}(W)$ is the character of W. □

The character of $\mathscr{H}$ can be determined from any G-orbit which spans $\mathscr{H}$.

Proposition 13.4. *Let $P = M(\nu)$ be the canonical Gramian of a G-frame for $\mathscr{H}$. Then the character χ of the representation is given by*

$$\chi(g) = \sum_{h \in G} \nu(h^{-1}gh).$$

In particular, if the G-frame is central, then $\chi = |G|\nu$.

Proof. By Corollary 10.3, we can assume that the G-frame is $(Pe_g)_{g \in G}$, with

$$g(Pe_h) = e_{gh}, \qquad \mathscr{H} = \mathrm{ran}(P).$$

Since $(Pe_h)_{h \in G}$ is a normalised tight frame, the trace formula (Exercise 2.13) gives

$$\chi(g) = \sum_{h \in G} \langle gPe_h, Pe_h \rangle = \sum_{h \in G} \langle Pe_{gh}, e_h \rangle = \sum_{h \in G} P_{h,gh} = \sum_{h \in G} \nu(h^{-1}gh).$$

If ν is a class function, then $\nu(h^{-1}gh) = \nu(g)$, and we get $\chi(g) = |G|\nu(g)$. □

For the orthogonal projections P_χ of (13.12) given by irreducible characters χ, the condition $P_\chi^2 = P_\chi$ and the orthogonality relation

$$P_{\chi_j} P_{\chi_k} = 0, \qquad \chi_j \neq \chi_k,$$

can be expressed using (13.4) as

$$\chi * \chi = \frac{|G|}{\chi(1)} \chi, \tag{13.16}$$

$$(\chi_j * \chi_k)(g) = \sum_{h \in G} \chi_j(gh)\chi_k(h^{-1}) = 0, \qquad \forall g \in G \quad (\chi_j \neq \chi_k). \tag{13.17}$$

These formulas for the convolution of irreducible characters are well known. The special case $g = 1$ in (13.17) gives the orthogonality of characters (11.5), i.e.,

$$\langle \chi_j, \chi_k \rangle := \frac{1}{|G|} \sum_{h \in G} \chi_j(h)\overline{\chi_k(h)} = 0, \qquad = \delta_{jk}. \tag{13.18}$$

13.6 The idempotents and the homogeneous decomposition

The (orthogonal) idempotents p_χ of (13.13), or, more precisely, the idempotents

$$q_\chi := p_{\overline{\chi}} = \frac{\chi(1)}{|G|} \sum_{g\in G} \chi(g^{-1})g \in Z(\mathbb{C}G). \tag{13.19}$$

play a very special role in the homogeneous decomposition of a $\mathbb{C}G$-module V (Lemma 10.2). For W an irreducible $\mathbb{C}G$-module with character χ, we recall the *homogeneous component* of V is

$$H_V(\chi) = H_V(W) := \sum_{\substack{X\subset V\\ X\cong W}} X.$$

Let $\hat{G}$ denote the irreducible characters of G. We now state (and prove) the well-known formula for the homogeneous components.

Theorem 13.2. *Let G be a finite group, and V be a $\mathbb{C}G$-module. Then the direct sum decomposition of V into it homogeneous components is given by*

$$V = \bigoplus_{\chi\in\hat{G}} q_\chi V, \tag{13.20}$$

i.e., $H_V(\chi) = q_\chi V$, where the sum is orthogonal if the action of G on V is unitary.

Proof. Let W be a d-dimensional irreducible with character χ. We will first show that

$$q_\chi w = w, \qquad \forall w \in W. \tag{13.21}$$

For this, we can assume that the action of G is unitary (Corollary 10.1). As in the proof of Proposition 13.3, choose $w_1, \ldots, w_d \in W$ so that $(gv)_{g\in G}$, $v = (w_1, \ldots, w_d)$ is a central normalised tight G-frame for W^d, i.e., $\langle gv, v\rangle = \frac{\chi(1)}{|G|}\chi(g)$. Here w_1 can be *any* nonzero element of W, up to a scalar multiple. We calculate

$$q_\chi v = \sum_{g\in G} \frac{\chi(1)}{|G|}\chi(g^{-1})gv = \sum_{g\in G} \langle g^{-1}v, v\rangle gv = \sum_{g\in G} \langle v, gv\rangle gv = v.$$

Taking the first component gives $q_\chi w_1 = w_1$, which gives (13.21).

As in Lemma 10.2, write V as a direct sum

$$V = V_1 \oplus V_2 \oplus \cdots \oplus V_m,$$

of irreducible G-invariant subspaces (which are orthogonal if the action is unitary). Using the properties $q_\chi^2 = q_\chi$ and $q_{\chi_j} q_{\chi_k} = 0$, $\chi_j \neq \chi_k$ it follows that $\sum_{\chi\in\hat{G}} q_\chi = 1$ (apply $\sum_\chi q_\chi$ to the irreducible submodules W of $\mathbb{C}G$), which gives (13.20). □

13.7 An illustrative example

We now return to our motivating example of the nonabelian group $G = D_3 = \langle a,b \rangle$ of order 6 acting on $\mathbb{R}^2$ via (13.1), i.e.,

$$a := \begin{pmatrix} -\frac{1}{2} & \frac{1}{2}\sqrt{3} \\ -\frac{1}{2}\sqrt{3} & -\frac{1}{2} \end{pmatrix}, \qquad b := \begin{pmatrix} 1 & 0 \\ 0 & -1 \end{pmatrix}. \tag{13.22}$$

The conjugacy classes of G are $\{1\}, \{a,a^2\}, \{b,ab,a^2b\}$. Since the action of G is irreducible, the orbit $\Phi = (gw)_{g\in G}$ of any vector

$$w := \frac{1}{3}\begin{pmatrix} x \\ y \end{pmatrix}, \qquad x^2+y^2 = 1$$

is a normalised tight frame, with Gramian the G-matrix $P = M(\nu)$, given by

$$\nu(1) = \tfrac{1}{3}, \qquad \nu(a) = -\tfrac{1}{6}, \qquad \nu(a^2) = -\tfrac{1}{6}, \qquad \nu(b) = \tfrac{1}{3}x^2 - \tfrac{1}{3}y^2,$$

$$\nu(ab) = -\tfrac{1}{6}x^2 - \tfrac{1}{\sqrt{3}}xy + \tfrac{1}{6}y^2, \qquad \nu(a^2b) = -\tfrac{1}{6}x^2 + \tfrac{1}{\sqrt{3}}xy + \tfrac{1}{6}y^2.$$

It is easy to verify (13.7) holds, e.g., for $g = 1$, we have

$$\nu(1)^2 + \nu(a)\nu(a^2) + \nu(a^2)\nu(a) + \nu(b)^2 + \nu(ab)^2 + \nu(a^2b)^2 = \frac{1}{3}(x^2+y^2)^2 = \nu(1).$$

From Proposition 13.4, we can determine the character χ of the representation

$$\chi(1) = 6\nu(1) = 2, \qquad \chi(a) = \chi(a^2) = 3\nu(a) + 3\nu(a^2) = -1,$$

$$\chi(b) = \chi(ab) = \chi(a^2b) = 2\nu(b) + 2\nu(ab) + 2\nu(a^2b) = 0.$$

With the order $1, a, a^2, b, ab, a^2b$, the characters of G are

$$\chi_1 = \begin{bmatrix} 1 \\ 1 \\ 1 \\ 1 \\ 1 \\ 1 \end{bmatrix}, \qquad \chi_2 \begin{bmatrix} 1 \\ 1 \\ 1 \\ -1 \\ -1 \\ -1 \end{bmatrix}, \qquad \chi_3 = \begin{bmatrix} 2 \\ -1 \\ -1 \\ 0 \\ 0 \\ 0 \end{bmatrix}. \tag{13.23}$$

Thus we can deduce from P that the action of G is irreducible with $\chi = \chi_3$. With q_{χ_j} the idempotents of (13.19), Theorem 13.2 gives $q_{\chi_1}V = q_{\chi_2}V = 0$, $q_{\chi_3}V = V$, i.e.,

$$I + a + a^2 \pm (b + ab + a^2b) = 0, \qquad \frac{2}{6}(2I - a - a^2 + 0(b + ab + a^2b)) = I.$$

where a and b are defined by (13.22).

13.8 The highly symmetric tight frames

For the irreducible action of $G = D_3$ on $\mathbb{R}^2$ given by (13.1) there are uncountably many inequivalent G-frames (see Figure 13.1). None of these are central, since the only central D_3-frame for $\mathbb{R}^2$ is the one obtained by taking the characters χ_1 and χ_2 of (13.23), which results in three copies of an orthogonal basis. We would like to think of the six equally spaced unit vectors ($\theta = \frac{\pi}{6}$), which has a larger symmetry group than the others, as being a G-frame worth singling out. This leads to the notion of a *highly symmetric* frame.

Definition 13.3. A finite frame Φ of distinct vectors is **highly symmetric** if the action of its symmetry group $\mathrm{Sym}(\Phi)$ is irreducible, transitive, and the stabiliser of any one vector (and hence all) is a nontrivial subgroup which fixes a subspace of dimension exactly one.

We recall that the action of the symmetry group of a finite frame is given by (9.1). If Φ is a highly symmetric frame, then (by the orbit size theorem)

$$|\mathrm{Sym}(\Phi)| > |\Phi|.$$

As defined, a highly symmetric frame has *distinct* vectors, and so it may not be a group frame (Theorem 10.4) unless the vectors are repeated some fixed number of times, e.g., it is naturally a $\mathrm{Sym}(\Phi)$-frame.

Since a frame is highly symmetric if and only if the canonical tight frame is, it suffices to consider only the highly symmetric tight frames. The key features of the class of *highly symmetric tight frames* are:

- There is a *finite* number of highly symmetric tight frames of n vectors for $\mathbb{C}^d$.
- They can be computed from the representations of abstract groups.
- It is possible to determine whether or not a given tight frame is highly symmetric.
- Some harmonic frames are highly symmetric tight frames.
- The vertices of the regular complex polytopes are highly symmetric tight frames.
- All finite reflection groups give highly symmetric tight frames.

There are no highly symmetric frames for $\mathbb{C}^1$ (by definition), so we let $d > 1$.

Example 13.4. (Equally spaced vectors) The three equally spaced unit vectors in $\mathbb{R}^2$ are a highly symmetric tight C_3-frame (each vector is fixed by the reflection through the line it lies on). The six equally spaced unit vectors are a highly symmetric tight D_3-frame, as discussed above. More generally, the n equally space unit vectors are a highly symmetric tight C_n-frame, and $D_{n/2}$-frame also, when n is even.

Example 13.5. (Harmonic frames) The standard orthonormal basis $\{e_j\}$ for $\mathbb{F}^d$ is not a highly symmetric tight frame, since its symmetry group fixes $e_1 + \cdots + e_d$ (and so its action is not irreducible). On the other hand, the vertices of the regular d-simplex always are (the three equally spaced vectors is the case $d = 2$). Since both of these frames are harmonic, we conclude that a highly symmetric tight frame may or may not be harmonic.

13.9 The construction of highly symmetric tight frames

We now show that up to similarity:

> There is a finite number of highly symmetric frames of n vectors for $\mathbb{F}^d$.

Theorem 13.3. *(Finiteness) Fix $n \geq d$. There is a finite number of highly symmetric normalised tight frames of n vectors for $\mathbb{F}^d$ (up to unitary equivalence).*

Proof. Suppose Φ is a highly symmetric normalised tight frame of n vectors for $\mathbb{F}^d$. Then it is determined, up to unitary equivalence, by the representation induced by $\mathrm{Sym}(\Phi)$, and a subgroup H which fixes only the one-dimensional subspace spanned by some vector in Φ. There is a finite number of choices for $\mathrm{Sym}(\Phi)$ since its order is at most $n(n-1)\cdots(n-d+1)$ (Exercise 9.2), and hence (by Maschke's theorem) a finite number of possible irreducible representations. As there is only a finite number of choices for H, it follows that the class of such frames is finite. □

This yields the following algorithm, which can be implemented in a symbolic algebra package such as `Magma` (see the worked example that follows). We denote the *stabiliser* of v by $\mathrm{Stab}(v) = \mathrm{Stab}_G(v) := \{g \in G : gv = v\}$.

Algorithm: To construct all highly symmetric tight frames Φ of n vectors in $\mathbb{F}^d$.

1. Start with an abstract group G. This corresponds to $\mathrm{Sym}(\Phi)$ or an appropriate subgroup, so that $|G|$ divides $n!$ and $n < |G| \leq n(n-1)\cdots(n-d+1)$.
2. Take all faithful irreducible representations $\rho : G \to GL_d(\mathbb{F})$. There is a finite number of these, and they can be computed.
3. *Find (up to conjugacy) all subgroups H of $\rho(G)$ which fix a subspace* $\mathrm{span}\{v\}$, $v \neq 0$. Then $\{gv\}_{g\in G}$ is a highly symmetric tight frame of $|G|/\mathrm{Stab}(v)$ vectors. No other subgroups of $\mathrm{Stab}(v)$ need be considered.
4. *Determine which of the highly symmetric tight frames obtained are unitarily equivalent (up to a reordering).*

Example 13.6. (*Magma* calculation) Let G be the solvable group `<18, 3>`, for which `Magma` gives the presentation

$$G = \langle g_1, g_2, g_3 : g_1^2 = g_2^3 = g_3^3 = 1, g_1^{-1} g_3 g_1 = g_3^2 \rangle.$$

The representations of G over $\mathbb{C}$ can be computed:

```
G:=SmallGroup(18,3);
r:=AbsolutelyIrreducibleModules(G,Rationals());
```

There are six of dimension 1 and three of dimension 2, the first given by

```
rho:=Representation(r[7]); rG:=ActionGroup(r[7]);
a:=rG.1=rho(G.1); b:=rG.2; c:=rG.3; sg:=Subgroups(rG);
```

$$a=\rho(g_1)=\begin{pmatrix}0&1\\1&0\end{pmatrix},\quad b=\rho(g_2)=\begin{pmatrix}\omega^2&0\\0&\omega^2\end{pmatrix},\quad c=\rho(g_3)=\begin{pmatrix}\omega^2&0\\0&\omega\end{pmatrix},$$

where $\omega := e^{\frac{2\pi i}{3}}$. The subspace fixed by a (nontrivial) subgroup H given by `sg` can be found by the command `NullspaceMatrix(M-Id)`, where `M` is a block matrix of generators for H and `Id` is the corresponding identity block matrix. Thus, we obtain two highly symmetric tight frames:

$$\begin{aligned} &6\,\text{vectors:} && v=v_1=(1,0), && \mathrm{Stab}(v_1)=\langle bc\rangle,\\ &9\,\text{vectors:} && v=v_2=(1,1), && \mathrm{Stab}(v_2)=\langle a\rangle, \end{aligned}$$

which are a cross and a cube (see Example 13.10). These are the only highly symmetric tight frames we obtain, since the eighth representation is not faithful, and $\rho(G)$ is the same for the seventh and ninth.

Example 13.7. There are no highly symmetric tight frames of five vectors in $\mathbb{C}^3$. Such a tight frame would have a symmetry group of order a multiple of 5, which is at most $5\cdot4\cdot3=60$. A computer search over all groups in this range shows there is no such frame. By way of contrast, the tight frame of five vectors in $\mathbb{C}^3$ with the largest symmetry group is the vertices of a *trigonal bipyramid*, which has symmetry group of order 12 (see Example 9.12).

13.10 Complex polytopes and finite reflection groups

The n equally spaced unit vectors can be viewed as the vertices of the regular n-gon. We now investigate the highly symmetric tight frames which come as the vertices of (regular) complex polytopes (which include the n-gon and the Platonic solids). The main idea is that by imposing enough regularity (symmetries must map flags to flags), the symmetry group is generated by (complex) reflections, which leads to a complete classification via the symmetry group.

A transformation $g\in GL_d(\mathbb{F})$ is a **(complex) reflection** (or **pseudoreflection**) if it has finite order m and $\mathrm{rank}(g-I)=1$, i.e., g fixes a hyperplane H and maps some $v\mapsto\omega v$ where $v\notin H$ is nonzero and ω is a primitive m-th root of unity. The terminology and geometric motivation comes from $\mathbb{R}^d$ with $\omega=-1$. A finite subgroup of $GL_d(\mathbb{F})$ is a **reflection group** if it is generated by its reflections.

Frames are sequences of vectors (points), whereas polytopes, such as the Platonic solids, have points, lines (through points), and faces, etc. The technical definition (to follow) specifies these *j-faces* ($j=0,1,\ldots$) as affine subspaces of $\mathbb{F}^d$, together with some combinatorial properties motivated by the case $\mathbb{R}^3$. Of course, such a face is the affine hull of the vertices it contains, and it is convenient to move between the two. For complex spaces, a line (1-face) may contain *more* than two points, which challenges one's intuition.

Definition 13.4. (see [Sch04]) A **d-polytope–configuration** is a finite family $\mathscr{P}$ of affine subspaces of $\mathbb{F}^d$ of dimensions $j = -1, 0, 1, \ldots, d$, called **elements** or ***j*-faces**, ordered by inclusion $\subset$, which form lattice with the properties

(i) If $F_{j-1} \subset F_{j+1}$ are $j-1$ and $j+1$ faces, then there are *at least* two j-faces contained between them. (Modified diamond condition)
(ii) If $F \subset G$ are faces, then there is a sequence of faces $F = H_0 \subset H_1 \subset \cdots \subset H_k = G$ with $\dim(H_j) = \dim(F) + j$, $\forall j$. (Connectedness)

For brevity, we call such a $\mathscr{P}$ a **complex polytope**. We now follow the usual practice and translate $\mathscr{P}$ so that the barycentre (average of the vertices) is zero. This allows the vertices to be thought of as vectors and ensures that the affine maps of the vertices to themselves are linear (and ultimately unitary).

Definition 13.5. The **symmetry group** $\mathrm{Sym}(\mathscr{P})$ of a d-polytope–configuration $\mathscr{P}$ (with barycentre 0) is the group of $g \in GL_d(\mathbb{F})$, which map the elements of $\mathscr{P}$ to themselves.

In particular, if $\Phi_{\mathscr{P}}$ is the points (vectors) of $\mathscr{P}$, then $\mathrm{Sym}(\mathscr{P})$ is a subgroup of $\mathrm{Sym}(\Phi_{\mathscr{P}})$ (viewed as linear transformations of $\mathbb{F}^d$).

Definition 13.6. A **flag** of d-polytope–configuration $\mathscr{P}$ is a sequence F of faces with

$$F = (F_{-1}, F_0, F_1, \ldots, F_d), \qquad F_{-1} \subset F_0 \subset F_1 \subset \cdots \subset F_d, \qquad \dim(F_j) = j, \quad \forall j,$$

and $\mathscr{P}$ is **regular** if $\mathrm{Sym}(\mathscr{P})$ is transitive on the flags of $\mathscr{P}$.

Shephard [She52], [She53] showed the symmetry group of a regular complex polytope is an irreducible reflection group and classified all such polytopes via their symmetry groups. More precisely, let F be a flag of a regular complex polytope $\mathscr{P}$, and c_j be the centre of the j-face F_j, i.e., the average of its vertices. Then there are generating reflections $r_0, \ldots, r_{d-1}$ for $\mathrm{Sym}(\mathscr{P})$ where r_j fixes $c_0, \ldots, c_{j-1}, c_{j+1}, \ldots, c_d$ and maps F_j to another j-face, i.e., r_j maps F to a flag which differs only in the j-face.

A **parabolic subgroup** of a finite reflection group $G \subset GL_d(\mathbb{F})$ is the pointwise stabiliser of a subset $V \subset \mathbb{F}^d$. Steinberg's fixed point theorem ([Ste64]) says that a parabolic subgroup is a finite reflection subgroup.

Theorem 13.4. *The vertices of the regular complex polytopes are highly symmetric tight frames. In particular, the vertices of the regular complex polytopes can be constructed from their abstract symmetry groups (which contains the corresponding reflection group).*

Proof. Let $\mathscr{P}$ be a regular complex polytope, and $\Phi = \Phi_{\mathscr{P}}$ be its vertices. View $G = \mathrm{Sym}(\Phi)$ as a subgroup of $GL_d(\mathbb{F})$. Then $H = \mathrm{Sym}(\mathscr{P})$ is a subgroup of G, which is irreducible and transitive on the flags, and in particular is transitive on the vertices Φ. Thus, Φ will be a highly symmetric (tight) frame provided that $\mathrm{Stab}_H(v) \subset \mathrm{Stab}_G(v)$ fixes a space of dimension exactly one for each $v \in \Phi$.

Fix a vertex $v \in \Phi$. Since H is a reflection group, Steinberg's fixed point theorem implies that $\mathrm{Stab}_H(v)$ is the group generated by all the reflections which fix v. If F is a flag with $F_0 = \{v\}$, then the $d-1$ generating reflections $r_1, \ldots, r_{d-1}$ fix v, and so the subspace fixed by them all is one-dimensional (and equal to $\mathrm{span}\{v\}$). Thus $\mathrm{Stab}_G(v)$ fixes only $\mathrm{span}\{v\}$. □

Corollary 13.2. *If $G \subset GL_d(\mathbb{F})$ is an irreducible finite reflection group, then $(gv)_{g\in G}$ is a highly symmetric tight frame for $\mathbb{F}^d$ if and only if $H = \mathrm{Stab}(v)$ is a maximal proper parabolic subgroup.*

Proof. Since the parabolic subgroups are generated by reflections, and reflections fix a hyperplane, the set V fixed by a maximal proper parabolic subgroup must be a one-dimensional subspace $V = \mathrm{span}\{v\}$, $v \neq 0$. □

In [ST54] (cf. [LT09]) all finite reflection groups were classified. Essentially, they appear as the symmetry groups of "semi-regular" complex polytopes. In the next subsections we outline the highly symmetric tight frames which can be obtained from the (imprimitive and primitive) finite reflection groups.

13.10.1 Imprimitive groups (ST 1-3)

A representation of G on $\mathbb{F}^d$ is **imprimitive** if $\mathbb{F}^d$ is a direct sum $\mathbb{F}^d = V_1 \oplus \cdots \oplus V_m$ of nonzero subspaces, such that the action of G on $\mathbb{F}^d$ permutes the V_j, otherwise it is **primitive**. The Shephard–Todd classification of the *imprimitive* irreducible complex reflection groups consists of three infinite families (ST 1–3) given by the groups $G(m,p,d)$, where $m > 1$, $p \mid m$, and

$$|G(m,p,d)| = m^d d!/p.$$

These are available in Magma via `ImprimitiveReflectionGroup(m,p,d)`, and can be constructed (cf. [LT09]) as a group of unitary transformations

$$\begin{aligned} G(1,1,d) &= \langle r_1, r_2, \ldots, r_{d-1} \rangle, \\ G(m,m,d) &= \langle s, r_1, r_2, \ldots, r_{d-1} \rangle, \\ G(m,1,d) &= \langle t, r_1, r_2, \ldots, r_{d-1} \rangle, \\ G(m,p,d) &= \langle s, t^p, r_1, r_2, \ldots, r_{d-1} \rangle, \quad 1 < p < m, \ p \mid m \end{aligned}$$

where r_j swaps e_j and e_{j+1}, t is the reflection $e_1 \mapsto \omega e_1$, $\omega = e^{\frac{2\pi i}{m}}$, and $s = t^{-1} r_1 t$, i.e.,

$$r_1 = \begin{bmatrix} 0 & 1 & \\ 1 & 0 & \\ & & I \end{bmatrix}, \quad t = \begin{bmatrix} \omega & & \\ & 1 & \\ & & I \end{bmatrix}, \quad s = \begin{bmatrix} 0 & \overline{\omega} & \\ \omega & 0 & \\ & & I \end{bmatrix}, \qquad (13.24)$$

where I is the identity matrix of size $d-2$. The three infinite families are:

ST 1: $G(1,1,d+1) \cong S_{d+1}$ acting on the d-dimensional subspace of vectors in $\mathbb{F}^{d+1}$ which are orthogonal to $e_1+\cdots+e_{d+1}$.
ST 2: $G(m,p,d)$, $m,d>1$, $p \mid m$, $(m,p,d) \neq (2,2,2)$ acting on $\mathbb{C}^d$.
ST 3: $G(m,1,1) \cong \mathbb{Z}_m$ acting on $\mathbb{C}$.

There are no highly symmetric tight frames for $\mathbb{C}$, so only ST 2 and ST 3 can give highly symmetric tight frames. We now give some indicative examples.

Example 13.8. (m-distance tight frame) Let $G = G(1,1,d+1) \cong S_{d+1}$ act on the d-dimensional subspace $\mathscr{H} = (e_1+\cdots+e_{d+1})^\perp$ of $\mathbb{F}^{d+1}$ via $\sigma e_j = e_{\sigma j}$, and

$$w_m := e_1+\cdots+e_m - \frac{m}{d+1-m}(e_{m+1}+\cdots+e_{d+1}), \qquad 1 \leq m \leq d.$$

Then $|\mathrm{Stab}(w_m)| = m!(d+1-m)!$, so $\Phi_m := \{gw_m\}_{g\in G}$ is a highly symmetric tight frame of $\binom{d+1}{m}$ vectors for the d-dimensional space $\mathscr{H}$, with $S_{d+1} \subset \mathrm{Sym}(\Phi_m)$. These are the only possibilities. This frame Φ_m is the *standard m-distance tight frame* of §12.16. For $m=1$, we obtain the simplex with vertices given by Φ_1. For the other cases, the vectors of Φ_m are the barycentres of the $(m-1)$-faces of this simplex, and so Φ_d is also a simplex. The special case $m=2$, $d=3$ gives the six vertices of the octahedron (which have symmetry group $S_4 \times \mathbb{Z}_2$).

Example 13.9. (28 equiangular lines in $\mathbb{R}^7$) The special case of Example 13.8 where $G = G(1,1,8)$ acts on the vector

$$v = 3w_2 = (3,3,-1,-1,-1,-1,-1,-1).$$

gives an orbit of 28 vectors in $\mathbb{R}^7$ which are an equiangular tight frame.

Example 13.10. (The generalised cross and cube) Let $G = G(m,1,d)$, $|G| = d!m^d$, and

$$v_k := e_1+\cdots+e_k, \qquad 1 \leq k \leq d.$$

Then v_k has $|\mathrm{Stab}(v_k)| = k!(d-k)!m^{d-k}$, and so its orbit gives a highly symmetric tight frame of $\binom{d}{k}m^k$ vectors for $\mathbb{R}^d$. These are the only possibilities.

The extreme cases are the **(generalised) cross** ($k=1$) and **cube** ($k=d$), which are regular complex polytopes. These terms originate from the case $m=2$, $d=3$, where we have the *octahedron* (6 vertices), the *cuboctahedron* (12 vertices) and the *cube* (8 vertices), respectively, and $G = G(2,1,3)$ is O_h (the *full octahedral group*). For $m=2$, $d=4$ (see Figure 13.2), the polytopes are the *hexadecachoron* (*16-cell*) (8 vertices), *octaplex* (*24-cell*) (24 vertices), *rectified tesseract* (32 vertices) and *tesseract* (16 vertices).

The cross and cube are harmonic frames, generated by the cyclic subgroups

$$\langle r_1 r_2 \cdots r_{d-1} t \rangle, \qquad \langle q_1,\ldots,q_d \rangle,$$

where $q_j = (r_1 r_2 \cdots r_{j-1})^{-1} t (r_1 r_2 \cdots r_{j-1})$ is the reflection $e_j \mapsto \omega e_j$.

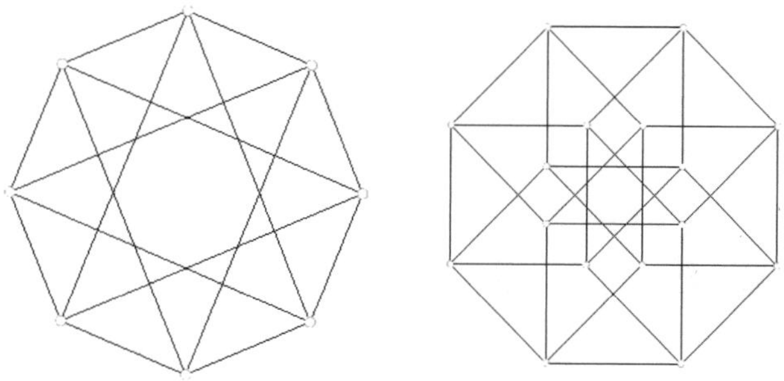

Fig. 13.2: Symbolic projections of the cross (hexadecachoron) and cube (tesseract) in $\mathbb{R}^4$.

The imprimitive reflection groups of the ST 2 family can be nested, (see [LT09]), e.g.,

$$G(m,p,d) \triangleleft G(m,1,d), \qquad G(m,p,2) \triangleleft G(2m,2,2).$$

Hence, a highly symmetric tight frame obtained from an imprimitive reflection group G may be a *subset* of one obtained for a larger imprimitive reflection group.

Example 13.11. (Nested irreducible reflection groups) Let

$$G = G(2,2,d), \quad d > 2, \qquad |G| = 2^{d-1}d! \qquad (\text{Coxeter group}\, D_d).$$

There are highly symmetric tight frames given by the orbits of e_1 and $e_1 + \cdots + e_d$. The first of these is the cross, which has a symmetry group larger than G, namely $G(2,1,d)$. The second is the *demicube*, a subset of half the vertices of the cube, which has symmetry group $G(2,1,d)$.

13.10.2 Primitive reflection groups (ST 4-37)

There are 34 (exceptional) finite reflection groups in the Shephard–Todd classification. Their numbers and rank (the dimension of the space they act on) are

ST 4–22 (rank 2), ST 23–27 (rank 3), ST 28–32 (rank 4),
ST 33 (rank 5), ST 34–35 (rank 6), ST 36 (rank 7), ST 37 (rank 8).

`Magma` calculations (see Tables 13.1 and 13.2) indicate the following behaviour:

- There are highly symmetric tight frames given by each primitive reflection group (Theorem 13.4). Some are not the vertices of a regular complex polytope.
- These highly symmetric tight frames are not harmonic.
- They may or may not be G-frames (of distinct vectors).
- They have a small number of angle moduli.

We now highlight a few examples (with indicative `Magma` code).

Example 13.12. (ST 23). All highly symmetric tight frames obtained from rank 2 reflection groups are group frames (of distinct vectors). This is not the case in higher dimensions. Let G be the Shephard–Todd group 23, $|G| = 120$, for which `Magma` gives the generators

$$g_1 = \begin{pmatrix} -1 & 0 & 0 \\ \frac{1}{2}(\sqrt{5}+1) & 1 & 0 \\ 0 & 0 & 1 \end{pmatrix}, \qquad g_2 = \begin{pmatrix} 1 & \frac{1}{2}(\sqrt{5}+1) & 0 \\ 0 & -1 & 0 \\ 0 & 1 & 1 \end{pmatrix}, \qquad g_3 = \begin{pmatrix} 1 & 0 & 0 \\ 0 & 1 & 1 \\ 0 & 0 & -1 \end{pmatrix},$$

which are *not* unitary matrices. We obtain three highly symmetric tight frames:

$$\begin{aligned} &12\,\text{vectors:} \qquad v = (\sqrt{5}-1,0,2), \\ &20\,\text{vectors:} \qquad v = (\sqrt{5}+3,0,2), \\ &30\,\text{vectors:} \qquad v = (1,1,1), \end{aligned}$$

which are the vertices of the *icosahedron*, *dodecahedron* and *icosidodecahedron*. The first of these is a group frame (for `<12, 3>`), and the other two are not.

Example 13.13. (24 vectors in $\mathbb{C}^2$). There are five regular complex polygons with 24 vertices. Their (flag to flag) symmetry groups are

$$\begin{aligned} &\texttt{ShephardTodd}(6) = <48,33>, \qquad \texttt{ShephardTodd}(6) = <48,33>, \\ &\texttt{ShephardTodd}(5) = <72,25>, \qquad \texttt{ShephardTodd}(8) = <96,67>, \\ &\texttt{ImprimitiveReflectionGroup}(12,1,2) = <288,239>. \end{aligned}$$

The four obtained from the primitive groups are *not* harmonic. The fifth frame is a generalised cross, which is harmonic.

In addition to these, there is a highly symmetric tight frame of 24 vectors (which is not a polygon) that can be obtained from the group

$$\texttt{G} := \texttt{ShephardTodd}(12) =< 48,29 >, \qquad G = \langle g_1, g_2, g_3 \rangle,$$
$$g_1 := \frac{1}{2}\begin{pmatrix} \omega^3-\omega & -\omega^3+\omega \\ -\omega^3+\omega & -\omega^3+\omega \end{pmatrix}, \quad g_2 := \frac{1}{2}\begin{pmatrix} \omega^3-\omega & \omega^3-\omega \\ \omega^3-\omega & -\omega^3+\omega \end{pmatrix}, \quad g_3 := \begin{pmatrix} 0 & -\omega \\ \omega^3 & 0 \end{pmatrix}.$$

and the vector $v = (1,\omega^3)$, where $\omega = e^{\frac{2\pi i}{8}}$. Similarly, this frame is not harmonic.

Let n be the number of vectors in a frame Φ with s angles $\mathrm{Ang}(\Phi)$, and k be the order of the group of scalar matrices which map Φ to Φ. Then, the estimate of Theorem 12.20 implies that

$$n \le b := k \begin{cases} \binom{d+s-1}{s}\binom{d+s-2}{s-1}, & 0 \in \mathrm{Ang}(\Phi); \\ \binom{d+s-1}{s}^2, & 0 \notin \mathrm{Ang}(\Phi). \end{cases} \tag{13.25}$$

The following tables list the highly symmetric tight frames of n vectors for $\mathbb{C}^d$ given by the complex reflection groups ST 4–37, as calculated in [BW13]. In these we list the "small group library" number (when possible), the bound b of (13.25) on the number of vectors (here (ℓ) denotes the number of lines when it is sharp), the number of angles s and the groups for which it is group frame with distinct vectors (when possible).

Table 13.1: The highly symmetric tight frames of n vectors in $\mathbb{C}^2$ given by the primitive reflection groups ST 4–22. Here (P) denotes a nonstarry regular complex polytope.

ST	d	order	n	b	s	group frame
4	2	$\langle 24,3\rangle$	8 (P)	8 (4)	1	$\langle 8,4\rangle$
5		$\langle 72,25\rangle$	24 (P)	24 (4)	1	$\langle 24,3\rangle$, $\langle 24,11\rangle$
6		$\langle 48,33\rangle$	16 (P)	16 (4)	1	$\langle 16,13\rangle$
			24 (P)	24 (6)	2	$\langle 24,3\rangle$
7		$\langle 144,157\rangle$	48	48 (4)	1	$\langle 48,47\rangle$, $\langle 48,33\rangle$
			72	72 (6)	2	$\langle 72,25\rangle$
8		$\langle 96,67\rangle$	24 (P)	24 (6)	2	$\langle 24,3\rangle$, $\langle 24,1\rangle$
9		$\langle 192,963\rangle$	48 (P)	48 (6)	2	$\langle 48,4\rangle$, $\langle 48,28\rangle$, $\langle 48,29\rangle$
			96 (P)	160	4	$\langle 96,67\rangle$, $\langle 96,74\rangle$
10		$\langle 288,400\rangle$	72 (P)	72 (6)	2	$\langle 72,12\rangle$, $\langle 72,25\rangle$
			96 (P)	144	3	$\langle 96,54\rangle$, $\langle 96,67\rangle$
11		$\langle 576,5472\rangle$	144	144 (6)	2	$\langle 144,69\rangle$, $\langle 144,121\rangle$, $\langle 144,122\rangle$
			192	288	3	$\langle 192,876\rangle$, $\langle 192,963\rangle$
			288	480	4	$\langle 288,400\rangle$, $\langle 288,638\rangle$
12		$\langle 48,29\rangle$	24	40	4	$\langle 24,3\rangle$
13		$\langle 96,192\rangle$	48	80	4	$\langle 48,28\rangle$, $\langle 48,29\rangle$
			48	48 (6)	2	$\langle 48,28\rangle$, $\langle 48,33\rangle$
14		$\langle 144,122\rangle$	48 (P)	72	3	$\langle 48,26\rangle$, $\langle 48,29\rangle$
			72 (P)	120	4	$\langle 72,25\rangle$
15		$\langle 288,903\rangle$	96	144	3	$\langle 96,182\rangle$, $\langle 96,192\rangle$
			144	240	4	$\langle 144,121\rangle$, $\langle 144,122\rangle$
			144	144 (6)	2	$\langle 144,121\rangle$, $\langle 144,157\rangle$
16		$\langle 600,54\rangle$	120 (P)	120 (12)	3	$\langle 120,5\rangle$, $\langle 120,15\rangle$
17		$\langle 1200,483\rangle$	240 (P)	240 (12)	3	$\langle 240,93\rangle$, $\langle 240,154\rangle$
			600 (P)	1440	8	$\langle 600,54\rangle$
18		$\langle 1800,328\rangle$	360 (P)	360 (12)	3	$\langle 360,51\rangle$, $\langle 360,89\rangle$
			600 (P)	900	5	$\langle 600,54\rangle$
19		$\langle 3600,*\rangle$	720	720 (12)	3	$\langle 720,420\rangle$, $\langle 720,708\rangle$
			1200	1800	5	$\langle 1200,483\rangle$
			1800	4320	8	$\langle 1800,328\rangle$
20		$\langle 360,51\rangle$	120 (P)	180	5	$\langle 120,5\rangle$
21		$\langle 720,420\rangle$	240 (P)	360	5	$\langle 240,93\rangle$
			360 (P)	864	8	$\langle 360,51\rangle$
22		$\langle 240,93\rangle$	120	288	8	$\langle 120,5\rangle$

Table 13.2: The highly symmetric tight frames of n vectors in $\mathbb{C}^d$, $3 \leq d \leq 8$, given by the primitive reflection groups ST 23–37. Here (P) denotes a nonstarry regular complex polytope.

ST	d	order	n	b	s	group frame
23	3	120	12 (P)	18	1	$\langle 12,3\rangle$
			20 (P)	72	2	–
			30	300	4	–
24		336	42	120	3	$\langle 42,2\rangle$
			56	450	4	–
25		648	27 (P)	27 (9)	1	$\langle 27,3\rangle$, $\langle 27,4\rangle$
			72	108	2	–
26		1296	54 (P)	54 (9)	1	$\langle 54,8\rangle$, $\langle 54,10\rangle$, $\langle 54,11\rangle$
			72 (P)	108	2	–
			216	1350	4	$\langle 216,88\rangle$
27		2160	216	1350	4	–
			270	1890	5	–
			360	9720	8	–
28	4	1152	24 (P)	80	2	$\langle 24,1\rangle$, $\langle 24,3\rangle$, $\langle 24,11\rangle$
			96	9408	6	$\langle 96,67\rangle$, $\langle 96,201\rangle$, $\langle 96,204\rangle$
29		7680	80	160	2	$\langle 80,30\rangle$
			160	800	3	–
			320	7840	5	$\langle 320,1581\rangle$, $\langle 320,1586\rangle$
			640	251680	10	–
30		14400	120 (P)	1400	4	$\langle 120,5\rangle$, $\langle 120,15\rangle$
			600 (P)	1109760	15	$\langle 600,54\rangle$
			720	3032400	18	–
			1200	78330560	32	–
31		46080	240	800	3	–
			1920	145200	9	$\langle 1920,*\rangle$
			3840	3162816	16	–
32		155520	240 (P)	240 (40)	2	–
			2160	28224	6	–
33	5	51840	80	450	2	–
			270	450	2	–
			432	31752	5	–
			1080	138600	7	–
34	6	39191040	756	*	*	*
			*	*	*	*
35		51840	27	441	2	$\langle 27,3\rangle$,$\langle 27,4\rangle$
			72	252	2	–
			216	213444	6	$\langle 216,86\rangle$, $\langle 216,88\rangle$
			720	232848	6	–
36	7	2903040	126	*	*	*
			*	*	*	*
37	8	696729600	240	*	*	*
			*	*	*	*

13.11 Projective representations

The four equiangular vectors $\Phi = (v, Sv, \Omega v, S\Omega v)$ in $\mathbb{C}^2$ of (1.7) are *not* the orbit of a group of order four. Indeed, if this was so, then they would be a harmonic frame (as all groups of order four are abelian), and hence have inner products in $\mathbb{Q}(i)$.

Nevertheless, since $\Omega S = -S\Omega$, the group H generated by S and Ω contains the scalar matrix $-I$, and so is $H = \{\pm I, \pm\Omega, \pm S, \pm S\Omega\}$. Thus, we can think of Φ as a H-frame (with vectors repeated), or as the orbit of the projective action of $G = H/\langle -I\rangle \cong \mathbb{Z}_2 \times \mathbb{Z}_2$ (for which the vectors are only defined up to a unit scalar multiple). The natural way to describe these equivalent viewpoints is via *projective representations* of the *index group* G. We will see that:

- For a given projective representation of G there is a canonical choice for H.
- A projective representation of G can be calculated from the representations of H.
- Many sets of equiangular lines come as projective orbits of irreducible actions.

Let G be a finite abstract group. A **projective representation** of G on a finite dimensional vector space $\mathscr{V}$ (over $\mathbb{F}$) is a group homomorphism

$$\rho_P : G \to PGL(\mathscr{V}) := \frac{GL(\mathscr{V})}{C}, \qquad C := \{c \in \mathbb{F} : c \neq 0\}. \tag{13.26}$$

Here $PGL(\mathscr{V})$ is the **projective linear group**, i.e., the group invertible linear transformations up to a nonzero scalar. The theory of representations (see §10.1) extends in the obvious way. In particular, by an appropriate choice of inner product on $\mathscr{V}$, the action of a projective representation can be taken to unitary.

Example 13.14. The matrices of (1.7) give a projective unitary representation of $\mathbb{Z}_2 \times \mathbb{Z}_2$ via

$$G = \mathbb{Z}_2 \times \mathbb{Z}_2 \mapsto PU(\mathbb{C}^2) : (j,k) \mapsto [S^j \Omega^k],$$

where $PU(\mathbb{F}^d)$ is the **projective unitary group**, i.e., the quotient of the unitary group by the subgroup C of unit modulus scalar matrices, and $[S^j\Omega^k] = S^j\Omega^k C$.

We now explain how a projective representation of a finite group G on $\mathbb{F}^d$ can be associated with a finite group $H \in GL(\mathbb{F}^d)$. Let E_g be any matrix in $\rho_P(g)$. We recall that $\rho_P(g)$ is the set of all nonzero scalar multiples of E. The fact $\rho_P(g)$ has finite order, i.e., $E_g^k = cI$ for some nonzero scalar c, does not imply that E_g has finite order. This can be rectified by using the key observation:

There are *exactly* d scalings of $E_g \in GL(\mathbb{C}^d)$ which have determinant 1, i.e.,

$$\hat{E}_g = \frac{\omega^j}{\det(E_g)^{1/d}} E_g, \qquad j = 0, 1, \ldots, d-1, \tag{13.27}$$

where $\det(E_g)^{1/d}$ is any fixed d-th root of $\det(E_g)$, and $\omega := e^{\frac{2\pi i}{d}}$.

13.12 The canonical abstract (error) group

Let $\hat{E}_g$ denote any of the d scalings of (13.27), so that

$$\det(\hat{E}_g) = 1, \qquad \forall g \in G.$$

Then the $d|G|$ matrices $H := \{\omega^j \hat{E}_g : j = 0, \ldots, d-1, g \in G\}$ are distinct. Moreover, they form a group, since

$$\hat{E}_g \hat{E}_h = \hat{w}_{g,h} \hat{E}_{gh}. \qquad \forall g, h \in G,$$

for some $\hat{w}_{g,h} \in \mathbb{C}$, and taking determinants of this gives

$$1 = \hat{w}_{g,h}^d \quad \Longrightarrow \quad \hat{w}_{g,h} \in \{1, \omega, \omega^2, \ldots \omega^{d-1}\}, \quad \forall g, h \in G.$$

Thus, we arrive at the following definition.

Definition 13.7. Let ρ_P be an irreducible projective representation of G on $\mathbb{C}^d$, and $E_g \in \rho_P(g)$, $g \in G$. Then associated **canonical error group** is

$$H := \{\omega^j \hat{E}_g : j = 0, \ldots, d-1, \, g \in G\},$$

and the abstract version of this group is called the **canonical abstract error group**. We will call G the **index group** of H, and d is **rank**.

Since conjugation preserves the determinant, it follows that equivalent projective representations have the same canonical abstract error group.

The action ρ of the canonical error group H on $\mathbb{C}^d$ is irreducible. Hence if this action is taken to be unitary, then (by Theorem 10.6) the unitary matrices $(E_g)_{g \in G}$ are an equal-norm tight frame for $M_d(\mathbb{C})$, i.e.,

$$A = \frac{d}{|G|} \sum_{g \in G} \langle A, E_g \rangle E_g, \qquad \forall A \in M_d(\mathbb{C}). \tag{13.28}$$

Moreover, by (13.10), the tight frame $(\rho(h))_{h \in H}$ is a *central group frame*.

The canonical error group H is (particular) *central extension* of G. It has centre

$$Z(H) = \langle \omega I \rangle \cong \mathbb{Z}_d,$$

since if a matrix commutes with the spanning sequence $(E_g)_{g \in G}$ for $M_d(\mathbb{C})$, then it commutes with all of $M_d(\mathbb{C})$ and therefore is a scalar matrix (Schur's lemma). Thus a group H can be a canonical (abstract) error group for at most one dimension d. Further, the index group G of a canonical (abstract) error group H is given by

$$G = \frac{H}{Z(H)}.$$

13.13 Nice error frames

In applications, it is convenient to describe a projective representation ρ_P of G in terms of unitary matrices $(E_g)_{g\in G}$, where $E_g \in \rho_P(g)$. A sequence $(E_g)_{g\in G}$ of unitary matrices gives a faithful unitary projective representation of G if and only if

1. E_1 is a scalar multiple of the identity I, and no other E_g is.
2. $E_g E_h = w_{g,h} E_{gh}$, $\forall g,h \in G$, where $w_{g,h} \in \mathbb{C}$.

Moreover, by Theorem 10.6, this projective representation is irreducible if and only if (13.28) holds. By Theorem 6.1, this irreducibilty condition can be written as

3. $\sum\limits_{g\in G} |\text{trace}(E_g)|^2 = |G|$.

For G a group (of order $\geq d^2$), a sequence $(E_g)_{g\in G}$ of unitary matrices in $M_d(\mathbb{C})$ is a **nice (unitary) error frame** with **index group** G if it satisfies 1, 2 and 3 (above). Our discussion gives:

Proposition 13.5. *Let G be a group and $(E_g)_{g\in G}$ be unitary matrices in $M_d(\mathbb{C})$. Then the following are equivalent*

1. *$(E_g)_{g\in G}$ is a nice error frame for $M_d(\mathbb{C})$.*
2. *$g \mapsto E_g$ is a faithful irreducible projective representation of G of degree d.*

In this case, the action of the canonical error group H on $\mathbb{C}^d$ is a special unitary faithful irreducible ordinary representation of the canonical abstract error group of degree d.

Above (and to follow) we write the projective representation ρ_P of (13.26) as $g \mapsto E_g$, with the understanding that $E_g \in \rho_P(g)$.

Example 13.15. (Nice error bases) When $|G| = d^2$, the condition 3 of a nice error frame reduces to

3. $\text{trace}(E_g) = 0$, $g \neq 1$, $g \in G$.

i.e., (13.28) is an orthogonal expansion. In the field of quantum error correcting codes, an orthonormal expansion for $M_d(\mathbb{C})$ is called a **error operator basis** and a **nice (unitary) error basis**, if the matrices come from a projective representation (and are unitary) [Kni96a]. These were then generalised to nice error frames.

Example 13.16. The **Pauli** matrices (used to study spin in quantum mechanics)

$$\sigma_1 = \sigma_x := \begin{pmatrix} 0 & 1 \\ 1 & 0 \end{pmatrix}, \quad \sigma_2 = \sigma_y := \begin{pmatrix} 0 & -i \\ i & 0 \end{pmatrix}, \quad \sigma_3 = \sigma_z := \begin{pmatrix} 1 & 0 \\ 0 & -1 \end{pmatrix}, \tag{13.29}$$

together with the identity are a nice error basis for the 2×2 matrices, with index group $\mathbb{Z}_2 \times \mathbb{Z}_2$. They have determinant -1, and generate the group `<16, 13>` of order 16. The group generated by just the reflections σ_1 and σ_3 contains $\pm i\sigma_2$ and is the *dihedral group* <8, 3>. The canonical error group for the nice error basis $\{I,\sigma_1,\sigma_2,\sigma_3\}$ is

$$H = \langle i\sigma_1, i\sigma_2, i\sigma_3\rangle,$$

which is the *quaternian group* <8, 4>.

We now specify which nice error frames are considered to be "equivalent".

Definition 13.8. Nice error frames $(E_g)_{g\in G}$ and $(F_h)_{h\in H}$ for $M_d(\mathbb{C})$ are **equivalent** if there is bijection $\sigma : G \to H$ between their index groups, scalars $(c_g)_{g\in G}$ and an invertible $T \in M_d(\mathbb{C})$, such that

$$F_{\sigma g} = c_g T^{-1} E_g T, \qquad \forall g \in G. \tag{13.30}$$

This is more general than the *equivalence* of projective representations, where $G = H$, and reindexing of the elements of $(E_g)_{g\in G}$ is not allowed.

Proposition 13.6. *Equivalent nice error frames have the* same *canonical abstract error group, and (in particular) the same index group.*

Proof. Suppose nice error frames $(E_g)_{g\in G}$ and $(F_h)_{h\in H}$ for $M_d(\mathbb{C})$ are equivalent. Then (13.30) scales to

$$\hat{F}_{\sigma g} = T^{-1}(\hat{c}_g E_g)T, \qquad \forall g \in G,$$

where $\hat{c}_g \in \{1,\omega,\omega^2,\ldots,\omega^{d-1}\}$ (by considering determinants). Thus the canonical error groups are conjugate via T and so are isomorphic. Since the index group is the abstract error group factored by its centre, the nice error frames also have the same index groups. □

Example 13.17. For $d = 1$, the only canonical abstract error group is $H = 1$.

Example 13.18. A nice error basis (projective representation of $\mathbb{Z}_d \times \mathbb{Z}_d$) is given by

$$G = \mathbb{Z}_d \times \mathbb{Z}_d \mapsto M_d(\mathbb{C}) : (j,k) \mapsto E_{(j,k)} = S^j \Omega^k,$$

where S is the *cyclic shift matrix*, and Ω is the *modulation matrix*, given by

$$(S)_{jk} := \delta_{j,k+1}, \qquad (\Omega)_{jk} = \omega^j \delta_{j,k}, \quad \omega := e^{\frac{2\pi i}{d}}.$$

This is the only nice error basis (up to equivalence) for $M_d(\mathbb{C})$ with index group $G = \mathbb{Z}_d \times \mathbb{Z}_d$ (cf. [BK73]). This nice error basis (which generalises the Pauli matrices) plays a key role in the construction of SICs (see Chapter 14). It is known as the *Heisenberg nice error basis*.

In §13.15, we outline how *all* nice error frames can be constructed using `Magma`.

13.14 Tensor products of nice error frames

Nice error frames can be constructed (and also deconstructed) via tensor products.

Proposition 13.7. *Let* $(E_{g_1})_{g_1 \in G_1}, (F_{g_2})_{g_2 \in G_2}$ *be nice error frames for* $M_{d_1}(\mathbb{C})$, $M_{d_2}(\mathbb{C})$. *Then their tensor product*

$$(E_{g_1} \otimes F_{g_2})_{(g_1,g_2) \in G_1 \times G_2}$$

is a nice error frame for $M_{d_1 d_2}(\mathbb{C})$. *In particular, a product of index groups is an index group. Moreover, the canonical error group is*

$$H = \{\omega^j (h_1 \otimes h_2) : 0 \leq j < d-1, h_1 \in H_1, h_2 \in H_2\}, \qquad \omega := e^{\frac{2\pi i}{d}}, \quad d := d_1 d_2.$$

where H_1, H_2 *are the canonical error groups of the nice error frames.*

Proof. In view of Proposition 13.5, the first part follows from the basic theory of (projective) representations. Alternatively, it can be verified directly, e.g., the tensor product satisfies condition 3 of the definition of a nice error basis since

$$\sum_{(g_1,g_2) \in G_1 \times G_2} |\mathrm{trace}(E_{g_1} \otimes F_{g_2})|^2 = \sum_{g_1} \sum_{g_2} |\mathrm{trace}(E_{g_1}) \mathrm{trace}(F_{g_2})|^2$$
$$= \Big(\sum_{g_1} |\mathrm{trace}(E_{g_1})|^2 \Big) \Big(\sum_{g_2} |\mathrm{trace}(F_{g_2})|^2 \Big) = |G|\,|H| = |G \times H|.$$

The tensor product group $H_1 \otimes H_2$ consists of scalar multiples of each $E_{g_1} \otimes E_{g_2}$, with determinant 1, but may not contain all d-roots of unity (if d_1 and d_2 are not coprime), and so we add these. □

Corollary 13.3. *A product of index groups is an index group, and in particular, a product of index groups for nice error bases is an index group for a nice error basis.*

Example 13.19. Let K be a finite abelian group of order d. Since K is a product of cyclic groups, it follows by taking tensor products of the Heisenberg nice error basis (Example 13.18) that $G = K \times K$ is the index group of a nice error basis for $M_d(\mathbb{C})$.

It can be shown (Theorem 13.5) that:

A nice error frame can have an abelian index group only if it is a nice error basis.

Example 13.20. Taking the tensor product of the two nonabelian index groups for $d = 4$, with the (abelian) index group for $d = 2$, gives two nonabelian index groups for $d = 8$, i.e.,

$$< 16,3 > \times < 4,2 > = < 64,193 >, \qquad < 16,11 > \times < 4,2 > = < 64,261 >.$$

13.15 Computing all nice error frames

Finding the centre of a finite group H and its irreducible representations are fast calculations, and a representation can always be made unitary. Thus the following characterisation of abstract error groups gives rise to a practical algorithm for their calculation, and hence that of the nice error frames they correspond to.

Proposition 13.8. *A group H is a canonical abstract error group if and only if*

1. *Its centre $Z(H)$ is cyclic of order d.*
2. *It has a faithful irreducible ordinary representation ρ of degree d, which is special, i.e.,* $\det(h) = 1$, $\forall h \in H$.

In particular, for $d > 1$ all canonical abstract error groups are nonabelian.

The nice error frame given by such a representation is $(E_g)_{g\in G}$, where

$$G := \frac{H}{Z(H)}, \qquad E_g \in \rho(g).$$

It remains only to determine which of these are equivalent. In this regard, we have:

Proposition 13.9. *(Equivalence) If $\rho : H \to M_d(\mathbb{C})$ is a faithful irreducible special unitary ordinary representation of H, then so is*

$$\rho_\sigma : h \mapsto \rho(\sigma h), \qquad \sigma \in \mathrm{Aut}(H),$$

where $\mathrm{Aut}(H)$ *denotes the automorphisms of H. These give equivalent nice error frames, even though the representations may not be equivalent when σ is an outer automorphism.*

Proof. Since any given automorphism σ of H fixes the centre $Z(H)$, it induces an automorphism $\sigma_G \in \mathrm{Aut}(G)$ on the index group $G = H/Z(H)$. Thus a nice error frame $(F_g)_{g\in G}$ for ρ_σ is reindexing of one for ρ, since

$$F_g := \rho_\sigma(g) = \rho(\sigma g) = \rho(\sigma_G(g)), \qquad \forall g \in G.$$

If σ is an inner automorphism, i.e., $\sigma h = k^{-1}hk$, for some $k \in H$, then ρ and ρ_σ are equivalent ordinary representations of H, since

$$\rho_\sigma(h) = \rho(k^{-1}hk)) = \rho(k)^{-1}\rho(h)\rho(k).$$

□

In practice, the action groups $\rho(H)$ of the ordinary representations ρ of a given H calculated in `magma` with the command

```
AbsolutelyIrreducibleModules(H,Rationals());
```

are often the same monomial group.

13.16 Examples of nice error frames

In [WC15], lists of nice error frames were produced using Proposition 13.8, i.e.,

Algorithm: To construct all rank d abstract error groups H of a given order.

1. Find all groups H (of the given order) with a cyclic centre of order d.
2. Find the faithful irreducible (ordinary) representations ρ of H of degree d.
3. Determine whether ρ is special, i.e., $\det(h) = 1$, $\forall h \in H$.

Table 13.3: The canonical abstract error groups H and index groups G for the first few nice error frames which are not bases, $2 \leq d \leq 4$.

$d=2$		$d=3$		$d=4$	
H	G	H	G	H	G
$\langle 12,1\rangle$	$\langle 6,1\rangle$	$\langle 36,11\rangle$	$\langle 12,3\rangle$	$\langle 80,28\rangle$	$\langle 20,3\rangle$
$\langle 16,9\rangle$	$\langle 8,3\rangle$	$\langle 54,8\rangle$	$\langle 18,4\rangle$	$\langle 96,157\rangle$	$\langle 24,8\rangle$
$\langle 20,1\rangle$	$\langle 10,1\rangle$	$\langle 63,3\rangle$	$\langle 21,1\rangle$	$\langle 96,215\rangle$	$\langle 24,14\rangle$
$\langle 24,3\rangle$	$\langle 12,3\rangle$	$\langle 72,42\rangle$	$\langle 24,12\rangle$	$\langle 128,523\rangle$	$\langle 32,27\rangle$
$\langle 24,4\rangle$	$\langle 12,4\rangle$	$\langle 81,9\rangle$	$\langle 27,3\rangle$	$\langle 128,545\rangle$	$\langle 32,24\rangle$
$\langle 28,1\rangle$	$\langle 14,1\rangle$	$\langle 108,15\rangle$	$\langle 36,9\rangle$	$\langle 128,749\rangle$	$\langle 32,34\rangle$
$\langle 32,20\rangle$	$\langle 16,7\rangle$	$\langle 108,22\rangle$	$\langle 36,11\rangle$	$\langle 128,782\rangle$	$\langle 32,31\rangle$
$\langle 36,1\rangle$	$\langle 18,1\rangle$	$\langle 117,3\rangle$	$\langle 39,1\rangle$	$\langle 128,864\rangle$	$\langle 32,6\rangle$
$\langle 40,4\rangle$	$\langle 20,4\rangle$	$\langle 144,68\rangle$	$\langle 48,3\rangle$	$\langle 128,880\rangle$	$\langle 32,9\rangle$
$\langle 44,1\rangle$	$\langle 22,1\rangle$	$\langle 162,14\rangle$	$\langle 54,5\rangle$	$\langle 128,1750\rangle$	$\langle 32,27\rangle$
$\langle 48,8\rangle$	$\langle 24,6\rangle$	$\langle 171,4\rangle$	$\langle 57,1\rangle$	$\langle 128,1799\rangle$	$\langle 32,28\rangle$
$\langle 48,28\rangle$	$\langle 24,12\rangle$	$\langle 189,8\rangle$	$\langle 63,3\rangle$	$\langle 128,2146\rangle$	$\langle 32,39\rangle$

Example 13.21. ($d = 2$) In view of Proposition 13.5, all canonical abstract error groups for $d = 2$ are given by the *ADE classification* of the finite subgroups of $SL_2(\mathbb{C})$ [Ste85]. The **generalised quaternian group** or **dicyclic group** of order $4n$ ($n > 1$), which is generated by the matrices

$$\begin{pmatrix} \omega_{2n} & 0 \\ 0 & \omega_{2n}^{-1} \end{pmatrix}, \quad \begin{pmatrix} 0 & -1 \\ 1 & 0 \end{pmatrix}, \qquad \omega_{2n} := e^{\frac{2\pi i}{2n}},$$

gives an infinite family of rank 2 of abstract canonical error groups H. These account for all rank the 2 abstract canonical error groups in Table 13.3, except for

$$H = < 24,3 >,\ G = < 12,3 >, \qquad H = < 48,28 >,\ G = < 24,12 >.$$

which come from the Shephard–Todd reflection groups with numbers 4 and 8.

From Table 13.3, we observe that index groups may be repeated in different dimensions.

Example 13.22. (Repeated index groups) A group G may be the index group for nice error frames in more than one dimension d, e.g., $G = <12,3>$ is the index group for a nice error frame for $M_2(\mathbb{C})$ ($H = <24,3>$), and also one for $M_3(\mathbb{C})$ ($H = <36,11>$).

None of the index groups in Table 13.3 are abelian. Indeed, basic results from character theory imply the following.

Theorem 13.5. *(Abelian index group) A nice error frame can have an abelian index group only if it is a nice error basis.*

Proof. Suppose that H is the canonical abstract error group of a nice error frame, and that $\chi : H \to \mathbb{C}$ is the character of a faithful irreducible representation with $\deg(\chi) = \chi(1) = d$. Recall that the **centre** of a character $\chi : H \to \mathbb{C}$ is the subgroup

$$Z(\chi) := \{h \in H : |\chi(h)| = \chi(1)\},$$

and that if χ is irreducible then

$$\frac{Z(\chi)}{\ker(\chi)} = Z\Big(\frac{H}{\ker(\chi)}\Big), \qquad \ker(\chi) := \{h \in H : \chi(h) = \chi(1)\}.$$

Since the representation is faithful, $\ker(\chi) = 1$, and so this becomes

$$Z(\chi) = Z(H).$$

Thus the index group is $G = H/Z(\chi)$, by Theorem 2.13 of [Isa06], if G is abelian, then

$$|G| = [H : Z(\chi)] = \chi(1)^2 = d^2.$$

□

The nice error bases (see Tables 13.4 and 14.1) play a prominent role in the construction of SICs. These may or may not have an abelian index group. It follows from Example 13.19, that $G = K \times K$ is the index group of a nice error basis for every abelian group K.

Example 13.23. (Inequivalent nice error bases) For $d = 8$, there are 47 canonical abstract error groups and only 42 index groups (see Table 14.1). In particular, there are three canonical abstract error groups for $G = <64, 67>$, and hence at least three inequivalent nice error bases with this index group. Moreover, two of these give rise to SICs, and one does not.

Table 13.4: Nice error bases for $d < 14$, $d \neq 8$. Here H is the canonical abstract error group, G is the index group and `sic` indicates that a SIC exists numerically.

d	H	G
1	$\langle 1,1\rangle$	$\langle 1,1\rangle = \mathbb{Z}_1$ sic
2	$\langle 8,4\rangle$	$\langle 4,2\rangle = \mathbb{Z}_2^2$ sic
3	$\langle 27,3\rangle$	$\langle 9,2\rangle = \mathbb{Z}_3^2$ sic
4	$\langle 64,19\rangle$	$\langle 16,2\rangle = \mathbb{Z}_4^2$ sic
	$\langle 64,94\rangle$	$\langle 16,3\rangle$
	$\langle 64,256\rangle$	$\langle 16,11\rangle$
	$\langle 64,266\rangle$	$\langle 16,14\rangle$
5	$\langle 125,3\rangle$	$\langle 25,2\rangle = \mathbb{Z}_5^2$ sic
6	$\langle 216,42\rangle$	$\langle 36,11\rangle = \mathbb{Z}_3 \times A_4$ sic
	$\langle 216,66\rangle$	$\langle 36,13\rangle$
	$\langle 216,80\rangle$	$\langle 36,14\rangle = \mathbb{Z}_6^2$ sic
7	$\langle 343,3\rangle$	$\langle 49,2\rangle = \mathbb{Z}_7^2$ sic
9	$\langle 729,24\rangle$	$\langle 81,2\rangle = \mathbb{Z}_9^2$ sic
	$\langle 729,30\rangle$	$\langle 81,4\rangle$
	$\langle 729,405\rangle$	$\langle 81,9\rangle$ sic
	$\langle 729,489\rangle$	$\langle 81,12\rangle$
	$\langle 729,503\rangle$	$\langle 81,15\rangle = (\mathbb{Z}_3 \times \mathbb{Z}_3)^2$
10	$\langle 1000,70\rangle$	$\langle 100,15\rangle$
	$\langle 1000,84\rangle$	$\langle 100,16\rangle = \mathbb{Z}_{10}^2$ sic
11	$\langle 1331,3\rangle$	$\langle 121,2\rangle = \mathbb{Z}_{11}^2$ sic

d	H	G
12	$\langle 1728,1294\rangle$	$\langle 144,68\rangle$ sic
	$\langle 1728,2011\rangle$	$\langle 144,92\rangle$
	$\langle 1728,2079\rangle$	$\langle 144,101\rangle = \mathbb{Z}_{12}^2$ sic
	$\langle 1728,2983\rangle$	$\langle 144,132\rangle$
	$\langle 1728,10718\rangle$	$\langle 144,95\rangle$
	$\langle 1728,10926\rangle$	$\langle 144,100\rangle$
	$\langle 1728,11061\rangle$	$\langle 144,102\rangle$
	$\langle 1728,13457\rangle$	$\langle 144,136\rangle$
	$\langle 1728,20393\rangle$	$\langle 144,170\rangle$
	$\langle 1728,20436\rangle$	$\langle 144,172\rangle$
	$\langle 1728,20556\rangle$	$\langle 144,177\rangle$
	$\langle 1728,20771\rangle$	$\langle 144,179\rangle$
	$\langle 1728,30353\rangle$	$\langle 144,184\rangle$
	$\langle 1728,30562\rangle$	$\langle 144,189\rangle$
	$\langle 1728,30928\rangle$	$\langle 144,193\rangle$
	$\langle 1728,30953\rangle$	$\langle 144,194\rangle$
	$\langle 1728,31061\rangle$	$\langle 144,196\rangle$
	$\langle 1728,31093\rangle$	$\langle 144,197\rangle = (\mathbb{Z}_2 \times \mathbb{Z}_6)^2$
13	$\langle 2197,3\rangle$	$\langle 169,2\rangle = \mathbb{Z}_{13}^2$ sic

Notes

Relaxing the condition of irreducibility gives a **stable isogon** (see [MVW16]).

Chapter 14
Weyl–Heisenberg SICs

The maximal number of equiangular vectors (lines) in $\mathbb{C}^d$ is less than or equal to d^2 (Theorem 12.2). The corresponding bound of $\frac{1}{2}d(d+1)$ for real equiangular lines in $\mathbb{R}^d$ is rarely attained (only for $d = 2, 3, 7, 23$). There is *compelling evidence* for

Zauner's conjecture (1999): There are d^2 equiangular lines in $\mathbb{C}^d$ (for all d).

This was given in Zauner's 1999 thesis (see [Zau10]), and is often stated in stronger forms which give a specific structure for the lines (as a group frame). These lines will be called *SICs* (see §14.2), and their existence is also known as the *SIC problem*. The evidence for this conjecture (which is steadily increasing) includes

- SICs have been constructed *numerically* for all $d \leq 121$ (to 8000 digit precision), and for a handful of other dimensions up to $d = 323$.
- SICs have been constructed *analytically* for $d = 2, \ldots, 16, 19, 24, 28, 35, 48$ [SG10], and most recently for $d = 17, 18, 20, 21, 30, 31, 37, 39, 43$ [ACFW17].

We now outline the detailed structure of SICs, using the following road map:

- SICs are *group covariant*, i.e., are G-frames for a discrete Heisenberg group. This reduces their construction to finding a *fiducial vector* (generating vector).
- The *Clifford group* (normaliser of the Heisenberg group) maps fiducial vectors to fiducial vectors.
- Fiducial vectors have certain symmetries, in particular they are eigenvectors of the Zauner matrix Z of order 3 (or M_1 in some special cases).
- The field given by the triple products defining a SIC has a solvable Galois group. It induces *Galois symmetries* of SICs (this is part proved and part conjecture).

Before reading the remainder of this chapter, please be warned:

The SIC problem is very addictive.

S.F.D. Waldron, *An Introduction to Finite Tight Frames*, Applied and Numerical Harmonic Analysis, https://doi.org/10.1007/978-0-8176-4815-2_14

14.1 Maximal sets of complex equiangular lines and vectors

Suppose that (v_j) is sequence of $n = d^2$ unit vectors in $\mathbb{C}^d$ which gives equality in the lower bound (6.22) for the second frame potential ($t = 2$), i.e.,

$$\sum_j \sum_k |\langle v_j, v_k\rangle|^4 = \frac{2d^3}{d+1}, \quad \text{or} \quad \sum_{j\neq k} |\langle v_j, v_k\rangle|^4 = \frac{2d^3}{d+1} - d^2 = \frac{d^2(d^2-1)}{(d+1)^2}.$$

By the variational characterisation (Theorem 6.1),

$$\sum_{j\neq k} |\langle v_j, v_k\rangle|^2 \geq \frac{1}{d}(d^2)^2 - d^2 = d^3 - d^2, \tag{14.1}$$

with equality if and only if (v_j) is a tight frame. By Cauchy–Schwarz,

$$\sum_{j\neq k} |\langle v_j, v_k\rangle|^2 \leq \sqrt{d^2(d^2-1)}\sqrt{\sum_{j\neq k} |\langle v_j, v_k\rangle|^4} = \frac{d^2(d^2-1)}{d+1} = d^3 - d^2,$$

so by (14.1) there must be equality, i.e., $|\langle v_j, v_k\rangle|^2 = C$, where $C > 0$ is constant. Thus (v_j) is an equiangular tight frame for $\mathbb{C}^d$ with the maximal possible number of vectors (see Theorem 12.2).

This condition on the vectors v_j can be expressed in terms of the associated lines $\mathbb{C}v_j$, and the rank one orthogonal projections $P_j = v_j v_j^*$, which satisfy

$$\langle P_j, P_k\rangle = \operatorname{trace}(P_j P_k^*) = \operatorname{trace}(v_j v_j^* v_k v_k^*) = \operatorname{trace}(v_j^* v_k v_k^* v_j) = |\langle v_j, v_k\rangle|^2.$$

We now give some of these characterisations.

Proposition 14.1. *Suppose that (v_j) is d^2 unit vectors in $\mathbb{C}^d$, and $P_j := v_j v_j^*$. Then the following are equivalent*

1. *(v_j) is a equiangular tight frame for $\mathbb{C}^d$.*
2. *The lines $(\mathbb{C}v_j)$ are equiangular.*
3. *$|\langle v_j, v_k\rangle|^2 = \frac{1}{d+1}$, $j \neq k$.*
4. *$\langle P_j, P_k\rangle = \frac{1}{d+1}$, $j \neq k$.*
5. *$\sum_j \sum_k |\langle v_j, v_k\rangle|^4 = \frac{2d^3}{d+1}$.*
6. *$\sum_j \sum_k \langle P_j, P_k\rangle^2 = \frac{2d^3}{d+1}$.*
7. *(v_j) is a complex $(2,2)$-design.*
8. *(P_j) is a complex projective $(2,2)$-design.*

Proof. If (v_j) is an equiangular tight frame with $|\langle v_j, v_k\rangle|^2 = C$, $j \neq k$, then (2.10) gives

$$(d^4 - d^2)C = d^3 - d^2) \quad \Longrightarrow \quad C = \frac{1}{d+1}.$$

Thus the equivalences follow from the previous discussion, and the definition of $(2,2)$-designs (see §6.9). □

14.2 SICs

The "rank one quantum measurements" (P_j) corresponding to d^2 equiangular lines in $\mathbb{C}^d$ have applications in quantum mechanics, e.g., in quantum state tomography.

Definition 14.1. A **SIC**[1], **SIC-POVM**, or **symmetric informationally complete positive operator valued measure** for $\mathbb{C}^d$ is d^2 rank one orthogonal projections (P_j) on $\mathbb{C}^d$, which satisfy the condition

$$\langle P_j, P_k \rangle = \operatorname{trace}(P_j P_k^*) = \frac{1}{d+1}, \qquad j \neq k. \tag{14.2}$$

For an explanation of the quantum physic underlying this definition see §14.3. The space of $d \times d$ matrices (linear operators on the Hilbert space $\mathscr{H} = \mathbb{C}^d$) can be expressed in terms of these rank one Hermitian matrices:

Proposition 14.2. *If (P_j) is a SIC for $\mathbb{C}^d$, then*

1. *(P_j) is a basis for the real vector space of Hermitian matrices.*
2. *(P_j) is a basis for $\mathbb{C}^{d \times d}$.*
3. *$(P_j - \frac{1}{d}I)$ is an equiangular tight frame for the traceless Hermitian matrices.*
4. *$(P_j - \frac{1}{d}I)$ is an equiangular tight frame for the traceless matrices in $\mathbb{C}^{d \times d}$.*

Proof. By Theorem 12.2 the d^2 matrices (P_j) are linearly independent (over $\mathbb{C}$ and hence $\mathbb{R}$), and so form a basis for the Hermitian matrices and $\mathbb{C}^{d \times d}$. The matrices $P_j - \frac{1}{d}I$ are traceless, with $\langle P_j - \frac{1}{d}I, P_k - \frac{1}{d}I \rangle = \langle P_j, P_k \rangle - \frac{1}{d}$, and so are equiangular. In addition, they are a tight frame for the (d^2-1)–dimensional spaces of traceless matrices asserted, as the variational characterisation (6.4) for being a tight frame holds, via the calculation

$$d^2\Big(1 - \frac{1}{d}\Big)^2 + (d^4 - d^2)\Big(\frac{1}{d+1} - \frac{1}{d}\Big)^2 = \frac{(d-1)d^2}{d+1} = \frac{1}{d^2-1}\Big(d^2(1 - \frac{1}{d})\Big)^2.$$

Here we used $\langle P_j, P_k \rangle = \frac{1}{d+1}$, $j \neq k$ and $\langle P_j, P_j \rangle = 1$. □

Remark 14.1. The equiangular tight frame $(P_j - \frac{1}{d}I)$ above is the d^2 vertices of a regular simplex in a (d^2-1)-dimensional real space. Indeed (see Exer. 14.2), the rank one orthogonal projections vv^* (or unit vectors $v \in \mathbb{C}^d$) giving a SIC are mapped to the traceless Hermitian matrices with unit norm (called the **Bloch sphere** when $d = 2$) by the map

$$vv^* \mapsto \sqrt{\frac{d}{d-1}}(vv^* - \frac{1}{d}I). \tag{14.3}$$

Since their image (the vertices of a regular simplex) has symmetry group S_{d^2}, it is expected that SICs should also have symmetries. This is borne out by the known constructions. Unfortunately, SICs can only be found by taking the preimage of *certain* regular simplices under (14.3), as it is not onto for $d > 2$.

---

[1] Vectors (v_j) satisfying 1, 2, 3, 5, 7 of Proposition 14.1 are also known as SICs.

14.3 Quantum measurements

Much of the literature on SICs comes from quantum mechanics (which uses Dirac notation). I now outline the physical interpretation of SICs (as kindly explained to me by Marcus Appleby).

The **Dirac** or **bra–ket notation** consists of the **bra** $\langle w|$ and the **ket** $|v\rangle$ which combine to give the **bra–ket** $\langle w|v\rangle$. We illustrate this notation:

$$\begin{array}{llllll} \textit{mathematics} & w^* = \langle \cdot, w\rangle & v & \langle v, w\rangle = w^* v & vw^* & e_j \\ \textit{bra-ket} & \langle w| & |v\rangle & \langle w|v\rangle = (\langle w|)(|v\rangle) & |v\rangle\langle w| & |e_j\rangle = |j\rangle \end{array}$$

In particular, our inner products are linear in the first variable, and the bra–ket is linear in the second variable.

The *state* of a *quantum system* of dimension d is described by a $d \times d$ positive semidefinite matrix ρ with $\operatorname{trace}(\rho) = 1$ called a **density matrix**. A **measurement** is described by positive semidefinite matrices $E_1, \ldots, E_n$ with $\sum_j E_j = I$. The (E_j) are called a **positive operator valued measure (POVM)**. The **probability distribution** for obtaining the measurement outcome j for a state ρ is $\operatorname{trace}(\rho E_j)$.

- For many measurements (E_j), e.g. spin and energy, there are infinitely many states which give some fixed probability distribution.
- There are *some* measurements for which the probability distribution fixes the state – these measurements are said to be **informationally complete**.
- Informationally complete measurements can be used to infer the state statistically (quantum tomography).
- An informationally complete POVM must have at least d^2 operators. It is said to be **minimally informationally complete** if it has exactly d^2 operators.

If (P_j) is a SIC for $\mathbb{C}^d$, say $P_j = v_j v_j^*$, then (v_j) is an equiangular tight frame for $\mathbb{C}^d$, and the frame expansion can be written as

$$I = \frac{1}{d} \sum_j P_j.$$

Thus the operators

$$E_j := \frac{1}{d} P_j$$

are a

Symmetric Informationally Complete Positive Operator Valued Measure,

where the term *symmetric* refers to the fact that $\langle E_j, E_k\rangle$ takes just two values, i.e.,

$$\langle E_j, E_k\rangle = \operatorname{trace}(E_j E_k) = \begin{cases} \frac{1}{d^2(d+1)}, & j \neq k; \\ \frac{1}{d^2}, & j = k. \end{cases}$$

This explains the origin of the term SIC-POVM (SIC).

14.4 Group covariance

All known SICs for $\mathbb{C}^d$ have a group *covariance property*, i.e., they are an H-frame.

- The group H can be the discrete (Weyl–)Heisenberg group, except when $d = 8$, where it can also be a product of such groups for the *Hoggar lines* (see §14.6).
- The Heisenberg group is a projective representation of $G = \mathbb{Z}_d \times \mathbb{Z}_d$, i.e., a nice error basis with index group G. Other nice error bases (with possibly nonabelian index groups) give rise to SICs, but these SICs are group covariant with respect to the Heisenberg group (or are the Hoggar lines).
- For d prime, the only group covariant SICs are for the Heisenberg group [Zhu10] (there might be SICs which are not group covariant).
- The *Clifford group* (normaliser of the Heisenberg group) maps Weyl–Heisenberg SICs to Weyl–Heisenberg SICs.

By definition, a SIC or set of equiangular lines is a projective object. To describe the SICs for $\mathbb{C}^d$ which are the projective orbit of a given vector (line) under

$$\rho : G \to PU(\mathbb{C}^d) \qquad \text{(projective unitary group)},$$

it is often convenient to deal with a generating unit vector $v \in \mathbb{C}^d$ (any in the line) and unitary matrices $E_g \in \rho(G)$, i.e., the corresponding nice error basis (see §13.13). The matrices $(E_g)_{g\in G}$ can be replaced by a finite group H which consists of unit scalar multiples of them, e.g., the canonical abstract error group, and the vector v by the orthogonal projection $P_v = vv^*$ onto the line it defines. In this way, a SIC can be equivalently presented as

- $(E_g v)_{g\in G}$ (each line appears once)
- An H-frame $(hv)_{h\in H}$ (the lines are repeated)
- A projective G-frame $(g \cdot P_v)_{g\in G}$, where $g \cdot A := \tilde{g}A\tilde{g}^{-1}$, $\tilde{g} \in \rho(g)$.

The vector v and the rank one orthogonal projection $P_v = vv^*$ are said to be **fiducial**.

By Corollary 8.2, the equiangular lines given by $(hv)_{h\in H}$ are determined up to projective unitary equivalence by their triple products

$$\langle h_1 v, h_2 v\rangle \langle h_2 v, h_3 v\rangle \langle h_3 v, h_1 v\rangle = \langle gv, v\rangle \langle hv, v\rangle \langle g^{-1}h^{-1}v, v\rangle, \quad g := h_2^{-1}h_1,\ h := h_3^{-1}h_2.$$

The field associated with a given covariant SIC is the extension of $\mathbb{Q}$ by these triple products, in which the fiducial projector $P_v = vv^*$ lies. The structure of this field plays a crucial role in the construction of Weyl–Heisenberg SICs (see §14.20).

Example 14.1. The SIC $(v, Sv, \Omega v, S\Omega v)$ of (1.7) is covariant with respect to $\mathbb{Z}_2 \times \mathbb{Z}_2$ (see Example 13.14). The fiducial vector and fiducial projector are

$$v = \frac{1}{\sqrt{6}} \begin{pmatrix} \sqrt{3+\sqrt{3}} \\ e^{\frac{\pi}{4}i}\sqrt{3-\sqrt{3}} \end{pmatrix}, \qquad P_v = vv^* = \frac{1}{6} \begin{pmatrix} 3+\sqrt{3} & \sqrt{3}-\sqrt{3}i \\ \sqrt{3}+\sqrt{3}i & 3-\sqrt{3} \end{pmatrix}.$$

We observe that here the field required to present v is larger than that for $P_v = vv^*$.

14.5 The discrete Heisenberg group and Weyl–Heisenberg SICs

We now generalise the Pauli matrices and the group (nice error basis) they generate to higher dimensions (see Examples 13.16 and 13.18).

Throughout, let ω and μ be the primitive d-th and $2d$-th roots of unity

$$\omega := e^{\frac{2\pi i}{d}}, \qquad \mu := e^{\frac{2\pi i}{2d}},$$

and take the indices for elements of $\mathbb{C}^d$ and $\mathbb{C}^{d\times d}$ from $\mathbb{Z}_d = \{0,1,\ldots,d-1\}$. Let $S \in \mathbb{C}^{d\times d}$ be the **cyclic shift matrix**, and $\Omega \in \mathbb{C}^{d\times d}$ be the **modulation matrix** given by

$$(S)_{jk} := \delta_{j,k+1}, \qquad (\Omega)_{jk} := \omega^j \delta_{j,k}. \tag{14.4}$$

For $d=2$, these are the Pauli matrices $\sigma_1 = \sigma_x$ and $\sigma_3 = \sigma_z$, and for $d=3$ they are

$$S = \begin{pmatrix} 0 & 0 & 1 \\ 1 & 0 & 0 \\ 0 & 1 & 0 \end{pmatrix}, \qquad \Omega = \begin{pmatrix} 1 & & \\ & \omega & \\ & & \omega^2 \end{pmatrix}.$$

These have order d, and satisfy the **commutativity relation**

$$\Omega^k S^j = \omega^{jk} S^j \Omega^k. \tag{14.5}$$

Thus the group generated by the unitary matrices S and Ω is

$$H := \langle S, \Omega \rangle = \{\omega^r S^j \Omega^k : r,j,k \in \mathbb{Z}_d\}. \tag{14.6}$$

This is called the **Heisenberg group**[2] (for $\mathbb{Z}_d$), as is the group

$$\hat{H} := \{ch : c \in \mathbb{T}, h \in H\} \subset \mathscr{U}(\mathbb{C}^d), \qquad \mathbb{T} := \{c \in \mathbb{C} : |c| = 1\}. \tag{14.7}$$

Since $\operatorname{trace}(S^j\Omega^k) = 0$, $(j,k) \neq (0,0)$, we have (see Proposition 13.5)

- The unitary matrices $(S^j\Omega)_{(j,k)\in\mathbb{Z}_d\times\mathbb{Z}_d}$ are a *nice error basis* for $\mathbb{C}^{d\times d}$.
- $(j,k) \mapsto S^j\Omega^k$ is a faithful irreducible projective representation of $\mathbb{Z}_d \times \mathbb{Z}_d$.

(see Example 13.18). In particular, the unitary action of H on $\mathbb{C}^d$ is irreducible, and so $(hv)_{h\in H}$ and $(S^j\Omega^k v)_{j,k\in\mathbb{Z}_d}$ are tight frames for $\mathbb{C}^d$ for any $v \neq 0$ (Theorem 10.5).

Definition 14.2. A SIC (equiangular tight frame of d^2 vectors) for $\mathbb{C}^d$ of the form $\Phi_v = (S^j\Omega^k v)_{j,k\in\mathbb{Z}_d}$ is said to be a **(Weyl-)Heisenberg SIC** (or **Heisenberg frame**).

A *Heisenberg frame* for $\mathbb{C}^d$ is generated from a single vector v by applying S (translation) and Ω (frequency shift). Thus, it is a discrete analogue of a Gabor system (Weyl–Heisenberg frame), and has good time–frequency localisation. In this analogy the fiducial vector v corresponds to the *mother wavelet*.

---

[2] It is also known as the **generalised Pauli** or **Weyl–Heisenberg group**.

14.6 The Hoggar lines

All the known SICs are group covariant, and all are Weyl–Heisenberg SICs, except for the *Hoggar lines* ($d = 8$), which we now describe. The original presentation of these lines [Hog98] was in the 4-dimensional space $\mathbb{H}^4$ over the quaternians $\mathbb{H}$. We follow the description of [GR09]. Consider the nice error basis given by the tensor product (see §13.14) of the Heisenberg nice error basis for $\mathbb{C}^2$ (see Example 13.18) with itself three times, i.e.,

$$\{E_g\}_{g\in(\mathbb{Z}_2^2)^3} = \{S^{j_1}\Omega^{j_2})\otimes(S^{k_1}\Omega^{k_2})\otimes(S^{\ell_1}\Omega^{\ell_2})\}_{(j,k,\ell)\in(\mathbb{Z}_2^2)^3},$$

where $S = \sigma_1 = \sigma_x = \begin{pmatrix} 0 & 1 \\ 1 & 0 \end{pmatrix}$, $\Omega = \sigma_3 = \sigma_z = \begin{pmatrix} 1 & 0 \\ 0 & -1 \end{pmatrix}$. Then the **Hoggar lines** are the SIC of 64 (equiangular) lines in $\mathbb{C}^8$ given by $\{E_g v\}_{g\in(\mathbb{Z}_2^2)^3}$, where

$$v = (0,0,1+i,1-i,1+i,-1-i,0,2).$$

The Hoggar lines are coinvariant with respect to other groups (see Table 14.1), many of which can be obtained as a subgroup of the Clifford group (see §14.7).

Table 14.1: The nice error bases for $d = 8$. Those which occur as subgroups of the Clifford group are labelled with an $\star$, and those which give rise to SICs are labelled SIC. All these SICs are the Hoggar lines (up to projective unitary equivalence), except for $H = <$ `512,451` $>$, $G = \mathbb{Z}_8^2$.

H	G
$\langle 512,451\rangle$	$\langle 64,2\rangle = \mathbb{Z}_8^2$ SIC*
$\langle 512,452\rangle$	$\langle 64,3\rangle$ SIC*
$\langle 512,35969\rangle$	$\langle 64,8\rangle$ SIC*
$\langle 512,36083\rangle$	$\langle 64,10\rangle$
$\langle 512,59117\rangle$	$\langle 64,34\rangle$ *
$\langle 512,59133\rangle$	$\langle 64,35\rangle$ *
$\langle 512,260804\rangle$	$\langle 64,58\rangle$ *
$\langle 512,261506\rangle$	$\langle 64,67\rangle$ SIC*
$\langle 512,261511\rangle$	$\langle 64,67\rangle$ SIC*
$\langle 512,261518\rangle$	$\langle 64,67\rangle$ *
$\langle 512,262018\rangle$	$\langle 64,60\rangle$ SIC*
$\langle 512,262052\rangle$	$\langle 64,62\rangle$ SIC*
$\langle 512,265618\rangle$	$\langle 64,69\rangle$ SIC*
$\langle 512,265839\rangle$	$\langle 64,68\rangle$ SIC*
$\langle 512,265911\rangle$	$\langle 64,71\rangle$ SIC*
$\langle 512,266014\rangle$	$\langle 64,72\rangle$ *
$\langle 512,266267\rangle$	$\langle 64,73\rangle$
$\langle 512,266357\rangle$	$\langle 64,75\rangle$ SIC
$\langle 512,266373\rangle$	$\langle 64,74\rangle$ SIC
$\langle 512,266477\rangle$	$\langle 64,78\rangle$ SIC
$\langle 512,266583\rangle$	$\langle 64,77\rangle$ SIC
$\langle 512,266616\rangle$	$\langle 64,82\rangle$
$\langle 512,400195\rangle$	$\langle 64,90\rangle$ SIC *
$\langle 512,400223\rangle$	$\langle 64,90\rangle$ SIC

H	G
$\langle 512,400443\rangle$	$\langle 64,123\rangle$ *
$\langle 512,401215\rangle$	$\langle 64,91\rangle$ SIC*
$\langle 512,402896\rangle$	$\langle 64,128\rangle$ *
$\langle 512,402951\rangle$	$\langle 64,138\rangle$ SIC
$\langle 512,402963\rangle$	$\langle 64,138\rangle$
$\langle 512,403139\rangle$	$\langle 64,162\rangle$ *
$\langle 512,406850\rangle$	$\langle 64,174\rangle$ *
$\langle 512,406879\rangle$	$\langle 64,167\rangle$ *
$\langle 512,406902\rangle$	$\langle 64,179\rangle$ *
$\langle 512,6276980\rangle$	$\langle 64,192\rangle = (\mathbb{Z}_2\times\mathbb{Z}_4)^2$ *
$\langle 512,6277027\rangle$	$\langle 64,193\rangle$ SIC*
$\langle 512,6278298\rangle$	$\langle 64,195\rangle$ SIC*
$\langle 512,6279917\rangle$	$\langle 64,202\rangle$ SIC
$\langle 512,6279938\rangle$	$\langle 64,202\rangle$ SIC
$\langle 512,6280116\rangle$	$\langle 64,203\rangle$
$\langle 512,6291080\rangle$	$\langle 64,226\rangle$
$\langle 512,6339777\rangle$	$\langle 64,211\rangle$
$\langle 512,6339869\rangle$	$\langle 64,207\rangle$
$\langle 512,6375318\rangle$	$\langle 64,236\rangle$
$\langle 512,6376278\rangle$	$\langle 64,216\rangle$
$\langle 512,7421157\rangle$	$\langle 64,242\rangle$
$\langle 512,10481364\rangle$	$\langle 64,261\rangle$
$\langle 512,10494180\rangle$	$\langle 64,267\rangle = (\mathbb{Z}_2^3)^2$ SIC

14.7 The Clifford group (normaliser of the Heisenberg group)

Here we study the *Clifford group* (normaliser of the Heisenberg group), which plays a key role in the construction (and counting) of Heisenberg SICs.

Proposition 14.3. *If U is a unitary matrix which normalises $\hat{H}$ of (14.7), i.e.,*

$$US^j\Omega^kU^{-1}\in\hat{H},\qquad \forall j,k\in\mathbb{Z}_d,$$

and $\Phi_v:=(S^j\Omega^kv)_{j,k\in\mathbb{Z}_d}$ is a SIC (i.e., is equiangular), then so is Φ_{Uv}.

Proof. Since $|\langle S^{j_1}\Omega^{k_1}v,S^{j_2}\Omega^{k_2}v\rangle|=|\langle S^{j_1-j_2}\Omega^{k_1-k_2}v,v\rangle|$, the angles between distinct vectors in Φ_v are

$$|\langle S^j\Omega^kv,v\rangle|=r,\qquad (j,k)\neq(0,0).$$

Since $U^{-1}S^j\Omega^kU=ch$, $c\in\mathbb{T}$, $h\in H$, with h not a scalar for $(j,k)\neq(0,0)$,

$$|\langle S^j\Omega^k(Uv),Uv\rangle|=|\langle U^{-1}S^j\Omega^kUv,v\rangle|=|\langle chv,v\rangle|=r,\qquad (j,k)\neq(0,0),$$

and so the angles in the second frame Φ_{Uv} are equal. □

Let $[U]:=\{cU:c\in\mathbb{T}\}=\{e^{it}U:t\in\mathbb{R}\}$, so that $[I]$ is the unitary scalar matrices.

Definition 14.3. The normaliser of the Heisenberg group $\hat{H}$ in the group of unitary matrices is called the **Clifford group**, and it is denoted by $\mathrm{C}(d)$. The **projective Clifford group** is $\mathrm{PC}(d):=\mathrm{C}(d)/[I]$ (its elements are called **Clifford operations**).

Since $H\subset \mathrm{C}(d)$, the action of $\mathrm{C}(d)$ on $\mathbb{C}^d$ is irreducible, and $Z(C(d))=[I]$. Since

$$S^*=S^T=S^{-1},\qquad \Omega^*=\Omega^{-1},\quad \Omega^T=\Omega, \tag{14.8}$$

the Heisenberg group and Clifford group are closed under taking the transpose and Hermitian transpose, and hence also entrywise conjugation $\overline{A}=(A^*)^T$, we have

Entrywise conjugation maps a given Heisenberg SIC fiducial to another.

This motivates the following.

Definition 14.4. The normaliser of the Heisenberg group $\hat{H}$ in $EU(\mathbb{C}^d)$ (the group of unitary and antiunitary maps $\mathbb{C}\to\mathbb{C}$) is the **extended Clifford group** $\mathrm{EC}(d)$. The **extended projective Clifford group** is $\mathrm{PEC}(d):=\mathrm{EC}(d)/[I]$.

Combining the above observation with Proposition 14.3 gives:

The extended Clifford group maps a given Heisenberg SIC fiducial to another.

Some elements of the Clifford group include the **Fourier matrix**[3] F, the diagonal matrix R, $M = RF$, and the permutation matrices P_σ, $\sigma \in \mathbb{Z}_d^*$, which are given by

$$(F)_{jk} := \frac{1}{\sqrt{d}}\omega^{jk}, \tag{14.9}$$

$$(R)_{jk} := \mu^{j(j+d)}\delta_{jk}, \tag{14.10}$$

$$(M)_{jk} := \frac{1}{\sqrt{d}}\mu^{j(j+d)+2jk}, \tag{14.11}$$

$$(P_\sigma)_{jk} := \delta_{j,\sigma k}, \quad \sigma \in \mathbb{Z}_d^*. \tag{14.12}$$

We observe that R, M are well defined, i.e., the value of $j(j+d)$ depends only on the integer j mod d. The entry $\mu^{j(j+d)}$ has many alternative descriptions, e.g.,

$$\mu^{j(j+d)} = \mu^{j^2}(-1)^j = \mu^{j^2}(-1)^{j^2} = (-\mu)^{j^2} = \mu^{(d+1)j^2}.$$

Elementary computations (see Exer. 14.5) give:

Lemma 14.1. *The unitary matrices F, R, M, P_σ normalise $\hat{H}$. Indeed*

$$F(S^j\Omega^k)F^{-1} = \omega^{-jk}S^{-k}\Omega^j, \tag{14.13}$$

$$R(S^j\Omega^k)R^{-1} = \mu^{j(j+d)}S^j\Omega^{j+k}, \tag{14.14}$$

$$M(S^j\Omega^k)M^{-1} = \mu^{k(k-2j+d)}S^{-k}\Omega^{j-k}, \tag{14.15}$$

$$P_\sigma(S^j\Omega^k)P_\sigma^{-1} = S^{\sigma j}\Omega^{\sigma^{-1}k}, \tag{14.16}$$

where σ^{-1} is the multiplicative inverse of $\sigma \in \mathbb{Z}_d^*$. If d is odd, then all the powers of μ above are even, and so can be expressed as powers of ω.

We will see (Theorem 14.1) that R (or M) along with F and H generate the Clifford group. The appearance of R (and hence M) can be explained. It appears in a direct search for diagonal matrices in the normaliser of $\hat{H}$.

Proposition 14.4. *The group of diagonal unitary matrices which normalise the Heisenberg group $\hat{H}$ is generated by the scalar matrices, Ω, and the matrix R given by*

$$R_{jk} := \mu^{j(j+d)}\delta_{jk}.$$

Proof. Suppose that $\Lambda = \operatorname{diag}(\lambda_j)$ normalises $\hat{H}$, and $\lambda_d := \lambda_0$. Then

$$\Lambda S \Lambda^{-1} = \begin{pmatrix} 0 & 0 & \cdots & \frac{\lambda_d}{\lambda_{d-1}} \\ \frac{\lambda_1}{\lambda_0} & 0 & \cdots & 0 \\ 0 & \frac{\lambda_2}{\lambda_1} & \cdots & 0 \\ \vdots & & & \vdots \end{pmatrix} = cS\Omega^k, \qquad \text{i.e.,} \quad \frac{\lambda_{j+1}}{\lambda_j} = c\,\omega^{jk}, \ \forall j,$$

---

[3] The matrix $F^* = F^{-1} = \overline{F}$ is also commonly referred to as the *Fourier matrix*, or as the *DFT (discrete Fourier transform) matrix.*

where $c \in \mathbb{C}$, $k \in \mathbb{Z}$. Solving this recurrence gives

$$\lambda_j = \lambda_0 c^j \omega^{\frac{1}{2}j(j-1)k} = \lambda_0 c^j \mu^{j(j-1)k} = \lambda_0 \mu^{j(j+d)k} (c\mu^{-k(d+1)})^j.$$

Since $\lambda_d = \lambda_0$, this gives

$$(c\mu^{-k(d+1)})^d = (c\mu^{-k(d+1)})^0 = 1 \quad \Longrightarrow \quad c\mu^{-k(d+1)} = \omega^m,$$

and so $\Lambda = \lambda_0 R^k \Omega^m$. □

We now consider the structure of $\mathrm{C}(d)$. Let

$$U_{(j,k)} := S^j \Omega^k, \qquad (j,k) \in \mathbb{Z}_d^2. \tag{14.17}$$

If $a \in \mathrm{C}(d)$, then

$$aU_\lambda a^{-1} = z_a(\lambda) U_{\psi_a(\lambda)}, \qquad \forall \lambda \in \mathbb{Z}_d^2, \tag{14.18}$$

which defines functions $\psi_a : \mathbb{Z}_d^2 \to \mathbb{Z}_d^2$ and $z_a : \mathbb{Z}_d^2 \to \mathbb{T}$, since no U_λ is a scalar multiple of another. For example, (14.13), (14.14) give

$$\psi_F \begin{pmatrix} j \\ k \end{pmatrix} = \begin{pmatrix} -k \\ j \end{pmatrix}, \qquad z_F(j,k) = \omega^{-jk}, \tag{14.19}$$

$$\psi_R \begin{pmatrix} j \\ k \end{pmatrix} = \begin{pmatrix} j \\ j+k \end{pmatrix}, \qquad z_R(j,k) = \mu^{j(j+d)}. \tag{14.20}$$

We now show the elements of the Clifford group factored by $\hat{H}$ can be indexed by the elements of $SL_2(\mathbb{Z}_d)$ (the 2×2 matrices over $\mathbb{Z}_d$ with determinant 1).

For a 2×2 matrix A, we define a symmetric matrix σ_A by

$$\sigma_A := \begin{pmatrix} \alpha\gamma & \beta\gamma \\ \beta\gamma & \beta\delta \end{pmatrix}, \qquad A = \begin{pmatrix} \alpha & \beta \\ \gamma & \delta \end{pmatrix}. \tag{14.21}$$

Lemma 14.2. *Let ψ_a and z_a be given by* (14.18). *Then the map*

$$\psi : \mathrm{C}(d) \to SL_2(\mathbb{Z}_d) : a \mapsto \psi_a \tag{14.22}$$

is a group homomorphism with kernel $\hat{H}$, and z_a satisfies

$$z_a(p+q) = \omega^{p^T \sigma_A q} z_a(p) z_a(q), \qquad p,q \in \mathbb{Z}_d^2 \tag{14.23}$$

where $A = \psi_a$ and σ_A is given by (14.21).

Proof. By (14.5), we have $U_p U_q = \omega^{p_2 q_1} U_{p+q}$. and so

$$\omega^{p_2 q_1} (aU_{p+q} a^{-1}) = aU_p U_q a^{-1} = (aU_p a^{-1})(aU_q a^{-1}),$$

which gives

$$\omega^{p_2q_1} z_a(p+q) U_{\psi_a(p+q)} = z_a(p) U_{\psi_a(p)} z_a(q) U_{\psi_a(q)}$$
$$= z_a(p) z_a(q) \omega^{\psi_a(p)_2 \psi_a(q)_1} U_{\psi_a(p)+\psi_a(q)}.$$

and hence

$$\psi_a(p+q) = \psi_a(p) + \psi_a(q), \tag{14.24}$$

$$\omega^{p_2q_1} z_a(p+q) = z_a(p) z_a(q) \omega^{\psi_a(p)_2 \psi_a(q)_1}. \tag{14.25}$$

For $p = p_1e_1 + p_2e_2 \in \mathbb{Z}_d^2$, from (14.24) we obtain

$$\psi_a(p) = p_1\psi_a(e_1) + p_2\psi_a(e_2) = [\psi_a(e_1), \psi_a(e_2)]p,$$

i.e., ψ_a can be represented by the 2×2 matrix $[\psi_a(e_1), \psi_a(e_2)]$.

Let $[p', q'] = [\psi_a(p), \psi_a(q)] = \psi[p, q]$, so that $\det([p', q']) = \det(\psi_a)\det([p, q])$. Since the quotient $z_a(p)z_a(q)/z_a(p+q)$ is symmetric in p and q, (14.25) gives

$$\begin{aligned} \omega^{p_2q_1 - p_2'q_1'} = \omega^{q_2p_1 - q_2'p_1'} \quad &\Longrightarrow \quad p_1'q_2' - q_1'p_2' = p_1q_2 - q_1p_2 \\ &\Longrightarrow \quad \det([p', q']) = \det([p, q]), \\ &\Longrightarrow \quad \det(\psi_a) = 1, \end{aligned} \tag{14.26}$$

i.e., $\psi_a \in SL_2(\mathbb{Z}_d)$. Using this (see Exer. 14.8), (14.25) can be written as (14.23).

Since $(ab)U_\lambda(ab)^{-1} = a(bU_\lambda b^{-1})a^{-1}$, we have

$$z_{ab}(\lambda) U_{\psi_{ab}(\lambda)} = a(z_b(\lambda) U_{\psi_b(\lambda)}) a^{-1} = z_b(\lambda) z_a(\psi_b(\lambda)) U_{\psi_a(\psi_b(\lambda))}, \tag{14.27}$$

so that $\psi_{ab}(\lambda) = \psi_a(\psi_b(\lambda))$, i.e., $a \mapsto \psi_a$ is a homomorphism.

We now determine the kernel of ψ. By (14.5), $\hat{H} \subset \ker\psi$. Suppose $\psi_a = I$, so that $aSa^{-1} = z_a(1,0)S$ and $a\Omega a^{-1} = z_a(0,1)\Omega$. Since $S^d = \Omega^d = I$, this implies that $z_a(1,0)$ and $z_a(0,1)$ are d-th roots of unity, say

$$aSa^{-1} = \omega^\alpha S, \qquad a\Omega a^{-1} = \omega^\beta \Omega. \tag{14.28}$$

If $a \in \hat{H}$, then (14.28) implies that a is a scalar multiple of $S^{-\beta}\Omega^\alpha$. Hence, we consider the unitary matrix $b = (S^{-\beta}\Omega^\alpha)^{-1}a$. By (14.28) and repeated application of (14.5), we have that

$$b(S^j\Omega^k)b^{-1} = \Omega^{-\alpha}S^\beta (aSa^{-1})^j (a\Omega a^{-1})^k S^{-\beta}\Omega^\alpha$$
$$= \Omega^{-\alpha}S^\beta (\omega^\alpha S)^j (\omega^\beta \Omega)^k S^{-\beta}\Omega^\alpha = S^j\Omega^k.$$

Since b commutes with the basis $(S^j\Omega^k)_{j,k\in\mathbb{Z}_d}$ for $\mathbb{C}^{d\times d}$, Schur's lemma implies that b must be a (unit) scalar matrix cI, and hence $a = cS^{-\beta}\Omega^\alpha \in \hat{H}$. □

Example 14.2. From Lemma 14.1, we have the following $\psi_a \in SL_2(\mathbb{Z}_d)$,

$$\psi_F = \begin{pmatrix} 0 & -1 \\ 1 & 0 \end{pmatrix}, \quad \psi_R = \begin{pmatrix} 1 & 0 \\ 1 & 1 \end{pmatrix}, \quad \psi_M = \begin{pmatrix} 0 & -1 \\ 1 & -1 \end{pmatrix}, \quad \psi_{P_\sigma} = \begin{pmatrix} \sigma^{-1} & \\ & \sigma \end{pmatrix}.$$

14.8 Generators for the Clifford group

The subgroup of the Clifford group $\mathrm{C}(d)$ generated by F,R (and the scalars) is the **symplectic unitaries**[4]

$$\mathrm{C}_{\mathrm{Sp}}(d) := \langle F, R, [I] \rangle.$$

The elements of $\mathrm{C}_{\mathrm{Sp}}(d)/[I]$ are called **symplectic operations**, and elements of the Heisenberg group $\hat{H}$ (or $\hat{H}/[I]$) are referred to as **Heisenberg operations**, **(Weyl) displacements** or **time–frequency shifts**. For d even (see Exer. 14.6),

$$\Omega^{\frac{d}{2}} = R^d, \quad S^{\frac{d}{2}} = F^{-1}\Omega^{\frac{d}{2}}F = F^{-1}R^dF, \quad S^{\frac{d}{2}}\Omega^{\frac{d}{2}} = F^{-1}R^dFR^d. \qquad (14.29)$$

Thus there are nontrivial symplectic operations which are also displacements.

It turns out that (14.29) are the only cases (see Corollary 14.2). This makes the description of the Clifford group more technical for d even (here R has order $2d$).

We now show that $\mathrm{C}(d)$ is generated by the normal subgroup $\hat{H}$ and F,R, i.e.,

Every Clifford operation is the product of a displacement operation and a symplectic operation.

Theorem 14.1. *(Clifford group generators) The homomorphism*

$$\psi : \mathrm{C}(d) \to SL_2(\mathbb{Z}_d) : a \mapsto \psi_a$$

maps F and R to generators for $SL_2(\mathbb{Z}_d)$, and hence is onto. Therefore $\mathrm{C}(d)$ is generated by the unitary scalar matrices, and

$$S, \ \Omega, \ F, \ R.$$

The Clifford group $\mathrm{C}(d)$ is closed under taking the transpose, Hermitian transpose, and entrywise conjugation $\overline{A} = (A^*)^T$.

Proof. By Lemma 14.2, the kernel of $a \mapsto \psi_a$ is $\hat{H}$. Since $\hat{H}$ is generated by the unitary scalar matrices and S,Ω, it suffices to show that $SL_2(\mathbb{Z}_d)$ is generated by

$$\psi_F = \begin{pmatrix} 0 & -1 \\ 1 & 0 \end{pmatrix}, \qquad \psi_R = \begin{pmatrix} 1 & 0 \\ 1 & 1 \end{pmatrix}. \qquad (14.30)$$

It is well known these matrices generate $SL_2(\mathbb{Z})$. Since the map of taking the entries of $A \in SL_2(\mathbb{Z})$ modulo d is a homomorphism onto $SL_2(\mathbb{Z}_d)$, they generate $SL_2(\mathbb{Z}_d)$. We observed the closure properties are a consequence of (14.8). Alternatively, for R and F, we have $R^* = \overline{R} = R^{-1}$, $R^T = R$, $F^* = \overline{F} = F^{-1}$, $F^T = F$. □

[4] This is because their action on $\hat{H}$ is given by a symplectic matrix (see Remark 14.2).

The order of $SL_2(\mathbb{Z}_d)$ is known (see [Gun62] Theorem 3, Chapter I)

$$|SL_2(\mathbb{Z}_d)| = d^3 \prod_{p|d} \Bigl(1 - \frac{1}{p^2}\Bigr), \qquad (p \text{ the prime factors of } d).$$

Hence, by Theorem 14.1, the number of Clifford operations is

$$\Bigl|\frac{\mathrm{C}(d)}{[I]}\Bigr| = \Bigl|\frac{\hat{H}}{[I]}\Bigr| \Bigl|\frac{\mathrm{C}(d)}{\hat{H}}\Bigr| = d^2 |SL_2(\mathbb{Z}_d)| = d^5 \prod_{p|d} \Bigl(1 - \frac{1}{p^2}\Bigr).$$

The extended Clifford group $\mathrm{EC}(d)$ is generated by the antiunitary map $C : v \mapsto \overline{v}$ of conjugation together with any set of generators for $\mathrm{C}(d)$ (see Corollary 14.3).

14.9 Indexing the Clifford operations

We now show each Clifford operation is uniquely determined by the pair (ψ_a, z_a). Define the semidirect product $SL_2(\mathbb{Z}_d) \ltimes \mathbb{T}^{\mathbb{Z}_d^2}$ via the multiplication

$$(A, z_A)(B, z_B) := (AB, (z_A \circ B) z_B), \tag{14.31}$$

where functions $\mathbb{Z}_d^2 \to \mathbb{T}$ are multiplied pointwise.

Corollary 14.1. *With the multiplication (14.31), the map*

$$\mathrm{C}(d) \to SL_2(\mathbb{Z}_d) \ltimes \mathbb{T}^{\mathbb{Z}_d^2} : a \mapsto (\psi_a, z_a) \tag{14.32}$$

is a homomorphism with kernel $[I]$*. Thus every Clifford operation* $[a] \in= \mathrm{C}(d)/[I]$ *has a unique index* (ψ_a, z_a)*, and these satisfy*

$$\psi_{ab} = \psi_a \psi_b, \qquad z_{ab} = (z_a \circ \psi_b) z_b, \tag{14.33}$$

$$\psi_{a^*} = \psi_{a^{-1}} = \psi_a^{-1}, \qquad z_{a^*} = z_{a^{-1}} = z_A^{-1} \circ \psi_a^{-1} = \overline{z_a} \circ \psi_{a^*}, \tag{14.34}$$

Further, if $\psi_a = \psi_b$*, then* z_a / z_b *is a character.*

Proof. It is easy to check (see Exer. 14.7) that $SL_2(\mathbb{Z}_d) \ltimes \mathbb{T}^{\mathbb{Z}_d^2}$ is a group with the multiplication (14.31), identity $(I, 1)$, and inverse $(A, z_A)^{-1} = (A^{-1}, z_A^{-1} \circ A^{-1})$. By (14.27), we have

$$\psi_{ab} = \psi_a \psi_b, \qquad z_{ab} = (z_a \circ \psi_b) z_b,$$

i.e., the map $a \mapsto (\psi_a, z_a)$ is a homomorphism. Thus (14.33) holds, as does (14.34) by the calculation $(\psi_{a^{-1}}, z_{a^{-1}}) = (\psi_a, z_a)^{-1} = (\psi_a^{-1}, z_a^{-1} \circ \psi_a^{-1})$.

Now suppose that a is in the kernel, i.e., $\psi_a = I$, $z_a = 1$. By Lemma 14.2, we have $a = cS^j \Omega^k \in \hat{H}$. Using (14.5), we therefore obtain (see Exer. 14.4)

$$a S^{p_1} \Omega^{p_2} a^{-1} = S^j \Omega^k S^{p_1} \Omega^{p_2} \Omega^{-k} S^{-j} = \omega^{k p_1 - j p_2} S^{p_1} \Omega^{p_2},$$

so that $z_a(p) = \omega^{kp_1 - jp_2} = 1, \forall p \in \mathbb{Z}_d^2$. Thus $j = k = 0$ and $a = cI \in [I]$, as supposed. For $\psi_a = \psi_b = A$, it follows from (14.23) or (14.25) that z_a/z_b is a character. □

We call the subgroup of $SL_2(\mathbb{Z}_d) \ltimes \mathbb{T}^{\mathbb{Z}_d^2}$ given by

$$\mathrm{Ind}(d) := \{(\psi_a, z_a) : a \in \mathrm{C}(d)\}$$

the **index group** of the Clifford operations, and the **index map** is the isomorphism

$$\mathrm{C}(d)/[I] \to \mathrm{Ind}(d) : [a] \mapsto (\psi_a, z_a). \tag{14.35}$$

Example 14.3. For $M = RF$, from (14.19) and (14.20), we calculate

$$\psi_M = \psi_R \psi_F = \begin{pmatrix} 1 & 0 \\ 1 & 1 \end{pmatrix} \begin{pmatrix} 0 & -1 \\ 1 & 0 \end{pmatrix} = \begin{pmatrix} 0 & -1 \\ 1 & -1 \end{pmatrix},$$

$$z_M = (z_R \circ \psi_F) z_F \quad \Longrightarrow \quad z_M(j,k) = \mu^{(-k)(-k+d)} \omega^{-jk} = \mu^{k(k+d)+2jk}.$$

The Clifford group $\mathrm{C}(d)$ and the index map $a \mapsto (\psi_a, z_a)$ can easily be set up in a computer algebra package such as `magma` by taking the matrix group generated by the matrices S, Ω, F, R (defined over a suitable cyclotomic field).

By Theorem 14.1, $\psi^{-1}(A) = \{ha : h \in \hat{H}\}, A \in SL_2(\mathbb{Z}_d)$, for any a with $\psi_a = A$. Hence to describe the elements of the Clifford group $\mathrm{C}(d)$, it suffices to know an a in each coset $\psi^{-1}(A)$. We now show this representative can be a symplectic matrix.

14.10 Appleby indexing

If d is odd, then $-\mu = \omega^{\frac{d+1}{2}}$, and it follows from (14.23) that

$$z_a(p) = (-\mu)^{p^T \sigma_A p} \hat{z}_a(p), \qquad \forall p \in \mathbb{Z}_d^2, \tag{14.36}$$

where $A = \psi_a$, and $\hat{z}_a$ is a character (see Exer. 14.8).

If d is even, then the factor $(-\mu)^{p^T \sigma_A p}$ above is not well defined. To obtain an analogue of (14.36), it is necessary "lift" A to a $B \in SL_2(\mathbb{Z}_{2d}^2)$. This "doubling" works, but the corresponding (Appleby) index $[B, \chi]$ is not unique. We now give the details as described by [App05].

Define displacement operators by

$$\hat{D}_p := (-\mu)^{p_1 p_2} S^{p_1} \Omega^{p_2}, \qquad p \in \mathbb{Z}^2. \tag{14.37}$$

These satisfy $\det(D_p) = 1$, and (see Exer. 14.15)

$$\hat{D}_p^{-1} = \hat{D}_{-p}, \qquad \hat{D}_p \hat{D}_q = (-\mu)^{\langle\langle p,q \rangle\rangle} \hat{D}_{p+q} = \omega^{\langle\langle p,q \rangle\rangle} \hat{D}_q \hat{D}_p, \tag{14.38}$$

and

$$\hat{D}_{p+dq} = \begin{cases} \hat{D}_p, & d \text{ odd;} \\ (-1)^{\langle\langle p,q \rangle\rangle}\hat{D}_p, & d \text{ even,} \end{cases} \tag{14.39}$$

where $\langle\langle \cdot,\cdot \rangle\rangle$ is the *symplectic form*

$$\langle\langle p,q \rangle\rangle := p_2q_1 - p_1q_2 = p^T \begin{pmatrix} 0 & -1 \\ 1 & 0 \end{pmatrix} q.$$

It follows from (14.39) that $\hat{D}_p$ depends only on p mod d', where

$$d' := \begin{cases} d, & d \text{ odd;} \\ 2d, & d \text{ even.} \end{cases}$$

We observe that $\langle\langle p,q \rangle\rangle$ has the property

$$\langle\langle Ap,Aq \rangle\rangle = \det(A)\langle\langle p,q \rangle\rangle, \qquad \forall p,q. \tag{14.40}$$

We now generalise (14.36), to show that for each $[a] \in \mathrm{C}(d)/[I]$ there exists a $B \in SL_2(\mathbb{Z}_{d'})$ and $\chi \in \mathbb{Z}_d^2$, such that

$$a\hat{D}_p a^{-1} = \omega^{\langle\langle \chi,Bp \rangle\rangle}\hat{D}_{Bp}, \qquad \forall p \in \mathbb{Z}_{d'}^2.$$

Here $\langle\langle \chi,Bp \rangle\rangle$ is interpreted as $\langle\langle \chi,Ap \rangle\rangle$, $A := B$ mod d, when d is even. We will write the pair (B,χ) as $[B,\chi]$, and call it an **Appleby index**.

Theorem 14.2. *Define the semidirect product* $SL_2(\mathbb{Z}_{d'}) \ltimes \mathbb{Z}_d^2$ *via the multiplication*

$$[B_1,\chi_1][B_2,\chi_2] := [B_1B_2,\chi_1 + A_1\chi_2], \quad A_1 := B_1 \bmod d. \tag{14.41}$$

Then there is a unique surjective homomorphism onto the Clifford operations

$$f : SL_2(\mathbb{Z}_{d'}) \ltimes \mathbb{Z}_d^2 \to \mathrm{C}(d)/[I], \tag{14.42}$$

with the property that for $[a] = f([B,\chi])$

$$a\hat{D}_p a^{-1} = \omega^{\langle\langle \chi,Bp \rangle\rangle}\hat{D}_{Bp}, \qquad \forall p \in \mathbb{Z}_{d'}^2, \tag{14.43}$$

i.e.,

$$A := \psi_a = B \bmod d, \qquad z_a(p) = \omega^{\langle\langle \chi,Ap \rangle\rangle}(-\mu)^{p^T\sigma_{Bp}}, \quad \forall p \in \mathbb{Z}_d^2. \tag{14.44}$$

This f is an isomorphism for d odd (i.e., $d' = d$), and for d even it has kernel

$$\ker f = \left\{ \left[\begin{pmatrix} 1+rd & sd \\ td & 1+rd \end{pmatrix}, \begin{pmatrix} s\frac{d}{2} \\ t\frac{d}{2} \end{pmatrix} \right] : r,s,t \in \{0,1\} \right\}. \tag{14.45}$$

Proof. If $a \in \mathrm{C}(d)$ satisfies (14.43), then (14.44) follows (see Exer. 14.16). In view of the isomorphism (14.35), f is uniquely defined, and it suffices to show that

$$\theta : SL_2(\mathbb{Z}_{d'}) \ltimes \mathbb{Z}_d^2 \to \mathrm{Ind}(d) : [B,\chi] \mapsto (A, z_a),$$

given by (14.44) is a surjective homomorphism.

We first show it is a homomorphism (as a map to $SL_2(\mathbb{Z}_d) \ltimes \mathbb{T}^{\mathbb{Z}_d^2}$). Now

$$\theta\big([B_1,\chi_1][B_2,\chi_2]\big) = \theta\big([B_1B_2,\chi_1 + A_1\chi_2]\big) = (A_1A_2, z_{a_1a_2}),$$

$$A_j := B_j \bmod d, \qquad z_{a_1a_2}(p) := \omega^{\langle\!\langle \chi_1 + A_1\chi_2, A_1A_2p\rangle\!\rangle}(-\mu)^{p^T\sigma_{B_1B_2}p},$$

and

$$\theta\big([B_1,\chi_1]\big)\theta\big([B_2,\chi_2]\big) = (A_1,z_{a_1})(A_2,z_{a_2}) = (A_1A_2,(z_{a_1}\circ A_2)z_{a_2}),$$

$$\big((z_{a_1}\circ A_2)z_{a_2}\big)(p) = \omega^{\langle\!\langle \chi_1, A_1A_2p\rangle\!\rangle}(-\mu)^{(B_2p)^T\sigma_{B_1}B_2p}\omega^{\langle\!\langle \chi_2, A_2p\rangle\!\rangle}(-\mu)^{p^T\sigma_{B_2}p},$$

so that θ is a homomorphism provided that

$$\langle\!\langle \chi_1 + A_1\chi_2, A_1A_2p\rangle\!\rangle = \langle\!\langle \chi_1, A_1A_2p\rangle\!\rangle + \langle\!\langle \chi_2, A_2p\rangle\!\rangle,$$

$$p^T\sigma_{B_1B_2}p = (B_2p)^T\sigma_{B_1}B_2p + p^T\sigma_{B_2}p.$$

The first follows since (14.40) gives

$$\langle\!\langle \chi_2, A_2p\rangle\!\rangle = \langle\!\langle A_1\chi_2, A_1A_2p\rangle\!\rangle,$$

and the second follows by the identity

$$\sigma_{B_1B_2} = B_2^T\sigma_{B_1}B_2 + \det(B_1)\sigma_{B_2}.$$

We calculate (as in Example 14.4)

$$\theta\Big(\Big[\begin{pmatrix}0 & -1\\ 1 & 0\end{pmatrix},0\Big]\Big) = (\psi_F, z_F), \qquad \theta\Big(\Big[\begin{pmatrix}1 & 0\\ 1 & 1\end{pmatrix},0\Big]\Big) = (\psi_R, z_R), \tag{14.46}$$

and

$$\theta\Big(\Big[I,\begin{pmatrix}\alpha\\ \beta\end{pmatrix}\Big]\Big) = (z_{S^\alpha\Omega^\beta}, \psi_{S^\alpha\Omega^\beta}),$$

so that θ maps generators for $SL_2(\mathbb{Z}_{d'}) \ltimes \mathbb{Z}_d^2$ to generators for $\mathrm{Ind}(d)$, and hence is a surjective homomorphism.

Finally, we determine $\ker f = \ker\theta$. By (14.44), we have $[B,\chi] \in \ker f$ if

$$A := \psi_a = B \bmod d = I, \qquad z_a(p) = \omega^{\langle\!\langle \chi, Ap\rangle\!\rangle}(-\mu)^{p^T\sigma_B p} = 1, \quad \forall p.$$

For d odd, $d' = d$, and so $B = I$ and $z_a(p) = \omega^{\chi_2p_1 - \chi_1p_2} = 1, \forall p$. Thus $[B,\chi] = [I,0]$, and f is an isomorphism. For d even, $B \bmod d = I$ gives

$$B = \begin{pmatrix} 1+rd & sd \\ td & 1+ud \end{pmatrix}, \qquad r,s,t,u \in \{0,1\},$$

and the condition $\det(B) = 1$ gives

$$\det(B) = (1+rd)(1+ud) - std^2 \equiv 1 + (r+u)d \bmod d' \quad \Longrightarrow \quad r = u,$$

so that

$$B = \begin{pmatrix} 1+rd & sd \\ td & 1+rd \end{pmatrix}, \quad \sigma_B = \begin{pmatrix} td(1+rd) & tdsd \\ sdtd & sd(1+rd) \end{pmatrix} \equiv \begin{pmatrix} td & \\ & sd \end{pmatrix} \bmod d'.$$

Hence $z_a(p) = \omega^{\langle\langle \chi, p \rangle\rangle}(-\mu)^{p^T \sigma_B p} = \omega^{\chi_2 p_1 - \chi_1 p_2}(-\mu)^{tdp_1^2 + sdp_2^2} = 1$, which gives

$$\omega^{\chi_1 p_2 - \chi_2 p_1} = (-1)^{tp_1^2 + sp_2^2} = (-1)^{tp_1 + sp_2} \omega^{\frac{d}{2}(tp_1 + sp_2)}, \quad \forall p.$$

Thus, $\chi_1 = \frac{d}{2}s$, $\chi_2 = -\frac{d}{2}t = \frac{d}{2}t$, and we obtain (14.45). □

In other words:

Each Clifford operation has an Appleby index $[B, \chi] \in SL_2(\mathbb{Z}_{d'}) \times \mathbb{Z}_d^2$.

- This is unique for d odd.
- There are eight choices (each differing by an element of $\ker f$) for d even.

Table 14.2: The index (ψ_a, z_a) and Appleby index $[B, \chi]$ for generators of the Clifford operations.

a	ψ_a	$z_a(j,k)$	$[B,\chi]$
$S^\alpha \Omega^\beta$	I	$\omega^{\beta j - \alpha k}$	$[I, \begin{pmatrix} \alpha \\ \beta \end{pmatrix}]$
F	$\begin{pmatrix} 0 & -1 \\ 1 & 0 \end{pmatrix}$	ω^{-jk}	$[\begin{pmatrix} 0 & -1 \\ 1 & 0 \end{pmatrix}, 0]$
R	$\begin{pmatrix} 1 & 0 \\ 1 & 1 \end{pmatrix}$	$\mu^{j(j+d)}$	$[\begin{pmatrix} 1 & 0 \\ 1 & 1 \end{pmatrix}, 0]$

Example 14.4. For the Appleby index $[B, \chi] = [\begin{pmatrix} 0 & -1 \\ 1 & 0 \end{pmatrix}, 0]$, we have $\sigma_B = \begin{pmatrix} 0 & -1 \\ -1 & 0 \end{pmatrix}$,

$$\omega^{\langle\langle \chi, Bp \rangle\rangle}(-\mu)^{p^T \sigma_B p} = (-\mu)^{-2jk} = \omega^{-jk} = z_F(j,k), \qquad p = \begin{pmatrix} j \\ k \end{pmatrix},$$

so that $\theta([\begin{pmatrix} 0 & -1 \\ 1 & 0 \end{pmatrix}, 0]) = (\psi_F, z_F)$ and $[\begin{pmatrix} 0 & -1 \\ 1 & 0 \end{pmatrix}, 0]$ is an Appleby index for $[F]$.

14.11 Symplectic unitaries

By (14.29), for d even, there are nontrivial *symplectic unitaries* (those generated by F, R and the scalars) which are in the Heisenberg group. We now characterise these.

Let m_d be the surjective homomorphism

$$m_d : SL_2(\mathbb{Z}_{d'}) \to SL_2(\mathbb{Z}_d) : B \mapsto A := B \pmod{d},$$

which is the identity for d odd, and for d even has kernel (see Theorem 14.2)

$$K := \left\{ \begin{pmatrix} 1+rd & sd \\ td & 1+rd \end{pmatrix} : r,s,t \in \{0,1\} \right\}, \qquad |K| = 8. \tag{14.47}$$

Corollary 14.2. *A matrix $a \in \mathrm{C}(d)$ is a symplectic unitary if and only if it has an Appleby index of the form $[B,0]$. Indeed, the map*

$$\alpha : SL_2(\mathbb{Z}_{d'}) \to \mathrm{C_{Sp}}(d)/[I] : B \mapsto f([B,0]) \tag{14.48}$$

is a surjective homomorphism, which is an isomorphism for d odd. For d even, $\ker\alpha = \{I, (d+1)I\}$*, and so the only nontrivial Heisenberg operations which are symplectic are given by*

$$S^{\frac{d}{2}},\ \Omega^{\frac{d}{2}},\ S^{\frac{d}{2}}\Omega^{\frac{d}{2}} \qquad (d \text{ even}).$$

Proof. By (14.41), we have

$$[B_1B_2,0] = [B_1,0][B_2,0],$$

and so α is a homomorphism. It is onto, since by (14.46), its image contains

$$\alpha\left(\begin{pmatrix} 0 & -1 \\ 1 & 0 \end{pmatrix}\right) = [F], \qquad \alpha\left(\begin{pmatrix} 1 & 0 \\ 1 & 1 \end{pmatrix}\right) = [R], \tag{14.49}$$

which are generators for $\mathrm{C_{Sp}}(d)/[I]$. Since ψ has kernel $\hat{H}$ (Lemma 14.2), it induces a well defined homomorphism $\hat{\psi} : \mathrm{C_{Sp}}(d)/[I] \to SL_2(\mathbb{Z}_d)$, with

$$\hat{\psi}([F]) = \psi(F) = \begin{pmatrix} 0 & -1 \\ 1 & 0 \end{pmatrix}, \qquad \hat{\psi}([R]) = \psi(R) = \begin{pmatrix} 1 & 0 \\ 1 & 1 \end{pmatrix}. \tag{14.50}$$

By (14.49) and (14.50), we conclude that

$$m_d = \hat{\psi} \circ \alpha,$$

since it holds for the generators (14.30) of $SL_2(\mathbb{Z}_{d'})$. The kernel of $\hat{\psi}$ consists of the symplectic operations which are also Heisenberg operations, i.e.,

$$\ker\hat{\psi} = \mathrm{C_{Sp}}(d)/[I] \cap \hat{H}/[I].$$

For d odd, m_d is an isomorphism, so that $\ker \hat{\psi} = \{[I]\}$. For d even,

$$m_d = \hat{\psi} \circ \alpha \quad \Longrightarrow \quad |\ker \hat{\psi}|\,|\ker \alpha| = |\ker m_d| = |K| = 8,$$

and (14.39) gives

$$\hat{D}_{(d+1)Ip} = (-1)^{\langle\!\langle p,p \rangle\!\rangle} \hat{D}_p = \hat{D}_{Ip} \quad \Longrightarrow \quad (d+1)I \in \ker \alpha \quad \Longrightarrow \quad |\ker \alpha| \geq 2.$$

In view of (14.29), we must have

$$\ker \alpha = \{I, (d+1)I\}, \qquad \ker \hat{\psi} = \{[I], [S^{\frac{d}{2}}], [\Omega^{\frac{d}{2}}], [S^{\frac{d}{2}}\Omega^{\frac{d}{2}}]\},$$

as claimed. □

In other words:

Each symplectic operation $[a]$ has an Appleby index of the form $[B, 0]$.

- This is unique for d odd.
- There are two choices ($[B, 0]$ and $[(d+1)B, 0]$) for d even.

We call $B \in SL_2(\mathbb{Z}_{d'})$ a **symplectic index** for $[a]$.

The following commutative diagram summarises Corollary 14.2.

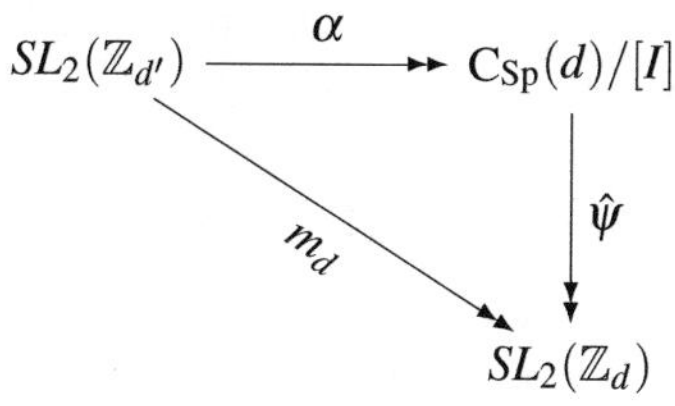

(14.51)

In particular, we have the following 1–1 indexing of the symplectic operations

$$\frac{C_{Sp}(d)}{[I]} \cong \begin{cases} SL_2(\mathbb{Z}_d), & d \text{ odd}; \\ \frac{SL_2(\mathbb{Z}_{2d})}{\langle (d+1)I \rangle}, & d \text{ even}. \end{cases}$$

Remark 14.2. Matrices in $SL_2(\mathbb{Z}_{d'})$ are said to be *symplectic* (see Exer. 14.10). If a is a symplectic unitary, with symplectic index B, then (14.43) gives

$$a\hat{D}_p a^{-1} = \hat{D}_{Bp}, \qquad \forall p (\in \mathbb{Z}_{d'}),$$

i.e., the conjugation action of a on the displacement $\hat{D}_p$ is given by multiplication of p by the symplectic matrix B. This is the origin of the term *symplectic unitary*.

The group $C_{Sp}(d)$ of symplectic unitaries is *not irreducible* for $d > 2$, since its centre contains the nondiagonal matrix $P_{-1} = F^2$. Calculations in $C_{Sp}(d)/[I]$ can be done in the group generated by F and R, which is finite (see Exer. 14.19 for details).

14.12 Permutation matrices

The permutation matrices are a subgroup of the symplectic unitaries.

Proposition 14.5. *The permutation matrices P_b, $b \in \mathbb{Z}_d^*$, are symplectic. Indeed, with $1 \le b < d$, we have*

$$P_b = (c_{b,d})^{-1} R^{b^{-1}} F R^b F R^{b^{-1}} F, \tag{14.52}$$

where b^{-1} is the inverse of b in $\mathbb{Z}_{d'}^*$, and $c_{b,d} = c_{b^{-1},d}$ is the Gauss sum

$$c_{b,d} := \frac{1}{\sqrt{d}} \sum_{j\in\mathbb{Z}_d} \mu^{bj(j+d)} = \frac{1}{2\sqrt{d}} G\big(b(d+1),2d\big).$$

Proof. Let $B = \begin{pmatrix} b & 0 \\ 0 & b^{-1} \end{pmatrix} \in SL_2(\mathbb{Z}_{d'})$. Then $\sigma_B = 0$, and so (14.16) gives

$$A := \psi_{P_b} = B \bmod d, \qquad z_{P_b}(p) = 1 = \omega^{\langle\langle 0,Ap\rangle\rangle}(-\mu)^{p^T\sigma_B p}, \quad \forall p \in \mathbb{Z}_d^2.$$

By Theorem 14.2, this implies that $[B,0]$ is an Appleby index for P_b, which is therefore a symplectic unitary, with symplectic index B. Now B can be factored

$$B = \begin{pmatrix} b & \\ & b^{-1} \end{pmatrix} = \begin{pmatrix} 0 & -1 \\ 1 & -b^{-1} \end{pmatrix} \begin{pmatrix} 0 & -1 \\ 1 & -b \end{pmatrix} \begin{pmatrix} 0 & -1 \\ 1 & -b^{-1} \end{pmatrix}. \tag{14.53}$$

In view of (14.46), a symplectic index for $R^b F$ is given by

$$\begin{pmatrix} 0 & -1 \\ 1 & -b \end{pmatrix} = \begin{pmatrix} 1 & 0 \\ 1 & 1 \end{pmatrix}^b \begin{pmatrix} 0 & -1 \\ 1 & 0 \end{pmatrix},$$

and so applying the homomorphism α of Corollary 14.2 to (14.53) gives (14.52), for some scalar $c_{b,d}$, to be determined. From (14.9) and (14.10), we have

$$(R^b F)_{jk} = \frac{1}{\sqrt{d}} \mu^{bj(j+d)+2jk}. \tag{14.54}$$

Hence, equating the $(0,0)$-entries of $c_{b,d} P_b (R^{b^{-1}} F)^{-1} = R^b F R^{b^{-1}} F$, gives

$$\frac{1}{\sqrt{d}} c_{b,d} = \sum_{j\in\mathbb{Z}_d} (R^b F)_{0j} (R^{b^{-1}} F)_{j0} = \frac{1}{d} \sum_{j\in\mathbb{Z}_d} \mu^{b^{-1} j(j+d)}.$$

We recall that $\mu^{j(j+d)}$ depends only on j modulo d, and $\mu^{jd} = \mu^{dj^2}$, so that

$$c_{b,d} = \frac{1}{2}\frac{1}{\sqrt{d}}\sum_{j=0}^{2d-1}\mu^{b^{-1}j(j+d)} = \frac{1}{2\sqrt{d}}\sum_{j=0}^{2d-1}\mu^{b^{-1}(d+1)j^2} = \frac{1}{2\sqrt{d}}G\big(b^{-1}(d+1),2d\big).$$

Evaluating the $(0,0)$-entries of (14.52), using (14.54), gives

$$c_{b,d} = \frac{1}{d\sqrt{d}}\sum_{j\in\mathbb{Z}_d}\sum_{k\in\mathbb{Z}_d}\mu^{bj(j+d)+2jk+b^{-1}k(k+d)} = c_{b^{-1},d}.$$

□

The formulas for evaluating Gauss sums imply that $c_{b,d}$ is an 8-th root of unity, e.g., if b has odd order, then $c_{b,d} = (\sqrt{i})^{1-d}$ (see Exer. 14.11).

Example 14.5. When $b=1$, (14.52) gives

$$M^3 = (RF)^3 = c_{1,d}P_1 = e^{-\frac{2\pi i}{8}(d-1)}I.$$

The map $\mathbb{Z}_d^* \to \mathrm{C}(d) : \sigma \to P_\sigma$ is a group homomorphism, since

$$(P_{\sigma_1\sigma_2})_{jk} = \sum_r (P_{\sigma_1})_{jr}(P_{\sigma_2})_{rk} = \delta_{j,\sigma_1 r}\delta_{r,\sigma_2 k} = \sigma_{j,\sigma_1\sigma_2 k} = (P_{\sigma_1\sigma_2})_{jk}.$$

From Corollary 14.2, we observe that

If a_1 and a_2 are symplectic matrices with symplectic indices B_1 and B_2, then B_1B_2 is a symplectic index for the symplectic matrix a_1a_2.

Using the symplectic index for P_β given by Proposition (14.52), and those for F and R given by (14.49), one obtains the Table 14.3.

Table 14.3: Some symplectic unitaries $[a]\in \mathrm{C}_{\mathrm{Sp}}(d)$, with an Appleby index $[B,0]$ (B is a symplectic index), the index (ψ_a, z_a). Here β is a unit in $\mathbb{Z}_d$ (and hence in $\mathbb{Z}_{d'}$). For the cases when a is antiunitary, or the off diagonal entries of $A=\psi_a$ are nonzero and not units, see §14.13.

Symplectic matrix $a\in f([B,0])$	$A=\psi_a\in SL(\mathbb{Z}_d)$ $B\in SL(\mathbb{Z}'_d)$	$z_a(j,k)$	index description
P_β	$\begin{pmatrix}\beta & 0\\ 0 & \beta^{-1}\end{pmatrix}$	1	diagonal
$R^\alpha P_\beta$	$\begin{pmatrix}\beta & 0\\ \alpha\beta & \beta^{-1}\end{pmatrix}$	$(-\mu)^{\alpha\beta^2j^2} = \mu^{\alpha\beta^2 j(j+d)}$	lower triangular
$F^{-1}R^\alpha P_\beta F$	$\begin{pmatrix}\beta^{-1} & -\alpha\beta\\ 0 & \beta\end{pmatrix}, \alpha\neq 0$	$(-\mu)^{-\alpha\beta^2k^2}$	upper triangular
$R^\alpha P_\beta F^{-1}R^\gamma$	$\begin{pmatrix}\beta\gamma & \beta\\ \alpha\beta\gamma-\beta^{-1} & \alpha\beta\end{pmatrix}$	$(-\mu)^{\gamma(\alpha\beta^2\gamma-1)j^2+2(\alpha\beta^2\gamma-1)jk+\alpha\beta^2k^2}$	b_{12} is a unit
$F^{-1}R^\alpha P_\beta F^{-1}R^\gamma F$	$\begin{pmatrix}\alpha\beta & \beta^{-1}-\alpha\beta\gamma\\ -\beta & \beta\gamma\end{pmatrix}$	$(-\mu)^{-\alpha\beta^2j^2+2(\alpha\beta^2\gamma-1)jk-\gamma(\alpha\beta^2\gamma-1)k^2}$	b_{21} is a unit

14.13 Calculating a symplectic matrix from its symplectic index

Using Table 14.3, a symplectic matrix a can be determined from its symplectic index B (or $\psi_a = B \bmod d$), except for when the off diagonal entries of B are nonzero and not units. We now consider this case, and also when a is antiunitary.

The antilinear map

$$C : \mathbb{C}^d \to \mathbb{C}^d; z \mapsto \overline{z}$$

of entrywise complex conjugation is in the extended Clifford group $\mathrm{EC}(d)$, i.e., it normalises $\hat{H}$, since

$$C(S^j \Omega^k)C^{-1} = S^j \Omega^{-k}. \tag{14.55}$$

Thus the antiunitary elements of $\mathrm{EC}(d)$ are precisely aC, $a \in \mathrm{C}(d)$. Let

$$\psi_C := J := \begin{pmatrix} 1 & 0 \\ 0 & -1 \end{pmatrix}.$$

We note $\det(J) = -1$. With U_λ defined by (14.17), from (14.55) we obtain

$$(aC)U_\lambda (aC)^{-1} = a(CU_\lambda C^{-1})a^{-1} = a(U_{\psi_C(\lambda)})a^{-1} = z_a(\psi_C(\lambda))U_{\psi_a \psi_C(\lambda)}.$$

In this way, we can extend the index (z_a, ψ_a) to $a \in \mathrm{EC}(d)$, where

$$\psi_a \in ESL_2(\mathbb{Z}_d) := \{A \in SL_2(\mathbb{Z}_d); \det(A) = \pm 1\} = SL_2(\mathbb{Z}_d) \cup \{AJ : A \in SL_2(\mathbb{Z}_d)\},$$

and

$$z_{aC} = z_a \circ \psi_C, \qquad \psi_{aC} = \psi_a \psi_C, \qquad a \in \mathrm{E}(d).$$

Similarly, the Appleby indexing extends. With $\hat{D}_p$ defined by (14.37), (14.55) gives

$$\begin{aligned} C\hat{D}_p Cv &= C(-\mu)^{p_1 p_2} S^{p_1} \Omega^{p_2} \overline{v} = (-\mu)^{-p_1 p_2} S^{p_1} \Omega^{-p_2} v \\ &= (-\mu)^{(Jp)_1 (Jp)_2} S^{(Jp)_1} \Omega^{(Jp)_2} = \hat{D}_{Jp} v, \end{aligned}$$

so that if $a \in \mathrm{C}(d)$ has Appleby index $[B, \chi]$, then

$$(aC)\hat{D}_p (aC)^{-1} = a(\hat{D}_{Jp})a^{-1} = \omega^{\langle\langle \chi, BJp \rangle\rangle} \hat{D}_{BJp}, \qquad \forall p \in \mathbb{Z}_{d'}^2, \tag{14.56}$$

i.e., aC has **extended Appleby index** $[BJ, \chi]$. Thus, the surjective homomorphism f of Theorem 14.2 extends to

$$f_E : ESL_2(\mathbb{Z}_{d'}) \ltimes \mathbb{Z}_d^2 \to \mathrm{EC}(d)/[I]. \tag{14.57}$$

We say that an antiunitary element $aC \in \mathrm{EC}(d)$ is **symplectic** if it has an (extended) Appleby index of the form $[BJ, 0]$, $B \in SL_2(\mathbb{Z}_{d'})$.

We now illustrate how a symplectic Clifford operation can be constructed from its (extended) Appleby index. This allows a matrix $a \in \mathrm{EC}(d)$ to be constructed with any given extended Clifford index $[B, \chi]$, $B \in SL_2(\mathbb{Z}_{d'})$, $\chi \in \mathbb{Z}_{d'}^2$.

Example 14.6. (Antiunitary symplectic) For $d = 8$, consider the symplectic index

$$\begin{pmatrix} 6 & 11 \\ 5 & 1 \end{pmatrix} \in ESL_2(\mathbb{Z}_{16})$$

from the Table 14.4. This matrix has determinant -1, and so given an antiunitary symplectic operation aC, $a \in C(d)$, where a symplectic index of a is given by

$$B = \begin{pmatrix} 6 & 11 \\ 5 & 1 \end{pmatrix} J^{-1} = \begin{pmatrix} 6 & 11 \\ 5 & 1 \end{pmatrix} \begin{pmatrix} 1 & 0 \\ 0 & -1 \end{pmatrix} = \begin{pmatrix} 6 & 5 \\ 5 & 15 \end{pmatrix}.$$

Since 5 is a unit in $\mathbb{Z}_{16}$ with inverse 13, we have

$$B = \begin{pmatrix} 1 & 0 \\ 3 & 1 \end{pmatrix} \begin{pmatrix} 5 & 0 \\ 0 & 13 \end{pmatrix} \begin{pmatrix} 0 & 1 \\ -1 & 0 \end{pmatrix} \begin{pmatrix} 1 & 0 \\ 14 & 1 \end{pmatrix},$$

and so we can take $aC = R^3 P_5 F^{-1} R^{14} C$ (see the fourth row of Table 14.3).

Lemma 14.3. *Let $B \in SL_2(\mathbb{Z}_{d'})$, then for $x \in \mathbb{Z}_{d'}$, we have*

$$B = \begin{pmatrix} \alpha & \beta \\ \gamma & \delta \end{pmatrix} = \begin{pmatrix} 1 & 0 \\ -x & 1 \end{pmatrix} \begin{pmatrix} 0 & -1 \\ 1 & 0 \end{pmatrix} \begin{pmatrix} \gamma + x\alpha & \delta + x\beta \\ -\alpha & -\beta \end{pmatrix}, \tag{14.58}$$

where x can be chosen so that $\delta + x\beta \in \mathbb{Z}_{d'}^*$. One such choice for x is the product of the primes which divide d but not $\delta := b_{22} \in \mathbb{Z}_{d'}$. In this way, a symplectic matrix $a \in \mathrm{C}(d)$ with symplectic index B can be constructed by using Table 14.3.

Proof. The formula (14.58) holds by multiplying out. Since $\det(B) = \alpha\delta - \beta\gamma = 1$, a prime p dividing d (and hence d') cannot divide both δ and β. We have two cases

$$\begin{aligned} p \mid \delta \quad &\Longrightarrow \quad p \nmid \beta,\ p \nmid x \quad \Longrightarrow \quad p \nmid \delta + x\beta, \\ p \nmid \delta \quad &\Longrightarrow \quad p \mid x \quad \Longrightarrow \quad p \nmid \delta + x\beta. \end{aligned} \tag{14.59}$$

In both of these cases, $p \nmid \delta + x\beta$, and so $\delta + x\beta$ is a unit in $\mathbb{Z}_{d'}$. □

Example 14.7. For $d = 6$, $d' = 12$, let B be the order 4 matrix

$$B := \begin{pmatrix} 3 & 2 \\ 4 & 3 \end{pmatrix} \in SL_2(\mathbb{Z}_{12}).$$

The prime divisors of d are $p = 2, 3$ and $\delta = 3$, so we can take $x = 2$, and (14.58) gives

$$B = \begin{pmatrix} 3 & 2 \\ 4 & 3 \end{pmatrix} = \begin{pmatrix} 1 & 0 \\ -2 & 1 \end{pmatrix} \begin{pmatrix} 0 & -1 \\ 1 & 0 \end{pmatrix} \begin{pmatrix} 10 & 7 \\ -3 & -2 \end{pmatrix}.$$

Thus $a = R^{-2} F R^{-2} F^{-1} R^{-2}$ is a symplectic unitary, with Appleby index $[B, 0]$. This matrix a has three zero entries in each column and satisfies $a^4 = I$.

14.14 The Zauner matrix

It follows from (14.46) that $M := RF$ has the symplectic index

$$B_M = \begin{pmatrix} 1 & 0 \\ 1 & 1 \end{pmatrix} \begin{pmatrix} 0 & 1 \\ -1 & 0 \end{pmatrix}^{-1} = \begin{pmatrix} 0 & -1 \\ 1 & -1 \end{pmatrix} \in SL_2(\mathbb{Z}_{d'}). \tag{14.60}$$

From $B_M^3 = I$, or Example 14.5, it follows the symplectic operation $[M]$ has order 3. This also follows from the *Gauss sum*

$$\sum_{j=0}^{d-1} \mu^{j(j+d)} = \sum_{j=0}^{d-1} (-\mu)^{j^2} = \frac{1}{2}\sum_{j=0}^{2d-1} e^{\frac{2\pi i}{2d}(d+1)j^2} = \frac{1}{2}G(d+1,2d) = \frac{\sqrt{d}}{e^{\frac{2\pi i}{8}(d-1)}}, \tag{14.61}$$

by a direct calculation

$$\begin{aligned}(M^2)_{jk} &= \frac{1}{d}\sum_r (-\mu)^{j^2+2jr}(-\mu)^{r^2+2rk} = \frac{1}{d}\sum_r (-\mu)^{(r+j+k)^2-k^2-2kj} \\ &= \frac{1}{d}\frac{\sqrt{d}}{e^{\frac{2\pi i}{8}(d-1)}}(-\mu)^{-k^2-2kj} = (e^{\frac{2\pi i}{8}})^{-(d-1)}(M^{-1})_{jk}. \end{aligned} \tag{14.62}$$

Thus the **Zauner matrix**[5] (see [Zau99], [Zau10]) given by the normalisation

$$Z := \zeta^{d-1}M = \zeta^{d-1}RF, \qquad \zeta := e^{\frac{2\pi i}{24}}, \tag{14.63}$$

has order 3, as do $Z^2, \overline{Z}, \overline{Z}^2 \in \mathrm{C}(d)$. By (14.15),

$$Z(S^j\Omega^k)Z^{-1} = \mu^{k(k-2j+d)}S^{-k}\Omega^{j-k}, \tag{14.64}$$

and by (14.33), we have

$$\psi_Z = \begin{pmatrix} 0 & -1 \\ 1 & -1 \end{pmatrix}, \quad \psi_{Z^2} = \begin{pmatrix} -1 & 1 \\ -1 & 0 \end{pmatrix}, \quad \psi_{\overline{Z}} = \begin{pmatrix} 0 & 1 \\ -1 & -1 \end{pmatrix}, \quad \psi_{\overline{Z}^2} = \begin{pmatrix} -1 & -1 \\ 1 & 0 \end{pmatrix}.$$

The Zauner matrix satisfies

$$R^{-1}ZR = \overline{Z}^2, \qquad R^{-1}Z^2R = \overline{Z}.$$

We have the following corollary of Theorem 14.1.

Corollary 14.3. *The extended Clifford group is generated by* $\hat{H}$, *and*

$$C \text{ (order 2)}, \qquad Z \text{ (order 3)}, \qquad F \text{ (order 4)}.$$

Proof. The extended Clifford group is generated by C, and a set of generators for $\mathrm{C}(d)$. Here, the generator R of Theorem 14.1 is replaced by $Z = \zeta^{d-1}RF$. □

---

5 The $\mathbf{Z} = \zeta^{d-1}\mathbf{FG}$ of [Zau99] is $\overline{Z}^2$, with $\mathbf{F} = F^{-1}$, $\mathbf{G} = R$, $\mathbf{U} = \Omega$, $\mathbf{V} = S^{-1}$.

14.15 The Scott–Grassl numerical SICs

A stronger form of Zauner's conjecture asserts the existence of a Weyl–Heisenberg SIC with a specific structure, i.e.,

Zauner's conjecture (Strong): For every dimension d, a Weyl–Heisenberg SIC (set of d^2 equiangular vectors) in $\mathbb{C}^d$ of the form

$$(S^j\Omega^k v)_{j,k\in\mathbb{Z}_d},$$

can be constructed, where v is an eigenvector of the Zauner matrix Z.

Following from the work of Renes, et al. [RBKSC04], an extensive (and ongoing) search led by Andrew Scott[6] [SG10] found *numerical* Weyl–Heisenberg SICs (for $d \leq 67$) and counted and indexed them (for $d \leq 50$). This has now been extended to a putative list of all Weyl–Heisenberg SICs with certain symmetries for $d \leq 90$ and at least one for $d \leq 120$ and $d = 124, 143, 147, 168, 172, 195, 199, 228, 259, 323$.

These **(Scott–Grassl) numerical SICs** were obtained by finding unit vectors $v \in \mathbb{C}^d$ that minimise the second frame potential of $(S^j\Omega^k v)_{(j,k)\in\mathbb{Z}_d^2}$, i.e., by Proposition 14.1 and $(S^j\Omega^k)^* = \Omega^{-k}S^{-j} = \omega^{jk}S^{-j}\Omega^{-k}$,

$$\sum_{(j,k)\in\mathbb{Z}_d^2} |\langle S^j\Omega^k v, v\rangle|^4 = \frac{1}{|\mathbb{Z}_d^2|}\sum_p\sum_q |\langle S^{p_1}\Omega^{p_2}v, S^{q_1}\Omega^{q_2}v\rangle|^4 \geq \frac{2d}{d+1},$$

with equality if and only v is a fiducial vector for a Weyl–Heisenberg SIC. These SIC fiducials can be presented to high accuracy, e.g., 1000 decimal places (see [Chi15]).

The numerical SICs given by a fiducial vector v have been invaluable in the study and analytic construction of SICs [ACFW17]. We now detail some of their properties which are summarised in Table 14.4 (kindly provided by Scott and Grassl).

The **Clifford action** (of the extended Clifford group) on the projectors $P_v = vv^*$ given by fiducial vectors v is $g\cdot P_v = (gv)(gv)^* = gP_vg^{-1}$, which gives orbits

$$\mathrm{orb}(v) := \{g(vv^*)g^{-1}\}_{g\in \mathrm{EC}(d)}.$$

The **symmetries** of a fiducial vector v is the stabiliser of the induced action of the projective group $\mathrm{PEC}(d)$, i.e.,

$$S = S(v) := \{[a] \in \mathrm{PEC}(d) : av = \lambda v \text{ for some (unit) scalar} \lambda \in \mathbb{C}\}. \qquad (14.65)$$

The symmetries $S = S(v)$ simplify the construction of a fiducial vector v, e.g., if v is an eigenvector of the Zauner matrix Z, then v lies in a subspace of $\mathbb{C}^d$ of dimension approximately $d/3$ (see Table 14.7). The symmetries of numerical SICs all appear

[6] Using Andrew Scott's code, C. A. Fuchs, M. C. Hoang and B. C. Stacey have found numerical Weyl–Heisenberg SICs for $d \leq 151$ (and counting) using a Chimera supercomputer.

to be symplectic operations, which give a cyclic group of order a multiple of three for $d > 3$. These include the operations with symplectic indices $F_z, F_a, F_b, F_c, F_d, F_e$ (see §14.18). The number of SICs modulo the action of the extended projective Clifford group $\mathrm{PEC}(d)$ is given in the # column of Table 14.4.

Table 14.4: The Weyl-Heisenberg covariant numerical SIC-POVMs of [Sco17]. Gaps in the classification are marked by ?'s, which either indicate an unexplored dimension (to be filled in later) or note the likely presence of an unknown general symmetry.

d	PEC(d) orbits				labels
	#	stabiliser			
		$\lvert S\rvert$	S	notes	
2	1	6	$\left\langle \begin{pmatrix} 0 & -1 \\ -1 & 0 \end{pmatrix}, F_z \right\rangle$		a
3	∞	6	$\left\langle \begin{pmatrix} 0 & -1 \\ -1 & 0 \end{pmatrix}, F_z \right\rangle$		a
	1	12	$\left\langle \begin{pmatrix} 0 & -1 \\ -1 & 0 \end{pmatrix}, -F_z \right\rangle$		b
	1	48	$\mathrm{ESL}_2(\mathbb{Z}_3)$		c
4	1	6	$\langle F_c F_z \rangle = \langle F_c \rangle \langle F_z \rangle$		a
5	1	3	$\langle F_z \rangle$		a
6	1	3	$\langle F_z \rangle$		a
7	1	3	$\langle F_z \rangle$		a
	1	6	$\langle F_c F_z \rangle = \langle F_c \rangle \langle F_z \rangle$		b
8	1	3	$\langle F_z \rangle$		a
	1	12	$\left\langle \begin{pmatrix} 6 & 11 \\ 5 & 1 \end{pmatrix} \right\rangle = \langle F_z \rangle \left\langle \begin{pmatrix} 3 & 6 \\ 10 & 9 \end{pmatrix} \right\rangle \ni F_b$	?	b
9	2	3	$\langle F_z \rangle$		a, b
10	1	3	$\langle F_z \rangle$		a
11	3	3	$\langle F_z \rangle$		a–c
12	1	3	$\langle F_z \rangle$		a
	1	6	$\left\langle \begin{pmatrix} 0 & 17 \\ 17 & 15 \end{pmatrix} \right\rangle \ni F_a$	$\begin{pmatrix} 0 & 17 \\ 17 & 15 \end{pmatrix} \sim F_e$	b
13	2	3	$\langle F_z \rangle$		a, b
14	2	3	$\langle F_z \rangle$		a, b
15	3	3	$\langle F_z \rangle$		a–c
	1	6	$\langle F_b F_z \rangle = \langle F_b \rangle \langle F_z \rangle$		d
16	2	3	$\langle F_z \rangle$		a, b
17	3	3	$\langle F_z \rangle$		a–c
18	2	3	$\langle F_z \rangle$		a, b
19	3	3	$\langle F_z \rangle$		a–c
	1	6	$\langle F_c F_z \rangle = \langle F_c \rangle \langle F_z \rangle$		d
	1	18	$\left\langle \begin{pmatrix} 3 & 12 \\ 7 & 15 \end{pmatrix} \right\rangle = \langle F_c \rangle \left\langle \begin{pmatrix} 7 & 14 \\ 5 & 2 \end{pmatrix} \right\rangle \ni F_z$	$\begin{pmatrix} 7 & 14 \\ 5 & 2 \end{pmatrix} \sim \mathbb{F}^d$	e
20	2	3	$\langle F_z \rangle$		a, b
21	4	3	$\langle F_z \rangle$		a–d
	1	3	$\langle F_a \rangle$		e
22	1	3	$\langle F_z \rangle$		a
23	6	3	$\langle F_z \rangle$		a–f
24	2	3	$\langle F_z \rangle$		a, b
	1	6	$\langle F_b F_z \rangle = \langle F_b \rangle \langle F_z \rangle$		c
25	2	3	$\langle F_z \rangle$		a, b

d	PEC(d) orbits				
	#	stabiliser			labels
		$\lvert S\rvert$	S	notes	
26	4	3	$\langle F_z\rangle$		a–d
27	6	3	$\langle F_z\rangle$		a–f
28	2	3	$\langle F_z\rangle$		a,b
	1	6	$\langle F_cF_z\rangle = \langle F_c\rangle\langle F_z\rangle$		c
29	4	3	$\langle F_z\rangle$		a–d
30	3	3	$\langle F_z\rangle$		a–c
	1	3	$\langle F_a\rangle$		d
31	7	3	$\langle F_z\rangle$		a–g
32	2	3	$\langle F_z\rangle$		a,b
33	4	3	$\langle F_z\rangle$		a–d
34	2	3	$\langle F_z\rangle$		a,b
35	8	3	$\langle F_z\rangle$		a–h
	1	6	$\langle F_bF_z\rangle = \langle F_b\rangle\langle F_z\rangle$		i
	1	12	$\left\langle\begin{pmatrix}15 & 3\\ 32 & 18\end{pmatrix}\right\rangle = \langle F_z\rangle\left\langle\begin{pmatrix}3 & 15\\ 20 & 18\end{pmatrix}\right\rangle \ni F_b$	?	j
36	4	3	$\langle F_z\rangle$		a–d
37	4	3	$\langle F_z\rangle$		a–d
38	4	3	$\langle F_z\rangle$		a–d
39	6	3	$\langle F_z\rangle$		a–f
	2	3	$\langle F_a\rangle$		g,h
	2	6	$\left\langle\begin{pmatrix}0 & 7\\ 28 & 6\end{pmatrix}\right\rangle \ni F_a$	$\begin{pmatrix}0 & 7\\ 28 & 6\end{pmatrix} \sim F_e$	i,j
40	2	3	$\langle F_z\rangle$		a,b
41	8	3	$\langle F_z\rangle$		a–h
42	4	3	$\langle F_z\rangle$		a–d
43	6	3	$\langle F_z\rangle$		a–f
44	6	3	$\langle F_z\rangle$		a–f
45	4	3	$\langle F_z\rangle$		a–d
46	3	3	$\langle F_z\rangle$		a–c
47	8	3	$\langle F_z\rangle$		a–h
48	4	3	$\langle F_z\rangle$		a–d
	1	3	$\langle F_a\rangle$		e
	1	6	$\langle F_bF_z\rangle = \langle F_b\rangle\langle F_z\rangle$		f
	1	24	$\left\langle\begin{pmatrix}4 & 37\\ 25 & 63\end{pmatrix}\right\rangle \ni F_a, F_b$	?	g
49	7	3	$\langle F_z\rangle$		a–g
50	2	3	$\langle F_z\rangle$		a,b
51	14	3	$\langle F_z\rangle$		a–n
52	3	3	$\langle F_z\rangle$		a–c
	1	6	$\langle F_cF_z\rangle = \langle F_c\rangle\langle F_z\rangle$		d
53	9	3	$\langle F_z\rangle$		a–i
	1	9	$\left\langle\begin{pmatrix}7 & 21\\ 32 & 28\end{pmatrix}\right\rangle \ni F_z$	$\begin{pmatrix}7 & 21\\ 32 & 28\end{pmatrix} \sim \mathbb{F}^d$	j
54	4	3	$\langle F_z\rangle$		a–d
55	6	3	$\langle F_z\rangle$		a–f
56	6	3	$\langle F_z\rangle$		a–f

d	PEC(d) orbits				
	#	stabiliser			labels
		$\|S\|$	S	notes	
57	6	3	$\langle F_z\rangle$		a–f
	2	3	$\langle F_a\rangle$		g,h
58	4	3	$\langle F_z\rangle$		a–d
59	12	3	$\langle F_z\rangle$		a–l
60	4	3	$\langle F_z\rangle$		a–d
61	6	3	$\langle F_z\rangle$		a–f
62	5	3	$\langle F_z\rangle$		a–e
63	14	3	$\langle F_z\rangle$		a,d–p
	2	6	$\langle F_bF_z\rangle = \langle F_b\rangle\langle F_z\rangle$		b,c
64	4	3	$\langle F_z\rangle$		a–d
65	8	3	$\langle F_z\rangle$		a–h
66	6	3	$\langle F_z\rangle$		d–i
	3	3	$\langle F_a\rangle$		a–c
67	7	3	$\langle F_z\rangle$		c–i
	2	6	$\langle F_cF_z\rangle = \langle F_c\rangle\langle F_z\rangle$		a,b
68	4	3	$\langle F_z\rangle$		a–d
69	8	3	$\langle F_z\rangle$		a–h
70	5	3	$\langle F_z\rangle$		a–e
71	18	3	$\langle F_z\rangle$		a–r
72	4	3	$\langle F_z\rangle$		a–d
73	4	3	$\langle F_z\rangle$		a–d
74	7	3	$\langle F_z\rangle$		a–g
75	12	3	$\langle F_z\rangle$		a–l
	3	3	$\langle F_a\rangle$		m–o
76	6	3	$\langle F_z\rangle$		a–f
77	8	3	$\langle F_z\rangle$		a–h
78	7	3	$\langle F_z\rangle$		a–g
79	14	3	$\langle F_z\rangle$		a–n
80	8	3	$\langle F_z\rangle$		a–h
	1	6	$\langle F_bF_z\rangle = \langle F_b\rangle\langle F_z\rangle$		i
81	12	3	$\langle F_z\rangle$		a–l
82	3	3	$\langle F_z\rangle$		a–c
83	16	3	$\langle F_z\rangle$		a–p
84	6	3	$\langle F_z\rangle$		a–f
	2	3	$\langle F_a\rangle$		g,h
	2	6	$\langle F_e\rangle$	${F_e}^2 \sim F_a$	i,j
85	4	3	$\langle F_z\rangle$		a–d
86	10	3	$\langle F_z\rangle$		a–j
87	12	3	$\langle F_z\rangle$		a–l
88	4	3	$\langle F_z\rangle$		a–d
89	10	3	$\langle F_z\rangle$		a–j
90	4	3	$\langle F_z\rangle$		a–d
91–121	≥ 1	≥ 3	$? \ni F_z$		a
99	≥ 3	≥ 6	$? \ni F_z, F_b$		b–d
111	≥ 1	≥ 9	$? \ni \mathbb{F}^d$	$\mathbb{F}^{d^3} \sim F_a$	
120	≥ 1	≥ 6	$? \ni F_z, F_b$		
	≥ 1	≥ 6	$? \ni F_a, F_b$		

d	PEC(d) orbits				
	#	stabiliser			labels
		|S|	S	notes	
124	≥ 1	≥ 6	$? \ni F_z, F_c$		a
143	≥ 1	≥ 6	$? \ni F_z, F_b$		a
147	≥ 1	≥ 6	$? \ni F_e$	$F_e^2 \sim F_a$	a
168	≥ 1	≥ 6	$? \ni F_z, F_b$		a
172	≥ 1	≥ 6	$? \ni F_z, F_c$		a
195	≥ 1	≥ 6	$? \ni F_z, F_b$		a
199	≥ 1	≥ 9	$? \ni \mathbb{F}^d$		a
228	≥ 1	≥ 6	$? \ni F_e$	$F_e^2 \sim F_a$	a
259	≥ 1	≥ 6	$? \ni F_z, F_c$		a
323	≥ 1	≥ 9	$? \ni \mathbb{F}^d$		a

14.16 Symmetries of the Weyl–Heisenberg SICs

The symmetries of a fiducial v (under the Clifford group action) given in Table 14.4 include the symplectic (anti)unitaries with symplectic indices (see Exer. 14.16)

$$F_z := \begin{pmatrix} 0 & d-1 \\ d+1 & d-1 \end{pmatrix} = (d+1)\begin{pmatrix} 0 & -1 \\ 1 & -1 \end{pmatrix}, \quad d \geq 2,$$

$$F_a := \begin{pmatrix} 1 & d+3 \\ \frac{4d-3}{3} & d-2 \end{pmatrix} = (d+1)\begin{pmatrix} d+1 & 3 \\ \frac{d-3}{3} & d-2 \end{pmatrix}, \quad d \equiv 3 \bmod 9, \quad d \neq 3,$$

$$F_b := \begin{pmatrix} -\beta & d \\ d & d-\beta \end{pmatrix}, \quad d = \beta^2 - 1, \quad \beta \geq 3,$$

$$F_c := \begin{pmatrix} \kappa & d-2\kappa \\ d+2\kappa & d-\kappa \end{pmatrix} = \begin{pmatrix} \kappa & d+2\kappa \\ d+2\kappa & d+\kappa \end{pmatrix}\begin{pmatrix} 1 & 0 \\ 0 & -1 \end{pmatrix}, \quad \begin{matrix} d = (3k\pm 1)^2 + 3, \\ \kappa = 3k^2 \pm k + 1, \end{matrix} \quad k \geq 0,$$

$$F_d := \begin{pmatrix} 0 & 1 \\ -1 & -k(k+3) \end{pmatrix}, \quad d = k^2(k+3) - 1,$$

$$F_e := \begin{pmatrix} 0 & 1 \\ 1 & d+3k \end{pmatrix} = \begin{pmatrix} 0 & -1 \\ 1 & -d-3k \end{pmatrix}\begin{pmatrix} 1 & 0 \\ 0 & -1 \end{pmatrix}, \quad d = 9k^2 + 3.$$

The first two are order three symplectic unitaries, namely the Zauner matrix Z, and

$$M_1 = a_{F_a} := (-1)^{d-1} R^{d-1} F^{-1} R^{-3} F R^{\frac{d}{3}}, \qquad d \equiv 3 \bmod 9, \quad d \neq 3. \qquad (14.66)$$

These have symplectic indices with trace -1 (see §14.18), and (see Table 14.4)

- For every dimension d there is a fiducial with the Zauner symmetry.
- In addition, for $d \equiv 3 \bmod 9$, $d \neq 3$, there are *exceptional* fiducials which are eigenvectors of M_1 (but not of a conjugate of Z), namely those with the labels

$$12b,\ 21e,\ 30d,\ 39ghij,\ 48eg,\ 57gh,\ 66abc,\ 75mno,\ 84ghij,\ \ldots$$

For all known Weyl–Heisenberg numerical SICs, the symmetries $S(v)$ of the fiducial vector v is a cyclic group of order a multiple of three for $d>3$.

In addition to a symmetry Z or M_1, the tabulated SICs have symmetries:

- The order 2 symplectic unitary
$$a_{F_b} = S^{\frac{1}{2}d(d+1)}\Omega^{\frac{1}{2}d(d+1)}P_{-k}, \qquad d=k^2-1,$$
for $d = 8b, 15d, 24c, 35ij, 48f, 63bc, 80i, 99, 120, 143a, 168a, 195a, 224,\ldots$
- The order 2 symplectic antiunitary
$$a_{F_c} = R^{-1}F^{-1}R^{\kappa(2\kappa-d)}P_{3\kappa}FR^{1-\kappa d}C, \qquad d=(3k\pm1)^2+3,$$
where $\kappa = 3k^2\pm k+1$, for $d = 4a, 7b, 19de, 28c, 52d, 67ab, 103, 124a, 172a, 199, 259a, 292,\ldots$
- The order 9 unitary
$$a_{F_d} = R^{-k(k+3)}F^{-1}, \qquad d=k^2(k+3)-3,$$
for $d = 19e, 53j, 111, 199a, 323a, 489,\ldots$
- The order 6 antiunitary
$$a_{F_e} = R^{d+3k}FC, \qquad d=9k^2+3,$$
for $d = 12b, 39ij, 84ij, 147a, 228a, 327,\ldots$

Example 14.8. ($d=4$) The Scott–Grassl SIC $4a$ for $\mathbb{C}^4$ is in the 1-eigenspace of Z (dimension 2) and has antiunitary symmetry aC, $a := R^{-1}F^{-1}R^{-2}P_3FR^{-3}$, where

$$Z=\frac{1}{2}\begin{pmatrix} \sqrt{i} & \sqrt{i} & \sqrt{i} & \sqrt{i} \\ -i & 1 & i & -1 \\ -\sqrt{i} & \sqrt{i} & -\sqrt{i} & \sqrt{i} \\ -i & -1 & i & 1 \end{pmatrix}, \qquad a=\frac{1}{2}\begin{pmatrix} 1-i & 0 & -1-i & 0 \\ 0 & -1-i & 0 & i-1 \\ -1-i & 0 & 1-i & 0 \\ 0 & i-1 & 0 & -1-i \end{pmatrix},$$

It therefore satisfies[7]
$$Zv=v, \qquad a\overline{v} = \frac{1}{\sqrt{2}}(1-i)v. \tag{14.67}$$

Each eigenspace of Z and M_1 (of order 3) has dimension $\approx \frac{d}{3}$, see Table 14.7 and (14.109). Equations such as (14.67) aid the search for exact fiducials. Further simplifications of the equations that determine a Weyl–Heisenberg SIC (see §14.27) can be obtained by choosing a conjugate of Z (or other symmetries) to be monomial (see §14.17 and §14.18).

---

[7] Here v is the numerical SIC $4a$, which has the (unnormalised) analytic form given by (14.83). The SIC chosen in (14.68) does not have the symmetry $a\overline{v}=cv$, $c\in\mathbb{T}$.

14.17 Monomial representations of the Clifford group

If d is a square, say $d=n^2$, then the commutativity relation (14.5) gives

$$\Omega^n S^n = \omega^{n^2} S^n \Omega^n = S^n \Omega^n,$$

so the Weyl–Heisenberg group has an abelian subgroup generated by S^n and Ω^n. By diagonalising this subgroup, [ABB+12] were led to a monomial representation of the Clifford group, i.e., one in which the matrices have exactly one nonzero entry in each row and column. We now summarise (without proof) this **phase-permutation representation**. Let $(e_k)_{k\in\mathbb{Z}_d}$ is the standard basis for $\mathbb{C}^d$.

In the basis (u_j) for $\mathbb{C}^d$ given by

$$u_j := \sum_{t=0}^{n-1} \omega^{-ntj_1} e_{nt+j_2}, \qquad j=(j_1,j_2)\in\mathbb{Z}_n^2,$$

the Clifford group elements a transform to monomial matrices

$$\hat{a} := U^{-1}aU, \qquad U:=[u_j],$$

where

$$\hat{S}e_j := \begin{cases} e_{(j_1,j_2+1)}, & j_2+1\neq 0; \\ \sigma^{j_1} e_{(j_1,0)} & j_2+1=0, \end{cases} \qquad \hat{\Omega}e_j := \omega^{j_2} e_{(j_1-1,j_2)}, \qquad \sigma := e^{\frac{2\pi i}{n}}.$$

Example 14.9. ($d=4$) For $n=2$, with the order $(0,0),(0,1),(1,0),(1,1)$ on the indices from $\mathbb{Z}_2^2$, we have

$$\hat{S} = \begin{pmatrix} 0&1&0&0\\ 1&0&0&0\\ 0&0&0&-1\\ 0&0&1&0 \end{pmatrix}, \quad \hat{\Omega} = \begin{pmatrix} 0&0&1&0\\ 0&0&0&i\\ 1&0&0&0\\ 0&i&0&0 \end{pmatrix}, \quad U = \begin{pmatrix} 1&0&1&0\\ 0&1&0&1\\ 1&0&-1&0\\ 0&1&0&-1 \end{pmatrix},$$

$$\hat{F} = \begin{pmatrix} 1&0&0&0\\ 0&0&1&0\\ 0&1&0&0\\ 0&0&0&-i \end{pmatrix}, \quad \hat{R} = \begin{pmatrix} 0&0&1&0\\ 0&-\sqrt{i}&0&0\\ 1&0&0&0\\ 0&0&0&-\sqrt{i} \end{pmatrix}, \quad \hat{Z} = \begin{pmatrix} 0&\sqrt{i}&0&0\\ 0&0&-i&0\\ \sqrt{i}&0&0&0\\ 0&0&0&1 \end{pmatrix}.$$

In the basis (u_j) the condition $Zv=v$ of (14.67) can be written as $\hat{Z}\,w=w$, $w=U^{-1}v$, i.e., $\sqrt{i}w_2=w_1$, $-iw_3=w_2$ and $\sqrt{i}w_1=w_3$. Thus we seek a fiducial of the form $w=(-i\sqrt{i},-i,1,a)$, $a\in\mathbb{C}$. It is easy to check that $a=\sqrt{2+\sqrt{5}}\sqrt{i}$ gives a fiducial vector v, which is a 1–eigenvector of Z, namely

$$v = \frac{1}{\sqrt{10+2\sqrt{5}}} \begin{pmatrix} 1-i\sqrt{i} \\ -i+\sqrt{2+\sqrt{5}}\sqrt{i} \\ -1-i\sqrt{i} \\ -i-\sqrt{2+\sqrt{5}}\sqrt{i} \end{pmatrix}, \qquad \sqrt{i}=\frac{1}{\sqrt{2}}(1+i). \tag{14.68}$$

14.18 The Clifford trace and symplectic unitaries of order 3

Each known Weyl–Heisenberg SIC has an order 3 symmetry Z or M_1 (see §14.16). The symplectic indices of these symplectic unitaries satisfy

$$\mathrm{trace}(F_z) = \mathrm{trace}(F_a) = -1.$$

We now classify such matrices up to conjugation by a symplectic operation.

The **Clifford trace** is the map

$$\mathrm{tr}_{\mathrm{C}} : \mathrm{C}(d) \to \mathbb{Z}_d : a \mapsto \mathrm{trace}(\psi_a).$$

Since $a \mapsto \psi_a$ is a homomorphism with kernel $\hat{H}$ (Lemma 14.2), this satisfies

$$\mathrm{tr}_{\mathrm{C}}(ab) = \mathrm{tr}_{\mathrm{C}}(ba), \qquad \forall a,b \in \mathrm{C}(d), \tag{14.69}$$

$$\mathrm{tr}_{\mathrm{C}}(ah) = \mathrm{tr}_{\mathrm{C}}(a), \qquad \forall a \in \mathrm{C}(d),\ \forall h \in \hat{H}. \tag{14.70}$$

In particular, the Clifford trace of any conjugate of Z or $Z^{-1} = Z^2$ is -1, e.g.,

$$\mathrm{tr}_{\mathrm{C}}(gZg^{-1}) = \mathrm{tr}_{\mathrm{C}}(Zg^{-1}g) = \mathrm{tr}_{\mathrm{C}}(Z) = \mathrm{trace}(\psi_Z) = -1,$$

and the Clifford trace is well defined on the Clifford operations, i.e.,

$$\mathrm{tr}_{\mathrm{C}}([a]) := \mathrm{tr}_{\mathrm{C}}(a), \qquad \forall [a] \in \mathrm{PC}(d).$$

The order of a Clifford operation is related to its Clifford trace, since

$$A^2 = \mathrm{trace}(A)A - I, \qquad \forall A \in SL_2(\mathbb{Z}_d). \tag{14.71}$$

Lemma 14.4. *A nonidentity extended Clifford operation* $[a] \in \mathrm{EC}(d)/[I]$ *with index* (A, z_a) *and Clifford trace* $t = \mathrm{trace}(A)$ *has order* 3 *if and only if*

$$(t^2-1)A = (t+1)I, \qquad z_a((t+1)Ap) = \omega^{(t+1)p^T M_A p}, \quad \forall p \in \mathbb{Z}_d^2, \tag{14.72}$$

where $M_A = \begin{pmatrix} \gamma(\alpha^3+2\alpha^2\delta+\alpha\delta^2-2\alpha-\delta) & \beta\gamma(\alpha+\delta-1)(\alpha+\delta+1) \\ \beta\gamma(\alpha+\delta-1)(\alpha+\delta+1) & \beta(\delta^3+2\alpha\delta^2+\alpha^2\delta-2\delta-\alpha) \end{pmatrix}$, $A = \begin{pmatrix} \alpha & \beta \\ \gamma & \delta \end{pmatrix}$.

Proof. Since a product of three antiunitaries is a unitary matrix, we have $a \in \mathrm{C}(d)$. In view of the isomorphism (14.35), $[a]$ has order 3 if and only if

$$(A, z_a)^3 = (A^3, (z_a \circ A^2)(z_a \circ A)z_a) = (I, 1).$$

From (14.71), we obtain

$$A^3 = A(tA - I) = t(tA - I) - A = (t^2-1)A - tI,$$

so that the condition $A^3 = I$ can be written as the first condition of (14.72).

We now consider the condition $(z_a \circ A^2)(z_a \circ A)z_a = 1$. By (14.23), we calculate

$$\begin{aligned} z_a(p)z_a(Ap)z_a(A^2p) &= \omega^{-p^T\sigma_A(Ap)}z_a(p+Ap)z_a(A^2p) \\ &= \omega^{-p^T\sigma_A(Ap)}\omega^{-(p+Ap)^T\sigma_A(A^2p)}z_a(p+Ap+A^2p) \\ &= \omega^{-p^T(\sigma_A A+\sigma_A A^2+A^T\sigma_A A^2)p}z_a(p+Ap+A^2p). \end{aligned}$$

By (14.71), we have

$$I+A+A^2 = I+A+tA-I = (1+t)A.$$

Using $\det(A) = \alpha\delta - \beta\gamma = 1$, a calculation gives

$$\begin{aligned} &\sigma_A A + \sigma_A A^2 + A^T\sigma_A A^2 \\ &= (\alpha+\delta+1)\begin{pmatrix} \gamma(\alpha^3+2\alpha^2\delta+\alpha\delta^2-2\alpha-\delta) & \beta\gamma(\alpha+\delta-1)(\alpha+\delta+1) \\ \beta\gamma(\alpha+\delta-1)(\alpha+\delta+1) & \beta(\delta^3+2\alpha\delta^2+\alpha^2\delta-2\delta-\alpha) \end{pmatrix}. \end{aligned}$$

Thus we may rewrite the condition $(z_a \circ A^2)(z_a \circ A)z_a = 1$, to obtain the result. □

Since $z_a(0) = 1$, $\forall a \in \mathrm{EC}(d)$, and $\mathrm{tr}_{\mathrm{C}}(I) = 2 = -1$ if and only if $d = 3$, we have

If $a \in \mathrm{C}(d)$ has Clifford trace -1 and $d \neq 3$, then $[a]$ has order 3.

Further, by taking the trace of the condition $(t^2-1)A = (t+1)I$, we have

The Clifford trace t of a Clifford operation of order 3 satisfies

$$(t-2)(t+1)^2 = 0. \tag{14.73}$$

For d prime, Clifford operators of order 3 must have Clifford trace -1.

Proposition 14.6. *Suppose that $d \neq 3$ and $a \in \mathrm{C}(d)$. Then*

1. *If a has Clifford trace -1, then $[a]$ has order 3.*
2. *If d is prime, then $[a]$ has order 3 if and only if a has Clifford trace -1.*

Proof. Since we have already proved 1, it suffices to prove for $d \neq 3$ prime and $[a]$ of order 3 that the Clifford trace $t = \mathrm{tr}_{\mathrm{C}}(a)$ is -1. We recall that t is a root of (14.73).

If $t \neq -1$, then $t+1$ is a unit (all nonzero elements of $\mathbb{Z}_d$ are units for d prime), so that $t = 2$. But, if $t = 2$, then (14.72) gives $3A = 3I$, and hence $A = I$ ($3 \in \mathbb{Z}_d^*$ for $d \neq 3$ prime), so that $a \in \hat{H}$ (by Lemma 14.2). Since $(S^j\Omega^k)^3 = \omega^3 S^{rj}\Omega^{rk}$ (see Exer. 14.4) and S, Ω have order d, the order of $[a]$ cannot be 3 (since 3 does not divide d). Thus $t = \mathrm{tr}_{\mathrm{C}}(a) - 1$ (when $[a]$ has order 3 and $d \neq 3$ is prime). □

A Clifford operation of order 3 is said to be **canonical order** 3 if it has Clifford trace -1 (see [App05]), e.g., the Zauner matrix Z and M_1 are canonical order 3.

Example 14.10. It follows from (14.69) and (14.70) that left or right multiplication of a canonical order 3 Clifford operation by a displacement operation gives another canonical order 3 operation, e.g., $[h_1 Z h_2]$ is canonical order 3 for any $h_1, h_2 \in \hat{H}$.

There are Clifford operations of order 3 with Clifford trace 2.

Example 14.11. If 3 divides d, then the symplectic unitary $R^{\frac{d'}{3}}$ (and its inverse) has order 3 and Clifford trace

$$\mathrm{tr}_{\mathbb{C}}(R^{\frac{d'}{3}}) = \mathrm{trace}\left(\begin{pmatrix} 1 & 0 \\ \frac{d'}{3} & 1 \end{pmatrix}\right) = 2,$$

as do the Weyl displacement operators $S^{\frac{d}{3}}, \Omega^{\frac{d}{3}}, S^{\frac{d}{3}}\Omega^{\frac{d}{3}}$.

There are Clifford operations of order 3 with Clifford trace $t \neq -1, 2$, i.e., for which (14.73) holds with $t-2$ and $t+1$ not units in $\mathbb{Z}_d$.

Example 14.12. For $d = 10$, $SL_2(\mathbb{Z}_{10})$ has a single conjugacy class of elements of order 3 and trace 4 and 7. These have representatives

$$A = \begin{pmatrix} 3 & 2 \\ 6 & 1 \end{pmatrix} \quad (\text{trace } 4), \qquad B = \begin{pmatrix} 6 & 5 \\ 5 & 1 \end{pmatrix} \quad (\text{trace } 7).$$

These can be lifted to symplectic indices which give symplectic unitaries of order 3 and Clifford trace 4 and 7 (see §14.13), e.g., $a = R^{10} F R^8 F^{-1} R^6$, $b = R^{10} F R^5 F^{-1} R^{15}$.

We need the following technical lemma of [BW17].

Lemma 14.5. *Suppose that $d \geq 2$, and let*

$$z := \psi_Z = \begin{pmatrix} 0 & -1 \\ 1 & -1 \end{pmatrix}, \qquad z^2 = \begin{pmatrix} -1 & 1 \\ -1 & 0 \end{pmatrix}, \tag{14.74}$$

$$m_1 := \begin{pmatrix} 1 & 3 \\ \frac{d-3}{3} & -2 \end{pmatrix}, \quad d \equiv 3 \bmod 9, \tag{14.75}$$

$$m_2 := \begin{pmatrix} 1 & 3 \\ \frac{2d-3}{3} & -2 \end{pmatrix}, \quad d \equiv 6 \bmod 9. \tag{14.76}$$

Then the conjugacy classes of elements of order 3 and trace -1 in $SL_2(\mathbb{Z}_d)$ have representatives

$$\{z\}, \quad d \not\equiv 0 \bmod 3, \tag{14.77}$$

$$\{z, z^2\}, \quad d \equiv 0 \bmod 9 \text{ or } d = 3, \tag{14.78}$$

$$\{z, z^2, m_1\}, \quad d \equiv 3 \bmod 9,\ d \neq 3, \tag{14.79}$$

$$\{z, z^2, m_2\}, \quad d \equiv 6 \bmod 9. \tag{14.80}$$

Lemma 14.6. *Let $\varphi : G \to H$ be a homomorphism of G onto H, with $|\ker\varphi| = 2^k$. If $h \in H$ has order 3, then there is an element $g \in G$ of order 3 with $\varphi(g) = h$.*

Proof. By the first isomorphism theorem for groups, we may assume that $H = G/K$, where $K = \ker\varphi$. Suppose that $h = aK \in G/K$ has order 3, i.e., $a^3 = x \in K$, where $a \notin K$. By Bézout's identity (the Euclidean algorithm) choose integers α, β with $1 = -3\alpha + 2^k\beta$. Let $g = ax^\alpha \in \langle a \rangle$. Then $\varphi(g) = ax^\alpha K = aK$, and

$$g^3 = (ax^\alpha)^3 = a^3x^{3\alpha} = x^{3\alpha+1} = x^{2^k\beta} = 1.$$

□

Lemma 14.7. *For d even, $SL_2(\mathbb{Z}_{2d})$ has no elements of of order 3 and trace $d-1$.*

Proof. If $A \in SL_2(\mathbb{Z}_{2d})$ has order 3, and $t = \text{trace}(A)$, then, by (14.71), we have

$$A^3 = A(tA - I) = t(tA - I) - A = (t^2-1)A - tI = I \quad \Longrightarrow \quad (t^2-1)A = (t+1)I.$$

For $t = d-1$, this gives $(d^2-2d)A = 0 = dI \pmod{2d}$, which not possible. □

We now characterise all symplectic unitaries of canonical order 3.

Theorem 14.3. *(Characterisation) The symplectic operations of canonical order 3 are conjugate in $C_{Sp}(d)/[I]$ to $[a]$, where $a \in C_{Sp}(d)$ is one of the following*

$$\begin{aligned} \{Z\}, &\quad d \not\equiv 0 \bmod 3, \\ \{Z, Z^2\}, &\quad d \equiv 0 \bmod 9 \text{ or } d = 3, \\ \{Z, Z^2, W_1\}, &\quad d \equiv 3 \bmod 9, d \neq 3, \\ \{Z, Z^2, W_2\}, &\quad d \equiv 6 \bmod 9, \end{aligned}$$

where

$$\begin{aligned} Z &:= e^{\frac{2\pi i}{24}(d-1)} RF^{-1}, \\ W_a &:= (-1)^{d-1} R^{\frac{2d}{3}a} F^{-1} R^3 FR, \end{aligned} \tag{14.81}$$

have order 3 in $C_{Sp}(d)$.

Proof. The key idea is to apply the fact that group homomorphisms map conjugacy classes to conjugacy classes to the commutative diagram (14.51) of §14.11, i.e.,

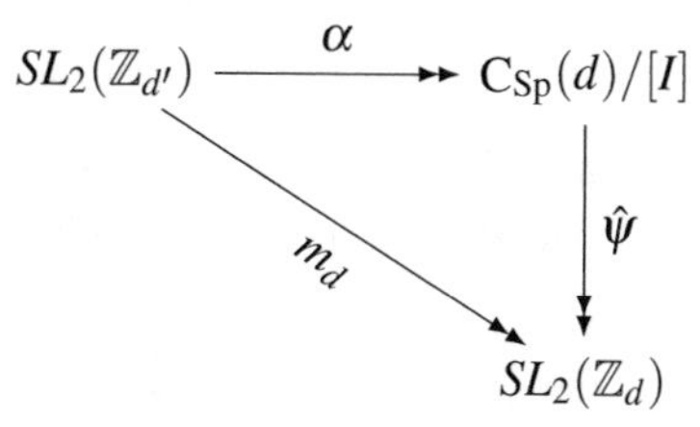

We observe that

- Since the kernel of $\hat{\Psi}$ has order 1 or 4 (d odd or even), the conjugacy classes of elements of order 3 and Clifford trace -1 in $C_{Sp}(d)/[I]$ map onto the conjugacy classes of elements of order 3 and trace -1 in $SL_2(\mathbb{Z}_d)$ (by Lemma 14.6).
- Since the kernel of α has 1 or 2 (d odd or even), each conjugacy class of an element of order 3 and Clifford trace -1 in $C_{Sp}(d)/[I]$ is the image under α of the conjugacy class of an element of order 3 in $SL_2(\mathbb{Z}_{d'})$ (by Lemma 14.6) and of trace -1 (by Lemma 14.7).

Thus the conjugacy classes of elements of order 3 and trace -1 in $SL_2(\mathbb{Z}_{d'})$ map onto the conjugacy classes of elements of canonical order 3 in $C_{Sp}(d)/[I]$, which in turn map onto the conjugacy classes of elements of order 3 and trace -1 in $SL_2(\mathbb{Z}_d)$. A count of the conjugacy classes in $SL_2(\mathbb{Z}_{d'})$ and $SL_2(\mathbb{Z}_d)$ (for d even) shows that these maps are 1–1, i.e., representatives of the conjugacy classes of elements of order 3 and trace -1 in $SL_2(\mathbb{Z}_{d'})$ give symplectic indices for representatives of the conjugacy classes of the symplectic operations of canonical order 3.

We now use Lemma 14.5 to calculate these symplectic indices (and show the injectivity asserted above) for the various cases.

For $d \not\equiv 0 \pmod 3$, we have $2d \not\equiv 0 \pmod 3$, and so there is a single conjugacy class with symplectic index z.

For $d \equiv 0 \pmod 9$, $d \neq 3$, we have $2d \equiv 0 \pmod 9$, and so there is are two conjugacy classes given by the symplectic indices z, z^2. For $d = 3$, we have $d' = d$, and there are two conjugacy classes given by the symplectic indices z, z^2.

For $d \equiv 3 \pmod 9$, $d \neq 3$, we have $2d \equiv 6 \pmod 9$, so that there are three conjugacy classes given by the symplectic indices z, z^2, and

$$\begin{pmatrix} 1 & 3 \\ \frac{d-3}{3} & -2 \end{pmatrix} \in SL_2(\mathbb{Z}_d) \quad (d \text{ odd}), \qquad \begin{pmatrix} 1 & 3 \\ \frac{2(2d)-3}{3} & -2 \end{pmatrix} \in SL_2(\mathbb{Z}_{2d}) \quad (d \text{ even}).$$

The second formula gives the first for d odd, and so works in both cases.

For $d \equiv 6 \pmod 9$, we have $2d \equiv 3 \pmod 9$, so that there are three conjugacy classes given by the symplectic indices z, z^2, and

$$\begin{pmatrix} 1 & 3 \\ \frac{2d-3}{3} & -2 \end{pmatrix} \in SL_2(\mathbb{Z}_d) \quad (d \text{ odd}), \qquad \begin{pmatrix} 1 & 3 \\ \frac{2d-3}{3} & -2 \end{pmatrix} \in SL_2(\mathbb{Z}_{2d}) \quad (d \text{ even}).$$

In the last two cases, the third conjugacy class is given by the symplectic indices m_1 and m_2 (respectively), where

$$m_j := \begin{pmatrix} 1 & 3 \\ \frac{4dj-3}{3} & -2 \end{pmatrix} \in SL_2(\mathbb{Z}_{d'}).$$

and m_j^2 is conjugate to m_j (since otherwise there would be four conjugacy classes). For convenience of presentation, we take the representative with symplectic index

$$w_j := m_j^2 = \begin{pmatrix} -2 & -3 \\ 1+\frac{2d}{3}j & 1 \end{pmatrix} = \begin{pmatrix} 1 & 0 \\ 1 & 1 \end{pmatrix}^{\frac{2d}{3}j} \begin{pmatrix} 0 & -1 \\ 1 & 0 \end{pmatrix}^{-1} \begin{pmatrix} 1 & 0 \\ 1 & 1 \end{pmatrix}^{3} \begin{pmatrix} 0 & -1 \\ 1 & 0 \end{pmatrix} \begin{pmatrix} 1 & 0 \\ 1 & 1 \end{pmatrix}.$$

By taking the symplectic operations corresponding to the representatives z, z^2, w_1, w_2 in the above conjugacy classes, i.e., Z, Z^2, W_1, W_2, we obtain representatives for the conjugacy classes of canonical order 3 symplectic operations. The normalisation of W_a in its definition (14.81) ensures that it has order 3 (see Exer. 14.14). □

From the above proof, we have:

The conjugacy classes of order 3 and trace -1 elements in $SL(\mathbb{Z}_{d'})$ are in 1–1 correspondence with the conjugacy classes of canonical order 3 symplectic operations.

The matrices W_0, W_1, W_2 given by (14.81) are defined for $d \mid 3$, and are sparse, i.e.,

$$(W_a)_{jk} = \sqrt{\frac{3}{d}}(-\sqrt{i})^{1-3d} \begin{cases} (-\mu)^{-\frac{1}{3}(j-k)^2+\frac{2d}{3}aj^2+k^2}, & j-k \equiv 0 \mod 3; \\ 0, & j-k \not\equiv 0 \mod 3. \end{cases}$$

Further W_0 is defined for all d, and is conjugate to Z via $\tau^{1-d}W_0 = (R^{-1}F)^{-1}$ $Z(R^{-1}F)$.

$$\tau^{1-d}W_0 = (R^{-1}F)^{-1}Z(R^{-1}F).$$

Example 14.13. The canonical order 3 unitary given by the symplectic index

$$(d+1)F_a = \begin{pmatrix} d+1 & 3 \\ \frac{d-3}{3} & d-2 \end{pmatrix} = \begin{pmatrix} 1 & 0 \\ d & 1 \end{pmatrix}^{-1} w_1^2 \begin{pmatrix} 1 & 0 \\ d & 1 \end{pmatrix}, \quad w_1^2 = \begin{pmatrix} 1 & 3 \\ \frac{4d-3}{3} & -2 \end{pmatrix}$$

is conjugate to $W_1^2 = W_1^{-1}$, and so is given by the order 3 symplectic unitary

$$a_{F_a} = R^d W_1^{-1} R^d = (-1)^{d-1} R^{d-1} F^{-1} R^{-3} F R^{\frac{d}{3}}.$$

All the known Weyl–Heisenberg SICs appear as eigenvectors of the canonical order 3 symplectic unitaries Z or $M_1 = a_{Fa}$ (equivalently W_1). As yet, no SIC fiducials have been found which are eigenvectors of W_2 (for $d \equiv 6 \mod 9$).

Since the symmetries appear to form a cyclic group, it is natural to look for roots of the symplectic indices for z and m_1 as the symplectic indices of extra symmetries.

Example 14.14. For $d = 8$, the matrix

$$b = \begin{pmatrix} 11 & 13 \\ 3 & 8 \end{pmatrix} = \begin{pmatrix} 6 & 11 \\ 5 & 1 \end{pmatrix}^2 \in SL_2(\mathbb{Z}_{16})$$

is a square root of $(d+1)z$. It is the symplectic index for the square root of Z given by $B = \zeta^{17} R^8 P_{13} F^{-1} R^7$, which is a symmetry of the SIC 8(c) of Table 14.4.

14.19 Conjugates of the canonical order 3 symplectic unitaries

For fiducial vectors which are eigenvectors of Z or M_1 (all those known do date) the equations characterising the SIC can be simplified (as in the Example 14.9) by conjugating Z (or M_1) by an element of $\mathrm{EC}(d)$ to obtain a monomial matrix (the conjugates of $S^j\Omega^k$ will continue to be monomial matrices).

We now use Theorem 14.3 to determine when the conjugate of Z or M_1 (or M_2 for that matter) by a symplectic operation is a monomial matrix of the form $R^\alpha P_\sigma$.

Since the permutation matrices in $C(d)$ are symplectic (see §14.12), there exists a permutation matrix $P_\sigma \in \mathrm{C}(d)$, $\sigma \in \mathbb{Z}_d^*$ of canonical order 3 if and only if

$$P_\sigma^3 = P_{\sigma^3} = I, \qquad \mathrm{tr}_\mathrm{C}(P_\sigma) = \sigma + \sigma^{-1} = -1,$$

i.e., the existence of an integer σ (for $d \neq 3$) with

$$\sigma^3 \equiv 1 \bmod d, \qquad \sigma^2 + \sigma + 1 \equiv 0 \bmod d. \tag{14.82}$$

For such a σ, we have $\begin{pmatrix} \sigma & 0 \\ \alpha & \sigma^{-1} \end{pmatrix}^3 = \begin{pmatrix} \sigma^3 & 0 \\ \alpha(1+\sigma+\sigma^2) & \sigma^{-3} \end{pmatrix} = \begin{pmatrix} 1 & 0 \\ 0 & 1 \end{pmatrix}$, so that:

If σ satisfies (14.82), then $[R^\alpha P_\sigma]$ is a canonical order 3 symplectic operation.

By the Chinese remainder theorem, it follows (see [App05]) that the condition (14.82) is equivalent to d satisfying:

(i) d has at least one prime divisor $\equiv 1 \bmod 3$.
(ii) d has no prime divisors $\equiv 2 \bmod 3$ (so that d is odd).
(iii) d is not divisible by 9.

The first few such d are

$$d = 7, 13, 19, 21, 31, 37, 39, 43, 49, 57, 61, 67, 73, 79, 91, 93, 97, \ldots$$

By Theorem 14.3, the monomial operation $[R^\alpha P_\sigma]$ is conjugate (via a symplectic operation) to one of $[Z]$, $[Z]^2$, $[M_1] = [W_1]$, $[W_2]$.

- For d not a multiple of 3 ($d \not\equiv 0 \bmod 3$), i.e.,

$$d = 7, 13, 19, 31, 37, 43, 49, 61, 67, 73, 79, 91, 97, \ldots$$

  there is single conjugacy class, and so all $[R^\alpha P_\sigma]$ are conjugate to $[Z]$.
- For d a multiple of 3, i.e.,

$$d = 21, 39, 57, 93, 111, 129, 147, 183, 201, 219, 237, \ldots$$

  we have $\frac{d}{3} \equiv 1 \bmod 3$, i.e., $d \equiv 3 \bmod 9$, and so the conjugacy classes are given by $[Z], [Z]^2, [M_1]$.

For a σ satisfying $\sigma^3 = 1$, $1+\sigma+\sigma^2$, the symplectic index calculations

$$gzg^{-1} = \begin{pmatrix} \sigma & 0 \\ 1 & \sigma^2 \end{pmatrix}, \qquad gz^2g^{-1} = \begin{pmatrix} \sigma^2 & 0 \\ -1 & \sigma \end{pmatrix}, \qquad g := \begin{pmatrix} 1 & \sigma \\ 0 & 1 \end{pmatrix}, \quad z = \begin{pmatrix} 0 & -1 \\ 1 & -1 \end{pmatrix},$$

give the following:

For any d, if σ satisfies (14.82), then

1. The monomial operation $[RP_\sigma]$ is a symplectic conjugate of $[Z]$.
2. The monomial operation $[R^{-1}P_\sigma]$ is a symplectic conjugate of $[Z^2]$.

Whenever d is a multiple of 3, i.e., $d \equiv 3 \bmod 9$, it appears (for the d listed above) that $[P_\sigma]$ is always a symplectic conjugate of $[M_1]$.

Example 14.15. For $d = 21$, no symplectic conjugate of Z is a permutation matrix, but many conjugates are monomial, e.g., the symplectic index calculation

$$\begin{pmatrix} 1 & 4 \\ 0 & 1 \end{pmatrix} \begin{pmatrix} 0 & -1 \\ 1 & -1 \end{pmatrix} \begin{pmatrix} 1 & 4 \\ 0 & 1 \end{pmatrix}^{-1} = \begin{pmatrix} 4 & 0 \\ 1 & 16 \end{pmatrix},$$

together with Table 14.3, gives

$$gZg^{-1} = \omega^7 R^{16} P_4, \qquad g := F^{-1} R^{-4} F.$$

Example 14.16. For $d = 19$, Appleby [App05] constructed an exact SIC fiducial which was an eigenvector of the order 18 antiunitary $P_{10}C$ with symplectic index

$$\begin{pmatrix} -9 & 0 \\ 0 & -2 \end{pmatrix} = \begin{pmatrix} -9 & 0 \\ 0 & 2 \end{pmatrix} J = g^{-1} \begin{pmatrix} 3 & 12 \\ 7 & 15 \end{pmatrix}^5 g, \quad g := \begin{pmatrix} 5 & 1 \\ 3 & 8 \end{pmatrix}.$$

In view of Table 14.4, this is the SIC $19e$. The exact fiducial is

$$v = b_0 e_0 + \sum_{r=1}^{18} b_1 e^{i\ell_r \theta} e_r, \quad b_0 = \sqrt{\frac{5+9\sqrt{5}}{95}}, \; b_1 = \sqrt{\frac{10-\sqrt{5}}{190}}, \; \theta = \cos^{-1}\Bigl(\sqrt{\frac{\sqrt{5}-1}{8}}\Bigr),$$

where $\ell_r = (\frac{r}{19}) \in \{-1,1\}$ is the Legendre symbol.

The extra symmetries given by $P_{10}C$ immediately imply that the fiducial vector v has the very simple structural form above. At the time of its construction, the only known exact SICs were for $d = 2,\ldots,7$ and $d = 8$ (the Hoggar lines). By using similar techniques, the exact Scott–Grassl SICs $24c$, $35j$ and $48g$ were constructed [SG10]. In subsequent constructions of exact SICs, the symmetries deduced from numerical SICs have been exploited.

14.20 The SIC field of a Weyl–Heisenberg SIC

The Weyl–Heisenberg SICs for $d \neq 3$ that have been found exactly, e.g., for $d = 4$

$$\begin{pmatrix} 8 \\ ((\sqrt{10}+\sqrt{2}-2\sqrt{5}-2)\sqrt{\sqrt{5}+1}-4)i+(\sqrt{10}+\sqrt{2})\sqrt{\sqrt{5}+1}+4\sqrt{2}-4 \\ -(8\sqrt{2}-8)i \\ -((\sqrt{10}+\sqrt{2}-2\sqrt{5}-2)\sqrt{\sqrt{5}+1}+4)i-(\sqrt{10}+\sqrt{2})\sqrt{\sqrt{5}+1}+4\sqrt{2}-4 \end{pmatrix} \tag{14.83}$$

have the following features:

- They are very complicated to present (in general), e.g., the three exact fiducial vectors for $d = 11$ given in [SG10] take up 22 pages.
- They are expressible by radicals (nested roots).

Since the equations defining a SIC involve quartic polynomials in *several* variables (see §14.27), it does not follow (from the fact a univariate quadratic can be solved by radicals) that the components of a SIC should be expressible by radicals.

We now consider the field defined by a Weyl–Heisenberg SIC. We will see that some of its Galois automorphisms map SICs to SICs. This leads to methods for the construction of exact SICs from numerical SICs. The presentation (to follow) assumes a basic familiarity with Galois theory.

In theory, the natural field in which to define a SIC is that generated by the triple products of its vectors (see §8.6). Since the Weyl–Heisenberg SICs are given by the orbit of a fiducial projector $\Pi = vv^*$ under the action of the Heisenberg group, and the Clifford group maps SIC fiducials to SIC fiducials (see §14.7), it is convenient and effective to (possibly) enlarge[8] the field to contain the entries of Π and μ.

Definition 14.5. The **SIC field** $\mathbb{E}$ of a Weyl–Heisenberg SIC with fiducial $\Pi = vv^*$ is the smallest extension of $\mathbb{Q}$ containing the entries of Π and $\mu = e^{\frac{\pi i}{d}}$.

The inclusion of μ ensures (see Exercises 14.22, 14.23) that:

The SIC field $\mathbb{E} = \mathbb{Q}(\Pi, \mu)$ depends only on the extended Clifford orbit of Π.

Since $\overline{\Pi} = (\Pi^*)^T = \Pi^T$ and $\overline{\mu} = \mu^{-1}$, it follows

The SIC field is closed under complex conjugation (denoted by g_c).

We have the (inclusion reversing) Galois correspondence between subfields of $\mathbb{E}$ and subgroups of the Galois group $\mathscr{G}$ (of automorphisms of the field $\mathbb{E}$ which fix $\mathbb{Q}$).

[8] In earlier work [AYAZ13], the SIC field $\mathbb{E}$ was defined to be the smallest normal extension of $\mathbb{Q}$ containing the entries of Π, $\sqrt{d}$, and μ. Here we use the more recent definition of [AFMY17].

14.21 The Galois group of a generic SIC

For $d > 1$, a SIC is said to be **generic** if it is a Weyl–Heisenberg SIC for $d \geq 4$, otherwise it is **sporadic** SIC. The sporadic SICs have special properties not shared by the generic SICs (see [Zhu15], [Sta17], [AFMY17], Exer. 14.21), e.g.,

The symmetry groups of the known sporadic SICs, i.e., the Hoggar lines and the $d = 2, 3$ Weyl–Heisenberg SICs are nonabelian (and doubly transitive on the SIC fiducial projectors [Zhu15]), whereas the known generic SICs have cyclic symmetry groups.

The generic SICs seem to have many special properties (in addition the Clifford group action), that were observed in the early constructions. These were formalised as a set of *conjectures* about the SIC field $\mathbb{E}$ (and the action of its Galois group) in [AYAZ13], and a refined version of these *facts* is given in [AFMY17]. We now outline these **facts** (conjectures) about the generic SICs. From now on, we assume that SICs are generic, and the **known** SICs are those reported in [ACFW17].

Fact 1. In every known case, the SIC field $\mathbb{E}$ is normal over $\mathbb{Q}$.

We recall $\mathbb{E}$ is *normal* over $\mathbb{Q}$ means that every irreducible polynomial over $\mathbb{Q}$ which has a root in $\mathbb{E}$ splits over $\mathbb{E}$ (i.e., all its roots are in $\mathbb{E}$).

Fact 2. In every known case, $\mathbb{E}$ is an extension of $\mathbb{K} := \mathbb{Q}(\sqrt{(d-3)(d+1)})$. Here $\sqrt{(d-3)(d+1)}$ is never an integer.

Fact 3. In every known case, $\mathrm{Gal}(\mathbb{E}/\mathbb{K})$ is a finite abelian group.

In particular, the Galois group of $\mathbb{E}$ over $\mathbb{Q}$ is solvable, and so a generic SIC is expressible by radicals.

We recall the *Kronecker–Weber theorem*, that the finite abelian extensions of $\mathbb{Q}$ are subfields of some cyclotomic field $\mathbb{Q}(e^{2\pi i/n})$. Finding a similar characterisation for the finite abelian extensions of a quadratic field (such as the extension $\mathbb{E}$ of $\mathbb{K}$ above) is an instance of *Hilbert's 12-th problem*. This was solved for imaginary quadratic fields $\mathbb{Q}(i\sqrt{n})$, n a positive integer, where such extensions are a subfield of a field generated by the torsion points of certain elliptic curves. For abelian extensions of $\mathbb{Q}(\sqrt{n})$, such as the conjectured $\mathbb{E}$, there is currently *no such construction*.

Even if the SIC field $\mathbb{E}$ is known (there are conjectures that it is a ray class field over $\mathbb{K}$), then it is not immediately obvious how to go from a Scott–Grassl numerical SIC (of high precision) to an exact SIC with entries in $\mathbb{E}$ (since $\mathbb{E}$ is dense in $\mathbb{C}$). This is part of the intrigue of the SIC problem, i.e., given the high precision numerical SICs, it is tempting to imagine that exact SICs cannot be too far away.

14.22 The Galois action on a fiducial projector

The extended Clifford group permutes the SIC fiducials via the action (Exer. 14.22)

$$a \cdot \Pi := (av)(av)^* = a\Pi a^{-1}, \quad \Pi = vv^*, \qquad a \in \mathrm{EC}(d).$$

There is a natural action of the Galois group $\mathscr{G} = \mathrm{Gal}(\mathbb{E}/\mathbb{Q})$ of the SIC field $\mathbb{E}$ on matrices over $\mathbb{E}$ given by

$$g(A) = g([a_{jk}]) := [g(a_{jk})], \qquad g \in \mathscr{G}, \quad A \in \mathbb{E}^{m \times n}.$$

One might hope that this Galois action also maps SIC fiducials to SIC fiducials. Indeed we have already seen (§14.7) that this is the case for complex conjugation, which we now denote by $g_c \in \mathscr{G}$. We will show that this is true for a large (index 2) subgroup $\mathscr{G}_c$ (consisting of the elements which commute with g_c).

If $g \in \mathscr{G} = \mathrm{Gal}(\mathbb{E}/\mathbb{Q})$ and $\Pi = [a_{jk}]$ is a fiducial projector, then a necessary condition for $g(\Pi)$ to be a fiducial projector is $g(\Pi)^* = g(\Pi) = g(\Pi^*)$, i.e.,

$$\overline{g(a_{kj})} = g(\overline{a_{kj}}) \quad \Longleftrightarrow \quad g_c g(a_{kj}) = g g_c(a_{kj}).$$

Since the automorphism g maps μ to another $2d$–th root of unity, this implies that $g \in \mathscr{G}$ must commute with complex conjugation $g_c \in \mathscr{G}$ (see Exer. 14.20).

For $g \in \mathscr{G} = \mathrm{Gal}(\mathbb{E}/\mathbb{Q})$ and Π a SIC fiducial, $g(\Pi)$ is a rank one orthogonal projection if and only if g is in the centraliser of complex conjugation $g_c \in \mathscr{G}$.

Let $\mathscr{G}_c$ be the centraliser of g_c in $\mathscr{G}$. We now show that $\mathscr{G}_c$ maps SIC fiducials to SIC fiducials. Let $\Pi = vv^*$ be a rank one orthogonal projection. Since $(S^j\Omega^k)_{j,k\in\mathbb{Z}_d}$ is an orthogonal (nice error) basis, we have

$$\Pi = \frac{1}{d} \sum_{j,k\in\mathbb{Z}_d} \langle \Pi, S^j\Omega^k \rangle S^j\Omega^k,$$

i.e., Π is determined by its *overlaps* (or scalar multiples of them)

$$\langle \Pi, S^j\Omega^k \rangle = \mathrm{trace}(vv^*(S^j\Omega^k)^*) = \langle v, S^j\Omega^k v \rangle \qquad (j,k) \neq (0,0),$$

and Π is a SIC fiducial if and only if these have constant modulus. We define the **overlaps** $\chi_p^\Pi := \mathrm{trace}(\Pi \hat{D}_p)$, $p \in \mathbb{Z}_{d'}^2$, where the $\hat{D}_p$ are the displacement operators of (14.37). We note that Π is a SIC fiducial if and only if its overlaps satisfy

$$\chi_p^\Pi := \mathrm{trace}(\Pi \hat{D}_p) = \begin{cases} 1, & p \equiv 0 \bmod d; \\ \frac{e^{i\theta_p}}{\sqrt{d}}, & p \not\equiv 0 \bmod d, \end{cases} \qquad p \in \mathbb{Z}_{d'}. \tag{14.84}$$

We refer to the θ_p as the **overlap phases** of the SIC fiducial Π.

Let $g \in \mathcal{G} = \mathrm{Gal}(\mathbb{E}/\mathbb{Q})$. Since $(-\mu)$ is a primitive d'–th root of unity, there is a unique integer $0 \le k_g < d'$, with

$$g(-\mu) = (-\mu)^{k_g}, \qquad k_g \in \mathbb{Z}_{d'}^*. \tag{14.85}$$

so that

$$g(\hat{D}_p) = g((-\mu)^{p_1 p_2} S^{p_1} \Omega^{p_2}) = (-\mu)^{k_g p_1 p_2} S^{p_1} \Omega^{k_g p_2} = \hat{D}_{H_g p}, \tag{14.86}$$

where

$$H_g := \begin{pmatrix} 1 & 0 \\ 0 & k_g \end{pmatrix} \in GL_2(\mathbb{Z}_{d'}), \qquad g \in \mathcal{G}. \tag{14.87}$$

Lemma 14.8. *Let $g \in \mathcal{G}_c$. If $\Pi = vv^*$ is a fiducial projector with SIC field $\mathbb{E}$, then so is $g(\Pi)$, i.e., $\mathcal{G}_c$ maps SICs to SICs (possibly on different extended Clifford orbits).*

Proof. Since $g \in \mathcal{G}_c$, it follows $g(\Pi)$ is a rank one orthogonal projection. By (14.86),

$$g(\chi_p^{\Pi}) = g(\mathrm{trace}(\Pi \hat{D}_p)) = \mathrm{trace}(g(\Pi)\hat{D}_{H_g p}) = \chi_{H_g p}^{g(\Pi)}, \qquad \forall p \in \mathbb{Z}_{d'}^2.$$

Since $g \in \mathcal{G}_c$, i.e., it commutes with conjugation, we have $g(|z|^2) = |g(z)|^2$, $z \in \mathbb{E}$, and so

$$|\chi_p^{g(\Pi)}|^2 = |g(\chi_{H_g^{-1}p}^{\Pi})|^2 = |\chi_{H_g^{-1}p}^{\Pi}|^2.$$

In view of (14.84), we conclude that $g(\Pi)$ gives a SIC, with SIC field $\mathbb{E}$. □

The subgroup $\mathcal{G}_c$ of $\mathrm{Gal}(\mathbb{E}/\mathbb{Q})$ maps SICs to SICs (with the same SIC field), with the Galois action on the overlaps given by

$$g(\chi_p^{\Pi}) = \chi_{H_g p}^{g(\Pi)}, \qquad \forall p \in \mathbb{Z}_{d'}^2. \tag{14.88}$$

It appears that $\mathcal{G}_c$ is a large (index 2) subgroup of $\mathcal{G}$.

Fact 4. In every known case,

$$\mathrm{Gal}(\mathbb{E}/\mathbb{K}) = \mathcal{G}_c = \text{the centraliser of complex conjugation.}$$

By (14.88), the Galois action of $\mathcal{G}_c$ maps SICs to SICs with the *same* SIC field, but on *possibly different* extended Clifford orbits. We will call the extended Clifford orbits with the same SIC field a **Galois multiplet**, with the number of orbits being its **size**, e.g., 4a is a Galois singlet, 9ab is a Galois doublet, 30abc is a Galois triplet, and 21abcd is a Galois quartet. It appears that orbit of a SIC fiducial under the combined Galois and Clifford actions is (all of) a multiplet.

14.23 Constructing exact SICs from numerical SICs

Before giving more detail about the Galois action on SICs, we explain why it has been so pivotal in the construction of exact SICs from numerical SICs.

Suppose now, for a moment, that

- The $\mathscr{G}_c$–orbits $\{\mathscr{O}_j\}$ of the overlaps of a SIC fiducial are known.
- $\mathbb{K} = \mathbb{Q}(\sqrt{(d-3)(d+1)})$ is fixed by $\mathscr{G}_c$ (Fact 4).
- $\mathscr{G}_c$ is abelian (Facts 3 and 4).

Then the $\mathscr{G}_c$–invariant polynomials

$$f_j(x) := \prod_{\chi \in \mathscr{O}_j} (x - \chi),$$

would have coefficients in the field $\mathbb{K}$. By using the Scott–Grassl numerical SICs, these coefficients could be calculated to a high degree of accuracy. If it were then possible to "guess" these coefficients in $\mathbb{K}$ exactly, then one could find the overlaps exactly by factoring the exact f_j ($\mathscr{G}_c$ is solvable), and so convert a numerical SIC into an exact SIC. This general method is called **precision bumping** (also see §9.8), and has been successful in constructing many exact SICs (see [Chi15], [ACFW17]), by using the PSLQ algorithm or LLL (Lenstra–Lenstra–Lovász) algorithm to "guess" elements of $\mathbb{K}$ (or some small extension of it, such as the $\mathbb{E}_0$ of §14.24).

14.24 The Galois action on a centred fiducial

The precision bumping algorithm of §14.23 for constructing exact SIC fiducials supposes that one know the $\mathscr{G}_c$–orbit of the overlaps. At this point, only the action of the subgroup $\mathscr{G}_0$ which maps a SIC to one on the same extended Clifford orbit is understood. We now summarise these results (see [AYAZ13], [AFMY17]).

Let $\mathscr{G}_0$ be the group which fixes the extended Clifford orbit of a fiducial Π, i.e.,

$$\mathscr{G}_0 = \mathscr{G}_0^\Pi := \{g \in \mathscr{G}_c : g(\Pi) \text{is on the same extended Clifford orbit as} \Pi\},$$

and $\mathbb{E}_0 = \mathbb{E}_0^\Pi$ be the fixed field of $\mathscr{G}_0$. Clearly, $\mathscr{G}_0$ is a subgroup of $\mathscr{G}_c$, which depends only on the extended Clifford orbit of Π (or it seems the multiplet).

Fact 5. In every known case, $\mathscr{G}_0 = \mathscr{G}_0^\Pi$ depends only on the multiplet of Π, and the index $[\mathbb{E}_0 : \mathbb{K}]$ is the size of the multiplet.

The group $S(v)$ of the symmetries (14.65) of a SIC fiducial vector v, are the stabiliser of the projector $\Pi = vv^*$ under the Clifford action (see Exer. 14.22), since

$$S(v) = \{[a] \in \mathrm{PEC}(d) : (av)(av)^* = vv^*\} = \{[a] \in \mathrm{PEC}(d) : a \cdot \Pi = \Pi\} =: S_\Pi.$$

By a simple calculation,

$$S_{[a]\cdot\Pi} = [a]S_\Pi[a]^{-1},$$

Hence it is natural to seek a fiducial $[a]\cdot\Pi$ on the extended Clifford orbit of Π for which the symmetries $[a]S_\Pi[a]^{-1}$ have a simple form. Now (see Exer. 14.24)

The Clifford action of an $[a] \in \mathrm{PEC}(d)$, with extended Appleby index $[B,b]$, on the overlaps is given by

$$\chi_p^\Pi = \omega^{\langle\!\langle b,\det(B)Bp\rangle\!\rangle}\chi_{\det(B)Bp}^{[a]\cdot\Pi}, \qquad \forall p \in \mathbb{Z}_{d'}^2. \tag{14.89}$$

In particular, if $[a]$ is symplectic, i.e., $b=0$, then the overlaps of Π and $a\Pi a^{-1}$ are a permutation of each other. We say that a fiducial Π is **centred** if its symmetries S_Π are (extended) symplectic operations. In particular, all the Scott-Grassl SICs are centred (see Table 14.4). In every known case, the symmetry group S_Π of a SIC fiducial can always be conjugated to be symplectic, i.e.,

On every extended Clifford orbit there is a *centred* SIC fiducial.

For a given centred SIC fiducial Π, we denote by $S_0(\Pi)$ the group of (extended) symplectic indices for S_Π, i.e., with f_E given by (14.57),

$$S_0(\Pi) := \{B \in ESL_2(\mathbb{Z}_{d'}) : \alpha_E(B) \in S_\Pi\}, \qquad \alpha_E(B) := f_E([B,0]).$$

Since $[a]\cdot\Pi = \Pi$ for $[a] \in S_\Pi$, it follows from (14.89) that

If Π is a centred SIC fiducial, then

$$B \in S_0(\Pi) \iff \chi_p^\Pi = \chi_{\det(B)Bp}^\Pi.$$

This clearly reduces the number of overlaps that must be found to construct a SIC.

We now consider a further refinement of the SIC field $\mathbb{E}$. Let D be the square-free part of $(d-3)(d+1)$, so that $\mathbb{K} := \mathbb{Q}(\sqrt{(d-3)(d+1)}) = \mathbb{Q}(\sqrt{D})$.

Let $g \in \mathrm{Gal}(\mathbb{E}/\mathbb{Q})$ be any element with $g(\sqrt{D}) = -\sqrt{D}$ (these exist, by the Galois correspondence). Then $g_1 := gg_cg^{-1}$ is an element of order 2, which fixes $\mathbb{K}$, since

$$g_1(\sqrt{D}) = g(\overline{-\sqrt{D}}) = g(-\sqrt{D}) = \sqrt{D},$$

and is independent of the choice of g, since

$$gg_cg^{-1} = \tilde{g}g_c\tilde{g}^{-1} \iff g^{-1}\tilde{g} \in \mathscr{G}_c = \mathrm{Gal}(\mathbb{E}/\mathbb{K}) \iff g^{-1}\tilde{g}\,\text{fixes}\,\mathbb{K},$$

with the last equivalence by: $g^{-1}\tilde{g}(\sqrt{D}) = g^{-1}(-\sqrt{D}) = \sqrt{D}$. In every known case, $g_1 \in \mathscr{G}_0^\Pi$, i.e., g_1 maps fiducials to fiducials on the same extended Clifford orbit.

From (14.85), we have $g^{-1}(-\mu)=(-\mu)^{k_g^{-1}}$, so that

$$g_1(-\mu)=gg_cg^{-1}(-\mu)=((-\mu)^{-k_g^{-1}})^{k_g}=(-\mu)^{-1} \quad\Longrightarrow\quad H_{g_1}=J:=\begin{pmatrix}1&\\&-1\end{pmatrix}.$$

Let $\mathbb{E}_1$ be the fixed field of the order 2 subgroup $\mathscr{G}_1=\langle g_1\rangle$ of $\mathscr{G}_c$. By the Galois correspondence, $\mathbb{Q}\subset\mathbb{K}\subset\mathbb{E}_0\subset\mathbb{E}_1\subset\mathbb{E}$, with the index $[\mathbb{E}:\mathbb{E}_1]=2$.

Fact 6. In every known case, $\mathbb{E}=\mathbb{E}_1(i\sqrt{d'})$.

We now seek to understand the extension of $\mathbb{E}_0$ to $\mathbb{E}_1$.

Fact 7. On every known extended Clifford orbit, there is a fiducial Π with its overlaps χ_p^Π in $\mathbb{E}_1$, i.e., by (14.88),

$$g_1(\chi_p^\Pi)=\chi_{Jp}^{g_1(\Pi)}=\chi_p^\Pi, \qquad \forall p\in\mathbb{Z}_{d'}. \tag{14.90}$$

A centred SIC fiducial satisfying (14.90) is said to be **strongly centred**. For $d\not\equiv 0 \bmod 3$, every centred fiducial is strongly centred (see Exer. 14.27). Whilst for $d\equiv 0 \bmod 3$, some, but not all, centred fiducials are strongly centred. Hence

On every extended Clifford orbit there is a *strongly centred* SIC fiducial.

We now assume that the SIC fiducial Π is *strongly centred*. Let $\mathscr{O}_1,\ldots,\mathscr{O}_\ell$ be the extended Clifford orbits of the multiplet for $\Pi=\Pi_1$, and Π_j be a strongly centred fiducial on $\mathscr{O}_j$. Then for every $g\in\mathscr{G}_c$, there is a symplectic operation $a=a_B\in \mathrm{EC}(d)$, with symplectic index $B=F_{g,j}\in ESL_2(\mathbb{Z}_{d'})$, for which

$$g(\Pi_j)=[a]\cdot\Pi_k=a\Pi_k a^{-1}, \qquad [a]=f_E([F_{g,j},0]).$$

The fact that an $a\in\mathrm{EC}(d)$ with $g(\Pi_j)=[a]\cdot\Pi_k$ can be chosen to be symplectic is a consequence of Π_j and Π_k being strongly centred (see Exer. 14.27).

By (14.88) and (14.89), we calculate

$$g(\chi_p^{\Pi_j})=\chi_{H_gp}^{g(\Pi_j)}=\chi_{H_gp}^{[a]\cdot\Pi_k}=\chi_{\det(B)B^{-1}H_gp}^{\Pi_k}=\chi_{G_{g,j}p}^{\Pi_k}, \qquad G_{g,j}:=\det(F_{g,j})F_{g,j}^{-1}H_g. \tag{14.91}$$

For $g\in\mathscr{G}_0$, we have $j=k$, and, in particular, for g_1, we have (see Exer. 14.27)

$$F_{g_1,j}=-J, \qquad G_{g_1,j}=\det(-J)(-J)^{-1}J=I.$$

For a given strongly centred $\Pi=\Pi_j$, the matrix $F_{g,j}\in ESL_2(\mathbb{Z}_{d'})$ is not unique, as it may be replaced by any element of the left coset $F_{g,j}S_0(\Pi)$. Correspondingly, the matrix $G_{g,j}\in GL_2(\mathbb{Z}_{d'})$ may be replaced by any element of the form

$$\det(F_{g,j}A^{-1})(F_{g,j}A^{-1})^{-1}H_g = \det(A^{-1})A\det(F_{g,j})F_{g,j}^{-1}H_g, \qquad A \in S_0(\Pi).$$

Since $\det(A^{-1}) = \det(A)$, we may write this set as the right coset $S(\Pi)G_{g,j}$, for the subgroup

$$S(\Pi) := \{\det(A)A : A \in S_0(\Pi)\}.$$

We now investigate whether the mapping of g to these cosets is a homomorphism.

In view of the homomorphism α_E, S_Π is a homomorphic image of $S_0(\Pi)$, i.e.,

$$S_\Pi \cong S_0(\Pi) \quad (d\,\text{odd}), \qquad S_\Pi \cong S_0(\Pi)/\langle dI\rangle \quad (d\,\text{even}).$$

It is easy to see that the map

$$\Theta : GL_2(\mathbb{Z}_{d'}) \to GL_2(\mathbb{Z}_{d'}) : A \mapsto \det(A)A \tag{14.92}$$

is a homomorphism, which maps $S_0(\Pi)$ onto $S(\Pi)$, and has kernel

$$\ker\Theta = \{cI : c^3 = 1, c \in \mathbb{Z}_{d'}^*\}.$$

If $cI \in S_0(\Pi)$ has order 3, then from $\det(cI) = c^2 = \pm 1$, we have that $c = 1$, i.e.,

The groups $S_0(\Pi)$ and $S(\Pi)$ are isomorphic (via θ).

Let $N(S_0(\Pi))$ and $N(S(\Pi))$ be the normaliser of $S_0(\Pi)$ and $S(\Pi)$ in $GL_2(\mathbb{Z}_{d'})$. If $gMg^{-1} = L$, then $\det(M) = \det(gMg^{-1}) = \det(L)$, and so

$$gMg^{-1} = L \quad \Longleftrightarrow \quad g(\det(M)M)g^{-1} = \det(L)L.$$

Hence $S_0(\Pi)$ and $S(\Pi)$ have the same normaliser in $GL_2(\mathbb{Z}_{d'})$, which we denote by

$$N(\Pi) = N(S_0(\Pi)) = N(S(\Pi)).$$

Since $g(\Pi) = [a]\cdot\Pi$, $a = a_{F_{g,j}}$, we have (see Exer. 14.27)

$$S_{g(\Pi)} = S_{[a]\cdot\Pi} \quad \Longrightarrow \quad g(S_\Pi) = [a]\cdot S_\Pi = aS_\Pi a^{-1}.$$

Hence, for $[a_L] = \alpha_E(L) \in S_\Pi$ there is $[a_M] = \alpha_E(M) \in S_\Pi$ with (see Exer. 14.26)

$$g([a_L]) = [a_{H_g L H_g^{-1}}] = [a a_M a^{-1}] = [a_{F_{g,j}} a_M a_{F_{g,j}^{-1}}],$$

i.e., $H_g L H_g^{-1} = F_{g,j} M F_{g,j}^{-1}$, up to equivalence of symplectic indices, and we have

$$H_g S_0(\Pi) H_g^{-1} = F_{g,j} S_0(\Pi) F_{g,j}^{-1}. \tag{14.93}$$

We observe that $g \mapsto k_g$ and $g \mapsto H_g$ of (14.85) and (14.86) are homomorphisms.

Lemma 14.9. *Let Π be a strongly centred fiducial. Then for a fixed j, the maps*

$$\mathscr{G}_0 \to \frac{N(\Pi)}{S_0(\Pi)} : g \mapsto F_{g,j}^{-1} H_g S_0(\Pi), \tag{14.94}$$

$$\mathscr{G}_0 \to \frac{N(\Pi)}{S(\Pi)} : g \mapsto G_{g,j} S(\Pi), \quad G_{g,j} := \det(F_{g,j}) F_{g,j}^{-1} H_g, \tag{14.95}$$

are homomorphisms, which have kernels $\langle g_c \rangle$ and $\langle g_1 \rangle$, respectively.

Proof. Let $F_g = F_{g,j}$, $G_g = G_{g,j}$. By (14.93), $F_g^{-1} H_g S_0(\Pi)(F_g^{-1} H_g)^{-1} = S_0(\Pi)$, so that $F_g^{-1} H_g \in N(S_0(\Pi))$, and the first map is well defined. With a_B chosen in $\mathbb{E}^{d\times d}$, we calculate

$$g_1 g_2(\Pi) = a_{F_{g_1 g_2}} \Pi a_{F_{g_1 g_2}}^{-1},$$

$$g_1 g_2(\Pi) = g_1(a_{F_{g_2}} \Pi a_{F_{g_2}}^{-1}) = a_{H_{g_1} F_{g_2} H_{g_1}^{-1}} a_{F_{g_1}} \Pi a_{F_{g_1}}^{-1} a_{H_{g_1} F_{g_2} H_{g_1}^{-1}}^{-1}.$$

By equating, rearranging, and taking symplectic indices, we obtain

$$F_{g_1 g_2}^{-1} H_{g_1 g_2} (F_{g_2}^{-1} H_{g_2})^{-1} (F_{g_1}^{-1} H_{g_1})^{-1} = F_{g_1 g_2}^{-1} H_{g_1} F_{g_2} H_{g_1}^{-1} F_{g_1} \in S_0(\Pi), \tag{14.96}$$

i.e., the first map is a group homomorphism. Above we used $H_{g_1 g_2} = H_{g_1} H_{g_2}$.

For the second map, we apply the homomorphism Θ of (14.92), for which $\Theta(S_0(\Pi)) = S(\Pi)$. It is well defined, since $\det(H_g)\det(H_g^{-1}) = 1$, gives

$$\det(F_g^{-1}) F_g^{-1} H_g S(\Pi) (\det(F_g^{-1}) F_g^{-1} H_g)^{-1} = S(\Pi).$$

Applying Θ to (14.96), and using $\det(H_{g_1 g_2}) = \det(H_{g_1})\det(H_{g_2})$ to simplify, gives $G_{g_1 g_2} G_{g_2}^{-1} G_{g_1}^{-1} \in S(\Pi)$, and so it is a homomorphism.

For the kernel of the first homomorphism, we have

$$F_g^{-1} H_g \in S_0(\Pi) \quad \Longrightarrow \quad \det(F_g^{-1})\det(H_g) = \pm 1 \quad \Longrightarrow \quad k_g = \det(H_g) = \pm 1.$$

For $k_g = 1$, we have g fixes μ and $H_g = I$, so that

$$F_g \in S_0(\Pi) \quad \Longrightarrow \quad g(\Pi) = \Pi \quad \Longrightarrow \quad g \text{ fixes } \mathbb{E} = \mathbb{Q}(\Pi, \mu) \quad \Longrightarrow \quad g = I.$$

For $k_g = -1$, we have $g(\mu) = \mu^{-1} = g_c(\mu)$ and $H_g = J$, so that

$$F_g \in J S_0(\Pi) \quad \Longrightarrow \quad g(\Pi) = [C] \cdot \Pi = g_c(\Pi) \quad \Longrightarrow \quad g = g_c \text{ on } \mathbb{E} \quad \Longrightarrow \quad g = g_c.$$

For the kernel of the second, we have a similar argument. As before

$$\det(F_g) F_g^{-1} H_g \in S(\Pi) \quad \Longrightarrow \quad k_g = \det(H_g) = \pm 1.$$

For $k_g = 1$, we have $\det(F_g^{-1}) F_g^{-1} \in S(\Pi)$ so that $F_g \in S_0(\Pi)$ and $g = I$. For $k_g = -1$, we have

$$\det(F_g^{-1}) F_g^{-1} J = \det(-J F_g^{-1})(-J F_g^{-1}) \in S(\Pi) \quad \Longrightarrow \quad -J F_g \in S_0(\Pi),$$

so that $F_g \in -JS_0(\Pi)$. Hence

$$g(\Pi) = [a_{-J}] \cdot \Pi = g_1(\Pi), \qquad g(\mu) = \mu^{-1} = g_1(\mu) \quad \Longrightarrow \quad g = g_1. \qquad \square$$

By (14.91), the action of $g \in \mathscr{G}_0$ on the overlaps of a strongly centred fiducial $\Pi = \Pi_j$ is given by

$$g(\chi_p^\Pi) = \chi_{G_g p}^\Pi, \qquad G_g = G_{g,j}. \tag{14.97}$$

Thus the second homomorphism (14.95) provides an explicit way to compute the $\mathscr{G}_0$–orbits of the overlaps, given that the image of $\mathscr{G}_0$ in $N(\Pi)/S(\Pi)$ is known. This makes the precision bumping method of §14.23 feasible.

Factoring out (14.95) by its kernel $\mathscr{G}_1 := \langle g_1 \rangle$ gives an injective homomorphism

$$\mathscr{G}_0/\mathscr{G}_1 = \mathrm{Gal}(\mathbb{E}_1/\mathbb{E}_0) \to \frac{N(\Pi)}{S(\Pi)} : g\langle g_1 \rangle \mapsto G_g S(\Pi). \tag{14.98}$$

In view of Fact 3, the image of this map must be an *abelian* group.

Key fact. In every known case, the map $g \mapsto G_g S(\Pi)$ of (14.98) defines an isomorphism

$$\mathrm{Gal}(\mathbb{E}_1/\mathbb{E}_0) \cong M(\Pi)/S(\Pi), \tag{14.99}$$

where $M(\Pi)$ is a maximal abelian subgroup of $GL_2(\mathbb{Z}_{d'})$ containing $S(\Pi)$, and g permutes the overlaps via $g(\chi_p^\Pi) = \chi_{G_g p}^\Pi$, $\forall p$.

Let $C(X)$ denote the centraliser in $GL_2(\mathbb{Z}_{d'})$ of a set X of matrices.

Example 14.17. (Type-z orbits) For a strongly centred fiducial Π which is given by an eigenvector of the Zauner matrix (or a conjugate of it) with symplectic index B, a calculation (see Exer. 14.25) shows the centraliser of B in $GL_2(\mathbb{Z}_{d'})$ is abelian, so that

$$\mathrm{Gal}(\mathbb{E}_1/\mathbb{E}_0) \cong C(\Pi)/S(\Pi), \qquad C(\Pi) := C(S(\Pi)) = C(B).$$

This is the original conjecture of [AYAZ13].

Example 14.18. (Type-a orbits) For a strongly centred fiducial Π which given by an eigenvector of the canonical order 3 operation with symplectic index F_a, the centraliser of F_a in $GL_2(\mathbb{Z}_{d'})$ is not abelian, and there are three maximal abelian subgroups containing $S(\Pi)$ (see [ACFW17]). In this case, the isomorphism (14.99) holds for an appropriate choice of $M(\Pi)$ in the known cases: 12b, 21e, 48g.

14.25 Minimal and maximal SIC fields

The facts and their consequences for the SIC field of §14.24 can be summarised:

The SIC field $\mathbb{E}$ of a generic strongly centred fiducial Π which is an n-let (multiplet of size n) is conjectured to have the Galois correspondence

$$\begin{array}{ccccccccc} 1 & \lhd & \mathscr{G}_1 & \lhd & \mathscr{G}_0 & \lhd & \mathscr{G}_c & \lhd & \mathscr{G} \\ \mathbb{E} & \rhd & \mathbb{E}_1 & \rhd & \mathbb{E}_0 & \rhd & \mathbb{K} & \rhd & \mathbb{Q} \\ & 2 & & \left|\frac{M(\Pi)}{S(\Pi)}\right| & & n & & 2 & \end{array} \tag{14.100}$$

where $\mathscr{G}_c$ is abelian, $\mathbb{K} = \mathbb{Q}(\sqrt{(d-3)(d+1)})$, and all of the subgroups and subfields shown are normal (with degree of the field extension given below).

Under these assumptions the *precision bumping* algorithm outlined in §14.20 proceeds as follows (see [ACFW17]). Let $\mathscr{G}_0/\mathscr{G}_1 \cong M(\Pi)/S(\Pi)$ act on the overlaps of a centred fiducial via (14.97) to obtain orbits $\{\mathscr{O}_j\}$. If $d \equiv 0 \bmod 3$, then Π may not be strongly centred, and it is convenient to replace the overlaps by their third powers (which are permuted by $\mathscr{G}_0/\mathscr{G}_1$). If Π is an n-let, then the polynomials

$$f_j(x) := \prod_{\chi \in \mathscr{O}_j} (x - \chi), \tag{14.101}$$

have coefficients in the field $\mathbb{E}_0$ (a degree n extension of $\mathbb{K}$). For a singlet ($n = 1$), i.e., $\mathscr{G}_c = \mathscr{G}_0$, the algorithm proceeds as already outlined. For an n-let, the $\mathscr{G}_c$-orbit of an overlap will be the union of n of the $\mathscr{G}_0/\mathscr{G}_1$-orbits of overlaps from centred fiducials lying on different extended Clifford orbits. Either the $\mathscr{G}_c$-orbit can be guessed in this way (e.g., if $n = 2$ and $g_c \notin \mathscr{G}_0$ then one could conjugate the Clifford orbit), or the polynomials f_j can be dealt with directly. In practice, so far, it has been possible to "guess" the next to leading coefficient of f_j exactly (since $\mathbb{E}_0$ is a small degree extension of $\mathbb{K}$), thereby determining $\mathbb{E}_0$, and applying the algorithm as before.

For a given SIC multiplet, there is a SIC field $\mathbb{E}$ (the **SIC field of the multiplet**). For different multiplets, the SIC field (and the Galois group $\mathscr{G}$) might be different. Calculations of [ACFW17] (for fixed d) suggest that

- Each SIC multiplet has a different SIC field.
- There is *minimal* SIC field $\mathbb{E}_{\min}$, which is contained in all SIC fields.
- There is *maximal* SIC field $\mathbb{E}_{\max}$, which contains in all SIC fields.

A SIC field which is not minimal or maximal is called an **intermediate SIC field**. We say that a multiplet (or any SIC in it) is **minimal**, **intermediate** or **maximal** if its SIC field is. See Figure 14.1 and Table 14.5 for the a summary of the SIC fields obtained in [ACFW17].

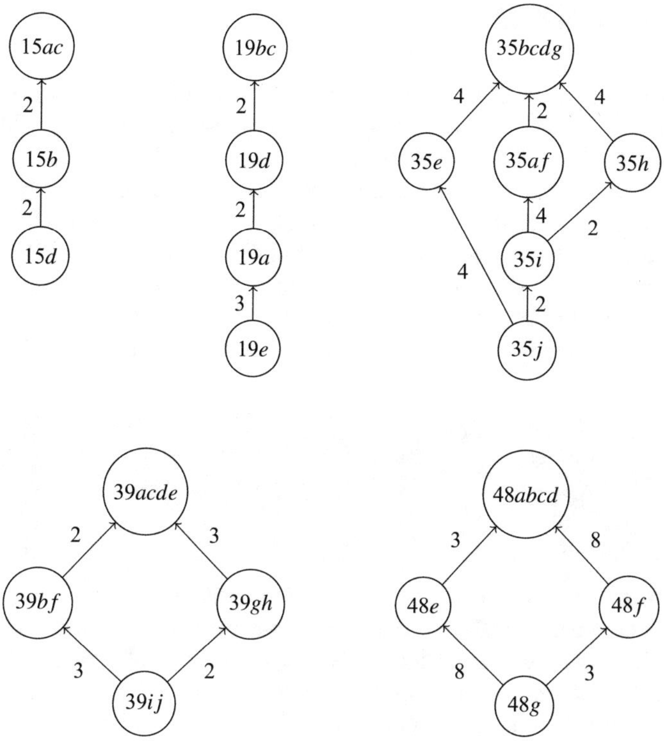

Fig. 14.1: The SIC field inclusions (with $\mathbb{E}_{\min}$ at the bottom) for cases where there are one or more intermediate fields, together with the degrees of the extensions.

Table 14.5: Minimal, maximal and intermediate multiplets

d	minimal multiplet	intermediate multiplet(s)	maximal multiplet	$[\mathbb{E}_{\max} : \mathbb{E}_{\min}]$
15	15*d*	15*b*	15*ac*	4
17	17*c*		17*ab*	2
18	18*ab*			
19	19*e*	19*a*, 19*d*	19*bc*	12
20	20*ab*			
21	21*e*		21*abcd*	3
24	24*c*		24*ab*	4
28	28*c*		28*ab*	4
30	30*d*		30*abc*	3
35	35*j*	35*i*, 35*e*, 35*h*, 35*af*	35*bcdg*	16
39	39*ij*	39*bf*, 39*gh*	39*acde*	6
48	48*g*	48*e*, 48*f*	48*abcd*	24

14.26 Ray class fields

A **ray class field** is an abelian extension of a global field (such as the algebraic number field $\mathbb{K} = \mathbb{Q}(\sqrt{D})$) associated with a **ray class group**. Every finite abelian extension of a number field is contained in one of its ray class fields.

The motivating example is the abelian extensions of $\mathbb{Q}$. Here the ray class field over $\mathbb{Q}$ associated with the ray class group $\mathbb{Z}_n^*$ is the field generated by the n-th roots of unity. Every finite abelian extension of $\mathbb{Q}$ is a subfield of such a ray class (cyclotomic) field (the Kronecker–Weber theorem). In general, the role of n above is played by the *conductor* of a ray class field. Most importantly, there are algorithms for calculating in ray class fields. Calculations (see [AFMY16], [ACFW17]) suggest

Ray class conjecture: The minimal SIC field $\mathbb{E} = \mathbb{E}_{\min}$ is the ray class field over $\mathbb{K} = \mathbb{Q}(\sqrt{(d-3)(d+1)}) = \mathbb{Q}(\sqrt{D})$ with conductor d', and ramification allowed at both infinite places[9].

Given this, the minimal SIC field (and extensions of it) can be constructed in `Magma` using `RayClassField(m)`. This makes calculations in the precision bumping algorithm, such as factoring the polynomials f_j of (14.101), faster than when the ray class field is constructed as a tower of extensions.

There are further conjectures about the SIC field (see [ACFW17]), e.g.,

Let $\mathbb{E} = \mathbb{E}_{\min}$ be the minimal SIC field in dimension d. Then in every known case the tower of fields (14.100) satisfies:

- $\mathbb{E}$ and $\mathbb{E}_1$ are ray class fields over $\mathbb{Q}(\sqrt{D})$ for which the finite part of the conductor is d'.
- $\mathbb{E}_1$ is the class field with ramification only allowed at the infinite place taking $\sqrt{D}$ to a positive real number.
- $\mathbb{E}_0$ is the Hilbert class field over $\mathbb{Q}(\sqrt{D})$.

We recall that D is the square-free part of $(d-3)(d+1)$. If D is a square-free positive integer, then it is the square-free part of $(d-3)(d+1)$ if and only if

$$(d-3)(d+1) = m^2 D \quad \Longleftrightarrow \quad (d-1)^2 - m^2 D = 4.$$

for some positive integer m. This is a modified version of *Pell's equation*, and has infinitely many solutions $d_1 < d_2 < d_3 < \cdots$ (see [AFMY17]), given by

$$d_j = 1 + 2T_j\Big(\frac{d_1 - 1}{2}\Big), \qquad j > 1,$$

[9] There are four ray class fields over $\mathbb{K} = \mathbb{Q}(\sqrt{(d-3)(d+1)})$ with conductor d'. The SIC field $\mathbb{E}$ is the largest of these, and the three others, which include $\mathbb{E}_1$, are subfields of $\mathbb{E}$.

where the T_j are the *Chebyshev polynomials of the first kind.*

If Zauner's conjecture holds with the SIC field structure conjectured, then

For every square-free positive integer D, there are infinitely many minimal SIC fields which are ray class fields over $\mathbb{Q}(\sqrt{D})$.

Example 14.19. The first few sequences of dimensions with D fixed are

$$d = 4,\ 8,\ 19,\ 48,\ 124,\ 323,\ 844,\ 2208,\ 5799,\ 15128,\ \ldots \qquad D = 5,$$
$$d = 5,\ 15,\ 53,\ 195,\ 725,\ 2703,\ 10085,\ 37635,\ 140453,\ \ldots \qquad D = 3,$$
$$d = 6,\ 24,\ 111,\ 528,\ 2526,\ 12099,\ 57966,\ 277728,\ \ldots \qquad D = 21,$$
$$d = 7,\ 35,\ 199,\ 1155,\ 6727,\ 39203,\ 228487,\ 1331715,\ \ldots \qquad D = 2,$$
$$d = 9,\ 63,\ 489,\ 3843,\ 30249,\ 238143,\ 1874889,\ 14760963,\ \ldots \qquad D = 15.$$

It can be shown [AFMY16] that for each one of these sequences, there are infinitely many distinct subsequences for which each dimension divides the following one. These subsequences are called **dimension towers** (the corresponding minimal SIC fields are nested).

For the $D = 5$ sequence $(d_j) = (4, 8.19, 48, 124, \ldots)$, it has been conjectured in [GS17] that there is a SIC with an antiunitary symmetry of order $6j$ given by the symplectic index

$$F_f := \begin{pmatrix} 0 & 1 \\ 1 & 1 \end{pmatrix} = \begin{pmatrix} 0 & -1 \\ 1 & -1 \end{pmatrix} \begin{pmatrix} 1 & 0 \\ 0 & -1 \end{pmatrix} = AJ,$$

i.e., the symplectic operation ZC. Such a SIC is called a *Lucas–Fibonacci* SIC, on account of the fact that the entries of the powers of $A = F_f J$ are the Fibonacci numbers. By directly solving (14.102) using Groebner basis methods, [GS17] have constructed an exact Lucas–Fibonacci SIC for $d = 124$ (labelled 124a), and have found numerical SICs for $d_6 = 323$, $d_7 = 844$, by searching in eigenspaces of ZC.

It is natural to speculate about similar infinite families of SICs for other values of D. In this regard, we list the first few dimensions for $D \leq 19$

D	2	3	5	6	7	10	11	13	14	15	17	19
d_1	7	5	4	11	17	39	21	12	31	9	67	341
d_2	35	15	8	99	255	1443	399	120	899	63	4355	115599
d_3	199	53	19	971	4049	54759	7941	1299	26911	489	287299	39302981

and observe from the symmetries given in Table 14.4, that for $D = 2$, there appears to be a family with $|S| = 6j$, $d = d_j$, and for $D = 3$ a family with $|S| = 3j$, $d = d_j$. There much on going work in this direction.

14.27 Equivalent equations for SIC fiducial vectors

The vectors $v \in \mathbb{C}^d$ giving SIC fiducials are determined by the d^2 quartic equations

$$|\langle S^j \Omega^k v, v\rangle|^2 = \frac{1}{d+1}, \quad (j,k) \neq (0,0), \qquad \|v\|^2 = 1, \tag{14.102}$$

in $v_1, \ldots, v_d, \overline{v_1}, \ldots, \overline{v_d}$ with coefficients from $\mathbb{Q}(\omega)$. The variational characterisation of $(2,2)$-designs (see Exer. 10.6) gives two equations (the first has degree 8)

$$\frac{1}{d^2} \sum_{(j,k)\in\mathbb{Z}_d^2} |\langle S^j \Omega^k v, v\rangle|^4 = \frac{2}{d(d+1)} \|v\|^4, \qquad \|v\|^2 = 1. \tag{14.103}$$

The algebraic variety of SIC fiducials (see §7) can be viewed as being real (by taking real variables $\Re v_j$ and $\Im v_j$ and polynomials with complex coefficients) or complex. In any case, it appears that:

The algebraic variety of SIC fiducials is zero dimensional (and *nonempty*), except for $d = 3$, where it is one dimensional.

The direct solution of equations such as above led to the earliest exact SICs, e.g., [SG10] used Groebner basis methods on (14.102), with v in an eigenspace of Z.

We observe that

$$\langle S^j \Omega^k v, v\rangle = (S^{-j} v)^* \Omega^k v = \sum_r \overline{v_{r+j}} v_r \omega^{kr}, \tag{14.104}$$

which is a polynomial in ω, with coefficients $U_j = (v_r \overline{v_{r+j}})_{r\in\mathbb{Z}_d}$. By using finite Fourier methods, this leads to a third set of simplified equations (see [ADF14], [BW07], [Kha08]), which we now present, and solve in a couple of cases.

Theorem 14.4. *A vector $v \in \mathbb{C}^d$ is a Weyl–Heisenberg SIC fiducial if and only if*

$$\langle U_s, S^{-t} U_s\rangle = \sum_{r\in\mathbb{Z}_d} v_r \overline{v}_{r+s} \overline{v}_{r+t} v_{r+s+t} = \begin{cases} 0, & s,t \neq 0; \\ \frac{1}{d+1}, & s \neq 0, t = 0, \quad s = 0, t \neq 0; \\ \frac{2}{d+1}, & (s,t) = (0,0). \end{cases} \tag{14.105}$$

Proof. Using (14.104), we may write $|\langle S^j \Omega^k v, v\rangle|^2$ as a polynomial of degree $d-1$ evaluated at a d-th root of unity

$$\begin{aligned} |\langle S^j \Omega^k v, v\rangle|^2 &= \Big(\sum_r v_r \overline{v_{r+j}} \omega^{kr}\Big)\Big(\sum_s \overline{v_s} v_{s+j} \omega^{-ks}\Big) \\ &= \sum_t \sum_r v_r \overline{v_{r+j}} \overline{v_{r-t}} v_{r-t+j} \omega^{kt} =: f_j(\omega^k). \end{aligned}$$

For (14.102) to hold for $j \neq 0$, we must have f_j be constant, equal to $\frac{1}{d+1}$ (since f_j is uniquely determined by its values at the d-th roots of unity), i.e.,

$$\sum_r v_r \overline{v_{r+j}} \overline{v_{r-t}} v_{r-t+j} = \begin{cases} 0, & t \neq 0; \\ \frac{1}{d+1}, & t = 0, \end{cases} \qquad j \neq 0,$$

which (after a change of variables) gives the first two equations in (14.105). Given that this holds, we have

$$f_0(\omega^k) = \frac{1}{d+1} \sum_{t=1}^{d-1} \omega^{kt} + \sum_r |v_r|^4.$$

For (14.102) to hold for $j = 0$, $k \neq 0$, we must have

$$f_0(\omega^k) = -\frac{1}{d+1} + \sum_r |v_r|^4 = \frac{1}{d+1} \iff \sum_r |v_r|^4 = \sum_r v_r \overline{v_r} \overline{v_r} v_r = \frac{2}{d+1},$$

i.e., the third equation in (14.105) must hold. This then ensures (14.102) holds for $j = 0$ and $k = 0$, i.e.,

$$f_0(1) = \frac{d-1}{d+1} + \frac{2}{d+1} = 1.$$

Thus for v to be fiducial it is necessary and sufficient that the subset of the equations (14.105) considered above holds. Given the symmetry in s and t of the left hand side, it follows that they all hold for a fiducial. □

From the proof, it is sufficient to require that (14.105) hold for $0 \leq s \leq t \leq \lfloor \frac{d}{2} \rfloor$.

Example 14.20. With $r_j := |v_j|$, the last two conditions in (14.105) become

$$\sum_j r_j^2 r_{j+s}^2 = \frac{1}{d+1}, \quad s \neq 0, \qquad \sum_j r_j^4 = \frac{2}{d+1}.$$

Example 14.21. For $d = 2$ and $v_j := r_j e^{\theta_j}$, the equations of (14.105) are

$$v_0^2 \overline{v_1}^2 + \overline{v_0}^2 v_1^2 = 0, \qquad 2 r_0^2 r_1^2 = \frac{1}{3}, \qquad r_0^4 + r_1^4 = \frac{2}{3}.$$

Solving the last two (using $r_0^2 + r_1^2 = 1$ to simplify the algebra) gives

$$r_0^2 = \frac{3 \pm \sqrt{3}}{6}, \quad r_1^2 = \frac{3 \mp \sqrt{3}}{6}.$$

The first then reduces to $\cos(2(\theta_1 - \theta_0)) = 0$, which has solution

$$2(\theta_1 - \theta_0) = \frac{\pi}{2} + \pi n \iff e^{i\theta_1} = e^{i\theta_0} e^{\frac{\pi}{4} i} i^n, \quad n = 0, 1, 2, 3.$$

This gives the unique SIC for $\mathbb{C}^2$ (up to projective unitary equivalence).

Example 14.22. We consider the sporadic SICs for $d = 3$, which is the only known case where there are *infinitely* many inequivalent SICs. Here (14.105) gives

$$\overline{v_2}\overline{v_1}v_0^2 + \overline{v_0}\overline{v_2}v_1^2 + \overline{v_1}\overline{v_0}v_2^2 = 0, \qquad r_2^2r_1^2 + r_0^2r_2^2 + r_1^2r_0^2 = \frac{1}{4}, \qquad r_0^4 + r_1^4 + r_2^4 = \frac{1}{2}.$$

From $r_0^4 + r_1^4 + r_2^4 = 1/2 = 2(1/4) = 2(r_0^2r_1^2 + r_1^2r_2^2 + r_2^2r_0^2)$, we obtain

$$(r_0^2 - (r_1^2 + r_2^2))^2 = r_0^4 + (r_1^4 + 2r_1^2r_2^2 + r_2^4) - 2r_0^2(r_1^2 + r_2^2) = 4r_1^2r_2^2.$$

If $r_0^2 \geq r_1^2 + r_2^2$, then taking square roots gives

$$r_0^2 - (r_1^2 + r_2^2) = 2r_1r_2 \quad \Longrightarrow \quad r_0^2 = r_1^2 + 2r_1r_2 + r_2^2 = (r_1 + r_2)^2 \quad \Longrightarrow \quad r_0 = r_1 + r_2,$$

and if $r_0^2 < r_1^2 + r_2^2$, then

$$r_0^2 - (r_1^2 + r_2^2) = -2r_1r_2 \quad \Longrightarrow \quad r_0^2 = r_1^2 - 2r_1r_2 + r_2^2 = (r_1 - r_2)^2,$$

so that $r_0 = r_1 - r_2$ for $r_1 \geq r_2$, and $r_0 = r_2 - r_1$ for $r_2 \geq r_1$. Thus we have three cases

$$r_0 = r_1 + r_2, \qquad r_1 = r_0 + r_2, \qquad r_2 = r_0 + r_1.$$

Since $r_0^2 + r_1^2 + r_2^2 = 1$ and $r_j \geq 0$, these three cases describe the three sides of a *spherical triangle* in the first octant on the unit sphere. The vertices of this triangle are given by the intersections of the three great circles, e.g.,

$$r_0 = r_1 + r_2 \quad \text{and} \quad r_1 = r_0 + r_2 \quad \Longrightarrow \quad (r_0, r_1, r_2) = (\frac{1}{\sqrt{2}}, \frac{1}{\sqrt{2}}, 0),$$

with the other vertices being $(\frac{1}{\sqrt{2}}, 0, \frac{1}{\sqrt{2}})$ and $(0, \frac{1}{\sqrt{2}}, \frac{1}{\sqrt{2}})$.

It remains only to satisfy the first equation, i.e.,

$$\overline{v_2}\overline{v_1}v_0^2 + \overline{v_0}\overline{v_2}v_1^2 + \overline{v_1}\overline{v_0}v_2^2 = 0. \tag{14.106}$$

We consider a representative case when $r = (r_0, r_1, r_2)$ is on a vertex, and on an edge of the spherical triangle. The other fiducials come from a symmetry of the triangle, i.e., by permuting the entries of v.

The *vertex* $(r_0, r_1, r_2) = (\frac{1}{\sqrt{2}}, \frac{1}{\sqrt{2}}, 0)$. Since $r_2 = |v_2| = 0$, the equation (14.106) is trivially satisfied, and as there are no other conditions, we have fiducial vectors

$$v = \frac{1}{\sqrt{2}} \begin{pmatrix} e^{i\theta} \\ e^{i\phi} \\ 0 \end{pmatrix}, \qquad \theta, \phi \in \mathbb{R}. \tag{14.107}$$

The *triangle edge* $r_0 = r_1 + r_2$, $r_1, r_2 > 0$. For convenience, let $z_j := v_j^2/\overline{v}_j$, so that $|z_j| = |v_j| = r_j$. Then dividing (14.106) by $\overline{v_0}\overline{v_1}\overline{v_2}$ yields the equivalent equation

$$z_0 + z_1 + z_2 = 0.$$

Since $z_0 = -z_1 - z_2$ and $|z_0| = r_0 = r_1 + r_2 = |-z_1| + |-z_2|$, the complex numbers $z_0, -z_1, -z_2$ must have the same argument, i.e.,

$$z_0 = r_0 e^{i\phi}, \quad z_1 = -r_1 e^{i\phi}, \quad z_2 = -r_2 e^{i\phi}, \qquad \phi \in \mathbb{R}.$$

Hence, writing $v_j = r_j e^{i\theta_j}$, so that $z_j = r_j e^{3i\theta_j}$, we calculate

$$\begin{aligned}
r_0 e^{3i\theta_0} &= r_0 e^{i\phi} &&\Longrightarrow \quad \theta_0 = \frac{\phi}{3} + \frac{2\pi}{3}k_0, \quad k_0 = 0,1,2,\\
r_1 e^{3i\theta_1} &= -r_1 e^{i\phi} &&\Longrightarrow \quad \theta_1 = \frac{\phi}{3} + \frac{\pi}{3} + \frac{2\pi}{3}k_1, \quad k_1 = 0,1,2,\\
r_2 e^{3i\theta_2} &= -r_2 e^{i\phi} &&\Longrightarrow \quad \theta_2 = \frac{\phi}{3} + \frac{\pi}{3} + \frac{2\pi}{3}k_2, \quad k_2 = 0,1,2.
\end{aligned}$$

After the change of variables $\theta := (\phi + 2\pi k_0)/3$, we can describe the fiducial vectors corresponding to the point (r_0, r_1, r_2) on this edge of the triangle by

$$v = e^{i\theta} \begin{pmatrix} r_0 \\ r_1 \mu \omega^{j_1} \\ r_2 \mu \omega^{j_2} \end{pmatrix}, \qquad \theta \in \mathbb{R}, \quad j_1, j_2 \in \{0,1,2\},$$

where $\mu = e^{\frac{2\pi i}{6}}$, $\omega = \mu^2$. We can give a more precise description of this edge by solving $r_0^2 + r_1^2 + r_2^2 = 1$ and $r_0 = r_1 + r_2$ for r_1, r_2 in terms of r_0, which gives

$$r_1 = \frac{r_0 \pm \sqrt{2 - 3r_0^2}}{2} \quad \text{and} \quad r_2 = \frac{r_0 \mp \sqrt{2 - 3r_0^2}}{2}.$$

Here the half of the edge where $r_1 \geq r_2$ is given by the 'top' choice in the formulas, and the other choice gives the half with $r_1 \leq r_2$. For these to give $r_1, r_2 > 0$, we must have

$$2 - 3r_0^2 \geq 0, \quad r_0 - \sqrt{2 - 3r_0^2} > 0 \quad \Longleftrightarrow \quad r_0^2 \leq \frac{2}{3}, \quad r_0^2 > \frac{1}{2}.$$

Hence $\frac{1}{2} < r_0^2 \leq \frac{2}{3}$ along this edge, with r_1 and r_2 given by the formulas above.

For the *vertex* SIC fiducial given by (14.107), we have the triple product

$$\Delta(v, Sv, S^2v) = \langle v, Sv\rangle \langle Sv, S^2v\rangle \langle S^2v, v\rangle = \frac{1}{8}e^{it}, \qquad t = 3(\phi - \theta).$$

and so there are *uncountably* many projectively unitarily inequivalent vertex SIC fiducials parameterised by $t = 3(\phi - \theta)$. In §8.5 of [Zhu12], the symmetries $S(v)$ of the SICs in three dimensions were calculated. For these, $|S(v)|$ can be 6 (the infinite family 3a), 12, or 28 (types 3b and 3c).

Example 14.23. Direct solution of the equations (14.105) shows that there exist *real* fiducial vectors $v \in \mathbb{R}^d$ for $d = 3, 7, 19$ (see [BW07], [Kha08]).

The solution of systems of equations such as those presented in this section, still play a vital role in the analytic construction of exact SICs.

Notes

Many of the Gauss sums that appear when multiplying elements of the Clifford group can be calculated by using the following *quadratic reciprocity law*

For $a,b,c \in \mathbb{Z}$, with $ac \neq 0$ and $ac+b$ even, one has the *quadratic reciprocity*

$$\sum_{j=0}^{|c|-1} e^{\pi i(aj^2+bj)/c} = \sqrt{\frac{|c|}{|a|}} e^{\pi i(|ac|-b^2)/4ac} \sum_{k=0}^{|a|-1} e^{-\pi i(ck^2+bk)/a}. \tag{14.108}$$

The Clifford group can be viewed from the point of view of Gabor analyis for the finite abelian group $\mathbb{Z}_d$ [FKL09], where the Clifford operations are referred to as *metaplectic operations* [FHK+08].

There is currently much activity on Zauner's conjecture, due to

- The recent conjectures about the structure of the SIC field and the action of its Galois group on fiducials, e.g., the ray class conjecture.
- The recent construction of many more exact SICs, and consequently SIC fields.
- The recent construction of many more numerical SICs.
- Efficienct methods for calculating the symmetry group from numerical SICs, and consequently conjectures about families of SICs with large symmetry groups, such as the Lucas–Fibonnacci SICs.

Some history and an extensive collection of references is given in [FHS17].

I wish to thank Marcus Appleby, Len Bos, Tuan Chien, Steve Flammia, Chris Fuchs, Gary Mcconnell, Markus Grassl, Andrew Scott, Jon Yard, Blake Stacey and Hjuangjun Zhu for sharing their results and insights on the SIC problem with me.

Exercises

14.1. Let (P_j) be a SIC for $\mathbb{C}^d$, and $\mathbb{C}^{d\times d}$ have the Frobenius inner product.
(a) Determine the condition on $c_1, c_2 \in \mathbb{R}$ which ensures that $P_j - c_1 I$ and $P_k - c_2 I$ are orthogonal for $j \neq k$.
(b) Find the dual basis to (P_j).
(c) Show that $(P_j - cI)$ is a basis if and only if $c \neq \frac{1}{d}$.
(d) Show that $(P_j - cI)$, $c = \frac{1}{d}(1 \pm \frac{1}{\sqrt{d+1}})$ is orthogonal.
(e) Find the projection of $(P_j - cI)$, $c \in \mathbb{R}$ onto the traceless matrices.

14.2. *The Bloch sphere.* Each unit vector $v \in \mathbb{C}^2$ can be scaled so that

$$v = \begin{pmatrix} \cos\frac{\theta}{2} \\ \sin\frac{\theta}{2} e^{i\phi} \end{pmatrix}, \qquad 0 \leq \theta \leq \pi, \quad 0 \leq \phi < 2\pi.$$

(a) Express the rank one orthogonal projection matrix vv^* in terms of

$$a = \sin\theta\cos\phi, \quad b = \sin\theta\sin\phi, \quad c = \cos\theta.$$

Here $(a,b,c) \in \mathbb{R}^3$, with $a^2 + b^2 + c^2 = 1$.
(b) Show that the map

$$vv^* \mapsto \sqrt{2}\big(vv^* - \frac{1}{2}I\big)$$

maps the rank one orthogonal projections on $\mathbb{C}^2$ (equivalently, unit vectors in $\mathbb{C}^2$) onto the traceless Hermitian matrices with unit Frobenius norm.
Remark: This identification of the rank one orthogonal projections (pure states of a two-level quantum mechanical system) with the unit sphere in the 2×2 traceless Hermitian matrices is called the **Bloch sphere**, with (a,b,c) a **Bloch vector**.
(c) Show that the map

$$vv^* \mapsto A = \sqrt{\frac{d}{d-1}}\big(vv^* - \frac{1}{d}I\big)$$

maps the rank one orthogonal projections on $\mathbb{C}^d$ (unit vectors in $\mathbb{C}^d$) to the traceless Hermitian matrices with unit Frobenius norm, but is not onto for $d > 2$.

14.3. The Bloch vector $(a,b,c) \in \mathbb{R}^3$ for a rank one orthogonal projection vv^* on $\mathbb{C}^2$ is given by

$$2vv^* - I = \begin{pmatrix} c & a - ib \\ a + ib & -c \end{pmatrix}.$$

(a) Calculate the the Bloch vectors of the SIC for $\mathbb{C}^2$ given by (1.7).
(b) The vectors $(\pm 1, 0, -\frac{1}{\sqrt{2}})$, $(0, \pm 1, \frac{1}{\sqrt{2}})$ are the vertices of a regular tetrahedron. Find the SIC corresponding to the Bloch vectors these give when normalised.

14.4. Show that the displacement operators $S^j \Omega^k$ given by (14.4) satisfy
(a) $\Omega^k S^j = \omega^{jk} S^j \Omega^k$.

(b) $(S^j\Omega^k)^r = \omega^{\frac{1}{2}r(r-1)jk}S^{rj}\Omega^{rk}$.
(c) For $h = cS^a\Omega^b \in \hat{H}$, one has $hS^j\Omega^k h^{-1} = \omega^{bj-ak}S^j\Omega^k$.

14.5. Show that the matrices F, R, M, P_σ of §14.5 satisfy
(a) $F(S^j\Omega^k)F^{-1} = \omega^{-jk}S^{-k}\Omega^j$.
(b) $R(S^j\Omega^k)R^{-1} = \mu^{j(j+d)}S^j\Omega^{j+k}$.
(c) $M(S^j\Omega^k)M^{-1} = \mu^{k(k-2j+d)}S^{-k}\Omega^{j-k}$.
(d) $P_\sigma(S^j\Omega^k)P_\sigma^{-1} = S^{\sigma j}\Omega^{\sigma^{-1}k}$.

14.6. For d even, show that $S^{\frac{d}{2}}$ and $\Omega^{\frac{d}{2}}$ are symplectic unitaries, i.e., belong to the subgroup of the Clifford group generated by F and R, with

$$\Omega^{\frac{d}{2}} = R^d, \qquad S^{\frac{d}{2}} = F^{-1}\Omega^{\frac{d}{2}}F = F^{-1}R^dF.$$

14.7. Show that $SL_2(\mathbb{Z}_d) \times \mathbb{T}^{\mathbb{Z}_d^2}$ is a group with the multiplication

$$(A, z_A)(B, z_B) := (AB, (z_A \circ B)z_B),$$

where functions $\mathbb{Z}_d^2 \to \mathbb{T}$ are multiplied pointwise.

14.8. The function $z_a : \mathbb{Z}_d^2 \to \mathbb{T}$ of (14.18) satisfies (14.25), i.e.,

$$z_a(p+q) = \omega^{c_A(p,q)}\, z_a(p)z_a(q),$$

where the symmetric function $c_A : \mathbb{Z}_d^2 \times \mathbb{Z}_d^2 \to \mathbb{Z}_d$ is given by

$$c_A(p,q) := (Ap)_2(Aq)_1 - p_2q_1, \qquad A := \psi_a.$$

(a) Show that $c_A(p,q)$ can be written as a quadratic form

$$c_A(p,q) = p^T\sigma_A q, \qquad \sigma_A := \begin{pmatrix} \alpha\gamma & \beta\gamma \\ \beta\gamma & \beta\delta \end{pmatrix}, \quad A = \psi_a = \begin{pmatrix} \alpha & \beta \\ \gamma & \delta \end{pmatrix}.$$

(b) Suppose that d is odd. Then $-\mu = \mu^{d+1} = \omega^{\frac{d+1}{2}}$, so that $\mathbb{Z}_d \to \mathbb{T} : p \mapsto (-\mu)^p$ is well defined. Show that $\hat{z}_a : \mathbb{Z}_d^2 \to \mathbb{T}$ given by

$$\hat{z}_a(p) := (-\mu)^{-c_A(p,p)}\, z_a(p) = (-\mu)^{-p^T\sigma_A p}\, z_a(p) = (-\mu)^{p_1p_2-(Ap)_1(Ap)_2}\, z_a(p),$$

is a character, i.e., $\hat{z}_a(p+q) = \hat{z}_a(p)\hat{z}_a(q)$, $\forall p, q$, and that

$$\hat{z}_{ab} = (\hat{z}_a \circ \psi_b)\hat{z}_b,$$

$$\hat{z}_F = \hat{z}_R = \hat{z}_M = 1, \qquad \hat{z}_{S^j\Omega^k}(p) = z_{S^j\Omega^k}(p) = \omega^{kp_1-jp_2}.$$

(c) Suppose that d is even. Choose a $B \in SL_2(\mathbb{Z}_{2d})$ with $B = A \bmod d$ (there are eight choices). Show that $(-\mu)^{p^T\sigma_B p}$ depends only on $p \bmod 2d$, so that

$$\hat{z}_{a,B} : \mathbb{Z}_d^2 \to \mathbb{T} : p \mapsto (-\mu)^{-p^T\sigma_B p}\, z_a(p)$$

is well defined. Show that $\hat{z}_{a,B}$ is a character, and so has the form

$$\hat{z}_{a,B}(p) = \omega^{\langle\langle \chi, Ap \rangle\rangle}, \qquad \text{where } \chi \in \mathbb{Z}_d^2.$$

14.9. Here we consider how the indexing of §14.9 depends on the normalisation of S and Ω. Let

$$\hat{S} := c_1 S, \quad \hat{\Omega} := c_2 \Omega, \qquad c_1, c_2 \in \mathbb{T},$$

and $\hat{U}_{(j,k)} := \hat{S}^j \hat{\Omega}^k$. As in (14.18), define a corresponding $\hat{z}_a : \mathbb{Z}_d^2 \to \mathbb{T}$ by

$$a\hat{U}_p a^{-1} = \hat{z}_a(p)\hat{U}_{\psi_a(p)} \quad \Longrightarrow \quad \hat{z}_a(p) = \frac{c_1^{p_1} c_2^{p_2}}{c_1^{\psi_a(p)_1} c_2^{\psi_a(p)_2}} z_a(p).$$

(a) Show that $\hat{z}_a$ satisfies

$$\hat{z}_{ab} = \hat{z}_b(\hat{z}_a \circ \psi_b), \qquad \hat{z}_{S^a\Omega^b}(j,k) = \omega^{bj-ak},$$

and so Corollary 14.1 holds with z_a replaced by $\hat{z}_a$.
(b) For the choice $c_1 = c_2 = -\mu$, show that $\det(\hat{U}_p) = 1$ and $\hat{z}_a(p)$ is a power of ω. $\hat{z}_a(j,k)$ is a power of ω, and calculate $\hat{z}_a(j,k)$ explicitly for $a = F, R, M, P_\sigma$

14.10. A $2n \times 2n$ matrix M over a field $\mathbb{F}$ is said to be **symplectic** if

$$M^T A M = A, \qquad A := \begin{pmatrix} 0 & I \\ -I & 0 \end{pmatrix}.$$

The symplectic matrices form a group $\mathrm{Sp}(n) = \mathrm{Sp}(2n, \mathbb{F})$.
(a) Show that for $n = 1$ the symplectic matrices are the matrices with determinant 1, i.e., $\mathrm{Sp}(2) = SL_2(\mathbb{F})$.
(b) Show $SL_2(\mathbb{Z})$ is a subgroup of $SL_2(\mathbb{R})$.
(c) Show that $SL_2(\mathbb{Z}_d)$ is a group.
Remark: The elements of $SL_2(\mathbb{Z}_d)$ are called *symplectic matrices.*

14.11. Show that if $b \in \mathbb{Z}_{d'}^*$ has odd order, then

$$c_{b,d} := \frac{1}{\sqrt{d}} \sum_{j \in \mathbb{Z}_d} \mu^{bj(j+d)} = (\sqrt{i})^{1-d}.$$

14.12. Let A be a normal matrix with eigenvalues $\lambda_1, \ldots, \lambda_k$, and P_j be the projection onto the λ_j–eigenspace. Since normal matrices are unitarily diagonalisable, we have

$$A^j = \lambda_1^j P_1 + \lambda_2^j P_2 + \cdots + \lambda_k^j P_k, \qquad j = 0, 1, 2, \ldots.$$

(a) Show that P_j can be written as a linear combination of $I, A, A^2, \ldots, A^{k-1}$.
(b) Suppose A has order k, so that its eigenvalues are $1, \omega, \omega^2, \ldots, \omega^{k-1}$, $\omega := e^{\frac{2\pi i}{k}}$ (possibly with zero multiplicity). Show that the projection onto the λ-eigenspace is

$$P_\lambda = \frac{1}{k}\big(I + (\overline{\lambda}A) + (\overline{\lambda}A)^2 + \cdots + (\overline{\lambda}A)^{k-1}\big).$$

(c) Show that if A is unitary and of order k, then the multiplicities of its eigenvalues (dimension of it eigenspaces) can be determined from trace(A^ℓ), $0 \le \ell \le \frac{k}{2}$.
(d) Calculate the multiplicities of the eigenvalues of the Fourier matrix F, i.e.,

Table 14.6: The multiplicities of the eigenvalues $1, i, i^2, i^3$ of F.

d	1	i	-1	$-i$
$4m$	$m+1$	m	m	$m-1$
$4m+1$	$m+1$	m	m	m
$4m+2$	$m+1$	m	$m+1$	m
$4m+3$	$m+1$	$m+1$	$m+1$	m

(e) Calculate the multiplicities of the eigenvalues of the the Zauner matrix Z, i.e.,

Table 14.7: The multiplicities of the eigenvalues $1, \tau, \tau^2$ of Z and $\overline{Z}^2$.

d	1	τ	τ^2
$3m$	$m+1$	m	$m-1$
$3m+1$	$m+1$	m	m
$3m+2$	$m+1$	$m+1$	m

14.13. Suppose that $\ell \mid d$. Let $\sqrt{i} = e^{\frac{2\pi i}{8}}$.
(a) Show that R^ℓ, and hence $F^{-1}R^\ell F = (F^{-1}RF)^\ell$, has order $\frac{d'}{\ell}$.
(b) For d odd, and for d even and ℓ odd, show that $F^{-1}R^\ell F$ is sparse, i.e.,

$$(F^{-1}R^\ell F)_{jk} = \sqrt{\frac{\ell}{d}}(\sqrt{i})^{1-\ell d}\begin{cases}(-1)^{j-k}\mu^{-\frac{1}{\ell}(j-k)^2}, & j-k \equiv 0 \bmod \ell;\\ 0, & j-k \not\equiv 0 \bmod \ell.\end{cases}$$

(c) For d even and ℓ even, show that $F^{-1}R^\ell F$ is sparse, i.e.,

$$(F^{-1}R^\ell F)_{jk} = \sqrt{\frac{\ell}{d}}(\sqrt{i})^{1-\ell d}\begin{cases}(-1)^{j-k}\mu^{-\frac{1}{\ell}(j-k)^2}, & j-k \equiv \frac{d}{2} \bmod \ell;\\ 0, & j-k \not\equiv \frac{d}{2} \bmod \ell.\end{cases}$$

14.14. Suppose that $3 \mid d$. Let $\tau = e^{\frac{2\pi i}{3}}$, $\sqrt{i} = e^{\frac{2\pi i}{8}}$, and

$$W_a := (-1)^{d-1}R^{\frac{2d}{3}a}F^{-1}R^3FR, \qquad a = 0,1,2.$$

(a) Show that W_a is sparse, i.e.,

$$(W_a)_{jk} = \sqrt{\frac{3}{d}}(-\sqrt{i})^{1-3d}\begin{cases}(-\mu)^{-\frac{1}{3}(j-k)^2+\frac{2d}{3}aj^2+k^2}, & j-k \equiv 0 \mod 3;\\ 0, & j-k \not\equiv 0 \mod 3.\end{cases}$$

(b) Show that W_a has order 3.
(c) Show that

$$\operatorname{trace}(W_a) = -\frac{i}{\sqrt{3}}\left(1+2\tau^{\frac{d}{3}}+2\tau^a(1-\tau^{\frac{d}{3}})\right).$$

(d) Find the multiplicities of the eigenvalues $1, \tau, \tau^2$ of

(i) W_1 for $d \equiv 3 \bmod 9$, $d \neq 3$.
(ii) W_2 for $d \equiv 6 \bmod 9$.

Remark: The first implies that the multiplicities of the eigenvalues m_λ of $M_1 = W_1^2$ are

$$m_\lambda = \frac{1}{3}\left(d+3(\lambda^2+\lambda)\right) = \begin{cases} \frac{d+6}{3}, & \lambda = 1; \\ \frac{d-3}{3}, & \lambda = \tau, \tau^2. \end{cases} \qquad (14.109)$$

(e) The matrix W_0 is well defined for all d. Show that it is conjugate to the Zauner matrix via

$$\tau^{1-d} W_0 = (R^{-1}F)^{-1} Z (R^{-1}F).$$

14.15. Show that the displacement operators $\hat{D}_p = (-\mu)^{p_1 p_2} S^{p_1} \Omega^{p_2}$, $p \in \mathbb{Z}^2$, satisfy

$$\hat{D}_p^{-1} = \hat{D}_{-p}, \qquad \hat{D}_p \hat{D}_q = (-\mu)^{\langle\langle p,q \rangle\rangle} \hat{D}_{p+q} = \omega^{\langle\langle p,q \rangle\rangle} \hat{D}_q \hat{D}_p,$$

and

$$\hat{D}_{p+dq} = \begin{cases} \hat{D}_p, & d \text{ odd}; \\ (-1)^{\langle\langle p,q \rangle\rangle} \hat{D}_p, & d \text{ even}, \end{cases}$$

where $\langle\langle \cdot,\cdot \rangle\rangle$ is the *symplectic form* $\langle\langle p,q \rangle\rangle := p_2 q_1 - p_1 q_2$.

14.16. *Appleby indexing*. We consider some details from §14.10.
(a) Suppose that $B \in SL_2(\mathbb{Z}_{d'})$ and $\chi \in \mathbb{Z}_d^2$. Show that if $a \in \mathrm{C}(d)$ satisfies

$$a \hat{D}_p a^{-1} = \omega^{\langle\langle \chi, Bp \rangle\rangle} \hat{D}_{Bp}, \qquad \forall p \in \mathbb{Z}_{d'}^2,$$

then the unique index (ψ_a, z_a) is given by

$$\psi_a = A := B \bmod d, \qquad z_a(p) = \omega^{\langle\langle \chi, Ap \rangle\rangle} (-\mu)^{p^T \sigma_B p}.$$

(b) Calculate the kernel of the Appleby homomorphism (14.42).
(c) Find the index (ψ_a, z_a) for the Clifford operations with Appleby index

(i) $[F_z, 0]$, $F_z := \begin{pmatrix} 0 & d-1 \\ d+1 & d-1 \end{pmatrix} = (d+1)\begin{pmatrix} 0 & -1 \\ 1 & -1 \end{pmatrix}$.
(ii) $[F_a, 0]$, $F_a := \begin{pmatrix} 1 & d+3 \\ \frac{4d-3}{3} & d-2 \end{pmatrix} = (d+1)\begin{pmatrix} d+1 & 3 \\ \frac{d-3}{3} & d-2 \end{pmatrix}$, $d \equiv 3 \bmod 9$, $d \neq 3$.
(iii) $[F_b, 0]$, $F_b := \begin{pmatrix} -\sqrt{d+1} & d \\ d & d-\sqrt{d+1} \end{pmatrix}$, $\sqrt{d+1} \in \mathbb{Z}$, $d \geq 8$.
(iv) $[F_c, 0]$, $F_c := \begin{pmatrix} \kappa & d-2\kappa \\ d+2\kappa & d-\kappa \end{pmatrix}$, $d = (3k \pm 1)^2 + 3$, $\kappa = 3k^2 \pm k + 1$, $k \geq 0$.

and determine the Clifford operation $[a]$ if it is obvious.

14.17. We show the eigenvectors of $S\Omega^\ell$, $\ell \in \mathbb{Z}_d$, give the MUBs of Theorem 12.22.
(a) Show that the matrices $\{S\Omega^\ell\}_{\ell\in\mathbb{Z}_d}$ commute up to a scalar.
(b) Show that $S\Omega^\ell$ and $S\Omega^m$ are (Frobenius) orthogonal when $\ell - m \in \mathbb{Z}_d^*$.
(c) Show that the columns $\mathscr{B}_\ell := \{R^\ell F e_j\}$ of $R^\ell F$ are eigenvectors of $S\Omega^\ell$.
(d) For any d, the d^2-1 matrices $\{S^j\Omega^k : 0 \le j \ne k < d\}$ can be partitioned into $d+1$ subsets of $d-1$ matrices

$$\{\Omega,\Omega^2,\ldots,\Omega^{d-1}\}, \qquad \mathscr{M}_\ell := \{S\Omega^\ell, S^2\Omega^{2\ell},\ldots,S^{d-1}\Omega^{(d-1)\ell}\}, \quad \ell \in \mathbb{Z}_d.$$

Show that matrices in the same set have the same eigenvectors, i.e., $\mathscr{E} = \{e_j\}$ and $\mathscr{B}_\ell$, respectively.

14.18. Here we consider the canonical abstract error group given by Clifford group.
(a) Find the determinant of the generators S, Ω, F, R for $\mathrm{C}(d)$.
Hint: First find the determinant of Z.
(b) Explain how each of the generators M above can be replaced by a scalar multiple $\hat{M} = c^{-1}M$ with determinant 1.
(c) Use `Magma` to investigate the canonical abstract error groups which appear as subgroups of $\langle \hat{S},\hat{\Omega},\hat{F},\hat{R}\rangle$ for $d = 2,3,4$.

14.19. Let $G = \langle F,R\rangle$ be the group generated by F and R.
(a) Show that G is finite, i.e., it contains only finitely many unitary scalar matrices.
(b) Show that for $d > 2$, the nondiagonal matrix $P_{-1} = F^2$ is in the centre of G.
Remark: By Schur's lemma, this implies that $\langle F,R\rangle$ and $\mathrm{C}_{\mathrm{Sp}}(d)$ are not irreducible.

14.20. Let $A = [a_{jk}] \in \mathbb{C}^{m\times n}$ with $\mathbb{E} := \mathbb{Q}(A) = Q(\{a_{jk}\})$, and $g \in \mathrm{Gal}(\mathbb{E}/\mathbb{Q})$.
(a) Show that if A is Hermitian, then $g(A)$ is Hermitian if and only if g commutes with complex conjugation.
(b) Show that if $A = vv^*$ is a rank one orthogonal projection, then $g(A)$ is also if and only if g commutes with complex conjugation.
(c) Now let $\mathbb{E}$ be the SIC field of a fiducial Π. Show that if $g(\Pi)$ is a SIC fiducial for some $g \in \mathrm{Gal}(\mathbb{E}/\mathbb{Q})$, then g must commute with complex conjugation on $\mathbb{E}$.

14.21. The symmetry group $S(v)$ of (14.65) for a generic SIC fiducial is *abelian*. Here we show that it is *nonabelian* for the SICs for $d = 2,3$ (listed in Table 14.4).
(a) Show that the order two antiunitary symplectic operation $b = F^{-1}C$ does not commute with the Zauner matrix Z.
(b) Show that the SICs for $d = 2,3$ have nonabelian symmetry groups.
(c) Show that if $b = F^{-1}C$ is a symmetry of a SIC (with a symmetry Z or M_1), then its symmetry group is nonabelian.
Remark: The Hoggar lines are the only other known SIC with a nonabelian (in fact noncyclic) symmetry group. Because of this (and other reasons), [Sta17] refers to the Hoggar lines and $d = 2,3$ SICs as **sporadic** SICs, and all other SICs as **generic**.

14.22. If v is a Weyl–Heisenberg SIC fiducial, then so is av, $a \in \mathrm{EC}(d)$, (see §14.7). Show that

$$[a]\cdot\Pi := (av)(av)^*, \quad \Pi = vv^*, \qquad [a] \in \mathrm{PEC}(d)$$

defines an action of the extended Clifford group on the SIC fiducial projectors.

14.23. Here we show the SIC field $\mathbb{E} = \mathbb{Q}(\Pi, \mu)$ of a Weyl–Heisenberg SIC fiducial projector $\Pi = vv^*$ depends only its extended Clifford orbit.
(a) Show that $\mathbb{E}$ is closed under complex conjugation.
(b) Let $\Pi' = (av)(av)^*$, $[a] \in \mathrm{PEC}(d)$, be a another fiducial in the extended Clifford orbit of Π. Show that $\mathbb{Q}(\Pi', \mu) = \mathbb{Q}(\Pi, \mu)$, i.e., the SIC field $\mathbb{E} = \mathbb{Q}(\Pi, \mu)$ depends only on the orbit of Π, and not the particular choice of the fiducial projector.

14.24. We consider the extended Clifford action on the overlaps χ_p^Π of a SIC Π.
(a) Show that if $a \in \mathrm{C}(d)$ has Appleby index $[B, b]$, then

$$\chi_p^\Pi = \omega^{\langle\langle b, Bp \rangle\rangle} \chi_{Bp}^{[a]\cdot\Pi}, \qquad \forall p \in \mathbb{Z}_{d'}^2.$$

(b) Show that the overlaps of $\overline{\Pi} = \Pi^T$ satisfy

$$\chi_p^{\overline{\Pi}} = \chi_{-Jp}^\Pi, \qquad \forall p \in \mathbb{Z}_{d'}^2, \qquad -J = \begin{pmatrix} -1 & \\ & 1 \end{pmatrix}.$$

(c) Show that if $[a] \in \mathrm{PEC}(d)$ has (extended) Appleby index $[B, b]$, then

$$\chi_p^\Pi = \omega^{\langle\langle b, \det(B)Bp \rangle\rangle} \chi_{\det(B)Bp}^{[a]\cdot\Pi}, \qquad \forall p \in \mathbb{Z}_{d'}^2.$$

14.25. Let Π be a centred fiducial for which $S_0(\Pi) \cong S(\Pi)$ is abelian (all the known cases). Show that if $B \in S_0(\Pi)$ is a conjugate of F_z, then there is a unique maximal abelian subgroup of $GL_2(\mathbb{Z}_{d'})$ containing $S_0(\Pi)$ and $S(\Pi)$, which is given by

$$C(\Pi) := C(S(\Pi)) = C(S_0(\Pi)) = C(B) = \{\alpha I + \beta B : \alpha, \beta \in \mathbb{Z}_{d'}\} \cap GL_2(\mathbb{Z}_{d'}),$$

where $C(X)$ denotes the centraliser of X in $GL_2(\mathbb{Z}_{d'})$.

14.26. Let $\mathscr{G} = \mathrm{Gal}(\mathbb{E}/\mathbb{Q})$ be the Galois group of the SIC field $\mathbb{E}$ of a fiducial Π, and H_g be given by (14.87). It can be shown (see []) that

$$\frac{\sqrt{i}^{d-1}}{\sqrt{d}} \in \mathbb{Q}(\mu).$$

(a) Show that for every Clifford operation $[a]$, we can choose $a \in \mathbb{E}^{d\times d}$.
(b) Show that if $a \in \mathrm{C}(d) \cap \mathbb{E}^{d\times d}$ and $g \in \mathscr{G}_c$ then $g(a) \in \mathrm{C}(d)$.
(c) Show that

$$g([a]) := [g(a)], \qquad g \in \mathscr{G}_c, \quad a \in \mathbb{E}^{d\times d},$$

defines an action of the group $\mathscr{G}_c$ on the Clifford operations.
(d) Show that action of $\mathscr{G}_c$ on $\mathrm{PC}(d)$ in terms of Appleby indexes is given by

$$g(f([B,b])) = f([H_g B H_g^{-1}, H_g b]).$$

In particular, $g \in \mathscr{G}_c$ maps the symplectic operation with symplectic index B to the symplectic operation with symplectic index $H_g B H_g^{-1}$.

14.27. Let Π be a SIC fiducial. We recall that $g(\Pi)$, $g \in \mathscr{G}_c$, is a fiducial (on the same multiplet), with $g(\Pi)$ on the same extended Clifford orbit when $g \in \mathscr{G}_0$.
(a) Show that the Clifford action commutes with the symmetries of a fiducial, i.e.,

$$S_{[a]\cdot \Pi} = [a] \cdot S_\Pi := a S_\Pi a^{-1}, \qquad \forall [a] \in \mathrm{PEC}(d).$$

(b) Show that the Galois action commutes with symmetries of a fiducial, i.e.,

$$S_{g(\Pi)} = g(S_\Pi), \qquad \forall g \in \mathscr{G}_c.$$

(c) Let $\mathscr{O}_1, \ldots, \mathscr{O}_\ell$ be the extended Clifford orbits of a SIC fiducial multiplet, and suppose that on every orbit $\mathscr{O}_j$ there is a centred fiducial Π_j. Fix Π_j. Show that for every $g \in \mathscr{G}_c$ there is a Π_k and $[a] \in \mathrm{PEC}(d)$ with Appleby index $[B,q]$, such that

$$g(\Pi_j) = [a] \cdot \Pi_k, \qquad 3q = \begin{cases} 0 \bmod d, & d \text{ is odd;} \\ 0 \bmod \frac{d}{2}, & d \text{ is even.} \end{cases}$$

(d) Show that if $d \not\equiv 0 \bmod 3$, then we can chose $q = 0$ above, so that a is symplectic, and $\mathscr{G}_c$ maps centred fiducials to centred fiducials (on the same multiplet).
(e) Show that if $d \not\equiv 0 \bmod 3$, then all centred fiducials are strongly centred.
(f) Show that if all the Π_j above are strongly centred, then we can take $[a]$ to be symplectic, and in this case

$$g_1(\Pi_j) = P_{-1} \overline{\Pi_j} P_{-1}.$$

14.28. Here we consider the group generated by the Clifford operations and the Galois symmetries of the SIC field $\mathbb{E}$, and its action on SIC fiducials. For $g \in \mathscr{G}$, and $a \in \mathrm{C}(d) \cap \mathbb{E}^{d\times d}$, we write ag for the map $\mathbb{E}^d \to \mathbb{E}^d$ given by

$$(ag)v := a(g(v)), \qquad \forall v \in \mathbb{E}^d.$$

(a) Show that $X = \{ag : a \in \mathrm{C}(d) \cap \mathbb{E}^{d\times d}, g \in \mathscr{G}_c\}$ is a group under composition, i.e., the multiplication

$$(a_1 g_1)(a_2 g_2) = a_1 g_1(a_2) g_1 g_2,$$

with inverse $(ag)^{-1} = g^{-1}(a^{-1}) g^{-1}$.
Remark: The map ag is said to be a g-**unitary**, since a is unitary and

$$(ag)(\alpha v + \beta w) = g(\alpha)(ag)(v) + g(\beta)(ag)(v), \qquad \alpha, \beta \in \mathbb{E}, \quad v, w \in \mathbb{E}^d.$$

A 1-unitary map is unitary, and a g_c-unitary map is antiunitary.
(b) Show that $\mathrm{PC}(d) \times \mathscr{G}_c = \{[a]g : [a] \in \mathrm{PC}(d), g \in \mathscr{G}_c\}$ is group with the induced multiplication

$$([a_1]g_1)([a_2]g_2) := [a_1 g_1(a_2)] g_1 g_2, \qquad a_1, a_2 \in \mathrm{C}(d) \cap \mathbb{E}^{d\times d},$$

which contains $\mathrm{PC}(d)$, $\mathscr{G}_c$ and $\mathrm{PEC}(d)$ as subgroups. We call $[a]g$ a g-unitary.
(c) Show there is a (natural) action of $\mathrm{PC}(d) \times \mathscr{G}_c =$ on fiducials $\Pi = vv^*$ given by

$$[a]g \cdot \Pi := (agv)(agv)^* = ag(\Pi)a^{-1} = [a] \cdot g(\Pi), \qquad a \in \mathrm{C}(d) \cap \mathbb{E}^{d \times d}.$$

Show that if $[a]g$, $a \in \mathrm{C}(d) \cap \mathbb{E}^{d \times d}$, stabilises Π, then v is an "eigenvector" of ag, i.e.,

$$(ag)v = \lambda v, \qquad \exists \lambda \in \mathbb{E}, \quad |\lambda| = 1.$$

Remark: The stabiliser of $\Pi = vv^*$ in the subgroup $\mathrm{PEC}(d) = \mathrm{PC}(d) \times \langle g_c \rangle$ is the group $S_\Pi = S(v)$. The stabiliser is in general larger, since for a $g \in \mathscr{G}_0$, we can choose $[a] \in \mathrm{PEC}(d)$, with $g(\Pi) = [a] \cdot \Pi$, and so the g-unitary $[a^{-1}]g$ stabilises Π.

(d) For a g-unitary $[a]g \in \mathrm{PC}(d) \times \mathscr{G}_c$, one can define an extended index (ψ_a, z_a, g), or an extended Appleby index $[B, b, g]$, $[a] = f([B, b])$. Show that

$$f_{\mathscr{G}_c} : SL_2(\mathbb{Z}_{d'}) \times \mathbb{Z}_d^2 \times \mathscr{G}_c \to \mathrm{PC}(d) \times \mathscr{G}_c : [B, b, g] \mapsto f([B, b])g,$$

is a homomorphism, where $SL_2(\mathbb{Z}_{d'}) \times \mathbb{Z}_d^2 \times \mathscr{G}_c$ is equipped with the multiplication

$$[B_1, b_1, g_1][B_2, b_2, g_1] = [B_1 H_{g_1} B_2 H_{g_1}^{-1}, b_1 + B_1 H_{g_1} b_2, g_1 g_2].$$

(e) The subgroup $SL_2(\mathbb{Z}_{d'}) \times \mathbb{Z}_d^2 \times \{1, g_c\}$ gives the extended Appleby triple indices for $\mathrm{PEC}(d)$. Let $J = H_{g_c}$. Show that the map

$$\Theta : SL_2(\mathbb{Z}_{d'}) \times \mathbb{Z}_d^2 \times \{1, g_c\} \to ESL_2(\mathbb{Z}_{d'}) \times \mathbb{Z}_d^2 : [B, b, g_c^j] \mapsto [BJ^j, b]$$

is an isomorphism between indices.
Remark: The map $f_{\mathscr{G}_c} \circ \Theta^{-1} : ESL_2(\mathbb{Z}_{d'}) \times \mathbb{Z}_d^2 \to \mathrm{PEC}(d)$ is the f_E of (14.57).

Chapter 15
Tight frames of orthogonal polynomials on the simplex

The orthogonal polynomials of degree k on the triangle are a finite dimensional inner product space which is invariant under the unitary action of the symmetry group G of the triangle (the dihedral group of order 6) given by

$$g \cdot f = f \circ g^{-1},$$

i.e., is a G-invariant space (see §10.10). It is natural to seek an orthogonal expansion for this space which is invariant under these symmetries. The polynomials in such an orthogonal expansion have a simple form, i.e., they consist of a small number of polynomials, together with those obtained by the changes of variables given by the action of G. In a few cases it is possible to find a G-invariant orthonormal basis, e.g., for the 3-dimensional space of quadratic Legendre polynomials (Example 10.20), but in general it is not. Here we present a natural G-invariant tight frame for this space (and more generally the Jacobi polynomials on the simplex). This is a great illustration of the usefulness of tight frames–the redundancy of a tight frame allows us to find an expansion with the symmetries of the space, which is not possible for a basis. Key aspects of this construction include:

- The Jacobi polynomials in the tight frame are given explicitly in terms of the Bernstein basis by using a multivariate generalisation of the ${}_2F_1$ hypergeometric function (the Lauricella function of type A).
- The tight frame considerably improves upon the previously known expansions (Appell's biorthogonal system which has some of the symmetries, and Proriol's orthogonal basis which has no symmetries and is given by a recursive formula).
- A polynomial is a Jacobi polynomial if and only if its Bernstein basis coefficients give a Hahn polynomial. The Bernstein coefficients are characterised by certain linear dependencies which can be expressed in terms of the adjoint of the degree elevation operator (for the Bernstein form).
- The proof that the frame is tight uses the fact that the Jacobi polynomials are eigenspaces of the Bernstein–Durrmeyer operator.

S.F.D. Waldron, *An Introduction to Finite Tight Frames*, Applied and Numerical Harmonic Analysis, https://doi.org/10.1007/978-0-8176-4815-2_15

15.1 Jacobi polynomials and their Bernstein coefficients

Throughout, let $\xi = (\xi_0, \xi_1, \ldots, \xi_d)$ be the barycentric coordinates of a d-simplex $T \subset \mathbb{R}^d$ with vertices $V = \{v_0, v_1, \ldots, v_d\}$, and volume $\mathrm{vol}_d(T)$. We recall (see §4.7) that the barycentric coordinates are the unique linear polynomials with

$$x = \sum_j \xi_j(x) v_j, \quad \sum_j \xi_j(x) = 1, \qquad \forall x \in \mathbb{R}^d.$$

In particular, the barycentric coordinate ξ_j is 1 at v_j and zero at all the other vertices. It is sometimes convenient to index the barycentric coordinates by the vertices that they correspond to (rather than labels for them). We will use standard multi-index notation as outlined in §4.9, e.g., $\Gamma(\nu) = \prod_j \Gamma(\nu_j)$, where Γ is the *Gamma function*.

Example 15.1. The barycentric coordinates of the interval (1-simplex) $[-1,1]$ with vertices $\{-1,1\}$ are

$$\xi_0(x) = \frac{1-x}{2}, \qquad \xi_1(x) = \frac{1+x}{2}, \tag{15.1}$$

and for the triangle (2-simplex) with vertices $\{0, e_1, e_2\}$ they are

$$\xi_0(x,y) = 1-x-y, \qquad \xi_1(x,y) = x, \qquad \xi_2(x,y) = y.$$

The **univariate Jacobi polynomials** are the orthogonal polynomials given by the inner product

$$\langle f, g \rangle_\nu := \frac{\Gamma(\nu_0+\nu_1)}{\Gamma(\nu_0)\Gamma(\nu_1)} \frac{1}{2} \int_{-1}^{1} f(x) g(x) \left(\frac{1-x}{2}\right)^{\nu_0-1} \left(\frac{1+x}{2}\right)^{\nu_1-1} dx, \qquad \nu_0, \nu_1 > 0.$$

The normalisation above is chosen so $\langle 1, 1 \rangle_\nu = 1$. The parameters are usually written as $\nu = (\nu_0, \nu_1) = (\alpha+1, \beta+1)$, with the Jacobi polynomial of degree n denoted by $P_n^{(\alpha,\beta)}$. The condition $\nu_j > 0$ ensures that the weight function is integrable.

Example 15.2. Well-known Jacobi polynomials include the **Legendre polynomials** ($\nu_j = 1$) and the **Chebyshev polynomials** ($\nu_j = \frac{1}{2}$), which are given by

$$\langle f, g \rangle_{(1,1)} = \frac{1}{2} \int_{-1}^{1} f(x) g(x)\, dx, \qquad \langle f, g \rangle_{(\frac{1}{2},\frac{1}{2})} = \frac{1}{\pi} \int_{-1}^{1} f(x) g(x) \frac{dx}{\sqrt{1-x^2}}.$$

In view of (15.1), the weight function on the interval $[-1,1]$ used to define the Jacobi polynomials is a product of powers of the barycentric coordinates of $[-1,1]$. By replacing $[-1,1]$ by a d-simplex T (a triangle for $d=2$), we obtain the following multivariate generalisation of the univariate Jacobi inner product.

Definition 15.1. Let T be a d-simplex in $\mathbb{R}^d$ with barycentric coordinates ξ. Then **(multivariate) Jacobi inner product** on T with parameters ν is given by

$$\langle f,g\rangle_\nu := \frac{\Gamma(|\nu|)}{\Gamma(\nu)}\frac{1}{d!\mathrm{vol}_d(T)}\int_T fg\,\xi^{\nu-1}, \qquad \nu\in\mathbb{R}^{d+1}, \quad \nu_j>0. \tag{15.2}$$

We call the Lebesgue measure on T weighted by $\xi^{\nu-1}$ the **Jacobi measure**.

A simple calculation shows that the condition $\nu_j>0$, $\forall j$ ensures the weight function $\xi^{\nu-1}$ is integrable on T, so the Jacobi inner product is well defined, and that

$$\langle \xi^\alpha,\xi^\beta\rangle_\nu = \frac{(\nu)_{\alpha+\beta}}{(|\nu|)_{|\alpha|+|\beta|}}, \qquad \alpha,\beta\in\mathbb{Z}_+^{d+1}, \tag{15.3}$$

where $(\nu)_\alpha := \prod_j(\nu_j)_{\alpha_j}$, with the **Pochhammer symbol** $(x)_n$ given by

$$(x)_n := x(x+1)\cdots(x+n-1).$$

The orthogonal polynomials of degree k (see §10.10) for the Jacobi measure are called the (**multivariate**) **Jacobi polynomials**, and we denote them by $\mathscr{P}_k^\nu$, i.e.,

$$\mathscr{P}_k^\nu := \{f\in\Pi_k(\mathbb{R}^d) : \langle f,h\rangle_\nu = 0, \forall h\in\Pi_{k-1}(\mathbb{R}^d)\},$$

where $\Pi_n(\mathbb{R}^d)$ is the space of polynomials of degree $\le n$ on $\mathbb{R}^d$. Since

$$\dim(\mathscr{P}_k^\nu) = \binom{k+d-1}{d-1},$$

the space $\mathscr{P}_k^\nu$ has dimension greater than 1 when $k>0$ and $d>1$.

Each polynomial $f\in\Pi_n(\mathbb{R}^d)$ can be expressed in terms of the Bernstein basis

$$f = \sum_{|\alpha|=n} c_\alpha(f)B_\alpha = \sum_{|\alpha|=n} c_\alpha B_\alpha,$$

where the **Bernstein polynomials** of degree n are defined by

$$B_\alpha := \binom{|\alpha|}{\alpha}\xi^\alpha = \frac{|\alpha|!}{\alpha!}\xi^\alpha = \frac{n!}{\alpha!}\xi^\alpha, \qquad |\alpha|=n, \quad \alpha\in\mathbb{Z}_+^{d+1}.$$

This basis for $\Pi_n(\mathbb{R}^d)$ is well suited to representing polynomials on the simplex T (see §4.9 for a generalisation to a frame for other polytopes). The coefficients

$$c(f) = c^n(f) = c = (c_\alpha)_{|\alpha|=n}$$

are referred to as the **Bernstein(–Bézier)** coefficients. By the multinomial theorem

$$f = \sum_{|\alpha|=n} c_\alpha B_\alpha = \sum_{|\alpha|=n} c_\alpha B_\alpha\Big(\sum_{i=0}^d \xi_i\Big)^j = \sum_{|\alpha|=n+j}(R^j c)_\alpha B_\alpha,$$

where the powers of the **degree raising operator** R are given by

$$(R^j c)_\alpha = \sum_{|\gamma|=j} \binom{j}{\gamma} \frac{(-\alpha)_\gamma}{(-|\alpha|)_j} c_{\alpha-\gamma}, \qquad j = 0,1,2,\ldots. \tag{15.4}$$

Here, we think of the Bernstein coefficients as a function $c : \alpha \mapsto c_\alpha$ defined on the simplex points

$$\Delta_n := \{\alpha \in \mathbb{Z}_+^{d+1} : |\alpha| = n\}, \qquad \#\Delta_n = \dim(\Pi_n(\mathbb{R}^d)).$$

We view such functions $c : \Delta_n \to \mathbb{R}$ as polynomials of degree n in d-variables by identifying c with the unique polynomial of degree n on the d-dimensional affine subspace $\{x \in \mathbb{R}^{d+1} : x_0 + x_1 + \cdots + x_d = n\}$ which takes the value c_α at $\alpha \in \Delta_n$. For example, by the multinomial theorem $f = 1 = \sum_{|\alpha|=n} B_\alpha$, and so 1 corresponds to the constant polynomial $c : \alpha \mapsto 1$. More generally, we have:

Proposition 15.1. *Suppose that $f = \sum_{|\alpha|=n} c_\alpha B_\alpha \in \Pi_n(\mathbb{R}^d)$ and $0 \le s \le n$. Then f has degree s if and only if $c : \alpha \mapsto c_\alpha$ is a polynomial of degree s.*

Proof. The polynomials $(B_\beta)_{|\beta|=s}$ are a basis for $\Pi_s(\mathbb{R}^d)$, and can be expressed

$$B_\beta = \sum_{|\alpha|=s} b_\alpha B_\alpha, \qquad b_\alpha := \begin{cases} 1, & \alpha = \beta; \\ 0, & \text{otherwise}. \end{cases}$$

Let $j := n - s$. Then by (15.4), the Bernstein coefficients of $B_\beta = \sum_{|\alpha|=n} c_\alpha B_\alpha$ are

$$c_\alpha = (R^j b)_\alpha = \sum_{|\gamma|=j} \binom{j}{\gamma} \frac{(-\alpha)_\gamma}{(-|\alpha|)_j} b_{\alpha-\gamma} = \frac{j!}{(\alpha-\beta)!} \frac{(-\alpha)_{\alpha-\beta}}{(-n)_j} = \frac{j!(-1)^j}{(-n)_j} \frac{(-\alpha)_\beta}{(-\beta)_\beta}.$$

Since $\Delta_n \to \mathbb{R} : \alpha \mapsto (-\alpha)_\beta$, $|\beta| = s$, is a polynomial of degree s, we obtain the correspondence. □

We define an inner product on the space of polynomials $\Delta_n \to \mathbb{R}$ of degree n by

$$\langle f, g \rangle_{\nu,n} := \sum_{|\alpha|=n} \frac{(\nu)_\alpha}{\alpha!} f(\alpha) g(\alpha). \tag{15.5}$$

The orthogonal polynomials of degree s corresponding to the discrete measure above are called the **Hahn polynomials**, and we denote them by $\mathscr{P}_s^{\nu,n}$, $0 \le s \le n$.

The **adjoint** R_ν^* of the degree raising operator R with respect to (15.5) is defined by

$$\langle Rc, b \rangle_{\nu,n} = \langle c, R_\nu^* b \rangle_{\nu,n-1}, \qquad c : \Delta_{n-1} \to \mathbb{R}, \quad b : \Delta_n \to \mathbb{R}. \tag{15.6}$$

A simple calculation (Exer. 15.1) shows the powers of R_ν^* are given by

$$((R_\nu^*)^j b)_\beta = \sum_{|\gamma|=j} \frac{(\beta+\nu)_\gamma}{(|\beta|+1)_j} \binom{j}{\gamma} b_{\beta+\gamma}, \qquad b : \Delta_n \to \mathbb{R}, \quad 0 \le j \le n. \tag{15.7}$$

We can now show that Jacobi polynomials are characterised by the fact that their Bernstein coefficients are a Hahn polynomial (and vice versa). This is a relatively new result (see [Cie87] for the univariate case, and [Wal06]).

Theorem 15.1. *Fix $\nu > 0$. Let $f = \sum_{|\alpha|=n} c_\alpha B_\alpha \in \Pi_n(\mathbb{R}^d)$, $c = (c_\alpha)$, and $0 \le s \le n$. Then the following are equivalent*

1. $f \in \mathscr{P}_s^\nu$ *(Jacobi polynomials).*
2. $c \in \mathscr{P}_s^{\nu,n}$ *(Hahn polynomials).*
3. $(R_\nu^*)^{n-s+1} c = 0$.

Proof. ($1 \Longleftrightarrow 2$) We have $f \in \mathscr{P}_s^\nu$ if and only if it is orthogonal to the spanning set $\{\xi^\beta\}_{|\beta|<s}$ for $\Pi_{s-1}(\mathbb{R}^d)$, i.e., by (15.3), we have

$$\langle f, \xi^\beta \rangle_\nu = \sum_{|\alpha|=n} c_\alpha \frac{n!}{\alpha!} \frac{(\nu)_{\alpha+\beta}}{(|\nu|)_{|\alpha|+|\beta|}} = \frac{n!}{(|\nu|)_{s+|\beta|}} \sum_{|\alpha|=n} \frac{(\nu)_\alpha}{\alpha!} c_\alpha (\nu+\alpha)_\beta = 0.$$

With $p_\beta : \Delta_n \to \mathbb{R} : \alpha \mapsto (\nu+\alpha)_\beta$, this orthogonality condition can be written as

$$\langle c, p_\beta \rangle_{\nu,n} = 0, \qquad |\beta| < s.$$

Since $\{p_\beta\}_{|\beta|<s}$ spans the space of polynomials of degree $< s$, this says $c \in \mathscr{P}_s^{\nu,n}$.

($2 \Longleftrightarrow 3$) Let $q_\beta : \Delta_n \to \mathbb{R} : \alpha \mapsto (-\alpha)_\beta$, $|\beta| = s-1$. With $k := n - |\beta| = n - s + 1$, using (15.7), we calculate

$$\begin{aligned}
\langle c, q_\beta \rangle_{\nu,n} &= \sum_{\substack{|\alpha|=n \\ \alpha \ge \beta}} \frac{(\nu)_\alpha}{\alpha!} c_\alpha (-\alpha)_\beta = \sum_{|\gamma|=k} \frac{(\nu)_{\beta+\gamma}}{(\beta+\gamma)!} c_{\beta+\gamma} (-\beta-\gamma)_\beta \\
&= (\nu)_\beta \sum_{|\gamma|=k} (\nu+\beta)_\gamma c_{\beta+\gamma} \frac{(-\beta-\gamma)_\beta}{(\beta+\gamma)!} = (\nu)_\beta \sum_{|\gamma|=k} (\nu+\beta)_\gamma c_{\beta+\gamma} \frac{(-1)^{|\beta|}}{\gamma!} \\
&= (\nu)_\beta \frac{(-1)^{s-1}}{k!} (s)_k \sum_{|\gamma|=k} \frac{(\nu+\beta)_\gamma}{(|\beta|+1)_k} c_{\beta+\gamma} \frac{k!}{\gamma!} \\
&= (\nu)_\beta (-1)^{s-1} \binom{n}{s-1} ((R_\nu^*)^k c)_\beta.
\end{aligned}$$

Since $\{q_\beta\}_{|\beta|=s-1}$ is a basis for the polynomials of degree $< s$, we have $c \in \mathscr{P}_s^{\nu,n}$ if and only if $\langle c, q_\beta \rangle_{\nu,n} = 0$, $|\beta| = s-1$, which is equivalent to $(R_\nu^*)^k c = 0$, by the calculation above. □

This association between Jacobi and Hahn polynomials preserves inner products:

Theorem 15.2. *([Wal06]) Suppose that $f = \sum_{|\alpha|=n} c_\alpha(f) B_\alpha$ and $g = \sum_{|\alpha|=n} c_\alpha(g) B_\alpha$. If f or g belongs to $\mathscr{P}_s^\nu$, $0 \le s \le n$, then*

$$\langle f, g \rangle_\nu = \frac{(n!)^2}{(n-s)!(|\nu|)_{n+s}} \sum_{|\alpha|=n} \frac{(\nu)_\alpha}{\alpha!} c_\alpha(f) c_\alpha(g) = \frac{(n!)^2}{(n-s)!(|\nu|)_{n+s}} \langle c(f), c(g) \rangle_{\nu,n}.$$

15.2 The Bernstein–Durrmeyer operator

To motivate the tight frame of Jacobi polynomials on a simplex in §15.3, we consider the **Bernstein–Durrmeyer** operator M_n^ν (see [Der85], [BX91]). This is defined on the continuous functions on the simplex T with the Jacobi inner product (15.2) by

$$M_n^\nu f := \sum_{|\alpha|=n} \frac{\langle f, \xi^\alpha \rangle_\nu}{\langle 1, \xi^\alpha \rangle_\nu} B_\alpha = \sum_{|\alpha|=n} \langle f, \xi^\alpha \rangle_\nu \frac{(|\nu|)_n}{(\nu)_\alpha} \frac{n!}{\alpha!} \xi^\alpha.$$

This self-adjoint operator is a natural generalisation of the Bernstein operator (see §4.10). It is *positive*, i.e.,

$$f \geq 0 \quad \Longrightarrow \quad M_n^\nu f \geq 0,$$

but does not reproduce the nonconstant linear polynomials. It is *degree reducing*, and satisfies an analogue of Theorem 4.6. It can be viewed as a *de la Vallée–Poussin mean* of the orthogonal projections onto $\mathscr{P}_s^\nu$ $0 \leq s \leq n$. From this, or by a simple calculation (see [Der85]), it follows that the eigenvalues of M_n^ν are

$$\lambda_s(M_n^\nu) = \frac{n!}{(n-s)!} \frac{1}{(n+|\nu|)_s}, \qquad 0 \leq s \leq n,$$

with corresponding eigenspace the Jacobi polynomials $\mathscr{P}_s^\nu$, i.e., for $0 \leq s \leq n$

$$f = (n-s)!(|\nu|)_{n+s} \sum_{|\alpha|=n} \frac{1}{\alpha!} \langle f, \xi^\alpha \rangle_\nu \frac{\xi^\alpha}{(\nu)_\alpha}, \qquad \forall f \in \mathscr{P}_s^\nu. \tag{15.8}$$

Let Q_s be the orthogonal projection onto $\mathscr{P}_s^\nu$. Then for $f \in \mathscr{P}_s^\nu$,

$$\langle f, \xi^\alpha \rangle_\nu = \langle Q_s f, \xi^\alpha \rangle_\nu = \langle f, Q_s(\xi^\alpha) \rangle_\nu,$$

and so from (15.8), we obtain

$$\begin{aligned} f &= (n-s)!(|\nu|)_{n+s} \sum_{|\alpha|=n} \frac{1}{\alpha!} \langle f, Q_s(\xi^\alpha) \rangle_\nu \frac{\xi^\alpha}{(\nu)_\alpha} \\ &= (n-s)!(|\nu|)_{n+s} \sum_{|\alpha|=n} \frac{(\nu)_\alpha}{\alpha!} \Big\langle f, \frac{Q_s(\xi^\alpha)}{(\nu)_\alpha} \Big\rangle_\nu \frac{Q_s(\xi^\alpha)}{(\nu)_\alpha}, \qquad \forall f \in \mathscr{P}_s^\nu. \end{aligned} \tag{15.9}$$

In other words, (15.9) gives a *tight frame* expansion for $\mathscr{P}_s^\nu$. In the next section, we obtain an explicit formula for $Q_s(\xi^\alpha)$ when $n = s$ (the case with the fewest vectors).

This result was found independently by [Ros99] (not presented in terms of tight frames) and by [XW01], [PW02]. The presentation in terms of the Bernstein–Durrmeyer operator M_n^ν given here is adapted from [Wal06].

15.3 Tight frames of Jacobi polynomials with symmetries

We now give the main result.

Theorem 15.3. *A tight frame expansion for the Jacobi polynomials $\mathscr{P}_n^\nu$ is given by*

$$f = (|\nu|)_{2n} \sum_{|\alpha|=n} \frac{(\nu)_\alpha}{\alpha!} \langle f, \phi_\alpha^\nu \rangle_\nu \phi_\alpha^\nu, \qquad \forall f \in \mathscr{P}_n^\nu, \tag{15.10}$$

where

$$\phi_\alpha^\nu := \frac{(-1)^n}{(n+|\nu|-1)_n} \sum_{\beta \le \alpha} \frac{(n+|\nu|-1)_{|\beta|}(-\alpha)_\beta}{(\nu)_\beta} \frac{\xi^\beta}{\beta!} \tag{15.11}$$

is the orthogonal projection of $\xi^\alpha/(\nu)_\alpha$ onto $\mathscr{P}_s^\nu$.

Proof. In view of the tight frame expansion (15.9) for $\mathscr{P}_n^\nu$ (take $s=n$), it suffices to prove that ϕ_α^ν is the orthogonal projection of $\xi^\alpha/(\nu)_\alpha$ onto $\mathscr{P}_n^\nu$. Since

$$\phi_\alpha^\nu \in \frac{\xi^\alpha}{(\nu)_\alpha} + \Pi_{n-1}(\mathbb{R}^d),$$

this reduces to showing that ϕ_α^ν is orthogonal to the basis $(\xi^\gamma)_{|\gamma|=n-1}$ for $\Pi_{n-1}(\mathbb{R}^d)$. Suppose that $|\gamma| = n-1$. Then

$$(|\nu|)_{|\beta|+|\gamma|} = (|\nu|)_{n-1}(|\nu|+n-1)_{|\beta|}, \qquad (\nu)_{\beta+\gamma} = (\nu)_\gamma(\nu+\gamma)_\beta,$$

and so, by (15.3), we have

$$\begin{aligned}
\langle \phi_\alpha^\nu, \xi^\gamma \rangle_\nu &= \frac{(-1)^n}{(n+|\nu|-1)_n} \sum_{\beta \le \alpha} \frac{(n+|\nu|-1)_{|\beta|}(-\alpha)_\beta}{(\nu)_\beta \beta!} \frac{(\nu)_{\beta+\gamma}}{(|\nu|)_{|\beta|+|\gamma|}} \\
&= \frac{(-1)^n}{(n+|\nu|-1)_n} \frac{(\nu)_\gamma}{(|\nu|)_{n-1}} \sum_{\beta \le \alpha} \frac{(-\alpha)_\beta}{(\nu)_\beta \beta!} (\nu+\gamma)_\beta.
\end{aligned}$$

By the Chu–Vandermonde identity, the last sum above simplifies to

$$\sum_{\beta \le \alpha} \frac{(-\alpha)_\beta(\nu+\gamma)_\beta}{(\nu)_\beta \beta!}(\nu+\gamma)_\beta = \sum_\beta \frac{(-\alpha)_\beta(\nu+\gamma)_\beta}{(\nu)_\beta \beta!} = \frac{(\nu-(\nu+\gamma))_\alpha}{(\nu)_\alpha} = \frac{(-\gamma)_\alpha}{(\nu)_\alpha}.$$

Since $|\gamma| < |\alpha| = n$, we must have $\gamma_j < \alpha_j$ for some j, i.e., $(-\gamma_j)_{\alpha_j} = 0$, and so $(-\gamma)_\alpha = 0$, which completes the proof. □

The Jacobi polynomials ϕ_α^ν in the tight frame for $\mathscr{P}_n^\nu$ can be written

$$\phi_\alpha^\nu = \frac{(-1)^n}{(n+|\nu|-1)_n} F_A(n+|\nu|-1, -\alpha, \nu; \xi),$$

where F_A is the **Lauricella function** of type A, which is given by

$$F_A(a,b;c;x) := \sum_{\alpha\in\mathbb{Z}_+^{d+1}} (a)_{|\alpha|} \frac{(b)_\alpha}{(c)_\alpha} \frac{x^\alpha}{\alpha!}, \qquad a\in\mathbb{R}, \quad b,c,x\in\mathbb{R}^{d+1}.$$

This is a multivariate generalisation of the hypergeometric function ${}_2F_1(a,b;c;x)$. We observe $F_A(a,-\beta,c;x)$ is a polynomial of degree $|\beta|$ in x when β is a multi-index.

Example 15.3. (Univariate Jacobi polynomials). The univariate Jacobi polynomials $P_n^{(\alpha,\beta)}$ for the weight $v=(\alpha+1,\beta+1)$ on $[-1,1]$ and barycentric coordinates ξ given by (15.1), can be written

$$P_n^{\alpha,\beta}(x) = \frac{(\alpha+1)_n}{n!} {}_2F_1\bigl(-n,1+\alpha+\beta+n;\alpha+1;\frac{1}{2}(1-x)\bigr). \tag{15.12}$$

The Jacobi polynomial for the multi-index $(n,0)$ given by (15.11) is

$$\begin{aligned}
\phi_{(n,0)}^v(x) &= \frac{(-1)^n}{(n+|v|-1)_n} \sum_{j=0}^n \frac{(n+|v|-1)_j(-n)_j}{(v_0)_j} \frac{\xi_0(x)^j}{j!} \\
&= \frac{(-1)^n}{(n+\alpha+\beta+1)_n} \sum_{j=0}^n \frac{(-n)_j(n+\alpha+\beta+1)_j}{(\alpha+1)_j} \frac{(\frac{1}{2}(1-x))^j}{j!} \\
&= \frac{(-1)^n}{(n+\alpha+\beta+1)_n} {}_2F_1(-n,1+\alpha+\beta+n;\alpha+1;\frac{1}{2}(1-x)) \\
&= \frac{(-1)^n}{(n+\alpha+\beta+1)_n} \frac{n!}{(\alpha+1)_n} P_n^{(\alpha,\beta)}(x)
\end{aligned}$$

Since $\mathscr{P}_n^v$ is one dimensional, each $\phi_{(n-k,k)}^v$, $0\le k\le n$, is a scalar multiple of $P_n^{\alpha,\beta}$.

The tight frame expansion for the Jacobi polynomials $\mathscr{P}_s^v$ given by (15.9) for $0\le s<n$ has more vectors than that of (15.10). By calculating $Q_s(\xi^\alpha)$ explicitly (see [Wal06]), one can generalise Theorem 15.3 as follows.

Theorem 15.4. *A tight frame for the Jacobi polynomials $\mathscr{P}_s^v$, $s\le n$ is given by*

$$f = (n-s)!(|v|)_{n+s} \sum_{|\alpha|=n} \frac{(v)_\alpha}{\alpha!} \langle f,\phi_\alpha^{v,s}\rangle_v \phi_\alpha^{v,s}, \qquad \forall f\in\mathscr{P}_s^v, \tag{15.13}$$

where

$$\phi_\alpha^{v,s} := \frac{(-1)^s}{(s+|v|-1)_s} \frac{\binom{n}{s}}{(|v|+2s)_{n-s}} \sum_{\substack{\beta\le\alpha \\ |\beta|\le s}} \frac{(s+|v|-1)_{|\beta|}(-\alpha)_\beta(-s)_{|\beta|}}{(v)_\beta(-n)_{|\beta|}} \frac{\xi^\beta}{\beta!} \tag{15.14}$$

is the orthogonal projection of $\xi^\alpha/(v)_\alpha$ onto $\mathscr{P}_s^v$. We also have

$$f = \frac{(n-s)!}{n!}(|v|)_{n+s} \sum_{|\alpha|=n} \langle f,\phi_\alpha^{v,s}\rangle_v B_\alpha, \qquad \forall f\in\mathscr{P}_s^v. \tag{15.15}$$

15.4 The orthogonal polynomials of Appell and Proriol

We now consider the symmetries of tight frames for $\mathscr{P}_n^\nu$ given by Theorem 15.4 and compare them with the previously known expansions.

The affine maps $g \in \mathrm{Aff}(\mathbb{R}^d)$ of the simplex T onto itself are uniquely determined by their action on the vertices $\{v_0, v_1, \ldots, v_d\}$ or the barycentric coordinates, i.e.,

$$g v_j = v_{\sigma j}, \qquad g \cdot \xi_j = \xi_{\sigma j},$$

where $\sigma \in S_{d+1}$ is a permutation (of the vertices). Under this identification, the symmetry group of the Jacobi measure is

$$G := \{\sigma \in S_{d+1} : \nu_{\sigma j} = \nu_j, \forall j\},$$

and the Jacobi polynomials $\mathscr{P}_n^\nu$ are a G-invariant space, where the unitary action of G is given by

$$\sigma \cdot f := f \circ g^{-1}, \qquad f \in \Pi(\mathbb{R}^d).$$

With $\sigma\alpha := (\sigma\alpha_0, \ldots, \sigma\alpha_d)$, applying $\sigma \cdot \xi_j = \xi_{\sigma j}$ to (15.14) gives

$$\sigma \cdot \phi_\alpha^{\nu,s} = \phi_{\sigma\alpha}^{\sigma\nu,s} = \phi_{\sigma\alpha}^{\nu,s}, \qquad \forall \sigma \in G,$$

i.e., the tight frames of Theorem given by Theorem 15.4 are G-invariant, i.e., the G-orbit of some smaller number of vectors.

Example 15.4. (Legendre polynomials) Suppose that all the ν_j are equal, e.g., the **Legendre polynomials** given by $\nu_j = 1$ (giving Lebesgue measure on T). Then the symmetry group is $G = S_{d+1}$, and the tight frame $(\phi_\alpha^\nu)_{|\alpha|=n}$ of Theorem 15.3 is the orbit of $p(n)$ polynomials, where $p(n)$ is the partition function (the number of partitions of n). For example, $3, 2+1, 1+1+1$ are the three partitions of $n = 3$, and so the tight frame for the cubic Jacobi polynomials $\mathscr{P}_3^\nu$ is the orbit of three polynomials for *any* dimension d.

Example 15.5. (Quadratic Jacobi polynomials) From (15.11), we have

$$\phi_{(2,0,\ldots,0)}^\nu = \frac{\xi_0^2}{\nu_0(\nu_0+1)} - \frac{2}{(2+|\nu|)}\frac{\xi_0}{\nu_0} + \frac{1}{(1+|\nu|)(2+|\nu|)},$$

$$\phi_{(1,1,,0,\ldots,0)}^\nu = \frac{\xi_0\xi_1}{\nu_0\nu_1} - \frac{1}{(2+|\nu|)}\left(\frac{\xi_0}{\nu_0} + \frac{\xi_1}{\nu_1}\right) + \frac{1}{(1+|\nu|)(2+|\nu|)}.$$

The remaining quadratic Jacobi polynomials in the tight frame $\Phi = (\phi_\alpha^\nu)_{|\alpha|=2}$ for $\mathscr{P}_2^\nu$ are obtained from these by making the substitution

$$\nu_0 \mapsto \nu_j, \qquad \nu_1 \mapsto \nu_k, \quad k \neq j.$$

When $\nu_0 = \nu_j$, $\nu_1 = \nu_k$, then this corresponds to a symmetry of $\mathscr{P}_2^\nu$. In any case, formulas for all the polynomials in Φ can be obtained from those two above, by substitutions, independent of the dimension d.

The **Appell polynomials** (introduced in [AKdF26]) are the nonorthogonal basis for $\mathscr{P}_n^\nu$ given by the subset of the tight frame $\Phi = (\phi_\alpha^\nu)_{|\alpha|=n}$ for which the k-th component of α is zero, i.e.,

$$\Phi_k := \{\phi_\alpha^\nu : |\alpha| = n, \alpha_k = 0\}.$$

There is an explicit formula for the polynomials in the dual basis (see [AKdF26], [FL74], [KMT91]). The Appell basis and its dual basis are invariant under the action of subgroup H of G (the symmetry group of the measure) given by

$$H := \{\sigma \in G : \nu_{\sigma k} = \nu_k\}.$$

The **Proriol polynomials** (introduced in [Pro57]) are an orthonormal basis for $\mathscr{P}_n^\nu$ given explicitly, but by complicated formulas (see [DX01]). These polynomials are not invariant under any of the (nonidentity) symmetries of the Jacobi weight.

The basis $\Phi = (\phi_\alpha^\nu)_{|\alpha|=n}$ for $\mathscr{P}_n^\nu$ has the following desirable properties:

- It is a tight frame.
- Its polynomials are given explicitly in the Bernstein form by using a multivariate version of the ${}_1F_2$ hypergeometric function (the Lauricella function F_A). These formulas do not become more complicated as d becomes large.
- It is invariant under all of the symmetries of the Jacobi weight.

As discussed, the bases of Appell and Proriol do not share all of these properties.

Example 15.6. Consider the three-dimensional space $\mathscr{P}_2^\nu$ of quadratic Legendre polynomials on a triangle. The symmetry group of the weight is S_3 (order 6), i.e., all permutations of the vertices give symmetries. The tight frame Φ is given by the orthogonal projections of

$$\xi_0^2, \quad \xi_1^2, \quad \xi_2^2, \quad \xi_1\xi_2, \quad \xi_0\xi_2, \quad \xi_0\xi_1$$

onto $\mathscr{P}_2^\nu$. This is the S_3-orbit of two polynomials. The Appell basis is the orthogonal projection of

$$\xi_0^2, \quad \xi_1^2, \quad \xi_0\xi_1,$$

which is invariant under the subgroup of order 2 generated by the permutation $\begin{pmatrix}0 & 1\end{pmatrix}$. The Proriol basis is given by the orthogonal projection of

$$\xi_0^2, \quad \xi_0(\xi_0 + 2\xi_1), \quad 4\xi_1^2 + 2\xi_0\xi_1 - \xi_0^2.$$

Notes

The idea of using finite tight frame expansions for spaces of *multivariate orthogonal polynomials* (not to be confused with *multiple orthogonal polynomials* [MFVA16]) appeared independently in [Ros99] and [XW01], [PW02]. A detailed account of the multivariate orthogonal polynomials, which includes the systems of Appell and Prorial, is given in [DX01]. The presentation in terms of the Bernstein–Durrmeyer operator M_n^ν that is given here is adapted from [RW04], [Wal06]. There are similar expansions for the multivariate *Hahn* and *continuous Hahn polynomials* [RW04], and for the multivariate orthogonal polynomials for a radially symmetric weight (see Chapter 16).

Tight frames allow for optimal expansions for spaces of multivariate orthogonal polynomials for specific weights, e.g., see [Dun87] and §10.10, §10.14.

Exercises

15.1. Let R be the degree raising operator given by (15.4), and R_ν^* be its adjoint as given by (15.6). Show that the j-th power of R_ν^* is given by (15.7), i.e.,

$$((R_\nu^*)^j b)_\beta = \sum_{|\gamma|=j} \frac{(\beta+\nu)_\gamma}{(|\beta|+1)_j} \binom{j}{\gamma} b_{\beta+\gamma}, \qquad b : \Delta_n \to \mathbb{R}, \quad 0 \le j \le n.$$

Chapter 16
Continuous tight frames for finite dimensional spaces

The tight frame expansion for n equally spaced unit vectors in $\mathbb{R}^2$ is

$$f = \frac{2}{n}\sum_{j=1}^{n}\langle f,u_j\rangle u_j, \qquad \forall f\in\mathbb{R}^2, \qquad u_j := \begin{pmatrix}\cos\frac{2\pi j}{n}\\ \sin\frac{2\pi j}{n}\end{pmatrix}. \tag{16.1}$$

We may take the limit of this, as $n\to\infty$, to obtain

$$f = \frac{2}{2\pi}\int_0^{2\pi}\langle f,u_\theta\rangle u_\theta\,d\theta, \qquad \forall f\in\mathbb{R}^2, \qquad u_\theta := \begin{pmatrix}\cos\theta\\ \sin\theta\end{pmatrix}, \tag{16.2}$$

which is the prototypical example of a *continuous* tight frame expansion.

A key feature of this expansion is that the vectors $(u_\theta)_{0\le\theta\le 2\pi}$ are invariant under the symmetries $O(2)$ of $\mathbb{R}^2$ (see the comments at the start of §9). Indeed, one can argue that this is the natural representation for the space $\mathcal{H}=\mathbb{R}^2$, which follows directly from its symmetries (apply $O(2)$ to any unit vector), and that expansions like (16.1) then follow by a process of discretisation (called *sampling*).

16.1 Continuous and discrete frames

The sum $\sum_{j\in J}$ in the definition of a frame can be a more general integral.

Definition 16.1. Let $\mathcal{H}$ be a Hilbert space and $(J,\mathcal{S},\mu)$ be a measure space. A **(generalised) frame** for $\mathcal{H}$ with respect to μ is a family $(f_j)_{j\in J}$ for which

1. For each $f\in\mathcal{H}$, $J\to\mathbb{F}: j\mapsto\langle f,f_j\rangle$ is $\mathcal{S}$-measurable on J.
2. There exist (frame bounds) $A,B>0$ such that

$$A\|f\|^2 \le \int_J |\langle f,f_j\rangle|^2\,d\mu(j) \le B\|f\|^2, \qquad \forall f\in\mathcal{H}.$$

S.F.D. Waldron, *An Introduction to Finite Tight Frames*, Applied and Numerical Harmonic Analysis, https://doi.org/10.1007/978-0-8176-4815-2_16

It is **tight** if one can choose $A = B$. Further, we will refer to the frame as being **discrete** if $\mu(\{j\}) > 0, \forall j \in J$, and **continuous** if $\mu(\{j\}) = 0, \forall j \in J$.

Example 16.1. If μ is the counting measure on a set J, then a generalised frame is precisely a frame.

Example 16.2. The vectors $(u_\theta)_{0\le\theta\le 2\pi}$ of (16.2) form a continuous tight frame for $\mathbb{R}^2$ (with respect to the Lebesgue measure), since

$$\int_0^{2\pi} |\langle x, u_\theta\rangle|^2 \, d\theta = \int_0^{2\pi} (x_1 \cos\theta + x_2 \sin\theta)^2 \, d\theta, = \pi \|x\|^2, \qquad \forall x \in \mathbb{R}^2.$$

The basic results on frames extend in the obvious fashion, i.e., replace the sum over J by an integral. In particular (see Exercises 16.1 and 16.2), we have:

Proposition 16.1. *Let $\Phi = (f_j)_{j\in J}$ be a generalised frame with respect to μ for $\mathcal{H}$. Then*

$$Sf := \int_J \langle f, f_j\rangle f_j \, d\mu(j)$$

defines a bounded invertible self-adjoint operator $S = S_{\Phi,\mu} : \mathcal{H} \to \mathcal{H}$, for which

$$f = \int_J \langle f, S^{-1} f_j\rangle f_j \, d\mu(j) = \int_J \langle f, f_j\rangle S^{-1} f_j \, d\mu(j)$$
$$= \int_J \langle S^{-\frac{1}{2}} f, f_j\rangle S^{-\frac{1}{2}} f_j \, d\mu(j), \qquad \forall f \in \mathcal{H}.$$

We call $(\tilde{f}_j) = (S^{-1} f_j)$ the **dual frame**, and $(S^{-\frac{1}{2}} f_j)$ the **canonical tight frame**.

Proposition 16.2. *(Variational characterisation) Let $(f_j)_{j\in J}$ be a generalised frame with respect to μ for a d-dimensional space $\mathcal{H}$. Then*

$$\int_J \int_J |\langle f_j, f_k\rangle|^2 \, d\mu(j) \, d\mu(k) \ge \frac{1}{d} \Big(\int_J \|f_j\|^2 \, d\mu(j) \Big)^2,$$

with equality if and only if (f_j) is tight.

Example 16.3. Let $\Pi_n^\circ(\mathbb{R}^d)$ be the space of homogeneous polynomials of degree n on $\mathbb{R}^d$, i.e., those polynomials f satisfying

$$f(x) = \|x\|^n f\Big(\frac{x}{\|x\|}\Big), \qquad x \ne 0.$$

In view of this, these polynomials are determined by their values on the unit sphere $\mathbb{S} := \{x \in \mathbb{R}^d : \|x\| = 1\}$, and so we can define an inner product on them by $\langle f, g\rangle = \int_{\mathbb{S}} fg$ (Lebesgue integration on $\mathbb{S}$). The *ridge polynomials* $\Phi = (\langle \cdot, v\rangle^n)_{v\in\mathbb{S}}$ give a continuous frame for $\Pi_n^\circ(\mathbb{R}^d)$ (it is well known that they span). A calculation shows that Φ is tight for $n = 0, 1$ (but not for $n \ge 2$). In §16.5, we will consider continuous tight frames for the this space (and the space of complex homogeneous polynomials).

16.2 The analysis and synthesis operators

Let $\Phi = (f_J)_{j \in J}$ be a generalised tight frame for $\mathscr{H}$ with respect to μ. Then the **synthesis operator** can be defined in the natural way (see Exer. 16.1)

$$V = V_\Phi : L_2(\mu) \to \mathscr{H} : a \mapsto \int_J a_j f_j \, d\mu(j).$$

From this, one can define the **Gramian** and the **canonical Gramian**

$$\mathrm{Gram}(\Phi) = V^*V : L_2(\mu) \to L_2(\mu), \qquad P_\Phi = V^* S^{-1} V : L_2(\mu) \to L_2(\mu),$$

where $S = VV^* : \mathscr{H} \to \mathscr{H}$ is the frame operator. The canonical Gramian P_Φ is an orthogonal projection. It can be represented by the "matrix" $[P_\Phi]=[\langle f_k, S^{-1} f_j \rangle]_{j,k \in J}$ (see Exer. 16.3), where

$$P_\Phi a = [P_\Phi] \cdot_\mu a := \int_J [P_\Phi]_{jk} a_k \, d\mu(k) = \int_J \langle f_k, S^{-1} f_j \rangle a_k \, d\mu(k).$$

Example 16.4. (Short-time Fourier transform) Let $\mathscr{H} = L_2(\mathbb{R}^d)$, $g \in L_2(\mathbb{R}^d)$ be nonzero, and $M_\omega T_x g(t) := e^{2\pi i \omega \cdot t} g(t-x)$. Then

$$\int_{\mathbb{R}^d} \int_{\mathbb{R}^d} |\langle f, M_\omega T_x g \rangle|^2 \, dx \, d\omega = \|g\|^2 \|f\|^2, \qquad \forall f \in L_2(\mathbb{R}^d),$$

so that $(M_\omega T_x g)_{(x,\omega) \in \mathbb{R}^d \times \mathbb{R}^d}$ is a tight continuous frame with respect to Lebesgue measure on $\mathbb{R}^d \times \mathbb{R}^d$. The analysis operator $V^* : L_2(\mathbb{R}^d) \to L_2(\mathbb{R}^d \times \mathbb{R}^d)$ given by

$$(V^* f)(x, \omega) = \langle f, M_\omega T_x g \rangle \int_{\mathbb{R}^d} f(t) \overline{g(t-x)} e^{-2\pi i \omega \cdot t} \, dt, \qquad x, \omega \in \mathbb{R}^d$$

is called the **short-time Fourier transform (STFT)** of f with respect to g (the *window function*). See [Grö01] for detail.

In principle, the definition of a continuous tight frame could be extended, so as to include many important integral transforms which behave in a similar way. For example, the **Fourier transform** $\hat{f}$ of $f \in L_2(\mathbb{R})$,

$$\hat{f}(\omega) := \int_{-\infty}^{\infty} f(t) e^{-2\pi i \omega} \, dt = \langle f, e^{2\pi i \omega \cdot} \rangle, \qquad \langle f, g \rangle := \int_{-\infty}^{\infty} f(t) \overline{g(t)} \, dt$$

gives the reconstruction formula

$$f(t) = \frac{1}{2\pi} \int_{-\infty}^{\infty} \hat{f}(\omega) e^{2\pi i \omega t} \, d\omega = \frac{1}{2\pi} \int_{-\infty}^{\infty} \langle f, e^{2\pi i \omega \cdot} \rangle e^{2\pi i \omega t} \, d\omega, \qquad \forall f \in L_2(\mathbb{R}),$$

which would be a continuous tight frame expansion if $e^{2\pi i \omega \cdot} \in L_2(\mathbb{R})$.

16.3 Reproducing kernels

Many interesting continuous tight frames come from reproducing kernels. Some, such as the family of zonal harmonics predate the theory of frames.

Definition 16.2. A Hilbert space $\mathscr{H}$, of functions defined on some set X, is called a **reproducing kernel Hilbert space** if each of the point evaluations $x \mapsto f(x)$, $x \in X$ is continuous, and hence has a Riesz representer $K_x \in \mathscr{H}$. For such a space, the function $K : X \times X \to \mathbb{F}$ defined by

$$K(x,y) := \langle K_y, K_x \rangle = K_y(x), \qquad \forall x,y \in X$$

is called the **reproducing kernel**.

The reproducing kernel is Hermitian, i.e.,

$$\overline{K_x(y)} = \overline{\langle K_x, K_y \rangle} = \langle K_y, K_x \rangle = K_y(x).$$

Example 16.5. Let $\mathscr{H}$ be a reproducing kernel Hilbert space of functions $X \to \mathbb{F}$, with the inner product given by a measure μ, i.e.,

$$\langle f,g \rangle := \int_X f(y)\overline{g(y)}\, d\mu, \tag{16.3}$$

If K is the reproducing kernel, then

$$f(x) = \langle f, K_x \rangle = \int_X f(y)\overline{K_x(y)}\, d\mu(y) = \int_X \langle f, K_y \rangle K_y(x)\, d\mu(y),$$

i.e.,

$$f = \int_X \langle f, K_y \rangle K_y\, d\mu(y), \qquad \forall f \in \mathscr{H}, \tag{16.4}$$

and so $(K_y)_{y \in X}$ is a normalised generalised tight frame with respect to the measure μ, which we will refer to as the **reproducing kernel tight frame**.

Clearly, every finite dimensional space with an inner product of the form (16.3) has a reproducing kernel, and hence a natural generalised tight frame.

Example 16.6. Let $\mathscr{H} = F_d$, a space of functions on $J = \{1,2,\ldots,n\}$, with the inner product given by the counting measure. The Riesz representer of the point evaluation $j \mapsto x_j = \langle x, e_j \rangle$ is the standard basis vector e_j, and so the reproducing kernel is

$$K(j,k) = e_k(j) = \delta_{jk}, \qquad K_j = e_j,$$

and the corresponding generalised tight frame is the standard orthonormal basis $(e_j)_{j=1}^n$. More generally, if $\mathscr{H}$ is a subspace of $\mathbb{F}^n$, then the reproducing kernel tight frame is $(Pe_j)_{j=1}^n$, where P is the orthogonal projection onto $\mathscr{H}$ (see Exer. 16.5).

A formula for the reproducing kernel can be computed from any frame.

Proposition 16.3. *Suppose that K is the reproducing kernel for $\mathcal{H}$, and (f_j) is a finite frame for $\mathcal{H}$, with alternate dual frame (g_j), e.g., $g_j = \tilde{f}_j$, then*

$$K(x,y) = \langle K_y, K_x\rangle = \sum_j g_j(x)\overline{f_j(y)}.$$

Proof. Expand $K_x \in \mathcal{H}$ in terms of this frame

$$K_x = \sum_j \langle K_x, f_j\rangle g_j = \sum_j \overline{\langle f_j, K_x\rangle} g_j = \sum_j \overline{f_j(x)} g_j.$$

Using this, and the frame expansion, we obtain

$$\langle K_y, K_x\rangle = \langle \sum_j \overline{f_j(y)} g_j, \sum_k \overline{f_k(x)} g_k\rangle = \sum_j \overline{f_j(y)} \sum_k \langle g_j, g_k\rangle f_k(x) = \sum_j \overline{f_j(y)} g_j(x),$$

as claimed. □

16.4 Zonal harmonics

The extension of (16.2) to $\mathbb{R}^d$ is most easily obtained by using *zonal harmonics*.

We denote the **(unit) sphere** in $\mathbb{R}^d$ by

$$\mathbb{S} = \mathbb{S}^{d-1} := \{x \in \mathbb{R}^d : \|x\| = 1\}.$$

A function $f : X \to \mathbb{R}$, $X \subset \mathbb{R}^d$ is **harmonic** if it satisfies *Laplace's equation*, i.e.,

$$\Delta f = 0, \qquad \Delta := D_1^2 + \cdots D_d^2.$$

Let $\mathcal{H}_k = \mathcal{H}_k(\mathbb{R}^d)$ be the space of homogeneous harmonic polynomials of degree k. The map $f \mapsto f|_{\mathbb{S}}$, of restriction of a function to the sphere, applied to $\mathcal{H}_k$ has trivial kernel, so that

$$\dim(\mathcal{H}_k) = \dim(\mathcal{H}_k(\mathbb{S})), \qquad \mathcal{H}_k(\mathbb{S}) := \{f|_{\mathbb{S}} : f \in \mathcal{H}_k\}.$$

The spaces $\mathcal{H}_k(\mathbb{R}^d)$ and $\mathcal{H}_k(\mathbb{S})$ are the (**solid** and **surface**) **spherical harmonics** of degree k. They are invariant under the action of $O(d)$ and $SO(d)$ (which are absolutely irreducible), and have dimension

$$\dim(\mathcal{H}_k) = \binom{k+d-1}{d-1} - \binom{k+d-3}{d-1}. \tag{16.5}$$

Spherical harmonics of different degrees are orthogonal to each other with respect to the inner product

$$\langle f,g\rangle_{\mathbb{S}} := \int_{\mathbb{S}} fg\,d\sigma = \frac{1}{\operatorname{area}(\mathbb{S})}\int_{\mathbb{S}} f(\xi)g(\xi)\,d\xi, \tag{16.6}$$

where σ is normalised surface-area measure on the sphere $\mathbb{S}$, $d\xi$ denotes Lebesgue measure on the sphere, and the area of the sphere is

$$\operatorname{area}(\mathbb{S}) := \int_{\mathbb{S}} 1\,d\xi = \frac{2\pi^{\frac{d}{2}}}{\Gamma(\frac{d}{2})}. \tag{16.7}$$

The spherical harmonics give the orthogonal decomposition

$$L_2(\mathbb{S}) = \bigoplus_{k=0}^{\infty} \mathscr{H}_k, \tag{16.8}$$

of $L_2(\mathbb{S})$ into absolutely irreducible $SO(d)$-invariant subspaces.

The spherical harmonics of a degree k with the inner product of (16.6) form a reproducing kernel Hilbert space, and so have a natural continuous tight frame.

Definition 16.3. The **zonal harmonic**[1] of degree k with **pole** $\xi \in \mathbb{S}$ is the Riesz representer of point evaluation at ξ, i.e., the unique $Z = Z_\xi^{(k)} \in \mathscr{H}_k$ with

$$f(\xi) = \langle f, Z_\xi^{(k)}\rangle_{\mathbb{S}} = \int_{\mathbb{S}} f Z_\xi^{(k)}\,d\sigma(\xi), \qquad \forall f \in \mathscr{H}_k. \tag{16.9}$$

Thus (16.4) gives the following reproducing kernel tight frame:

Example 16.7. The zonal harmonics $(Z_\xi^{(k)})_{\xi\in\mathbb{S}}$ are a continuous tight frame for the space $\mathscr{H}_k$ of spherical harmonics of degree k, i.e.,

$$f = \int_{\mathbb{S}} \langle f, Z_\xi^{(k)}\rangle_{\mathbb{S}} Z_\xi^{(k)}\,d\sigma(\xi), \qquad \forall f \in \mathscr{H}_k. \tag{16.10}$$

We recall some basic facts about zonal harmonics (cf [SW71] and [ABR01]).

Definition 16.4. A function f defined on a $O(d)$-invariant subspace of $\mathbb{R}^d$ (such as $\mathbb{S}$ or $\mathbb{R}^d$) is **zonal** with **pole** $\xi \in \mathbb{S}$ if it can be written in the form

$$f(x) = g(\langle x, \xi\rangle, \|x\|).$$

This definition is equivalent to:

- f is invariant under the action of the subgroup of $O(d)$ which fixes ξ.
- f constant on parallels of the sphere, i.e., f is constant on $H \cap S$, where H is any hyperplane in $\mathbb{R}^d$ which is orthogonal to the vector ξ, and S is a sphere (on which f is defined).

---

[1] The zonal harmonic $Z_\xi^{(k)}$ is also commonly defined for unnormalised surface-area measure, which adds a scaling factor to the formulas for it presented here.

Zonal functions generalise *ridge functions* and *radial functions*, which have the form

$$f(x)=g(\langle x,\xi\rangle)\quad \text{(ridge function)},\qquad f(x)=g(\|x\|)\quad \text{(radial function)}.$$

The zonal harmonic $Z_\xi^{(k)}\in\mathscr{H}_k$ is a *zonal* function, as can be seen from the following explicit formula (see Exer. 16.8)

$$Z_\xi^{(k)}(x)=\|x\|^k C_k^{(\frac{d}{2})}\Bigl(\frac{\langle x,\xi\rangle}{\|x\|}\Bigr)-\|x\|^k C_{k-2}^{(\frac{d}{2})}\Bigl(\frac{\langle x,\xi\rangle}{\|x\|}\Bigr), \tag{16.11}$$

where $C_k^{(\lambda)}$ are the *ultraspherical (Gegenbauer) polynomials*, with $C_k^\lambda:=0$, $k<0$.

If f_η is zonal with pole η, then we can move the pole of f_η to be ξ by applying any $g\in O(d)$ with $\xi=g\eta$. We use the notation f_ξ to denote the corresponding (well defined) zonal function $f_\xi:=gf_\eta:=f_\eta\circ g^{-1}$, $g\in O(d)$, $\xi=g\eta$. In particular,

$$Z_{g\xi}^{(k)}=gZ_\xi^{(k)}:=Z_\xi^{(k)}\circ g^{-1},\qquad \forall g\in O(d). \tag{16.12}$$

The zonal function $Z_\xi^{(k)}$ is *peaked* at ξ (see Exer. 16.13), i.e.,

$$\|Z_\xi^{(k)}\|_{\mathbb{S}}^2=\langle Z_\xi^{(k)},Z_\xi^{(k)}\rangle_{\mathbb{S}}=Z_\xi^{(k)}(\xi)=\dim(\mathscr{H}_k), \tag{16.13}$$

$$|Z_\xi^{(k)}(\eta)|\le\dim(\mathscr{H}_k),\qquad \forall\eta\in\mathbb{S}. \tag{16.14}$$

Example 16.8. (Homogeneous linear polynomials). For $k=1$, $\mathscr{H}_k$ is the space of homogeneous linear polynomials in d real variables, and (16.11) gives

$$Z_\xi^{(1)}(x)=d\langle x,\xi\rangle,$$

and (16.10) becomes

$$f=\int_{\mathbb{S}}\langle f,Z_\xi^{(1)}\rangle_{\mathbb{S}}Z_\xi^{(1)}\,d\sigma=\frac{d}{\operatorname{area}(\mathbb{S})}\int_{\mathbb{S}}f(\xi)\langle\cdot,\xi\rangle\,d\xi,\qquad \forall f\in\mathscr{H}_1.$$

Let f be homogeneous linear polynomial $\langle\cdot,x\rangle$, to obtain

$$\langle\cdot,x\rangle=\frac{d}{\operatorname{area}(\mathbb{S})}\int_{\mathbb{S}}\langle x,\xi\rangle\langle\cdot,\xi\rangle\,d\xi=\Bigl\langle\cdot,\frac{d}{\operatorname{area}(\mathbb{S})}\int_{\mathbb{S}}\langle x,\xi\rangle\xi\,d\xi\Bigr\rangle,$$

and we deduce the following generalisation of (16.2) to $\mathbb{R}^d$.

Proposition 16.4. *The vectors $(\xi)_{\xi\in\mathbb{S}}$ of the unit sphere are a continuous tight frame for $\mathbb{R}^d$, i.e.,*

$$x=\frac{d}{\operatorname{area}(\mathbb{S})}\int_{\mathbb{S}}\langle x,\xi\rangle\xi\,d\xi,\qquad \forall x\in\mathbb{R}^d. \tag{16.15}$$

By linearity, the expansion (16.15) also holds for $x\in\mathbb{C}^d$, and an analogue exists for $\mathbb{S}$ replaced by the complex unit sphere $\mathbb{S}_{\mathbb{C}}$ (see Exer. 16.12).

16.5 Homogeneous polynomials

Let $\Pi_n^\circ(\mathbb{R}^d)$ be the space of homogeneous polynomials on $\mathbb{R}^d$ of degree n, with the inner product $\langle f,g\rangle_{\mathbb{S}}$ defined by (16.6). We will find the reproducing kernel tight frame for this space. Every homogeneous polynomial $p \in \Pi_n^\circ(\mathbb{R}^d)$ can be written uniquely

$$p(x) = \sum_{0\le j\le \frac{n}{2}} \|x\|^{2j} p_{n-2j}(x) = \sum_{j=0}^{[\frac{n}{2}]} \|x\|^{2j} p_{n-2j}(x), \tag{16.16}$$

where $p_{n-2j} \in \mathscr{H}_{n-2j}(\mathbb{R}^d)$. This and the density of the polynomials in $L_2(\mathbb{S})$ gives the orthogonal decomposition of (16.8). Since the restriction map

$$\Pi_n^\circ(\mathbb{R}^d) \to L_2(\mathbb{S}) : f \mapsto f|_{\mathbb{S}}$$

is injective (see Example 16.3), from (16.16) we obtain the natural identification

$$\Pi_n^\circ(\mathbb{R}^d) \approx \Pi_n^\circ(\mathbb{S}) = \bigoplus_{0\le j\le \frac{n}{2}} \mathscr{H}_{n-2j}(\mathbb{S}). \tag{16.17}$$

We say that a space is **rotationally invariant** if it is $SO(d)$-invariant. The desired tight frame for the rotationally invariant subspace $\Pi_k^\circ(\mathbb{R}^d)$ of $L_2(\mathbb{S})$ is special case of the following general result.

Theorem 16.1. *Let $\mathscr{H}$ be a rotationally invariant subspace of $L_2(\mathbb{S})$. Then*

$$\mathscr{H} = \bigoplus_{j\in J} \mathscr{H}_j, \tag{16.18}$$

for some subset J of $\mathbb{N}$, and we have the generalised tight frame expansion

$$f = \sum_{j\in J} \int_{\mathbb{S}} \langle f, Z_\xi^{(j)}\rangle_{\mathbb{S}} Z_\xi^{(j)}\, d\xi = \int_{\mathbb{S}} \sum_{j\in J} \langle f, Z_\xi^{(j)}\rangle_{\mathbb{S}} Z_\xi^{(j)}\, d\xi, \qquad \forall f \in \mathscr{H}. \tag{16.19}$$

Moreover, for $\mathscr{H}$ finite dimensional, i.e., J finite, let

$$Z_\xi = Z_\xi^J := \sum_{j\in J} Z_\xi^{(j)} \in \mathscr{H}. \tag{16.20}$$

Then $(Z_\xi)_{\xi\in\mathbb{S}}$ is the reproducing kernel tight frame for $\mathscr{H}$, i.e.,

$$f = \int_{\mathbb{S}} f(\xi) Z_\xi\, d\xi = \int_{\mathbb{S}} \langle f, Z_\xi\rangle_{\mathbb{S}} Z_\xi\, d\xi, \qquad \forall f \in \mathscr{H}. \tag{16.21}$$

Proof. The action of $SO(d)$ on $\mathscr{H}_j$ is absolutely irreducible, and so the orthogonal projection of $\mathscr{H}$ onto $\mathscr{H}_j$ is either 0 or $\mathscr{H}_j$, and (16.18) holds (see Exer. 16.7).

Recall (see Example 16.7) that $(Z_\xi^{(j)})_{\xi\in\mathbb{S}}$ is a continuous tight frame for $\mathscr{H}_k$, i.e.,

$$\int_{\mathbb{S}} |\langle f, Z_\xi^{(j)}\rangle|^2 \, d\sigma(\xi) = \|f\|^2, \qquad \forall f \in \mathscr{H}_j.$$

Let $f \in \mathscr{H}$, and P_j be the orthogonal projection onto $\mathscr{H}_j$. Then

$$\langle f, Z_\xi^{(j)}\rangle_{\mathbb{S}} = \langle f, P_j Z_\xi^{(j)}\rangle_{\mathbb{S}} = \langle P_j f, Z_\xi^{(j)}\rangle_{\mathbb{S}},$$

so that

$$\sum_{j\in J} \int_{\mathbb{S}} |\langle f, Z_\xi^{(j)}\rangle|^2 \, d\xi = \sum_{j\in J} \int_{\mathbb{S}} |\langle P_j f, Z_\xi^{(j)}\rangle|^2 \, d\xi = \sum_{j\in J} \|P_j f\|^2 = \|f\|^2.$$

By Fubini, we can exchange the sum and integral above, and so we have (16.19).

Now suppose that J is finite. Since $\langle f, Z_\xi^{(j)}\rangle = \langle f, Z_\xi\rangle$, from (16.19) we obtain

$$f = \int_{\mathbb{S}} \sum_{j\in J} \langle f, Z_\xi^{(j)}\rangle_{\mathbb{S}} Z_\xi^{(j)} \, d\sigma(\xi) = \int_{\mathbb{S}} \sum_{j\in J} \langle f, Z_\xi\rangle_{\mathbb{S}} Z_\xi^{(j)} \, d\sigma(\xi) = \int_{\mathbb{S}} \langle f, Z_\xi\rangle_{\mathbb{S}} Z_\xi \, d\sigma(\xi),$$

which is (16.21). By Exer. 16.6 the reproducing kernel for $\mathscr{H}$ is the sum of the reproducing kernels for $\mathscr{H}_j$, $j \in J$, and so we recognise $(Z_\xi)_{\xi\in\mathbb{S}}$ as the reproducing kernel tight frame for $\mathscr{H}$. □

This generalises Example 16.7, which is the special case $J = \{k\}$, $\mathscr{H} = \mathscr{H}_k$. In addition to (16.11), the *solid* zonal harmonics are given by the formulas

$$\begin{aligned} Z_\xi^{(k)}(x) &= \|x\|^k Z_\xi^{(k)}\Big(\frac{x}{\|x\|}\Big) \\ &= (d+2k-2) \sum_{j=0}^{[k/2]} (-1)^j \frac{d(d+2)\cdots(d+2k-2j-4)}{2^j j!(k-2j)!} \langle x, \xi\rangle^{k-2j} \|x\|^{2j} \\ &= \frac{2k+d-2}{d-2} \|x\|^k C_k^{(\frac{d-2}{2})}\Big(\frac{\langle x, \xi\rangle}{\|x\|}\Big) \quad \text{(Funk–Hecke formula).} \end{aligned} \tag{16.22}$$

We now consider $\Pi_n^\circ(\mathbb{R}^d)$, i.e., the case $J = \{n-2j : 0 \le j \le \frac{n}{2}\}$.

Corollary 16.1. *(Homogeneous polynomials) The reproducing kernel tight frame expansion for $\Pi_n^\circ(\mathbb{R}^d)$ is given by*

$$f = \int_{\mathbb{S}} \langle f, Z_\xi\rangle_{\mathbb{S}} Z_\xi \, d\sigma(\xi), \qquad \forall f \in \Pi_n^\circ(\mathbb{R}^d), \tag{16.23}$$

where

$$Z_\xi(x) := \sum_{0\le j\le \frac{n}{2}} \|x\|^{2j} Z_\xi^{(n-2j)}(x) = \|x\|^n C_n^{(\frac{d}{2})}\Big(\frac{\langle x, \xi\rangle}{\|x\|}\Big), \qquad x \in \mathbb{R}^d. \tag{16.24}$$

Proof. The reproducing kernel tight frame $(p_\xi)_{\xi\in\mathbb{S}}$ for $\bigoplus_{0\le j\le \frac{n}{2}} \mathscr{H}_{n-2j}(\mathbb{S})$ is given by

$$p_\xi = \sum_{0\le j\le \frac{n}{2}} Z_\xi^{(n-2j)}|_{\mathbb{S}}.$$

A polynomial in the form (16.16) is mapped by the identification as follows:

$$\Pi_n^\circ(\mathbb{R}^d) \to \bigoplus_{0\le j\le \frac{n}{2}} \mathscr{H}_{n-2j}(\mathbb{S}) : p = \sum_{0\le j\le \frac{n}{2}} \|\cdot\|^{2j} p_{n-2j} \mapsto \sum_{0\le j\le \frac{n}{2}} p_{n-2j}|_{\mathbb{S}}.$$

The inverse of p_ξ under this identification map is Z_ξ, which establishes (16.23). The second formula in (16.24) follows from (16.11), which gives the telescoping sum

$$Z_\xi(x) = \sum_{0\le j\le \frac{n}{2}} \|x\|^{2j}\Big\{\|x\|^{n-2j} C_{n-2j}^{(\frac{d}{2})}\Big(\frac{\langle x,\xi\rangle}{\|x\|}\Big) - \|x\|^{n-2j} C_{n-2j-2}^{(\frac{d}{2})}\Big(\frac{\langle x,\xi\rangle}{\|x\|}\Big)\Big\},$$

which simplifies to (16.24), since $C_k^{(\lambda)} := 0$, $k<0$. □

Example 16.9. (Spherical polynomials) In view of (16.16), the space polynomials of degree n on the sphere is given by the choice $J = \{0,1,\ldots,n\}$, i.e.,

$$\Pi_n(\mathbb{S}) = \mathscr{H}_0 \oplus \mathscr{H}_1 \oplus \cdots \oplus \mathscr{H}_n = \Pi_n^\circ(\mathbb{S}) \oplus \Pi_{n-1}^\circ(\mathbb{S}).$$

From (16.24), we obtain

$$Z_\xi(x) = C_n^{(\frac{d}{2})}(\langle x,\xi\rangle) + C_{n-1}^{(\frac{d}{2})}(\langle x,\xi\rangle), \qquad x\in\mathbb{S}.$$

Example 16.10. (Polynomial wavelets) For $J = \{n+1, n+2, \ldots, n+s\}$, we have

$$\mathscr{H} = \Pi_{n+s}(\mathbb{S}) \ominus \Pi_n(\mathbb{S}) = \mathscr{H}_{n+1} \oplus \mathscr{H}_{n+2} \oplus \cdots \oplus \mathscr{H}_{n+s},$$

and from (16.11) and (16.22) we obtain

$$\begin{aligned} Z_\xi(x) &= C_{n+s}^{(\frac{d}{2})}(\langle x,\xi\rangle) + C_{n+s-1}^{(\frac{d}{2})}(\langle x,\xi\rangle) - C_n^{(\frac{d}{2})}(\langle x,\xi\rangle) - C_{n-1}^{(\frac{d}{2})}(\langle x,\xi\rangle) \\ &= \sum_{k=n+1}^{n+s} \frac{2k+d-2}{d-2} C_k^{(\frac{d-2}{2})}(\langle x,\xi\rangle). \end{aligned}$$

This zonal function Z_ξ is localised in space near ξ. The coefficients in the second formula can be modified to obtain a zonal *polynomial wavelet* for $\Pi_{n+s}(\mathbb{S}) \ominus \Pi_n(\mathbb{S})$ which has good space–frequency localisation [Fer07].

Example 16.11. (Poisson kernel) It is natural to take $J = \mathbb{N}$ to obtain a reproducing kernel for $\mathscr{H} = L_2(\mathbb{S})$. In this case, the series for Z_ξ does not converge in $L_2(\mathbb{S})$. Nevertheless, it sums to give the **Poisson kernel** (see Exer. 16.8)

$$P(x,\xi) = \sum_{j=0}^{\infty} Z_\xi^{(j)}(x) = \frac{1-\|x\|^2}{\|x-\xi\|^d}.$$

16.6 Orthogonal polynomials for a radially symmetric weight

Here we consider the multivariate orthogonal polynomials $\mathscr{P}_n$ (see §10.10) for a radially symmetric measure μ on $\mathbb{R}^d$, i.e., one for which the symmetry group of the measure is $O(d)$, and so $\mathscr{P}_n$ is $O(d)$-invariant.

For simplicity, we suppose μ is Lebesgue integration with a nonnegative radial weight function $w : [0,R) \to \mathbb{R}$ on the ball $B_R := \{x \in \mathbb{R}^d : \|x\| < R\}$, $0 < R \leq \infty$, (which defines an inner product on Π_n), i.e.,

$$\langle f,g \rangle = \langle f,g \rangle_w := \int_{B_R} f(x)g(x)\, w(\|x\|)\, dx. \tag{16.25}$$

By using the characterisation of G-frames (Theorem 10.9) extended to the infinite compact group $G = SO(d)$, we find a single polynomial $p \in \mathscr{P}_n$ of unit norm, for which $(gp)_{g\in G}$ is a continuous tight frame for $\mathscr{P}_n$, i.e.,

$$f = \dim(\mathscr{P}_n) \int_{g\in SO(d)} \langle f, gp \rangle gp \, d\nu_d(g), \qquad \forall f \in \mathscr{P}_n, \tag{16.26}$$

where ν_d is the *normalised Haar measure* on $SO(d)$. By choosing $p = p_\xi$ to be zonal, we obtain

$$f = \dim(\mathscr{P}_n) \int_{\mathbb{S}} \langle f, p_\xi \rangle p_\xi \, d\sigma(\xi), \qquad \forall f \in \mathscr{P}_n,$$

with one choice for p_ξ giving the reproducing kernel tight frame. Both of these expansions can be discretised to obtain finite tight frames.

16.6.1 The $O(d)$-invariant subspaces of $\mathscr{P}_n$

To find a $p \in \mathscr{P}_n$ giving (16.26), we need the $SO(d)$-invariant subspaces V_j of $\mathscr{P}_n$, i.e., the $O(d)$-invariant subspaces (we use the subgroup $SO(d)$ for convenience).

We will repeatedly use the fact that if a function f defined on B_R can be factored into a *radial* and *angular* part

$$f(x) = \mathscr{R}(\|x\|)\Theta\Big(\frac{x}{\|x\|}\Big),$$

then (by Fubini's theorem) it can be integrated

$$\int_{B_R} f(x)\, dx = \int_{\mathbb{S}} \int_0^R \mathscr{R}(r)\Theta(\xi)\, r^{d-1} dr\, d\xi = \Big(\int_0^R \mathscr{R}(r)\, r^{d-1} dr\Big)\Big(\int_{\mathbb{S}} \Theta(\xi)\, d\xi\Big).$$

In particular, for $j \neq k$, polynomials in $\mathscr{H}_j$ and $\mathscr{H}_k$ multiplied by radial polynomials are orthogonal with respect to the inner product (16.25) for $\mathscr{P}_n$. This leads to the decomposition of $\mathscr{P}_n$ into its $O(d)$-invariant subspaces.

Lemma 16.1. *For $0 \leq j \leq \frac{n}{2}$, let $P_j = P_j^{(n)}$ be an orthogonal polynomial of degree j for the univariate weight on $[0,R^2)$ given by*

$$t \mapsto t^{n-2j+\frac{d-2}{2}} w(\sqrt{t}). \tag{16.27}$$

Then $\mathscr{P}_n$ is the orthogonal direct sum of the absolutely irreducible $O(d)$-invariant subspaces

$$\mathscr{P}_n = \bigoplus_{0\leq j\leq \frac{n}{2}} V_j^{(n)}, \qquad V_j^{(n)} := P_j^{(n)}(\|\cdot\|^2)\mathscr{H}_{n-2j}, \tag{16.28}$$

where the inner product on $V_j = V_j^{(n)}$ is given by

$$\langle h_1 P_j(\|\cdot\|^2), h_2 P_j(\|\cdot\|^2)\rangle = \operatorname{area}(\mathbb{S})\langle h_1,h_2\rangle_{\mathbb{S}} \|P_j\|_{w,n-2j}^2, \qquad \forall h_1,h_2 \in \mathscr{H}_{n-2j}, \tag{16.29}$$

where

$$\|p\|_{w,k}^2 := \frac{1}{2}\int_0^{R^2} |p(t)|^2 t^{k+\frac{d-2}{2}} w(\sqrt{t})\,dt. \tag{16.30}$$

Proof. We have already observed that the $V_j = V_j^{(n)}$ are orthogonal to each other. Since $\mathscr{H}_{n-2j}$ is an absolutely irreducible $O(d)$-invariant subspace, it follows that V_j is also. Moreover, by (16.17), we have

$$\sum_{0\leq j\leq \frac{n}{2}} \dim(V_j) = \sum_{0\leq j\leq \frac{n}{2}} \dim(\mathscr{H}_{n-2j}) = \dim(\Pi_n^\circ(\mathbb{R}^d) = \dim(\mathscr{P}_n).$$

Hence to prove (16.28), it suffices to show that $V_j \subset \mathscr{P}_n$, which we now do.

Let $h_\alpha \in \mathscr{H}_\alpha$, $h_\beta \in \mathscr{H}_\beta$ and p and q be polynomials, then

$$\begin{aligned}
&\langle h_\alpha p(\|\cdot\|^2), h_\beta q(\|\cdot\|^2)\rangle \\
&= \int_{B_R} \|x\|^\alpha h_\alpha\Bigl(\frac{x}{\|x\|}\Bigr) p(\|x\|^2)\|x\|^\beta h_\beta\Bigl(\frac{x}{\|x\|}\Bigr) q(\|x\|^2)\, w(\|x\|)\,dx \\
&= \Bigl(\int_0^R p(r^2)q(r^2)r^{\alpha+\beta+d-2} w(r)\,r\,dr\Bigr)\Bigl(\int_{\mathbb{S}} h_\alpha(\xi)h_\beta(\xi)\,d\xi\Bigr) \\
&= \operatorname{area}(\mathbb{S})\Bigl(\int_0^{R^2} p(t)q(t)t^{\frac{\alpha+\beta}{2}+\frac{d-2}{2}} w(\sqrt{t})\,\frac{dt}{2}\Bigr)\langle h_\alpha,h_\beta\rangle_{\mathbb{S}}.
\end{aligned} \tag{16.31}$$

Let $h_{n-2j} \in \mathscr{H}_{n-2j}$ and p_j be a univariate polynomial of degree j. In view of (16.16), the polynomial $p_j(\|\cdot\|^2)h_{n-2j}$ is in $\mathscr{P}_n$ if and only if

$$\langle p_j(\|\cdot\|^2)h_{n-2j}, \|\cdot\|^{2\ell} h_{k-2\ell}\rangle = 0, \qquad h_{k-2\ell} \in \mathscr{H}_{k-2\ell}, \quad 0\leq k<n, \quad 0\leq \ell \leq \frac{k}{2}.$$

By (16.31) and orthogonality of spherical harmonics of different degrees, this holds except for when $n-2j = k-2\ell$, in which case $\ell = j - \frac{1}{2}(n-k) < j$, and we require

$$\int_0^{R^2} p_j(t) t^\ell t^{n-2j+\frac{d-2}{2}} w(\sqrt{t})\, dt = 0,$$

which is satisfied by the choice $p_j = P_j$.

The formulas (16.29) and (16.30) are special case of (16.31). □

Example 16.12. (Gegenbauer polynomials). The orthogonal polynomials on the unit ball B_1 corresponding to the weight

$$w(r) := (1-r^2)^\alpha r^{2\beta}, \qquad \alpha > -1,\ \beta > -\frac{d}{2}, \tag{16.32}$$

are the **generalised Gegenbauer polynomials, Gegenbauer polynomials** (when $\beta = 0$) and **Legendre polynomials** (when $\alpha = \beta = 0$). For this weight

$$t^{n-2j+\frac{d-2}{2}} w(\sqrt{t}) = t^{n-2j+\frac{d-2}{2}} (1-t)^\alpha t^\beta = (1-t)^\alpha t^{n-2j+\frac{d-2}{2}+\beta},$$

so that

$$P_j = P_j^{(\alpha, n-2j+\frac{d-2}{2}+\beta)}(2(\cdot)-1),$$
$$\|P_j\|^2_{w,n-2j} = \frac{1}{2} \frac{\Gamma(j+\alpha+1)\Gamma(n-j+\frac{d}{2}+\beta)}{(\alpha+n+\frac{d}{2}+\beta) j!\,\Gamma(\alpha+n-j+\frac{d}{2}+\beta)}, \tag{16.33}$$

where $P_j^{(\alpha,\beta)}$ are the univariate Jacobi polynomials given by (15.12).

Example 16.13. (Hermite polynomials). The orthogonal polynomials on $\mathbb{R}^d = B_\infty$ corresponding to the weight

$$w(r) = r^{2\beta} e^{-r^2}, \qquad \beta \ge 0, \tag{16.34}$$

are called **generalised Hermite polynomials** and **Hermite polynomials** ($\beta = 0$). For these, we can take

$$P_j = L_j^{(n-2j+\frac{d-2}{2}+\beta)}, \qquad \|P_j\|^2_{w,n-2j} = \frac{1}{2} \frac{\Gamma(n-j+\frac{d}{2}+\beta)}{j!}, \tag{16.35}$$

where $L_n^{(\alpha)}$ are the *generalised Laguerre polynomials* given by

$$L_n^{(\alpha)}(x) := \frac{1}{n!} x^{-\alpha} e^x \frac{d^n}{dx^n}(x^{n+\alpha} e^{-x}), \qquad \int_0^\infty (L_n^{(\alpha)}(x))^2 x^\alpha e^{-x}\, dx = \frac{\Gamma(\alpha+n+1)}{n!}.$$

Example 16.14. In view of (16.28), an orthonormal basis for $V_j^{(n)}$ (and hence $\mathscr{P}_n$) can be obtained from an orthonormal basis $(Y_\beta) = (S_{n-2j,\beta})$ for $\mathscr{H}_{n-2j}$ (see [DX01] for detail). This basis is not $O(d)$-invariant (none can be).

We now find a continuous tight frame for $\mathscr{P}_n$ which is $O(d)$-invariant.

16.6.2 Continuous tight frames for $\mathscr{P}_n$

We now give a continuous tight frame $(p_\xi)_{\xi\in\mathbb{S}}$ for the orthogonal polynomials $\mathscr{P}_n$, which is $O(d)$-invariant. This is obtained by considering all group frames $(gp)_{g\in G}$ for $\mathscr{P}_n$, where G is the continuous group $G = SO(d)$.

We assume the decomposition of $\mathscr{P}_n = \bigoplus V_j$ given by Lemma 16.1 and suppose that P_j is normalised, so that its leading term has a positive coefficient. Let $\nu = \nu_d$ be the normalised Haar measure on $SO(d)$.

Theorem 16.2. *Let $p \in \mathscr{P}_n$ be any unit-norm polynomial of the form*

$$p = \sum_{0\le j\le \frac{n}{2}} \sqrt{\frac{\dim(\mathscr{H}_{n-2j})}{\dim(\mathscr{P}_n)}}\, p_j, \qquad p_j \in V_j^{(n)}, \quad \|p_j\| = 1. \tag{16.36}$$

Then $\{gp\}_{g\in SO(d)}$ is an equal-norm continuous tight frame for $\mathscr{P}_n$, i.e.,

$$f = \dim(\mathscr{P}_n) \int_{SO(d)} \langle f, gp\rangle gp\, d\nu(g), \qquad \forall f \in \mathscr{P}_n, \tag{16.37}$$

and these are all such $p \in \mathscr{P}_n$. Moreover, p can be chosen to be zonal, in which case

$$f = \dim(\mathscr{P}_n) \int_{\mathbb{S}} \langle f, p_\xi\rangle p_\xi\, d\sigma(\xi), \qquad \forall f \in \mathscr{P}_n. \tag{16.38}$$

There are a finite number of such $p = p_\xi$ with a given pole ξ. Of these, we call

$$p_\xi := \frac{1}{\sqrt{\dim(\mathscr{P}_n)}} \sum_{0\le j\le \frac{n}{2}} \frac{Z_\xi^{(n-2j)}}{\sqrt{\operatorname{area}(\mathbb{S})}} \frac{P_j(\|\cdot\|^2)}{\|P_j\|_{w,n-2j}}. \tag{16.39}$$

the canonical choice of p (the leading term of P_j has a positive coefficient).

Proof. Let $G = SO(d)$. We observe that Theorem 10.9 extends to this G (or any infinite compact group), with the finite sum $\frac{1}{|G|}\sum_{g\in G}$ replaced by integration with respect to the normalised Haar measure on G.

The action of G on $\mathscr{P}_n$ is unitary and is absolutely irreducible on the subspaces $V_j = V_j^{(n)}$. Moreover, none of the V_j are $\mathbb{C}G$-isomorphic to each other (see [FH87]). This is easily seen for $d \ge 3$ where (16.5) implies the V_j have different dimensions, and for $d = 2$ from following the explicit description of V_j

$$V_j = \operatorname{span}\{(x,y) \mapsto \Re((a+ib)(x+iy)^{n-2j})P_j(x^2+y^2) : a,b \in \mathbb{R}\}.$$

Hence, by Theorem 10.9, the choice (16.36) gives (16.37). Here we write $\dim(V_j)$ as $\dim(\mathscr{H}_{n-2j})$ and normalise p so that it is a unit vector. A simple calculation shows that p is zonal with a pole at ξ if and only if each p_j is. The space of zonal functions in $\mathscr{H}_{n-2j}$ with pole ξ is one-dimensional and spanned by $Z_\xi^{(n-2j)}$. Thus, by (16.29) and (16.13)

$$p_j = \pm\frac{Z_\xi^{(n-2j)}P_j(\|\cdot\|^2)}{\|Z_\xi^{(n-2j)}P_j(\|\cdot\|^2)\|} = \pm\frac{Z_\xi^{(n-2j)}P_j(\|\cdot\|^2)}{\sqrt{\operatorname{area}(\mathbb{S})}\sqrt{\dim(\mathscr{H}_{n-2j})}\|P_j\|_w},$$

where the "+" choice gives the canonical choice of p_ξ (which is most peaked at ξ).

Finally, suppose p is zonal, so that p_ξ is zonal with pole $\xi \in S$. Let G_ξ be the subgroup of $G = SO(d)$ which fixes the point ξ, i.e.,

$$G_\xi := \{g \in SO(d) : g\xi = \xi\} \cong SO(d-1).$$

Then (see [SD80] for detail) the integral of (16.37) can be computed

$$\begin{aligned}\int_{SO(d)} \langle f, gp\rangle gp\, d\nu_d(g) &= \int_{\mathbb{S}}\int_{SO(d-1)} \langle f, gp\rangle gp\, d\nu_{d-1}(g)\, d\sigma(\xi)\\ &= \int_{\mathbb{S}}\int_{G_\xi} \langle f, p_\xi\rangle p_\xi\, d\nu_{d-1}(g)\, d\sigma(\xi) = \int_{\mathbb{S}} \langle f, p_\xi\rangle p_\xi\, d\sigma(\xi),\end{aligned}$$

which gives (16.38). □

The first few zonal harmonics on $\mathbb{R}^d$ are given by

$$\begin{aligned}Z_\xi^{(0)} &= 1, \qquad Z_\xi^{(1)} = d\langle x,\xi\rangle, \qquad Z_\xi^{(2)} = \frac{d+2}{2}\Big(d\langle x,\xi\rangle^2 - \|x\|^2\Big),\\ Z_\xi^{(3)} &= \frac{d(d+4)}{6}\langle x,\xi\rangle\Big((d+2)\langle x,\xi\rangle^2 - 3\|x\|^2\Big), \qquad (16.40)\\ Z_\xi^{(4)} &= \frac{d(d+6)}{24}\Big((d^2+6d+8)\langle x,\xi\rangle^4 - (6d+12)\langle x,\xi\rangle^2\|x\|^2 + 3\|x\|^4\Big).\end{aligned}$$

Example 16.15. Let $\mathscr{P}_4$ be the quartic Legendre polynomials on the unit disc. Then by (16.33) and (16.40), the summands of the canonical choice p_ξ are

$$\begin{aligned}p_0(x) &= \frac{1}{\sqrt{\pi}}\Big(16\langle x,\xi\rangle^4 - 16\|x\|^2\langle x,\xi\rangle^2 + 2\|x\|^4\Big),\\ p_1(x) &= \frac{1}{\sqrt{\pi}}\Big(4\langle x,\xi\rangle^2 - 2\|x\|^2\Big)\Big(4\|x\|^2 - 3\Big),\\ p_2(x) &= \frac{1}{\sqrt{\pi}}\Big(6\|x\|^4 - 6\|x\|^2 + 1\Big),\end{aligned}$$

and so the canonical choice is a ridge polynomial

$$p_\xi(x) = p_0(x) + p_1(x) + p_2(x) = \frac{1}{\sqrt{\pi}}\Big(16\langle x,\xi\rangle^4 - 12\langle x,\xi\rangle^2 + 1\Big).$$

The choice $q_\xi := p_0 + p_1 - p_2$ is not a ridge polynomial (see Fig. 16.1).

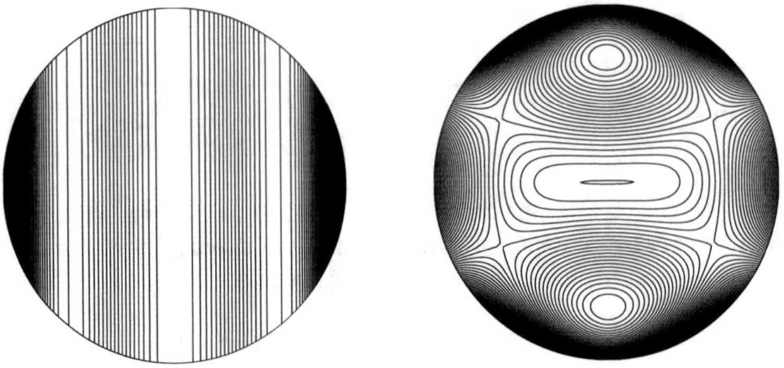

Fig. 16.1: Contour plots of the quartic Legendre polynomials p_ξ and q_ξ from Example 16.15 for $\xi=(1,0)$. Clearly the canonical choice p_ξ is a ridge function.

For the Legendre polynomials (constant weight) the canonical choice for p_ξ is always a ridge polynomial. This is not the case for the Gegenbauer polynomials for a nonconstant weight (see [Wal09]). The corresponding continuous tight frame of ridge polynomials for the Legendre polynomials was used by Petrushev [Pet99] to study approximation by ridge functions and neural networks on the unit ball.

Corollary 16.2. *(Legendre polynomials). For the constant weight* 1 *on the unit ball, the canonical choice for p in Theorem 16.2 is the ridge polynomial given by*

$$p_\xi(x)=\frac{\sqrt{2n+d}}{\sqrt{\operatorname{area}(S)}\sqrt{\dim(\mathscr{P}_n)}}C_n^{(\frac{d}{2})}(\langle x,\xi\rangle). \tag{16.41}$$

Example 16.16. For the Legendre polynomials on the disc in $\mathbb{R}^2$, (16.41) gives

$$p_\xi(x)=\frac{\sqrt{2n+2}}{\sqrt{2\pi}\sqrt{n+1}}C_n^{(1)}(\langle x,\xi\rangle)=\frac{1}{\sqrt{\pi}}U_n(\langle x,\xi\rangle), \tag{16.42}$$

where U_n are the *Chebyshev polynomials of the second kind* (see Example 16.15). In this case, both (16.37) and (16.38) reduce to

$$f=\frac{n+1}{2\pi}\int_0^{2\pi}\langle f,R_\theta p\rangle R_\theta p\,d\theta, \qquad \forall f\in\mathscr{P}_n, \tag{16.43}$$

where R_θ is rotation by θ and $p=p_\xi$ (for any ξ).

For the Legendre polynomials on the unit ball in $\mathbb{R}^3$, (16.41) gives

$$p_\xi(x)=\frac{\sqrt{2n+3}}{\sqrt{4\pi}\sqrt{\binom{n+2}{2}}}C_n^{(\frac{3}{2})}(\langle x,\xi\rangle).$$

In this case, the integral in (16.37) is over the manifold $SO(3)$ of dimension 3, and the integral in (16.38) is over $\mathbb{S}$ which has dimension 2.

16.6.3 The reproducing kernel for $\mathscr{P}_n$

We now consider the reproducing kernel tight frame for $\mathscr{P}_n$ (and associated spaces).

Lemma 16.2. *Let $p \neq 0$ be a univariate polynomial. Then the reproducing kernel for the absolutely irreducible $SO(d)$-invariant subspace $p(\|\cdot\|^2)\mathscr{H}_k$ of $L_2(\mu)$ is*

$$K(x,y) = \frac{p(\|x\|^2)p(\|y\|^2)}{\operatorname{area}(\mathbb{S})\|p\|_{w,k}^2} Z^{(k)}(x,y),$$

where $\|p\|_{w.k}$ is given by (16.30) and $Z^{(k)}$ is the polynomial given by

$$Z^{(k)}(x,y) := \|x\|^k\|y\|^k Z^{(k)}_{\frac{x}{\|x\|}}\Big(\frac{y}{\|y\|}\Big) = \|x\|^k\|y\|^k Z^{(k)}_{\frac{y}{\|y\|}}\Big(\frac{x}{\|x\|}\Big). \tag{16.44}$$

Proof. Let $K_y(x) := K(x,y)$, so that

$$K_y = \frac{p(\|\cdot\|^2)p(\|y\|^2)}{\operatorname{area}(\mathbb{S})\|p\|_{w,k}^2}\|y\|^k Z^{(k)}_{\frac{y}{\|y\|}}, \qquad y \neq 0.$$

Then, for $f \in \mathscr{H}_k$, using (16.31) and (16.9), we calculate

$$\begin{aligned}\langle p(\|\cdot\|^2)f, K_y\rangle &= \frac{p(\|y\|^2)}{\operatorname{area}(\mathbb{S})\|p\|_{w,k}^2}\|y\|^k\langle p(\|\cdot\|^2)f, p(\|\cdot\|^2)Z^{(k)}_{\frac{y}{\|y\|}}\rangle \\ &= p(\|y\|^2)\|y\|^k\langle f, Z^{(k)}_{\frac{y}{\|y\|}}\rangle_{\mathbb{S}} \\ &= p(\|y\|^2)\|y\|^k f\Big(\frac{y}{\|y\|}\Big) = p(\|y\|)^2 f(y),\end{aligned}$$

so that $K(x,y)$ is the reproducing kernel for $p(\|\cdot\|^2)\mathscr{H}_k$. □

An explicit formula for $Z^{(k)}(x,y)$ is given by (16.22), i.e.,

$$Z^{(k)}(x,y) = (d+2k-2)\sum_{0\le j\le \frac{k}{2}}(-1)^j\frac{d(d+2)\cdots(d+2k-2j-4)}{2^j j!(k-2j)!}\langle x,y\rangle^{k-2j}\|x\|^{2j}\|y\|^{2j}.$$

Example 16.17. The reproducing kernel for the subspace $V_j^{(n)} = P_j^{(n)}(\|\cdot\|^2)\mathscr{H}_{n-2j}$ of $\mathscr{P}_n$ given in Lemma 16.1 is

$$K_j^{(n)}(x,y) = \frac{P_j^{(n)}(\|x\|^2)P_j^{(n)}(\|y\|^2)}{\operatorname{area}(\mathbb{S})\|P_j^{(n)}\|_{w,n-2j}^2} Z^{(n-2j)}(x,y),$$

where $Z^{(k)}$ is the polynomial given by (16.44).

We now give two $SO(d)$-invariant formulas for the reproducing kernel for $\mathscr{P}_n$, which follow from Theorem 16.2.

Theorem 16.3. *The reproducing kernel for $\mathscr{P}_n$ is given by the formulas*

$$K_n(x,y) = \dim(\mathscr{P}_n) \int_{\mathbb{S}} p_\xi(x) p_\xi(y)\, d\sigma(\xi)$$
$$= \sum_{0\le j\le \frac{n}{2}} \frac{P_j^{(n)}(\|x\|^2) P_j^{(n)}(\|y\|^2)}{\operatorname{area}(\mathbb{S}) \|P_j^{(n)}\|_{w,n-2j}^2} Z^{(n-2j)}(x,y), \qquad (16.45)$$

where p_ξ is the canonical choice and $Z^{(k)}$ is the polynomial given by (16.44).

Proof. Expanding (16.38) gives

$$f(y) = \dim(\mathscr{P}_n) \int_{\mathbb{S}} \Big(\int_{B_R} f(x) p_\xi(x)\, d\mu(x) \Big) p_\xi(y)\, d\sigma(\xi)$$
$$= \int_{B_R} f(x) \Big(\dim(\mathscr{P}_n) \int_{\mathbb{S}} p_\xi(x) p_\xi(y)\, d\sigma(\xi) \Big)\, d\mu(x),$$

so that the reproducing kernel for $\mathscr{P}_n$ is given by the first formula.

Since $\mathscr{P}_n$ is the orthogonal direct sum (16.28), i.e.,

$$\mathscr{P}_n = \bigoplus_{0\le j\le \frac{n}{2}} V_j^{(n)},$$

its reproducing kernel is the sum of the reproducing kernels for the $V_j^{(n)}$ given in Example 16.17 (see Exer. 16.6), which gives the second formula. □

The reproducing kernel for a finite dimensional rotationally invariant subspace of $L_2(\mu)$ can be calculated by using the above techniques. This is a little more involved than for $L_2(\mathbb{S})$ (see Theorem 16.1), since the homogeneous components of the $\mathbb{C}SO(d)$-module $L_2(\mu)$ contain more than a single copy of each irreducible. Indeed, the homogeneous components, as defined by (10.14), are

$$\sum_{j=0}^{\infty} \|\cdot\|^{2j} \mathscr{H}_k, \qquad k = 0, 1, \ldots.$$

Example 16.18. Since the polynomials of degree k are the orthogonal direct sum

$$\Pi_k(\mathbb{R}^d) = \mathscr{P}_0 \oplus \mathscr{P}_1 \oplus \cdots \oplus \mathscr{P}_k = \bigoplus_{n=0}^{k} \bigoplus_{0\le j\le \frac{n}{2}} V_j^{(n)}, \qquad (16.46)$$

the reproducing kernel of $\Pi_k(\mathbb{R}^d) \subset L_2(\mu)$ is

$$K(x,y) = \sum_{n=0}^{k} \sum_{0\le j\le \frac{n}{2}} \frac{P_j^{(n)}(\|x\|^2) P_j^{(n)}(\|y\|^2)}{\operatorname{area}(\mathbb{S}) \|P_j^{(n)}\|_{w,n-2j}^2} Z^{(n-2j)}(x,y). \qquad (16.47)$$

We now consider the reproducing kernel for the homogeneous polynomials.

Example 16.19. By (16.16) and (16.31), the homogeneous polynomials of degree n are the orthogonal direct sum

$$\Pi_n^\circ(\mathbb{R}^d) = \bigoplus_{0 \le j \le \frac{n}{2}} \|\cdot\|^{2j} \mathscr{H}_{n-2j}.$$

Hence, by Lemma 16.2, the reproducing kernel of $\Pi_n^\circ(\mathbb{R}^d)$ is

$$K(x,y) = \sum_{0 \le j \le \frac{n}{2}} \frac{\|x\|^{2j}\|y\|^{2j}}{\operatorname{area}(\mathbb{S})\|(\cdot)^j\|_{w,n-2j}} Z^{(n-2j)}(x,y).$$

16.6.4 Finite tight frames for $\mathscr{P}_n$

We now outline how the continuous tight frame expansions for $\mathscr{P}_n$ of §16.6.2 can be discretised (sampled) to obtain a finite tight frame expansion.

For a fixed $x \in \mathbb{R}^d$ and $f \in \mathscr{P}_n$, (16.38) gives

$$f(x) = \dim(\mathscr{P}_n) \int_{\mathbb{S}} \langle f, p_\xi \rangle p_\xi(x)\, d\xi. \tag{16.48}$$

The above integral of the polynomial $\xi \mapsto \langle f, p_\xi \rangle p_\xi(x)$ of degree $2n$ can be replaced by a quadrature rule (spherical design) to obtain a discrete form of (16.38).

Definition 16.5. A finite subset Θ of $\mathbb{S}$ together with weights $c_\xi \in \mathbb{R}$, $\xi \in \Theta$ is called a **quadrature** (or **cubature**) **rule** of **degree** k for the sphere if

$$\int_S f\, d\sigma(\xi) = \sum_{\xi \in \Theta} c_\xi f(\xi), \qquad \forall f \in \Pi_k(\mathbb{R}^d).$$

An equal weight quadrature rule, i.e., one with $c_\xi = 1/|\Theta|$, $\forall \xi \in \Theta$ is known as a *spherical k-design* (see §6.4). There is an extensive literature on quadrature rules for the sphere (see [Str71], [CR93], [Coo99]).

Theorem 16.4. *(Finite tight frame). Let $\Theta \subset \mathbb{S}$ be a cubature rule of degree $2n$ for the sphere $\mathbb{S}$ with weights $(c_\xi)_{\xi\in\Theta}$, and p_ξ be the canonical choice (16.39). Then*

$$f = \dim(\mathscr{P}_n) \sum_{\xi \in \Theta} c_\xi \langle f, p_\xi \rangle p_\xi, \qquad \forall f \in \mathscr{P}_n. \tag{16.49}$$

Proof. Apply the quadrature rule of degree $2n$ to (16.48) to obtain

$$f(x) = \frac{\dim(\mathscr{P}_n)}{\operatorname{area}(\mathbb{S})} \int_{\mathbb{S}} \langle f, p_\xi \rangle p_\xi(x)\, d\xi = \dim(\mathscr{P}_n) \sum_{\xi \in \Theta} c_\xi \langle f, p_\xi \rangle p_\xi(x),$$

which is (16.49). □

Example 16.20. For equal weight quadrature rules, i.e., spherical designs, (16.49) reduces to

$$f=\frac{\dim(\mathscr{P}_n)}{|\Theta|}\sum_{\xi\in\Theta}\langle f,p_\xi\rangle p_\xi,\qquad \forall f\in\mathscr{P}_n. \tag{16.50}$$

We now consider the bivariate polynomials, i.e., $d=2$. Here

$$SO(2)=\{R_\theta:0\le\theta<2\pi\},\qquad R_\theta:=\begin{pmatrix}\cos\theta & -\sin\theta\\ \sin\theta & \cos\theta\end{pmatrix},$$

where R_θ is rotation through θ.

Example 16.21. (The circle) Let Θ_k be any set of k equally spaced points on the unit circle $\mathbb{S}$ ($d=2$). These give an equal weight quadrature rule of degree $k-1$ for $\mathbb{S}$ (see Exer. 6.10). Hence (16.50) holds for $\Theta=\Theta_k$, $k\ge 2n+1$. This also extends to when $k\ge n+1$ and k is odd by writing $\Theta_{2k}=\Theta_k\cup R_\pi\Theta_k$, and using $R_\pi p_\xi=(-1)^n p_\xi$.

Example 16.22. (Logan–Shepp) Taking $k=2(n+1)$ in Example 16.21 gives two copies (up to a scalar ±1) of the orthonormal basis $(R_{\frac{j\pi}{n+1}}p_0)_{j=0}^n$ for $\mathscr{P}_n$.

In particular, for the Legendre polynomials on the unit disc (constant weight), the continuous tight frame expansion (16.43) can be discretised to the orthogonal expansion

$$f=\frac{n+1}{2\pi}\int_0^{2\pi}\langle f,R_{\frac{j\pi}{n+1}}p\rangle R_{\frac{j\pi}{n+1}}p,\quad \forall f\in\mathscr{P}_n,\qquad p(x,y):=\frac{1}{\sqrt{\pi}}U_n(x),$$

of Logan and Shepp [LS75].

In a similar vein, one can obtain discrete versions of (16.37).

Definition 16.6. A finite subgroup G of $\mathrm{SO}(d)$ **generates** a spherical t-design if the set $\Theta=\{g\eta\}_{g\in G}$ is a spherical t-design for some (and hence every) $\eta\in S$.

Such groups are said to be *t-homogeneous* (see [Ban84], [dlHP04]).

Corollary 16.3. *Let G be a finite subgroup of* $\mathrm{SO}(d)$ *which generates a spherical $2n$-design, and $p=p_\xi$ the canonical choice (16.39). Then*

$$f=\frac{\dim(\mathscr{P}_n)}{|G|}\sum_{g\in G}\langle f,gp\rangle gp,\qquad \forall f\in\mathscr{P}_n. \tag{16.51}$$

Proof. Let $\Theta=\{g\xi\}_{g\in G}$ in (16.50), and use $gp_\xi=p_{g\xi}$. □

Example 16.23. (The circle) Similarly to Example 16.21, (16.51) holds for

$$G=\langle R_{\frac{2\pi}{k}}\rangle\subset\mathrm{SO}(2),$$

the cyclic group of rotations through multiples of $2\pi/k$ (of order k), where $k\ge n+1$ and k is odd, or $k\ge 2n+1$ and k is even.

16.7 Functions on the complex sphere

We now outline how the orthogonal decomposition (16.8) of $L_2(\mathbb{S})$ into spherical harmonics can be extended to complex-valued functions on the complex sphere

$$\mathbb{S}_\mathbb{C} = \tilde{\mathbb{S}} := \{x \in \mathbb{C}^d : \|x\| = 1\}.$$

For more detail see [Rud80]. The inner product (16.6) is replaced by

$$\langle f, g \rangle_{\mathbb{S}_\mathbb{C}} := \int_{\mathbb{S}_\mathbb{C}} f \overline{g} \, d\sigma, \tag{16.52}$$

where σ is normalised surface-area measure on the sphere in $\mathbb{R}^{2d}$. From (16.8), we have the orthogonal decomposition into absolutely irreducible $O(2d)$-subspaces

$$L_2(\mathbb{S}_\mathbb{C}) = \bigoplus_{k=0}^{\infty} \mathscr{H}_k(\mathbb{S}_\mathbb{C}), \tag{16.53}$$

where $\mathscr{H}_k(\mathbb{S}_\mathbb{C}) = \mathscr{H}_k(\mathbb{C}^d)$ is the complex vector space of all *harmonic* (as functions on $\mathbb{R}^{2d}$) homogeneous (with respect to real scalars) polynomials of degree k on $\mathbb{R}^{2d}$ (which is identified with $\mathbb{C}^d$). The monomials $z \mapsto z^\alpha \overline{z}^\beta$, $|\alpha| + |\beta| = k$, are an orthogonal basis for $\mathscr{H}_k(\mathbb{S}_\mathbb{C}) = \mathscr{H}_k(\mathbb{C}^d)$, and a calculation shows

$$\int_{\mathbb{S}} |z^\alpha|^2 \, d\sigma(z) = \frac{(d-1)!\alpha!}{(d-1+|\alpha|)!}. \tag{16.54}$$

Let $\mathscr{U} = \mathscr{U}(d)$ be the group of all unitary operators on $\mathbb{C}^d$, which is a compact subgroup of $O(2d)$. The $\mathscr{U}$-invariant subspaces $\mathscr{H}_k(\mathbb{S}_\mathbb{C})$, $k \neq 0$, are not irreducible. Let $H(p,q)$ be the subspace of $\mathscr{H}_k(\mathbb{S}_\mathbb{C})$, $k = p+q$, consisting of all polynomials on $\mathbb{C}^d$ that have *bidegree* (p,q), i.e.,

$$H(p,q) := \operatorname{span}\{z \mapsto z^\alpha \overline{z}^\beta : |\alpha| = p, |\beta| = q\}.$$

This is the space $\Pi^\circ_{p,q}(\mathbb{C}^d)$ considered in Exer. 6.17. Then $\mathscr{H}_k(\mathbb{S}_\mathbb{C})$ is the orthogonal direct sum

$$\mathscr{H}_k(\mathbb{S}_\mathbb{C}) = \bigoplus_{p+q=k} H(p,q),$$

of absolutely irreducible $\mathscr{U}$-invariant subspaces, so that

$$L_2(\mathbb{S}_\mathbb{C}) = \bigoplus_{k=0}^{\infty} \mathscr{H}_k(\mathbb{S}_\mathbb{C}) = \bigoplus_{k=0}^{\infty} \bigoplus_{p+q=k} H(p,q), \tag{16.55}$$

where none of the $H(p,q)$ in this orthogonal direct sum are $\mathbb{C}\mathscr{U}$-isomorphic.

We now consider the reproducing kernel tight frames for various $\mathscr{U}$-invariant subspaces of $L_2(\mathbb{S}_\mathbb{C})$. Here we write $z^\alpha \overline{z}^\beta$ for the monomial $z \mapsto z^\alpha \overline{z}^\beta$.

Lemma 16.3. *The reproducing kernel for the irreducible $\mathscr{U}$-invariant subspace $H(p,q)$ of $L_2(\mathbb{S}_\mathbb{C})$ is*

$$K_{pq}(z,w) = \frac{(d-1+p+q)!}{(d-1)!} \sum_{|\alpha|=p} \sum_{|\beta|=q} \frac{z^\alpha \overline{z}^\beta \overline{w}^\alpha w^\beta}{(\alpha+\beta)!}. \tag{16.56}$$

Proof. We first observe that by (16.54)

$$\langle z^\alpha \overline{z}^\beta, z^\alpha \overline{z}^\beta \rangle_{\mathbb{S}_\mathbb{C}} = \int_{\mathbb{S}_\mathbb{C}} z^\alpha \overline{z}^\beta \overline{z}^\alpha z^\beta \, d\sigma(z) = \int_{\mathbb{S}_\mathbb{C}} |z^{\alpha+\beta}|^2 \, d\sigma(z) = \frac{(d-1)!(\alpha+\beta)!}{(d-1+|\alpha+\beta|)!}.$$

Since $(z^\alpha \overline{z}^\beta)_{|\alpha|=p,|\beta|=q}$ is an orthogonal basis for $H(p,q)$, Proposition 16.3 gives

$$K_{pq}(z,w) = \sum_{|\alpha|=p} \sum_{|\beta|=q} z^\alpha \overline{z}^\beta \overline{w}^\alpha w^\beta \frac{(d-1+p+q)!}{(d-1)!(\alpha+\beta)!},$$

which is (16.56). □

Example 16.24. (Holomorphic polynomials). The space $H(p,0)$ is the holomorphic homogeneous polynomials of degree p. Its reproducing kernel is

$$K_{p0}(z,w) = \binom{d-1+p}{d-1} \sum_{|\alpha|=p} z^\alpha \overline{w}^\alpha \binom{p}{\alpha} = \binom{d-1+p}{d-1} \langle z,w \rangle^p. \tag{16.57}$$

In view of (16.55), the reproducing kernel for the space of holomorphic polynomials of degree $\leq n$ is

$$K(z,w) = \sum_{p=0}^{n} \binom{d-1+p}{d-1} \langle z,w \rangle^p.$$

Summing over all p gives the *Szegö kernel* for the holomorphic functions in $L_2(\mathbb{S}_\mathbb{C})$

$$S(z,w) = \sum_{p=0}^{\infty} K_{p0}(z,w) = \sum_{p=0}^{\infty} \binom{d-1+p}{d-1} \langle z,w \rangle^p = \frac{1}{(1-\langle z,w \rangle)^d}.$$

Example 16.25. (Spherical harmonics) The reproducing kernel for $\mathscr{H}_k(\mathbb{S}_\mathbb{C}) = \mathscr{H}_k(\mathbb{C}^d)$ is given by

$$\sum_{p+q=k} K_{pq}(z,w) = \|z\|^k C_k^{(d)}\Big(\frac{\langle z,w \rangle + \langle w,z \rangle}{2\|z\|}\Big) - \|z\|^k C_{k-2}^{(d)}\Big(\frac{\langle z,w \rangle + \langle w,z \rangle}{2\|z\|}\Big).$$

Summing over all p and q gives the Poisson kernel for $\mathbb{S}_\mathbb{C}$

$$P(z,w) = \frac{1-\|z\|^2}{\|w-z\|^{2d}}, \qquad z \in B, \quad w \in \mathbb{S}.$$

The details of these examples are given in Exer. 16.10.

16.8 G-frames for infinite groups

The previous constructions of tight G-frames for the continuous group $G = SO(d)$ relied on the extension of Theorem 10.8 to infinite groups G. Motivated by these, we now briefly discuss such extensions in generality. We first suppose that:

- G is a *locally compact (topological) group*, and μ is the associated *(left) Haar measure*, which is normalised so that $\mu(G) = 1$ when G is compact.
- There is a *unitary continuous representation* of G on a Hilbert space $\mathscr{H}$, i.e., a unitary representation ρ for which the map $(g, v) \mapsto gv := \rho(g)v$ is continuous.

It is natural the role played by (algebraically) irreducible subspaces is replaced by *topologically irreducible subspaces*. A continuous representation of G on V is said to be **topologically irreducible** if V does not contain a proper closed G-invariant subspace (for V finite dimensional, there is no distinction between the two).

The closure of a topologically irreducible subspace is a topologically irreducible subspace. For a continuous unitary action on $\mathscr{H}$, the orthogonal complement of a G-invariant subspace is G-invariant.

If G is *compact*, then all topologically irreducible unitary representations are finite dimensional. This leads to an orthogonal decomposition of $\mathscr{H}$ into finite dimensional irreducible G-invariant subspaces (Peter–Weyl theorem).

If G is *not compact*, e.g., $G = \mathbb{Z}$, then the situation is far more complicated:

- Irreducible representations may not exist, and they can be infinite dimensional.
- The frame operator for a G-orbit does not obviously converge.

Here is an example of G noncompact (see Exer. 16.11 for detail).

Example 16.26. Let $G = \mathbb{Z}$ act on $\mathscr{H} = \ell_2(\mathbb{Z})$ via the *bilateral shift* $j \cdot v = S^j v$, where $Se_k := e_{k+1}$. This gives a unitary continuous action. Elementary calculations show that $(S^j v)_{j \in J}$ is a tight frame if and only if it is an orthogonal basis, and the only v with finite support giving a tight frame are the standard basis vectors $v = e_k$ (and multiples of them). There exist vectors with infinite support giving tight frames, e.g.,

$$v = (\ldots, -\frac{1}{5}, 0, -\frac{1}{3}, 0, -1, 0, 1, 0, \frac{1}{3}, 0, \frac{1}{5}, 0, \frac{1}{7}, \ldots). \tag{16.58}$$

The vectors v for which $(S^j v)_{j \in \mathbb{Z}}$ is a tight frame (orthogonal basis) can be better understood by identifying them as the Fourier coefficients of a 2π-periodic function $f \in L_2(\mathbb{T})$, where $v_j = \langle f(z), z^j \rangle_{L_2(\mathbb{T})}$ ($\mathbb{T} = [0, 2\pi]$ has the normalised Haar measure). Here the shift S corresponds to multiplication of $f(z)$ by $z = e^{it}$. The condition for v to give a tight frame is that f have constant modulus. The examples $v = e_k$ correspond to $f(z) = z^k$ (the only trigonometric polynomials with modulus 1), and the v of (16.58) to the function $f = 1$ on $[0, \pi]$ and $f = -1$ on $[\pi, 2\pi]$ (up to a scalar).

We now consider the case when G is compact, e.g., $G = SO(d)$. The Peter–Weyl theorem ensures:

- A continuous action of G (compact) on $\mathscr{H}$ can be taken to be unitary.
- For a continuous unitary action of G on $\mathscr{H}$, there is an orthogonal decomposition of $\mathscr{H}$ into finite dimensional absolutely irreducible subspaces.

Theorem 16.5. *Let G be a compact group with a continuous unitary action on $\mathscr{H}$. Then the orbit $(gv)_{g\in G}$ is a (generalised) tight frame for $\mathscr{H}$ if and only if $\mathscr{H}$ is finite dimensional, and the conditions of Theorem 10.8 hold, where (10.16) is replaced by*

$$\int_G \langle v_j, gv_j\rangle gv_k \, d\mu(g) = 0. \tag{16.59}$$

Proof. We first suppose that $\mathscr{H}$ is finite dimensional, and so can be written as an orthogonal direct sum $\mathscr{H} = V_1 \oplus \cdots \oplus V_m$ of irreducible G-invariant spaces. Since

$$\int_G |\langle f, gv\rangle|^2 \, d\mu(g) \le \int_G \|f\|^2 \|gv\|^2 \, d\mu(g) = \mu(G)\|f\|^2\|v\|^2,$$

the frame operator for $\Phi = (gv)_{g\in G}$ is well defined (see Exer. 16.1) by

$$S_\Phi(f) = \int_G \langle f, gv\rangle gv \, d\mu(g), \qquad \forall f \in \mathscr{H}.$$

The argument of Theorem 10.8 then follows with $\sum_{g\in G}$ replaced by integration with respect to the Haar measure. In particular, we observe $\operatorname{trace}(S_\Phi) = \mu(G)\|v\|^2$ and Schur's lemma (Lemma 10.4) also holds for G infinite (and V_j finite dimensional).

We now suppose that $\mathscr{H}$ is infinite dimensional. Since the orthogonal projection of a tight frame onto a closed subspace is a tight frame (for the subspace), we can assume without loss of generality that

$$\mathscr{H} = V_1 \oplus V_2 \oplus V_3 \oplus \cdots,$$

a countable direct sum of (finite dimensional) irreducible G-invariant subspaces. Suppose that $(gv)_{g\in G}$ is tight frame for $\mathscr{H}$, and write $v = v_1 + v_2 + \cdots$, $v_j \in V_j$. Then $(gw)_{g\in G}$, $w := v_1 + \cdots + v_m$, is a tight frame for the finite dimensional space $V_1 \oplus \cdots \oplus V_m$, and so the conditions of Theorem 10.8 hold. In particular,

$$v_j \ne 0, \quad \forall j, \qquad \frac{\|v_j\|^2}{\|v_k\|^2} = \frac{\dim(V_j)}{\dim(V_k)}.$$

This implies

$$\|w\|^2 = \sum_j \|v_j\|^2 = \frac{\|v_1\|^2}{\dim(V_1)} \sum_{j=1}^{\infty} \dim(V_j) = \infty,$$

which is not possible. Thus there can be a tight G-frame $(gw)_{g\in G}$ for $\mathscr{H}$ only when $\mathscr{H}$ has finite dimension. □

The considerations of §10.13 on when a G-frame exists extend, e.g.,

For G compact and $\mathscr{H}$ finite dimensional, there is a tight G-frame for $\mathscr{H}$ if and only if each irreducible $\mathbb{C}G$-module W in $\mathscr{H}$ has multiplicity $\leq \dim(W)$.

This result is closely connected with the result of [GM71] (for G compact) on the existence of **cyclic vectors**, i.e., v for which the closed span of $(gv)_{g\in G}$ is $\mathscr{H}$. The condition for $V = \mathscr{H}$ to have cyclic vectors is that it has countable dimension, and

$$\mathrm{mult}(W,V) \leq \dim(W), \qquad \text{for every irreducble } \mathbb{C}G\text{-module } W.$$

By considering the projection v_W of a cyclic vector $v \in V$ onto its homogeneous components $H_V(W)$, it is easy enough to see that this is a necessary condition. Conversely, by choosing v_W so that $(gv_W)_{g\in G}$ is a tight frame for $H_V(W)$, and $\sum_W \|v_W\|^2 < \infty$, one can construct a cyclic vector $v = \sum_W v_W$.

For G a countable group, G-frames (for the counting measure) are studied up to unitary equivalence in [HL00] (using Von Neumann algebras), and for G compact, they are studied in [Ive15] (using the Zak transform). Both approaches make use of the left regular representation of G, which plays the same role as the group algebra $\mathbb{C}G$ in the case when G is finite dimensional.

Example 16.27. Let $G = SO(d)$ act on the space of polynomials $\Pi_k(\mathbb{R}^d)$, with a radially symmetric measure. From the orthogonal direct sum (16.46), we have

$$\Pi_2(\mathbb{R}^d) = \mathscr{H}_0 \oplus \mathscr{H}_1 \oplus P_1^{(2)}(\|\cdot\|^2)\mathscr{H}_0 \oplus \mathscr{H}_2.$$

Since $\dim(\mathscr{H}_0) = 1$ and the multiplicity of the irreducible $\mathscr{H}_0$ in $\Pi_2(\mathbb{R}^d)$ is two, there is no G-frame for the quadratic polynomials $\Pi_2(\mathbb{R}^d)$ (or $\Pi_k(\mathbb{R}^d)$, $k \geq 2$). Since $\Pi_k(\mathbb{R}^d)$ is finite dimensional, it is possible to construct a reproducing kernel tight frame (see Example 16.18).

Notes

Continuous tight frames were introduced and studied in detail in [AAG93]. A good account of zonal harmonics is given in [ABR01]. The section on the orthogonal polynomials for a radially symmetric weight was adapted from [Wal09]. I thank Tom Ter Elst and Joey Iverson for many useful discussions about this chapter.

Exercises

16.1. Let $(f_j)_{j\in J}$ be a generalised frame with respect to μ for a Hilbert space $\mathscr{H}$ (possibly infinite dimensional).
(a) Show that

$$g \mapsto \int_J \overline{\langle f, f_j\rangle}\langle g, f_j\rangle \, d\mu(j) =: \langle g, Sf\rangle$$

defines a bounded linear functional on $\mathscr{H}$ and denote its Riesz representer by

$$Sf = \int_J \langle f, f_j\rangle f_j \, d\mu(j).$$

(b) Show this defines a linear map $S : \mathscr{H} \to \mathscr{H}$, with

$$A\|f\|^2 \le \langle Sf, f\rangle \le B\|f\|^2, \qquad \forall f \in \mathscr{H}.$$

(c) Show that S has a bounded inverse, and

$$\begin{aligned} f &= \int_J \langle f, S^{-1} f_j\rangle f_j \, d\mu(j) = \int_J \langle f, f_j\rangle S^{-1} f_j \, d\mu(j) \\ &= \int_J \langle f, S^{-1/2} f_j\rangle S^{-1/2} f_j \, d\mu(j), \qquad \forall f \in \mathscr{H}. \end{aligned}$$

(d) Show that the *synthesis* and *analysis* operators (see §2.4) can be generalised to

$$V : \ell_2(\mu) \to \mathscr{H} : a \mapsto \int_J a_j f_j \, d\mu(j),$$

$$V^* : \mathscr{H} \to \ell_2(\mu) : f \mapsto (\langle f, f_j\rangle)_{j\in J},$$

where Va is defined as the Riesz representer of

$$f \mapsto \int_J \langle f, a_j f_j\rangle \, d\mu(j) =: \langle f, Va\rangle.$$

Remark: Here $S = VV^*$, and one can define the Gramian as $V^*V : \ell_2(\mu) \to \ell_2(\mu)$.

16.2. Let $(f_j)_{j\in J}$ be a generalised frame with respect to μ for a d-dimensional Hilbert space $\mathscr{H}$. Show that the variational characterisation (Theorem 6.1) extends in the obvious way (Proposition 16.2), i.e.,

$$\int_J \int_J |\langle f_j, f_k\rangle|^2 \, d\mu(j)\, d\mu(k) \ge \frac{1}{d}\Big(\int_J \|f_j\|^2 \, d\mu(j)\Big)^2,$$

with equality if and only if $(f_j)_{j\in J}$ is tight.

16.3. The *canonical Gramian* of a generalised frame $\Phi = (f_j)_{j\in J}$ with respect to μ is given by

$$P_\Phi = V^* S^{-1} V : L_2(\mu) \to L_2(\mu),$$

where $S = VV^*$ is the frame operator and V is the synthesis operator of Φ.
(a) Show that P_Φ is an orthogonal projection.
(b) Show that the "matrix" $[P_\Phi] := [\langle f_k, S^{-1} f_j\rangle]_{j,k\in J}$ represents P_Φ in the sense

$$P_\Phi a = [P_\Phi] \cdot_\mu a := \Big(\int_J [P_\Phi]_{jk} a_k \, d\mu(k)\Big)_{j\in J} = \Big(\int_J \langle f_k, S^{-1} f_j\rangle a_k \, d\mu(k)\Big)_{j\in J}.$$

(c) Let $v_k := (\langle f_k, S^{-1} f_j\rangle)_{j\in J}$ be the k-th "column" of $[P_\Phi]$. Show that $v_k \in L_2(\mu)$, and $(v_j)_{j\in J}$ gives a copy of the canonical tight frame $(S^{-1/2} f_j)_{j\in J}$, i.e.,

$$\langle v_r, v_s\rangle_{L_2(\mu)} = \langle f_r, f_s\rangle, \qquad \forall r,s \in J.$$

16.4. Suppose $(f_j)_{j\in J}$ is a unit-norm generalised tight frame for a d-dimensional space $\mathscr{H}$, i.e., $\|f_j\| = 1$, $\forall j \in J$. Show that μ is finite, and

$$f = \frac{d}{\mu(J)} \int_J \langle f, f_j\rangle f_j \, d\mu(j), \qquad \forall f \in \mathscr{H}.$$

16.5. Let $\mathscr{H}$ be a subspace of $\mathbb{F}^n$, and P be the orthogonal projection onto $\mathscr{H}$.
(a) Show that the tight frame $(K_j)_{j=1}^n$ corresponding to the reproducing kernel for $\mathscr{H}$ is given by $K_j = Pe_j$.
(b) Find this normalised tight frame explicitly for

$$\mathscr{H} = \{x \in \mathbb{F}^n : x_1 + \cdots + x_n = 0\}.$$

16.6. Suppose that $\mathscr{H}$ is a reproducing kernel Hilbert space.
(a) Show that any subspace of $\mathscr{H}$ is again a reproducing kernel Hilbert space.
(b) Suppose $\mathscr{H} = \bigoplus_j \mathscr{H}_j$, an orthogonal direct sum of subspaces. Show that the reproducing kernel of $\mathscr{H}$ is $K = \sum_j K_j$, where K_j is the reproducing kernel of $\mathscr{H}_j$.

16.7. Use the orthogonal decomposition $L_2(\mathbb{S}) = \bigoplus_j \mathscr{H}_j$ of $L_2(\mathbb{S})$ into absolutely irreducible rotationally invariant subspaces to show that the rotationally invariant subspaces of $L_2(\mathbb{S})$ have the form

$$\mathscr{H} = \bigoplus_{j\in J} \mathscr{H}_j, \qquad \text{for some } J \subset \mathbb{N}.$$

16.8. The **Poisson kernel** for the unit ball $B = \{x \in \mathbb{R}^d : \|x\| < 1\}$ is given by

$$P(x,\xi) = \frac{1 - \|x\|^2}{\|x - \xi\|^d} = \frac{1 - \|x\|^2}{(1 - 2\langle x, \xi\rangle + \|x\|^2)^{\frac{d}{2}}}, \qquad x \in B, \ \xi \in \mathbb{S}.$$

It has the property that for every u which is harmonic on the closed unit ball

$$u(x) = \int_{\mathbb{S}} u(\xi) P(x,\xi) \, d\xi.$$

From this, and (16.8), (16.9), (16.7), it follows that

$$P(x,\xi) = \sum_{k=0}^{\infty} Z_\xi^{(k)}(x), \qquad x \in B,\ \xi \in \mathbb{S},$$

where the series converges absolutely and locally uniformally. Use the generating function for the Gegenbauer polynomials

$$\frac{1}{(1-2yt+t^2)^\lambda} = \sum_{k=0}^{\infty} C_k^{(\lambda)}(y)t^k$$

to expand the Poisson kernel in terms of the zonal harmonics, to obtain the formula

$$Z_\xi^{(k)}(x) = \|x\|^k C_k^{\frac{d}{2}}\Bigl(\frac{\langle x,\xi\rangle}{\|x\|}\Bigr) - \|x\|^k C_{k-2}^{\frac{d}{2}}\Bigl(\frac{\langle x,\xi\rangle}{\|x\|}\Bigr).$$

16.9. (Linear polynomials on the sphere). The spaces $\Pi_1(\mathbb{S})$ and $\Pi_1(\mathbb{R}^d)$ of linear polynomials on the sphere and on $\mathbb{R}^d$ have dimension $d+1$.
(a) Find the reproducing kernel tight frame (Z_ξ) for these spaces, with the norms $\|\cdot\|_{\mathbb{S}}$ and $\|\cdot\|_w$, respectively.
(b) Let $\xi_1,\ldots,\xi_4$ be four points on the sphere in $\mathbb{R}^3$. Show that the zonal functions $(Z_{\xi_j})_{j=1}^4$ are a basis for $\Pi_1(\mathbb{S})$ if and only if the points $\{\xi_j\}$ do not lie on a circle.
(c) Show that $(Z_{\xi_j})_{j=1}^4$ is an orthogonal basis for $\Pi_1(\mathbb{S})$ if and only if the points $\{\xi_j\}$ are the vertices of a regular tetrahedron.

16.10. The $\mathscr{U}$-invariant subspaces of $L_2(\mathbb{S}_\mathbb{C})$ are given by the subsums of (16.55). Thus the reproducing kernel of such a space is the sum of the reproducing kernels of its summands (cf Theorem 16.1).
(a) Show that the reproducing kernel of $H(p,0)$ is zonal.
(b) Find the reproducing kernel for the holomorphic polynomials of degree $\le n$.
(c) Show that the sum of the reproducing kernels of the homogeneous holomorphic functions of all degrees (these are orthogonal) is the the **Szegö kernel**

$$S(z,w) := \frac{1}{(1-\langle z,w\rangle)^d}, \qquad z \in B, \quad w \in \mathbb{S}.$$

It has the property that for every f which is holomorphic on the closed unit ball

$$f(z) = \int_{\mathbb{S}_\mathbb{C}} f(w)S(z,w)\,d\sigma(w).$$

(d) Expand the Poisson kernel

$$P(z,w) = \sum_{p=0}^{\infty}\sum_{q=0}^{\infty} K_{pq}(z,w) = \frac{1-\|z\|^2}{\|w-z\|^{2d}},$$

to find a formula for the reproducing kernel for $\mathscr{H}_k(\mathbb{S}_\mathbb{C}) = \mathscr{H}_k(\mathbb{C}^d)$.

16.11. Let S be the *bilateral shift* on $\ell_2(\mathbb{Z})$, given by $Se_j := e_{j+1}$. Then $j \cdot v = S^j v$ defines unitary continuous action of the noncompact group $G = \mathbb{Z}$ on $\mathscr{H} = \ell_2(\mathbb{Z})$.
(a) Let $v \in \ell_2(\mathbb{Z})$ be nonzero. Show that if $(S^j v)_{j\in\mathbb{Z}}$ is a tight frame for $\ell_2(\mathbb{Z})$, then the frame bound is $A = \|v\|^2$, i.e., $(S^j v)$ is orthogonal basis.
(b) Let $v = v_a e_a + v_b e_b \neq 0$, $a \neq b$. Show

$$\sum_j |\langle x, S^j v\rangle|^2 = \|v\|^2\|x\|^2 + 2\Re\bigl(v_a \overline{v_b}\langle S^{a-b}x, x\rangle\bigr),$$

and conclude that $(S^j v)_{j\in\mathbb{Z}}$ is a tight frame for $\ell_2(\mathbb{Z})$ if and only if $v_a = 0$ or $v_b = 0$.
(c) Find all finitely supported vectors v for which $(S^j v)_{j\in\mathbb{Z}}$ is a tight frame for $\ell_2(\mathbb{Z})$.
(d) Let $v = (\ldots, -\frac{1}{5}, 0, -\frac{1}{3}, 0, -1, 0, 1, 0, \frac{1}{3}, 0, \frac{1}{5}, 0, \frac{1}{7}, \ldots)$. Show that $(S^j v)_{j\in\mathbb{Z}}$ is a tight frame for $\ell_2(\mathbb{Z})$.
(e) Determine all vectors v for which $(S^j v)_{j\in\mathbb{Z}}$ is a tight frame for $\ell_2(\mathbb{Z})$.

16.12. Let $\mathbb{S}_\mathbb{C} := \{z \in \mathbb{C}^d : \|z\| = 1\}$ be the complex unit sphere in $\mathbb{C}^d \approx \mathbb{R}^{2d}$, and σ be Lebesgue surface-area measure on $\mathbb{S}_\mathbb{C}$ viewed as a unit sphere in $\mathbb{R}^{2d}$. Deduce the analogue of (16.15), i.e.,

$$z = \frac{d}{\operatorname{area}(\mathbb{S}_\mathbb{C})} \int_{\mathbb{S}_\mathbb{C}} \langle z, \xi\rangle \xi \, d\sigma(\xi), \qquad \forall z \in \mathbb{C}^d.$$

16.13. Show that the zonal harmonic $Z_\xi^{(k)}$ is localised at $\xi \in \mathbb{S}$ in the following sense.
(a) $|Z_\xi^{(k)}(x)| < Z_\xi^{(k)}(\xi)$, $\forall x \neq \xi$, $x \in \mathbb{S}$.
(b) $Z_\xi^{(k)}(\xi) = \|Z_\xi^{(k)}\|_\mathbb{S}^2 = \dim(\mathscr{H}_k)$.
(c) The maximum

$$\max\{p(\xi) : \|p\|_\mathbb{S} = 1, p \in \mathscr{H}_k(\mathbb{S}^{d-1})\} = \sqrt{\dim(\mathscr{H}_k)} = O(k^{\frac{d}{2}}), \qquad k \to \infty$$

is attained if and only if $p = \frac{1}{\sqrt{\dim(\mathscr{H}_k)}} Z_\xi^{(k)}$.

16.14. The variational condition (6.5) for (f_j) to be a finite tight frame for F_d is

$$\sum_j \sum_k |\langle f_j, f_k\rangle|^2 = \frac{1}{d}\Bigl(\sum_j \langle f_j, f_j\rangle\Bigr)^2.$$

Show that the following analogous condition holds

$$\int_\mathbb{S}\int_\mathbb{S} |\langle x, y\rangle|^2 \, d\sigma(x)\, d\sigma(y) = \frac{1}{d}\Bigl(\int_\mathbb{S} \langle x, x\rangle \, d\sigma(x)\Bigr)^2.$$

Remark: This can be interpreted as saying that although a generic frame of n unit vectors for F_d is not tight, for $n \gg d$ it is close to being tight (also see Exer. 6.3).

16.15. Let $\langle \cdot, \cdot \rangle_{\circ,k}$ denote the *apolar* inner product of (6.19) on $H(k,0)$ (the space of holomorphic homogeneous polynomials of degree k).

(a) Use (6.21), to show that

$$\langle f,g\rangle_{\circ,k} = \binom{k+d-1}{d-1}\langle f,g\rangle_{\mathbb{S}_{\mathbb{C}}}.$$

(b) Express the reproducing kernel tight frame given by (16.57) in terms of $\langle\cdot,\cdot\rangle_{\circ,k}$.
(c) Use the formula

$$\int_{\mathbb{S}_{\mathbb{C}}} f(\langle\xi,\eta\rangle)\,d\sigma(\xi) = \frac{d-1}{\pi}\int_0^{2\pi}\int_0^1 (1-r^2)^{d-2} f(re^{i\theta})\,r\,dr\,d\theta, \quad \eta\in\mathbb{S}_{\mathbb{C}},$$

to calculate $\|\langle\cdot,\eta\rangle^k\|_{\mathbb{S}_{\mathbb{C}}}$.

Solutions

Exercises of Chapter 1

1.1 Since $\sum_j \langle f, u_j \rangle u_j = \sum_j (u_j^* f) u_j = V(V^* f) = (VV^*) f$, we obtain the matrix form. Verify that (1.1) holds for a particular choice of u_1, u_2, u_3. If R is a rotation, then $[Ru_1, Ru_2, Ru_3] = RV$, and $(RV)(RV)^* = R(VV^*)R^* = \frac{3}{2}RR^* = \frac{3}{2}I$. Thus (1.1) holds for Ru_1, Ru_2, Ru_3.

1.2 (a) Suppose $f = \langle f, u_1 \rangle v_1 + \langle f, u_2 \rangle v_2$, $\forall f \in \mathbb{R}^2$. Since the RHS is linear in f, it suffices that this holds for the basis $\{u_1, u_2\}$ i.e., $u_1 = v_1 - \frac{1}{2}v_2$, $u_2 = -\frac{1}{2}v_1 + v_2$, which leads to $v_1 = \frac{2}{3}(2u_1 + u_2)$, $v_2 = \frac{2}{3}(u_1 + 2u_2)$.
(b) Yes. Since $u_1 + u_2 + u_3 = 0$, the coefficients satisfy $\sum_j \langle f, u_j \rangle = 0$. Hence if one is changed, then this sum will no longer be zero.
(c) No. Two coefficients can be changed while preserving $\sum_j \langle f, u_j \rangle = 0$.

1.3 *Equiangularity/equispacing.*
(a) Verify that these are a tight frame by using the matrix form of Exer. 1.1.
(b), (c) Direct computation.
(d) Though equally spaced from each other, these vectors do not "fill up" the space $\mathbb{C}^2$, e.g., if $w = (-1, 0)$, then $\|v_j - w\| = \sqrt{2 + \sqrt{2}} > \sqrt{3}$, $\forall j$.

1.4 *Gabor and Wavelet systems.*
(a) $\mathscr{T}_a \mathscr{M}_b f(x) = (\mathscr{M}_b f)(x-a) = e^{2\pi i b \cdot x} e^{-2\pi i b \cdot a} f(x-a) = e^{-2\pi i a \cdot b} \mathscr{M}_b \mathscr{T}_a f(x)$.
(b) Since $\mathscr{T}_a \mathscr{T}_b = \mathscr{T}_{a+b}$ and $\mathscr{M}_a \mathscr{M}_b = \mathscr{M}_{a+b}$, $\forall a, b$, the set G is closed under multiplication and inversion and so forms a group (generated by any set of generators for the subgroups $\{\mathscr{T}_a\}_{a \in A}$ and $\{\mathscr{M}_b\}_{b \in B}$).
(c) Since $\mathscr{D}_{2^j} = \mathscr{D}_2^j$, $\mathscr{T}_k = \mathscr{T}_1^k$, and

$$\mathscr{D}_{2^j} \mathscr{T}_k f = 2^{\frac{j}{2}} f(2^j \cdot - k) = 2^{\frac{j}{2}} f(2^j(\cdot - \frac{k}{2^j})) = \mathscr{T}_{\frac{k}{2^j}} \mathscr{D}_j f,$$

the group generated by $\mathscr{D}_2$ and $\mathscr{T}_1$ contains the translates of the diadic integers $\{\mathscr{T}_{\frac{k}{2^j}}\}_{j,k \in \mathbb{Z}}$, which are not contained in the set $\{\mathscr{D}_{2^j} \mathscr{T}_k\}_{j,k \in \mathbb{Z}}$.

S.F.D. Waldron, *An Introduction to Finite Tight Frames*, Applied and Numerical Harmonic Analysis, https://doi.org/10.1007/978-0-8176-4815-2

Exercises of Chapter 2

2.1 *Polarisation identity.*[1] Expand, and use $\langle g,f\rangle = \overline{\langle f,g\rangle}$, to get

$$\|f+g\|^2 - \|f-g\|^2 = 2(\langle f,g\rangle + \langle g,f\rangle) = 2(\langle f,g\rangle + \overline{\langle f,g\rangle}) = 4\Re\langle f,g\rangle.$$

If the inner product is complex, then this gives

$$\|f+ig\|^2 - \|f-ig\|^2 = 4\Re\langle f,ig\rangle = 4\Re(-i\langle f,g\rangle) = 4\Im\langle f,g\rangle.$$

2.2 Take the inner product of Parseval with g to obtain Plancherel. The reverse implication follows by the uniqueness of the Riesz representation and

$$\langle f,g\rangle = \frac{1}{A}\sum_{j\in J}\langle f,f_j\rangle\langle f_j,g\rangle = \langle \frac{1}{A}\sum_{j\in J}\langle f,f_j\rangle f_j, g\rangle, \qquad \forall g\in\mathscr{H}.$$

Taking $g=f$ in Plancherel gives the tight frame condition. Conversely, given a tight frame, the polarisation identity gives

$$\begin{aligned}4\Re\langle f,g\rangle &= \|f+g\|^2 - \|f-g\|^2 = \frac{1}{A}\sum_{j\in J}\Bigl(|\langle f+g,f_j\rangle|^2 - |\langle f-g,f_j\rangle|^2\Bigr)\\ &= \frac{1}{A}\sum_{j\in J}4\Re(\langle f,f_j\rangle\langle f_j,g\rangle) = 4\Re\Bigl(\frac{1}{A}\sum_{j\in J}\langle f,f_j\rangle\langle f_j,g\rangle\Bigr),\end{aligned}$$

and $4\Im\langle f,g\rangle = 4\Im(\frac{1}{A}\sum_{j\in J}\langle f,f_j\rangle\langle f_j,g\rangle)$, when $\mathscr{H}$ is complex.
Remark. This result extends to countably (or even uncountably) infinite tight frames, with the interchange of sums and inner products being justified by considering the appropriate limits of partial sums.

2.3 Clearly, $P=\frac{1}{A}VV^*$ is self-adjoint, i.e., $P^*=P$. Since (f_j) is a tight frame for $\mathscr{K}$, V is onto $\mathscr{K}$ and Parseval gives $VV^*|_{\mathscr{K}} = AI_{\mathscr{K}}$, so that

$$P^2 = (\frac{1}{A}VV^*|_{\mathscr{K}})(\frac{1}{A}VV^*) = I_{\mathscr{K}}\frac{1}{A}VV^* = P.$$

2.4 (a) Given the uniqueness of the coefficients in the orthogonal expansion

$$f = \sum_{j\in J}\frac{\langle f,f_j\rangle}{\langle f_j,f_j\rangle}f_j, \qquad \forall f\in\mathscr{H},$$

by Parseval, $(f_j)_{j\in J}$ is a tight frame if and only if $\|f_j\|^2 = \langle f_j,f_j\rangle = A$, $\forall j$, and this is normalised if and only if $\|f_j\| = \sqrt{A} = 1$, $\forall j$.
(b) Taking $f = f_j/\|f_j\|$ in the normalised tight frame condition gives

[1] A real or complex normed linear space $(X,\|\cdot\|)$ is an inner product space, with the inner product given by the polarisation identity, if and only if it satisfies the *parallelogram identity* $\|f+g\|^2 + \|f-g\|^2 = 2\|f\|^2 + 2\|g\|^2$, $\forall f,g\in X$.

$$1 = \|f\|^2 = \sum_{k\in J} |\langle \frac{f_j}{\|f_j\|}, f_k\rangle|^2 = \|f_j\|^2 + \frac{1}{\|f_j\|^2} \sum_{k\neq j} |\langle f_j, f_k\rangle|^2,$$

so we must have $\|f_j\| \leq 1$, with equality if and only if $|\langle f_j, f_k\rangle| = 0$, $\forall k \neq j$.

2.5 (a) Since unitary maps preserve norms: $\|U^* f\| = \|f\|$, and we obtain

$$A\|f\|^2 = A\|U^* f\|^2 = \sum_j |\langle U^* f, f_j\rangle|^2 = \sum_j |\langle f, Uf_j\rangle|^2, \quad \forall f.$$

(b) Since (Tf_j) is a tight frame for the finite dimensional space $\mathcal{H}$, the map T is onto, and hence invertible. Using the normalised tight frame property of (Tf_j), followed by that of (f_j), we obtain

$$T^{-1}f = \sum_j \langle f, Tf_j\rangle T^{-1}(Tf_j) = \sum_j \langle T^* f, f_j\rangle f_j = T^* f, \qquad \forall f \in \mathcal{H},$$

so that $T^{-1} = T^*$, and T is unitary.

2.6 Apply P to the tight frame expansion, and use $f = Pf$, $f \in \mathcal{H}$, to obtain

$$f = P\Bigl(\sum_j \langle f, f_j\rangle f_j\Bigr) = \sum_j \langle Pf, f_j\rangle Pf_j = \sum_j \langle f, Pf_j\rangle Pf_j, \quad f \in \mathcal{H}.$$

2.7 (a)$\Longrightarrow$(b) Suppose that Q is a partial isometry. Clearly, $(QQ^*)^* = QQ^*$. Let $Q| = Q|_{\mathrm{ran}(Q^*)}$, and $x \in \ker(Q^*) = \mathrm{ran}(Q)^\perp$, $y \in \mathrm{ran}(Q)$. Then

$$(QQ^*)^2(x+y) = QQ^*QQ^*y = Q(Q|^*Q|)Q^*y = QQ^*y,$$

so that QQ^* is the orthogonal projection onto $\mathrm{ran}(Q)$.
(b)$\Longrightarrow$(c) Suppose that QQ^* is an orthogonal projection. Clearly, Q^*Q is Hermitian. Let $x \in \ker(Q)$ and $y = Q^*z \in \ker(Q)^\perp = \mathrm{ran}(Q^*)$. Then

$$(Q^*Q)^2(x+y) = Q^*QQ^*QQ^*z = Q^*(QQ^*)^2 z = Q^*QQ^*z = Q^*Qy,$$

so that Q^*Q is the orthogonal projection onto $\mathrm{ran}(Q^*)$.
(c)$\Longrightarrow$(a) Suppose that Q^*Q is an orthogonal projection, so that $(Q^*Q)^* = Q^*Q$, $(Q^*Q)^2 = Q^*Q$. Let $y = Q^*z \in \ker(Q)^\perp = \mathrm{ran}(Q^*)$. Then

$$\|Qy\|^2 = \langle QQ^*z, QQ^*z\rangle = \langle (QQ^*)^2 z, z\rangle = \langle QQ^*z, z\rangle = \langle Q^*z, Q^*z\rangle = \|y\|^2,$$

so that Q is a partial isometry.
(a)$\Longrightarrow$(d) If Q is a partial isometry, then it can be factored $Q = UP$, where P is the orthogonal projection onto $(\ker Q)^\perp$ and $U : (\ker Q)^\perp \to Q(\mathcal{H})$ is unitary, so by Exer. 2.5 and 2.6, it follows that $Q\Phi$ is a normalised tight frame (for its span).
(d)$\Longrightarrow$(b) Suppose $\Phi = (f_j)$ and $Q\Phi = (Qf_j)$ are normalised tight frames. We will show that QQ^* is the orthogonal projection onto $\mathrm{ran}(Q)$. For $y \in \mathrm{ran}(Q)^\perp = \ker(Q^*)$, we have $QQ^*y = Q0 = 0$, and so it suffices to show $QQ^*y = y$, $\forall y = Qf \in \mathrm{ran}(Q)$.

This follows by first expanding in the normalised tight frame (Qf_j), and then in (f_j) (and using linearity):

$$y = Qf = \sum_j \langle Qf, Qf_j \rangle Qf_j = Q\sum_j \langle Q^*Qf, f_j \rangle f_j = Q(Q^*Qf) = QQ^*y.$$

2.8 (a) For a given U of size n, a random equal-norm tight frame for $\mathbb{C}^d$ is given by the MATLAB code: `J=randperm(n), J=J(1:d), V=U(J,:).`
(b) The Fourier matrix F is given the MATLAB code: `F=fft(eye(n))'/sqrt(n)` or `w=exp(2*pi*i/n), F=w.^([0:n-1]'*[0:n-1])/sqrt(n)`. Since $\omega^n = 1$ and $1+\omega+\omega^2+\cdots+\omega^{n-1} = 0$ (for any n–root of unity $\omega \neq 1$), we have

$$(F^*F)_{jk} = \sum_\ell (F^*)_{j\ell} F_{\ell k} = \frac{1}{d}\sum_\ell \omega^{-j\ell}\omega^{k\ell} = \frac{1}{d}\sum_{0\le\ell<n} (\omega^{k-j})^\ell = \frac{1}{d} d\delta_{jk}.$$

Similarly, $(F^2)_{jk} = \frac{1}{d}\sum_\ell (\omega^{j+k})^\ell = \delta_{j+k,0}$, with the indices mod n, and so

$$(F^4)_{jk} = (F^2F^2)_{jk} = \sum_\ell \delta_{j+\ell,0}\delta_{k+\ell,0} = \delta_{jk}.$$

2.9 (a) Let $z_j := x_j + iy_j$, so $w_j = z_j^2$, and the frame operator is

$$S = VV^* = \frac{1}{4}\begin{bmatrix} \sum_j (z_j+\overline{z_j})^2 & \frac{1}{i}\sum_j (z_j^2 - \overline{z_j}^2) \\ \frac{1}{i}\sum_j (z_j^2 - \overline{z_j}^2) & -\sum_j (z_j - \overline{z_j})^2 \end{bmatrix}, \quad V = \begin{bmatrix} \frac{z_1+\overline{z_1}}{2} & \cdots & \frac{z_n+\overline{z_n}}{2} \\ \frac{z_1-\overline{z_1}}{2i} & \cdots & \frac{z_n-\overline{z_n}}{2i} \end{bmatrix}.$$

This is diagonal, with equal diagonal entries if and only if

$$\sum_j z_j^2 - \sum_j \overline{z_j}^2 = 0, \quad \sum_j z_j^2 + \sum_j \overline{z_j}^2 = 0 \quad \Longleftrightarrow \quad \sum_j w_j = \sum_j z_j^2 = 0,$$

in which case $S = \frac{1}{2}\sum_j |z_j|^2 I$.
(b) Tight frames (f_j) and (g_j) for $\mathbb{R}^2$ are projectively unitarily equivalent if and only if $g_j = c\alpha_j U f_j$, $\forall j$, where U is unitary, $c > 0$, $\alpha_j = \pm 1$. Assume wlog that U is rotation by θ, so this becomes $w_j = ce^{i\theta}\alpha_j z_j$, $\forall j$, where $z_j := \Re f_j + i\Im f_j$, $w_j := \Re g_j + i\Im g_j$, i.e., the diagram vectors satisfy $w_j^2 = \zeta z_j^2$, $\forall j$, $\zeta = c^2 e^{2i\theta}$.
(c) Three unit vectors are a tight frame for $\mathbb{R}^2$ if and only if the sum of their diagram vectors is zero: $z_1^2 + z_2^2 + z_3^2 = 0$. By (b), we may assume that these diagram vectors are the third roots of unity, so z_1, z_2, z_3 are distinct sixth roots of unity, none of which is the negatives of another, and so the vectors in $\mathbb{R}^2$ they represent are projectively equivalent to three equally spaced vectors.
(d) Four unit vectors (v_j) are a tight frame if and only if the sum of their diagram vectors (z_j^2) is zero in $\mathbb{C}$, i.e., the Argand diagram of their sum is a parallelogram. Suppose wlog that $z_1^2 = -z_2^2 = (iz_2)^2$ and $z_3^2 = -z_4^2 = (iz_4)^2$. Then $z_1 = \pm iz_2$, $z_3 = \pm iz_4$, i.e., $v_1 \perp v_2$, $v_3 \perp v_4$, and so (v_j) is a union of two orthonormal bases for $\mathbb{R}^2$.
(e) By construction, the sum of the diagram vectors is zero, and so they give a tight frame (see Figure A.1). Moreover, none contains an orthonormal basis (these

correspond to diagram vectors which are negatives of each other). The inner product between the first and last has modulus $\cos\theta$, which gives the projective unitary inequivalence.

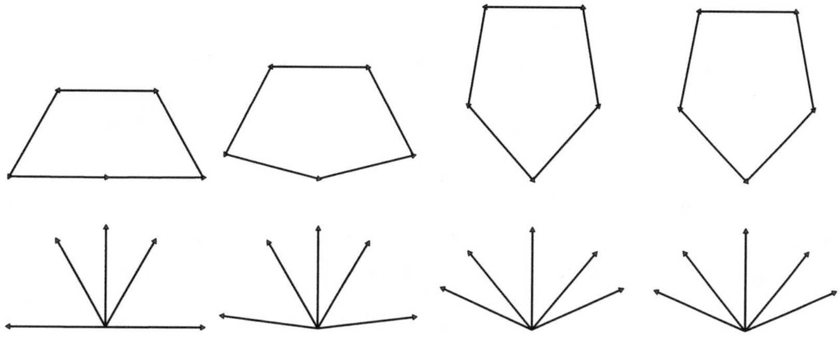

Fig. A.1: The diagram vectors, and their sum (above) for the tight frames of Exer. 2.9.

2.10 (a) The diagram vectors of a unit-norm tight frame for $\mathbb{R}^2$ can be ordered $z_j^2 = e^{i\theta_j}$, $0 \leq \theta_1 \leq \theta_2 \leq \cdots \leq \theta_n < 2\pi$, so the Argand diagram for their sum $z_1^2 + \cdots + z_n^2 = 0$ is a convex n-gon with unit length sides. This polygon has the same shape as that obtained from $(cz_{\sigma j}^2)$, where $c = \pm 1$ and σ is a permutation of the indices and so, by Exer. 2.9 (b), depends only on the equivalence class. Conversely, such a convex polygon is uniquely determined by the sequence of its exterior angles $\psi_1, \ldots, \psi_n$ (taken in either angular direction), and $z_j^2 := e^{i\theta_j}$, $\theta_j : \sum_{1 \leq \ell \leq j} \psi_j$ is a sequence of diagram vectors (since the exterior angles add to 2π) which correspond to this polygon.
(b) Since the angle between vectors is half the angle between their diagram vectors, two vectors are orthogonal to each other if and only if the angle between their diagram vectors is π, i.e., they are negatives of each other and hence correspond to a pair of parallel sides of the diagram vector polygon.
(c) Suppose wlog the angles of n equally spaced vectors are $\frac{2\pi}{n} j$, $1 \leq j \leq n$, so the angles of the diagram vectors are $\frac{4\pi}{n} j$, $1 \leq j \leq n$. For n odd, this is original angles (reordered) so the polygon is the regular n-gon. For $n \neq 4$ even, this is two copies of the angles $\frac{2\pi}{n/2} k$, $1 \leq k \leq \frac{n}{2}$, so that polygon is the regular $\frac{n}{2}$-gon of side length 2 (where each 'side' is two collinear unit length edges). For $n = 4$, the polygon is degenerate, a flat parallelogram.
(d) Yes. Suppose all vectors are nonzero (to avoid edges of zero length); then, as in (a), we can order the diagram vectors: $z_1^2, \ldots, z_n^2$, where $z_j = r_j e^{i\theta_j}$, and $0 \leq \theta_1 \leq \theta_2 \leq \cdots \leq \theta_n < 2\pi$. As before, the corresponding convex polygon (with sides of length $|z_j|^2$) depends only on the equivalence class.

2.11 (a) $C(\alpha v + \beta w) = \overline{\alpha v} + \overline{\beta w} = \overline{\alpha}\,\overline{v} + \overline{\beta}\,\overline{w} = \overline{\alpha} C(v) + \overline{\beta} C(w)$.
(b) $S_{\overline{\Phi}}(Cf) = \sum_j \langle \overline{f}, \overline{f_j} \rangle \overline{f_j} = \sum_j \overline{\langle f, f_j \rangle f_j} = C(S_\Phi f)$, and $\langle \overline{f_j}, \overline{f_k} \rangle = \overline{\langle f_j, f_k \rangle}$.
(c) With this definition of $\overline{v}$, we need only to check that the operations of (2.18) hold: $\overline{v+w} = \overline{(v_j + w_j)} = (\overline{v_j + w_j}) = (\overline{v_j} + \overline{w_j}) = (\overline{v_j}) + (\overline{w_j}) = \overline{v} + \overline{w}$, etc.

2.12 (a) $\text{trace}(S) = \text{trace}(VV^*) = \text{trace}(V^*V) = \sum_j \langle f_j, f_j \rangle$.
(b) Since S is self-adjoint, $(\|S\|_F)^2 = \text{trace}(SS^*) = \text{trace}(S^2)$, and we get

$$\text{trace}(S^2) = \text{trace}(V^*V(V^*V)^*) = \|V^*V\|_F^2 = \sum_j \sum_k |\langle f_j, f_k \rangle|^2.$$

2.13 Let $V = [f_j]$. By Parseval: $VV^* = I_{\mathscr{H}}$, so $L = LVV^*$, and we have

$$\text{trace}(L) = \text{trace}(V^*LV) = \sum_j e_j^*(V^*LV)e_j = \sum_j e_j^*(V^*Lf_j) = \sum_j \langle Lf_j, f_j \rangle.$$

2.14 Let(e_k) be an orthonormal basis.
(a) $\{\frac{1}{\sqrt{2}}e_1, (\frac{1}{\sqrt{2}})^2 e_1, (\frac{1}{\sqrt{2}})^3 e_1, \ldots\} \cup \{e_2, \ldots, e_d\}$ is a normalised tight frame for $\mathscr{H}$.
(b) The trace formula (2.9) holds for infinite frames, since (by Plancherel)

$$\text{trace}(S) = \sum_k \langle Se_k, e_k \rangle = \sum_k \langle \sum_j \langle e_k, f_j \rangle f_j, e_k \rangle = \sum_j \Big(\sum_k \langle f_j, e_k \rangle \langle e_k, f_j \rangle \Big) = \sum_j \langle f_j, f_j \rangle.$$

Therefore $\sum_j \|f_j\|^2 = dA < \infty$, and in particular $\|f_j\| \to 0$.
(c) Since $\|f_j\| \to 0$ for an infinite frame, it cannot have equal (nonzero) norms.

2.15 If $\|f_j\| = c$, $\forall j$, then the trace formula (2.9) gives

$$\sum_j \|f_j\|^2 = nc^2 = dA \quad \Longrightarrow \quad c = \sqrt{\frac{dA}{n}}.$$

2.16 Since (f_j) is equiangular, Exer. 2.15 gives $\langle f_j, f_j \rangle = \|f_j\|^2 = \frac{dA}{n}$, $\forall j$. Suppose that $|\langle f_j, f_k \rangle| = c$, $\forall j \neq k$, then the variational formula (2.10) gives

$$(n^2 - n)c^2 + n\Big(\frac{dA}{n}\Big)^2 = \frac{1}{d}\Big(n\frac{dA}{n}\Big)^2 \quad \Longrightarrow \quad c = \frac{A}{n}\sqrt{\frac{d(n-d)}{n-1}}.$$

2.17 (a) By Theorem 2.1, Φ is a normalised tight frame for $\mathscr{H}$ if and only if the Hermitian matrix $P = \text{Gram}(\Phi)$ is an orthogonal projection of rank d, i.e., its eigenvalues are 0 and 1, with exactly d being nonzero.
(b) If all eigenvalues of S_Φ are 1, it is the identity, giving the Parseval identity.
(c), (d) The singular values of V and V^* are precisely the (nonzero) eigenvalues of the frame operator $S = VV^*$, or, equivalently the Gramian $P = V^*V$.

2.18 (a)$\Longrightarrow$(b) Since $VV^* = I$, $\langle V^*f, V^*g \rangle = \langle VV^*f, g \rangle = \langle f, g \rangle$, $\forall f, g$.
(b)$\Longrightarrow$(c) Take $f = g$, so that $\|V^*f\|^2 = \langle V^*f, V^*f \rangle = \langle f, f \rangle = \|f\|^2$, $\forall f$.
(c)$\Longrightarrow$(a) Expand $\|V^*f\|^2 = (\|(\langle f, f_j \rangle)\|_2)^2 = \sum_j |\langle f, f_j \rangle|^2 = \|f\|^2$, $\forall f$.

2.19 If there is such a unitary U, then clearly $\langle g_j, g_k \rangle = \langle Uf_j, Uf_k \rangle = \langle f_j, f_k \rangle$, $\forall j, k$. Now suppose $\langle g_j, g_k \rangle = \langle f_j, f_k \rangle$, $\forall j, k$ holds. Assume, wlog, that $\{f_1, \ldots, f_d\}$ is a basis for $\mathscr{H}$, and define a linear map U by

$$U : \mathscr{H} \to \mathscr{K}, \qquad Uf_j := g_j, \quad 1 \le j \le d.$$

Then U is unitary, since for $h_1 = \sum_{j=1}^d \alpha_j f_j$, $h_2 = \sum_{k=1}^d \beta_k f_k$,

$$\langle Uh_1, Uh_2 \rangle = \sum_{j,k} \alpha_j \overline{\beta_k} \langle g_j, g_k \rangle = \sum_{j,k} \alpha_j \overline{\beta_k} \langle f_j, f_k \rangle = \langle h_1, h_2 \rangle.$$

For $k > d$ and $1 \le j \le d$, we have

$$\begin{aligned} \langle Uf_k - g_k, g_j \rangle &= \langle Uf_k, g_j \rangle - \langle g_k, g_j \rangle = \langle Uf_k, Uf_j \rangle - \langle g_k, g_j \rangle \\ &= \langle f_k, f_j \rangle - \langle g_k, g_j \rangle = 0, \end{aligned}$$

which implies $Uf_k = g_k$ since $\{g_1, \ldots, g_d\}$ is a basis for $\mathscr{K}$.

2.20 (a) $\Phi = (f_j)_{j \in J}$ and $\Psi = (g_k)_{k \in K}$ are unitarily equivalent up to reordering if and only if there is a bijection $\sigma : J \to K$ for which Φ and $(g_{\sigma j})_{j \in J}$ are unitarily equivalent, i.e., by Corollary 2.1,

$$\operatorname{Gram}(\Phi) = \operatorname{Gram}((g_{\sigma j})_{j \in J}) = ([g_k]Q)^*[g_k]Q = Q^* \operatorname{Gram}(\Psi) Q,$$

where $Q = Q_\sigma \in \mathbb{F}^{K \times J}$ is the permutation matrix $Q := [e_{\sigma j}]_{j \in J}$.
(b) They are projectively unitarily equivalent up to reordering if and only if $g_{\sigma j} = \alpha_j U f_j$, $\forall j$, with $\sigma : J \to K$ a bijection, $|\alpha_j| = 1$, $\forall j$, and U unitary, i.e., $(\alpha_j f_j)_{j \in J}$ and $(g_{\sigma j})_{j \in J}$ are unitarily equivalent, i.e., with $\Lambda := \operatorname{diag}(\alpha_j)_{j \in J}$ and Q as above

$$\Lambda^* \operatorname{Gram}(\Phi) \Lambda = ([f_j]\Lambda)^*[f_j]\Lambda = ([g_k]Q)^*[g_k]Q = Q^* \operatorname{Gram}(\Psi) Q.$$

(c) If these are projectively unitarily equivalent as above, then

$$\langle g_{\sigma j}, g_{\sigma k} \rangle = \langle \alpha_j U f_j, \alpha_k U f_k \rangle = \alpha_j \overline{\alpha_k} \langle f_j, f_k \rangle \quad \Longrightarrow \quad |\langle g_{\sigma j}, g_{\sigma k} \rangle| = |\langle f_j, f_k \rangle|.$$

The equiangular harmonic frames with the Gramians of (2.12) and (2.17), i.e.,

$$\begin{bmatrix} \frac{2}{3} & -\frac{1}{3} & -\frac{1}{3} \\ -\frac{1}{3} & \frac{2}{3} & -\frac{1}{3} \\ -\frac{1}{3} & -\frac{1}{3} & \frac{2}{3} \end{bmatrix}, \qquad \begin{bmatrix} \frac{2}{3} & \frac{-\omega^2}{3} & -\frac{\omega}{3} \\ \frac{-\omega}{3} & \frac{2}{3} & -\frac{\omega^2}{3} \\ \frac{-\omega^2}{3} & -\frac{\omega}{3} & \frac{2}{3} \end{bmatrix}, \qquad \omega := e^{\frac{2\pi}{3}}$$

satisfy this condition, but are not projectively unitarily equivalent.
Remark: Projective unitary equivalence is determined by the m products (see §8).

2.21 (a) By Exer. 2.20, normalised tight frames Φ and Ψ are projectively unitarily equivalent (up to reordering) if and only if their Gramians P_Φ and P_Ψ satisfy

$$\Lambda^* P_\Phi \Lambda = Q^* P_\Psi Q \quad \Longleftrightarrow \quad \Lambda^*(I - P_\Phi)\Lambda = Q^*(I - P_\Psi)Q,$$

i.e., the complementary normalised tight frames are projectively unitarily equivalent (up to reordering).

(b) The complementary normalised tight frame of an equal-norm tight frame of $n = d+1$ vectors for $\mathbb{F}^d$ consists of $d+1$ scalars with modulus $1/\sqrt{d+1}$. Since all such frames are projectively unitarily equivalent, the result follows by (a).
(c) We observe that up to a scalar multiple (α_j) is given by

$$\alpha_j \overline{\alpha_k} \langle f_j, f_k \rangle = \langle \alpha_j U f_j, \alpha_k U f_k \rangle = \langle g_j, g_k \rangle.$$

Taking $\alpha_1 = 1$, we have

$$\alpha_2 = \frac{\langle g_2, g_1 \rangle}{\langle f_2, f_1 \rangle} = \frac{1+\omega}{-1} = \omega^2, \quad \alpha_3 = \frac{\langle g_3, g_1 \rangle}{\langle f_3, f_1 \rangle} = \frac{1+\omega^2}{-1} = \omega,$$

with U given by $U(\alpha_j f_j) = g_j$, and so $U = [\alpha_1 f_1, \alpha_2 f_2]^{-1}[g_1, g_2] = I$.

2.22 Let $\{a,b,c\} \in \mathbb{C}$ be the complementary normalised tight frame. Since unitary maps preserve the Gramian, we can assume that $c \geq 0$, and since it is normalised $|a|^2 + |b|^2 + |c|^2 = 1$, thus the complementary frame has frame operator

$$V = [a, b, \sqrt{1 - |a|^2 - |b|^2}], \qquad |a|^2 + |b|^2 \leq 1.$$

Thus all Gramians for normalised tight frames of three vectors for $\mathbb{C}^2$ are given by

$$I - V^*V = \begin{bmatrix} 1 - |a|^2 & -\overline{a}b & -\overline{a}\sqrt{|a|^2+|b|^2} \\ -\overline{b}a & 1 - |b|^2 & -\overline{b}\sqrt{|a|^2+|b|^2} \\ -a\sqrt{|a|^2+|b|^2} & -b\sqrt{|a|^2+|b|^2} & |a|^2+|b|^2 \end{bmatrix}.$$

2.23 Let P be the Gramian of a normalised tight frame. Then $I - P$ is the Gramian of the complementary tight frame, and these frames are unitarily equivalent if and only if their Gramians are equal (Corollary 2.1), i.e., $P = I - P$. This implies $P = \frac{1}{2}I$, which is *not* an orthogonal projection matrix.

The projectively unitarily equivalent normalised tight frames of two vectors for $\mathbb{F}^1$ given by $V = [\frac{1}{\sqrt{2}}, \frac{1}{\sqrt{2}}b]$ and $W = [\frac{1}{\sqrt{2}}, -\frac{1}{\sqrt{2}}b]$, $|b| = 1$, are complements of each other. Further examples might be given by equiangular tight frames of $2d$ vectors for $\mathbb{F}^d$ (a calculation excludes the case $d = 2$).

2.24 The complementary frame to $V = [f_j]$ is `null(V)'`

2.25 (a) In terms of the synthesis matrix $V = [f_j]_{j=1^n} \in \mathbb{C}^{d \times n}$, we seek a $d \times n$ matrix V with columns of equal (nonzero) length, say d, and rows of equal length which are orthogonal. For $n \geq d = 1$, we can take V to be any $1 \times n$ matrix with entries of unit modulus. Suppose $n \geq d \geq 2$, and write $n = kd + d + r$, $0 \leq r < d$. We seek a V of the form $V = d[V_1, V_2, \ldots, V_k, W]$, where each V_j is a unitary matrix (say I). This V has the desired properties provided the $d \times (d+r)$ matrix W has columns of length 1 (the same as those of V_j) and rows of equal length, i.e., there exists an equal-norm tight frame of $d+r$ vectors in $\mathbb{F}^d$. Such a frame is given by an orthonormal basis when $r = 0$, or by the complement of an equal-norm tight frame of $d+r$ vectors in $\mathbb{F}^r$, which we can construct by (strong) induction since $r < d$.

(b) The following `matlab` code gives such a function

```
function V = ENTF(n,d)
  if d==1, V=ones(1,n); end;
  if n==d, V=d*eye(d); end;
  if d<n & n <2*d, V=sqrt(n*d)*null(ENTF(n,n-d))'; end;
  if n>=2*d & d>1, V=[d*eye(d) ENTF(n-d,d)]; end;
```

Here `eye(d)` can be replaced by any $d \times d$ unitary matrix and `ones(1,n)` any $1 \times n$ matrix with unit modulus entries.
(c) Now an example

$$[1,1,1] \overset{\text{take complement}}{\longrightarrow} \begin{bmatrix} \sqrt{3} & -\sqrt{3} & 0 \\ 1 & 1 & -2 \end{bmatrix} \overset{\text{add columns}}{\longrightarrow} \begin{bmatrix} 2 & 0 & \sqrt{3} & -\sqrt{3} & 0 \\ 0 & 2 & 1 & 1 & -2 \end{bmatrix}$$

$$\overset{\text{take complement}}{\longrightarrow} \begin{bmatrix} 3 & 0 & -\sqrt{3} & \sqrt{3} & 0 \\ 0 & 3 & -1 & -1 & 2 \\ 0 & 0 & \sqrt{5} & \sqrt{5} & \sqrt{5} \end{bmatrix} \overset{\text{add columns}}{\longrightarrow} V = \begin{bmatrix} 3 & 0 & 0 & 3 & 0 & -\sqrt{3} & \sqrt{3} & 0 \\ 0 & 3 & 0 & 0 & 3 & -1 & -1 & 2 \\ 0 & 0 & 3 & 0 & 0 & \sqrt{5} & \sqrt{5} & \sqrt{5} \end{bmatrix},$$

2.26 (a) Each of the maps $A_j : f \mapsto \langle f, f_j \rangle f_j$ is a (bounded) positive operator since $\langle A_j f, f \rangle = |\langle f, f_j \rangle|^2 \geq 0$, and hence so are the differences $F_t - F_s$, $s < t$ (which are finite sums of these). Clearly, $F_0 = O_{\mathscr{H}}$ and $F_n = I_{\mathscr{H}}$.
(b) We have the telescoping sum $\sum_{j=1}^n (F_j - F_{j-1}) = F_n - F_0 = I_{\mathscr{H}}$, and $Q_j := E_j - E_{j-1}$ (which satisfies $Q_j^* = Q_j$) is an orthogonal projection, since

$$(E_j - E_{j-1})^2 = E_j^2 - E_j E_{j-1} - E_{j-1} E_j + E_{j-1}^2 = E_j - E_{j-1} - E_{j-1} + E_{j-1} = Q_j.$$

Similarly, if $j < k$, then $Q_j Q_k = Q_k Q_j = 0$.
(c) Let $\tilde{\mathscr{K}} \subset \mathscr{K}$ be the orthogonal complement of the kernel of the linear map

$$L : \mathscr{K} \to \mathbb{C}^n : f \mapsto (\langle f, f_j \rangle)_{j=1}^n.$$

Since $f \in \ker(L) \cap \mathscr{H} \implies f = \sum_j \langle f, f_j \rangle f_j = 0$, we have $\mathscr{H} \subset \tilde{\mathscr{K}}$. Further, $\operatorname{ran}(Q_k) \cap \mathscr{H} \neq \{0\}$, since otherwise, $\operatorname{ran}(Q_k) \perp \mathscr{H}$, so that $PQ_k = 0$, and

$$\langle f_k, f_k \rangle f_k = (F_k - F_{k-1}) f_k = PQ_k f_k = 0,$$

which contradicts $f_k \neq 0$. Thus, we may choose $f \in \operatorname{ran}(Q_k)$, with $Pf \neq 0$, so that $Lf = LPf$ is a nonzero scalar multiple of the standard basis vector e_k for $\mathbb{C}^n$, and hence L maps onto $\mathbb{C}^n$. A dimension count gives $\dim(\tilde{\mathscr{K}}) = \operatorname{rank}(L) = n$, and so we may replace $\mathscr{K}$ by $\tilde{\mathscr{K}}$.
(d) Suppose that (f_j) is a normalised tight frame for $\mathbb{F}^d$, i.e., the rows of $V = [f_1, \ldots, f_n] \in \mathbb{F}^{d \times n}$ are orthonormal. Extend V by adding a further $n-d$ orthonormal rows, to obtain a unitary matrix U. The columns of U are an orthonormal basis for $\mathbb{F}^n$, and their orthogonal projection onto the first d components gives (f_j).

2.27 Since $F := [f_j] = U + iV$, $U := [u_j]$, $V := [v_j]$, the normalised tight frame condition for (f_j) is

$$FF^* = (U+iV)(U+iV)^* = UU^* + VV^* + i(VU^* - UV^*) = I,$$

i.e., $UU^* + VV^* = I$, and $VU^* = UV^*$. The sequence of $2n$ real vectors has synthesis operator $S = [U,V]$, which satisfies $SS^* = UU^* + VV^* = I$, and so is a tight frame.

2.28 Let (f_j) be a normalised tight frame for $\mathbb{R}^d$. Write $f \in \mathbb{C}^d$ as $f = u + iv$, $u, v \in \mathbb{R}^d$. Then

$$\sum_j |\langle f, f_j\rangle|^2 = \sum_j |\langle u+iv, f_j\rangle|^2 = \sum_j \Big(|\langle u, f_j\rangle|^2 + |\langle v, f_j\rangle|^2\Big) = \|u\|^2 + \|v\|^2 = \|f\|^2.$$

2.29 (a) Let $V = [f_j]$, $W = [g_k]$. As in Exer. 2.13, we calculate

$$\operatorname{trace}(M^*L) = \operatorname{trace}(M^*LVV^*) = \operatorname{trace}((MV)^*LV) = \textstyle\sum_j \langle Lf_j, Mf_j\rangle,$$
$$\operatorname{trace}(M^*L) = \operatorname{trace}(M^*WW^*L) = \operatorname{trace}((L^*W)^*M^*W) = \textstyle\sum_k \langle M^*g_k, L^*g_k\rangle.$$

(b) Let $L \in \mathscr{L}(\mathscr{H},\mathscr{K})$. We verify the Parseval identity for L (applied to f). Observe

$$\langle L, g_k f_j^*\rangle_{HS} = \operatorname{trace}((g_k f_j^*)^* L) = \operatorname{trace}(f_j g_k^* L) = \operatorname{trace}(g_k^* L f_j) = \langle Lf_j, g_k\rangle,$$

and so, by linearity, we have

$$\begin{aligned}\sum_{j,k} \langle L, g_k f_j^*\rangle_{HS}(g_k f_j^*) f &= \sum_{j,k} \langle Lf_j, g_k\rangle g_k \langle f, f_j\rangle = \sum_j \langle f, f_j\rangle \Big(\sum_k \langle Lf_j, g_k\rangle g_k\Big) \\ &= \sum_j \langle f, f_j\rangle Lf_j = L\Big(\sum_j \langle f, f_j\rangle f_j\Big) = Lf.\end{aligned}$$

2.30 (a) $VV^* = I$ gives $Vx = V(V^*f) = f$, $WAV^* = W(W^*LV)V^* = L$.
(b) We have $[\alpha L + \beta M] = W^*(\alpha L + \beta M)V = \alpha W^*LV + \beta W^*MV = \alpha[L] + \beta[M]$, and $[L^*] = V^*L^*W = (W^*LV)^* = [L]^*$.
(c) Suppose that $M : \mathscr{K} \to \mathscr{L}$, $(h_\ell)_{\ell \in L}$ is a normalised tight frame for $\mathscr{L}$, and let $X = [h_\ell]_{\ell \in L}$. Then $[ML] = X^*LMV = (X^*MW)(W^*LV) = [M][L]$.
(d) With $x = [f]$, part (a) gives $Ax = [L][f] = [Lf]$ and $\lambda x = \lambda[f] = [\lambda f]$. But $f \mapsto [f]$ is 1–1, so that $Ax = \lambda x$ if and only if $Lf = \lambda f$.
(e) The singular values of A are the positive square roots of the nonzero eigenvalues of $A^*A = (W^*LV)^*W^*LV = V^*L^*WW^*LV = V^*L^*LV$, with the eigenspaces giving the corresponding singular vectors. If x is an eigenvector of A^*A for $\lambda \neq 0$, then $f = Vx \neq 0$, and we may apply (d).

Exercises of Chapter 3

3.1 Let g_j^* denote $f \mapsto \langle f, g_j\rangle$, so that $A_j = f_j g_j^*$.
(a) $\operatorname{trace}(A_j) = \operatorname{trace}(f_j g_j^*) = \operatorname{trace}(g_j^* f_j) = \langle f_j, g_j\rangle$.
(b) Since $A^2 f = f_j(g_j^* f_j) g_j^* f = \langle f_j, g_j\rangle Af$, take $c = \langle f_j, g_j\rangle$, $P : f \mapsto \frac{\langle f, g_j\rangle}{\langle f_j, g_j\rangle} f_j$.

(c) Since $A_j^* = g_j f_j^*$, we have $P^* f = \frac{\langle f, f_j \rangle}{\langle g_j, f_j \rangle} g_j$, so $P = P^*$ if and only if $\langle f, g_j \rangle f_j = \langle f, f_j \rangle g_j$, $\forall f$, i.e., f_j and g_j are scalar multiples of each other.
(d) $\operatorname{trace}(A_j A_k^*) = \operatorname{trace}(f_j (g_j^* g_k) f_k^*) = \langle g_k, g_j \rangle \operatorname{trace}(f_k^* f_j) = \langle g_k, g_j \rangle \langle f_j, f_k \rangle$.

3.2 *Least squares solution.*
(a) $V(\lambda a + (1-\lambda) b) = \lambda V a + (1-\lambda) V b = \lambda f + (1-\lambda) f = f$.
(b) $V(V^* S^{-1} f) = VV^*(VV^*)^{-1} f = f$.
(c) Since $V^* S^{-1} f$ is a solution, $\mathscr{A} = V^* S^{-1} f + \ker(V)$. Since $\operatorname{ran}(V^*)$ is orthogonal to $\ker(V)$, for any $a = V^* S^{-1} f + b \in \mathscr{A}$, $b \in \ker(V)$, Pythagorus gives

$$\sum_j |c_j|^2 = \|c\|^2 = \|V^* S^{-1} f\|^2 + \|b\|^2,$$

so the unique solution of minimal ℓ_2-norm is obtained by choosing $b = 0$.

3.3 (a) With $A^\dagger := A^*(AA^*)^{-1}$, we have $AA^\dagger = AA^*(AA^*)^{-1} = I$, $A^\dagger A = A^*(AA^*)^{-1} A$ are Hermitian, and $AA^\dagger A = IA = A$, $A^\dagger A A^\dagger = A^\dagger I = A^\dagger$.
(b) With $P^\dagger := P$, $PP^\dagger = P^\dagger P = P$ is Hermitian, $PP^\dagger P = P$, $P^\dagger P P^\dagger = P^\dagger$.
(c) Let $A = \operatorname{Gram}(\Phi) = V^* V$, and $A^\dagger := \operatorname{Gram}(\tilde{\Phi}) = (S^{-1} V)^*(S^{-1} V) = V^* S^{-2} V$, $S = VV^*$. We verify that $A^\dagger$ is the pseudoinverse of A: $AA^\dagger = V^* V V^* S^{-2} V = V^* S^{-1} V$ and $A^\dagger A = V^* S^{-2} V V^* V = V^* S^{-1} V$ are Hermitian, and

$$AA^\dagger A = (AA^\dagger) A = (V^* S^{-1} V) V^* V = V^* V = A,$$

$$A^\dagger A A^\dagger = (V^* S^{-1} V) V^* S^{-2} V = V^* S^{-2} V = A^\dagger.$$

(d) The synthesis operator of Φ^{can} is $S^{-\frac{1}{2}} V$, $S = VV^*$, so by (a) and (c), we have

$$\begin{aligned}\operatorname{Gram}(\Phi^{\mathrm{can}}) &= (S^{-\frac{1}{2}} V)^* S^{-\frac{1}{2}} V = V^* S^{-1} V = V^*(VV^*)^{-1} V = V^\dagger V \\ &= V^* S^{-1} V = V^* V (V^* S^{-2}) = \operatorname{Gram}(\Phi) \operatorname{Gram}(\Phi)^\dagger.\end{aligned}$$

3.4 Expanding Tf in the normalised tight frame (f_j) gives

$$f = T^{-1}(Tf) = \sum_j \langle Tf, f_j \rangle T^{-1} f_j = \sum_j \langle f, T^* f_j \rangle T^{-1} f_j.$$

Replace T^* by T^{-1} to obtain the other equality. Let $V = [f_j]$. Then frame operator of $(T^* f_j)$ is $T^* V (T^* V)^* = T^*(VV^*) T = T^* T$, and so the dual frame is

$$(T^* T)^{-1} T^* f_j = T^{-1} f_j.$$

3.5 Since $\sum_j |\langle g, Q f_j \rangle|^2 = \sum_j |\langle Q^* g, f_j \rangle|^2 \le B_\Phi \|Q^* g\|^2 \le B_\Phi \|Q\|^2 \|g\|^2$, we have $B_\Phi \le B_\Phi \|Q\|^2$. Similarly, $\sum_j |\langle g, Q f_j \rangle|^2 \ge A_\Phi \|Q^* g\|$. If $g \in \operatorname{span} \Psi$, then $g = Qf$. But $Q = QQ^\dagger Q = (QQ^\dagger)^* Q = (Q^\dagger)^* Q^* Q$, and so we obtain $\|g\| = \|(Q^\dagger)^* Q^* Q f\| \le \|(Q^\dagger)^*\| \|Q^* g\|$. Thus $\sum_j |\langle g, Q f_j \rangle|^2 \ge A_\Phi \|Q^\dagger\|^{-2} \|g\|^2$, giving $A_\Psi \ge A_\Phi \|Q^\dagger\|^{-2}$.

3.6 (a)$\Longrightarrow$(b) $\sum_{j=1}^n \frac{|\langle f,f_j\rangle|^2}{\|f\|^2} \geq A$ gives the lower frame bound. For the upper bound: $\sum_{j=1}^n |\langle f,f_j\rangle|^2 \leq \sum_{j=1}^n \|f\|^2\|f_j\|^2 = B\|f\|^2$, $B := \sum_{j=1}^n \|f_j\|^2 < \infty$.
(b)$\Longrightarrow$(e)$\Longleftrightarrow$(f) V^* is 1–1 if and only if its kernel is zero. The lower frame bound gives: $\|V^*f\|^2 = \sum_j |\langle f,f_j\rangle|^2 \geq A\|f\|^2 > 0$, $\forall f \neq 0$, i.e., $\ker(V^*) = 0$.
(c)$\Longleftrightarrow$(d)$\Longleftrightarrow$(e) (f_j) spans $\mathscr{H}$ if and only if V is onto iff V^* is 1–1.
(f)$\Longrightarrow$(a) Since $\inf_{f\neq 0}\sum_{j=1}^n \frac{|\langle f,f_j\rangle|^2}{\|f\|^2} = \inf_{\|g\|=1} |\langle g,f_j\rangle|^2$ and $\{g : \|g\| = 1\}$ is compact ($\mathscr{H}$ is finite dimensional), the infimum is attained and so is nonzero.

3.7 Since S is positive definite, it is unitarily diagonalisable $S = U\Lambda U^*$, where $U = [u_1,\ldots,u_d]$ and $\Lambda = \mathrm{diag}(\lambda_1,\ldots,\lambda_d)$, with $\lambda_j > 0$, so that

$$\langle Sf,f\rangle = \langle U\Lambda U^*f,f\rangle = \langle \Lambda U^*f,U^*f\rangle = \sum_j \lambda_j |\langle f,u_j\rangle|^2.$$

Let $\lambda_{\min} = \min \lambda_j$ and $\lambda_{\max} = \max \lambda_j$. Then we have

$$\lambda_{\min}\|f\|^2 = \lambda_{\min}\sum_j |\langle f,u_j\rangle|^2 \leq \langle Sf,f\rangle \leq \lambda_{\max}\sum_j |\langle f,u_j\rangle|^2 = \lambda_{\max}\|f\|^2,$$

with equality if (and only if) f is an eigenvector for $\lambda_{\min}$, $\lambda_{\max}$, respectively.

3.8 Let $f = f_j$ in the frame bound inequality

$$A\|f_j\|^2 \leq \|f_j\|^4 + \sum_{k\neq j} |\langle f_j,f_k\rangle|^2 \leq B\|f_j\|^2.$$

(a) Immediate.
(b) Equality in (a) if and only if $\sum_{k\neq j} |\langle f_j,f_k\rangle|^2 = 0$, i.e., $f_j \perp f_k$, $\forall j \neq k$.
(c) Since $\|f_j\|^2 < A$ gives $\|f_j\|^4 < A\|f_j\|^2$, we must have

$$\sum_{k\neq j} |\langle f_j,f_k\rangle|^2 > 0 \quad \Longrightarrow \quad f_j \text{ is not orthogonal to } \mathrm{span}_{k\neq j} f_k \quad \Longrightarrow$$

3.9 Let $S = VV^*$. The commutativity relations of Exer. 3.11 give

$$[\tilde{f}_j] = S^{-1}V = V\mathrm{Gram}(\Phi)^\dagger, \qquad [f_j^{\mathrm{can}}] = S^{-\frac{1}{2}}V = V(\mathrm{Gram}(\Phi)^\dagger)^{\frac{1}{2}}.$$

Expanding gives

$$\begin{aligned}
[\tilde{f}_j] &= S^{-1}V = (U_1\Sigma U_2^* U_2\Sigma^* U_1^*)^{-1}V = U_1(\Sigma\Sigma^*)^{-1}U_1^*(U_1\Sigma U_2^*) \\
&= U_1\mathrm{diag}(1/\sigma_1^2,\ldots)\mathrm{diag}(\sigma_1,\ldots)U_2^* = U_1\mathrm{diag}(1/\sigma_1,\ldots)U_2^*, \\
[f_j^{\mathrm{can}}] &= S^{-\frac{1}{2}}V = (U_1(\Sigma\Sigma^*)^{-1}U_1^*)^{\frac{1}{2}}(U_1\Sigma U_2^*) \\
&= U_1\mathrm{diag}(1/\sigma_1,\ldots)\mathrm{diag}(\sigma_1,\ldots)\Sigma U_2^* = U_1\mathrm{diag}(1,\ldots)U_2^*.
\end{aligned}$$

3.10 Let $S = S_\Phi = VV^*$, where $V = [f_j]_{j\in J}$, so that $S_{\tilde{\Phi}} = S^{-1}$, and we have
(a) $[\tilde{f}_j^{\text{can}}] = (S^{-1})^{-\frac{1}{2}} S^{-1} V = S^{-\frac{1}{2}} V = [f_j^{\text{can}}]$.
(b) $G_\Phi = \text{Gram}(\Phi) = V^*V$, $G_{\tilde{\Phi}} = \text{Gram}(\tilde{\Phi}) = (S^{-1}V)^* S^{-1} V = V^* S^{-2} V$, and $G_{\Phi^{\text{can}}} = \text{Gram}(\Phi^{\text{can}}) = (S^{-\frac{1}{2}}V)^* S^{-\frac{1}{2}} V = V^* S^{-1} V$, so we obtain

$$G_\Phi G_{\tilde{\Phi}} = V^*VV^*S^{-2}V = V^*S^{-1}V = G_{\Phi^{\text{can}}} = V^*S^{-2}VV^*V = G_{\tilde{\Phi}} G_\Phi.$$

(c) Use $\tilde{f}_j = S_\Phi^{-1} f_j$ and $S_\Phi^{-1} = S_{\tilde{\Phi}}$.

3.11 (a) Observe $SV = (VV^*)V = V(V^*V) = VG$, and use induction.
(b) By (a), $p(S)V = Vp(G)$ for all polynomials p. Now use Weierstrass density.
(c) Similarly, approximate $A^\dagger$ via the *Tikhonov regularisation* formula

$$A^\dagger = \lim_{\delta\to 0} (A^*A + \delta I)^{-1} A^* = \lim_{\delta\to 0} A^*(AA^* + \delta I)^{-1}. \tag{.1}$$

3.12 $P := VW^* = WV^* = P^*$, and $P|_{\mathscr{K}} = I$, gives $P^2 = I|_{\mathscr{K}}(VW^*) = P$.

3.13 Let S be the frame operator of (f_j). Since S^{-1} is a positive definite operator, $\langle f_j, \tilde{f}_j\rangle = \langle f_j, S^{-1} f_j\rangle = \langle S^{-\frac{1}{2}} f_j, S^{-\frac{1}{2}} f_j\rangle = \langle f_j^{\text{can}}, f_j^{\text{can}}\rangle = \|f_j^{\text{can}}\|^2 \geq 0$. Since (f_j^{can}) is a normalised tight frame, it satifies (2.9), i.e., $\sum_j \|f_j^{\text{can}}\|^2 = d$.

3.14 Let T be the frame operator for (f_j), i.e., $T = S|_{\mathscr{K}}$. The assertions amount to the claim $T^{-1} = S^\dagger|_{\mathscr{K}}$. Now $\text{ran}(S^\dagger) = \text{ran}(S^*) = \text{ran}(S) = \mathscr{K}$, so for $y = Sx \in \mathscr{K}$, we have $(TS^\dagger|_{\mathscr{K}})y = SS^\dagger Sx = Sx = y$, i.e., $TS^\dagger|_{\mathscr{K}} = I_{\mathscr{K}}$. Also see Corollary 3.5.

3.15 The (j,k)-entry of Gram(Φ) is $\langle f_k, f_j\rangle = v_j^* v_k = \langle v_k, v_j\rangle$ (so the columns of L give a copy of Φ). Now apply Exer. 3.14 with $V = L$.

3.16 Let $S = VV^*$.
(a) Ψ is a frame since its synthesis operator QV is onto $\mathscr{H}$.
(b) $\tilde{g}_j = (QVV^*Q^*)^{-1} Q f_j = (Q^*)^{-1} S^{-1} Q^{-1} Q f_j = (Q^*)^{-1} S^{-1} f_j = (Q^*)^{-1} \tilde{f}_j$.
(c) $g_j^{\text{can}} = (QVV^*Q^*)^{-\frac{1}{2}} Q f_j = (QSQ^*)^{-\frac{1}{2}} QS^{\frac{1}{2}} (S^{-\frac{1}{2}} f_j) = U f_j^{\text{can}}$, where the matrix $U = (QSQ^*)^{-\frac{1}{2}} QS^{\frac{1}{2}}$ is unitary, since $UU^* = (QSQ^*)^{-\frac{1}{2}} QS^{\frac{1}{2}} S^{\frac{1}{2}} Q^* (QSQ^*)^{-\frac{1}{2}} = I$.
(d) Let $Q = cU$, then $(QSQ^*)^{-\frac{1}{2}} QS^{\frac{1}{2}} = (c^2USU^*)^{-\frac{1}{2}} cUS^{\frac{1}{2}} = US^{-\frac{1}{2}} U^* US^{\frac{1}{2}} = U$, with $US^{\frac{1}{2}}U^* = (USU^*)^{\frac{1}{2}}$ following from $US^jU^* = (USU^*)^j$, $j = 1, 2, \ldots$ and the Weierstrass density theorem.

3.17 Define $\tilde{W} \in \mathbb{F}^{d\times n}$ by $W = U_1 \tilde{W} U_2^*$. Then

$$\tilde{W}\tilde{W}^* = U_1^* W U_2 U_2^* W^* U_1 = U_1^*(WW^*)U_1 = AU_1^* U_1 = AI,$$

so that

$$\min_{\substack{W\in\mathbb{F}^{d\times n}, A>0\\ WW^*=AI}} \|V - W\|_F = \min_{\substack{\tilde{W}\in\mathbb{F}^{d\times n}, A>0\\ \tilde{W}\tilde{W}^*=AI}} \|U_1 \Sigma U_2^* - U_1 \tilde{W} U_2^*\|_F = \min_{\substack{\tilde{W}\in\mathbb{F}^{d\times n}, A>0\\ \tilde{W}\tilde{W}^*=AI}} \|\Sigma - \tilde{W}\|_F.$$

Let $\Sigma = \text{diag}(\sigma_1, \sigma_2, \ldots)$ and $\tilde{W} = [w_{jk}]$. Then

$$\begin{aligned}
\|\Sigma-\tilde{W}\|_F &= \mathrm{trace}((\Sigma-\tilde{W})(\Sigma-\tilde{W})^*) = \mathrm{trace}(\Sigma\Sigma^*-\tilde{W}\Sigma^*-\Sigma\tilde{W}*+W\tilde{W}) \\
&= \mathrm{trace}(VV^*-\tilde{W}\Sigma^*-\Sigma\tilde{W}^*+AI) \\
&= \mathrm{trace}(VV^*)+dA-\mathrm{trace}(\tilde{W}\Sigma^*+\Sigma\tilde{W}^*),
\end{aligned}$$

where

$$\mathrm{trace}(\tilde{W}\Sigma^*+\Sigma\tilde{W}^*) = 2\sum_{j=1}^{d}\sigma_j\Re(\tilde{w}_{jj}).$$

The condition $\tilde{W}\tilde{W}^* = AI$ says that $\tilde{W}$ has orthogonal rows of length $\sqrt{A}$, so the unique choice maximising $\Re(\tilde{w}_{jj})$ is $\tilde{w}_{jj} = \sqrt{A}$, $\tilde{w}_{jk} = 0$, $j \neq k$. Thus $\|V-W\|_F$ is minimised, when

$$dA - 2\sum_j \sigma_j\sqrt{A} = \sum_j (A - 2\sigma_j\sqrt{A})$$

is maximised, which occurs when $\sqrt{A} = \frac{1}{d}\sum_j \sigma_j$. Thus, with A defined as above, the unique minimiser W is given by

$$W = U_1[\sqrt{A}I, 0]U_2^* = \sqrt{A}U_2^*.$$

This gives Corollary 3.3, since the singular values σ_j of V are the square roots of the eigenvalues λ_j of the frame operator $S = VV^*$, and the canonical tight frame for $V = [f_j]$ is given by (3.7).

3.18 It suffices to show that $\Re\sum_j\langle f_j, g_j\rangle \leq \sum_{k_1}\sum_{k_2}\sqrt{\lambda_{k_1}}\sqrt{\mu_{k_2}}|\langle u_{k_1}, v_{k_2}\rangle|$. By Cauchy–Schwarz,

$$\begin{aligned}
\sum_j |\langle f_j, u_{k_1}\rangle\langle v_{k_2}, g_j\rangle| &\leq \Big(\sum_j |\langle f_j, u_{k_1}\rangle|^2\Big)^{\frac{1}{2}}\Big(\sum_j |\langle v_{k_2}, g_j\rangle|^2\Big)^{\frac{1}{2}} \\
&= \sqrt{\langle S_\Phi u_{k_1}, u_{k_1}\rangle}\sqrt{\langle S_\Psi v_{k_2}, v_{k_2}\rangle} = \sqrt{\lambda_{k_1}}\sqrt{\mu_{k_2}},
\end{aligned}$$

which gives

$$\begin{aligned}
\Re\sum_j\langle f_j, g_j\rangle \leq \Big|\sum_j\langle f_j, g_j\rangle\Big| &= \Big|\sum_j\sum_{k_1}\sum_{k_2}\langle f_j, u_{k_1}\rangle\langle u_{k_1}, v_{k_2}\rangle\langle v_{k_2}, g_j\rangle\Big| \\
&\leq \sum_{k_1}\sum_{k_2}|\langle u_{k_1}, v_{k_2}\rangle|\sum_j|\langle f_j, u_{k_1}\rangle\langle v_{k_2}, g_j\rangle| \\
&\leq \sum_{k_1}\sum_{k_2}|\langle u_{k_1}, v_{k_2}\rangle|\sqrt{\lambda_{k_1}}\sqrt{\mu_{k_2}}.
\end{aligned}$$

3.19 (a) Let $e = (1, \ldots, 1)$. It is easy to check that $Me = (na - a + 1)e$, and so e is an eigenvector for $na - a + 1$. If $v \perp e$, i.e., $\langle v, e\rangle = v_1 + \cdots + v_d = 0$, then $(Mv)_j = av_1 + \cdots + av_d - av_j + v_j = (1-a)v_j$, so that v is an eigenvector for $1-a$.
(b) The matrix M is positive semidefinite if and only if its eigenvalues are nonnegative: $na - a + 1, 1 - a \geq 0$, i.e., $\frac{-1}{n-1} \leq a \leq 1$. Its rank is the number of nonzero eigenvalues which is either 1 ($a = 1$), $n - 1$ ($a = \frac{-1}{n-1}$), or n (otherwise).

(c) If $\operatorname{rank}(M) = n-1$, then $a = \frac{-1}{n-1}$, and so M is the Gramian of the $n = d+1$ vertices of a simplex in $\mathbb{R}^d$, see (2.14).
(d) If $\operatorname{rank}(M) = n$, then M is invertible, and so the Gramian of $(\tilde{u}_j)$ is $M^\dagger = M^{-1}$. Let B be the $n \times n$ matrix with diagonal entries b, and off diagonal entries c. Then

$$(MB)_{jk} = \begin{cases} b+(n-1)ac, & j=k; \\ c+ab+(n-2)ac, & j \neq k \end{cases}$$

It is easy to check that $MB = I$, i.e., $B = \operatorname{Gram}((\tilde{v}_j))$, when

$$b = \frac{na-2a+1}{(1-a)(na-a+1)}, \qquad c = \frac{-a}{(1-a)(na-a+1)}.$$

Thus $(\tilde{v}_j)$ is (a scalar multiple) of an isogonal configuration of vectors.

3.20 Recall a frame Φ is real if and only if $\operatorname{Gram}(\Phi)$ has real entries.
(a) Use (3.4), and observe that the Tikhonov regularisation formula (.1) implies the pseudoinverse of a matrix is real if and only the matrix itself is.
(b) Use (3.9), i.e., $\operatorname{Gram}(\Phi^{\mathrm{can}}) = \operatorname{Gram}(\Phi)\operatorname{Gram}(\tilde{\Phi})$.
(c) No. Suppose that Φ is a basis for which the Gramian has complex entry (easily constructed), then Φ^{can} is an orthonormal basis, which is real.

3.21 Let $V = [f_j]$. The dual frame of (Pf_j) and (Pg_j) are equal if and only their synthesis operators are, i.e., $(PV(PV)^*)^{-1}PV = P(V^*)^{-1}$, which gives
(a) $PV(V^*) = (PVV^*P)P \iff PS = PSP$.
(b) Taking the adjoint of (a) gives $SP = (PS)^* = (PSP)^* = PSP$.
(c) By (a) and (b), we have $PS = SP$, and conversely, given this, we may obtain (a) and (b) by right and left multiplication by P. Thus (a), (b) and (c) are equivalent.
(d) Since $\operatorname{ran}(P) = \mathscr{H}$, (d) follows from any of (a), (b) and (c). Conversely, suppose that $S\mathscr{H} \subset \mathscr{H}$. Then $SPf \in \mathscr{H}$, so that $SPf = PSPf$, $\forall f$, which is (b).

3.22 Let $L = U\Sigma V^*$ be a singular value decomposition, where the diagonal entries of Σ are $\sigma_1, \ldots, \sigma_m$, and $V = [v_1, \ldots, v_m]$. Since U is unitary, we have

$$\|Lx\| = \|U\Sigma V^* x\| = \|\Sigma V^* x\| = \|(\sigma_j \langle x, v_j\rangle)_{j=1}^m\| = \Big(\sum_{j=1}^m \sigma_j^2 |\langle x, v_j\rangle|^2\Big)^{\frac{1}{2}}.$$

Hence (since V is unitary), we obtain

$$\|Lx\| \le M\Big(\sum_{j=1}^m |\langle x, v_j\rangle|^2\Big)^{\frac{1}{2}} = M\|V^* x\| = M\|x\|, \qquad M := \max \sigma_j,$$

with equality if and only if $\langle x, v_j\rangle = 0$ for every singular value σ_j less than the maximum, i.e., x is right-singular vector for the maximum singular value. Similarly, since $(\ker L)^\perp$ is the orthogonal direct sum of the right singular vector subspaces for the nonzero singular values, for a nonzero $x \in (\ker L)^\perp$

$$\|Lx\| \geq m\Big(\sum_{j=1}^{m} |\langle x, v_j\rangle|^2\Big)^{\frac{1}{2}} = m\|V^*x\| = m\|x\|, \qquad m := \min_{\sigma_j \neq 0} \sigma_j,$$

with equality if and only if $\langle x, v_j\rangle = 0$ for every singular value σ_j greater than m, the minimum nonzero singular value, i.e., x is right singular vector for m.

3.23 Since the norm is unitarily invariant, it suffices to consider the first inequality. Let m and M be the smallest and largest singular values of L, so, by Exer. 3.22,

$$m\|x\| \leq \|Lx\| \leq M\|x\|, \qquad \forall x.$$

Then, $\|(L-U)x\| \geq \big|\|Lx\| - \|Ux\|\big| = \big|\|Lx\| - \|x\|\big|$, so that

$$\|L-U\| \geq \max_{x\neq 0}\left|\frac{\|Lx\|}{\|x\|} - 1\right| \geq \max\{|m-1|, |M-1|\} = \|L-I\|.$$

3.24 (a) If $L : \mathscr{H} \to \mathscr{K}$ is a linear map between finite dimensional Hilbert spaces, with singular values $\sigma_1 \geq \sigma_2 \geq \cdots \geq \sigma_m$, then there are the sharp inequalities

$$\big(\min_{\sigma_j\neq 0} \sigma_j\big)\|x\| \leq \|Lx\| \leq \big(\max_j \sigma_j\big)\|x\|, \qquad \forall (\ker L)^\perp = \operatorname{ran}(L^*).$$

(see Exer. 3.22 for details). Suppose that $V = U_1 \Sigma U_2^*$ is a singular value decomposition of V, where $\Sigma = \big[\operatorname{diag}(\sqrt{\lambda_1}, \ldots, \sqrt{\lambda_d})\ 0\big]$ and $\lambda_1, \ldots, \lambda_d > 0$ are the eigenvalues of $S = VV^*$. The inequalities (and their sharpness) then follow by taking L to be V, V^*, $S = VV^*$ and $G = \operatorname{Gram}(\Phi) = V^*V$, and observing that these have singular value decompositions

$$V = U_1\Sigma U_2^*, \quad V^* = U_2\Sigma^* U_1^*, \quad S = U_1\Sigma\Sigma^* U_1^*, \quad G = U_2\Sigma^*\Sigma U_2^*.$$

The equivalence of these inequalities with the frame bounds follows from Exer. 3.7.

3.25 Let $V = U_1\Sigma U_2^*$ be a singular value decomposition of $V = [f_j]$. Then

$$S_\Phi = VV^* = U_1\Sigma U_2^* U_2 \Sigma^* U_1^* = U_1 \Sigma\Sigma^* U_1^*,$$

which is a unitary diagonalisation of S_Φ. Let $\Lambda := \Sigma\Sigma^* = \operatorname{diag}(\lambda_1, \ldots, \lambda_d)$. Since $S_{\tilde{\Phi}} = S_\Phi^{-1} = (U_1 \Lambda U_1^*)^{-1} = U_1\Lambda^{-1}U_1^*$, we have

$$\begin{aligned} \|\alpha S_\Phi + \beta S_{\tilde{\Phi}} + \gamma I\| &= \|\alpha U_1\Lambda U_1^* + \beta U_1 \Lambda^{-1} U_1^* + \gamma U_1 U_1^*\| \\ &= \|\alpha\Lambda + \beta\Lambda^{-1} + \gamma I\| \quad (U_1 \text{ is unitary}) \\ &= \max_{1\leq j\leq d} \big|\alpha\lambda_j + \beta\lambda_j^{-1} + \gamma\big| \quad (\|\cdot\| \text{ is the spectral norm}). \end{aligned}$$

We have $\operatorname{Gram}(\Phi) = U_2\Sigma^*\Sigma U_2^*$, $\operatorname{Gram}(\tilde{\Phi}) = (U_2\Sigma^*\Sigma U_2^*)^\dagger = U_2(\Sigma^*\Sigma)^\dagger U_2^*$, and

$$\operatorname{Gram}(\Phi^{\text{can}}) = \operatorname{Gram}(\Phi)\operatorname{Gram}(\tilde{\Phi}) = U_2\Sigma^*\Sigma(\Sigma^*\Sigma)^\dagger U_2^*.$$

Since Σ has the block form $\Sigma = \begin{bmatrix}\Lambda^{1/2} & 0\end{bmatrix}$, $\Sigma^*\Sigma = \mathrm{diag}(\Lambda,0)$, $(\Sigma^*\Sigma)^\dagger = \mathrm{diag}(\Lambda^{-1},0)$, and we obtain

$$\begin{aligned}&\|\alpha\,\mathrm{Gram}(\Phi)+\beta\,\mathrm{Gram}(\tilde{\Phi})+\gamma\mathrm{Gram}(\Phi^{\mathrm{can}})\|\\ &= \|U_2\{\alpha\begin{pmatrix}\Lambda & \\ & 0\end{pmatrix}+\beta\begin{pmatrix}\Lambda^{-1} & \\ & 0\end{pmatrix}+\gamma\begin{pmatrix}I & \\ & 0\end{pmatrix}\}U_2^*\|\\ &= \|\alpha\begin{pmatrix}\Lambda & \\ & 0\end{pmatrix}+\beta\begin{pmatrix}\Lambda^{-1} & \\ & 0\end{pmatrix}+\gamma\begin{pmatrix}I & \\ & 0\end{pmatrix}\| = \|\alpha\Lambda+\beta\Lambda^{-1}+\gamma I\|.\end{aligned}$$

3.26 Since the function $\lambda\mapsto\lambda-\frac{1}{\lambda}$, $\lambda>0$, is increasing and changes sign at $\lambda=1$, the maximum is either $\frac{1}{A}-A$, where $A\le 1$, and we need to prove

$$\frac{1}{A}-A\ge\frac{B-A}{\sqrt{AB}} \iff (\sqrt{AB}-1)(\sqrt{AB}+A^2)\le 0,$$

or it is $B-\frac{1}{B}$, where $B\ge 1$, and we need to prove

$$B-\frac{1}{B}\ge\frac{B-A}{\sqrt{AB}} \iff (\sqrt{AB}-1)(\sqrt{AB}+B^2)\ge 0.$$

The first clearly holds for $B\le 1$, and the second for $A\ge 1$ (with equality iff $AB=1$). It therefore suffices to consider the case $A<1$, $B>1$. Here either

$$\frac{1}{A}-A=\frac{1}{A}-\frac{1}{1/A}\ge B-\frac{1}{B} \iff A\le\frac{1}{B} \iff \sqrt{AB}\le 1,$$

$$\frac{1}{A}-A=\frac{1}{A}-\frac{1}{1/A}\le B-\frac{1}{B} \iff \frac{1}{B}\le A \iff \sqrt{AB}\ge 1,$$

and again the corresponding inequality holds via

$$\frac{1}{A}-A\ge\frac{1}{A}-\frac{1}{B}\ge\left(\frac{1}{A}-\frac{1}{B}\right)\sqrt{AB}, \qquad B-\frac{1}{B}\ge B-A\ge\frac{B-A}{\sqrt{AB}},$$

(with equality iff $AB=1$).

Alternatively, suppose that the inequality is false, i.e.,

$$\left|A-\frac{1}{A}\right|,\left|B-\frac{1}{B}\right|<\frac{B-A}{\sqrt{AB}},$$

then one obtains the contradiction

$$\left(A-\frac{1}{A}\right)^2+\left(B-\frac{1}{B}\right)^2<2\frac{(B-A)^2}{AB} \iff \frac{(A^2+B^2)(AB-1)^2}{A^2B^2}<0.$$

3.27 (a) Consider the first minimisation. Assume $t>0$. From

$$|tB-1|^2-|tA-1|^2=t\frac{B-A}{A+B}\left(t-\frac{2}{A+B}\right),$$

we conclude that

$$|tB-1| \geq |tA-1|, \quad \forall t \geq \frac{2}{A+B}, \qquad |tB-1| \leq |tA-1|, \quad \forall t \leq \frac{2}{A+B}.$$

Hence

$$\min_{t\geq\frac{2}{A+B}} \max\{|tA-1|,|tB-1|\} = \min_{t\geq\frac{2}{A+B}} |tB-1| = \min_{t\geq\frac{2}{A+B}} (tB-1) = \frac{B-A}{A+B},$$

$$\min_{t\leq\frac{2}{A+B}} \max\{|tA-1|,|tB-1|\} = \min_{t\leq\frac{2}{A+B}} |tA-1| = \min_{t\leq\frac{2}{A+B}} (1-tA) = \frac{B-A}{A+B},$$

each of which is attained if and only if $t = \frac{2}{A+B}$, which gives the first minimum. For the second one, observe that $0 < \frac{1}{B} \leq \frac{1}{A}$, and apply the first to obtain

$$\min_{t>0} \max\Big\{\Big|\frac{1}{tA}-1\Big|,\Big|\frac{1}{tB}-1\Big|\Big\} = \frac{\frac{1}{A}-\frac{1}{B}}{\frac{1}{A}+\frac{1}{B}} = \frac{B-A}{A+B},$$

which is attained if and only if $\frac{1}{t} = \frac{2}{\frac{1}{A}+\frac{1}{B}}$, i.e., $t = \frac{A+B}{2AB}$.
(b) Since $[A,B] \to \mathbb{R} : \lambda \mapsto |c^2\lambda-1|$ and $[\frac{1}{B},\frac{1}{A}] \to \mathbb{R} : \lambda^{-1} \mapsto |\frac{1}{c^2}\lambda^{-1}-1|$ attain their maxima at an endpoint, from (3.31) and (3.32) we obtain

$$\operatorname{dist}(c\Phi, \Phi^{\text{can}}) = \max\{|c^2A-1|,|c^2B-1|\},$$

$$\operatorname{dist}(\frac{1}{c}\tilde{\Phi}, \Phi^{\text{can}}) = \max\Big\{\Big|\frac{1}{c^2A}-1\Big|,\Big|\frac{1}{c^2B}-1\Big|\Big\}.$$

By part (a), with $t = \sqrt{c}$, the minima of these distances over $c > 0$ is $\frac{B-A}{A+B}$, and this is attained as claimed.
(c) The four quantities in the min–max are continuous functions of t, which are monotone at every point except the value of t which makes them zero. Hence, for the minima to occur two of these functions must equal the maximum (if not, then the one nonzero value equal to the maximum could be reduced slightly, so as to reduce the maximum of all four values). Thus it suffices to consider only the values of t which make (at least) two of these functions equal, and the common value M (a potential min–max). Calculations (or a sketch) show the minimum M is given by

$$|tB-1| = \Big|\frac{1}{tA}-1\Big|, \qquad M = \sqrt{\frac{B}{A}}-1 \quad \Big(\text{at } t = \frac{1}{\sqrt{AB}}\Big).$$

(d) By the formulas in (b) above, we have

$$\max\big\{\operatorname{dist}(c\Phi, \Phi^{\text{can}}), \operatorname{dist}(\frac{1}{c}\tilde{\Phi}, \Phi^{\text{can}})\big\}$$
$$= \max\Big\{|c^2A-1|,|c^2B-1|,\Big|\frac{1}{c^2A}-1\Big|,\Big|\frac{1}{c^2B}-1\Big|\Big\}.$$

By part (c), with $t = c^2$, the minimum of this over $c > 0$ is $\sqrt{\frac{B}{A}} - 1$, and this is attained precisely when $c^2 = \frac{1}{\sqrt{AB}}$.

(e) We observe that $\mathrm{Gram}(cU\Phi^{\mathrm{can}}) = c^2\,\mathrm{Gram}(\Phi^{\mathrm{can}})$, and the eigenvalues of S_Φ satisfy $B = \frac{1}{A} = \lambda_1 \geq \lambda_2 \cdots \geq \lambda_d = A$, so Exer. 3.25 gives

$$\begin{aligned}\mathrm{dist}(\Phi, cU\Phi^{\mathrm{can}}) &= \|S_\Phi - cI\| = \|\mathrm{Gram}(\Phi) - c^2\,\mathrm{Gram}(\Phi^{\mathrm{can}})\| \\ &= \max_j |\lambda_j - c^2| = \max\Big\{|A - c^2|, \Big|\frac{1}{A} - c^2\Big|\Big\},\end{aligned}$$

$$\mathrm{dist}(\tilde{\Phi}, cU\Phi^{\mathrm{can}}) = \max_j |\lambda_j^{-1} - c^2| = \max\Big\{|A - c^2|, \Big|\frac{1}{A} - c^2\Big|\Big\}.$$

By the reasoning of (c), the minimum occurs when

$$|A - c^2| = \Big|\frac{1}{A} - c^2\Big| \quad \Longleftrightarrow \quad c^2 = \frac{A^2+1}{2A}.$$

3.28 Since $\frac{1}{B} \leq \frac{1}{A}$, we may assume that $B \geq 1$. Suppose that

$$\max\{|A-1|, |B-1|\} = \max\Big\{\Big|\frac{1}{A} - 1\Big|, \Big|\frac{1}{B} - 1\Big|\Big\} = \max\Big\{\frac{|A-1|}{A}, \frac{|B-1|}{B}\Big\}.$$

Consider cases. If $|A-1| \geq |B-1|$, then $A \leq 1$, which gives

$$|A-1| = \frac{|A-1|}{A} \quad \Longrightarrow \quad A = 1, B = 1.$$

If $|A-1| \leq |B-1|$, then there are two possibilities

$$|B-1| = \frac{|B-1|}{B} \quad \Longrightarrow \quad B = 1, A = 1,$$

$$|B-1| = \frac{|A-1|}{A} \quad \Longrightarrow \quad |AB - A| = |A-1| \quad \Longrightarrow \quad AB \in \{1, 2A-1\}.$$

When $AB = 2A - 1$, we have

$$A \leq B = \frac{2A-1}{A} \quad \Longrightarrow \quad (A-1)^2 = A^2 - 2A + 1 \leq 0 \quad \Longrightarrow \quad A = 1, B = 1,$$

and so $AB = 1$ in all cases.

3.29 Recall, from (3.13), that $\mathrm{ran}(V^*) = \mathrm{ran}(\mathrm{Gram}(\Phi^{\mathrm{can}}))$.

(a)$\Longrightarrow$(c) $W = QV$ gives $\mathrm{ran}(W^*) = \mathrm{ran}((QV)^*) = \mathrm{ran}(V^*Q^*) \subset \mathrm{ran}(V^*)$.

(c)$\Longrightarrow$(a) Suppose that $\mathrm{ran}(W^*) \subset \mathrm{ran}(V^*)$. Let $Q = WV^\dagger$. Since $V^\dagger$ maps $\mathrm{ran}(V)$ onto $\ker(V)^\perp = \mathrm{ran}(V^*)$, we can decompose x as $x = V^\dagger a + b$, $b \in \ker(V) \subset \ker(W)$. Then $QVx = W(V^\dagger V V^\dagger)a + WV^\dagger V b = WV^\dagger a + 0 = Wx$, $\forall x$, so that $Q\Phi = \Psi$.

(b)$\Longleftrightarrow$(c) The Gramians are orthogonal projections (determined by their ranges), and so their product PQ equals Q if and only if $\mathrm{ran}(Q) \subset \mathrm{ran}(P)$.

(c)$\Longrightarrow$ (*d*) W^* is 1–1 and $\mathrm{ran}(W^*) \subset \mathrm{ran}(V^*) = \ker(V)^\perp$, so that VW^* is 1–1.
(d)$\Longleftrightarrow$ (*e*) Immediate, since $VW^*g = \sum_j \langle g, g_j\rangle f_j$.

3.30 Let $V = [f_j]$, $W = [g_j]$. Then the closeness condition is

$$\|(W-V)c\| \le \lambda \|Vc\|, \quad \forall c.$$

(a) If $Vc = 0$, then this implies $(W-V)c = 0$, and so $Wc = (W-V)c + Vc = 0$. Thus $\ker(V) \subset \ker(W)$, and Exer. 3.29 gives $\Psi = Q\Phi$.
(b) In view of (a), $W = QV$ and the closeness condition is equivalent to

$$\|(Q-I)Vc\| \le \lambda \|Vc\|, \qquad \forall c \in \ker(V)^\perp = \mathrm{ran}(V^*).$$

Since Φ is a frame, $S = VV^*$ is onto, and so Vc above can be an arbitrary element of $\mathscr{H}$. Thus the closeness condition is equivalent to λ being an upper bound for $\|Q-I\|$, and $\mathrm{cl}(\Psi, \Phi) = \|Q-I\|$.
(c) Since $\|Vc\| = \|(V-W)c + Wc\| \le \|(V-W)c\| + \|Wc\|$, we have

$$\|(W-V)c\| \le \lambda \big(\|(V-W)c\| + \|Wc\|\big) \quad \Longrightarrow \quad \|(V-W)c\| \le \frac{\lambda}{1-\lambda}\|Wc\|,$$

i.e., Φ is close to Ψ, with $\mathrm{cl}(\Phi, \Psi) \le \frac{\lambda}{1-\lambda}$.

3.31 (a) We have $\|M\|\|Sx\| \ge \|MSx\| \ge \|x\| - \|(I-MS)x\| \ge (1 - \|I-MS\|)\,\|x\|$, i.e.,

$$\|Sx\| \ge \frac{1-\|I-MS\|}{\|M\|}\|x\|, \qquad \forall x \in X.$$

(b) $\|F(g_1) - F(g_2)\| = \|(I-MS)(g_1-g_2)\| \le \|I-MS\|\|I-MS\|$.
(c) By (a), MS is bounded below: $\|MSx\| \ge (1-\|I-MS\|)\|x\|$, and so

$$\|(MS)^{-1}\| \le \frac{1}{1-\|I-MS\|}.$$

3.32 (a) The eigenvalues of $S = S_\Phi$ satisfy $A \le \lambda_j \le B$, so that

$$\|I - \frac{2}{A+B}S\| = \max_j \big|1 - \frac{2\lambda_j}{A+B}\big| \le \max_{A\le\lambda\le B}\big|1 - \frac{2\lambda}{A+B}\big|.$$

The maximum of $\lambda \mapsto \big|1-\frac{2\lambda}{A+B}\big|$ above occurs at the endpoints, and so it is

$$\max\big\{\big|1-\frac{2A}{A+B}\big|, \big|1-\frac{2B}{A+B}\big|\big\} = \frac{B-A}{A+B}.$$

(b) The error in the fixed point iteration $g_{n+1} := F(g_n)$ for a Banach contraction F with constant κ and fixed point g can be estimated by either of

$$\|g_n - g\| \le \kappa^n \|g - g_0\|, \qquad \|g_n - g\| \le \frac{\kappa^n}{1-\kappa}\|g_1 - g_0\|.$$

Using the estimate $\kappa = \|I - \frac{2}{A+B}S\| \le \frac{B-A}{A+B} < 1$, from the second we obtain

$$\|g_n - g\| \le \frac{\left(\frac{B-A}{A+B}\right)^n}{1-\frac{B-A}{A+B}}\|\frac{2}{A+B}h\| = \frac{\|h\|}{A}\left(\frac{B-A}{A+B}\right)^n.$$

Remark. The choice $h = Sf$, i.e., $g = f$, gives the *frame algorithm* (3.37), for which the first estimate gives

$$\|g_n - f\| \le \|f\|\left(\frac{B-A}{A+B}\right)^n.$$

3.33 Let $S = U\Lambda U^*$ be a unitary diagonalisation of S, and $f(x) := x^{-\frac{1}{2}}, x > 0$. Since

$$f^{(j)}(x) = \frac{(-1)^j(2j)!}{2^{2j}j!}x^{-\frac{2j+1}{2}},$$

the Taylor series expansion of f about $c = \frac{A+B}{2}$ is

$$\begin{aligned}
x^{-\frac{1}{2}} &= \sum_{j=0}^{\infty}\frac{(-1)^j(2j)!}{2^{2j}(j!)^2}\left(\frac{A+B}{2}\right)^{-\frac{2j+1}{2}}\left(x-\frac{A+B}{2}\right)^j \\
&= \sqrt{\frac{2}{A+B}}\sum_{j=0}^{\infty}\frac{(-1)^j(2j)!}{2^{2j}(j!)^2}\left(1-\frac{2}{A+B}x\right)^j,
\end{aligned}$$

which (by the ratio test) is absolutely convergent for $0 < x < A+B$. By Exer. 3.32,

$$\|I - \frac{1}{A+B}S\| \le \frac{B-A}{A+B} = \left|1-\frac{2}{A+B}A\right| = \left|1-\frac{2}{A+B}B\right|$$

and so the series for $S^{-\frac{1}{2}}$ is absolutely convergent. Its partial sums satisfy

$$\sqrt{\frac{2}{A+B}}\sum_{j=0}^{n}\frac{(2j)!}{2^{2j}(j!)^2}\left(I-\frac{2}{A+B}S\right)^j = U\left\{\sqrt{\frac{2}{A+B}}\sum_{j=0}^{n}\frac{(2j)!}{2^{2j}(j!)^2}\left(I-\frac{2}{A+B}\Lambda\right)^j\right\}U^*.$$

Since the eigenvalues of S (diagonal entries of Λ) satisfy $0 < A \le \lambda_j \le B < A+B$, the partial sums in the $\{\quad\}$ converge to $\Lambda^{-\frac{1}{2}}$ (by considering entries), and so the series converges to $S^{-\frac{1}{2}} = U\Lambda^{-\frac{1}{2}}U^*$.

3.34 Let $V = [f_j]$, and use equivalence (f) of Proposition 3.4. We calculate

$$I - P_\Phi = I - V^*S^{-1}V = \frac{1}{3}\begin{pmatrix}1&1&1\\1&1&1\\1&1&1\end{pmatrix} = vv^*, \qquad v := \frac{1}{\sqrt{3}}(1,1,1),$$

Thus all possible 2×3 matrices $L(I-P_\Phi)$ are determined by $z = Lv$, i.e.,

$$L(I-P_\Phi) = L(vv^*) = (Lv)v^* = zv^* = \frac{1}{\sqrt{3}}[z,z,z],$$

and so all possible duals are given by $W = [\tilde{f}_1 + w, \tilde{f}_2 + w, \tilde{f}_3 + w]$, $w = \frac{1}{\sqrt{3}} z \in \mathbb{R}^2$.
(b) Let $W = [g_j]$. Then (g_j) is a pseudodual if and only if

$$(2/\sqrt{3})\det(VW^*) = w_{11}w_{22} - w_{21}w_{12} + w_{13}(w_{21} - w_{22}) + w_{23}(w_{12} - w_{11}) \neq 0.$$

The left-hand side above is zero if $g_1 = g_2$, i.e., $w_{11} = w_{12}$ and $w_{21} = w_{22}$, otherwise the coefficient of w_{13} or of w_{23} is nonzero, and $g_3 = (w_{13}, w_{23})$ can be chosen to make the left-hand side nonzero.

3.35 (a) This follows immediately from $WV^* = (WQ^{-1})(VQ^*)^*$.
(b) The synthesis operator of the canonical dual of the frame given by VQ^* is

$$U = (VQ^*QV^*)^{-1}VQ^*.$$

By (3.40), this dual frame uniquely minimises $\|U\|$ over all dual frames to VQ^*. Thus, by (a), $W = UQ = (VQ^*QV^*)^{-1}VQ^*Q$ is the unique dual frame of Φ which minimises $\|W\|_Q = \|WQ^{-1}\| = \|U\|$.

3.36 (a) To $(\langle f, \hat{f}_j \rangle) = (\langle f, f_j \rangle) - (\langle f, f_j - \hat{f}_j \rangle)$ apply the triangle inequality

$$\Big(\sum_j |\langle f, \hat{f}_j \rangle|^2\Big)^{\frac{1}{2}} \leq \Big(\sum_j |\langle f, f_j \rangle|^2\Big)^{\frac{1}{2}} + \Big(\sum_j |\langle f, f_j - \hat{f}_j \rangle|^2\Big)^{\frac{1}{2}} \leq \sqrt{B}\|f\| + \sqrt{R}\|f\|,$$

and the reverse triangle inequality

$$\sqrt{A}\|f\| - \sqrt{R}\|f\| \leq \Big(\sum_j |\langle f, f_j \rangle|^2\Big)^{\frac{1}{2}} - \Big(\sum_j |\langle f, f_j - \hat{f}_j \rangle|^2\Big)^{\frac{1}{2}} \leq \Big(\sum_j |\langle f, \hat{f}_j \rangle|^2\Big)^{\frac{1}{2}}.$$

Thus $(\sqrt{A} - \sqrt{R})^2$ and $(\sqrt{B} + \sqrt{R})^2$ are lower and upper frame bounds for $(\hat{f}_j)$.
(b) Let $V = [f_j]$, $W = [\hat{f}_j]$. Then the synthesis operator of the canonical dual of $\hat{\Phi}$ is $U = (WW^*)^{-1}W$, and so $\|UV^* - I\| = \|UV^* - UW^*\| \leq \|U\|\|V^* - W^*\|$. Now

$$\|V^* - W^*\| = \sup_{\|f\|=1} \|(V^* - W^*)f\| = \sup_{\|f\|=1} \Big(\sum_j |\langle f_j - \hat{f}_j, f \rangle|^2\Big)^{\frac{1}{2}} \leq \sqrt{R}.$$

Similarly, $\|U\| = \|U^*\|$ is bounded by the square root of an upper frame bound for the canonical dual of $\hat{\Phi}$ (see Exer. 3.24), which by (a) is $\leq 1/(\sqrt{A} - \sqrt{R})$. Thus $\|UV^* - I\| < 1$, so that the canonical dual of $\hat{\Phi}$ and Φ are approximate duals, if $\sqrt{R}/(\sqrt{A} - \sqrt{R}) < 1$, which holds if and only if $R < \frac{A}{4}$.

3.37 (a) Let V and $W = cV$ be the synthesis operators of Φ and $c\Phi$. Then

$$\|WV^* - I\| = \|cVV^* - I\| = \max_j |c\lambda_j - 1| = \max\{|cA - 1|, |cB - 1|\}$$

(see Exer. 3.27 for details). We consider cases: $c \leq 0$ gives $|cB - 1| = 1 - cB \geq 1$, $0 < c < \frac{2}{B}$ gives $-1 < cA - 1 \leq cB - 1 < 1$, and $c \geq \frac{2}{B}$ gives $cB - 1 \geq 1$. The minimum occurs when $|cA - 1| = |cB - 1|$, i.e., $c = \frac{2}{A+B}$.

3.38 By replacing $\mathscr{H}$ by $\mathscr{V}+\mathscr{W}$, we may assume without loss of generality that $\mathscr{H}$ is finite dimensional. Let $d_1=\dim(\mathscr{V})$, $d_2=\dim(\mathscr{W})$ and $n=\dim(\mathscr{H})$.
(a) Suppose that $\mathscr{V}\cap\mathscr{W}^\perp=\mathscr{W}\cap\mathscr{V}^\perp=0$. Then the algebraic direct sums $\mathscr{V}\oplus_a\mathscr{W}^\perp$ and $\mathscr{W}\oplus_a\mathscr{V}^\perp$ are subspaces $\mathscr{H}$, and so their dimensions satisfy

$$d_1+(n-d_2)\le n,\quad d_2+(n-d_1)\le n \quad\Longrightarrow\quad d_1=d_2.$$

Thus a dimension count gives (ii). Conversely, if (ii) holds, then (i) is immediate.
(b) The first part was proved in (a). Let $Q=P^*$, $P=P_{\mathscr{V},\mathscr{W}^\perp}$, and $w\in\mathscr{W}$, $v^\perp\in\mathscr{V}^\perp$. Then

$$\langle Qv^\perp,z\rangle=\langle v^\perp,Pz\rangle=0,\quad \forall z\in\mathscr{H}\quad\Longrightarrow\quad Qv^\perp=0,$$

$$\langle Qz,w^\perp\rangle=\langle z,Pw^\perp\rangle=\langle z,0\rangle=0,\quad \forall w^\perp\in\mathscr{W}^\perp\quad\Longrightarrow\quad Qz\in(\mathscr{W}^\perp)^\perp=\mathscr{W}.$$

Since P and Q have the same rank, we conclude that $Q|_\mathscr{W}$ is a bijection $\mathscr{W}\to\mathscr{W}$. Since $P^2=P$ implies $Q^2=Q$, we conclude that $Q|_\mathscr{W}=I_\mathscr{W}$. Thus $Q=P_{\mathscr{W},\mathscr{V}^\perp}$.

3.39 Let $V=[f_j]$ and W be the synthesis operator of a finite frame for $\mathscr{W}$. Then $[g_j]=W(V^*W)^\dagger$, and so the canonical oblique dual of (g_j) has synthesis operator

$$U=V((W(V^*W)^\dagger)^*V)^\dagger=V((W^*V)^\dagger W^*V)^\dagger=V(W^*V)^\dagger W^*V.$$

We observe that $\ker(W^*)=\mathscr{W}^\perp$, so that W^* is 1–1 on $\operatorname{ran}(V)$ (since $\mathscr{W}^\perp\cap\mathscr{V}=0$), and $\ker(W^*V)=\ker(V)$. Now right multiplication of the above by $(W^*V)^\dagger$, using $A^\dagger AA^\dagger=A^\dagger$, gives $U(W^*V)^\dagger=V(W^*V)^\dagger$. Thus it suffices to show that $U=V$ on

$$(\operatorname{ran}((W^*V)^\dagger)^\perp=\ker((V^*W)^\dagger)=\ker(W^*V)=\ker(V),$$

which follows immediately from the formula for U (which has right factor of V).

3.40 (a) $A^\dagger AA^\dagger=A^\dagger$ gives $P^2=V(\Lambda V)^\dagger\Lambda V(\Lambda V)^\dagger\Lambda=V(\Lambda V)^\dagger\Lambda=P$.
(b) $AA^\dagger A=A$ gives $\Lambda PV=\Lambda V(\Lambda V)^\dagger\Lambda V=\Lambda V$, i.e., $\Lambda(Pf)=\Lambda(f)$, $\forall f\in\mathscr{V}$.
(c) By (b), $\Lambda PV=\Lambda V$ so that $\operatorname{rank}(P)\ge\operatorname{rank}(\Lambda V)$, while $P=V(\Lambda V)^\dagger V$ gives $\operatorname{rank}(P)\le\operatorname{rank}((\Lambda V)^\dagger)=\operatorname{rank}(\Lambda V)$, so $\operatorname{rank}(P)=\operatorname{rank}(\Lambda V)$.
(d) If $\dim(\mathscr{L}|_\mathscr{V})=\dim(\Lambda|_\mathscr{V})\ge\dim(\mathscr{V})$, then $\Lambda|_\mathscr{V}$ is 1–1, so that ΛV maps $\ker(V)^\perp$ bijectively onto $\Lambda\mathscr{V}$, hence $(\Lambda V)^\dagger$ maps $\Lambda\mathscr{V}$ onto $\ker(V)^\perp$, and so P maps onto $\mathscr{V}=V(\ker(V)^\perp)$.
(e) If $\dim(\mathscr{L})\le\dim(\mathscr{L}|_\mathscr{V})$, then $\operatorname{ran}(\Lambda)=\operatorname{ran}(\Lambda|_\mathscr{V})$, so for $x\in X$ there is $v\in\mathscr{V}$ with $\Lambda x=\Lambda v$, which gives $\Lambda P(x-v)=\Lambda V(\Lambda V)^\dagger\Lambda(x-v)=0=\Lambda(x-v)$, while (b) gives $\Lambda Pv=\Lambda v$. Adding these gives $\Lambda Px=\Lambda P\{(x-v)+v\}=\Lambda x$.
(f) If $\lambda_k(f)=\langle f,g_k\rangle$, then $\Lambda=W^*$, where $W=[g_k]$, so $P=V(W^*V)^\dagger W^*$.

3.41 If $\Phi=(f_j)$ is a basis and $V=[f_j]$, then the orthonormal basis given by the canonical tight frame has synthesis operator $S^{-\frac12}V=VG^{-\frac12}$, where $G=\operatorname{Gram}(\Phi)$ (see Exer. 3.11). For the Legendre weight $w=1$ on $[-1,1]$, applying Gram–Schmidt and then Löwdin orthogonalisation to the first four normalised monomials gives following sequences of polynomials. As one would expect, for Löwdin orthogonalisation, different sequences of polynomials of degree $0,1,\ldots,n$ give different orthonor-

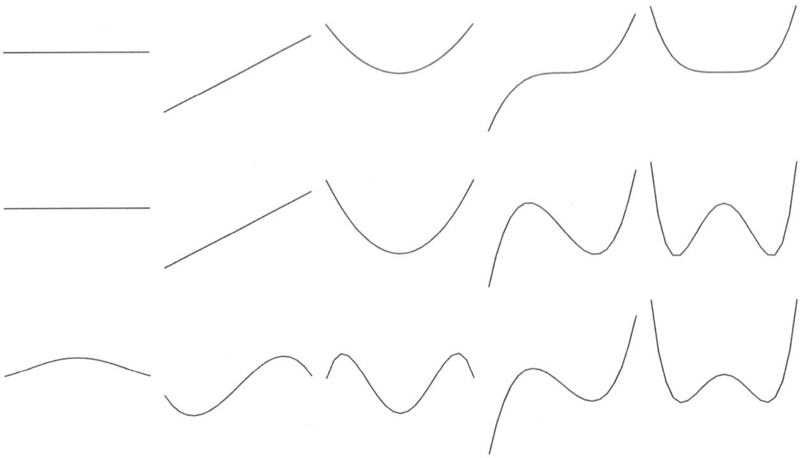

Fig. A.2: Gram–Schmidt and Löwdin orthogonalisation of the first five monomials

mal bases (in contrast to Gram–Schmidt). Starting instead with the Bernstein basis, one obtains.

Notice that here, Gram–Schmidt does not preserve the symmetries of the Bernstein basis, while Löwdin does. These latter polynomials presumably maintain good conditioning and have the advantage of being orthogonal.

They also seem to maintain the partition of unity property.

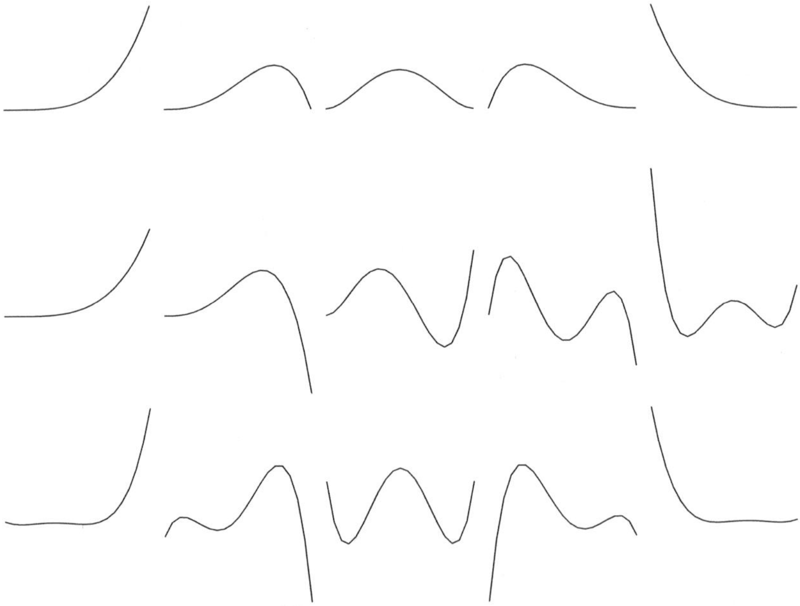

Fig. A.3: Gram–Schmidt and Löwdin orthogonalisation of the Bernstein basis

Exercises of Chapter 4

4.1 (a)$\iff$(b)$\iff$(c) Since $f \in X$ is determined by $(\lambda(f))_{\lambda \in X'}$,

$$
\begin{aligned}
V\Lambda = I_X \quad &\iff \quad f = I_X f = V\Lambda f = \sum_j \lambda_j(f) f_j, \quad \forall f \\
&\iff \quad \lambda(f) = \sum_j \lambda_j(f)\lambda(f_j) = \Big(\sum_j \lambda(f_j)\lambda_j\Big)(f), \quad \forall f \in X, \forall \lambda \\
&\iff \quad \lambda = \sum_j \lambda(f_j)\lambda_j, \quad \forall \lambda.
\end{aligned}
$$

(a)$\iff$(d)$\iff$(e)$\iff$(f) Since V is onto X, we can multiply and cancel by V on the right, and since Λ is 1–1, we can multiply and cancel by Λ on the left. Thus

$$
\begin{aligned}
V\Lambda = I_X \quad &\iff \quad G^2 = \Lambda(V\Lambda)V = \Lambda(I_X)V = G \quad (\text{multiply by}\, \Lambda \,\text{and}\, V) \\
&\iff \quad V = V(\Lambda V) = VG \quad (\text{cancel}\, \Lambda) \\
&\iff \quad \Lambda = (\Lambda V)\Lambda = G\Lambda \quad (\text{multiply by}\, \Lambda,\, \text{cancel V}).
\end{aligned}
$$

4.2 Since $c_j^\Phi(f) = \langle f, S^{-1} f_j\rangle = \langle S^{-\frac{1}{2}} f, S^{-\frac{1}{2}} f_j\rangle$, where $(S^{-\frac{1}{2}} f_j)$ is the canonical tight frame, we compute

$$
\begin{aligned}
\langle f, g\rangle_\Phi &= \langle (\langle S^{-\frac{1}{2}} f, S^{-\frac{1}{2}} f_j\rangle), (\langle S^{-\frac{1}{2}} g, S^{-\frac{1}{2}} f_j\rangle)\rangle \\
&= \sum_j \langle S^{-\frac{1}{2}} f, S^{-\frac{1}{2}} f_j\rangle \langle S^{-\frac{1}{2}} f_j, S^{-\frac{1}{2}} g\rangle = \langle S^{-\frac{1}{2}} f, S^{-\frac{1}{2}} g\rangle \qquad (\text{Plancherel}).
\end{aligned}
$$

4.3 By Proposition 4.9, $\xi_j = \xi_k$ if and only if $\psi_j = \psi_k = v$. In this case, write $\xi_j = \xi_k$ as f_v, to obtain

$$
x = \sum_j \xi_j(x)\psi_j = \sum_{v \in \Theta} m_v f_v(x) v.
$$

Since the barycentric coordinates for Θ are unique, we have that $m_v f_v = \ell_v$.

Exercises of Chapter 5

5.1 Suppose that (f_j) is a disjoint union $\cup_m \{f_j : j \in J_m\}$ of tight frames, with frame bounds A_m. Then for f_j, $j \in J_m$, we have

$$
Sf_j = \sum_k \langle f_j, f_k\rangle f_k = \sum_{k \in J_m} \langle f_j, f_k\rangle f_k = A_m f_j.
$$

Conversely, suppose that each f_j is an eigenvector of S. Let $\{\lambda_m\}$ be the eigenvalues of S, and $\Phi_m := \{j : Sf_j = \lambda_m f_j\}$. Since S is Hermitian, its eigenspaces span(Φ_m) are orthogonal, so that for $f_j \in \Phi_m$, we have

$$S_{\Phi_m} f_j = \sum_{f_k \in \Phi_m} \langle f_j, f_k \rangle f_k = \sum_k \langle f_j, f_k \rangle f_k = S f_j = \lambda_m f_j,$$

and so (by linearity) each Φ_m is a tight frame. Thus (f_j) is the disjoint union of the tight frames Φ_m.

5.2 (a) Since $\langle f+g, \phi_j \rangle = \langle f, \phi_j \rangle$ and $\langle f+g, \psi_k \rangle = \langle g, \psi_k \rangle$, we obtain

$$S(f+g) = \sum_j \langle f+g, \phi_j \rangle \phi_j + \sum_k \langle f+g, \psi_k \rangle \psi_k = S_\Phi(f) + S_\Psi(g).$$

(b) The dual vector to $\phi_j + 0 \in \Phi \cup \Psi$ is

$$S^{-1}(\phi_j + 0) = S_\Phi^{-1}(\phi_j) + S_\Psi^{-1}(0) = \tilde{\Phi}_j + 0,$$

and similarly for $0 + \psi_k \in \Phi \cup \Psi$.
(c) Since $\langle \phi_j + 0, 0 + \psi_k \rangle = 0$, the entries of the Gramian are zero if they correspond to a vector in Φ and one in Ψ.

5.3 Let P_1 and P_2 be the orthogonal projections onto $\mathscr{H}_1$ and $\mathscr{H}_2$. Then

$$(\langle f+g, \phi_j + \psi_j \rangle) = (\langle f, \phi_j \rangle) + (\langle g, \psi_j \rangle) = V^* f + W^* g,$$

so that

$$\begin{aligned} &(\phi_j + \psi_j) \text{ is a frame for } \mathscr{H}_1 \oplus \mathscr{H}_2 \\ \iff \quad & U = [\phi_j + \psi_j] : \mathbb{F}^J \to \mathscr{H}_1 \oplus \mathscr{H}_2 \text{ is onto} \\ \iff \quad & U^* = V^* P_1 + W^* P_2 : f + g \mapsto (\langle f+g, \phi_j + \psi_j \rangle) \text{ is } 1-1 \\ \iff \quad & \operatorname{ran}(V^*) \cap \operatorname{ran}(W^*) = 0 \quad (\text{since } V^* : \mathscr{H}_1 \to \mathbb{F}^J \text{ and } W^* : \mathscr{H}_2 \to \mathbb{F}^J \text{ are } 1-1). \end{aligned}$$

5.4 (a) Since $\|f+g\|^2 = \|f\|^2 + \|g\|^2$, $\forall f+g \in \mathscr{H}_1 \oplus \mathscr{H}_2$, we may add the frame bounds

$$A_\Phi \|f\|^2 \le \langle S_\Phi f, f \rangle \le B_\Phi \|f\|^2, \qquad A_\Psi \|g\|^2 \le \langle S_\Psi g, g \rangle \le B_\Psi \|g\|^2,$$

to obtain the sharp inequalities

$$\min\{A_\Phi, A_\Psi\} \|f+g\|^2 \le \langle S(f+g), f+g \rangle \le \max\{B_\Phi, B_\Psi\} \|f+g\|^2.$$

Thus the sequence is frame for $\mathscr{H}_1 \oplus \mathscr{H}_2$, with the above frame bounds.
(b) Suppose $S(f+g) = S_\Phi(f) + S_\Psi(g)$. Then S (and hence S^{-1}) maps $\mathscr{H}_j \to \mathscr{H}_j$, so that

$$\begin{aligned} & f+g = S^{-1}(S_\Phi(f) + S_\Psi(g)) = S^{-1} S_\Phi(f) + S^{-1} S_\Psi(g) \\ \implies \quad & f = S^{-1} S_\Phi(f), \; g = S^{-1} S_\Psi(g) \\ \implies \quad & S^{-1}(f) = S_\Phi^{-1}(f), \; S^{-1}(g) = S_\Psi^{-1}(g). \end{aligned}$$

Similarly, $S^{-1}(f+g) = S_\Phi^{-1}(f) + S_\Psi^{-1}(g)$ implies that $S(f) = S_\Phi(f)$, $S(g) = S_\Psi(g)$.

5.5 Suppose that (f_j) is a simple lift, say $f_j = \phi_j + \alpha\psi \in \mathcal{H} \oplus \mathcal{K}$, $\sum_j \phi_j = 0$. Then

$$\sum_j f_j = \alpha|J|\psi \neq 0, \qquad \langle \sum_j f_j, f_k\rangle = \langle \alpha|J|\psi, \phi_k + \alpha\psi\rangle = |\alpha|^2|J|\langle \psi, \psi\rangle.$$

Conversely, suppose that $\psi := \sum_j f_j \neq 0$ and $\langle \sum_j f_j, f_k\rangle = C$, $\forall k$. Let P be the orthogonal projection onto $\psi^\perp$. Then (f_j) is the direct sum

$$f_j = Pf_j + (I-P)f_j = Pf_j + \frac{\langle f_j, \psi\rangle}{\langle \psi, \psi\rangle}\psi = Pf_j + \frac{C}{\langle \psi, \psi\rangle}\psi,$$

which is a simple lift of (Pf_j).

Exercises of Chapter 6

6.1 Since A is Hermitian, it is unitarily diagonalisable, with real eigenvalues $\lambda_1, \ldots, \lambda_n$, and an orthonormal basis of eigenvectors $\{u_1, \ldots, u_n\}$. The inequality follows from the Cauchy–Schwartz inequality applied to $(1, \ldots, 1)$ and $(\lambda_1, \ldots, \lambda_n)$, with equality if and only if $\lambda_1 = \cdots = \lambda_r = c$. Thus

$$A = \sum_{j=1}^{d} \lambda_j u_j u_j^* = \sum_{j=1}^{r} c u_j u_j^* = c[u_1, \ldots, u_r][u_1, \ldots, u_r]^* = cUU^*.$$

6.2 With $K := \sum_j \langle f_j, f_j\rangle = \sum_j \|f_j\|^2 > 0$, we have

$$(n^2 - n)\max_{j\neq k} |\langle f_j, f_k\rangle|^2 + \sum_j |\langle f_j, f_j\rangle|^2 \geq \sum_j \sum_k |\langle f_j, f_k\rangle|^2 \geq \frac{K^2}{d}.$$

6.3 In `matlab`, we take a quotient, with value $\leq d$ (with equality for a tight frame)

```
c=[]; m=100; d=7;
for n=d:m,
V=rand(d,n)-1/2; c(n)=trace(V'*V)^2/norm(V'*V,'fro')^2;
end;
c(1)=0, c(m+1)=d, plot(c)
```

Large numbers of vectors do give frames which are close to tight. The fact that tight frames correspond to cubature rules gives a heuristic explanation for this. A more precise treatment of *random frames* is given in [Ehl12].

6.4 (a) For $\mathbb{F}^1 = \mathbb{F}$, the condition for perfect tightness becomes

$$2|v_j\overline{v_k}|^2 + \frac{|v_j|^4 + |v_k|^4}{n-1} = \frac{1}{d}\left(2|v_j|^2|v_k|^2 + \frac{|v_j|^4 + |v_k|^4}{n-1}\right),$$

which holds if and only if $d = 1$ (for $v_j \neq 0$).

(b) Suppose some vector v_k is zero, then the perfect tightness condition gives

$$\frac{\|v_j\|^4}{n-1} = \frac{1}{d}\left(\frac{\|v_j\|^4}{n-1}\right),$$

which can only hold (for $v_j \neq 0$) if $d=1$.
(c) An equal-norm frame is perfect if and only if it is an equiangular tight frame.
(d) This is an open question.

6.5 If (f_j) is orthogonal, i.e., $|\langle f_j, f_k\rangle| = 0$, $j \neq k$, then $\mathrm{FP}(f_1,\ldots,f_n) = n$, while any other choice gives a larger value.

6.6 The lower bound and equality follow directly from Theorem 6.1. For the upper bound, Cauchy–Schwarz gives

$$\frac{\sum_{j=1}^n \sum_{k=1}^n |\langle f_j, f_k\rangle|^2}{\left(\sum_{j=1}^n \|f_j\|^2\right)^2} \leq \frac{\sum_{j=1}^n \sum_{k=1}^n \|f_j\|^2 \|f_k\|^2}{\left(\sum_{j=1}^n \|f_j\|^2\right)^2} = 1,$$

with equality if and only if all f_j and f_k are scalar multiples of each other.

6.7 The integral over $\mathbb{S}$ is invariant under a unitary change of variables U.
(a) Choose U with $Ue_1 = y/\|y\|$, so that $\langle y, Ux\rangle = \|y\|\langle e_1, x\rangle$, which gives

$$\int_{\mathbb{S}} p_y \, d\sigma = \int_{\mathbb{S}} (p_y \circ U)\, d\sigma = \int_{\mathbb{S}} \|y\|^2 p_{e_1}\, d\sigma = \|y\|^2 \int_{\mathbb{S}} p_{e_1}\, d\sigma = c\|y\|^2.$$

Thus

$$\frac{1}{n}\sum_j |\langle y, \phi_j\rangle|^2 = \frac{1}{n}\sum_j p_y(\phi_j) = \int_{\mathbb{S}} p_y\, d\sigma = c\|y\|^2,$$

i.e., Φ is a tight frame.
(b) Since $\Pi_2^\circ(\mathbb{R}^d) = \mathrm{span}\{p_y : y \in \mathbb{R}^d\}$, it suffices to consider $p = p_y$

$$\int_{\mathbb{S}} p_y\, d\sigma = c\|y\|^2 = c\frac{1}{A}\sum_{j=1}^n |\langle y, \phi_j\rangle|^2 = \frac{c}{A}\sum_{j=1}^n p_y(\phi_j).$$

6.8 By Proposition 6.1, we seek a balanced equal-norm tight frame of n (distinct) vectors for $\mathbb{R}^d$. We must have $n > d$, since the sum of the vectors in an orthonormal basis is nonzero. Let $V = [v_1, \ldots, v_n]$ be a balanced normalised tight frame for $\mathbb{R}^d$. The condition V be balanced is that it is orthogonal to the normalised tight frame $U = [1/\sqrt{n}, \ldots, 1/\sqrt{n}]$ (see Lemma 5.1). Let $W = [w_1, \ldots, w_n]$ be the complement of the direct sum of U and V, which is a balanced tight frame for $\mathbb{R}^{n-1-d}$. Since there is no balanced tight frame of n vectors for $\mathbb{R}^1$ when n is odd, we cannot have $n-1-d = 1$ and n odd. In all the other cases, two designs can be obtained by taking an *unlifted real cyclic harmonic frame* of n vectors for $\mathbb{R}^d$ (see Corollary 11.3).

6.9 (a) For $t=1$ or $d=1$, we have $c_t(d,\mathbb{R})=c_t(d,\mathbb{C})$. For $t>1$, we have

$$\frac{c_t(d,\mathbb{C})}{c_t(d,\mathbb{R})}=\frac{c_{t-1}(d,\mathbb{C})\frac{t}{d+t-1}}{c_{t-1}(d,\mathbb{R})\frac{2t-1}{d+2(t-1)}}=\frac{c_{t-1}(d,\mathbb{R})}{c_{t-1}(d,\mathbb{C})}\frac{t(d+2t-2)}{t(d+2t-2)+(t-1)(d-1)},$$

and so, by induction, there is strict inequality for $t>1$, $d>1$.
(b) Using $c_{\frac{t}{2}}(d,\mathbb{R})=\frac{1\cdot3\cdot5\cdots(t-1)}{d(d+2)\cdots(d+t-2)}$, $x^{\otimes\frac{t}{2}}\otimes\overline{x}^{\otimes\frac{t}{2}}=x^{\otimes t}$, $\sum_k\|f_k\|^{2t}=n$, we get the following conditions, which are equivalent to being an order t spherical half-design,

$$\sum_{j=1}^n\sum_{k=1}^n\langle f_j,f_k\rangle^t=n^2\frac{1\cdot3\cdot5\cdots(t-1)}{d(d+2)\cdots(d+t-2)},$$

$$\|x\|^t=\langle x,x\rangle^{\frac{t}{2}}=\frac{1}{n}\frac{d(d+2)\cdots(d+t-2)}{1\cdot3\cdot5\cdots(t-1)}\sum_{j=1}^n|\langle x,f_j\rangle|^t,\qquad\forall x\in\mathbb{R}^d,$$

$$\int_{\mathbb{S}}x^{\otimes t}\,d\sigma(x)=\frac{1}{n}\sum_{j=1}^n f_j^{\otimes t}.$$

(c) For a monomial x^α, $(-x)^\alpha=(-1)^{|\alpha|}x^\alpha$, so the space of odd polynomials is $\Pi_1^\circ\oplus\Pi_3^\circ\oplus\Pi_5^\circ\oplus\cdots$, and by (6.26), we have

$$\int_{\mathbb{S}}x^\alpha\,d\sigma(x)=0=\frac{1}{n}\sum_{j=1}^{n/2}\{(f_j)^\alpha+(-f_j)^\alpha\},\qquad|\alpha|=1,3,5,\ldots.$$

Thus (6.33) integrates the odd monomials, and hence all odd polynomials.

6.10 (a) The claim is that for all bivariate polynomials $p\in\Pi_{n-1}(\mathbb{R}^2)$

$$\int_{\mathbb{S}(\mathbb{R}^2)}p(x,y)\,d\sigma(x,y)=\frac{1}{2\pi}\int_0^{2\pi}p(\cos\theta,\sin\theta)\,d\theta=\frac{1}{n}\sum_{j=1}^n p\Big(\cos\frac{2\pi}{n}j,\sin\frac{2\pi}{n}j\Big).$$

By identifing $(x,y)\in\mathbb{R}^2$ with $z=x+iy\in\mathbb{C}$, this can be written as

$$\int_{\mathbb{S}(\mathbb{C})}q(z,\overline{z})\,d\sigma(z)=\frac{1}{n}\sum_{j=1}^n q(\omega^j,\overline{\omega^j}),\qquad\omega:=e^{\frac{2\pi}{n}},\quad\forall q\in\Pi_{n-1}(\mathbb{R}^2).$$

Let $q(z,\overline{z}):=z^\alpha\overline{z}^\beta$, $\alpha+\beta<n$. Then $\alpha=\beta$ or $\alpha-\beta\not\equiv0\pmod n$, so that

$$\frac{1}{n}\sum_{j=1}^n q(z,\overline{z})=\frac{1}{n}\sum_{j=1}^n(\omega^j)^\alpha(\omega^{-j})^\beta=\frac{1}{n}\sum_{j=1}^n\omega^{(\alpha-\beta)j}=\begin{cases}0,&\alpha\neq\beta;\\1,&\alpha=\beta.\end{cases}$$

This equals the integral $\int_{\mathbb{S}(\mathbb{C})}z^\alpha\overline{z}^\beta\,d\sigma(z)$, which is given by taking $d=1$ in (6.25).
(b) It suffices to show (6.9), i.e., that (w_j) is a spherical (t,t)-design for $\mathbb{R}^2$. Using the trigonometric identity $\cos^2\theta=\frac{\cos2\theta+1}{2}$ and the cubature rule of (a), we have

$$\sum_j\sum_k |\langle w_j,w_k\rangle|^{2t} = \sum_j\sum_k \Big(\cos j\frac{\pi}{n}\cos k\frac{\pi}{n} + \sin j\frac{\pi}{n}\sin k\frac{\pi}{n}\Big)^{2t}$$
$$= \sum_j\sum_k \Big(\cos(j-k)\frac{\pi}{n}\Big)^{2t} = \sum_j\sum_k \Big(\frac{\cos\frac{2\pi}{n}(j-k)+1}{2}\Big)^t$$
$$= n^2\frac{1}{n}\sum_j \Big(\frac{\cos\frac{2\pi}{n}j+1}{2}\Big)^t = n^2\frac{1}{2\pi}\int_0^{2\pi}\Big(\frac{\cos\theta+1}{2}\Big)^t\,d\theta.$$

By (6.25), with $\alpha=(t,0)$ and $d=2$, the integral simplifies to

$$\frac{1}{2\pi}\int_0^{2\pi}\Big(\frac{\cos\theta+1}{2}\Big)^t\,d\theta = \frac{1}{\pi}\int_0^{2\pi}\Big(\cos\frac{\theta}{2}\Big)^{2t}\frac{d\theta}{2} = \frac{1}{\pi}\int_0^{\pi}(\cos x)^{2t}\,dx$$
$$= \frac{1}{2\pi}\int_0^{2\pi}(\cos x)^{2t}\,dx = \frac{(\frac12)_t}{(\frac22)_t} = \frac12\cdot\frac34\cdots\frac{2t-3}{2t-2}\cdot\frac{2t-1}{2t} = c_t(d,\mathbb{R}).$$

6.11 We verify (6.39), noting that $c_2(d,\mathbb{C}) = \frac{2}{d(d+1)}$, $c_3(d,\mathbb{C}) = \frac{6}{d(d+1)(d+2)}$.
(a) $d^2\Big\{(d^2-1)\Big(\frac{1}{\sqrt{d+1}}\Big)^4 + 1^4\Big\} = \frac{2d^3}{d+1} = \frac{2}{d(d+1)}(d^2)^2$.
(b) $md\Big\{(md-d)\Big(\frac{1}{\sqrt{d}}\Big)^4 + 1^4 + (d-1)0^4\Big\} = m(m+d-1) = \frac{2}{d(d+1)}(md)^2$ if and only if $m=d+1$ ($d\neq 1$).
(c) $d(d+1)\Big\{(d(d+1)-d)\Big(\frac{1}{\sqrt{d}}\Big)^6 + 1^6 + (d-1)0^6\Big\} = (d+1)^2 = \frac{6d^2(d+1)^2}{d(d+1)(d+2)}$ if and only if $d=2$.

6.12 Verify that $40(1+27(\frac{1}{\sqrt3})^{2t}) = c_t(4,\mathbb{C})40^2$ holds for $t=1,2,3$.

6.13 The common angle is $|\langle v_j,v_k\rangle| = \frac1d\sqrt{\frac{d(n-d)}{n-1}} = \frac{1}{\sqrt{d+2}}$, $j\neq k$ (see Exer. 2.16). Thus

$$\sum_j\sum_k |\langle v_j,v_k\rangle|^4 = (n^2-n)\Big(\frac1d\sqrt{\frac{d(n-d)}{n-1}}\Big)^4 + n,$$
$$c_2(d,\mathbb{R})\Big(\sum_\ell \|v_\ell\|^4\Big)^2 = \frac{1\cdot 3}{d(d+2)}n^2,$$

and a simple calculation shows that these are equal if and only if $n=\frac12 d(d+1)$.

6.14 (a) It is easy to see the vectors in $\Phi_0=(v_j)$ (64 of type 1 and 56 of type 2) satisfy $|\langle v_j,v_k\rangle|\in\{0,1,2\}$. For v_j fixed, a simple calculation shows that the number of times $|\langle v_j,v_k\rangle|$ takes the values $0,1,2$ is $63,56,1$. Thus we verify (6.39) as follows

$$\sum_j\sum_k |\langle v_j,v_k\rangle|^6 = 120(63\cdot 0^6 + 56\cdot 1^6 + 1\cdot 2^6)$$
$$= \frac{1\cdot3\cdot5}{8\cdot10\cdot12}(120\cdot 2^3)^2 = c_3(8,\mathbb{R})\Big(\sum_\ell \|v_\ell\|^6\Big)^2.$$

(b) By (a), Φ_0 and hence Φ integrates all homogeneous polynomials of degree 6, and therefore of degrees $0,2,4$ also. Since Φ is centrally symmetric, it integrates all homogeneous polynomials of odd order, and in particular those of orders $1,3,5,7$.

6.15 Take $f = x^{\otimes t}$, $g = y^{\otimes t}$ in the Plancherel identity (2.4) for the tight frame for $\mathrm{Sym}^t(\mathscr{H})$, using (2.9) and (6.16) to determine A

$$\langle x^{\otimes t}, y^{\otimes t}\rangle = \frac{1}{A}\sum_j \langle x^{\otimes t}, f_j^{\otimes t}\rangle\langle f_j^{\otimes t}, y^{\otimes t}\rangle, \qquad \binom{t+d-1}{t}A = \sum_\ell \|f_\ell^{\otimes t}\|^2.$$

After simplification by (6.17) this gives (a). Setting $y = x$ gives (b).

6.16 Recall (Exer. 2.11) that $\overline{\mathscr{H}}$ has inner product given by $\langle \overline{x}, \overline{y}\rangle := \overline{\langle x,y\rangle}$, $x,y \in \mathscr{H}$, and so the apolar inner product on $\mathrm{Sym}^t(\mathscr{H}) \otimes \mathrm{Sym}^t(\overline{\mathscr{H}})$ satisfies

$$\langle x^{\otimes t} \otimes \overline{y}^{\otimes t}, v^{\otimes t} \otimes \overline{w}^{\otimes t}\rangle_\circ = \langle x,v\rangle^t \langle \overline{y}, \overline{w}\rangle^t = (\langle x,v\rangle\overline{\langle y,w\rangle})^t, \qquad x,y,v,w \in \mathscr{H}.$$

Write $\xi = a - b$. Using (6.29), we obtain

$$\begin{aligned}
\langle a,a\rangle &= \int_{\mathbb{S}}\int_{\mathbb{S}} \langle x^{\otimes t} \otimes \overline{x}^{\otimes t}, y^{\otimes t} \otimes \overline{y}^{\otimes t}\rangle_\circ \, d\sigma(x)\, d\sigma(y) = \int_{\mathbb{S}}\int_{\mathbb{S}} |\langle x,y\rangle|^{2t}\, d\sigma(x)\, d\sigma(y) \\
&= \int_{\mathbb{S}} \|y\|^{2t} c_t(d,\mathbb{F})\, d\sigma(y) = c_t(d,\mathbb{F}), \\
\langle b,b\rangle &= \frac{1}{C^2}\sum_j\sum_k \langle f_j^{\otimes t} \otimes \overline{f_j}^{\otimes t}, f_k^{\otimes t} \otimes \overline{f_k}^{\otimes t}\rangle_\circ = \frac{1}{C^2}\sum_j\sum_k |\langle f_j, f_k\rangle|^{2t}, \\
\langle a,b\rangle &= \frac{1}{C}\sum_j \int_{\mathbb{S}} \langle x^{\otimes t} \otimes \overline{x}^{\otimes t}, f_j^{\otimes t} \otimes \overline{f_j}^{\otimes t}\rangle_\circ \, d\sigma(x) = \frac{1}{C}\sum_j \int_{\mathbb{S}} |\langle x, f_j\rangle|^{2t}\, d\sigma(x) \\
&= \frac{1}{C}\sum_j \|f_j\|^{2t} c_t(d,\mathbb{F}) = c_t(d,\mathbb{F}) = \langle b,a\rangle,
\end{aligned}$$

which combine to give the result for ξ. Let $P_x := \langle \cdot, x^{\otimes t}\rangle x^{\otimes t}$. The Frobeneous inner product between P_x and P_y is (cf. Exer. 3.1)

$$\langle P_x, P_y\rangle_F = \mathrm{trace}(P_x P_y^*) = \langle x^{\otimes t}, y^{\otimes t}\rangle\langle y^{\otimes t}, x^{\otimes t}\rangle = |\langle x,y\rangle|^{2t}.$$

Using this, a simple modification of the above argument gives the result for

$$Q = \int_{\mathbb{S}} P_x\, d\sigma(x) - \frac{1}{C}\sum_{j=1}^n P_{f_j}.$$

6.17 (a) The linear functionals $f \mapsto \partial^\alpha \overline{\partial}^\beta f(0)$, $|\alpha| = t, |\beta| = r$ are dual to these monomials. Thus counting gives the dimension.

(b) Recall (Exer. 2.11) that $\langle \overline{v}, \overline{w}\rangle := \langle w, v\rangle$. Thus the inner product induced by taking the apolar inner product on $\mathrm{Sym}^t(\mathscr{H}^*)$ and $\mathrm{Sym}^r(\overline{\mathscr{H}^*})$ is given by

$$\begin{aligned}\langle\langle\cdot,v\rangle^t\langle x,\cdot\rangle^r,\langle\cdot,w\rangle^t\langle y,\cdot\rangle^r\rangle_\circ &= \langle\langle\cdot,v\rangle^{\otimes t}\otimes\langle\bar{\cdot},\overline{x}\rangle^{\otimes t},\langle\cdot,w\rangle^{\otimes t}\otimes\langle\bar{\cdot},\overline{y}\rangle^{\otimes r}\rangle\\ &= \langle\langle\cdot,v\rangle,\langle\cdot,w\rangle\rangle^t\langle\langle\bar{\cdot},\overline{x}\rangle,\langle\bar{\cdot},\overline{y}\rangle\rangle^r = \langle w,v\rangle^t\langle\overline{y},\overline{x}\rangle^r. = \langle w,v\rangle^t\langle x,y\rangle^r.\end{aligned}$$

(c) Let $p=\langle\cdot,v\rangle^t\langle x,\cdot\rangle^r$ and $y=w$ in (6.71), to obtain

$$\langle p,\langle\cdot,w\rangle^t\langle w,\cdot\rangle^r\rangle_\circ = \langle w,v\rangle^t\langle x,w\rangle^r = p(w).$$

(d) If $p\in\Pi^\circ_{t,r}(\mathbb{C}^d)$ is orthogonal to P, then by (c) we have

$$p(w)=\langle p,\langle\cdot,w\rangle^t\langle w,\cdot\rangle^r\rangle=0,\quad\forall w\quad\Longrightarrow\quad P^\perp=0\quad\Longrightarrow\quad\Pi^\circ_{t,r}(\mathbb{C}^d)=P.$$

(e) It suffices to consider $p=\langle\cdot,v\rangle^t\langle x,\cdot\rangle^r$, $q=\langle\cdot,w\rangle^t\langle y,\cdot\rangle^r$. We have

$$\langle\partial,v\rangle(\langle z,\overline{w}\rangle)=(\textstyle\sum_j\overline{v}_j\partial_j)(\sum_k w_kz_k)=\overline{v}_1w_1+\cdots+\overline{v}_dw_d=\langle w,v\rangle,$$

$$\langle\overline{\partial},\overline{x}\rangle(\langle\overline{z},y\rangle)=(\textstyle\sum_j x_j\overline{\partial}_j)(\sum_k\overline{y}_k\overline{z}_k)=x_1\overline{y}_1+\cdots+x_d\overline{y}_d=\langle x,y\rangle.$$

Since $\tilde{q}(z)=\overline{\langle\overline{z},w\rangle^t\langle y,\overline{z}\rangle^r}=\langle z,\overline{w}\rangle^t\langle\overline{z},y\rangle^r$, and $p(\partial)\tilde{q}$ is a constant, we obtain

$$p(\partial)\tilde{q}(z)=\langle\partial,v\rangle^t\langle\overline{\partial},\overline{x}\rangle^r(\langle z,\overline{w}\rangle^t\langle\overline{z},y\rangle^r)=t!\langle w,v\rangle^t r!\langle x,y\rangle^r=t!r!\langle p,q\rangle_\circ.$$

6.18 Making the substitution (6.41) in Theorem 6.7 gives:

(a) $c_t(d,\mathbb{F})\|x\|^{2t}=\displaystyle\sum_{j=1}^n w_j|\langle x,\phi_j\rangle|^{2t},\quad x\in\mathbb{F}^d.$

(b) $c_t(d,\mathbb{F})\langle x,y\rangle^t=\displaystyle\sum_{j=1}^n w_j\langle x,\phi_j\rangle^t\langle\phi_j,y\rangle^t,\quad x\in\mathbb{F}^d.$

(c) $\displaystyle\int_{\mathbb{S}}p(x)\,d\sigma(x)=\sum_{j=1}^n w_j\,p(\phi_j),\quad\forall p\in\Pi^\circ_{t,t}(\mathbb{S}).$

(d) $\displaystyle\int_{\mathbb{S}}x^{\otimes t}\otimes\overline{x}^{\otimes t}\,d\sigma(x)=\sum_{j=1}^n w_j\,(\phi_j^{\otimes t}\otimes\overline{\phi_j}^{\otimes t}).$

(e) $\displaystyle\int_{\mathbb{S}}\langle\cdot,x^{\otimes t}\rangle x^{\otimes t}\,d\sigma(x)=\sum_{j=1}^n w_j\,\langle\cdot,\phi_j^{\otimes t}\rangle\phi_j^{\otimes t}.$

(f) $\displaystyle\int_{\mathbb{S}}\int_{\mathbb{S}}g(|\langle x,y\rangle|^2)\,d\sigma(y)\,d\sigma(x)=\sum_{j=1}^n\sum_{k=1}^n w_jw_k\,g(|\langle\phi_j,\phi_k\rangle|^2),\quad\forall g\in\Pi_t(\mathbb{R}).$

6.19 (a) Since $\|x\|^{2t}=(x_1+\cdots+x_d)^t$, $\langle x,v\rangle^{2t}=(v_1x_1+\cdots v_dx_d)^{2t}$, we have

$$\frac{\partial}{\partial x_j}\|x\|^{2t}=2tx_j\|x\|^{t-1},\quad\Big(\frac{\partial}{\partial x_j}\Big)^2\|x\|^{2t}=2t\|x\|^{t-1}+2tx_j(2(t-1)x_j\|x\|^{t-2}),$$

$$\frac{\partial}{\partial x_j}\langle x,v\rangle^{2t}=2tv_j\langle x,v\rangle^{2t-1},\quad\Big(\frac{\partial}{\partial x_j}\Big)^2\langle x,v\rangle^{2t}=2tv_j(2(t-1)v_j)\langle x,v\rangle^{2t-2}.$$

(b) Since $\|z\|^2=z_1\overline{z}_1+\cdots+z_d\overline{z}_d$, $|\langle z,v\rangle|^2=(z_1\overline{v}_1+\cdots+z_d\overline{v}_d)(\overline{z}_1v_1+\cdots+\overline{z}_dv_d)$

$$\partial_j(\|z\|^{2t}) = t\overline{z}_j\|z\|^{2(t-1)}, \quad \overline{\partial}_j\partial_j(\|z\|^{2t}) = t\|z\|^{2t-2} + t\overline{z}_j(t-1)z_j\|z\|^{2(t-2)},$$

$$\partial_j(|\langle z,v\rangle|^{2t}) = t|\langle z,v\rangle|^{2(t-1)}\overline{v}_j\langle\overline{z},\overline{v}\rangle,$$

$$\overline{\partial}_j\partial_j(|\langle z,v\rangle|^{2t}) = t|\langle z,v\rangle|^{2(t-1)}\overline{v}_jv_j + t(t-1)|\langle z,v\rangle|^{2(t-2)}\langle z,v\rangle v_j\overline{v}_j\langle\overline{z},\overline{v}\rangle.$$

(c) Applying the Laplacian in each case gives

$$c_t(d,\mathbb{R})2t(d+2t-2)\|x\|^{2(t-1)} = \frac{1}{\sum_\ell \|f_\ell\|^{2t}} \sum_{j=1}^n 2t(2t-1)|\langle x,f_j\rangle|^{2(t-1)}\|f_j\|^2,$$

$$c_t(d,\mathbb{C})4t(t+d-1)\|z\|^{2(t-1)} = \frac{1}{\sum_\ell \|f_\ell\|^{2t}} \sum_{j=1}^n 4t^2|\langle z,f_j\rangle|^{2(t-1)}\|f_j\|^2,$$

which is the Bessel identity for t replaced by $t-1$, since

$$c_t(d,\mathbb{R})\frac{d+2t-2}{2t-1} = c_{t-1}(d,\mathbb{R}), \qquad c_t(d,\mathbb{C})\frac{t+d-1}{t} = c_{t-1}(d,\mathbb{C}).$$

Remark. This (with induction) provides an alternative proof of Proposition 6.2.

6.20 We verify the generalised Bessel identity (6.31). Let $C = \sum_j \|v_j\|^{2t} = \sum_k \|w_k\|^{2t}$. Then

$$\begin{aligned} c_t(d,\mathbb{F})\|x\|^{2t} &= \frac{1}{2}\Big(\frac{1}{C}\sum_j |\langle x,v_j\rangle|^{2t} + \frac{1}{C}\sum_k |\langle x,w_k\rangle|^{2t}\Big) \\ &= \frac{1}{2C}\Big(\sum_j |\langle x,v_j\rangle|^{2t} + \sum_k |\langle x,w_k\rangle|^{2t}\Big), \quad \forall x \in \mathbb{F}^d, \end{aligned}$$

where $2C = \sum_j \|v_j\|^{2t} + \sum_k \|w_k\|^{2t}$.

6.21 Suppose that $(f_j)_{j=1}^n$ is a weighted (t,t)-design with the minimal number of vectors. We claim that the polynomials $x \mapsto |\langle x,f_j\rangle|^{2t}$ in $\Pi_{t,t}^\circ(\mathbb{F}^d)$ are linearly independent and hence obtain the result by a dimension count using (6.16) and (6.70). First scale (f_j) so that the generalised Bessel identity (6.31) becomes

$$\|x\|^{2t} = \sum_{j=1}^n |\langle x,f_j\rangle|^{2t}, \qquad \forall x \in \mathbb{F}^d.$$

Now, by way of contradiction, suppose that $\sum_j c_j|\langle x,f_j\rangle|^{2t} = 0$, where $c_j \in \mathbb{R}$, and, without loss of generality, $c_j \le c_n = 1$, $\forall j$. Subtracting this from the above gives

$$\|x\|^{2t} = \sum_{j=1}^{n-1}(1-c_j)|\langle x,f_j\rangle|^{2t} = \sum_{j=1}^{n-1} |\langle x,v_j\rangle|^{2t}, \qquad \forall x \in \mathbb{F}^d.$$

where $v_j := (1-c_j)^{\frac{1}{2t}} f_j$. Thus, by the generalised Bessel identity (6.31), $(v_j)_{j=1}^{n-1}$ is a (t,t)-design of $n-1$ vectors for $\mathbb{F}^d$, which contradicts the minimality of n.

6.22 The generalised Plancherel identity (6.32) implies the holomorphic polynomial $x \mapsto \langle x, y\rangle^t$ is in the span of the polynomials $x \mapsto \langle x, f_j\rangle^t$, $1 \le j \le n$. By Exer. 6.17, the polynomials $\{x \mapsto \langle x, y\rangle : y \in \mathbb{F}^d\}$ span $\Pi_t^\circ(\mathbb{C}^d) = \Pi_{t,0}^\circ(\mathbb{C}^d)$, with a similar result for $\mathbb{R}^d$, and so we have

$$n \ge \dim(\Pi_t^\circ(\mathbb{F}^d)) = \binom{t+d-1}{d-1}.$$

6.23 Suppose that (f_j) is the tight frame corresponding to a (t,t)-design (g_j), i.e., $f_j = \|g_j\|^{t-1} g_j$ (see Proposition 6.2). Taking norms gives $\|f_j\| = \|g_j\|^t$, and hence $g_j = f_j / \|f_j\|^{\frac{t-1}{t}}$. Substituting this into (6.39) gives the condition

$$\sum_{j=1}^n \sum_{k=1}^n \Big|\Big\langle \frac{f_j}{\|f_j\|^{\frac{t-1}{t}}}, \frac{f_k}{\|f_k\|^{\frac{t-1}{t}}}\Big\rangle\Big|^{2t} = c_t(d,\mathbb{F}) \Big(\sum_{\ell=1}^n \Big\| \frac{f_\ell}{\|f_\ell\|^{\frac{t-1}{t}}} \Big\|^{2t}\Big)^2.$$

6.24 (a) $\langle P_x, P_y\rangle = \operatorname{trace}(xx^*yy^*) = \operatorname{trace}(y^*xx^*y) = \langle x,y\rangle\langle y,x\rangle = |\langle x,y\rangle|^2$.
(b) By (a), we have $\rho(P_x,P_y)^2 = \langle P_x - P_y, P_x - P_y\rangle = 2 - 2\langle P_x, P_y\rangle$.

(c) By (a) and (b), we have $\rho(P_x,P_y)^2 = 2 - 2\langle P_x,P_y\rangle = 2 - 2|\langle \frac{x}{\|x\|}, \frac{y}{\|y\|}\rangle|^2$.
(d) Clearly $\operatorname{trace}(P_x - \frac{1}{d}I) = 1 - \frac{1}{d}d = 0$, and we calculate
$\|(P_x - \frac{1}{d}I) - (P_y - \frac{1}{d}I)\|^2 = \langle P_x - P_y, P_x - P_y\rangle = 2 - 2\langle P_x, P_y\rangle = \rho(P_x,P_y)^2$.

6.25 For $\mathbb{F} = \mathbb{R}, \mathbb{C}$, each $p \ne 0$ can be expressed as a finite sum of monomials:

$$p(x) = \sum_j c_j x^j \quad (c_j \ne 0), \qquad p(z) = \sum_{(\alpha,\beta)} c_{\alpha,\beta} z^\alpha \overline{z}^\beta \quad (c_{\alpha,gb} \ne 0).$$

Since these monomials are linearly independent, we have

$$p(ax) = \sum_j (a^j c_j) x^j = p(x), \quad \forall a \quad \Longrightarrow \quad c_j a^j = c_j, \ \forall a \quad \Longrightarrow \quad p \text{ is even},$$

$$p(az) = \sum_{(\alpha,\beta)} (a^{|\alpha|} \overline{a}^{|\beta|} c_{\alpha,\beta}) z^\alpha \overline{z}^\beta = p(z) \quad \Longrightarrow \quad c_{\alpha,\beta} a^{|\alpha|-|\beta|} = c_{\alpha,\beta}, \ \forall a$$

so that $|\alpha| = |\beta|$ for $c_{\alpha,\beta} \ne 0$. Hence in both cases $p \in \oplus_j \Pi_{j,j}$.

6.26 Suppose that $f(x+a) - f(a) = b$ has two solutions j and k (for some a and b). Then

$$y^{j+a} - y^j = b = y^{k+a} - y^k \quad \Longrightarrow \quad (y^j - y^k)(y^a - 1) = 0,$$

so that either $j = k$ or $a = 0$. Thus for $(a,b) \ne (0,0)$, there is at most one solution to $f(x+a) - f(a) = b$, and so f is 1–uniform.

6.27 The equations have $d + 2d(d-1) = d(2d-1)$ solutions of the form

$$(w,x,y,z) = (w,w,w,w), \qquad (w,x,y,z) = (w,x,w,x) \text{ or } (w,x,x,w).$$

We can rewrite the equations as

$$w-z=y-x=a, \qquad f(w)-f(z)=f(y)-f(x)=b,$$

for some $a \in G$, $b \in H$. We now treat cases. For $a=0$, we have $z=w$, $y=x$. For $a \neq 0$, we have $w=z+a$, $y=x+a$, so that

$$f(z+a)-f(z)=f(x+a)-f(x)=b$$

and the 1–uniformity of f implies $z=x$, and so $w=y$. Thus there are only the $d(2d-1)$ solutions first mentioned.

6.28 (a) Subtracting the component of the force in the direction a gives

$$\mathrm{EFF}(a,b)=v-\langle v,a\rangle a=\langle a,b\rangle(\langle b,a\rangle a-b).$$

(b) Since a_j is a λ–eigenvector of the frame operator S and $\mathrm{EFF}(a_j,a_j)=0$,

$$\sum_{k\neq j}\mathrm{EFF}(a_j,a_k)=\sum_k\langle a_j,a_k\rangle(\langle a_k,a_j\rangle a_j-a_k)=\langle Sa_j,a_j\rangle a_j-Sa_j=\lambda a_j-\lambda a_j=0.$$

6.29 A simple calculation, using $\frac{\partial}{\partial\overline{z}}z=0$, $\frac{\partial}{\partial\overline{z}}\overline{z}=1$, gives

$$\frac{\partial}{\partial\overline{v_{\alpha\beta}}}\langle v_j,v_k\rangle=\frac{\partial}{\partial\overline{v_{\alpha\beta}}}\sum_s v_{sj}\overline{v_{sk}}=\delta_{k\beta}v_{\alpha j}.$$

Therefore (using the chain rule), we calculate

$$\begin{aligned}\frac{\partial p}{\partial\overline{v_{\alpha\beta}}}(V)&=\sum_j\sum_k t|\langle v_j,v_k\rangle|^{2(t-1)}\frac{\partial}{\partial\overline{v_{\alpha\beta}}}\Big(\langle v_j,v_k\rangle\langle v_k,v_j\rangle\Big)\\&=\sum_j\sum_k t|\langle v_j,v_k\rangle|^{2(t-1)}\Big(\delta_{k\beta}v_{\alpha j}\langle v_k,v_j\rangle+\langle v_j,v_k\rangle\delta_{j\beta}v_{\alpha k}\Big)\\&=\sum_j t|\langle v_j,v_\beta\rangle|^{2(t-1)}v_{\alpha j}\langle v_\beta,v_j\rangle+\sum_k t|\langle v_\beta,v_k\rangle|^{2(t-1)}\langle v_\beta,v_k\rangle v_{\alpha k}\\&=2t\sum_j|\langle v_j,v_\beta\rangle|^{2(t-1)}\langle v_\beta,v_j\rangle v_{\alpha j},\end{aligned}$$

$$\frac{\partial g}{\partial\overline{v_{\alpha\beta}}}(V)=\sum_\ell t\|v_\ell\|^{2(t-1)}\frac{\partial}{\partial\overline{v_{\alpha\beta}}}\langle v_\ell,v_\ell\rangle=\sum_\ell t\|v_\ell\|^{2(t-1)}\delta_{\ell\beta}v_{\alpha\ell}=t\|v_\beta\|^{2(t-1)}v_{\alpha\beta}.$$

6.30 To find the entries of the Hessian, we use the Wirtinger calculus:

$$\frac{\partial}{\partial\overline{z}}=\frac{1}{2}\Big(\frac{\partial}{\partial x}+i\frac{\partial}{\partial x}\Big),$$

which gives

$$\frac{\partial^2 h}{\partial x_{ab}\partial x_{\alpha\beta}}=2\Re\Big(\frac{\partial^2 h}{\partial\overline{v_{ab}}\partial x_{\alpha\beta}}\Big),\qquad\frac{\partial^2 h}{\partial y_{ab}\partial x_{\alpha\beta}}=2\Im\Big(\frac{\partial^2 h}{\partial\overline{v_{ab}}\partial x_{\alpha\beta}}\Big),$$

$$\frac{\partial^2 h}{\partial x_{ab}\partial y_{\alpha\beta}} = 2\Re\Big(\frac{\partial^2 h}{\partial \overline{v_{ab}}\partial y_{\alpha\beta}}\Big), \qquad \frac{\partial^2 h}{\partial y_{ab}\partial y_{\alpha\beta}} = 2\Im\Big(\frac{\partial^2 h}{\partial \overline{v_{ab}}\partial y_{\alpha\beta}}\Big).$$

We now take $h=p$ and $h=g$. By Exer. 6.29, we have

$$\frac{\partial p}{\partial x_{\alpha\beta}}(V) = 2\Re\Big(\frac{\partial p}{\partial \overline{v_{\alpha\beta}}}(V)\Big) = 2t\sum_j |\langle v_j, v_\beta\rangle|^{2(t-1)}\big(v_{\alpha j}\langle v_\beta, v_j\rangle + \overline{v_{\alpha j}}\langle v_j, v_\beta\rangle\big),$$

$$\frac{\partial g}{\partial x_{\alpha\beta}}(V) = 2\Re\Big(\frac{\partial g}{\partial \overline{v_{\alpha\beta}}}(V)\Big) = t\|v_\beta\|^{2(t-1)}(v_{\alpha\beta}+\overline{v_{\alpha\beta}}).$$

Differentiating these gives

$$\begin{aligned}
\frac{\partial^2 p}{\partial \overline{v_{ab}}\partial x_{\alpha\beta}}(V) &= 2t\sum_j |\langle v_j, v_\beta\rangle|^{2(t-1)}\frac{\partial}{\partial \overline{v_{ab}}}\big(v_{\alpha j}\langle v_\beta, v_j\rangle + \overline{v_{\alpha j}}\langle v_j, v_\beta\rangle\big)\\
&\quad + 2t\sum_j 2\Re\big(v_{\alpha j}\langle v_\beta, v_j\rangle\big)(t-1)|\langle v_j, v_\beta\rangle|^{2(t-2)}\frac{\partial}{\partial \overline{v_{ab}}}\big(\langle v_j, v_\beta\rangle\langle v_\beta, v_j\rangle\big)\\
&= 2t\sum_j |\langle v_j, v_\beta\rangle|^{2(t-1)}\big(v_{\alpha j}\delta_{bj}v_{a\beta} + \delta_{a\alpha}\delta_{bj}\langle v_j, v_\beta\rangle + \overline{v_{\alpha j}}\delta_{b\beta}v_{aj}\big)\\
&\quad + 4t\sum_j \Re\big(v_{\alpha j}\langle v_\beta, v_j\rangle\big)(t-1)|\langle v_j, v_\beta\rangle|^{2(t-2)}\big(\langle v_j, v_\beta\rangle\delta_{bj}v_{a\beta} + \delta_{b\beta}v_{aj}\langle v_\beta, v_j\rangle\big)\\
&= 2t|\langle v_b, v_\beta\rangle|^{2(t-1)}\big(v_{\alpha b}v_{a\beta} + \delta_{a\alpha}\langle v_b, v_\beta\rangle\big) + 2t\delta_{b\beta}\sum_j |\langle v_j, v_\beta\rangle|^{2(t-1)}\overline{v_{\alpha j}}v_{aj}\\
&\quad + 4t(t-1)|\langle v_b, v_\beta\rangle|^{2(t-2)}\Re\big(v_{\alpha b}\langle v_\beta, v_b\rangle\big)\big(v_{a\beta}\langle v_b, v_\beta\rangle\big)\\
&\quad + 4t(t-1)\delta_{b\beta}\sum_j |\langle v_j, v_\beta\rangle|^{2(t-2)}\Re\big(v_{\alpha j}\langle v_\beta, v_j\rangle\big)v_{aj}\langle v_\beta, v_j\rangle,\\
\frac{\partial^2 g}{\partial \overline{v_{ab}}\partial x_{\alpha\beta}}(V) &= t\|v_\beta\|^{2(t-1)}\frac{\partial}{\partial \overline{v_{ab}}}(v_{\alpha\beta}+\overline{v_{\alpha\beta}})\\
&\qquad + t(t-1)\|v_\beta\|^{2(t-2)}(v_{\alpha\beta}+\overline{v_{\alpha\beta}})\frac{\partial}{\partial \overline{v_{ab}}}\langle v_\beta, v_\beta\rangle\\
&= t\|v_\beta\|^{2(t-1)}\delta_{a\alpha}\delta_{b\beta} + t(t-1)\|v_\beta\|^{2(t-2)}(v_{\alpha\beta}+\overline{v_{\alpha\beta}})\delta_{b\beta}v_{a\beta}.
\end{aligned}$$

Similarly, since $2\Im(z) = i(\overline{z}-z)$, we have

$$\frac{\partial p}{\partial y_{\alpha\beta}}(V) = 2\Im\Big(\frac{\partial p}{\partial \overline{v_{\alpha\beta}}}(V)\Big) = 2t\sum_j |\langle v_j, v_\beta\rangle|^{2(t-1)}i\big(\overline{v_{\alpha j}}\langle v_j, v_\beta\rangle - v_{\alpha j}\langle v_\beta, v_j\rangle\big),$$

$$\frac{\partial g}{\partial y_{\alpha\beta}}(V) = 2\Im\Big(\frac{\partial g}{\partial \overline{v_{\alpha\beta}}}(V)\Big) = t\|v_\beta\|^{2(t-1)}i(\overline{v_{\alpha\beta}}-v_{\alpha\beta}).$$

Differentiating these gives

$$\frac{\partial^2 p}{\partial \overline{v_{ab}} \partial y_{\alpha\beta}}(V) = 2t \sum_j |\langle v_j, v_\beta\rangle|^{2(t-1)} i\big(\delta_{a\alpha}\delta_{bj}\langle v_j, v_\beta\rangle + \overline{v_{\alpha j}}\delta_{b\beta}v_{aj} - v_{\alpha j}\delta_{bj}v_{a\beta}\big)$$
$$+4t\sum_j \Im\big(v_{\alpha j}\langle v_\beta, v_j\rangle\big)(t-1)|\langle v_j, v_\beta\rangle|^{2(t-2)}\big(\langle v_j, v_\beta\rangle\delta_{bj}v_{a\beta} + \delta_{b\beta}v_{aj}\langle v_\beta, v_j\rangle\big)$$
$$= 2ti|\langle v_b, v_\beta\rangle|^{2(t-1)}\big(\delta_{a\alpha}\langle v_b, v_\beta\rangle - v_{\alpha b}v_{a\beta}\big) + 2ti\delta_{b\beta}\sum_j |\langle v_j, v_\beta\rangle|^{2(t-1)}\overline{v_{\alpha j}}v_{aj}$$
$$+4t(t-1)|\langle v_b, v_\beta\rangle|^{2(t-2)}\Im\big(v_{\alpha b}\langle v_\beta, v_b\rangle\big)\big(v_{a\beta}\langle v_b, v_\beta\rangle\big)$$
$$+4t(t-1)\delta_{b\beta}\sum_j |\langle v_j, v_\beta\rangle|^{2(t-2)}\Im\big(v_{\alpha j}\langle v_\beta, v_j\rangle\big)v_{aj}\langle v_\beta, v_j\rangle,$$
$$\frac{\partial^2 g}{\partial \overline{v_{ab}} \partial y_{\alpha\beta}}(V) = t\delta_{a\alpha}\delta_{b\beta} i\|v_\beta\|^{2(t-1)} + 2t\Im(v_{\alpha\beta})(t-1)\|v_\beta\|^{2(t-2)}\delta_{b\beta}v_{a\beta}.$$

From these formulas, one can calculate the Hessian matrix of p and g. To find the Hessian matrix of $f := p - c_t(d, \mathbb{F}^d)g^2$, use the fact

$$\frac{\partial^2}{\partial r \partial s}(g^2) = \frac{\partial}{\partial r}\Big(2g\frac{\partial g}{\partial s}\Big) = 2g(V)\frac{\partial^2 g}{\partial r \partial s} + 2\frac{\partial g}{\partial r}\frac{\partial g}{\partial s}, \qquad r, s \in X.$$

6.31 (a) Let $f_1 = (1,0)$, $f_2 = (1,1)$. Then

$$g(x,y) = \frac{2}{3}\frac{2x^2+2xy+y^2}{x^2+y^2}, \qquad g(1,0) = \frac{4}{3}, \quad g(0,1) = \frac{2}{3}.$$

(b) Expand $\|\sum_j c_j f_j\|^2 = \langle \sum_j c_j f_j, \sum_k c_k f_k\rangle = \sum_j \sum_k c_j \overline{c_k}\langle f_j, f_k\rangle$, so that

$$\|\sum_j c_j f_j\|^2 \le \sum_j\sum_k |c_j||c_k||\langle f_j, f_k\rangle| \le \sum_j\sum_k |c_j|\sqrt{|\langle f_j, f_k\rangle|}c_k\sqrt{|\langle f_j, f_k\rangle|}.$$

By Cauchy–Schwarz, using the fact that both norms are equal, we have

$$\|\sum_j c_j f_j\|^2 \le \sum_j\sum_k |c_j|^2|\langle f_j, f_k\rangle| = \sum_j |c_j|^2 \sum_k |\langle f_j, f_k\rangle|.$$

(c) Take $c_j = \langle x, f_j\rangle / (\sum_\ell |\langle f_\ell, f_j\rangle|)$ in (b), and simplify.
(d) Use $\sum_\ell |\langle f_\ell, f_j\rangle| \le \max_k \sum_\ell |\langle f_\ell, f_k\rangle|$, $\forall j$.
(e) In the argument, apply Cauchy–Schwarz in the form

$$\|\sum_j c_j f_j\|^2 \le \sum_j\sum_k |c_j||c_k||\langle f_j, f_k\rangle| \le \Big(\sum_j\sum_k |c_j|^2|c_k|^2\Big)^{\frac{1}{2}}\Big(|\langle f_j, f_k\rangle|^2\Big)^{\frac{1}{2}}$$
$$= \Big(\sum_j |c_j|^2\Big)\Big(|\langle f_j, f_k\rangle|^2\Big)^{\frac{1}{2}}.$$

Now take $c_j = \langle x, f_j\rangle$ in $|\sum_j \overline{c_j}\langle x, f_j\rangle|^2 \le \|x\|^2(\sum_j |c_j|^2)(|\langle f_j, f_k\rangle|^2)^{\frac{1}{2}}$.

Exercises of Chapter 7

7.1 (a) A block matrix calculation gives

$$(I\ 0)\begin{pmatrix} U_{11} & U_{12} \\ U_{21} & U_{22} \end{pmatrix} = (U_{11}\ U_{12}) = (I\ 0) \quad \Longrightarrow \quad U = \begin{pmatrix} I & 0 \\ 0 & U_{22} \end{pmatrix}.$$

(b) Since $\dim(U(\mathbb{F}^n)) - \dim(U(\mathbb{F}^{n-d}))$, we calculate

$$\dim(\mathcal{N}_{n,\mathbb{R}^d}) = \frac{1}{2}n(n-1) - \frac{1}{2}(n-d)(n-d-1) = \frac{1}{2}d(2n-d-1),$$

$$\dim(\mathcal{N}_{n,\mathbb{C}^d}) = n^2 - (n-d)^2 = d(2n-d) = 2\dim(\mathcal{N}_{n,\mathbb{R}^d}) + d.$$

7.2 (a) The derivative

$$f'(t) = \frac{2}{\sqrt{1-t^2}}\Big(t\sqrt{1-t^2}(\|v\|^2 - \|w\|^2) + (1-2t^2)\alpha\Big), \qquad t \neq 1,$$

can be zero only when

$$t^2(1-t^2)\beta = (1-2t^2)^2\alpha^2, \qquad \beta := (\|v\|^2 - \|w\|^2)^2.$$

For $4\alpha^2 + \beta \neq 0$, this gives

$$t^2 = \frac{1 \pm r}{2}, \qquad r := \sqrt{\frac{\beta}{\beta + 4\alpha^2}}.$$

For this t, we obtain

$$f(t) = \frac{1}{2}(\|v\|^2 + \|w\|^2) \pm \frac{r}{2}(\|v\|^2 - \|w\|^2) + \sqrt{1-r^2}\alpha.$$

This is clearly maximised by the choice for which $\pm(\|v\|^2 - \|w\|^2) = \sqrt{\beta}$ (and is minimised by the other choice). Thus possible local maximum and minimum values are

$$f(t) = \frac{1}{2}(\|v\|^2 + \|w\|^2) \pm \frac{r}{2}\sqrt{\beta} + \sqrt{1-r^2}\alpha = \frac{1}{2}(\|v\|^2 + \|w\|^2) + \frac{1}{2}\frac{\pm\beta + 4|\alpha|\alpha}{\sqrt{\beta + 4\alpha^2}}.$$

(b) The choices of α that give the maximum and minimum value of the possible local maximum value and possible local minimum value from (a) are

$$\alpha = \overline{\sigma}\langle v, w\rangle = |\langle v, w\rangle|, \qquad \alpha = \overline{\sigma}\langle v, w\rangle = -|\langle v, w\rangle|,$$

which give the values

$$f(t) = \frac{1}{2}(\|v\|^2 + \|w\|^2) \pm \frac{1}{2}\sqrt{\beta + 4\alpha^2}.$$

(c) We have $f(0) = \|w\|^2$, $f(1) = \|v\|^2$. Since $\sqrt{\beta + 4\alpha^2} \geq \sqrt{\beta} = |\|v\|^2 - \|w\|^2|$, the maximum and minimum values of (b) are the global maximum and minimum.

7.3 (a),(b) Since the vectors of $V^{(k)}$ satisfy (7.7), the condition (i) gives

$$\sum_{j=k+1}^{n} \|v_j^{(k)}\|^2 = \sum_{j=k+1}^{n} a_j.$$

Suppose that $\|v_{k+1}^{(k)}\|^2 < a_{k+1}$ and (a) does not hold, i.e., $\|v_j^{(k)}\| = 0$, $j \geq k+2$. Then

$$\sum_{j=k+1}^{n} \|v_j^{(k)}\|^2 = \|v_{k+1}^{(k)}\|^2 < a_{k+1} \leq \sum_{j=k+1}^{n} a_j,$$

which is a contradiction. Now suppose that $\|v_{k+1}^{(k)}\|^2 > a_{k+1}$ and (b) does not hold, i.e., $\|v_j^{(k)}\|^2 > a_{k+1}$, $\forall j \geq k+2$. Then we again obtain a contradiction

$$\sum_{j=k+1}^{n} \|v_j^{(k)}\|^2 > \sum_{j=k+1}^{n} a_{k+1} \geq \sum_{j=k+1}^{n} a_j.$$

Exercises of Chapter 8

8.1 (a) The forward implication follows by taking the inner product with f, and the reverse follows from the polarisation identity (as in Exer. 2.2).
(b) It follows from (8.30) that set of $c = (c_j)$ satisfying (8.29) is an affine subspace of $\mathbb{F}^n$ and so has an element of minimal ℓ_2-norm. If $c = (c_j)$ satisfies (8.30), then so does $(\Re(c_j))$. Since $|\Re(c_j)| \leq |c_j|$, the c with minimal ℓ_2-norm must have real entries. Taking the trace of the linear operators in (8.29) gives

$$\operatorname{trace}(I_{\mathscr{H}}) = \dim(\mathscr{H}) = \sum_j c_j \operatorname{trace}(\phi_j \phi^*) = \sum_j c_j \|\phi_j\|^2.$$

8.2 (a) S is Hermitian, and (8.31) gives $AI_{\mathscr{H}} \leq S \leq BI_{\mathscr{H}}$. Thus S has eigenvalues $0 < A \leq \lambda \leq B$, and so is invertible, with $(1/B)I_{\mathscr{H}} \leq S^{-1} \leq (1/A)I_{\mathscr{H}}$.
(b) The signed frame operator of (f_j), σ, can be written $S = V\Lambda_\sigma V^*$, where $V = [f_j]$ and Λ_σ is diagonal with diagonal entries σ. Thus the signed frame operator of $(\tilde{f}_j)$, σ is

$$(S^{-1}V)\Lambda_\sigma(S^{-1}V)^* = S^{-1}(V\Lambda_\sigma V^*)S^{-1} = S^{-1}SS^{-1} = S^{-1}.$$

Expanding $I_{\mathscr{H}} = SS^{-1} = S^{-1}S$ gives the signed frame expansion, e.g.,

$$f = S(S^{-1}f) = \sum_j \sigma_j \langle S^{-1}f, f_j \rangle f_j = \sum_j \sigma_j \langle f, S^{-1}f_j \rangle f_j = \sum_j \sigma_j \langle f, \tilde{f}_j \rangle f_j.$$

Since S is positive definite, the canonical tight signed frame $f_j^{\rm can} := S^{-1/2} f_j$ is well defined. The signed frame operator of $(f_j^{\rm can})$ is

$$(S^{-1/2}V)\Lambda_\sigma(S^{-1/2}V)^* = S^{-1/2}(V\Lambda_\sigma V^*)S^{-1/2} = S^{-1/2}SS^{-1/2} = I_{\mathscr{H}},$$

i.e., $(f - j^{\rm can}, \sigma$ is a tight signed frame.

8.3 Suppose that $(f_j)_{j=1}^n$ is a tight fusion frame for $\mathbb{F}^d$ with signature

$$\sigma_1 = \cdots = \sigma_m = 1, \qquad \sigma_{m+1} = \cdots = \sigma_n = -1.$$

Then, by Cauchy–Schwarz, we have

$$\|f\|^2 = \sum_{j=1}^n \sigma_j |\langle f, f_j\rangle|^2 \le \sum_{j=1}^m |\langle f, f_j\rangle|^2 \le \Big(\sum_{j=1}^m \|f_j\|^2\Big)\|f\|^2, \qquad \forall f \in \mathbb{F}^d,$$

so that $(f_1, \ldots, f_m)$ is frame for $\mathbb{F}^d$, and hence we must have $m \ge d$.

Now let $(f_1, \ldots, f_m)$, $m \ge d$, be any normalised tight frame for $\mathbb{F}^d$ and σ be the signature above. Using the definitions and results of Exercise 8.2, we observe that $(f_1, \ldots, f_n)$ will be a signed frame for $\mathbb{F}^d$ with signature σ provided that

$$\sum_{j=1}^n \sigma_j |\langle f, f_j\rangle|^2 = \|f\|^2 - \sum_{j=m+1}^n |\langle f, f_j\rangle|^2 \ge \Big(1 - \sum_{j=m+1}^n \|f_j\|^2\Big)\|f\|^2 \ge A\|f\|^2, \quad \forall f,$$

for some $A > 0$. This is easy achieved by choosing $(f_{m+1}, \ldots, f_n)$ sufficiently small. The canonical tight frame for this signed frame is then a tight signed frame with the desired signature.

8.4 We consider the (j,k)-entries

$$\begin{aligned}\big((aa^*)\circ(bb^*)\big)_{jk} &= (aa^*)_{jk}(bb^*)_{jk} = (a_{j1}\overline{a_{k1}})(b_{j1}\overline{b_{k1}}) = (a_{j1}b_{j1})(\overline{a_{k1}}\overline{b_{k1}}) \\ &= (a\circ b)_{j1}\big((a\circ b)^*\big)_{1k} = \big((a\circ b)(a\circ b)^*\big)_{jk}.\end{aligned}$$

8.5 We recall (see Exer. 6.17) that $\Pi_{r,s}^\circ(\mathbb{C}^d)$ has dimension n, and p_j is the Riesz representer of the point evaluation δ_j (for the apolar inner product). Thus (a) and (b) are equivalent. Suppose that (a), and hence (b), holds. Then each $p \in \Pi_{r,s}^\circ(\mathbb{C}^d)$ has a unique expansion $p = \sum_j c_k p_k$, where the coefficients are uniquely determined by the linear system

$$\sum_k c_k p_k(v_j) = p(v_j), \quad 1 \le j \le n,$$

i.e., $Ac = [p(v_j)]$, where $a_{jk} = \delta_j(v_k) = \langle v_j, v_k\rangle^r \langle v_k, v_j\rangle^s = \langle v_j, v_k\rangle^r \overline{\langle v_j, v_k\rangle}^s$. Since $[p(v_j)] = [\delta_j(p)]$ can be any vector in $\mathbb{C}^n$, we conclude A is invertible. Conversely, suppose that $A = [\delta_j(v_k)]$ is invertible, then its rows are linearly independent. Since the rows of A are the linear functionals δ_j restricted to (v_k), we conclude that the δ_j are linearly independent and hence are basis for $\Pi_{r,s}^\circ(\mathbb{C}^d)'$.

Exercises of Chapter 9

9.1 (a) Let Φ and Ψ be orthonormal bases for $\mathbb{C}^{d_1}$ and $\mathbb{C}^{d_2}$, then $\Phi \cup \Psi$ is an orthonormal basis for $\mathbb{C}^{d_1+d_2}$, and we have $S_{d_1} \times S_{d_2} \subsetneq S_{d_1+d_2}$.
(b) If $\Psi = \Phi$, then there are additional symmetries which swap the two copies of Φ.
(c) Let Φ and Ψ be orthonormal bases for $\mathbb{C}^{d_1}$ and $\mathbb{C}^{d_2}$, then $\Phi \otimes \Psi$ is an orthonormal basis for $\mathbb{C}^{d_1 d_2}$, and we have $S_{d_1} \times S_{d_2} \subsetneq S_{d_1 d_2}$.
(d) Let Φ and Ψ be complements, so that $\Phi \oplus \Psi$ is a basis, and so every permutation gives a symmetry of $\Phi \oplus \Psi$.

9.2 (a) This is immediate, since $\mathrm{Sym}(\Phi)$ is a subgroup of S_n, which has order $n!$.
(b) If $\pi_\Phi(\sigma) = L_\sigma = I$, then $f_{\sigma j} = f_j$, $\forall j$ (since the vectors are distinct), so that σ must be the identity. Each $L_\sigma = \pi_\Phi(\sigma)$ is determined by its action on a basis of vectors for $\mathscr{H}$ taken from (f_j), which gives another basis of vectors in (f_j). If there are only m vectors in (f_j), then number of possible choices for the image of the first basis vector is m, for the second $m-1$, and so on (for the d elements of the basis).

9.3 (a) Let $\Phi = (f_j)$ be a finite frame for $\mathscr{H}$. Then for $\sigma, \tau \in \mathrm{Sym}(\Phi)$, we have

$$(L_\sigma L_\tau) f_j = L_\sigma (L_\tau f_j) = L_\sigma f_{\tau j} = f_{\sigma\tau j} = L_{\sigma\tau} f_j,$$

which implies $\pi_\Phi(\sigma\tau) = \pi_\Phi(\sigma)\pi_\Phi(\tau)$, since (f_j) spans $\mathscr{H}$.
(b) We calculate

$$S_\Phi(gf) = \sum_j \langle gf, f_j \rangle f_j = (g^*)^{-1} \sum_j \langle f, g^* f_j \rangle g^* f_j = (g^*)^{-1} S_{g^*\Phi}(f), \quad \forall f \in \mathscr{H}.$$

(c) If Φ is tight, then $g = L_\sigma$ is unitary, i.e., $g^* = g^{-1}$, and from (b) we have

$$S_\Phi(\sigma f) = S_\Phi(gf) = g S_{g^{-1}\Phi)}(f) = g S_{\sigma^{-1}\Phi}(f) = g S_\Phi(f) = \sigma S_\Phi(f).$$

9.4 In view of Theorem 4.3 and Example 9.1, we may assume that $\Phi = (v_j)$ is tight frame which is not a basis. Since each permutation is a symmetry $|\langle v_j, v_k \rangle| = R \neq 0$, $j \neq k$, i.e., Φ is equiangular, and its frame graph Γ is complete. Let $\mathscr{T}$ be the spanning tree for Γ with root v_1 and edges $\{v_1, v_j\}$, $j \neq 1$. By Theorem 8.2, we can assume that $\langle v_1, v_j \rangle = -r$, $j \neq 1$, where $r > 0$. Since symmetries preserve the triple products for distinct vectors, these must all be equal to some real C. Hence

$$\langle v_1, v_j \rangle \langle v_j, v_k \rangle \langle v_k, v_1 \rangle = C \quad \Longrightarrow \quad \langle v_j, v_k \rangle = \frac{C}{r^2}, \quad j \neq k,\ j,k \neq 1.$$

By Exer. 3.19, we have that if $C = r^3$, then $\mathrm{Gram}(\Phi)$ has rank n, which is not possible (Φ is not a basis), and so we must have $C/r^2 = -r$, so that $\langle v_j, v_k \rangle = -r$, $\forall j \neq k$, and therefore Φ is the vertices of the simplex.

9.5 Let Φ be an α-partition frame. Since a frame and it complement have the same symmetry group, it suffices to calculate the symmetry group of the complementary tight frame (2.15). Let M be the set values $\{\alpha_j\}$, and $m_\# := |\{j : \alpha_j = m\}|$, $m \in M$.

Since the action of the symmetry group is unitary, it must map the vectors $e_j/\sqrt{\alpha_j}$ with $\alpha_j = m$ to themselves. The subspaces $\mathscr{H}_m := \operatorname{span}\{e_j : \alpha_j = m\} \subset \mathbb{R}^k$ are orthogonal to each other, and so the symmetry group (and projective symmetry group) is the product of the symmetry groups for the equal-norm tight frames of the $m \cdot m_\#$ vectors contained in $\mathscr{H}_m$. There are $m_\#!$ unitary maps which map these vectors to themselves. For each of these maps, the image of the m copies of each of the $m_\#$ vectors $\frac{e_j}{\sqrt{m}} \in \mathscr{H}_m$ (which are equal) can be reordered in $m!$ ways, giving

$$|\mathrm{Sym}(\Phi)| = |\mathrm{Sym}_{\mathrm{P}}(\Phi)| = \prod_{m \in M} m_\#!(m!)^{m_\#}.$$

In particular, if the α_j are distinct, then the symmetry group is $S_{\alpha_1} \times S_{\alpha_2} \times \cdots \times S_{\alpha_k}$, and for a proper α–partition frame, i.e., $\alpha_j \geq 2$, $\forall j$, we have the lower estimate

$$|\mathrm{Sym}(\Phi)| = |\mathrm{Sym}_{\mathrm{P}}(\Phi)| \geq \prod_{m \in M} 2^{m_\#} = 2^n > n.$$

9.6 From the Gramians (see Example 2.8), it is easy to see that $\mathrm{Sym}(\Phi) = S_3$, and $\mathrm{Sym}(\Psi)$ has order 3 and is generated by the cycle $a = (123)$. The faithful actions of these symmetry groups are given by the generators

$$\pi_\Phi((12)) = \begin{pmatrix} 0 & 1 \\ 1 & 0 \end{pmatrix}, \quad \pi_\Phi((123)) = \begin{pmatrix} \omega & \\ & \omega^2 \end{pmatrix}, \quad \pi_\Psi((123)) = \begin{pmatrix} 1 & \\ & \omega \end{pmatrix}.$$

These tight frames are projectively equivalent (see Exer. 2.21, or observe that all the triple products are equal to -1). Thus $\mathrm{Sym}_{\mathrm{P}}(\Phi) = \mathrm{Sym}_{\mathrm{P}}(\Psi) = S_3$.

9.7 Since the action of $G = \langle a, b \rangle$ is unitary,

$$|\langle g_1 w_1, g_2 w_2 \rangle| = |\langle w_1, g_1^{-1} g_2 w_2 \rangle|, \qquad g_1, g_2 \in G, \quad w_1, w_2 \in \{v, e_1, e_2\}.$$

Using $1 + \omega + \omega^2 + \omega^3 + \omega^4 = 0$ to simplify, these angles can be computed as

$$|\langle v, av \rangle| = |\langle v, a^4 v \rangle| = \frac{\sqrt{5 - \sqrt{5}}}{\sqrt{10}}, \qquad |\langle v, a^2 v \rangle| = |\langle v, a^3 v \rangle| = \frac{\sqrt{5 + \sqrt{5}}}{\sqrt{10}},$$

$$|\langle v, a^j bv \rangle| = \frac{|\omega^{2j} + \zeta^2|}{\sqrt{5}}, \quad 0 \leq j \leq 4, \qquad \sum_{j=0}^{4} |\langle v, a^j bv \rangle|^8 = \frac{14}{25},$$

$$|\langle e_1, a^j bv \rangle| = |\langle e_2, a^j v \rangle| = \frac{2}{\sqrt{10 + 2\sqrt{5}}}, \qquad 0 \leq j \leq 4,$$

$$|\langle e_2, a^j bv \rangle| = |\langle e_1, a^j v \rangle| = \frac{1 + \sqrt{5}}{\sqrt{10 + 2\sqrt{5}}}, \qquad 0 \leq j \leq 4.$$

From these, we verify the condition for being a $(4,4)$-design, i.e.,

$$\sum_j\sum_k |\langle u_j,u_k\rangle|^8 = 12\cdot 1+10\Big\{2\cdot\Big(\frac{\sqrt{5-\sqrt{5}}}{\sqrt{10}}\Big)^8+2\cdot\Big(\frac{\sqrt{5+\sqrt{5}}}{\sqrt{10}}\Big)^8+\frac{14}{25}\Big\}+2\cdot 0$$
$$+20\Big\{\Big(\frac{2}{\sqrt{10+2\sqrt{5}}}\Big)^8+\Big(\frac{1+\sqrt{5}}{\sqrt{10+2\sqrt{5}}}\Big)^8\Big\}=\frac{144}{5}=c_4(2,\mathbb{C})(12)^2.$$

Exercises of Chapter 10

10.1 Since the action is unitary, $\langle hf,gv_j\rangle=\langle f,h^*gv_j\rangle=\langle f,h^{-1}gv_j\rangle$, and so we can repeat the argument of Lemma 10.1, which gives

$$S(hf)=\sum_g\langle hf,gw\rangle gv=h\sum_g\langle f,h^{-1}gw\rangle h^{-1}gv=h\sum_k\langle f,kw\rangle kv=h(Sf).$$

10.2 Consider the $\mathbb{F}G$-homomorphism $\sigma_kS:V_j\to W$ given by

$$\sigma_kSf_j=\sigma_k\left(\sum_{g\in G}\langle f_j,gv_j\rangle gv_k\right).$$

By Lemma 10.4,

$$\sigma_kS=c\sigma_j,\qquad c=\langle\sigma_kSv_j,\sigma_jv_j\rangle/\|\sigma_jv_j\|^2.$$

Using Theorem 10.5, we calculate

$$\begin{aligned}\langle\sigma_kSv_j,\sigma_jv_j\rangle&=\langle\sum_{g\in G}\langle v_j,gv_j\rangle g\sigma_kv_k,\sigma_jv_j\rangle=\sum_{g\in G}\langle v_j,gv_j\rangle\langle g\sigma_kv_k,\sigma_jv_j\rangle\\&=\sum_{g\in G}\langle g^{-1}v_j,v_j\rangle\langle\sigma_kv_k,\sigma_jg^{-1}v_j\rangle=\langle\sigma_kv_k,\sigma_j\sum_{g\in G}\langle v_j,g^{-1}v_j\rangle g^{-1}v_j\rangle\\&=\langle\sigma_kv_k,\sigma_j\frac{|G|\,\|v_j\|^2}{\dim(V_j)}v_j\rangle=\frac{|G|\,\|v_j\|^2}{\dim(V_j)}\langle\sigma_kv_k,\sigma_jv_j\rangle.\end{aligned}$$

Therefore, the condition (10.16), i.e., $Sv_j=0$, is

$$\frac{|G|\,\|v_j\|^2}{\dim(V_j)}\frac{\langle\sigma_kv_k,\sigma_jv_j\rangle}{\|\sigma_jv_j\|^2}\sigma_k^{-1}\sigma_jv_j=0\iff\langle\sigma_jv_j,\sigma_kv_k\rangle=0.$$

10.3 (a) We observe $\rho(\sigma):\mathbb{R}^8\to\mathbb{R}^8$ maps $\mathscr{H}$ to $\mathscr{H}$. This gives an action since

$$\rho(\sigma)\rho(\tau)x=\rho(\sigma)(x_{\tau j})=(x_{\sigma\tau j})=\rho(\sigma\tau)x\implies\rho(\sigma)\rho(\tau)=\rho(\sigma\tau),$$

which is unitary since $\langle\rho(\sigma)x,\rho(\sigma)y\rangle=\sum_j x_{\sigma j}y_{\sigma j}=x_jy_j=\langle x,y\rangle$.
(b) The stabiliser of v consists of all 2!6! permutations which map $\{1,2\}$ to itself. Thus, by the orbit size theorem, the orbit has $8!/(2!6!)=28$ vectors.

10.4 (a) We have that $\langle x,x\rangle_\rho = \frac{1}{|G|}\sum_g \langle \rho(g)x, \rho(g)x\rangle = \frac{1}{|G|}\sum_g \|\rho(g)x\|^2 \geq 0$, with equality possible if and only if each term is zero, i.e., $x=0$ (since each $\rho(g)$ is 1–1).
(b) Since $\rho(g)\rho(h)=\rho(gh)$, we obtain

$$\langle \rho(h)x, \rho(h)y\rangle_\rho = \frac{1}{|G|}\sum_g \langle \rho(g)\rho(h)x, \rho(g)\rho(h)y\rangle = \langle x,y\rangle_\rho.$$

(c) Clearly $\tilde{\rho}$ is a group homomorphism, and each $\tilde{\rho}(g)$ is unitary, since by (b)

$$\begin{aligned}\langle \tilde{\rho}(g)x, \tilde{\rho}(g)y\rangle &= \langle A\rho(g)B^{-1}x, \rho(g)B^{-1}y\rangle = \langle \rho(g)B^{-1}x, \rho(g)B^{-1}y\rangle_\rho \\ &= \langle B^{-1}x, B^{-1}y\rangle_\rho = \langle A^{-1}x, y\rangle_\rho = \langle A(A^{-1}x), y\rangle = \langle x,y\rangle.\end{aligned}$$

(d) We have $B^{-1}\Psi = B^{-1}(B\rho(g)B^{-1}Bv)_{g\in G} = (\rho(g)v)_{g\in G} = \Phi$, so that Φ is similar to Ψ. Since Ψ is a G-frame, Theorem 10.1 implies that Ψ^{can} is a G-frame.

10.5 $\langle\!\langle \cdot,\cdot\rangle\!\rangle$ is G-invariant if and only if for each $g\in G$

$$\begin{aligned}\langle\!\langle \rho(g)x, \rho(g)y\rangle\!\rangle = \langle\!\langle x,y\rangle\!\rangle \quad &\Longleftrightarrow \quad \langle M\rho(g)x, \rho(g)y\rangle_\rho = \langle Mx,y\rangle_\rho, \quad \forall x,y \\ &\Longleftrightarrow \quad \langle \rho(g)^{-1}M\rho(g)x, y\rangle_\rho = \langle Mx,y\rangle_\rho, \quad \forall x,y \\ &\Longleftrightarrow \quad \rho(g)^{-1}M\rho(g) = M.\end{aligned}$$

By Schur's lemma, the last condition implies that M is nonzero scalar multiple of the identity matrix when ρ is absolutely irreducible, i.e., in this case all G-invariant inner products are scalar multiples of each other.

10.6 (a) Since the action of G on $\mathscr{H}$ is unitary, we have

$$\begin{aligned}\sum_g\sum_h |\langle gv, hv\rangle|^2 &= \sum_g\sum_h |\langle v, g^{-1}hv\rangle|^2 = |G|\sum_g |\langle v, gv\rangle|^2 \\ &\geq \frac{1}{d}\Big(\sum_g \langle gv, gv\rangle\Big)^2 = \frac{1}{d}(|G|\|v\|^2)^2 = \frac{1}{d}|G|^2\|v\|^4.\end{aligned}$$

(b) As above, the (t,t)-design condition simplifies, since

$$\sum_g\sum_h |\langle gv, hv\rangle|^{2t} = |G|\sum_g |\langle v, gv\rangle|^{2t}, \qquad \Big(\sum_{g\in G}\|gv\|^{2t}\Big)^2 = (|G|\|v\|^{2t})^2 = |G|^2\|v\|^{4t}.$$

10.7 Use the notation $v\oplus w$ and $v\otimes w$ for elements of $\mathscr{H}_1\oplus\mathscr{H}_2$ and $\mathscr{H}_1\otimes\mathscr{H}_2$.
(a) If $\Phi=(gv)_{g\in G}$ and $\Psi=(gw)_{g\in G}$ are disjoint G-frames, then their direct sum is $\Phi\oplus\Psi = (g(v\oplus w))_{g\in G}$, where $g(x\oplus y) := gx\oplus gy$ is a unitary action.
(b),(c) If $\Phi_j = (g_jv_j)_{g_j\in G_j}$ are G-frames, then

$$\begin{aligned}\Phi_1 \hat{+} \Phi_2 &= \Big((g_1,g_2)\big(\frac{1}{\sqrt{|G_2|}}v_1 \oplus \frac{1}{\sqrt{|G_1|}}v_2\big)\Big)_{(g_1,g_2)\in G_1\times G_2} \\ \Phi_1\otimes\Phi_2 &= \big((g_1,g_2)(v_1\otimes v_2)\big)_{(g_1,g_2)\in G_1\times G_2}\end{aligned}$$

where $(g_1,g_2)(x\oplus y) := g_1x\oplus g_2y$ and $(g_1,g_2)(x\otimes y) := g_1x\otimes g_2y$ are unitary.
(d) Let Ψ be the complement of a normalised tight G-frame $\Phi = (\phi_g)_{g\in G}$. Since $\mathrm{Gram}(\Phi)+\mathrm{Gram}(\Psi)$ is the identity matrix, we have

$$\langle \psi_h,\psi_g\rangle = \begin{cases} -\langle \phi_h,\phi_g\rangle = -\langle g^{-1}h\phi_1,\phi_1\rangle, & h\neq g; \\ 1-\langle \phi_h,\phi_g\rangle = 1-\langle g^{-1}h\phi_1,\phi_1\rangle, & h=g \end{cases}$$

so that $\mathrm{Gram}(\Psi)$ is a G-matrix, and, by Theorem 10.3, Ψ is G-frame.

10.8 If the action was not irreducible, then there would be a plane $P\subset\mathbb{R}^3$ that was G-invariant (otherwise $\mathbb{R}^3$ would be a sum of 1-dimensional G-invariant subspaces, and so G would be abelian). The restriction of the action of G to this plane P would be a faithful representation of G, and so G would be isomorphic to a finite subgroup of $\mathscr{U}(\mathbb{R}^2)$. This is impossible, since the finite subgroups of $\mathscr{U}(\mathbb{R}^2)$ are cyclic or dihedral.

10.9 (a) Suppose V is an $\mathbb{F}G$-module. Since $v\mapsto gv$ is linear map, $\rho(g)v := gv$ defines a map $\rho: G\to GL(V)$, and since $(g,v)\mapsto gv$ is an action

$$\rho(g)\rho(h)v = g(hv) = (gh)v = \rho(gh)v \quad\Longrightarrow\quad \rho \text{ is a homomorphism.}$$

Conversely, suppose that ρ is representation, and let $gv := \rho(g)v$. Then

$$g(hv) = \rho(g)\rho(h)v = \rho(gh)v = (gh)v, \qquad 1v = \rho(1)v = v$$

so $(g,v)\mapsto gv$ is an action, and $\rho(g): v\mapsto gv$ is a linear map (by definition).
(b) By definition, W is a $\mathbb{F}G$-submodule of V if it is closed under vector addition and scalar multiplication, i.e., is a vector subspace, and it is closed under the action of G, i.e., is G-invariant.

10.10 First observe that τ^{-1} is an $\mathbb{F}G$-isomorphism, since

$$\tau g = g\tau \quad\Longrightarrow\quad g = \tau^{-1}g\tau \quad\Longrightarrow\quad g\tau^{-1} = \tau^{-1}g\tau\tau^{-1} = \tau^{-1}g.$$

The map $\sigma\tau^{-1}: V_j\to V_j$ is an $\mathbb{F}G$-isomorphism (check taking the inverse commutes with the action of G). Let $\lambda\in\mathbb{C}$ be an eigenvector (for a nonzero eigenvalue) of $\sigma\tau^{-1}$. Then $(gv)_{g\in G}$ spans V_j, and we commute

$$\sigma\tau^{-1}gv = g\sigma\tau^{-1}v = g(\lambda v) = \lambda(gv),$$

so that $\sigma\tau^{-1} = \lambda I_{V_j}$, as claimed.

10.11 We have the equivalences

$$g_1H = g_2H \quad\Longleftrightarrow\quad g_2^{-1}g_1\in H \quad\Longleftrightarrow\quad g_2^{-1}g_1v = v \quad\Longleftrightarrow\quad g_1v = g_2v.$$

(a) The forward direction implies $\mathscr{C}\to\mathscr{H}: gH\mapsto gv$ is a well-defined map onto the vectors in Φ, and the reverse direction that this is 1–1.

(b) Let N be the kernel of ρ, so that $N \subset H$. If $G/H \to \mathscr{U}(\mathscr{H}) : gH \mapsto \rho(g)$ defines a group homomorphism, then by the first isomorphism theorem the composition $G/N \to G/H \to \mathscr{U}(\mathscr{H}) : gN \mapsto gH \mapsto \rho(g)$ is a well-defined injective map, and so we must have $H = N$.
(c) If G is abelian, then $h(gv) = (hg)v = (gh)v = g(hv) = gv$, $\forall h \in H, g \in G$, and so H stabilises $\mathscr{H} = \operatorname{span}\{gv\}_{g\in G}$.

10.12 (a) By Exercise 10.6, the condition for $(gv)_{g\in G}$ to be a (t,t)-design is that

$$\frac{1}{|G|}\sum_{g\in G} |\langle v, gv\rangle|^{2t} = c_t(d,\mathbb{F})\|v\|^{4t}.$$

This will hold for every v if the above polynomials of degree $4t$ in the entries of v and $\overline{v}$ are equal. This can be verified in `Magma` for $t = 5$, but not $t = 6$. An easy way to do this is to check equality at a set of points v on which a polynomial in $\Pi^{\circ}_{2t,2t}(\mathbb{C}^d)$ is determined by its values.
(b) We observe that a has order 5 and $b^2 = -I$. Hence if v is an eigenvector of a, then $(gv)_{g\in G}$ consists of $120/10 = 12$ lines (from each of which we can select a vector to obtain $(5,5)$-design of 12 vectors).

10.13 Let $G = C_n = \langle a\rangle$ act on a θ-isogonal configuration $(x_j)_{j=1}^d$ via a cyclic shift: $ax_j := x_{j+1}$, $x_{d+1} := x_1$. This action is unitary, since

$$\begin{aligned}\langle a(\sum_j \alpha_j x_j), a(\sum_k \beta_k x_k)\rangle &= \langle \sum_j \alpha_j x_{j+1}, a\sum_k \beta_k x_{k+1}\rangle \\ &= \sum_j\sum_{k\neq j} \alpha_j\beta_k \cos\theta + \sum_j \alpha_j\beta_j = \langle \sum_j \alpha_j x_j, \sum_k \beta_k x_k\rangle,\end{aligned}$$

and so (x_j) is a C_n-frame.

Exercises of Chapter 11

11.1 (a) Let

$$(u_j) = \left(\begin{bmatrix}1\\1\\1\end{bmatrix}, \begin{bmatrix}\omega\\\omega^2\end{bmatrix}, \begin{bmatrix}\omega^2\\\omega\end{bmatrix}\right), \quad (v_j) = \left(\begin{bmatrix}1\\1\end{bmatrix}, \begin{bmatrix}1\\\omega\end{bmatrix}, \begin{bmatrix}1\\\omega^2\end{bmatrix}\right), \qquad \omega := e^{2\pi i/3}.$$

Then we calculate

$$\|u_j - u_k\| = \sqrt{6}, \quad j\neq k, \qquad \|v_j - v_k\| = \sqrt{3}, \quad j \neq k,$$

and $|\langle u_j, u_k\rangle| = |\langle v_j, v_k\rangle| = 1$, $j\neq k$.
(b) A calculation, see (c), gives $\|w_j - w_k\| = \sqrt{5}$, $j\neq k$. Here $|\langle w_j, w_k\rangle| = \frac{\sqrt{5}\pm 1}{2}$.
(c) Since $v_j = (1, \omega^j, \ldots, \omega^{(d-1)j})$, with $\omega := e^{2\pi i/(2d-1)}$, we calculate

$$\|v_j - v_k\|^2 = \sum_{a=0}^{d-1} |\omega^{aj} - \omega^{ak}|^2 = 2d-1 - \sum_{m=1}^{2d-1} \omega^{m(j-k)} = 2d-1, \quad j \neq k.$$

Since $w_j = (\omega^j, \omega^{2j}, \ldots, \omega^{dj})$, with $\omega := e^{2\pi i/(2d+1)}$, we calculate

$$\|w_j - w_k\|^2 = \sum_{a=1}^{d} |\omega^{aj} - \omega^{ak}|^2 = 2d - \sum_{m=1}^{2d} \omega^{m(j-k)} = 2d-(-1) = 2d+1, \quad j \neq k.$$

Remark: Notice that the separation between points in Ψ is larger than that for those in Φ (which has *fewer* points), and so equispacing is perhaps not the most useful notion of "equal spacing". The difficulty here is that though an equal distance from each other, the points are clumped in a particular cone. This phenomenon can be seen for the standard basis vectors in $\mathbb{C}^3$, which are equally spaced, and all lie in the first octant.

11.2 (a) The symmetry group has order 384 (it is the group $< 384, 5557 >$), which is the largest for the class of cyclic harmonic frames of this size. This frame consists of scalar multiples of an orthonormal basis.
(b) The symmetry groups are both $< 200, 31 >$, which is the largest for its class. The projective symmetry groups are also both $< 200, 31 >$.

11.3 Since $U^* = U^{-1}$, $Uf = \sum_j \langle Uf, U^j v\rangle U^j = U \sum_j \langle f, U^{j-1} v\rangle U^{j-1}$, which gives

$$f = \sum_{j=0}^{n-1} \langle f, U^{j-1} v\rangle U^{j-1} v = \sum_{j=0}^{n-1} \langle f, U^j v\rangle U^j v, \quad \forall f.$$

Cancelling terms in the above gives $\langle f, U^{-1} v\rangle U^{-1} v = \langle f, U^{n-1} v\rangle U^{n-1} v$, so that $U^n v = cv$, for some $c \in \mathbb{F}$. For $(U^j v)$ to be a frame, v must be nonzero, and so $\|v\| = \|U^n v\| = \|cv\| = |c|\|v\|$, implies that $|c| = 1$. On the spanning set $\{U^k v\}_{k=0}^{n-1}$, we have

$$U^n(U^k v) = U^n \sum_j \langle U^k, U^j v\rangle U^j v = \sum_j \langle U^k, U^j v\rangle U^j (cv) = c(U^k v),$$

so that $U^n = cI$. If $1/\alpha = \sqrt[n]{c}$ is an n th root of c, then $(\alpha U)^n = I$, and so $\big((\alpha U)^j v\big)_{j=0}^{n-1} = (v, \alpha U v, \alpha^2 U^2 v, \ldots, \alpha^{n-1} U^{n-1} v)$ is a cyclic harmonic frame.

11.4 Let Φ be the harmonic frame of n vectors given by $\{a\} \subset G$. For the vectors of Φ to be distinct, a must generate G, which implies that $G = \mathbb{Z}_n$ with a a unit. The corresponding harmonic frames are $(\omega^{aj})_{j \in \mathbb{Z}_n}$, where $\omega = e^{2\pi i/n}$, i.e., the nth roots of unity.

11.5 Let G be an abelian group of order n. If $J = \{a, b\} \subset G$ gives a real harmonic frame of n distinct vectors, then $b = -a$ (to be real) and a generates G (to have distinct vectors). Thus we must have $G = \mathbb{Z}_n$, with a a unit. For such a given subset $\{a, -a\} \subset \mathbb{Z}_n$, the corresponding cyclic harmonic frame is

$$\left(\begin{bmatrix}1\\1\end{bmatrix}, \begin{bmatrix}\omega^a\\\omega^{-a}\end{bmatrix}, \begin{bmatrix}\omega^{2a}\\\omega^{-2a}\end{bmatrix}, \ldots \begin{bmatrix}\omega^{(d-1)a}\\\omega^{-(d-1)a}\end{bmatrix}\right),$$

which is unique up to reordering. This harmonic frame must be the n equally spaced unit vectors in $\mathbb{R}^2$, since the equally spaced unit vectors are the orbit of a vector under the unitary action of the cyclic group of order n given by rotations through $\frac{2\pi}{n}$ (and therefore are a real cyclic harmonic frame).

11.6 Representatives of the six multiplicatively inequivalent two-element subsets of $\mathbb{Z}_6$ (the unique abelian group of order 6) together with the distances $\|v_j - v_k\|$ and angles $|\langle v_j, v_k\rangle|$, $j \neq k$ are

$$\begin{array}{llllll} \{0,1\} & 1,\sqrt{3},2, & 0,1,\sqrt{3}, & \{1,2\} & 2,\sqrt{6}, & 0,1,\sqrt{3}, \\ \{1,3\} & \sqrt{3},\sqrt{5},2\sqrt{2}, & 1,2, & \{1,4\} & 2,\sqrt{6}, & 0,2, \\ \{1,5\} & \sqrt{2},\sqrt{6},2\sqrt{2}, & 1,2, & \{3,4\} & \sqrt{3},2,\sqrt{7}, & 0,1,\sqrt{3}. \end{array}$$

Since the distances and angles are fixed under $v_j \mapsto \overline{v_j}$, we conclude that none of these harmonic frames are conjugates of each other.

11.7 Since matrix multiplication in G commutes, for $g_1, g_2 \in G$, we have

$$g_1\phi = g_2\phi \quad \Longrightarrow \quad g_1hv = g_2hv, \quad \forall h \in G \quad \Longrightarrow \quad g_1 = g_2.$$

If G is nonabelian, then this does not hold. For example, let G be the dihedral group of the square acting on $\mathbb{R}^2$ in the usual way, with a reflection fixing the x-axis. Then the orbit of $v = e_1$ consists of four equally spaced vectors, each repeated twice.

11.8 Suppose that $|G| = n$, $|J| = d$. Then the normalised tight frame given by J is $\frac{1}{\sqrt{n}}(\xi|_J)_{\xi\in\hat{G}}$, and similarly for the one given by $G \setminus J$. The (ξ,η)-entry of the Gramians for these normalised tight frames are

$$\langle \eta|_J, \xi|_J\rangle = \sum_{j\in J} \eta(j)\overline{\xi(j)}, \qquad \langle \eta|_{G\setminus J}, \xi|_{G\setminus J}\rangle = \sum_{j\in G\setminus J} \eta(j)\overline{\xi(j)}.$$

These sum to $\frac{1}{n}\langle \eta, \xi\rangle$, which is the identity matrix (orthogonality of the characters).

11.9 Let $n = |\Phi|$ and $v_j^a := v_j$, $1 \leq a \leq d$. Then (up to a scalar multiple) the vectors of Ψ are

$$(v_{j_1}^1, v_{j_2}^2, \ldots, v_{j_d}^d), \qquad 1 \leq j_1, j_2, \ldots, j_d \leq n.$$

We think of these vectors as being indexed by the pairs

$$\{(a, j_a)\}_{1\leq a\leq d, 1\leq j_a\leq n}.$$

There are symmetries given by

$$(a, j_a) \mapsto (\tau a, j_a), \quad \tau \in S_d, \qquad (a, j_a) \mapsto (a, \sigma_a j_a), \quad (\sigma_1, \ldots, \sigma_d) \in \mathrm{Sym}(\Phi)^d.$$

The subgroups of the symmetries of these two types have trivial intersection, and so $|\mathrm{Sym}(\Psi)| \geq |S_d||\mathrm{Sym}(\Phi)^d| = d!m^d$.

11.10 Let $\Psi_{\{\xi_1,\ldots,\xi_d\}} = (\psi_g)_{g\in G}$, $\psi_g := (\xi_j(g))_{j=1}^d$ denote the harmonic frame given by a choice of characters $\{\xi_1,\ldots,\xi_d\} \subset \hat{G}$.
(a) For $(v_g)_{g\in G} = \Psi_{\{\xi_1,\ldots,\xi_d\}}$, we calculate

$$\begin{aligned}\langle v_{a+j}, v_{a+k}\rangle &= \xi_1(a+j)\overline{\xi_1(a+k)} + \cdots + \xi_d(a+j)\overline{\xi_d(a+k)} \\ &= \xi_1(a)\xi_1(j)\overline{\xi_1(a)\xi_1(k)} + \cdots + \xi_d(a)\xi_d(j)\overline{\xi_d(a)\xi_d(k)} \\ &= \xi_1(j)\overline{\xi_1(k)} + \cdots + \xi_d(j)\overline{\xi_d(k)} = \langle v_j, v_k\rangle.\end{aligned}$$

(b) Let $(v_g)_{g\in G} = \Psi_{\{\xi_1,\ldots,\xi_d\}}$ and $(w_g)_{g\in G} = \Psi_{\{\eta_1,\ldots,\eta_d\}}$. If there is unitary U and a permutation $\sigma : G \to G$ with $v_g = Uw_{\sigma g}$, $\forall g \in G$, then, by (a), we calculate

$$\langle v_j, v_k\rangle = \langle Uw_{\sigma j}, Uw_{\sigma k}\rangle = \langle w_{\sigma j}, w_{\sigma k}\rangle = \langle w_{\sigma j - \sigma k + b}, w_b\rangle,$$

and so we may take $a = \sigma j - \sigma k + b$.
(c) Take $j = 1$, $k = 0$, $b = 0$ in (b), to obtain $\langle v_1, v_0\rangle = \langle w_a, w_0\rangle$, i.e.,

$$\sum_{j=1}^{d} \xi_j(1) = \sum_{j=1}^{d} \eta_j(a).$$

Moreover, if $G = \mathbb{Z}_n$ and σ is an automorphism, then $a = \sigma 1$ is a unit.

11.11 (a) The frame given by $J \subset \mathbb{Z}_p$ is real if and only if J is closed under taking inverses, i.e., $J = -J$. Since p is an odd prime, 2 is a unit, and so $j = -j$ if and only if $j = 0$. Thus J has the stated form. Since $\mathbb{Z}_p^* = \mathbb{Z}_p \setminus \{0\}$, K generates $\mathbb{Z}_p$ for $d > 1$.
(b) The unit group $\mathbb{Z}_p^*$ is cyclic of even order $p-1$. Let $a \in \mathbb{Z}_p^*$ have order j. The action of a on $\mathbb{Z}_p \setminus \{0\}$ gives $\frac{p-1}{j}$ orbits of size j. We now count the number of orbits of the d-element subsets of $\mathbb{Z}_p$ which give real harmonic frames (as above) under the action of $\mathbb{Z}_p^*$, which is the number real harmonic frames.

First consider the case when d is even. Suppose that a fixes some $J = K \cup -K$, $|J| = d$. If j is even, then $-1 = a^{\frac{j}{2}}$, so J consists of $\frac{d}{j}$ orbits under the action of a (half the orbit in K and the other half in $-K$), and we must have $j|d$. If j is odd, then K consists of $\frac{d/2}{j}$ orbits of size j (their negatives give the remaining $\frac{d/2}{j}$ orbits which make up J), and we must have $j|(d/2)$. Thus Burnside orbit counting gives

$$h_{p,d}^{\mathbb{R}} = \frac{1}{p-1}\Big\{ \sum_{\substack{j|\gcd(p-1,d) \\ j \text{ even}}} \binom{\frac{p-1}{j}}{\frac{d}{j}} \varphi(j) + \sum_{\substack{j|\gcd(p-1,\frac{d}{2}) \\ j \text{ odd}}} \binom{\frac{p-1}{2j}}{\frac{d}{2j}} \varphi(j) \Big\}.$$

When d is odd, the subsets J giving real frames have the form $J = \{0\} \cup K \cup -K$. These are multiplicatively equivalent if and only if the sets $K \cup -K$ are, and so we may apply the previous count (with d replaced by $d-1$). The formula so obtained also holds for $d = 1$.

Exercises of Chapter 12

12.1 If $A=[a_{jk}]\in\mathbb{F}^{d\times d}$ is Hermitian, i.e., $\overline{a_{jk}}=a_{kj}$, then it is determined by

$$a_{jj}\in\mathbb{R},\quad 1\le j\le d,\qquad a_{jk}\in\mathbb{F},\quad 1\le j<k\le d.$$

Thus the real vector space of Hermitian matrices has dimension

$$d+\frac{1}{2}d(d-1)=\frac{1}{2}d(d+1)\quad(\mathbb{F}=\mathbb{R}),\qquad d+2\frac{1}{2}d(d-1)=d^2\quad(\mathbb{F}=\mathbb{C}).$$

12.2 (a) Take the trace of $P_k=\sum_j c_jP_jP_k$ to obtain

$$1=\sum_{j\ne k}c_j\alpha^2+c_k=\sum_j c_j\alpha^2+c_k(1-\alpha^2),$$

so that the c_k are constant, hence equal to $\frac{d}{n}$ (by taking the trace of $I=\sum_j c_jP_j$).
(b) Substituting $c_j=\frac{d}{n}$ above, and multiplying by n gives

$$n=nd\alpha^2+d(1-\alpha^2).$$

Thus $n(1-d\alpha^2)=d(1-\alpha^2)>0$, so $\alpha^2<\frac{1}{d}$, and we may solve for n.

12.3 By (2.9), we have $\|f_j\|^2+\sum_{k\ne j}\|f_k\|^2=d$, and so

$$\|f_j\|^2=\sum_k|\langle f_j,f_k\rangle|^2=\|f_j\|^4+\sum_{k\ne j}C^2\|f_j\|^2\|f_k\|^2=\|f_j\|^4+C^2\|f_j\|^2(d-\|f_j\|^2).$$

Thus $1-C^2d=\|f_j\|^2(1-C^2)$. If $C=1$, then the vectors are all collinear (which is not possible), and so we conclude that $\|f_j\|^2$ is independent of j.

12.4 By construction, two of the x_j are 3 and the other six are -1. If two such vectors are not equal, then the number of times they can have a 3 in same position is 0 or 1. The corresponding inner products for these cases are

$$\frac{1}{24}(4\cdot 3(-1)+4\cdot(-1)^2)=-\frac{1}{3},\qquad \frac{1}{24}(1\cdot 3^2+2\cdot 3(-1)+5\cdot(-1)^2)=\frac{1}{3}.$$

12.5 Consider the n lines given by the f_j (some might be coincident). The acute angle between the lines given by f_j and f_k is $\theta_{jk}=\cos^{-1}(|\langle f_j,f_k\rangle|)$. Since $\cos^{-1}$ is strictly decreasing on $[0,1]$, the asserted inequality is equivalent to

$$\cos^{-1}\Bigl(\max_{j\ne k}|\langle f_j,f_k\rangle|\Bigr)\le\cos^{-1}\cos\Bigl(\frac{\pi}{n}\Bigr)\quad\Longleftrightarrow\quad\min_{j\ne k}\theta_{jk}\le\frac{\pi}{n}.$$

Rotate the (f_j) so that one line is the x-axis. Then, the lines partition the upper half plane into n sectors. The angle of the smallest sector gives $\min_{j\ne k}\theta_{jk}$. This must be $\le\frac{\pi}{n}$, since otherwise the sum of the angles of the n sectors would exceed π. There is equality if and only if all the n sectors have angle $\frac{\pi}{n}$. The strict inequality follows by calculus.

12.6 The normalisations of Φ and Ψ are $(\sqrt{\frac{d}{n}}f_j)$ and $(\sqrt{\frac{n-d}{n}}g_j)$. Thus

$$\langle\sqrt{\frac{n-d}{n}}g_j,\sqrt{\frac{n-d}{n}}g_k\rangle=-\langle\sqrt{\frac{d}{n}}f_j,\sqrt{\frac{d}{n}}f_k\rangle \quad\Longrightarrow\quad |\langle g_j,g_k\rangle|=\frac{d}{n-d}|\langle f_j,f_k\rangle|.$$

Thus if Φ minimises $\mathcal{M}_\infty(\Phi)$ over all unit-norm tight frames of n vectors for $\mathbb{F}^d$, then Ψ minimises $\mathcal{M}_\infty(\Psi)$ over all unit-norm tight frames of n vectors for $\mathbb{F}^{n-d}$.

12.7 (a) Let Σ be the Seidel matrix of the 5-cycle $(1,2,3,4,5)$, i.e.

$$\Sigma=\begin{pmatrix}0&-1&1&1&-1\\-1&0&-1&1&1\\1&-1&0&-1&1\\1&1&-1&0&-1\\-1&1&1&-1&0\end{pmatrix}.$$

This has eigenvalues $-\sqrt{5},-\sqrt{5},0,\sqrt{5},\sqrt{5}$. The Gramian of the corresponding *nontight* equiangular frame Φ is

$$\mathrm{Gram}(\Phi)=Q=I+\frac{1}{\sqrt{5}}\Sigma.$$

From this, we calculate (see Exer. 3.3)

$$\mathrm{Gram}(\tilde{\Phi})=Q^\dagger=\begin{pmatrix}\frac{2}{5}&a&b&b&a\\a&\frac{2}{5}&a&b&b\\b&a&\frac{2}{5}&a&b\\b&b&a&\frac{2}{5}&a\\a&b&b&a&\frac{2}{5}\end{pmatrix},\quad a:=\frac{3-\sqrt{5}}{20},\ b:=\frac{3+\sqrt{5}}{20},$$

$$\mathrm{Gram}(\Phi^{\mathrm{can}})=QQ^\dagger=\begin{pmatrix}\frac{3}{5}&-a&b&b&-a\\-a&\frac{3}{5}&-a&b&b\\b&-a&\frac{3}{5}&-a&b\\b&b&-a&\frac{3}{5}&-a\\-a&b&b&-a&\frac{3}{5}\end{pmatrix},\quad a:=\frac{\sqrt{5}-1}{10},\ b:=\frac{\sqrt{5}+1}{10}.$$

Thus the minimal angles between the lines given by Φ, $\tilde{\Phi}$ and Φ^{can} are

$$\cos^{-1}\frac{1}{\sqrt{5}}\approx 63.4349^\circ,\quad \cos^{-1}\frac{3+\sqrt{5}}{8}\approx 49.1176^\circ,\quad \cos^{-1}\frac{1+\sqrt{5}}{6}\approx 57.3610^\circ.$$

(b) It is easy to verify that $\Phi=(v_j)$ is given by vectors that lie in five of the six diagonals of the regular icosahedron, e.g., (using the hint) we may take

$$V=[v_j]=\begin{pmatrix}0&0&c&-1&c\\1&1&0&c&0\\c&-c&1&0&-1\end{pmatrix},\qquad c:=\frac{1+\sqrt{5}}{2}.$$

(c) It is easy to verify that $\Phi^{\text{can}}=(v_j)$ is the harmonic frame given by $\{0,2,3\}\subset\mathbb{Z}_5$, i.e.,

$$V=[v_j]=\begin{pmatrix}1&1&1&1&1\\1&\omega^2&\omega^4&\omega&\omega^3\\1&\omega^3&\omega&\omega^4&\omega^2\end{pmatrix},\qquad \omega:=e^{\frac{2\pi i}{5}},$$

which is the lifted fifth roots of unity (see Example 5.9).

12.8 This does not hold in general. For the four isogonal vectors in $\mathbb{R}^3$, the dual frame is equiangular (see Example 3.9), but for the five equiangular lines in $\mathbb{R}^3$ given by five of the six diagonals of the regular icosahedron, the dual frame has two angles $|\langle v_j,v_k\rangle|=\frac{1}{8}(3\pm\sqrt{5})$ (see Exer. 12.7).

12.9 (a) Adding the last d rows of A gives $(1,1,\ldots,1)$. Subtracting the multiple $\frac{1}{d}$ of this from the first n rows gives the row equivalent matrix

$$B=\begin{pmatrix}(\alpha^2-\frac{1}{d})J+(1-\alpha^2)I_n & 0\\ \frac{1}{d}J & I_d\end{pmatrix}.$$

(b) Since B is block upper triangular, its eigenvalues are those of the leading blocks: $(1-\alpha^2)$ with multiplicity $n-1$, $n(\alpha-\frac{1}{d})+(1-\alpha^2)$ with multiplicity 1 and 1 with multiplicity d. Thus, the rank of A is either $n+d-1$ if

$$n(\alpha-\frac{1}{d})+(1-\alpha^2)=0,$$

which is equivalent to Φ being tight, or it has rank $n+d$. In any case

$$n+d-1\le\operatorname{rank}(A)\le d^2.$$

12.10 The vectors have equal norms, since

$$\|\tilde{v}_{a,j}\|^2=\langle v_{a,j},v_{a,j}\rangle+2=r+2,\qquad \|\tilde{w}_\ell\|^2=\frac{1}{2}v+\frac{3}{2}=\frac{1}{2}(v+3)=r+2.$$

By (12.15) and (12.16), we have

$$\langle\tilde{v}_{a,j},\tilde{v}_{a,k}\rangle=-h_{0,j}^{(a)}\overline{h_{0,k}^{(a)}}+2h_{0,j}^{(a)}\overline{h_{0,k}^{(a)}}=h_{0,j}^{(a)}\overline{h_{0,k}^{(a)}},\qquad j\neq k,$$

$$\langle\tilde{v}_{a,j},\tilde{v}_{b,k}\rangle=h_{\beta,j}^{(a)}\overline{h_{\beta,k}^{(b)}},\qquad a\neq b.$$

Since $\langle w_\ell,w_m\rangle=-\alpha_\ell\overline{\alpha_m}$, $\ell\neq m$, $\|w_\ell\|^2=v$, we have

$$\langle\tilde{w}_\ell,\tilde{w}_m\rangle=\frac{1}{2}(-\alpha_\ell\overline{\alpha_m})+\frac{3}{2}\alpha_\ell\overline{\alpha_m}=\alpha_\ell\overline{\alpha_m},\qquad \ell\neq m.$$

Finally,

$$\langle \tilde{v}_{a,j}, \tilde{w}_\ell \rangle = \langle \sqrt{2} h_{0,j}^{(a)} e_a, \frac{1}{\sqrt{2}} w_j \rangle = h_{0,j}^{(a)} \overline{(w_j)_a},$$

and so the vectors are equiangular (Hadamard matrices have entries of modulus 1).

The space $\mathbb{C}^{\mathscr{B}} \oplus \mathbb{C}^{\mathscr{V}} \oplus \mathbb{C}$ has dimension

$$d := |\mathscr{B}| + |\mathscr{V}| + 1 = \frac{1}{6} v(v-1) + v + 1 = \frac{1}{6}(v+2)(v+3),$$

and the frame $(f_j) := (\tilde{v}_{a,j}) \cup (\tilde{w}_\ell)$ has

$$n := v(r+1) + (v+1) = \frac{1}{2}(v+1)(v+2)$$

vectors with $\|f_j\|^2 = r+2 = \frac{1}{2}(v+3)$, and $|\langle f_j, f_k \rangle| = 1$, $j \neq k$. This frame is a tight frame for $\mathbb{C}^d$, by the variational characterisation (Theorem 6.1), i.e.,

$$\begin{aligned} \sum_j \sum_k |\langle f_j, f_k \rangle|^2 &= n \cdot (\frac{1}{2}(v+3))^2 + (n^2 - n) \cdot 1 \\ &= \frac{3}{8}(v+1)^2(v+2)(v+3) = \frac{1}{d}\Big(n \cdot \frac{1}{2}(v+3)\Big)^2 = \Big(\sum_j \|f_j\|^2\Big)^2. \end{aligned}$$

The common angle is $|\langle f_j, f_k \rangle| / \|f_j\|^2 = \frac{1}{r+2}$, $j \neq k$.

12.11 (a) Since Σ is Hermitian, we have $-iC^T = \Sigma^* = \Sigma = iC$, so that $-C^T = C$.
(b) The condition of Theorem 12.7 for Σ to give an equiangular tight frame of n vectors for $\mathbb{C}^d$ is

$$C^T C = -C^2 = \Sigma^2 = (n-1)I + \mu \Sigma = (n-1)I + \mu i C,$$

which that implies $C^T C = (n-1)I$ (since μ is real), i.e., C is a conference matrix, where $\mu = 0$, i.e., $n = 2d$.

12.12 (a) Expanding out $C^T C = nI$, using $C^T = -C$ and $\mathbf{1}\mathbf{1}^* = J$, we have

$$\begin{pmatrix} 0 & -\mathbf{1}^* \\ \mathbf{1} & -A \end{pmatrix} \begin{pmatrix} 0 & \mathbf{1}^* \\ -\mathbf{1} & A \end{pmatrix} = \begin{pmatrix} n & -\mathbf{1}^* A \\ A\mathbf{1} & J - A^2 \end{pmatrix} = \begin{pmatrix} n & 0 \\ 0 & nI \end{pmatrix}.$$

Equating the blocks gives $J - A^2 = nI$ and $A\mathbf{1} = 0$. From $A\mathbf{1} = 0$, we have

$$AJ = (A\mathbf{1})\mathbf{1}^* = 0, \qquad JA = (-AJ)^* = 0.$$

(b) By construction, Σ is a signature matrix, and so it suffices to verify the second condition of Theorem 12.7. Using $J^2 = nJ$, $AJ = JA = 0$ and $A^2 = J - nI$, we have

$$\begin{aligned}
\Sigma^2 &= \frac{1}{n+1}\big((J-I)^2 - nA^2 \pm i\sqrt{n}(J-I)A \pm i\sqrt{n}A(J-I)\big) \\
&= \frac{1}{n+1}\big(nJ - 2J + I - n(J - nI) \mp 2i\sqrt{n}A\big) = \frac{1}{n+1}\big((n^2-1)I + 2(I - J \mp i\sqrt{n}A)\big) \\
&= (n-1)I \mp \frac{2}{n+1}\big(\pm(J-I) + i\sqrt{n}A\big) = (n-1)I \mp \frac{2}{\sqrt{n+1}}\Sigma.
\end{aligned}$$

Thus the second condition holds, with

$$\mu = \mp\frac{2}{\sqrt{n+1}}, \qquad d = \frac{n}{2} - \frac{n\mu}{2\sqrt{4(n-1)+\mu^2}} = \frac{n}{2} \pm \frac{1}{2}.$$

(c) Since $|\mu| \le 2$, we can choose $\zeta = \frac{1}{2}(-\mu + \sqrt{4 - |\mu|^2}i)$ in the third condition.

12.13 (a) Since -1 has no square root in $\mathrm{GF}(q)$, it follows that $N = -S$. Hence, if we add 0 to the difference set S, then the additional differences $0 - s, s - 0$, $s \in S$, give exactly one extra copy of each nonzero element of G.
(b) Since $G = N \cup \{0\} \cup S$ (disjoint union), N and $N \cup \{0\}$ are the complementary difference sets to $S \cup \{0\}$ and S (respectively). Since $x \mapsto -x$ is an automorphism of G, it follows that the harmonic frames given by S, N and by $S \cup \{0\}, N \cup \{0\}$ are unitarily equivalent up to reordering by an automorphism (Theorem 11.4).
(c) Let $z = x + iy = \langle \xi|_S, \eta|_S \rangle$ be the inner product between two distinct vectors of the harmonic frame $\Phi_S = (\xi|_S)_{\xi \in \hat{G}}$. The inner product between the corresponding vectors of $\Phi_{S\cup\{0\}}$ is $z+1$. Since each of these frames is equiangular (they are given by difference sets), $|z+1|^2 - |z|^2 = 2x+1$ is constant, i.e., with $d = \frac{n-1}{2}$, we have

$$2x + 1 = |z+1|^2 - |z|^2 = \frac{(d+1)(n-d-1)}{n-1} - \frac{d(n-d)}{n-1} = 0 \quad \Longrightarrow \quad x = -\frac{1}{2}.$$

Thus the real parts of the signature matrices for the equiangular harmonic frames given by S and $S \cup \{0\}$ are

$$\frac{-\frac{1}{2}}{d}\sqrt{\frac{d(n-1)}{n-d}} = -\frac{1}{\sqrt{n+1}}, \qquad \frac{1-\frac{1}{2}}{d+1}\sqrt{\frac{(d+1)(n-1)}{n-d-1}} = \frac{1}{\sqrt{n+1}}.$$

These signature matrices can be calculated explicitly by using the formula for quadratic Gauss sums (see [Kal06], [Ren07]), or by (12.63).
(d) Let $W = [\xi|_S]_{\xi \in \hat{G}}$ be the synthesis map for the equiangular harmonic frame $\Phi_S = (\xi|_S)_{\xi \in \hat{G}}$, and $\mathbf{1} = (1, \ldots, 1)$. By (c), the off diagonal entries of W^*W have the form

$$z = -\frac{1}{2} + iy, \qquad \frac{1}{4} + y^2 = \frac{\frac{q-1}{2}(q - \frac{q-1}{2})}{q-1} \quad \Longrightarrow \quad y^2 = \frac{q}{4}.$$

Since Φ_S is unlifted, we have $W\mathbf{1}^* = 0$. Hence the frame of $4m = q+1$ vectors with synthesis operator

$$V = \begin{pmatrix} \sqrt{q} & \mathbf{1}^* \\ 0 & \sqrt{2}W \end{pmatrix}$$

is a tight frame for $\mathbb{C}^{2m}$ (it has orthogonal rows of equal norm). Its Gramian is

$$V^*V = \begin{pmatrix} \sqrt{q} & 0 \\ \mathbf{1} & \sqrt{2}W^* \end{pmatrix} \begin{pmatrix} \sqrt{q} & \mathbf{1}^* \\ 0 & \sqrt{2}W \end{pmatrix} = \begin{pmatrix} q & \sqrt{q}\mathbf{1}^* \\ \sqrt{q}\mathbf{1} & \mathbf{1}\mathbf{1}^* + 2W^*W \end{pmatrix}.$$

The off diagonal entries of $\mathbf{1}\mathbf{1}^* + 2W^*W$ have the form

$$1+2\Big(-\frac{1}{2}+iy\Big) = 2iy, \qquad |2iy| = 2\sqrt{\frac{q}{4}} = \sqrt{q},$$

and so it is equiangular. This example is due to Zauner [Zau99]. For an equiangular tight frame of $n = 2d$ vectors for $\mathbb{C}^d$ whose signature matrix has off diagonal entries $\pm i$, it can be shown that the canonical tight frame of the frame obtained by removing a vector is an equiangular tight frame of $n-1$ vectors for $\mathbb{C}^d$ (see [Ren07], [Str08]).

12.14 By Theorem 12.6, the Gramian of the normalised equiangular tight frame is

$$P = \frac{d}{n}\big(I + C_{n,d}\hat{\Sigma}\big), \qquad C_{n,d} := \sqrt{\frac{n-d}{d(n-1)}}.$$

Since $\frac{d}{n-d}C_{n,d} = C_{n,n-d}$, the complementary orthogonal projection is

$$I - P = I - \frac{d}{n}I - \frac{d}{n}C_{n,d}\hat{\Sigma} = \frac{n-d}{n}\Big(I + C_{n,n-d}(-\hat{\Sigma})\Big),$$

so the complementary equiangular tight frame has reduced signature matrix $-\hat{\Sigma}$. The complement of a $\operatorname{srg}(v,k,\lambda,v)$ is a $\operatorname{srg}(v, v-k-1, v-2-2k+\mu, v-2k+\lambda)$, which gives the desired parameters. Alternatively, a count shows the vertex degree of the complementary graph is $k' = (n-1)-k-1$, and the other parameters follow from (12.35)

$$\lambda' = \frac{3(n-k-2)-n}{2}, \qquad \mu' = \frac{n-k-2}{2}.$$

12.15 Without loss of generality, we may assume that Q_{n-r} is the leading principal submatrix, so that

$$Q = \begin{pmatrix} Q_{n-r} & V \\ V^* & B \end{pmatrix}, \qquad V \in \mathbb{C}^{n-r \times r}.$$

Thus (by block multiplication), the leading block of of $dQ^2 - nQ$ gives

$$dQ_{n-r}^2 - nQ_{n-r} = -dVV^*.$$

Since $\operatorname{rank}(VV^*) = \operatorname{rank}(V^*V) \le \min\{r, n-r\}$, the result follows, and the equations are nontrivial only when $r < \frac{1}{2}n$.

12.16 (a) Since G is Hermitian with constant diagonal 1 and constant modulus off diagonal entries, we have all but the unitary property. Since the frame is tight, its canonical Gramian is an orthogonal projection, i.e. $(\frac{d}{n}G)^2 = \frac{d}{n}G$, and we calculate

$$U^*U = U^2 = I - 4\frac{d}{n}G + 4\frac{d}{n}G = I.$$

(b) By construction G is Hermitian, with constant diagonal 1 and constant modulus off diagonal entries. Thus it suffices to show that $\frac{d}{n}G$ is an orthogonal projection, i.e., $G^2 = \frac{n}{d}G$. Since $U^2 = U^*U = I$, we calculate

$$G^2 = \left(\frac{n}{2d}\right)^2 (I - 2U + U^2) = \left(\frac{n}{2d}\right)^2 2(I-U) = \frac{n}{d}G.$$

12.17 By the variational characterisation of tight frames, we have

$$\sum_j \sum_k |\langle v_j, v_k\rangle|^2 = \sum_{\substack{j,k \\ j\neq k}} |\langle v_j, v_k\rangle|^2 + n = \frac{n^2}{d}.$$

We recall that in Hölder's inequality with one vector being $(1,\ldots,1)$, there is equality if and only if the other vector has entries of constant modulus.
(a) In Hölder's inequality, take the conjugate exponents $\frac{2}{2-p}$ and $\frac{2}{p}$, to obtain

$$\begin{aligned}\sum_j \sum_k |\langle v_j, v_k\rangle|^p &= \sum_{\substack{j,k \\ j\neq k}} |\langle v_j, v_k\rangle|^p + n \le \Big(\sum_{\substack{j,k \\ j\neq k}} (1)^{\frac{2}{2-p}}\Big)^{1-\frac{p}{2}} \Big(\sum_{\substack{j,k \\ j\neq k}} (|\langle v_j, v_k\rangle|^p)^{\frac{2}{p}}\Big)^{\frac{p}{2}} + n \\ &= (n^2-n)^{1-\frac{p}{2}} \Big(\sum_{\substack{j,k \\ j\neq k}} |\langle v_j, v_k\rangle|^2\Big)^{\frac{p}{2}} + n = \frac{(\frac{n^2}{d}-n)^{\frac{p}{2}}}{(n^2-n)^{\frac{p}{2}-1}} + n,\end{aligned}$$

with equality if and only if (v_j) is equiangular.
(b) Take the conjugate exponents $\frac{p}{p-2}$ and $\frac{p}{2}$ in Hölder's inequality, to obtain

$$\begin{aligned}\frac{n^2}{d} &= \sum_{\substack{j,k \\ j\neq k}} |\langle v_j, v_k\rangle|^2 + n \le \Big(\sum_{\substack{j,k \\ j\neq k}} (1)^{\frac{p}{p-2}}\Big)^{1-\frac{2}{p}} \Big(\sum_{\substack{j,k \\ j\neq k}} (|\langle v_j, v_k\rangle|^2)^{\frac{p}{2}}\Big)^{\frac{2}{p}} + n \\ &= (n^2-n)^{1-\frac{2}{p}} \Big(\sum_{\substack{j,k \\ j\neq k}} |\langle v_j, v_k\rangle|^p\Big)^{\frac{2}{p}} + n,\end{aligned}$$

which gives the desired inequality, with equality if and only if (v_j) is equiangular.
(c) For $p=2$, equality in (a) and (b) is the variational characterisation of being tight.

12.18 (a) Since two frames are unitarily equivalent if and only if their Gramians are equal, the condition (12.65) is equivalent to

$$(U[V_1,W_1])^* U[V_1,W_1] = \left([V_2,W_2]\begin{pmatrix} P_\sigma\Lambda_1 & \\ & P_\tau\Lambda_2\end{pmatrix}\right)^* [V_2,W_2]\begin{pmatrix} P_\sigma\Lambda_1 & \\ & P_\tau\Lambda_2\end{pmatrix}.$$

Block multiplying out, using $V_j^*V_j = I$, gives

$$\begin{pmatrix} I & V_1^*W_1 \\ W_1^*V_1 & I\end{pmatrix} = \begin{pmatrix} I & (P_\sigma\Lambda_1)^*V_2^*W_2(P_\tau\Lambda_2) \\ (P_\tau\Lambda_2)^*W_1^*V_1(P_\sigma\Lambda_1) & I\end{pmatrix},$$

which is equivalent to $H_1 = \Lambda_1^{-1} P_\sigma^{-1} H_2 P_\tau \Lambda_2$ (Hadamard matrix equivalence).
(b) The nonzero 4-products for the vectors $(V_1 e_j)$ and $(W_1 e_j)$ of the first pair of MUBs are

$$\langle V_1 e_k, W_1 e_j\rangle\langle W_1 e_j, V_1 e_m\rangle\langle V_1 e_m, W_1 e_\ell\rangle\langle W_1 e_\ell, V_1 e_k\rangle = (H_1)_{jk}(H_1)_{mj}(H_1)_{\ell m}(H_1)_{k\ell}.$$

Since the vectors in the pairs of MUBs are projectively unitarily equivalent via $UV_1 e_j = \alpha_j V_2 e_{\sigma j}$, $UW_1 e_j = \beta_j W_2 e_{\tau j}$, the condition for equivalence of Hadamard matrices is

$$(H_1)_{jk}(H_1)_{k\ell}(H_1)_{\ell m}(H_1)_{mj} = (H_2)_{\sigma j,\tau k}(H_2)_{\sigma k,\tau \ell}(H_2)_{\sigma \ell,\tau m}(H_2)_{\sigma m,\tau j}.$$

It is enough to check this condition for j,k,l,m corresponding to a basis of 4-cycles for the cycle space of the frame graph (the complete bipartite graph $K_{d,d}$), which has dimension $d^2 - 2d + 1 = (d-1)^2$.

12.19 (a) Multiplying out, using $\mu^2 = \omega$, gives

$$(F^{-1}R^a F)_{jk} = \frac{1}{d}\sum_{s=0}^{d-1} \omega^{-js}\mu^{as(s+d)}\omega^{sk} = \frac{1}{2d}\sum_{s=0}^{2d-1} \mu^{as(s+d)+2s(k-j)}.$$

(b) Suppose d is odd. Since the $s=\beta$ and $s=\beta+d$ terms in the second sum above are equal, this implies that the sum above is twice the sum over the even terms, i.e.,

$$\begin{aligned} G(a, ad+2(k-j), 2d) &= 2\sum_{r=0}^{d-1} \mu^{a(2r)^2+(ad+2(k-j))(2r)} = 2\sum_{r=0}^{d-1} \omega^{2ar^2+2(k-j)r} \\ &= 2G(2a, 2(k-j), d). \end{aligned}$$

Since 2 is a unit, with $\overline{2} = \frac{1-d}{2}$, completing the square gives the desired formula

$$G(2a, 2(k-j), d) = \sum_{r=0}^{d-1} \omega^{2a\{(r+\overline{2}\overline{a}(k-j))^2 - (\overline{2}\overline{a}(k-j))^2\}} = \omega^{-\overline{2}\overline{a}(k-j)^2} G(2a, 0, d).$$

Suppose d is even. Then we can complete the square as follows.

$$\begin{aligned} G(a, ad+2(k-j), 2d) &= \sum_{s=0}^{2d-1} \mu^{a\{(s+\frac{d}{2}+\overline{a}(k-j))^2 - (\frac{d}{2}+\overline{a}(k-j))^2\}} \\ &= \mu^{-a(\frac{d}{2}+\overline{a}(k-j))^2} G(a, 0, 2d). \end{aligned}$$

(c) By (a) and (b), the modulus of the entries of $F^{-1}R^a F$ for d odd and even are

$$\frac{1}{2d}\cdot 2\cdot\sqrt{d} = \frac{1}{\sqrt{d}}, \qquad \frac{1}{2d}\cdot\sqrt{2}\sqrt{2d} = \frac{1}{\sqrt{d}}.$$

Exercises of Chapter 14

14.1 (a) Expanding gives $\langle P_j - c_1 I, P_k - c_2 I\rangle = \frac{1}{d+1} - c_1 - c_2 + dc_1c_2 = 0$.
(b) For $c_1 = 0$, we solve the above to get $c_2 = \frac{1}{d+1}$. Thus $(P_j - \frac{1}{d+1}I)$ is the dual basis to (P_j).
(c) Since the vectors giving (P_j) are a tight frame, we have $\sum_j P_j = dI$, i.e., the linear dependency $\sum_j (P_j - \frac{1}{d}I) = 0$. Suppose that $\sum_j a_j(P_j - cI) = 0$ for some scalars a_j, i.e., $\sum_j a_j P_j = c\sum_j a_j I$. Since (P_j) is a basis and $\sum_j P_j = dI$, the a_j must be constant, say $a_j = \alpha$, and the condition becomes $\alpha \sum_j P_j = \alpha dI = d^2 c\alpha I$. Thus (a_j) can be nonzero if and only if $c = \frac{1}{d}$.
(d) By (a) and (c), $(P_j - cI)$ is orthogonal if and only if $\frac{1}{d+1} - 2c + dc^2 = 0$, $c \neq \frac{1}{d}$, i.e., $c = \frac{1}{d}(1 \pm \frac{1}{\sqrt{d+1}})$.
(e) Since the traceless matrices are the orthogonal complement of I, the projection is

$$P_j - cI - \frac{\langle P_j - cI, I\rangle}{\langle I, I\rangle} I = P_j - cI - \frac{1}{d}(1 - cd)I = P_j - \frac{1}{d}I.$$

Remark: By (d), c can be chosen so $(P_j - cI)$ is an orthogonal basis, and so this shows that $(P_j - \frac{1}{d})$ is an (equiangular) tight frame for the traceless matrices.

14.2 (a) Expanding, and using the trigonometric identities $\sin 2x = 2\sin x \cos x$, $2\cos^2 x = 1 + \cos 2x$, we have

$$vv^* = \begin{pmatrix} \cos^2\frac{\theta}{2} & \sin\frac{\theta}{2}\cos\frac{\theta}{2}e^{-i\phi} \\ \sin\frac{\theta}{2}\cos\frac{\theta}{2}e^{i\phi} & \sin^2\frac{\theta}{2} \end{pmatrix} = \frac{1}{2}\begin{pmatrix} 1+c & a-ib \\ a+ib & 1-c \end{pmatrix}.$$

(b) Using (a), the image of vv^* is

$$A = \frac{\sqrt{2}}{2}\begin{pmatrix} c & a-ib \\ a+ib & -c \end{pmatrix},$$

where $\|A\|_F^2 = \frac{1}{2}(c^2 + (a^2+b^2) + (a^2+b^2) + c^2) = 1$. The form of A clearly shows that the map is onto.
(c) Since A is Hermitian, it suffices to calculate its trace and Frobenius norm:

$$\operatorname{trace}(A) = \sqrt{\frac{d}{d-1}}(vv^* - \frac{1}{d}I) = \sqrt{\frac{d}{d-1}}(1 - \frac{d}{d}) = 0,$$

$$\|A\|_F^2 = \operatorname{trace}(A^2) = \frac{d}{d-1}\operatorname{trace}\big(vv^* - \frac{2}{d}vv^* + \frac{1}{d^2}I\big) = \frac{d}{d-1}(1 - \frac{2}{d} + \frac{d}{d^2}) = 1.$$

If the map is onto, then $P = \sqrt{\frac{d-1}{d}}A + \frac{1}{d}I$ is a rank-one orthogonal projection on $\mathbb{C}^d$ for any traceless Hermitian matrix A with unit Frobenius norm. If A is zero except for the principal 2×2 submatrix B, then

$$P=\sqrt{\frac{d-1}{d}}\begin{pmatrix}B&0\\0&0\end{pmatrix}+\frac{1}{d}\begin{pmatrix}I_2&0\\0&I_{d-2}\end{pmatrix}=\begin{pmatrix}\sqrt{\frac{d-1}{d}}B+\frac{1}{d}I_2&0\\0&\frac{1}{d}I_{d-2}\end{pmatrix},$$

which clearly has rank larger than one for $d>3$.
See [SoS16] for more detail about the Bloch sphere and its relationship with SICs.

14.3 (a) The synthesis map of the Bloch vectors for $\Phi=(v,Sv,\Omega v,S\Omega v)$ is

$$\frac{1}{\sqrt{3}}\begin{pmatrix}1&1&-1&-1\\1&-1&-1&1\\1&-1&1&-1\end{pmatrix},$$

which clearly gives the vertices of a regular tetrahedron.
(b) Taking the first column of $2vv^*=\begin{pmatrix}1+c&a-ib\\a+ib&1-c\end{pmatrix}$ gives the SIC

$$\frac{1}{\sqrt{18-6\sqrt{3}}}\begin{pmatrix}3-\sqrt{3}\\\pm\sqrt{6}\end{pmatrix},\quad \frac{1}{\sqrt{18+6\sqrt{3}}}\begin{pmatrix}3+\sqrt{3}\\\pm i\sqrt{6}\end{pmatrix}.$$

14.4 (a) Since $(S^j\Omega^k)_{ab}=\sum_\alpha(S^j)_{a\alpha}(\Omega^k)_{\alpha b}=\sum_\alpha\delta_{a,\alpha+j}\omega^{kb}\delta_{\alpha b}=\omega^{kb}\delta_{a,b+j}$,

$$(\Omega^kS^j)_{ab}=\sum_\alpha(\Omega^k)_{a\alpha}(S^j)_{\alpha b}=\sum_\alpha\omega^{ka}\delta_{a\alpha}\delta_{\alpha,b+j}=\omega^{k(b+j)}\delta_{a,b+j}=(\omega^{jk}S^j\Omega^k)_{ab}.$$

(b) Use (a) and induction on r (it clearly holds for $r=0,1$):

$$\begin{aligned}(S^j\Omega)^r&=\omega^{\frac{1}{2}(r-1)(r-2)jk}S^{(r-1)j}\Omega^{(r-1)k}S^j\Omega^k\\&=\omega^{\frac{1}{2}(r-1)(r-2)jk}S^{(r-1)j}(\omega^{(r-1)jk}S^j\Omega^{(r-1)k})\Omega^k=\omega^{\frac{1}{2}r(r-1)jk}S^{rj}\Omega^{rk}.\end{aligned}$$

In particular, for d even, we have $(S\Omega)^d=-I$, and so $S\Omega$ has order $2d$.
(c) Since $c\overline{c}=1$, repeated application of (a) gives

$$\begin{aligned}h(S^j\Omega^k)h^{-1}&=S^a\Omega^bS^j(\Omega^k\Omega^{-b}S^{-a})=S^a\Omega^bS^j(\omega^{-a(k-b)}S^{-a}\Omega^{k-b})\\&=\omega^{-a(k-b)}S^a(\Omega^bS^{j-a})\Omega^{k-b}=\omega^{-a(k-b)+b(j-a)}S^j\Omega^k=\omega^{bj-ak}S^j\Omega^k.\end{aligned}$$

14.5 Using, $(S^j\Omega^k)_{\alpha\beta}=\omega^{\beta k}\delta_{\alpha,\beta+j}$, we have

$$\begin{aligned}(F(S^j\Omega^k)F^{-1})_{ab}&=\sum_\alpha\sum_\beta(F)_{a\alpha}(S^j\Omega^k)_{\alpha\beta}(F^*)_{\beta b}=\sum_\alpha\sum_\beta\frac{\omega^{a\alpha}}{\sqrt{d}}\omega^{\beta k}\delta_{\alpha,\beta+j}\frac{\omega^{-b\beta}}{\sqrt{d}}\\&=\frac{1}{d}\sum_\alpha\omega^{a\alpha}\omega^{(\alpha-j)k}\omega^{-b(\alpha-j)}=\omega^{-jk}\omega^{bj}\frac{1}{d}\sum_\alpha\omega^{(a+k-b)\alpha}\\&=\omega^{-jk}\omega^{bj}\delta_{a,b-k}=(\omega^{-jk}S^{-k}\Omega^j)_{ab},\end{aligned}$$

$$(R(S^j\Omega^k)R^{-1})_{ab}=\sum_\alpha\sum_\beta(R)_{a\alpha}(S^j\Omega^k)_{\alpha\beta}(R^*)_{\beta b}$$

$$
\begin{aligned}
&= \sum_\alpha \sum_\beta \mu^{a(a+d)} \delta_{a\alpha} \omega^{\beta k} \delta_{\alpha,\beta+j} \mu^{-b(b+d)} \delta_{\beta b} \\
&= \mu^{(b+j)(b+j+d)} \omega^{bk} \mu^{-b(b+d)} \delta_{a,b+j} = \mu^{j(j+d)+2bj} \omega^{bk} \delta_{a,b+j} \\
&= \mu^{j(j+d)} \omega^{b(j+k)} \delta_{a,b+j} = (\mu^{j(j+d)} S^j \Omega^{j+k})_{ab}, \\
M(S^j\Omega^k)M^{-1} &= R(F^{-1}S^j\Omega^k F)R^{-1} = R(\omega^{-jk}S^{-k}\Omega^j)R^{-1} \\
&= \omega^{-jk}\mu^{-k(-k+d)}S^{-k}\Omega^{j-k} = \mu^{k(k-2j+d)}S^{-k}\Omega^{j-k}, \\
(P_\sigma(S^j\Omega^k)P_\sigma^{-1})_{ab} &= \sum_\alpha \sum_\beta (P_\sigma)_{a\alpha}(S^j\Omega^k)_{\alpha\beta}(P_\sigma^*)_{\beta b} \\
&= \sum_\alpha \sum_\beta \delta_{a,\sigma\alpha} \omega^{\beta k} \delta_{\alpha,\beta+j} \delta_{b,\sigma\beta} = \omega^{\sigma^{-1}kb} \delta_{\sigma^{-1}a,\sigma^{-1}b+j} \\
&= \omega^{\sigma^{-1}bk} \delta_{a,b+\sigma j} = (S^{\sigma j}\Omega^{\sigma^{-1}k})_{ab}.
\end{aligned}
$$

14.6 The matrices R^d and $\Omega^{\frac{d}{2}}$ are diagonal, with

$$(R^d)_{jj} = \mu^{j(j+d)d} = (-1)^{j(j+d)} = (-1)^{j^2} = (-1)^j = (\Omega^{\frac{d}{2}})_{jj}.$$

From (14.13), we have $F(S^{\frac{d}{2}})F^{-1} = \Omega^{\frac{d}{2}}$.

14.7 We have associativity

$$\big((A,z_A)(B,z_B)\big)(C,z_C) = \big(AB,(z_A\circ B)z_B\big)(C,z_C) = \big(ABC,(z_A\circ BC)(z_B\circ C)z_C\big),$$

$$(A,z_A)\big((B,z_B)(C,z_C)\big) = (A,z_A)\big(BC,(z_B\circ C)z_C\big) = \big(ABC,(z_A\circ BC)(z_B\circ C)z_C\big),$$

and identity $(I,1)$

$$(I,1)(A,z_A) = (IA,(1\circ A)z_A) = (A,z_A), \qquad (A,z_A)(I,1) = (AI,(z_A\circ I)1) = (A,z_A),$$

with inverse $(A,Z_A)^{-1} = (A^{-1}, z_A^{-1}\circ A^{-1}) = (A^{-1},(Z_A\circ A^{-1})^{-1})$

$$
\begin{aligned}
(A,Z_A)^{-1}(A,Z_A) &= (A^{-1},Z_A^{-1}\circ A^{-1})(A,Z_A) = (A^{-1}A,(Z_A^{-1}\circ A^{-1}A)Z_A) = (I,1), \\
(A,Z_A)(A,Z_A)^{-1} &= (A,Z_A)(A^{-1},Z_A^{-1}\circ A^{-1}) = (AA^{-1},(Z_A\circ A^{-1})(Z_A^{-1}\circ A^{-1})) = (I,1).
\end{aligned}
$$

14.8 (a) Expanding, and using $\det(A) = \alpha\delta - \beta\gamma = 1$, gives

$$
\begin{aligned}
c_A(p,q) &= (\gamma p_1 + \delta p_2)(\alpha q_1 + \beta q_2) - p_2 q_1 \\
&= \alpha\gamma p_1 q_1 + \beta\gamma p_1 q_2 + (\alpha\delta - 1)p_2 q_1 + \beta\delta p_2 q_2 \\
&= \alpha\gamma p_1 q_1 + \beta\gamma p_1 q_2 + \beta\gamma p_2 q_1 + \beta\delta p_2 q_2 = p^T \sigma_A q.
\end{aligned}
$$

(b) Since $c_A(p+q,p+q) = c_A(p,p) + 2c_A(p,q) + c_A(q,q)$, we have

$$\frac{\hat{z}_a(p+q)}{\hat{z}_a(p)\hat{z}_a(q)} = \frac{z_a(p+q)}{z_a(p)z_a(q)} \frac{(-\mu)^{c_A(p,p)}(-\mu)^{c_A(q,q)}}{(-\mu)^{c_A(p+q,p+q)}} = \frac{\omega^{c_A(p,q)}}{(-\mu)^{2c_A(p,q)}} = 1.$$

Let $B := \psi_b$. Substituting into $z_{ab}(p) = z_a(Bp)z_b(p)$ gives

$$\hat{z}_{ab}(p)\frac{(-\mu)^{(ABp)_1(ABp)_2}}{(-\mu)^{p_1p_2}} = \hat{z}_a(Bp)\frac{(-\mu)^{(ABp)_1(ABp)_2}}{(-\mu)^{(Bp)_1(Bp)_2}}\hat{z}_b(p)\frac{(-\mu)^{(Bp)_1(Bp)_2}}{(-\mu)^{p_1p_2}},$$

and cancellation gives $\hat{z}_{ab}(p) = \hat{z}_a(Bp)\hat{z}_b(p)$. From (14.19), (14.20), we get

$$\hat{z}_F(j,k) = \omega^{-jk}(-\mu)^{jk-(-k)j} = \omega^{-jk}\omega^{jk} = 1,$$
$$\hat{z}_R(j,k) = \mu^{j(j+d)}(-\mu)^{jk-j(j+k)} = \mu^{j(j+d)}\mu^{-j^2}(-1)^{-j^2} = \mu^{jd}(-1)^j = 1.$$

Since $M = RF$, this gives $\hat{z}_M = (\hat{z}_R \circ \psi_F)\hat{z}_F = 1$. For $a = S^j\Omega^k$, $\psi_a = I$, so that $\hat{z}_{S^j\Omega^k} = z_{S^j\Omega^k}$. A simple calculation (see Exer. 14.4) gives $z_{S^j\Omega^k}$, i.e.,

$$(S^j\Omega^k)S^{p_1}\Omega^{p_2}(S^j\Omega^k)^{-1} = \omega^{kp_1-jp_2}S^{p_1}\Omega^{p_2} \quad \Longrightarrow \quad z_{S^j\Omega^k}(p) = \omega^{kp_1-jp_2}.$$

(c) For any $B = \begin{pmatrix} \alpha & \beta \\ \gamma & \delta \end{pmatrix} \in SL_2(\mathbb{Z}_{2d})$, expanding $(p+dj)^T\sigma_B(p+dj)$ gives

$$\alpha\gamma(p_1^2+2dj_1p_1+d^2j_1^2)+2\beta\gamma(p_1p_2+dj_1p_2+dj_2p_1+d^2j_1j_2)$$
$$+\beta\delta(p_2^2+2dj_2p_2+d^2j_2^2) \equiv \alpha\gamma p_1^2+2\beta\gamma p_1p_2+\beta\delta p_2^2 \equiv p^T\sigma_Bp \bmod 2d.$$

Hence the argument of (b) goes through.

14.9 (a) From $z_{ab}(p) = z_b(p)z_a(\psi_b(p))$ and $\psi_{ab}(p) = \psi_a(\psi_b(p))$, we obtain

$$\frac{c_1^{p_1}c_2^{p_2}}{c_1^{\psi_{ab}(p)_1}c_2^{\psi_{ab}(p)_2}}z_{ab}(p) = \frac{c_1^{p_1}c_2^{p_2}}{c_1^{\psi_b(p)_1}c_2^{\psi_b(p)_2}}z_b(p)\frac{c_1^{\psi_b(p)_1}c_2^{\psi_b(p)_2}}{c_1^{\psi_a(\psi_b(p))_1}c_2^{\psi_a(\psi_b(p))_2}}z_a(\psi_b(p)),$$

which is $\hat{z}_{ab} = \hat{z}_b(\hat{z}_a \circ \psi_b)$. Since $S^a\Omega^b(S^j\Omega^k)(S^a\Omega^b)^{-1} = \omega^{bj-ak}S^j\Omega^k$, we have

$$\hat{z}_{S^a\Omega^b}(j,k) = z_{S^a\Omega^b}(j,k) = \omega^{bj-ak}.$$

(b) Since $\det(S) = \det(\Omega) = (-1)^{d-1}$, and $(-\mu)^d = (-1)^d\mu^d = (-1)^{d+1}$, we have

$$\det(\hat{U}_p) = \det(c_1^jc_2^kS^j\Omega^k) = (-\mu)^{d(j+k)}(-1)^{(d-1)(j+k)} = (-1)^{2d(j+k)} = 1.$$

Hence, $z_a(p)^d = \det(z_a(p)\hat{U}_{\psi_a(p)}) = \det(a\hat{U}_pa^{-1}) = 1$, i.e., $\hat{z}_a(p)$ is a dth root of unity. From Lemma 14.1, we calculate

$$\hat{z}_F(j,k) = \frac{\mu^{(d+1)(j+k)}}{\mu^{(d+1)(k-j)}}\omega^{-jk} = \omega^{j-jk}.$$

$$\hat{z}_R(j,k) = \frac{\mu^{(d+1)(j+k)}}{\mu^{(d+1)(j+j+k)}}\mu^{j(j+d)} = \mu^{j(j-1)}.$$

$$\hat{z}_M(j,k) = \frac{\mu^{(d+1)(j+k)}}{\mu^{(d+1)(-k+j-k)}}\mu^{k(k-2j+d)} = \mu^{k(k-2j+3)}.$$

$$\hat{z}_M(j,k) = \frac{\mu^{(d+1)(j+k)}}{\mu^{(d+1)(\sigma^{-1}j+\sigma k)}} = \mu^{(d+1)((1-\sigma^{-1})j+(1-\sigma)k)}.$$

These are indeed powers of ω (for the last, if d is even then σ,σ^{-1} are odd).

14.10 (a) For $M = \begin{pmatrix}\alpha & \beta\\ \gamma & \delta\end{pmatrix}$, the condition $M^TAM = A$ gives

$$\begin{pmatrix}\alpha & \beta\\ \gamma & \delta\end{pmatrix}^T \begin{pmatrix}0 & 1\\ -1 & 0\end{pmatrix} \begin{pmatrix}\alpha & \beta\\ \gamma & \delta\end{pmatrix} = \begin{pmatrix}0 & \alpha\delta-\beta\gamma\\ -(\alpha\delta-\beta\gamma) & 0\end{pmatrix} = \begin{pmatrix}0 & 1\\ -1 & 0\end{pmatrix},$$

which is equivalent to $\det(M) = \alpha\delta - \beta\gamma = 1$.
(b) The inverse of $M \in SL_2(\mathbb{Z})$ is given by the formula

$$\begin{pmatrix}\alpha & \beta\\ \gamma & \delta\end{pmatrix} = \frac{1}{\alpha\delta-\beta\gamma}\begin{pmatrix}\delta & -\beta\\ -\gamma & \alpha\end{pmatrix} = \begin{pmatrix}\delta & -\beta\\ -\gamma & \alpha\end{pmatrix}.$$

(c) As in (b), the inverse in $SL_2(\mathbb{Z}_d)$ is given by

$$\begin{pmatrix}\alpha & \beta\\ \gamma & \delta\end{pmatrix} = \begin{pmatrix}\delta & -\beta\\ -\gamma & \alpha\end{pmatrix}.$$

14.11 Suppose that $b \in \mathbb{Z}_{d'}^*$ has odd order, say $2a+1$. Since b^a is a unit, we have

$$\sqrt{d}c_{b,d} = \sum_{j\in\mathbb{Z}_d} \mu^{bj(j+d)} = \sum_{j=0}^{d-1} \mu^{bb^a j(b^a j+d)} = \sum_{j=0}^{d-1} \mu^{j^2+b^{a+1}dj} = \sum_{j=0}^{d-1} e^{\frac{\pi i}{d}(j^2+b^{a+1}dj)}.$$

Evaluating the Gauss sum using the quadratic reciprocity law (14.108) gives

$$\sqrt{d}c_{b,d} = \sqrt{d}e^{\frac{\pi i}{4d}(d-(b^{a+1}d)^2)} = \sqrt{d}(\sqrt{i})^{1-b^{2a+2}d}.$$

For d even, b must be odd, and for d odd, we may assume that b is odd (by replacing it by $b+d$ if need be). Thus $b^2 \equiv 1 \bmod 8$, so that $c_{b,d} = (\sqrt{i})^{1-d}$.

14.12 (a) The coefficient matrix of the $k\times k$ linear system

$$\begin{pmatrix}1 & 1 & \cdots & 1\\ \lambda_1 & \lambda_2 & \cdots & \lambda_k\\ \lambda_1^2 & \lambda_2^2 & \cdots & \lambda_k^2\\ \vdots & \vdots & & \\ \lambda_1^{k-1} & \lambda_2^{k-1} & \cdots & \lambda_k^{k-1}\end{pmatrix} \begin{pmatrix}P_1\\ P_2\\ P_3\\ \vdots\\ P_k\end{pmatrix} = \begin{pmatrix}I\\ A\\ A^2\\ \vdots\\ A^{k-1}\end{pmatrix}$$

is the transpose of a Vandermonde matrix and so is invertible. Thus the system can be solved to give each P_j as a unique linear combination of $I,A,A^2,\ldots,A^{k-1}$.
(b) For the eigenvalues $1,\omega,\omega^2,\ldots,\omega^{k-1}$ the coefficient matrix is the $k\times k$ Fourier matrix F_k (multiplied by $\sqrt{k}$). Since F_k is unitary, we can compute the inverse of the coefficient matrix by taking its Hermitian transpose, to solve for P_{ω^j}

$$\begin{pmatrix} P_1 \\ P_\omega \\ P_{\omega^2} \\ \vdots \\ P_{\omega^{k-1}} \end{pmatrix} = \frac{1}{k} \begin{pmatrix} 1 & 1 & \cdots & 1 \\ 1 & \overline{\omega} & \cdots & \overline{\omega}^{k-1} \\ 1 & \overline{\omega}^2 & \cdots & \overline{\omega}^{2(k-1)} \\ \vdots & \vdots & & \\ 1 & \overline{\omega}^{k-1} & \cdots & \overline{\omega}^{(k-1)^2} \end{pmatrix} \begin{pmatrix} I \\ A \\ A^2 \\ \vdots \\ A^{k-1} \end{pmatrix}.$$

(c) If A is $d \times d$, then by (b) the multiplicity of the eigenvalue λ is

$$\operatorname{trace}(P_\lambda) = \frac{1}{k}\big(d + \overline{\lambda}\operatorname{trace}(A) + \overline{\lambda}^2 \operatorname{trace}(A^2) + \cdots + \overline{\lambda}^{k-1}\operatorname{trace}(A^{k-1})\big).$$

If A is unitary, i.e., $A^{-1} = A^*$, and $\frac{k}{2} < \ell \leq k-1$, then $A^\ell = A^{\ell-k} = (A^*)^{k-\ell}$ gives

$$\operatorname{trace}(A^\ell) = \operatorname{trace}((A^{k-\ell})^*) = \overline{\operatorname{trace}(A^{k-\ell})}, \qquad 1 \leq k-\ell < \frac{k}{2}.$$

(d) It follows from $F^2 = P_{-1}$ that F has order 4. Since $F^3 = F^{-1} = \overline{F}$, we have

$$m_\lambda = \frac{1}{4}\big(d + \overline{\lambda}\operatorname{trace}(F) + \overline{\lambda}^2 \operatorname{trace}(F^2) + + \overline{\lambda}^3 \overline{\operatorname{trace}(F)}\big),$$

where, by the quadratic reciprocity law (14.108),

$$\operatorname{trace}(F) = \frac{1}{\sqrt{d}} G(1,0,d) = \frac{1}{2}(1+i)\big(1+(-i)^d\big),$$

and

$$\operatorname{trace}(F^2) = \sum_j \frac{1}{d} \sum_\alpha (\omega^{2j})^\alpha = \frac{1}{d} \sum_{j:2j=0} \sum_\alpha 1^\alpha = \frac{1}{2}\big(3 + (-1)^d\big).$$

Thus m_λ depends on d modulo 4, as given by Table 14.6.
(e) Since $Z^3 = I$, $Z^2 = Z^{-1} = Z^*$, we have

$$m_\lambda = \frac{1}{3}\big(d + \overline{\lambda}\operatorname{trace}(Z) + \overline{\lambda}^2 \overline{\operatorname{trace}(Z)}\big).$$

With $\tau = e^{\frac{2\pi i}{3}}$, the quadratic reciprocity law (14.108), for $a = 3$, $b = c = d$, gives

$$\operatorname{trace}(Z) = \zeta^{d-1} \frac{1}{\sqrt{d}} \sum_j \mu^{j(j+d)+2j^2} = \zeta^{d-1} \frac{1}{\sqrt{d}} \sqrt{\frac{d}{3}} \zeta^{3-d} \sum_{k=0}^{2} \tau^{-2dk(k+1)}$$

$$= \frac{\zeta^2}{\sqrt{3}}(2+\tau^{-d}) = \frac{1}{3}(1-\tau^2)(2+\tau^{-d}) = \begin{cases} 2+\tau, & d \equiv 0 \bmod 3; \\ 1, & d \equiv 1 \bmod 3; \\ 1+\tau, & d \equiv 2 \bmod 3. \end{cases}$$

Thus m_λ depends on d modulo 3, as given by Table 14.7.

14.13 (a) It suffices to show that R has order d'. Since $R_{jj} = (-\mu)^{j^2}$ and $-\mu$ is a primitive d'th root of unity, the diagonal matrix R has order d' (consider $j = 1$).

(b),(c) Let $\tau = e^{\frac{2\pi i}{\ell}}$, a primitive ℓth root of unity. Take $a=\ell$, $b=\ell d+2(k-j)$, $c=d$ in the quadratic reciprocity law (14.108), to get

$$\begin{aligned}(F^{-1}R^{\ell}F)_{jk} &= \sum_{\alpha}\sum_{\beta}(F^{-1})_{\alpha j}(R^{\ell})_{\alpha\beta}(F)_{\beta k} = \frac{1}{d}\sum_{\alpha=0}^{d-1}\mu^{-2\alpha j}\mu^{\ell\alpha(\alpha+d)}\mu^{2\alpha k} \\ &= \frac{1}{d}\sqrt{\frac{d}{\ell}}e^{\frac{\pi i}{4\ell d}\left(\ell d-(\ell d+2(k-j))^2\right)}\sum_{\beta=0}^{\ell-1}e^{-\frac{\pi i}{\ell}(d\beta^2+(\ell d+2(k-j))\beta)} \\ &= \frac{1}{d}\sqrt{\frac{d}{\ell}}(\sqrt{i})^{1-\ell d}(-1)^{j-k}\mu^{-\frac{1}{\ell}(k-j)^2}\sum_{\beta=0}^{\ell-1}(-1)^{\frac{d}{\ell}\beta^2+d\beta}\tau^{(j-k)\beta}.\end{aligned}$$

For d odd, and for d even and ℓ odd, we have

$$(-1)^{\frac{d}{\ell}\beta^2+d\beta} = (-1)^{\frac{d}{\ell}\beta+d\beta} = (-1)^{\beta(\frac{d}{\ell}+d)} = 1,$$

so that the last sum becomes $\sum_{\beta}(\tau^{j-k})^{\beta}$, which is ℓ if $\tau^{j-k}=1$, i.e., $j-k\equiv 0 \bmod \ell$ and is zero otherwise, which gives (b). For d even and ℓ even, we have

$$(-1)^{\frac{d}{\ell}\beta^2+d\beta} = (-1)^{\frac{d}{\ell}\beta} = e^{-\frac{2\pi i}{\ell}\frac{d}{2}\beta} = \tau^{-\frac{d}{2}\beta},$$

so the sum becomes $\sum_{\beta}(\tau^{j-k-\frac{d}{2}})^{\beta}$, which is ℓ if $j-k-\frac{d}{2}\equiv 0 \bmod \ell$ and is zero otherwise, which gives (c).

14.14 (a) From Exer. 14.13 (b), we have

$$(F^{-1}R^3F)_{jk} = \sqrt{\frac{3}{d}}(\sqrt{i})^{1-3d}\begin{cases}(-\mu)^{-\frac{1}{3}(j-k)^2}, & j-k\equiv 0 \bmod 3;\\ 0, & j-k\not\equiv 0 \bmod 3,\end{cases}$$

since $(-1)^{j-k} = (-1)^{\frac{1}{3}(j-k)} = (-1)^{-\frac{1}{3}(j-k)^2}$ for $j-k\equiv 0 \pmod 3$. Multiplication by $(-1)^{d-1}=(-1)^{1-3d}$, followed by left and right multiplication by the diagonal matrices $R^{\frac{2d}{3}a}$ and R, gives the desired formula for $(W_a)_{jk}$.

(b) The symplectic operation $[M]$, $M := R^{\frac{2d}{3}a}F^{-1}R^3FR$, has order 3, i.e., $(cM)^3=I$, for some scalar $c\in\mathbb{T}$, since it has the following symplectic index of order 3

$$w_a = \begin{pmatrix}-2 & -3\\ 1+\frac{2d}{3}a & 1\end{pmatrix} = \begin{pmatrix}1&0\\1&1\end{pmatrix}^{\frac{2d}{3}a}\begin{pmatrix}0&-1\\1&0\end{pmatrix}^{-1}\begin{pmatrix}1&0\\1&1\end{pmatrix}^{3}\begin{pmatrix}0&-1\\1&0\end{pmatrix}\begin{pmatrix}1&0\\1&1\end{pmatrix}.$$

Since cM is unitary, we can determine c from $(cM)^2=(cM)^{-1}=c^{-1}M^*$ (at some nonzero entry). By the formula of (a) divided by $(-1)^{d-1}=(-1)^{1-3d}$, we calculate

$$(M^2)_{00} = \sum_{\alpha=0}^{d-1}M_{0\alpha}M_{\alpha 0} = \left(\sqrt{\frac{3}{d}}(\sqrt{i})^{1-3d}\right)^2\sum_{\alpha\equiv 0 \bmod 3}(-\mu)^{-\frac{1}{3}\alpha^2+\alpha^2-\frac{1}{3}\alpha^2+\frac{2d}{3}a\alpha^2}.$$

Since $(-1)^{3\beta^2} = (-1)^{\beta} = \mu^{d\beta}$ and $3 \mid \alpha$, the sum over $\alpha \equiv 0 \pmod 3$ evaluates to

$$\sum_{\alpha}(-\mu)^{\frac{1}{3}\alpha^2} = \sum_{\beta=0}^{\frac{d}{3}-1}(-\mu)^{3\beta^2} = \sum_{\beta=0}^{\frac{d}{3}-1}\mu^{3\beta^2+d\beta} = \sum_{\beta=0}^{\frac{d}{3}-1} e^{\frac{\pi i}{d/3}(\beta^2+\frac{d}{3}\beta)} = \sqrt{\frac{d}{3}}(\sqrt{i})^{1-\frac{d}{3}},$$

by the quadratic reciprocity law (14.108) Since $(M^*)_{00} = \sqrt{\frac{3}{d}}(\sqrt{i})^{-(1-3d)}$, we have

$$c^3 = \frac{(M^*)_{00}}{(M^2)_{00}} = \frac{(\sqrt{i})^{-(1-3d)}}{(\sqrt{i})^{2(1-3d)}(\sqrt{i})^{1-\frac{d}{3}}} = (\sqrt{i})^{4(\frac{d}{3}-1)} = (-1)^{\frac{d}{3}-1} = (-1)^{d-1},$$

so that $c = (-1)^{d-1}$, and W_a has order 3.

(c) Since $F^{-1}R^3F$ is circulant, with constant diagonal $\sqrt{\frac{3}{d}}(\sqrt{i})^{1-3d}$, we have

$$\operatorname{trace}(W_a) = (-1)^{d-1}\operatorname{trace}(R^{\frac{2d}{3}a+1}F^{-1}R^3F) = (-1)^{d-1}\sqrt{\frac{3}{d}}(\sqrt{i})^{1-3d}\operatorname{trace}(R^{\frac{2d}{3}a+1}).$$

Since $\tau^3 = 1$, $\mu^{\frac{2d}{3}} = \tau$, and $3 \mid d$, we calculate

$$\operatorname{trace}(R^{\frac{2d}{3}a+1}) = \sum_{j=0}^{d-1}\mu^{(\frac{2d}{3}a+1)j(j+d)} = \sum_{j=0}^{d-1}(\tau^a\mu)^{(\frac{2d}{3}a+1)j(j+d)} = \sum_{j=0}^{d-1}\tau^{aj^2}\mu^{j^2+dj}$$

$$= \sum_{c=0}^{2}\sum_{k=0}^{\frac{d}{3}-1}\tau^{a(3k+c)^2}\mu^{(3k+c)^2+d(3k+c)} = \sum_{c=0}^{2}\tau^{ac^2}\mu^{cd+c^2}\sum_{k=0}^{\frac{d}{3}-1}(\mu^3)^{3k^2+(d+2c)k}.$$

By the quadratic reciprocity law, the sum over k can be written as

$$\sum_{k=0}^{\frac{d}{3}-1} e^{\frac{\pi i}{d/3}(3k^2+(d+2c)k)} = \sqrt{\frac{d}{9}}e^{\frac{\pi i}{4d}(d-(d+2c)^2)}\sum_{\beta=0}^{2} e^{-\frac{\pi i}{3}(\frac{d}{3}\beta^2+(d+2c)\beta)}$$

$$= \sqrt{\frac{d}{9}}\mu^{-cd-c^2}(\sqrt{i})^{1-d}\sum_{\beta=0}^{2}(-\tau)^{\frac{d}{3}\beta^2+(d+2c)\beta},$$

where the sum over β simplifies to $1+\tau^{\frac{d}{3}-c}+\tau^{\frac{d}{3}+c}$. Thus

$$\operatorname{trace}(W_a) = (-1)^{d-1}\sqrt{\frac{3}{d}}(\sqrt{i})^{1-3d}\sqrt{\frac{d}{9}}(\sqrt{i})^{1-d}\sum_{c=0}^{2}\tau^{ac^2}\left(1+\tau^{\frac{d}{3}-c}+\tau^{\frac{d}{3}+c}\right)$$

$$= -\frac{i}{\sqrt{3}}\left(1+2\tau^{\frac{d}{3}}+2\tau^a(1-\tau^{\frac{d}{3}})\right).$$

(d) Since $\overline{\tau}^b = \tau^{-b} = \tau^{2b} = (\tau^b)^2$, the eigenvalues $\lambda = 1, \tau, \tau^2$ of W_a satisfy $\overline{\lambda} = \lambda^2$, and so, by Exer. 14.12, the multiplicity of the eigenvalue λ of W_a is

$$m_\lambda = \frac{1}{3}\big(d + \lambda^2 \operatorname{trace}(W_a) + \lambda \overline{\operatorname{trace}(W_a)}\big).$$

which depends on a and $\frac{d}{3}$ (mod 3).
(i) For $d \equiv 3 \bmod 9$, $d \neq 3$, $\frac{d}{3} \equiv 1 \pmod 3$, so that $\tau^{\frac{d}{3}} = \tau$, $\operatorname{trace}(W_1) = 3$, and

$$m_\lambda = \frac{1}{3}\big(d + 3(\lambda^2 + \lambda)\big) = \begin{cases} \frac{d+6}{3}, & \lambda = 1; \\ \frac{d-3}{3}, & \lambda = \tau, \tau^2. \end{cases}$$

(ii) For $d \equiv 6 \bmod 9$, $\frac{d}{3} \equiv 2 \pmod 3$, so that $\operatorname{trace}(W_2) = -3$, and

$$m_\lambda = \frac{1}{3}\big(d - 3(\lambda^2 + \lambda)\big) = \begin{cases} \frac{d-6}{3}, & \lambda = 1; \\ \frac{d+3}{3}, & \lambda = \tau, \tau^2. \end{cases}$$

(e) The fact $cW_0 = (R^{-1}F)^{-1}Z(R^{-1}F)$, with c a scalar, follows from the conjugacy

$$w_0 = g^{-1}zg, \qquad g := \begin{pmatrix} 0 & -1 \\ 1 & 1 \end{pmatrix} = \begin{pmatrix} 1 & 0 \\ 1 & 1 \end{pmatrix}^{-1} \begin{pmatrix} 0 & -1 \\ 1 & 0 \end{pmatrix}.$$

By matrix multiplication, $\tau^{1-d}W_0 = (R^{-1}F)^{-1}Z(R^{-1}F)$ is equivalent to

$$RFRF^{-1}RF^{-1} = (-\tau\zeta)^{d-1}I = (\sqrt{i})^{1-d}I.$$

From the calculation (14.62), we have $RFRFRF = (\sqrt{i})^{1-d}I$, and so it suffices to prove that $F^{-1}RF^{-1} = FRF$. By the quadratic reciprocity law, we have

$$\begin{aligned}(F^{-1}RF^{-1})_{jk} &= \frac{1}{d}\sum_{\alpha=0}^{d-1} \mu^{-2j\alpha}\mu^{\alpha(\alpha+d)}\mu^{-2k\alpha} = \frac{1}{d}\sum_{\alpha=0}^{d-1} e^{\frac{\pi i}{d}(\alpha^2+\alpha(d-2j-2k))} \\ &= \frac{1}{d}\sqrt{d}e^{\frac{\pi i}{4d}(d-(d-2j-2k)^2)} = \frac{1}{\sqrt{d}}(\sqrt{i})^{1-d}(-1)^{j+k}\mu^{-(j+k)^2}.\end{aligned}$$

From this, we have $(F^{-1}RF^{-1})_{jk} = (F^{-1}RF^{-1})_{-j,-k} = (FRF)_{jk}$.

14.15 By the commutativity relation (14.5), we have

$$\begin{aligned}\hat{D}_p\hat{D}_q &= (-\mu)^{p_1p_2+q_1q_2} S^{p_1}\Omega^{p_2}S^{q_1}\Omega^{q_2} = (-\mu)^{p_1p_2+q_1q_2}\omega^{p_2q_1}S^{p_1+q_1}\Omega^{p_2+q_2} \\ &= (-\mu)^{p_2q_1-p_1q_2}(-\mu)^{(p_1+p_2)(q_1+q_2)}S^{p_1+q_1}\Omega^{p_2+q_2} = (-\mu)^{\langle p,q\rangle}\hat{D}_{p+q}.\end{aligned}$$

From this, we obtain

$$\hat{D}_p\hat{D}_{-p} = (-\mu)^{\langle p,-p\rangle}\hat{D}_{p-p} = I \quad\Longrightarrow\quad \hat{D}_p^{-1} = \hat{D}_{-p},$$

$$\hat{D}_p\hat{D}_q = (-\mu)^{\langle p,q\rangle-\langle q,p\rangle}\hat{D}_q\hat{D}_p = (-\mu)^{2\langle p,q\rangle}\hat{D}_q\hat{D}_p = \omega^{\langle p,q\rangle}\hat{D}_q\hat{D}_p.$$

Now

$$\hat{D}_{p+dq} = (-\mu)^{(p_1+dq_1)(p_2+dq_2)} S^{p_1} \Omega^{p_2} = (-\mu)^{d(p_1q_2+p_2q_1+dp_1q_1)} \hat{D}_p.$$

Since $(-\mu)^d = (-1)^{d+1}$, for d odd the scalar on the left-hand side is 1, and for d even it is

$$(-1)^{p_1q_2+p_2q_1} = (-1)^{p_2q_1-p_1q_2} = (-1)^{\langle p,q\rangle}.$$

14.16 (a) Since $\hat{D}_p = (-\mu)^{p_1p_2}\hat{U}_p$, where $\hat{U}_p := U_{p \bmod d} = S^{p_1}\Omega^{p_2}$, we have

$$a(-\mu)^{p_1p_2}\hat{U}_p a^{-1} = \omega^{\langle \chi, Bp\rangle}(-\mu)^{(Bp)_1(Bp)_2}\hat{U}_{Bp}, \qquad \forall p \in \mathbb{Z}^2.$$

Since $Bp \equiv Ap \bmod d$, we have $\psi_a = A$, and

$$z_a(p) = \omega^{\langle \chi, Ap\rangle}(-\mu)^{(Bp)_1(Bp)_2 - p_1p_2}, \qquad p \in \mathbb{Z}_d^2,$$

where

$$c_B(p,p) := (Bp)_1(Bp)_2 - p_1p_2,$$

is calculated modulo d' (but depends only on p mod d). As in Exer. 14.8 (a), we have $c_B(p,p) = p^T\sigma_B p$.

(b) An element $[B,\chi] \in \ker f$ provided $\theta([B,\chi]) = (I,1)$. For d odd, this implies $B = \psi_a = I$, while $z_a(p) = \omega^{\langle \chi, p\rangle} = 1$ gives $\chi = 0$, and so f is 1–1. For d even, we obtain (14.5) since

$$\psi_a = B \bmod d = I, \quad \det(B) = 1 \quad \Longrightarrow \quad B = \begin{pmatrix} 1+rd & sd \\ td & 1+rd \end{pmatrix},$$

and using $\sigma_B = \begin{pmatrix} td & \\ & sd \end{pmatrix}$ and $(-\mu)^{dj^2} = \omega^{\frac{d}{2}j}$ gives

$$z_a(p) = \omega^{\langle \chi, p\rangle}(-\mu)^{dtp_1^2 + dsp_2^2} = 1 = \omega^{\chi_2p_1 - \chi_1p_2}\omega^{\frac{d}{2}tp_1 + \frac{d}{2}sp_2} \quad \Longrightarrow \quad \chi = \begin{pmatrix} s\frac{d}{2} \\ t\frac{d}{2} \end{pmatrix}.$$

(c) Each Appleby index has the form $[B,0]$, so that

$$z_a(p) = (-\mu)^m = \mu^{(d+1)m}, \qquad m := (Bp)_1(Bp)_2 - p_1p_2.$$

(i) Since $2d, d^2 \equiv 0 \bmod d'$, $dj^2 \equiv dj \bmod d'$, we calculate

$$m = (d-1)p_1((d+1)p_1 + (d-1)p_2) - p_1p_2 \equiv p_2^2 - 2p_1p_2 \bmod d',$$

$$(d+1)m \equiv p_2^2 - 2p_1p_2 + dp_2^2 - 2dp_1p_2 \equiv p_2^2 - 2p_1p_2 + dp_2 \bmod d',$$

so that

$$\psi_a = \begin{pmatrix} 0 & -1 \\ 1 & -1 \end{pmatrix}, \quad z_a(j,k) = \mu^{k(k-2j+d)}.$$

Thus $a = Z$ (up to a scalar).

(ii) We have

$$\begin{aligned} m &= (p_1+(d+3)p_2)((\frac{4}{3}d-1)p_1+(d-2)p_2)-p_1p_2 \\ &\equiv (\frac{4}{3}d-1)p_1^2-6p_1p_2+(d-6)p_2^2 \bmod d', \end{aligned}$$

which gives $(d+1)m \equiv (\frac{1}{3}d-1)p_1^2-6p_1p_2+(d-6)p_2^2 \bmod d'$, so that

$$\psi_a = \begin{pmatrix} 1 & 3 \\ 3\alpha & -2 \end{pmatrix}, \qquad z_a(j,k) = \mu^{(\frac{1}{3}d-1)j^2-6jk+(d-6)k^2}.$$

We have the factorisation

$$(d+1)F_a = \begin{pmatrix} d+1 & 3 \\ \frac{d-3}{3} & d-2 \end{pmatrix} = \begin{pmatrix} 1 & 0 \\ 1 & 1 \end{pmatrix}^{d-1} \begin{pmatrix} 0 & -1 \\ 1 & 0 \end{pmatrix}^{-1} \begin{pmatrix} 1 & 0 \\ 1 & 1 \end{pmatrix}^{-3} \begin{pmatrix} 0 & -1 \\ 1 & 0 \end{pmatrix} \begin{pmatrix} 1 & 0 \\ 1 & 1 \end{pmatrix}^{\frac{d}{3}},$$

and so, by Lemma 14.3, we have $a = R^{d-1}F^{-1}R^{-3}FR^{\frac{d}{3}}$ (up to a scalar multiple).
(iii) Let $\beta = \sqrt{d+1}$. We have

$$\begin{aligned} m &= (-\sqrt{d+1}p_1+dp_2)(dp_1+(d-\sqrt{d+1})p_2) \\ &\equiv -d\sqrt{d+1}(p_1^2+p_1p_2+p_2^2)+dp_1p_2 \bmod d', \end{aligned}$$

For d odd, $m(d+1) \equiv (d^2+d)p_1p_2 \equiv 0 \bmod 2d$, so that $z_a(p)=1$. For d even,

$$\begin{aligned} z_a(p) &= \mu^{(d+1)m} = \mu^m = (-1)^{-\sqrt{d+1}(p_1^2+p_1p_2+p_2^2)+p_1p_2} \\ &= (-1)^{p_1^2+p_1p_2+p_2^2+p_1p_2} = (-1)^{p_1+p_2} = \omega^{\frac{d}{2}(p_1+p_2)}. \end{aligned}$$

Thus

$$\psi_a = \begin{pmatrix} -\beta & \\ & -\beta \end{pmatrix}, \qquad z_a(j,k) = \omega^{\frac{d}{2}(d+1)(j+k)} = \begin{cases} 1, & d \text{ odd}; \\ (-1)^{j+k}, & d \text{ even}, \end{cases}$$

and we can take

$$a = S^{\frac{1}{2}d(d+1)}\Omega^{\frac{1}{2}d(d+1)}P_{-\beta} = \begin{cases} P_{-\beta}, & d \text{ odd}; \\ S^{\frac{d}{2}}\Omega^{\frac{d}{2}}P_{-\beta}, & d \text{ even}. \end{cases}$$

(iv) We observe that if d is even, then k is even and κ is odd, so that $\kappa d \equiv d \bmod d'$. Since $\det(F_c) = -1$, the operation is antiunitary. We have the factorisation

$$F_c = \begin{pmatrix} d-2\kappa & \kappa \\ d-\kappa & d+2\kappa \end{pmatrix} \begin{pmatrix} 0 & -1 \\ 1 & 0 \end{pmatrix} \begin{pmatrix} 1 & 0 \\ 0 & -1 \end{pmatrix},$$

where

$$\begin{pmatrix} d-2\kappa & \kappa \\ d-\kappa & d+2\kappa \end{pmatrix} = \begin{pmatrix} 1 & 0 \\ 1 & 1 \end{pmatrix}^{d+2} \begin{pmatrix} \kappa & 0 \\ 0 & d-3\kappa \end{pmatrix} \begin{pmatrix} 0 & -1 \\ 1 & 0 \end{pmatrix}^{-1} \begin{pmatrix} 1 & 0 \\ 1 & 1 \end{pmatrix}^{d-2}.$$

Since $\kappa(d-3\kappa) \equiv 1 \bmod d'$, κ is a unit in $\mathbb{Z}_{d'}$, and so we may take $a = R^{d+2}P_\kappa F^{-1} R^{d-2}FC$.

14.17 (a) The commutativity follows from (14.5), i.e.,

$$(S\Omega^\ell)(S\Omega^m) = S(\omega^\ell S\Omega^\ell)\Omega^m = \omega^\ell S^2\Omega^{\ell+m}.$$

(b) Since $\langle A,B\rangle = \operatorname{trace}(AB^*)$, we have

$$\langle S\Omega^\ell, S\Omega^m\rangle = \operatorname{trace}(S\Omega^\ell\Omega^{-m}S^{-1}) = \operatorname{trace}(\Omega^{\ell-m}) = 0, \quad \ell-m\in\mathbb{Z}_d^*.$$

(c) A symplectic index for $a = (R^\ell F)^{-1} = F^{-1}R^{-\ell}$ is given by

$$B = \begin{pmatrix} 0 & 1 \\ -1 & 0 \end{pmatrix} \begin{pmatrix} 1 & 0 \\ -\ell & 1 \end{pmatrix} = \begin{pmatrix} -\ell & 1 \\ -1 & 0 \end{pmatrix},$$

so that

$$z_a(j,k) = (-\mu)^{(-\ell j+k)(-j)-jk} = (-\mu)^{\ell j^2-2jk}.$$

Thus

$$(R^\ell F)^{-1}S\Omega^\ell(R^\ell F) = aS\Omega^\ell a^{-1} = z_a(1,\ell)S^{-\ell 1+\ell}\Omega^{-1} = (-\mu)^{-\ell}\Omega^{-1},$$

i.e., $S\Omega^\ell$ has distinct eigenvalues $(-\mu)^{-\ell}\omega^{-j} = \mu^{(d-1)\ell}\omega^{-j}$ with corresponding eigenvectors $R^\ell Fe_j$.

(d) The matrices in the first set are diagonal and so have eigenvectors $\{e_j\}$. Up to multiplication by a scalar, the matrices in $\mathscr{M}_\ell$ are powers of $S\Omega^\ell$ and so have the same eigenvectors. We can compute the diagonalisation by

$$(S\Omega^\ell)^r = \omega^{\frac{1}{2}r(r-1)\ell}S^r\Omega^{r\ell} = ((-\mu)^{-\ell}\Omega^{-1})^r = (-\mu)^{-r\ell}\Omega^{-r},$$

or

$$(R^\ell F)^{-1}S^r\Omega^{r\ell}(R^\ell F) = z_a(r,r\ell)S^{-\ell r+r\ell}\Omega^{-r} = (-\mu)^{-r^2\ell}\Omega^{-r}.$$

The eigenvalues will not all be distinct if r is not a unit.

14.18 (a) Recall the determinant is the product of the eigenvalues. We have

$$\det(S) = \det(FSF^{-1}) = \det(\Omega) = \prod_{j=0}^{d-1}\omega^j = \omega^{\sum_j j} = \omega^{\frac{1}{2}d(d-1)} = \mu^{d(d-1)} = (-1)^{d-1}.$$

From the formula for the sum of consecutive squares, we have

$$\sum_{j=1}^{d} j(j+d) = \frac{1}{6}d(d+1)(2d+1) + \frac{1}{2}d^2(d+1) = \frac{1}{6}d(d+1)(5d+1).$$

Since $\zeta = e^{\frac{2\pi i}{24}} = \tau^2 \sqrt{i}^3$, we have that $\det(R)$ depends on d mod 12, i.e.,

$$\det(R) = \sum_{j=1}^{d} \mu^{j(j+d)} = \mu^{\frac{1}{6}d(d+1)(5d+1)} = \zeta^{2(d+1)(5d+1)} = \tau^{1-d^2} i^{3(d+1)^2}.$$

The determinant of F and Z can be calculated from the Tables 14.6 and 14.7. Since $\det(F)$ depends on d mod 8, we first consider $\det(Z)$, for which the table gives

$$\det(Z) = \tau^{(d-1)^2} = \begin{cases} 1, & d \equiv 1 \bmod 3 \\ \tau & \text{otherwise.} \end{cases}$$

Since $Z = \zeta^{d-1} RF$, we have

$$\begin{aligned} \det(F) &= \det(\zeta^{d-1} I)^{-1} \det(R)^{-1} \det(Z) = \zeta^{-d(d-1)} \zeta^{-2(d+1)(5d+1)} \zeta^{8(d-1)^2} \\ &= \zeta^{-3d^2-3d+6} = (\sqrt{i}^3)^{-3d^2-3d+6} = \sqrt{i}^{-d^2-d+2} = i^{1-\frac{1}{2}d(d+1)}. \end{aligned}$$

(b) Since $\det(c^{-1}M) = c^{-d}\det(M)$, c is a dth root of $\det(M)$. Since μ is a dth root of -1, we can take

$$\hat{S} := \begin{cases} S, & d \text{ odd}; \\ \mu S, & d \text{ even}; \end{cases} \qquad \hat{\Omega} := \begin{cases} \Omega, & d \text{ odd}; \\ \mu\Omega, & d \text{ even}. \end{cases}$$

For F, we can take $c = i^a$ to be a fourth root of unity when $d \not\equiv 0 \bmod 4$, by solving

$$c^d = \sqrt{i}^{2ad} = \sqrt{i}^{-d^2-d+2} \quad \Longleftrightarrow \quad 2a \equiv -d^2 - d + 2 \bmod 8,$$

e.g., for $d \equiv 3 \bmod 8$, $6a \equiv 6$ gives $a = 1$, $\hat{F} = c^{-1}F = -iF$. For $d \equiv 0, 4 \bmod 8$, we must take c to be a $4d$th root of unity. By similar calculations for Z, we obtain

$$\hat{F} = \begin{cases} \mu^{-\frac{1}{2}} F, & d \equiv 0 \bmod 8; \\ F, & d \equiv 1, 6 \bmod 8; \\ iF, & d \equiv 2, 7 \bmod 8; \\ -iF, & d \equiv 3 \bmod 8; \\ \mu^{\frac{1}{2}} F, & d \equiv 4 \bmod 8; \\ -F, & d \equiv 5 \bmod 8; \end{cases} \qquad \hat{Z} = \begin{cases} \omega^{-\frac{1}{3}}, & d \equiv 0 \bmod 3; \\ Z, & d \equiv 1 \bmod 3; \\ \tau Z, & d \equiv 2 \bmod 3; \end{cases}$$

and one can take $\hat{R} = \hat{Z}\hat{F}^{-1}$.

(c) The subgroups of $\langle \hat{S}, \hat{\Omega}, \hat{F}, \hat{R} \rangle$ containing $\langle \hat{S}, \hat{\Omega} \rangle$ are canonical abstract error groups. For $d = 2$, the canonical abstract error group, index group pairs are

$$\langle 8,4 \rangle, \langle 4,2 \rangle \quad \langle 16,9 \rangle, \langle 8,3 \rangle, \quad \langle 24,3 \rangle, \langle 12,3 \rangle, \quad \langle 48,28 \rangle, \langle 24,12 \rangle.$$

These include the two which are not dicyclic groups (see Example 13.21). For $d=3$,

$$\langle 27,3\rangle,\ \langle 9,2\rangle,\quad \langle 54,8\rangle,\ \langle 18,4\rangle,\quad \langle 81,9\rangle,\ \langle 27,3\rangle,\quad \langle 108,15\rangle,\ \langle 36,9\rangle,$$
$$\langle 162,14\rangle,\ \langle 54,5\rangle,\quad \langle 216,88\rangle,\ \langle 72,41\rangle,\quad \langle 648,532\rangle,\ \langle 216,153\rangle.$$

For $d=4$, there are 19 canonical abstract error groups. The first few are

$$\langle 64,19\rangle,\ \langle 16,2\rangle,\quad \langle 128,749\rangle,\ \langle 32,34\rangle,\quad \langle 128,782\rangle,\ \langle 32,31\rangle,\quad \langle 128,545\rangle,\ \langle 32,24\rangle,$$
$$\langle 192,4\rangle,\ \langle 48,3\rangle,\quad \langle 256,24064\rangle,\ \langle 64,242\rangle,\quad \langle 256,21237\rangle,\ \langle 64,236\rangle,$$
$$\langle 256,17275\rangle,\ \langle 64,216\rangle,\quad \langle 256,217\rangle,\ \langle 64,18\rangle,\quad \langle 256,395\rangle,\ \langle 64,34\rangle.$$

The remaining ones have orders 384, 512, 768, 1024, 1536, 3072.

14.19 (a) Since F and R have finite orders (4 and d'), their eigenvalues, and hence determinants, are roots of unity. Thus F and R can be multiplied by appropriate roots of unity to obtain matrices $\hat{F}$ and $\hat{R}$ with determinant 1 (see Exer. 14.18). The group $\langle \hat{F},\hat{R}\rangle$ contains all elements of $\langle F,R\rangle$ up to multiplication by an nth root of unity (where n is fixed). Since $\langle \hat{F},\hat{R}\rangle \subset SL_d(\mathbb{C})$, the only scalar matrices that can belong to it are those given by dth roots of unity. Since

$$\frac{C_{Sp}(d)}{[I]} \cong \frac{\langle F,R\rangle}{[I]\cap\langle F,R\rangle} \cong \frac{\langle \hat{F},\hat{R},\omega I\rangle}{\langle \omega I\rangle},$$

and $C_{Sp}(d)/[I]$ is finite, we conclude that $\langle F,R\rangle$ must be finite.
(b) Since F commutes with $P_{-1}=(P_{-1})^{-1}=F^2$, it suffices to show that R does also

$$(P_{-1}RP_{-1})_{jk} = \sum_\alpha\sum_\beta \delta_{j,-\alpha}\mu^{\alpha(\alpha+d)}\delta_{\alpha,\beta}\delta_{\beta,-k} = \mu^{-j(-j+d)}\delta_{jk} = \mu^{j(j+d)}\delta_{jk} = (R)_{jk}.$$

14.20 (a) Since $A=A^*$, we have $g(A)^*=g(A)=g(A^*)$ if and only if

$$\overline{g(a_{kj})} = g(\overline{a_{kj}}), \qquad \forall j,k,$$

which is equivalent to g commuting with conjugation on $\mathbb{Q}(\{a_{jk}\})$ (since g fixes $\mathbb{Q}$).
(b) If g commutes with conjugation, then $g(v^*)=g(\overline{v})^T=\overline{g(v)}^T=g(v)^*$, so that

$$g(vv^*)=g(v)g(v^*)=g(v)g(v)^*, \qquad g(v)^*g(v)=g(v^*)g(v)=g(v^*v)=g(1)=1,$$

i.e., $g(vv^*)$ is a rank one orthogonal projection, with the necessity that g commutes with conjugation from (a).
(c) Since $g(\Pi)$ is an orthogonal projection, by (b), g must commute with complex conjugation on $\mathbb{Q}(\Pi)$. Since g is an automorphism of $\mathbb{E}$ fixing $1\in\mathbb{Q}$, it maps μ to another $2d$th root of unity, and so g commutes with complex conjugation on $\mathbb{Q}(\mu)$, and hence on the SIC field $\mathbb{E}=\mathbb{Q}(\Pi,\mu)$.

14.21 We observe that $b=F^{-1}C$ has the (extended) symplectic index $B=\begin{pmatrix}0&-1\\-1&0\end{pmatrix}$.

(a) It suffices to show the symplectic indices of b and Z do not commute:

$$Bz = \begin{pmatrix} 0 & -1 \\ -1 & 0 \end{pmatrix}\begin{pmatrix} 0 & -1 \\ 1 & -1 \end{pmatrix} = \begin{pmatrix} -1 & 1 \\ 0 & 1 \end{pmatrix} \neq \begin{pmatrix} 1 & 0 \\ 1 & -1 \end{pmatrix} = \begin{pmatrix} 0 & -1 \\ 1 & -1 \end{pmatrix}\begin{pmatrix} 0 & -1 \\ -1 & 0 \end{pmatrix} = zB.$$

From (a), it follows that the subgroup generated by b and Z is nonabelian (for $d \geq 2$).
(b) Since b and Z are symmetries of all the $d = 2,3$ SICs, their symmetry groups are nonabelian.
(c) If b and Z are symmetries of a SIC, then the symmetry group is nonabelian. Similarly, if b and M_1 are symmetries of a SIC, then the symmetry group is nonabelian, since their symplectic indices do not commute:

$$B\begin{pmatrix} d+1 & 3 \\ \frac{d-3}{3} & d-2 \end{pmatrix} = \begin{pmatrix} \frac{3-d}{3} & d+2 \\ d-1 & -3 \end{pmatrix} \neq \begin{pmatrix} -3 & d-1 \\ d+2 & \frac{3-d}{3} \end{pmatrix} = \begin{pmatrix} d+1 & 3 \\ \frac{d-3}{3} & d-2 \end{pmatrix} B.$$

14.22 Since $c\overline{c} = 1$, $c \in \mathbb{T}$, $[a]\cdot\Pi = [a]\cdot(vv^*)$ is well defined, and we observed that it maps (Weyl–Heisenberg) SIC fiducials to SIC fiducials. For a unitary, we have

$$[a]\cdot\Pi = (av)(av)^* = a(vv^*)a^* = a\Pi a^{-1}.$$

For aC antiunitary, we observe $(aC)^{-1} = C^{-1}a^{-1}$ and $\overline{\Pi} = C\Pi C^{-1}$, so that

$$[aC]\cdot\Pi = (aCv)(aCv)^* = (a\overline{v})(a\overline{v})^* = a\overline{vv^*}a^* = a\overline{\Pi}a^{-1} = (aC)\Pi(aC)^{-1}.$$

Thus, for $[a] \in \mathrm{PEC}(d)$, $[a]\cdot\Pi = a\Pi a^{-1}$ and so is an action on the SIC fiducials.

14.23 (a) It suffices to show that the conjugates of the generators of $\mathbb{E}$ are in $\mathbb{E}$. Clearly $\overline{\mu} = \mu^{-1} \in \mathbb{E}$. Since $\Pi^* = (vv^*)^* = vv^* = \Pi$, we have $\overline{\Pi} = (\Pi^*)^T = \Pi^T$, i.e., the conjugate of an entry of Π is an entry of Π.
(b) Since $[a]\cdot(vv^*) := (av)(av)^*$, $[a] \in \mathrm{PEC}(d)$ defines an action of the extended Clifford group on the Weyl–Heisenberg SIC fiducial projectors (see Exer. 14.22), it sufficies to show that the entries of $\Pi' = [a]\cdot\Pi$ belong to $\mathbb{E} = \mathbb{Q}(\Pi,\mu)$ for the nonscalar generators $a = S,\Omega,F,R,C$ (Theorem 14.1). For a unitary, $\Pi' = a\Pi a^{-1}$, so the entries of Π' belong to $\mathbb{E}$ provided that the entries of a do. This works for $a = S,\Omega,R$ (since $\mu \in \mathbb{E}$ by definition). For $a = F$, we modify this argument. We observe that $\Pi' = (\sqrt{d}F)\Pi(\sqrt{d}F)^{-1}$, where $\sqrt{d}F$ has entries in $\mathbb{E}$ (formerly the SIC field was defined to contain $\sqrt{d}$, so that F itself would have entries in the SIC field [AYAZ13]). Finally, for $a = C$, $\Pi' = \overline{\Pi}$, which has entries in $\mathbb{E}$ by (a).

14.24 (a) We have $\chi_p^{[a]^{-1}\cdot\Pi} = \mathrm{trace}(a^{-1}\Pi a\hat{D}_p) = \mathrm{trace}(\Pi a\hat{D}_p a^{-1})$, so (14.43) gives

$$\chi_p^{[a]^{-1}\cdot\Pi} = \mathrm{trace}(\Pi\omega^{\langle b,Bp\rangle}\hat{D}_{Bp}) = \omega^{\langle b,Bp\rangle}\chi_{Bp}^{\Pi}.$$

Replacing Π by $[a]\cdot\Pi$ gives the stated formula.
(b) Since $\Pi^* = \Pi$, we have $\overline{\Pi} = \Pi^T$. Using $\mathrm{trace}(A) = \mathrm{trace}(A^T)$, we calculate

$$\chi_p^{\overline{\Pi}} = \chi_p^{\Pi^T} = \mathrm{trace}(\Pi^T\hat{D}_p) = \mathrm{trace}(\hat{D}_p^T\Pi) = \mathrm{trace}(\Pi\hat{D}_p^T).$$

Since $\hat{D}_p = (-\mu)^{p_1 p_2} S^{p_1} \Omega^{p_2}$ and $S^T = S^{-1}$, $\Omega^T = \Omega$, (14.5) gives

$$\hat{D}_p^T = (-\mu)^{p_1 p_2} \Omega^{p_2} S^{-p_1} = (-\mu)^{p_1 p_2} \omega^{-p_1 p_2} S^{-p_1} \Omega^{p_2} = (-\mu)^{-p_1 p_2} S^{-p_1} \Omega^{p_2} = \hat{D}_{-Jp},$$

so that

$$\chi_p^{\overline{\Pi}} = \operatorname{trace}(\Pi \hat{D}_p^T) = \operatorname{trace}(\Pi \hat{D}_{-Jp}) = \chi_{-Jp}^{\Pi}.$$

(c) By part (a), the formula holds for an Appleby index with $\det(B) = 1$. Hence it suffices to consider an antiunitary $[aC]$, with Appleby index $[BJ, b] = [B, b][J, 0]$, where $\det(BJ) = -1$. Now $[aC] \cdot \Pi = aC\Pi Ca^{-1} = a\overline{\Pi}a^{-1} = [a] \cdot \overline{\Pi}$. Since $a \in \mathrm{C}(d)$ has Appleby index $[B, b]$, by (a) and (b), we have

$$\chi_{-Jp}^{\Pi} = \chi_p^{\overline{\Pi}} = \omega^{\langle b, Bp \rangle} \chi_{Bp}^{[a] \cdot \overline{\Pi}} = \omega^{\langle b, Bp \rangle} \chi_{Bp}^{[aC] \cdot \Pi}.$$

Replacing p by $-Jp$ above gives

$$\chi_p^{\Pi} = \chi_{(-J)^2 p}^{\Pi} = \omega^{\langle b, -BJp \rangle} \chi_{-BJp}^{[aC] \cdot \Pi} = \omega^{\langle b, \det(BJ) BJp \rangle} \chi_{\det(BJ) BJp}^{[aC] \cdot \Pi}.$$

14.25 First suppose that $B = F_z = \begin{pmatrix} 0 & -1 \\ 1 & -1 \end{pmatrix}$. Then a matrix $A = \begin{pmatrix} \alpha & \beta \\ \gamma & \delta \end{pmatrix} \in GL_2(\mathbb{Z}_{d'})$ is in $C(F_z)$ if and only if $AF_z = F_z A$, i.e., $\gamma = -\beta$, $\alpha + \beta = \delta$, which gives

$$A = \begin{pmatrix} \alpha & \beta \\ -\beta & \alpha + \beta \end{pmatrix} = \alpha \begin{pmatrix} 1 & 0 \\ 0 & 1 \end{pmatrix} + \beta \begin{pmatrix} 0 & 1 \\ -1 & 1 \end{pmatrix} = \alpha I - \beta F_z.$$

Therefore $C(F_z) = \{\alpha I + \beta F_z : \alpha, \beta \in \mathbb{Z}_{d'}\} \cap GL_2(\mathbb{Z}_{d'})$. By the Cayley–Hamilton theorem, we have $F_z^2 + F_z + I = 0$, so that

$$\begin{aligned}(\alpha_1 I + \beta_1 F_z)(\alpha_2 I + \beta_2 F_z) &= \alpha_1 \alpha_2 I + (\alpha_1 \beta_2 + \alpha_2 \beta_1) F_z + \beta_1 \beta_2 (-F_z - I) \\ &= (\alpha_2 I + \beta_2 F_z)(\alpha_1 I + \beta_1 F_z),\end{aligned}$$

and $C(F_z)$ is abelian. Thus $C(F_z)$ is the unique maximal abelian subgroup containing F_z, and hence containing $S_0(\Pi)$ and $S(\Pi)$ (which are abelian and contain F_z).

Now suppose that $B = AF_z A^{-1}$, for some $A \in ESL_2(\mathbb{Z}_{d'})$. Then

$$C(B) = AC(F_z)A^{-1} = \{\alpha I + \beta AF_z A^{-1} : \alpha, \beta \in \mathbb{Z}_{d'}\} \cap GL_2(\mathbb{Z}_{d'}),$$

which is abelian, since it is the conjugate of an abelian group. It is the maximal abelian subgroup containing B, and hence containing $S_0(\Pi)$ and $S(\Pi)$, as before.

14.26 (a) It suffices to show this for the generators $a = S, \Omega, F, R$ (Theorem 14.1). For these, we have $S, \Omega, R, \sqrt{i}^{d-1} F \in \mathbb{Q}(\mu)^{d \times d} \subset \mathbb{E}^{d \times d}$.

(b) It suffices to consider the generators of part (a). By (14.86), we have $g(S) = S$, $g(\Omega) = \Omega^{k_g}$, and similarly $g((-\mu)^{j^2} \delta_{jk}) = (-\mu)^{k_g j^2} \delta_{jk}$ gives $g(R) = R^{k_g}$. Since g commutes with conjugation, we have

$$(g(\sqrt{i}^{d-1} F))^* g(\sqrt{i}^{d-1} F) = g(F^* F) = g(I) = I,$$

so that $g(\sqrt{i}^{d-1}F)$ is unitary, and it is in the Clifford group, since

$$g(\sqrt{i}^{d-1}F)S^j\Omega^k g(\sqrt{i}^{d-1}F)^{-1} = g(FS^j\Omega^{k_g^{-1}k}F^{-1}) = g(\omega^{-k_g^{-1}jk}S^{k_g^{-1}k}\Omega^{-j})$$
$$= \omega^{-jk}S^{k_g^{-1}k}\Omega^{-k_g j}.$$

(c) We observe $g([a])$ is well defined, since if $b \in [a]$ with $b = za$, $z \in \mathbb{E}$, $|z| = 1$, then $|g(z)|^2 = g(z)\overline{g(z)} = g(z)g(\bar{z}) = g(z\bar{z}) = g(1) = 1$, so that $g(b) = g(z)g(a) \in [g(a)]$. This defines an action of $\mathscr{G}_c$, since $1([a]) = [1(a)] = [a]$ and

$$g(h([a])) = g([h(a)]) = [g(h(a))] = [(gh)(a)] = (gh)([a]).$$

(d) The Appleby index multiplication $[B_1, b_1][B_2, b_2] = [B_1B_2, b_1 + B_1b_2]$ satisfies

$$[H_gB_1H_g^{-1}, H_gb_1][H_gB_2H_g^{-1}, H_gb_2] = [H_g(B_1B_2)H_g^{-1}, H_g(b_1 + B_1b_2)],$$

and so it suffices to show the formula for the generators. From (b), we have

$$g(f([0,p])) = g([D_p]) = [D_{H_gp}] = f([0, H_gp]),$$

and with B_a denoting the symplectic index of a symplectic operation $[a]$,

$$B_{g(R)} = B_{R^{k_g}} = (B_R)^{k_g} = \begin{pmatrix} 1 & 0 \\ k_g & 1 \end{pmatrix} = \begin{pmatrix} 1 & 0 \\ 0 & k_g \end{pmatrix}\begin{pmatrix} 1 & 0 \\ 1 & 1 \end{pmatrix}\begin{pmatrix} 1 & 0 \\ 0 & k_g^{-1} \end{pmatrix} = H_gB_RH_g^{-1},$$

$$B_{g(F)} = \begin{pmatrix} 0 & k_g^{-1} \\ -k_g & 0 \end{pmatrix} = \begin{pmatrix} 1 & 0 \\ 0 & k_g \end{pmatrix}\begin{pmatrix} 0 & 1 \\ -1 & 0 \end{pmatrix}\begin{pmatrix} 1 & 0 \\ 0 & k_g^{-1} \end{pmatrix} = H_gB_FH_g^{-1}.$$

14.27 (a) Since $[a] \cdot \Pi = a\Pi a^{-1}$, we have

$$[b] \in S_\Pi \iff b\Pi b^{-1} = \Pi \iff aba^{-1}(a\Pi a^{-1})(aba^{-1})^{-1} = a\Pi a^{-1}$$
$$\iff [a][b][a]^{-1} = [aba^{-1}] \in S_{[a]\cdot\Pi}.$$

(b) By Exer. 14.26, for any $[a] \in \mathrm{PEC}(d)$, we may assume that $\mathbb{Q}(a) \subset \mathbb{E}$. Thus

$$[a] \in S_\Pi \iff [a]\cdot\Pi = \Pi \iff a\Pi a^{-1} = \Pi, \quad \text{where } \mathbb{Q}(a) \subset \mathbb{E}$$
$$\iff g(a)g(\Pi)g(a)^{-1} = g(\Pi) \iff [g(a)]\cdot g(\Pi) = g(\Pi)$$
$$\iff g([a]) = [g(a)] \in S_{g(\Pi)}.$$

(c) The fiducial $g(\Pi_j)$ is on some orbit $\mathscr{O}_k$, and so there is $[a] \in \mathrm{PEC}(d)$ with

$$g(\Pi_j) = [a]\cdot\Pi_k.$$

By (a) and (b), we have

$$S_{g(\Pi_j)} = S_{[a]\cdot\Pi_k} \implies g(S_{\Pi_j}) = [a]\cdot S_{\Pi_k} = [a]S_{\Pi_k}[a]^{-1}.$$

Let $[a_L] \in S_{\Pi_j}$ be canonical order 3, with symplectic index L. Then by the above, there is some $[a_M] \in S_{\Pi_k}$, with symplectic index M, and

$$g([a_L]) = [g(a_L)] = [a][a_M][a]^{-1}.$$

Let $[B,q]$ be an Appleby index of a. Since $H_g L H_g^{-1}$ is a symplectic index of $g([a_L])$, it follows from the above that the Appleby indices satisfy

$$[H_g L H_g^{-1}, 0] \equiv [B,q][H_g M H_g^{-1}, 0][B,q]^{-1} \bmod K,$$

where K is the kernel of the Appleby indexing homomorphism f_E of (14.57), which is the kernel of f (see Theorem 14.2), and $a \equiv b \pmod K$ means that $a^{-1}b \in K$. Expanding $[H_g L H_g^{-1}, 0][B,q] \equiv [B,q][H_g M H_g^{-1}, 0]$ gives

$$[H_g L H_g^{-1} B, H_g L H_g^{-1} q] \equiv [B H_g M H_g^{-1}, q] \bmod K.$$

For d odd, K is trivial, so that $H_g L H_g^{-1} q = q$. For d even, multiplication by an element of K adds $(s\frac{d}{2}, t\frac{d}{2})$, $s,t \in \{0,1\}$, to the second index, so that

$$(A-I)q = (H_g L H_g^{-1} - I)q = \begin{cases} 0 \bmod d, & d \text{ is odd;} \\ 0 \bmod \frac{d}{2}, & d \text{ is even,} \end{cases}$$

where $A := H_g L H_g^{-1} \bmod d$ has trace -1 (since a_L is canonical order 3). By (14.71), we have $A^2 = -A - I$, so that $(-A-2I)(A-I) = 3I$, and we have

$$3q = (-A-2I)(A-I)q = \begin{cases} 0 \bmod d, & d \text{ is odd;} \\ 0 \bmod \frac{d}{2}, & d \text{ is even.} \end{cases}$$

(d) If d is odd, then 3 is a unit, so that $3q = 0$ gives $q = 0$. If d is even, we have

$$q = \begin{pmatrix} s\frac{d}{2} \\ t\frac{d}{2} \end{pmatrix}, \qquad s,t \in \{0,1\}.$$

Multiplication of $[B,q]$ by $[\begin{pmatrix} 1+rd & sd \\ td & 1+rd \end{pmatrix}, \begin{pmatrix} s\frac{d}{2} \\ t\frac{d}{2} \end{pmatrix}] \in K$ gives a symplectic index. Hence, if $\Pi = \Pi_j$ is a centred fiducial, then $g(\Pi) = [a] \cdot \Pi_k$, with $[a]$ symplectic, so that

$$S_{g(\Pi)} = S_{[a]\cdot \Pi_k} = [a] S_{\Pi_k} [a]^{-1},$$

which consists of symplectic operations, i.e., $g(\Pi)$ is a centred fiducial.
Remark: If $d \equiv 0 \bmod 3$, then it can be shown that q is unique, with $q \equiv 0 \bmod \frac{d}{3}$.
(e) We show that if Π is strongly centred, then all centred fiducials on its extended Clifford orbit are *strongly* centred. By (c), every centred fiducial on the extended Clifford orbit of Π has the form $[a] \cdot \Pi$, where $[a]$ is symplectic, with symplectic index B. By (14.89), we have

$$\chi^{[a]\cdot\Pi}_{\det(B)Bp} = \chi^{\Pi}_{p} \in \mathbb{E}_1,$$

so that $[a]\cdot\Pi$ is strongly centred.
(f) For the first part, we need only to consider the case $d \equiv 0 \bmod 3$. Here, we have $g(\Pi_j) = [a]\cdot\Pi_k$, where $[a]$ has extended Appleby index $[B,q]$ with $q \equiv 0 \bmod \frac{d}{3}$, i.e., $q_j = \alpha_j \frac{d}{3}$, $\alpha_j \in \{0,1,2\}$. By (14.89),

$$\chi^{\Pi}_{p} = \omega^{\langle q,\det(B)Bp\rangle} \chi^{[a]\cdot\Pi_k}_{\det(B)Bp}, \qquad \forall p \in \mathbb{Z}^2_{d'},$$

which implies that the third roots of unity $\omega^{q_j} = (e^{\frac{2\pi i}{3}})^{\alpha_j}$ are in $\mathbb{E}_1$. Now

$$g_1(e^{\frac{2\pi i}{3}}) = g_1((-\mu)^{\frac{2d}{3}}) = (-\mu)^{-\frac{2d}{3}} = e^{-\frac{2\pi i}{3}},$$

so ω^{q_j} is fixed by g_1 (is in $\mathbb{E}_1$) if and only if $a_j = 0$, i.e., $q = 0$ and $[a]$ is symplectic. Now consider all d. Since $g_1 \in \mathscr{G}_0$, we have $g_1(\Pi_j) = [a]\cdot\Pi_j$, where $[a]$ is sympletic, with sympletic index B. Let $\Pi = \Pi_j$. Then by (14.90) and (14.89), we have

$$\chi^{g_1(\Pi)}_{Jp} = \chi^{\Pi}_{p}, \quad \chi^{\Pi}_{p} = \chi^{[a]\cdot\Pi}_{\det(B)Bp} = \chi^{g_1(\Pi)}_{\det(B)Bp}, \qquad \forall p,$$

so that $J = \det(B)B$, which gives $B = -J$. The symplectic operation with sympletic index $B = -J$ is $P_{-1}C$, so that

$$g_1(\Pi) = [P_{-1}C]\cdot\Pi = P_{-1}g_c(\Pi)P_{-1}^{-1} = P_{-1}\overline{\Pi}P_{-1}.$$

14.28 (a) The composition is given by the multiplication formula, since

$$(a_1g_1)_\circ(a_2g_2)(v) = (a_1g_1)(a_2g_2(v)) = a_1g_1(a_2)g_1(g_2(v)) = a_1g_1(a_2)g_1g_2(v),$$

and the product is in X (see Exer. 14.26). Since composition is associative, it suffices to verify the formula given for the inverse, i.e.,

$$(g^{-1}(a^{-1})g^{-1})(ag) = g^{-1}(a^{-1})g^{-1}(a)g^{-1}g = g^{-1}(a^{-1}a)1 = g^{-1}(I) = I,$$
$$(ag)(g^{-1}(a^{-1})g^{-1}) = ag(g^{-1}(a^{-1}))gg^{-1} = aa^{-1}1 = I.$$

(b) The multiplication is well defined, since for $[a] \in \mathrm{PC}(d)$ one can always choose $a \in \mathbb{E}^{d\times d}$ (see Exer. 14.26), and in its definition a_j is unique up to multiplication by a unit scalar $c_j \in \mathbb{E}$, and therefore

$$([c_1a_1]g_1)([c_2a_2]g_2) = [c_1a_1g_1(c_2a_2)]g_1g_2 = [c_1c_2a_1g_1(a_2)]g_1g_2 = [a_1g_1(a_2)]g_1g_2.$$

The group properties can verified by taking representatives in $\mathbb{E}^{d\times d}$, and using (a). It is clear $\mathrm{PC}(d)$ and $\mathscr{G}_c$ are subgroups. $\mathrm{PEC}(d)$ appears as the subgroup $\mathrm{EC}(d)\times\langle g_c\rangle$ (it is easily checked the multiplication is consistent).
(c) The formula for $[a]g\cdot\Pi$ is well defined (multiplying a by a unit scalar in $\mathbb{E}$ does not change its value). It gives a group action, since $[I]1\cdot\Pi = \Pi$, and

$$[a_1]g_1 \cdot ([a_2]g_2 \cdot \Pi) = [a_1]g_1 \cdot (a_2 g_2(\Pi) a_2^{-1}) = a_1 g_1(a_2) g_1(g_2(\Pi)) g_1(a_2^{-1}) a_1^{-1}$$
$$= (a_1 g_1(a_2))(g_1 g_2)(\Pi)(a_1 g_1(a_2))^{-1} = ([a_1]g_1)([a_2]g_2) \cdot \Pi.$$

If $\Pi = vv^*$ is stabilised by ag, i.e., $(agv)^*(agv) = vv^*$, then unit vectors agv and v give the same rank-one orthogonal projector and so must be unit scalar multiples of each other. Since $[a]g$ has finite order ($\mathscr{G}_c$ is assumed to be a finite abelian group), the scalar is a root of unity.

(d) It is easy to verify that this multiplication gives a group, with identity $[I,0,1]$, and inverse

$$[B,b,g]^{-1} = [H_g^{-1} B^{-1} H_g, -H_g^{-1} B^{-1} b, g^{-1}].$$

Further, we have

$$f_{\mathscr{G}_c}([B_1,b_1,g_1][B_2,b_2,g_2]) = f([B_1 H_{g_1} B_2 H_{g_1}^{-1}, b_1 + B_1 H_{g_1} b_2]) g_1 g_2,$$

and since f is a homomorphism

$$f_{\mathscr{G}_c}([B_1,b_1,g_1]) f_{\mathscr{G}_c}([B_2,b_2,g_2]) = f([B_1,b_1]) g_1 f([B_2,b_2]) g_2$$
$$= f([B_1,b_1]) g_1(f([B_2,b_2])) g_1 g_2 = f([B_1,b_1]) f([H_{g_1} B_2 H_{g_1}^{-1}, H_{g_1} b_2]) g_1 g_2$$
$$= f([B_1,b_1][H_{g_1} B_2 H_{g_1}^{-1}, H_{g_1} b_2]) g_1 g_2 = f([B_1 H_{g_1} B_2 H_{g_1}^{-1}, b_1 + B_1 H_{g_1} b_2]) g_1 g_2.$$

Thus $f_{\mathscr{G}_c}$ is a homomorphism, with kernel $\ker(f_{\mathscr{G}_c}) = \{[B,b,1] : [B,b] \in \ker(f)\}$.

(e) The map Θ is a homomorphism, since

$$\Theta([B_1,b_1,g_c^{j_1}][B_2,b_2,g_c^{j_2}]) = \Theta([B_1 J^{j_1} B_2 J^{-j_1}, b_1 + B_1 J^{j_1} b_2, g_c^{j_1+j_2}])$$
$$= [B_1 J^{j_1} B_2 J^{-j_1} J^{j_1+j_2}, b_1 + B_1 J^{j_1} b_2]$$
$$= [B_1 J^{j_1}, b_1][B_2 J^{j_2}, b_2] = \Theta([B_1,b_1,g_c^{j_1}])\Theta([B_2,b_2,g_c^{j_2}]).$$

An index $[B,b,g_c^j]$ is in its kernel if and only if $[BJ^j,b] = [I,0]$, i.e., $b=0$, and

$$BJ^j = I \quad \Longrightarrow \quad (-1)^j = \det(BJ^j) = \det(I) = 1 \quad \Longrightarrow \quad j=0, \quad B=I,$$

so that $[B,b,g_c^j] = [I,0,1]$, and therefore Θ is an isomorphism.

Exercises of Chapter 15

15.1 Suppose $0 \le j \le n$. Then for $c : \Delta_{n-j} \to \mathbb{R}$ and $b : \Delta_n \to \mathbb{R}$, we calculate

$$\langle R^j c, b \rangle_{\nu,n} = \sum_{|\alpha|=n} \frac{(\nu)_\alpha}{\alpha!} (R^j c)_\alpha b_\alpha = \sum_{|\alpha|=n} \frac{(\nu)_\alpha}{\alpha!} \sum_{|\gamma|=j} \binom{j}{\gamma} \frac{(-\alpha)_\gamma}{(-|\alpha|)_j} c_{\alpha-\gamma} b_\alpha$$
$$= \sum_{|\beta|=n-j} \sum_{|\gamma|=j} \frac{(\nu)_{\beta+\gamma}}{(\beta+\gamma)!} \binom{j}{\gamma} \frac{(-\beta-\gamma)_\gamma}{(-n)_j} c_\beta b_{\beta+\gamma}$$

$$= \sum_{|\beta|=n-j} \frac{(\nu)_\beta}{\beta!} c_\beta \sum_{|\gamma|=j} \beta! \frac{(\nu+\beta)_\gamma}{(\beta+\gamma)!} \binom{j}{\gamma} \frac{(-\beta-\gamma)_\gamma}{(-n)_j} b_{\beta+\gamma},$$

and so $(R_\nu^*)^j$ is given by

$$((R_\nu^*)^j b)_\beta = \sum_{|\gamma|=j} \beta! \frac{(\nu+\beta)_\gamma}{(\beta+\gamma)!} \binom{j}{\gamma} \frac{(-\beta-\gamma)_\gamma}{(-n)_j} b_{\beta+\gamma}$$

$$= \sum_{|\gamma|=j} (\nu+\beta)_\gamma \binom{j}{\gamma} \frac{(-1)^j}{(-n)_j} b_{\beta+\gamma} = \sum_{|\gamma|=j} \frac{(\beta+\nu)_\gamma}{(|\beta|+1)_j} \binom{j}{\gamma} b_{\beta+\gamma}.$$

Exercises of Chapter 16

16.1 (a) The integral converges and defines a bounded linear map, since

$$\Big|\int_J \overline{\langle f,f_j\rangle}\langle g,f_j\rangle\,d\mu(j)\Big|^2 \le \int_J |\langle f,f_j\rangle|^2\,d\mu(j) \int_J |\langle g,f_j\rangle|^2\,d\mu(j) \le B\|f\|^2\|g\|^2.$$

(b) The linearity of S and inequality follow from

$$\langle Sf,g\rangle = \overline{\langle g,Sf\rangle} = \int_J \langle f,f_j\rangle\overline{\langle g,f_j\rangle}\,d\mu(j), \qquad \langle Sf,f\rangle = \int_J |\langle f,f_j\rangle|^2\,d\mu(j).$$

(c) Expand $f = SS^{-1}f = S^{-1}Sf$, and use

$$\langle S^{-1}f,f_j\rangle = \langle f,S^{-1}f_j\rangle, \qquad S^{-1}\int_J \langle f,f_j\rangle f_j\,d\mu(j) = \int_J \langle f,f_j\rangle S^{-1}f_j\,d\mu(j).$$

(d) The linear functional defining Va is bounded, since (by Cauchy–Schwarz)

$$\left|\int_J \langle f,f_j\rangle\overline{a_j}\,d\mu(j)\right|^2 \le \int_J |\langle f,f_j\rangle|^2\,d\mu(j)\,\|a\|^2_{\ell_2(\mu)} \le B\|f\|^2\,\|a\|^2_{\ell_2(\mu)}.$$

The formula for V^*f follows from

$$\langle V^*f,a\rangle = \langle f,Va\rangle_{\ell_2(\mu)} = \int_J \langle f,f_j\rangle\overline{a_j}\,d\mu(j) = \langle((\langle f,f_j\rangle)_{j\in J},a\rangle_{\ell_2(\mu)}.$$

16.2 With S, the frame operator (see Exer. 16.1), the proof of Theorem 6.1 gives

$$\operatorname{trace}(S)^2 \le d\operatorname{trace}(S^2),$$

with equality if and only if $(f_j)_{j\in J}$ is tight. Let (e_ℓ) be an orthonormal basis, then the *trace formula* (see Exer. 2.13) and Plancherel gives

$$\text{trace}(S) = \sum_\ell \langle Se_\ell, e_\ell\rangle = \sum_\ell \int_J |\langle f_j, e_\ell\rangle|^2 \, d\mu(j) = \int_J \left(\sum_\ell |\langle f_j, e_\ell\rangle|^2 \right) d\mu(j)$$
$$= \int_J \|f_j\|^2 \, d\mu(j),$$

$$\text{trace}(S^2) = \sum_\ell \langle S(Se_\ell), e_\ell\rangle = \sum_\ell \int_J \left(\int_J \langle e_\ell, f_k\rangle \langle f_k, f_j\rangle \, d\mu(k) \right) \langle f_j, e_\ell\rangle \, d\mu(j)$$
$$= \int_J \int_J \left(\sum_\ell \langle e_\ell, f_k\rangle \langle f_j, e_\ell\rangle \right) \langle f_k, f_j\rangle \, d\mu(k) d\mu(j)$$
$$= \int_J \int_J |\langle f_k, f_j\rangle|^2 \, d\mu(k) d\mu(j).$$

Substituting these into the inequality above gives the result.

16.3 We recall that S (and hence S^{-1}) is Hermitian.
(a) $P_\Phi^* = (V^*S^{-1}V)^* = V^*S^{-1}V = P_\Phi$, $P_\Phi^2 = V^*S^{-1}VV^*S^{-1}V = V^*S^{-1}V = P_\Phi$.
(b) We have

$$(P_\Phi a)_j = \langle S^{-1}Va, f_j\rangle = \langle Va, S^{-1}f_j\rangle = \int_J \langle a_k f_k, S^{-1} f_j\rangle \, d\mu(k)$$
$$= \int_J \langle f_k, S^{-1} f_j\rangle a_k \, d\mu(k) = ([P_\Phi] \cdot_\mu a)_j.$$

(c) From the frame definition, we have

$$\|v_k\|^2_{L(\mu)} = \int_J |\langle f_k, S^{-1} f_j\rangle|^2 \, d\mu(j) = \int_J |\langle S^{-1} f_k, f_j\rangle|^2 \, d\mu(j) \le B \|S^{-1} f_k\|^2,$$

so that $v_k \in L_2(\mu)$. We calculate

$$\langle v_r, v_s\rangle_{L_2(\mu)} = \int_J \langle f_r, S^{-1} f_j\rangle \overline{\langle f_r, S^{-1} f_j\rangle} \, d\mu(j)$$
$$= \int_J \langle S^{-\frac{1}{2}} f_r, S^{-\frac{1}{2}} f_j\rangle \langle S^{-\frac{1}{2}} f_j, S^{-\frac{1}{2}} f_r\rangle \, d\mu(j) = \langle S^{-\frac{1}{2}} f_r, S^{-\frac{1}{2}} f_s\rangle,$$

by the Plancherel identity for the normalised tight frame $(S^{-\frac{1}{2}} f_j)$.

16.4 The frame operator $S = AI$ and so has finite trace:

$$\text{trace} S = \int_J \text{trace}\langle \cdot, f_j\rangle f_j \, d\mu(j) = \int_J \|f_j\|^2 \, d\mu(j) = \mu(J) = Ad,$$

which gives $\frac{1}{A} = \frac{d}{\mu(J)}$.

16.5 (a) Let $x \in \mathscr{H}$, so that $x = Px$, and point evaluation of x at j is given by

$$j \mapsto x_j = \langle x, e_j \rangle = \langle Px, e_j \rangle = \langle x, Pe_j \rangle,$$

which implies $K_j = Pe_j$.
(b) Since $\mathscr{H}$ is the orthogonal complement of the vector $(1, 1, \ldots, 1)$, one computes

$$K_j(k) = \begin{cases} \frac{n-1}{n}, & k = j; \\ -\frac{1}{n}, & \text{otherwise.} \end{cases}$$

This frame is the vertices of the simplex.

16.6 (a) Let P be the orthogonal projection onto a subspace $\mathscr{K}$ of $\mathscr{H}$, and K be the reproducing kernel for $\mathscr{H}$. For $f \in \mathscr{K}$, we have

$$f(x) = \langle f, K_x \rangle = \langle Pf, K_x \rangle = \langle f, PK_x \rangle,$$

so that $\mathscr{K}$ is reproducing kernel Hilbert space, with kernel $K_{\mathscr{K}}(x,) = \langle PK_y, PK_x \rangle$.
(b) Let P_j be the orthogonal projection onto $\mathscr{H}_j$, and write $f = \sum_j f_j$, $f_j \in \mathscr{H}_j$. Then

$$\begin{aligned} f(x) &= \sum_j f_j(x) = \sum_j \langle f_j, (K_j)_x \rangle = \sum_j \langle P_j f, (K_j)_x \rangle = \sum_j \langle f, P_j (K_j)_x \rangle = \sum_j \langle f, (K_j)_x \rangle \\ &= \langle f, (\sum_j K_j)_x \rangle, \qquad \forall f \in \mathscr{H}, \end{aligned}$$

so that $\sum_j K_j$ is the reproducing kernel of $\mathscr{H}$.

16.7 Every subspace $\mathscr{H}$ of $L_2(\mathbb{S})$ is the orthogonal direct sum of its projections onto the $\mathscr{H}_j$. If $\mathscr{H}$ is rotationally invariant, then so is its projection onto the rotationally invariant subspace $\mathscr{H}_k$, which is either $\mathscr{H}_k$ or 0 (by the irreducibility of $\mathscr{H}_k$). Thus $\mathscr{H}$ has the asserted form.

16.8 Substitute $\lambda = \frac{d}{2}$, $t = \|x\|$, $y = \frac{\langle x,y \rangle}{\|x\|}$, to obtain

$$\sum_{k=0}^{\infty} Z_\xi^{(k)}(x) = (1 - \|x\|^2) \sum_{k=0}^{\infty} C_k^{(\frac{d}{2})} \Big(\frac{\langle x, \xi \rangle}{\|x\|} \Big) \|x\|^k.$$

16.9 (a) For $\Pi_1(\mathbb{S}) = \mathscr{H}_0 \oplus \mathscr{H}_1$, (16.20) and (16.40) give

$$Z_\xi(x) = Z_\xi^{\{0,1\}}(x) = Z_\xi^{(0)}(x) + Z_\xi^{(1)}(x) = 1 + d\langle x.\xi \rangle.$$

For $\Pi_1(\mathbb{R}^d) = \mathscr{P}_0 \oplus \mathscr{P}_1 = V_0^{(0)} \oplus V_0^{(1)}$, (16.47) gives the reproducing kernel

$$K(x,y) = \frac{Z^{(0)}(x,y)}{\text{area}(\mathbb{S})\|1\|_{w,0}^2} + \frac{Z^{(1)}(x,y)}{\text{area}(\mathbb{S})\|1\|_{w,1}^2} = \frac{1}{\text{area}(\mathbb{S})} \Big(\frac{1}{\|1\|_{w,0}^2} + \frac{d\langle x, \xi \rangle}{\|1\|_{w,1}^2} \Big),$$

with the corresponding tight frame given by $Z_\xi = K(\cdot, \xi)$.

(b) Let $\xi_j = (x_j, y_j, z_j)$, then the Gramian of $(Z_{\xi_j})_{j=1}^4$ is

$$\begin{bmatrix} 1+3\langle\xi_1,\xi_1\rangle & 1+3\langle\xi_1,\xi_2\rangle & 1+3\langle\xi_1,\xi_3\rangle & 1+3\langle\xi_1,\xi_4\rangle \\ 1+3\langle\xi_2,\xi_1\rangle & 1+3\langle\xi_2,\xi_2\rangle & 1+3\langle\xi_2,\xi_3\rangle & 1+3\langle\xi_2,\xi_4\rangle \\ 1+3\langle\xi_3,\xi_1\rangle & 1+3\langle\xi_3,\xi_2\rangle & 1+3\langle\xi_3,\xi_3\rangle & 1+3\langle\xi_3,\xi_4\rangle \\ 1+3\langle\xi_4,\xi_1\rangle & 1+3\langle\xi_4,\xi_2\rangle & 1+3\langle\xi_4,\xi_3\rangle & 1+3\langle\xi_4,\xi_4\rangle \end{bmatrix}.$$

This factors as A^*A, with A column equivalent to B, where

$$A = \begin{bmatrix} 1 & 1 & 1 & 1 \\ \sqrt{3}x_1 & \sqrt{3}x_2 & \sqrt{3}x_3 & \sqrt{3}x_4 \\ \sqrt{3}y_1 & \sqrt{3}y_2 & \sqrt{3}y_3 & \sqrt{3}y_4 \\ \sqrt{3}z_1 & \sqrt{3}z_2 & \sqrt{3}z_3 & \sqrt{3}z_4 \end{bmatrix}, \qquad B = \begin{bmatrix} 1 & 0 & 0 & 0 \\ \sqrt{3}x_1 & x_2-x_1 & x_3-x_1 & x_4-x_1 \\ \sqrt{3}y_1 & y_2-y_1 & y_3-y_1 & y_4-y_1 \\ \sqrt{3}z_1 & z_2-z_1 & z_3-z_1 & z_4-z_1 \end{bmatrix}.$$

The polynomials $(Z_{\xi_j})_{j=1}^4$ are a basis if and only if their Gramian is invertible, i.e., B is invertible. By taking a cofactor expansion along the first row, it follows that B is invertible if and only if the vectors $\xi_2-\xi_1$, $\xi_3-\xi_1$, $\xi_4-\xi_1$ are not coplanar.
(c) The basis $(Z_{\xi_j})_{j=1}^4$ is orthogonal if the off diagonal entries of its Gramian are 0, i.e.,

$$1+3\langle\xi_j,\xi_k\rangle = 0, \quad j\neq k \quad \Longrightarrow \quad \langle\xi_j,\xi_k\rangle = -\frac{1}{3}, \quad j\neq k,$$

so that $(\xi_j)_{j=1}^4$ are the vertices of a regular simplex.

16.10 (a) By (16.56) and the multinomial identity, we have

$$K_{p0}(z,w) = \binom{d-1+p}{d-1} \sum_{|\alpha|=p} z^\alpha \overline{w}^\alpha \binom{p}{\alpha} = \binom{d-1+p}{d-1} \langle z,w\rangle^p.$$

(b) Since the holomorphic polynomials of degrees $0,1,\ldots,n$ are orthogonal, we add their reproducing kernels, to get

$$K(z,w) = \sum_{p=0}^{n} \binom{d-1+p}{d-1} \langle z,w\rangle^p.$$

(c) Differentiating the (absolutely convergent) geometric series

$$1+x+x^2+\cdots = \frac{1}{1-x}, \qquad |x|<1,$$

$d-1$ times gives

$$(d-1)! + \frac{d!}{1!}x + \frac{(d+1)!}{2!}x^2 + \cdots = \frac{(d-1)!}{(1-x)^d}, \qquad |x|<1,$$

i.e.,

$$\sum_{p=0}^{\infty} \binom{d-1+p}{p} x^p = \frac{1}{(1-x)^d}, \qquad |x|<1.$$

Thus, we have

$$\sum_{p=0}^{\infty} K_{p0}(z,w) = \binom{d-1+p}{d-1} \langle z,w\rangle^p = \frac{1}{(1-\langle z,w\rangle)^d}.$$

(d) Expanding the Poisson kernel, as in Exer. 16.8, gives

$$\begin{aligned} P(z,w) &= \frac{1-\|z\|^2}{\|w-z\|^{2d}} = \frac{1-\|z\|^2}{(1-\langle z,w\rangle - \langle w,z\rangle + \|z\|^2)^d} \\ &= \frac{1-\|z\|^2}{(1-2yt+t^2)^\lambda}, \qquad y = \frac{\langle z,w\rangle + \langle w,z\rangle}{2\|z\|}, \quad t = \|z\|, \quad \lambda = d, \\ &= (1-\|z\|^2)\sum_{k=0}^{\infty} C_k^{(d)}\Big(\frac{\langle z,w\rangle + \langle w,z\rangle}{2\|z\|}\Big)\|z\|^k. \end{aligned}$$

Equating the homogeneous polynomials of degree k in z gives

$$\sum_{p+q=k} K_{pq}(z,w) = \|z\|^k C_k^{(d)}\Big(\frac{\langle z,w\rangle + \langle w,z\rangle}{2\|z\|}\Big) - \|z\|^k C_{k-2}^{(d)}\Big(\frac{\langle z,w\rangle + \langle w,z\rangle}{2\|z\|}\Big).$$

16.11 (a) For $x = e_k$, we have

$$\sum_{j=-\infty}^{\infty} |\langle x, S^j v\rangle|^2 = \sum_j |v_j|^2 = \|v\|^2\|x\|^2.$$

Scaling so that $\|v\| = 1$ gives the orthogonality (see Exer. 2.4).
(b) Since $S^* = S^{-1}$, we calculate

$$\begin{aligned} \sum_j |\langle x, S^j v\rangle|^2 &= \sum_j |x_{a+j}\overline{v_a} + x_{b+j}\overline{v_b}|^2 \\ &= \sum_j |x_{a+j}|^2|v_a|^2 + \sum_j |x_{b+j}|^2|v_b|^2 + \sum_j \big(\overline{x_{a+j}}x_{b+j}v_a\overline{v_b} + x_{a+j}\overline{x_{b+j}}\,\overline{v_a}v_b\big) \\ &= \|v\|^2\|x\|^2 + v_a\overline{v_b}\langle S^{-b}x, S^{-a}x\rangle + \overline{v_a}v_b\langle S^{-a}x, S^{-b}x\rangle \\ &= \|v\|^2\|x\|^2 + 2\Re\big(v_a\overline{v_b}\langle S^{a-b}x, x\rangle\big). \end{aligned}$$

Here the absolute convergence of the terms in the bracketed sum follows by the Cauchy–Schwarz inequality. For $x = ze_a + e_b$, $z \in \mathbb{C}$, we have

$$\langle S^{a-b}x, x\rangle = \langle ze_{2a-b} + e_a, ze_a + e_b\rangle = \langle e_a, ze_a\rangle = \overline{z}.$$

Thus for $(S^j v)$ to be a tight frame (with frame bound $\|v\|^2$), we must have $v_a\overline{v_b} = 0$.
(c) The previous argument can be modified to show that a v with finite support gives a tight frame $(S^j v)$ for $\ell_2(\mathbb{Z})$ if and only if v is a nonzero scalar multiple of a standard basis vector. Here, a calculation gives

$$\sum_j |\langle x, S^j v\rangle|^2 = \|v\|^2\|x\|^2 + 2\sum_{\substack{\{a,b\}\subset \mathrm{supp}(v)\\ a\neq b}} \Re\big(v_a\overline{v_b}\langle S^{a-b}x, x\rangle\big).$$

Since $a-b$ could be equal for different pairs, we have to be careful with our choices for x. Let a and b be the minimum and the maximum of $\mathrm{supp}(v)$ (the support of v). If $\mathrm{supp}(v)$ has at least two elements, i.e., $a<b$, then choosing $x = ze_a + e_b$ above leads to the necessary condition $v_a\overline{v_b} = 0$ for a tight frame, which cannot be satisfied.
(d) By (a), it suffices to show $(S^j v)$ is orthogonal. Clearly, $\langle S^k v, S^\ell\rangle = \langle S^{k-\ell}v, v\rangle = 0$ for $k-\ell$ odd, and so it remains to consider the case $k-\ell = 2m$, $m \neq 0$. Here

$$\langle S^{2m}v, v\rangle = \sum_k v_k\overline{v_{k+2m}} = \sum_{n\in\mathbb{Z}} \frac{1}{2n+1}\frac{1}{2n+2m+1} = \frac{1}{2m}\sum_n\Big(\frac{1}{2n+1} - \frac{1}{2n+2m+1}\Big),$$

which is zero, since the series converges absolutely.
(e) To find all such $v \in \ell_2(\mathbb{Z})$, we identify each $v \in \ell_2(\mathbb{Z})$ with a $f \in L_2(\mathbb{T})$ via the Fourier transform, i.e., $f(z) = \sum_j v_j z^j$, $v_j = \langle f(z), z^j\rangle_{L_2(\mathbb{T})}$, where

$$\langle f(z), g(z)\rangle_{L_2(\mathbb{T})} := \frac{1}{2\pi}\int_0^{2\pi} f(e^{it})\overline{g(e^{it})}\,dt,$$

and $\|v\| = \|f\|_{L_2(\mathbb{T})}$. Here the shift S on $\ell_2(\mathbb{Z})$ corresponds to multiplication by z on $L_2(\mathbb{T})$, so that $(S^j v)$ gives a tight frame for $\ell_2(\mathbb{Z})$ if and only if $(z^j f(z))_{j\in\mathbb{Z}}$ gives a tight frame for $L_2(\mathbb{T})$, i.e.,

$$\sum_j |\langle g(z), z^j f(z)\rangle_{L_2(\mathbb{T})}|^2 = \|f\|^2_{L_2(\mathbb{T})}\|g\|^2_{L_2(\mathbb{T})}, \qquad \forall g \in L_2(\mathbb{T}).$$

Now

$$\sum_j |\langle g(z), z^j f(z)\rangle_{L_2(\mathbb{T})}|^2 = \sum_j |\langle g(z)\overline{f(z)}, z^j\rangle_{L_2(\mathbb{T})}|^2 = \|g(z)\overline{f(z)}\|^2_{L_2(\mathbb{T})},$$

so that f must satisfy

$$\|fg\|_{L_2(\mathbb{T})} = \|f\|_{L_2(\mathbb{T})}\|g\|_{L_2(\mathbb{T})}, \qquad \forall g \in L_2(\mathbb{T}),$$

i.e., f must have constant modulus. Thus the v for which $(S_j v)$ is a tight frame for $\ell_2(\mathbb{Z})$ are precisely those v which are the Fourier coefficients of a nonzero function with constant modulus. The example of (d) is (up to a scalar) given by $f = 1$ on $[0,\pi]$ and $f = -1$ on $[\pi, 2\pi]$. Since the only trigonometric polynomials with unit modulus on $\mathbb{T}$ are the monomials, we also recover (b) and (c).

16.12 Let $z = (z_1, \ldots, z_d) \in \mathbb{C}^d$, and fix j. We need to show that

$$z_j = \frac{d}{\mathrm{area}(\tilde{\mathbb{S}})}\int_{\tilde{\mathbb{S}}} \langle z, \xi\rangle \xi_j \, d\sigma(\xi) = \frac{d}{\mathrm{area}(\tilde{\mathbb{S}})}\int_{\tilde{\mathbb{S}}} (z_1\overline{\xi_1} + \cdots + z_d\overline{\xi_d})\xi_j\, d\sigma(\xi).$$

This follows from the simple calculations

$$\int_{\tilde{\mathbb{S}}} \overline{\xi_k}\xi_j \, d\sigma(\xi) = 0, \quad k \neq j,$$

$$\int_{\tilde{\mathbb{S}}} |\xi_j|^2 \, d\sigma(\xi) = \frac{1}{d}\int_{\tilde{\mathbb{S}}} 1 \, d\sigma(\xi) = \frac{\text{area}(\tilde{\mathbb{S}})}{d} \qquad (\text{since } \sum_j |\xi_j|^2 = 1).$$

16.13 For $p \in \mathscr{H}_k$, Cauchy–Schwarz gives

$$|p(\xi)| = |\langle p, Z_\xi^{(k)} \rangle_{\mathbb{S}}| \leq \|p\|_{\mathbb{S}} \|Z_\xi^{(k)}\|_{\mathbb{S}},$$

with equality if and only if p is a scalar multiple of $Z_\xi^{(k)}$. Taking $p = Z_x^{(k)}$ gives

$$|Z_\xi^{(k)}(x)| = |Z_x^{(k)}(\xi)|| \leq \|Z_x^{(k)}\|_{\mathbb{S}} \|Z_\xi^{(k)}\|_{\mathbb{S}} = \|Z_\xi^{(k)}\|_{\mathbb{S}}^2 = Z_\xi^{(k)}(\xi),$$

with equality if and only if $x = \xi$, which gives (a). Let (u_j) be an orthonormal basis for $\mathscr{H}_k$. Then taking the trace of linear operators in (16.10) gives (b), i.e.,

$$\begin{aligned}
\dim(\mathscr{H}_k) &= \sum_j \langle u_j, \int_{\mathbb{S}} \langle u_j, Z_\xi^{(k)} \rangle_{\mathbb{S}} Z_\xi^{(k)} \, d\sigma(\xi) \rangle_{\mathbb{S}} = \sum_j \int_{\mathbb{S}} \langle u_j, \langle u_j, Z_\xi^{(k)} \rangle_{\mathbb{S}} Z_\xi^{(k)} \rangle_{\mathbb{S}} \, d\sigma(\xi) \\
&= \int_{\mathbb{S}} \Big(\sum_j |\langle u_j, Z_\xi^{(k)} \rangle_{\mathbb{S}}|^2 \Big) \, d\sigma(\xi) = \int_{\mathbb{S}} \|Z_\xi^{(k)}\|_{\mathbb{S}}^2 \, d\sigma(\xi) = \|Z_\xi^{(k)}\|_{\mathbb{S}}^2 = Z_\xi^{(k)}(\xi).
\end{aligned}$$

By the argument for (a) and (b), we have that the maximum is $\|Z_\xi^{(k)}\|_{\mathbb{S}} = \sqrt{\dim(\mathscr{H}_k)}$, which is attained if and only if $p = Z_\xi^{(k)} / \|Z_\xi^{(k)}\|_{\mathbb{S}}$ (the unique scalar multiple of $Z_\xi^{(k)}$ which is positive at ξ and has unit norm).

16.14 This the variational characterisation (see Proposition 16.2) of $(\xi)_{\xi \in \mathbb{S}}$ being a continuous tight frame for $\mathbb{F}^d$ (Proposition 16.4, Exer. 16.12). It can be proved directly by using (6.29), i.e.,

$$\int_{\mathbb{S}} \int_{\mathbb{S}} |\langle x, y \rangle|^2 \, d\sigma(x) \, d\sigma(y) = \int_{\mathbb{S}} \|y\|^2 c_1(d, \mathbb{F}) \, d\sigma(y) = \frac{1}{d} = \frac{1}{d} \Big(\int_{\mathbb{S}} \langle x, x \rangle \, d\sigma(x) \Big)^2.$$

16.15 The monomials $(z^\alpha)_{|\alpha|=k}$ are orthogonal with respect to $\langle \cdot, \cdot \rangle_{\circ,k}$ and $\langle \cdot, \cdot \rangle_{\mathbb{S}_\mathbb{C}}$,

$$\langle z^\alpha, z^\alpha \rangle_{\circ,k} = \frac{\alpha!}{|\alpha|!}, \qquad \langle z^\alpha, z^\alpha \rangle_{\mathbb{S}_\mathbb{C}} = \frac{(d-1)!\alpha!}{(d-1+|\alpha|)!} = \frac{(d-1)!|\alpha|!}{(d-1+|\alpha|)!} \langle z^\alpha, z^\alpha \rangle_{\circ,k}.$$

(a) This follows from the above by linearity.
(b) By the reproducing property, (a) and (6.20), we have

$$
\begin{aligned}
f &= \int_{\mathbb{S}_\mathbb{C}} \langle f, \binom{k+d-1}{d-1} \langle \cdot, w\rangle^k \rangle_{\mathbb{S}_\mathbb{C}} \binom{k+d-1}{d-1} \langle \cdot, w\rangle^k \, d\sigma(w) \\
&= \binom{k+d-1}{d-1} \int_{\mathbb{S}_\mathbb{C}} \langle f, \langle \cdot, w\rangle^k \rangle_{\circ,k} \langle \cdot, w\rangle^k \, d\sigma(w) \\
&= \binom{k+d-1}{d-1} \int_{\mathbb{S}_\mathbb{C}} f(w) \langle \cdot, w\rangle^k \, d\sigma(w), \qquad \forall f \in H(k,0).
\end{aligned}
$$

(c) From the formula, we obtain a Beta integral

$$
\begin{aligned}
\|\langle \cdot, \eta\rangle^k\|_{\mathbb{S}_\mathbb{C}}^2 &= \int_{\mathbb{S}_\mathbb{C}} |\langle \xi, \eta\rangle|^{2k} \, d\sigma(\xi) = \frac{d-1}{\pi} \int_0^{2\pi} \int_0^1 (1-r^2)^{d-2} r^{2k} \, r \, dr \, d\theta \\
&= (d-1) \int_0^1 (1-r^2)^{d-2} r^{2k} \, 2r \, dr = (d-1) \int_0^1 (1-t)^{d-2} t^k \, dt \\
&= (d-1) \frac{(d-2)!k!}{(k+d-1)!} \binom{k+d-1}{d-1}^{-1}.
\end{aligned}
$$

References

[AAG93] Ali, S.T., Antoine, J.-P., Gazeau, J.-P.: Continuous frames in Hilbert space. Ann. Phys. **222**(1), 1–37 (1993)

[ABB+12] Appleby, D.M., Bengtsson, I., Brierley, S., Grassl, M., Gross, D., Larsson, J.-Å.: The monomial representations of the Clifford group. Quantum Inf. Comput. **12**(5–6), 404–431 (2012)

[ABR01] Axler, S., Bourdon, P., Ramey, W.: Harmonic Function Theory. Graduate Texts in Mathematics, vol. 137, 2nd edn. Springer, New York (2001)

[ACFW17] Appleby, M., Chien, T.-Y., Flammia, S., Waldron, S.: Constructing exact symmetric informationally complete measurements from numerical solutions. ArXiv e-prints (2017). arXiv:1703.05981. http://adsabs.harvard.edu/abs/2017arXiv170305981A

[ADF14] Appleby, D.M., Dang, H.B., Fuchs, C.A.: Symmetric informationally-complete quantum states as analogues to orthonormal bases and minimum-uncertainty states. Entropy **16**(3), 1484–1492 (2014)

[AEG80] Aiken, J.G., Erdos, J.A., Goldstein, J.A.: On löwdin orthogonalization. Int. J. Quantum Chem. **18**(4), 1101–1108 (1980)

[AFF11] Appleby, D.M., Flammia, S.T., Fuchs, C.A.: The Lie algebraic significance of symmetric informationally complete measurements. J. Math. Phys. **52**(2), 022202 (2011)

[AFMY16] Appleby, M., Flammia, S., McConnell, G., Yard, J.: Generating ray class fields of real quadratic fields via complex equiangular lines (2016)

[AFMY17] Appleby, M., Flammia, S., McConnell, G., Yard, J.: SICs and algebraic number theory. Found, Phys (2017)

[AG63] Akhiezer, N.I., Glazman, I.M.: Theory of Linear Operators in Hilbert Space. Translated from the Russian by Merlynd Nestell, vol. II. Frederick Ungar Publishing Co., New York (1963)

[AK05] Asgari, M.S., Khosravi, A.: Frames and bases of subspaces in Hilbert spaces. J. Math. Anal. Appl. **308**(2), 541–553 (2005)

[AKdF26] Appell, P., Kampé de Fériet, J.: Fonctions Hypergéométriqes et Hypersphériques-Polynomes d'Hermite. Gauthier-Villars, Paris (1926)

[AM15] Azarija, J., Marc, T.: There is no $(75,32,10,16)$ strongly regular graph (2015)

[AM16] Azarija, J., Marc, T.: There is no $(95,40,12,20)$ strongly regular graph (2016)

[AN13] Abdollahi, A., Najafi, H.: Frame graphs. Linear and Multilinear Algebra **0**(0), 1–15 (2017). Taylor & Francis. https://doi.org/10.1080/03081087.2017.1347135

[App05] Appleby, D.M.: Symmetric informationally complete-positive operator valued measures and the extended Clifford group. J. Math. Phys. **46**(5), 052107 (2005)

S.F.D. Waldron, *An Introduction to Finite Tight Frames*, Applied and Numerical Harmonic Analysis, https://doi.org/10.1007/978-0-8176-4815-2

[AYAZ13] Appleby, D.M., Yadsan-Appleby, H., Zauner, G.: Galois automorphisms of a symmetric measurement. Quantum Inf. Comput. **13**(7–8), 672–720 (2013)

[Bal99] Balan, R.: Equivalence relations and distances between Hilbert frames. Proc. Am. Math. Soc. **127**(8), 2353–2366 (1999)

[Ban84] Bannai, E.: Spherical designs and group representations. Combinatorics and Algebra (Boulder, Colo., 1983). Contemporary Mathematics, vol. 34, pp. 95–107. American Mathematical Society, Providence (1984)

[BB09] Bannai, E., Bannai, E.: A survey on spherical designs and algebraic combinatorics on spheres. Eur. J. Combin. **30**(6), 1392–1425 (2009)

[BBHS10] Bannai, E., Bannai, E., Hirao, M., Sawa, M.: Cubature formulas in numerical analysis and Euclidean tight designs. Eur. J. Combin. **31**(2), 423–441 (2010)

[BBRV02] Bandyopadhyay, S., Oscar Boykin, P., Roychowdhury, V., Vatan, F.: A new proof for the existence of mutually unbiased bases. Algorithmica **34**(4), 512–528 (2002). Quantum computation and quantum cryptography

[BC40] Brauer, R., Coxeter, H.S.M.: A generalization of theorems of Schönhardt and Mehmke on polytopes. Trans. Roy. Soc. Canada. Sect. III. **3**(34), 29–34 (1940)

[BDKS16] Balla, I., Dräxler, F., Keevash, P., Sudakov, B.: Equiangular lines and spherical codes in Euclidean space (2016)

[BE10] Bodmann, B.G., Elwood, H.J.: Complex equiangular Parseval frames and Seidel matrices containing pth roots of unity. Proc. Am. Math. Soc. **138**(12), 4387–4404 (2010)

[BE15] Bachoc, C., Ehler, M.: Signal reconstruction from the magnitude of subspace components. IEEE Trans. Inf. Theory **61**(7), 4015–4027 (2015)

[BF03] Benedetto, J.J., Fickus, M.: Finite normalized tight frames. Adv. Comput. Math. **18**(2–4), 357–385 (2003). Frames

[BG04] Baumert, L.D., Gordon, D.M.: On the existence of cyclic difference sets with small parameters. High Primes and Misdemeanours: Lectures in Honour of the 60th Birthday of Hugh Cowie Williams. Fields Institute Communications, vol. 41, pp. 61–68. American Mathematical Society, Providence (2004)

[BGOY14] Barg, A., Glazyrin, A., Okoudjou, K., Yu, W.-H.: Finite two-distance tight frames (2014)

[BH15] Bodmann, B.G., Haas, J.: Frame potentials and the geometry of frames. J. Fourier Anal. Appl. **21**(6), 1344–1383 (2015)

[BK73] Baggett, L., Kleppner, A.: Multiplier representations of abelian groups. J. Funct. Anal. **14**, 299–324 (1973)

[BK06] Benedetto, J.J., Kolesar, J.D.: Geometric properties of grassmannian frames for 2 and 3. EURASIP J. Adv. Signal Process. **2006** (2006)

[BLPY10] Blum, J., Lammers, M., Powell, A.M., Yılmaz, Ö.: Sobolev duals in frame theory and sigma-delta quantization. J. Fourier Anal. Appl. **16**(3), 365–381 (2010)

[BOT15] Bannai, E., Okuda, T., Tagami, M.: Spherical designs of harmonic index t. J. Approx. Theory **195**, 1–18 (2015)

[BP05] Bodmann, B.G., Paulsen, V.I.: Frames, graphs and erasures. Linear Algebr. Appl. **404**, 118–146 (2005)

[BPR14] Bondarenko, A.V., Prymak, A., Radchenko, D.: Non-existence of $(76,30,8,14)$ strongly regular graph and some structural tools (2014)

[BPT09] Bodmann, B.G., Paulsen, V.I., Tomforde, M.: Equiangular tight frames from complex Seidel matrices containing cube roots of unity. Linear Algebr. Appl. **430**(1), 396–417 (2009)

[BR04] Bukhshtaber, V.M., Ris, E.G.: Rings of continuous functions, symmetric products, and Frobenius algebras. Uspekhi Mat. Nauk **59**(1(355)), 125–144 (2004)

[Bra11] Bramwell, J.: On the existence of spherical (t,t)-designs. Honours Project, University of Auckland, 1 (2011)

[Bro07] Brouwer, A.E.: Strongly regular graphs. In: Colbourn, C.J., Dinitz, J.H. (eds.) Handbook of Combinatorial Designs, pp. 852–868. Chapman and Hall/CRC, Boca Raton (2007)

[Buk16] Bukh, B.: Bounds on equiangular lines and on related spherical codes. SIAM J. Discret. Math. **30**(1), 549–554 (2016)

[But62] Butson, A.T.: Generalized Hadamard matrices. Proc. Am. Math. Soc. **13**, 894–898 (1962)

[BvL84] Brouwer, A.E., van Lint, J.H.: Strongly regular graphs and partial geometries. Enumeration and Design (Waterloo, Ont., 1982), pp. 85–122. Academic Press, Toronto (1984)

[BW07] Bos, L., Waldron, S.: Some remarks on Heisenberg frames and sets of equiangular lines. N. Z. J. Math. **36**, 113–137 (2007)

[BW13] Broome, H., Waldron, S.: On the construction of highly symmetric tight frames and complex polytopes. Linear Algebr. Appl. **439**(12), 4135–4151 (2013)

[BW17] Bos, L., Waldron, S.: SICs and the elements of order 3 in the Clifford group. 1 (2017). preprint

[BWB10] Brierley, S., Weigert, S., Bengtsson, I.: All mutually unbiased bases in dimensions two to five. Quantum Inf. Comput. **10**(9–10), 803–820 (2010)

[BX91] Berens, H., Xu, Y.: On Bernstein–Durrmeyer polynomials with Jacobi weights. Approximation Theory and Functional Analysis (College Station. TX, 1990), pp. 25–46. Academic Press, Boston (1991)

[BY13] Barg, A., Wei-Hsuan, Y.: New bounds for spherical two-distance sets. Exp. Math. **22**(2), 187–194 (2013)

[BY14] Barg, A., Yu, W.-H.: New bounds for equiangular lines. Discrete Geometry and Algebraic Combinatorics. Contemporary Mathematics, vol. 625, pp. 111–121. American Mathematical Society, Providence (2014)

[Cas98] Casazza, P.G.: Every frame is a sum of three (but not two) orthonormal bases-and other frame representations. J. Fourier Anal. Appl. **4**(6), 727–732 (1998)

[Cas04] Casazza, P.G.: Custom building finite frames. Wavelets, Frames and Operator Theory. Contemporary Mathematics, vol. 345, pp. 61–86. American Mathematical Society, Providence (2004)

[CCKS97] Calderbank, A.R., Cameron, P.J., Kantor, W.M., Seidel, J.J.: Z_4-Kerdock codes, orthogonal spreads, and extremal Euclidean line-sets. Proc. London Math. Soc. (3) **75**(2), 436–480 (1997)

[CE04] Christensen, O., Eldar, Y.C.: Oblique dual frames and shift-invariant spaces. Appl. Comput. Harmon. Anal. **17**(1), 48–68 (2004)

[CFM12] Casazza, P.G., Fickus, M., Mixon, D.G.: Auto-tuning unit norm frames. Appl. Comput. Harmon. Anal. **32**(1), 1–15 (2012)

[CFM+13] Cahill, J., Fickus, M., Mixon, D.G., Poteet, M.J., Strawn, N.: Constructing finite frames of a given spectrum and set of lengths. Appl. Comput. Harmon. Anal. **35**(1), 52–73 (2013)

[CFW15] Chien, T.-Y., Flynn, V., Waldron, S.: Tight frames for cyclotomic fields and other rational vector spaces. Linear Algebr. Appl. **476**, 98–123 (2015)

[Chi10] Chien, T.-Y.: On the unitary equivalence between cyclic harmonic frames. Master's thesis, University of Auckland, 12 (2010)

[Chi15] Chien, T.-Y.: Equiangular lines, projective symmetries and nice error frames. Ph.D. thesis, University of Auckland, 1 (2015)

[Chr03] Christensen, O.: An Introduction to Frames and Riesz Bases. Applied and Numerical Harmonic Analysis. Birkhäuser Boston Inc., Boston (2003)

[CHS96] Conway, J.H., Hardin, R.H., Sloane, N.J.A.: Packing lines, planes, etc.: packings in Grassmannian spaces. Exp. Math. **5**(2), 139–159 (1996)

[Cie87] Ciesielski, Z.: Explicit formula relating the Jacobi, Hahn and Bernstein polynomials. SIAM J. Math. Anal. **18**(6), 1573–1575 (1987)

[CK03] Casazza, P.G., Kovačević, J.: Equal-norm tight frames with erasures. Adv. Comput. Math. **18**(2–4), 387–430 (2003)

[CK13] Casazza, P.G., Kutyniok, G. (eds.): Finite Frames - Theory and Applications. Applied and Numerical Harmonic Analysis. Birkhäuser/Springer, New York (2013)

[CKL08] Casazza, P.G., Kutyniok, G., Li, S.: Fusion frames and distributed processing. Appl. Comput. Harmon. Anal. **25**(1), 114–132 (2008)

[CL06] Casazza, P.G., Leon, M.T.: Existence and construction of finite tight frames. J. Concr. Appl. Math. **4**(3), 277–289 (2006)

[CL10] Christensen, O., Laugesen, R.S.: Approximately dual frames in Hilbert spaces and applications to Gabor frames. Sampl. Theory Signal Image Process. **9**(1–3), 77–89 (2010)

[CMS13] Cahill, J., Mixon, D.G., Strawn, N.: Connectivity and irreducibility of algebraic varieties of finite unit norm tight frames (2013)

[Coo99] Cools, R.: Monomial cubature rules since "Stroud": a compilation. II. J. Comput. Appl. Math. **112**(1–2), 21–27 (1999)

[Cox73] Coxeter, H.S.M.: Regular Polytopes, 3rd edn. Dover Publications Inc., New York (1973)

[CR93] Cools, R., Rabinowitz, P.: Monomial cubature rules since "Stroud": a compilation. J. Comput. Appl. Math. **48**(3), 309–326 (1993)

[CRT06] Candès, E.J., Romberg, J.K., Tao, T.: Stable signal recovery from incomplete and inaccurate measurements. Commun. Pure Appl. Math. **59**(8), 1207–1223 (2006)

[CW11] Chien, T.-Y., Waldron, S.: A classification of the harmonic frames up to unitary equivalence. Appl. Comput. Harmon. Anal. **30**(3), 307–318 (2011)

[CW16] Chien, T.-Y., Waldron, S.: A characterization of projective unitary equivalence of finite frames and applications. SIAM J. Discret. Math. **30**(2), 976–994 (2016)

[Dau92] Daubechies, I.: Ten Lectures on Wavelets. CBMS-NSF Regional Conference Series in Applied Mathematics, vol. 61. Society for Industrial and Applied Mathematics (SIAM), Philadelphia (1992)

[Dav79] Davis, P.J.: Circulant Matrices. A Wiley-Interscience Publication, New York, Pure and Applied Mathematics. Wiley (1979)

[dC00] de Caen, D.: Large equiangular sets of lines in Euclidean space. Electron. J. Combin. **7**. Research Paper **55**, 3 (2000)

[Deg07] Degraer, J.: Isomorph-free exhaustive generation algorithms for association schemes. Ph.D. thesis, Ghent University, 1 2007

[Der85] Derriennic, M.-M.: On multivariate approximation by Bernstein-type polynomials. J. Approx. Theory **45**(2), 155–166 (1985)

[DGS77] Delsarte, P., Goethals, J.M., Seidel, J.J.: Spherical codes and designs. Geom. Dedicata **6**(3), 363–388 (1977)

[DHC12] Datta, S., Howard, S., Cochran, D.: Geometry of the Welch bounds. Linear Algebr. Appl. **437**(10), 2455–2470 (2012)

[DHS10] Duncan, D.M., Hoffman, T.R., Solazzo, J.P.: Equiangular tight frames and fourth root Seidel matrices. Linear Algebr. Appl. **432**(11), 2816–2823 (2010)

[DHST08] Dhillon, I.S., Heath Jr., R.W., Strohmer, T., Tropp, J.A.: Constructing packings in Grassmannian manifolds via alternating projection. Exp. Math. **17**(1), 9–35 (2008)

[DJ96] Davis, J.A., Jedwab, J.: A survey of Hadamard difference sets. Groups, Difference Sets, and the Monster (Columbus, OH, 1993). Ohio State University Mathematics Research Institute Publications, vol. 4, pp. 145–156. de Gruyter, Berlin (1996)

[dlHP04] de la Harpe, P., Pache, C.: Spherical designs and finite group representations (some results of E. Bannai). Eur. J. Combin. **25**(2), 213–227 (2004)

[DS52] Duffin, R.J., Schaeffer, A.C.: A class of nonharmonic Fourier series. Trans. Am. Math. Soc. **72**, 341–366 (1952)

[DS06] Dykema, K., Strawn, N.: Manifold structure of spaces of spherical tight frames. Int. J. Pure Appl. Math. **28**(2), 217–256 (2006)

[Dun87] Dunkl, C.F.: Orthogonal polynomials on the hexagon. SIAM J. Appl. Math. **47**(2), 343–351 (1987)

[DV04] Digernes, T., Varadarajan, V.S.: Models for the irreducible representation of a Heisenberg group. Infin. Dimens. Anal. Quantum Probab. Relat. Top. **7**(4), 527–546 (2004)

[DW15] Duc Do, T., Waldron, S.: Multivariate Bernstein operators and redundant systems. J. Approx. Theory **192**, 215–233 (2015)

[DX01] Dunkl, C.F., Xu, Y.: Orthogonal Polynomials of Several Variables. Encyclopedia of Mathematics and its Applications, vol. 81. Cambridge University Press, Cambridge (2001)

[EB03] Eldar, Y.C., Bölcskei, H.: Geometrically uniform frames. IEEE Trans. Inform. Theory **49**(4), 993–1006 (2003)

[Ehl12] Ehler, M.: Random tight frames. J. Fourier Anal. Appl. **18**(1), 1–20 (2012)

[Eld03] Eldar, Y.C.: Sampling with arbitrary sampling and reconstruction spaces and oblique dual frame vectors. J. Fourier Anal. Appl. **9**(1), 77–96 (2003)

[ER93] Ehrenborg, R., Rota, G.-C.: Apolarity and canonical forms for homogeneous polynomials. Eur. J. Combin. **14**(3), 157–181 (1993)

[Fer07] Noemí Laín Fernández: Optimally space-localized band-limited wavelets on $\mathbb{S}^{q-1}$. J. Comput. Appl. Math. **199**(1), 68–79 (2007)

[FH87] Faraut, J., Harzallah, K.: Deux cours d'analyse harmonique. Progress in Mathematics, vol. 69. Birkhäuser, Boston (1987). Papers from the Tunis summer school held in Tunis, 27 August–15 September 1984

[FHK+08] Feichtinger, H.G., Hazewinkel, M., Kaiblinger, N., Matusiak, E., Neuhauser, M.: Metaplectic operators on $\mathbb{C}^n$. Q. J. Math. **59**(1), 15–28 (2008)

[FHS17] Fuchs, C.A., Hoang, M.C., Stacey, B.C.: The SIC Question: history and state of play (2017)

[Fic01] Fickus, M.C.: Finite normalized tight frames and spherical equidistribution. ProQuest LLC, Ann Arbor, MI, 2001. Thesis (Ph.D.)–University of Maryland, College Park

[FJMP16] Fickus, M., Jasper, J., Mixon, D.G., Peterson, J.: Tremain equiangular tight frames (2016)

[FKL09] Feichtinger, H.G., Kozek, W., Luef, F.: Gabor analysis over finite abelian groups. Appl. Comput. Harmon. Anal. **26**(2), 230–248 (2009)

[FL74] Fackerell, E.D., Littler, R.A.: Polynomials biorthogonal to Appell's polynomials. Bull. Aust. Math. Soc. **11**, 181–195 (1974)

[FM15] Fickus, M., Mixon, D.G.: Tables of the existence of equiangular tight frames (2015)

[FMP16] Fickus, M., Marks, J.D., Poteet, M.J.: A generalized Schur-Horn theorem and optimal frame completions. Appl. Comput. Harmon. Anal. **40**(3), 505–528 (2016)

[FMT12] Fickus, M., Mixon, D.G., Tremain, J.C.: Steiner equiangular tight frames. Linear Algebr. Appl. **436**(5), 1014–1027 (2012)

[Fol01] Folland, G.B.: How to integrate a polynomial over a sphere. Am. Math. Monthly **108**(5), 446–448 (2001)

[FPT02] Frank, M., Paulsen, V.I., Tiballi, T.R.: Symmetric approximation of frames and bases in Hilbert spaces. Trans. Am. Math. Soc. **354**(2), 777–793 (2002)

[Fru32] Frucht, R.: Über die Darstellung endlicher Abelscher Gruppen durch Kollineationen. J. Reine Angew. Math. **166**, 16–29 (1932)

[FS17] Foucart, S., Skrzypek, L.: On maximal relative projection constants. J. Math. Anal. Appl. **447**(1), 309–328 (2017)

[FWW06] Feng, D.-J., Wang, L., Wang, Y.: Generation of finite tight frames by Householder transformations. Adv. Comput. Math. **24**(1–4), 297–309 (2006)

[GKK01] Goyal, V.K., Kovačević, J., Kelner, J.A.: Quantized frame expansions with erasures. Appl. Comput. Harmon. Anal. **10**(3), 203–233 (2001)

[GKMS16] Greaves, G., Koolen, J.H., Munemasa, A., Szöllősi, F.: Equiangular lines in Euclidean spaces. J. Combin. Theory Ser. A **138**, 208–235 (2016)

[GM71] Greenleaf, F., Moskowitz, M.: Cyclic vectors for representations of locally compact groups. Math. Ann. **190**, 265–288 (1971)

[GR01] Godsil, C., Royle, G.: Algebraic Graph Theory. Graduate Texts in Mathematics, vol. 207. Springer, New York (2001)

[GR09] Godsil, C., Roy, A.: Equiangular lines, mutually unbiased bases, and spin models. Eur. J. Combin. **30**(1), 246–262 (2009)

[Gre16] Greaves, G.R.W.: Equiangular line systems and switching classes containing regular graphs (2016)

[Gro93] Grochenig, K.: Acceleration of the frame algorithm. IEEE Trans. Signal Process. **41**(12), 3331–3340 (1993)

[Grö01] Gröchenig, K.: Foundations of Time-Frequency Analysis. Applied and Numerical Harmonic Analysis. Birkhäuser Boston Inc., Boston (2001)

[GS79] Goethals, J.-M., Seidel, J.J.: Spherical designs. In: Relations between combinatorics and other parts of mathematics (Proceedings of Symposia in Pure Mathematics, Ohio State University, Columbus, Ohio, 1978), Proceedings of Symposia in Pure Mathematics, vol. XXXIV, pp. 255–272. American Mathematical Society, Providence (1979)

[GS87] Glover, H., Sjerve, D.: The genus of $\mathrm{PSl}_2(q)$. Reine Angew. Math. **380**, 59–86 (1987)

[GS17] Grassl, M., Scott, A.J.: Fibonacci-Lucas SIC-POVMs (2017)

[GT16] Goyeneche, D., Turek, O.: Equiangular tight frames and unistochastic matrices (2016)

[Gun62] Gunning, R.C.: Lectures on Modular Forms. Notes by Armand Brumer. Annals of Mathematics Studies, vol. 48. Princeton University Press, Princeton (1962)

[GVT98] Goyal, V.K., Vetterli, M., Thao, N.T.: Quantized overcomplete expansions in $\mathbf{R}^N$: analysis, synthesis, and algorithms. IEEE Trans. Inform. Theory **44**(1), 16–31 (1998)

[Haa48] Haantjes, J.: Equilateral point-sets in elliptic two- and three-dimensional spaces. Nieuw Arch. Wiskunde **2**(22), 355–362 (1948)

[Haa97] Haagerup, U.: Orthogonal maximal abelian $*$-subalgebras of the $n \times n$ matrices and cyclic n-roots. Operator Algebras and Quantum Field Theory (Rome, 1996), pp. 296–322. International Press, Cambridge (1997)

[Had40] Hadwiger, H.: Über ausgezeichnete Vektorsterne und reguläre Polytope. Comment. Math. Helv. **13**, 90–107 (1940)

[Hae93] Haemers, W.H.: There exists no $(76, 21, 2, 7)$ strongly regular graph. Finite Geometry and Combinatorics (Deinze 1992). London Mathematical Society Lecture Note Series, vol. 191, pp. 175–176. Cambridge University Press, Cambridge (1993)

[Hal98] Hall Jr., M.: Combinatorial Theory. Wiley Classics Library, 2nd edn. Wiley, A Wiley-Interscience Publication, New York (1998)

[Han07] Han, D.: Classification of finite group-frames and super-frames. Canad. Math. Bull. **50**(1), 85–96 (2007)

[HI82] Howlett, R.B., Martin Isaacs, I.: On groups of central type. Math. Z. **179**(4), 555–569 (1982)

[Hir10] Hirn, M.: The number of harmonic frames of prime order. Linear Algebr. Appl. **432**(5), 1105–1125 (2010)

[HJ90] Horn, R.A., Johnson, C.R.: Matrix Analysis. Cambridge University Press, Cambridge (1990). Corrected reprint of the 1985 original

[HJ91] Horn, R.A., Johnson, C.R.: Topics in Matrix Analysis. Cambridge University Press, Cambridge (1991)

[HK16] Hughes, D., Klish, Y.-S.: The existence and construction of spherical (t, t)-designs. Summer Project, University of Auckland, 12 (2016)

[HKLW07] Han, D., Kornelson, K., Larson, D., Weber, E.: Frames for Undergraduates. Student Mathematical Library, vol. 40. American Mathematical Society, Providence (2007)

[HL00] Han, D., Larson, D.R.: Frames, bases and group representations. Mem. Am. Math. Soc. **147**(697), x+94 (2000)

[HMR+00] Hochwald, B.M., Marzetta, T.L., Richardson, T.J., Sweldens, W., Urbanke, R.: Systematic design of unitary space-time constellations. IEEE Trans. Inf. Theory **46**(6), 1962–1973 (2000)

[Hog82] Hoggar, S.G.: t-designs in projective spaces. Eur. J. Combin. **3**(3), 233–254 (1982)

[Hog89] Hoggar, S.G.: Tight 4- and 5-designs in projective spaces. Graphs Combin. **5**(1), 87–94 (1989)

[Hog98] Hoggar, S.G.: 64 lines from a quaternionic polytope. Geom. Dedicata **69**(3), 287–289 (1998)

[HP04] Holmes, R.B., Paulsen, V.I.: Optimal frames for erasures. Linear Algebr. Appl. **377**, 31–51 (2004)

[HR07] Hillar, C.J., Rhea, D.L.: Automorphisms of finite Abelian groups. Am. Math. Monthly **114**(10), 917–923 (2007)

[HS96] Hardin, R.H., Sloane, N.J.A.: McLaren's improved snub cube and other new spherical designs in three dimensions. Discret. Comput. Geom. **15**(4), 429–441 (1996)

[Hug16] Hughes, D.: The existence and construction of spherical (t,t)-designs. Summer Project, University of Auckland, 12 (2016)

[Isa06] Martin Isaacs, I.: Character Theory of Finite Groups. AMS Chelsea Publishing, Providence (2006). Corrected reprint of the 1976 original [Academic Press, New York, MR0460423]

[Iva81] Ivanović, I.D.: Geometrical description of quantal state determination. J. Phys. A **14**(12), 3241–3245 (1981)

[Ive15] Iverson, J.W.: Frames generated by compact group actions (2015)

[JL01] James, G., Liebeck, M.: Representations and Characters of Groups, 2nd edn. Cambridge University Press, New York (2001)

[JMF14] Jasper, J., Mixon, D.G., Fickus, M.: Kirkman equiangular tight frames and codes. IEEE Trans. Inf. Theory **60**(1), 170–181 (2014)

[Joh07] Johnson, K.W.: Group representation theory via group matrices and group determinants. in preparation, 1 (2007)

[Kai94] Kaiser, G.: A Friendly Guide to Wavelets. Birkhäuser Boston Inc., Boston (1994)

[Kal06] Kalra, D.: Complex equiangular cyclic frames and erasures. Linear Algebr. Appl. **419**(2–3), 373–399 (2006)

[KC07a] Kovacevic, J., Chebira, A.: Life beyond bases: the advent of frames (part i). IEEE Signal Process. Mag. **24**(4), 86–104 (2007)

[KC07b] Kovacevic, J., Chebira, A.: Life beyond bases: the advent of frames (part ii). IEEE Signal Process. Mag. **24**(5), 115–125 (2007)

[KC08] Kovaevi, J., Chebira, A.: An introduction to frames. Foundations and Trends in Signal Processing **2**(1), 1–94 (2008)

[Kha08] Khatirinejad, M.: On Weyl-Heisenberg orbits of equiangular lines. J. Algebr. Combin. **28**(3), 333–349 (2008)

[KMT91] Kalnins, E.G., Miller Jr., W., Tratnik, M.V.: Families of orthogonal and biorthogonal polynomials on the N-sphere. SIAM J. Math. Anal. **22**(1), 272–294 (1991)

[Kni96a] Knill, E.: Group representations, error bases and quantum codes. arXiv:quant-ph/9608049 (1996)

[Kni96b] Knill, E.: Non-binary unitary error bases and quantum codes. arXiv:quant-ph/9608048 (1996)

[Kön99] König, H.: Cubature formulas on spheres. Advances in multivariate approximation (Witten-Bommerholz, 1998). Mathematical Research, vol. 107, pp. 201–211. Wiley, Berlin (1999)

[KOPT13] Kutyniok, G., Okoudjou, K.A., Philipp, F., Tuley, E.K.: Scalable frames. Linear Algebr. Appl. **438**(5), 2225–2238 (2013)

[KP11] Kotelina, N.O., Pevnyi, A.B.: The Venkov inequality with weights and weighted spherical half-designs. J. Math. Sci. (N. Y.) **173**(6), 674–682 (2011). Problems in mathematical analysis. No. 55

[KR02] Klappenecker, A., Rötteler, M.: Beyond stabilizer codes. I. Nice error bases. IEEE Trans. Inf. Theory **48**(8), 2392–2395 (2002)

[KR04] Klappenecker, A., Rötteler, M.: Constructions of mutually unbiased bases. Finite Fields and Applications. Lecture Notes in Computer Science, vol. 2948, pp. 137–144. Springer, Berlin (2004)

[KR05] Klappenecker, A., Roetteler, M.: Mutually unbiased bases are complex projective 2-designs. arXiv:quant-ph/0502031 (2005)

[Lev98] Levenshtein, V.: On designs in compact metric spaces and a universal bound on their size. Discret. Math. **192**(1–3), 251–271 (1998). Discrete metric spaces (Villeurbanne, 1996)

[LH10] Lopez, J., Han, D.: Optimal dual frames for erasures. Linear Algebr. Appl. **432**(1), 471–482 (2010)

[Li95] Li, S.: On general frame decompositions. Numer. Funct. Anal. Optim. **16**(9–10), 1181–1191 (1995)

[LM90] Leung, K.H., Ma, S.L.: Constructions of partial difference sets and relative difference sets on p-groups. Bull. Lond. Math. Soc. **22**(6), 533–539 (1990)

[LMO10] LeCompte, N., Martin, W.J., Owens, W.: On the equivalence between real mutually unbiased bases and a certain class of association schemes. Eur. J. Combin. **31**(6), 1499–1512 (2010)

[LMO14] Lemvig, J., Miller, C., Okoudjou, K.A.: Prime tight frames. Adv. Comput. Math. **40**(2), 315–334 (2014)

[LO04] Li, S., Ogawa, H.: Pseudoframes for subspaces with applications. J. Fourier Anal. Appl. **10**(4), 409–431 (2004)

[Löw70] Löwdin, P.-O.: On the nonorthogonality problem. Adv. Quantum Chem. **5**, 185–199 (1970)

[LS75] Logan, B.F., Shepp, L.A.: Optimal reconstruction of a function from its projections. Duke Math. J. **42**(4), 645–659 (1975)

[LS04] Lyubich, Y.I., Shatalova, O.A.: Isometric embeddings of finite-dimensional l_p-spaces over the quaternions. Algebra i Analiz **16**(1), 15–32 (2004)

[LT09] Lehrer, G.I., Taylor, D.E.: Unitary Reflection Groups. Australian Mathematical Society Lecture Series, vol. 20. Cambridge University Press, Cambridge (2009)

[LY09] Li, S., Yan, D.: Frame fundamental sensor modeling and stability of one-sided frame perturbation. Acta Appl. Math. **107**(1–3), 91–103 (2009)

[Ma94] Ma, S.L.: A survey of partial difference sets. Des. Codes Cryptogr. **4**(3), 221–261 (1994)

[MFVA16] Martínez-Finkelshtein, A., Van Assche, W.: What is...a multiple orthogonal polynomial? Not. Am. Math. Soc. **63**(9), 1029–1031 (2016)

[MM93] Massey, J.L., Mittelholzer, T.: Welch's bound and sequence sets for code-division multiple-access systems. Sequences. II (Positano, 1991), pp. 63–78. Springer, New York (1993)

[Mor04] Morandi, P.J.: Computing the symmetry groups of the platonic solids with the help of maple. Resonance **9**(8), 18–26 (2004)

[Mur93] Murdoch, T.A.: Isogonal configurations. Am. Math. Monthly **100**(4), 381–384 (1993)

[Mus09] Musin, O.R.: Spherical two-distance sets. J. Combin. Theory Ser. A **116**(4), 988–995 (2009)

[MVW16] Mixon, D.G., Villar, S., Ward, R.: Clustering subgaussian mixtures by semidefinite programming (2016)

[MW16] Marshall, S., Waldron, S.: On the number of harmonic frames. (2016). arXiv:1611.07121 [math.FA]

[Pet99] Petrushev, P.P.: Approximation by ridge functions and neural networks. SIAM J. Math. Anal. **30**(1), 155–189 (1999). (electronic)

[Pro57] Proriol, J.: Sur une famille de polynomes à deux variables orthogonaux dans un triangle. C. R. Acad. Sci. Paris **245**, 2459–2461 (1957)

[PW02] Peng, I., Waldron, S.: Signed frames and Hadamard products of Gram matrices. Linear Algebr. Appl. **347**, 131–157 (2002)

[RBKSC04] Renes, J.M., Blume-Kohout, R., Scott, A.J., Caves, C.M.: Symmetric informationally complete quantum measurements. J. Math. Phys. **45**(6), 2171–2180 (2004)

[Ren07] Renes, J.M.: Equiangular tight frames from Paley tournaments. Linear Algebr. Appl. **426**(2–3), 497–501 (2007)

[Rie88] Rieffel, M.A.: Projective modules over higher-dimensional noncommutative tori. Canad. J. Math. **40**(2), 257–338 (1988)

[Ros97] Rosenfeld, M.: In praise of the Gram matrix. The Mathematics of Paul Erdős, II. Algorithms and Combinatorics, vol. 14, pp. 318–323. Springer, Berlin (1997)

[Ros99] Rosengren, H.: Multivariable orthogonal polynomials and coupling coefficients for discrete series representations. SIAM J. Math. Anal. **30**(2), 232–272 (1999)

[RS07] Roy, A., Scott, A.J.: Weighted complex projective 2-designs from bases: optimal state determination by orthogonal measurements. J. Math. Phys. **48**(7), 072110 (2007)

[RS14] Roy, A., Suda, S.: Complex spherical designs and codes. J. Combin. Des. **22**(3), 105–148 (2014)

[Rud80] Rudin, W.: Function Theory in the Unit Ball of C^n. rundlehren der Mathematischen Wissenschaften [Fundamental Principles of Mathematical Science], vol. 241. Springer, New York (1980)

[Rud90] Rudin, W.: Fourier Analysis on Groups. Wiley Classics Library. Wiley, New York (1990). Reprint of the 1962 original, A Wiley-Interscience Publication

[RW02] Reams, R., Waldron, S.: Isometric tight frames. Electron. J. Linear Algebr. **9**, 122–128 (2002)

[RW04] Rosengren, H., Waldron, S.: Tight frames of Jacobi and Hahn polynomials on a simplex. preprint, 1 (2004)

[Sch37] Schönhardt, E.: über die summe der projektionen eines vektors. Deutsche Math. **2**, 446–451 (1937)

[Sch60] Schwinger, J.: Unitary operator bases. Proc. Natl. Acad. Sci. U.S.A. **46**, 570–579 (1960)

[Sch04] Schulte, E.: Symmetry of polytopes and polyhedra. Handbook of Discrete and Computational Geometry. CRC Press Series: Discrete Mathematics and Its Applications, 2nd edn, pp. 311–330. CRC, Boca Raton (2004)

[Sco17] Scott, A.J.: SICs: extending the list of solutions (2017)

[SD80] Schempp, W., Dreseler, B.: Einführung in die harmonische Analyse. B. G. Teubner, Stuttgart (1980). Mathematische Leitfäden. [Mathematical Textbooks]

[Sei01] Seidel, J.J.: Definitions for spherical designs. J. Stat. Plan. Inference **95**(1–2), 307–313 (2001). Special issue on design combinatorics: in honor of S. S. Shrikhande

[SG10] Scott, A.J., Grassl, M.: Symmetric informationally complete positive-operator-valued measures: a new computer study. J. Math. Phys. **51**(4), 042203 (2010)

[SH03] Strohmer, T., Heath Jr., R.W.: Grassmannian frames with applications to coding and communication. Appl. Comput. Harmon. Anal. **14**(3), 257–275 (2003)

[She52] Shephard, G.C.: Regular complex polytopes. Proc. Lond. Math. Soc. **3**(2), 82–97 (1952)

[She53] Shephard, G.C.: Unitary groups generated by reflections. Can. J. Math. **5**, 364–383 (1953)

[Sid74] Sidel'nikov, V.M.: New estimates for the closest packing of spheres in n-dimensional Euclidean space. Mat. Sb. (N.S.) **95**(137), 148–158 (1974)

[SK97] Saff, E.B., Kuijlaars, A.B.J.: Distributing many points on a sphere. Math. Intell. **19**(1), 5–11 (1997)

[SNS09] Skinner, A.J., Newell, V.A., Sanchez, R.: Unbiased bases (Hadamards) for six-level systems: four ways from Fourier. J. Math. Phys. **50**(1), 012107 (2009)

[SoS16] Słomczyński, W., Szymusiak, A.: Highly symmetric POVMs and their informational power. Quantum Inf. Process. **15**(1), 565–606 (2016)

[ST54] Shephard, G.C., Todd, J.A.: Finite unitary reflection groups. Can. J. Math. **6**, 274–304 (1954)

[Sta17] Stacey, B.C.: Sporadic SICs and the normed division algebras. Found, Phys (2017)

[STDH07] Sustik, M.A., Tropp, J.A., Dhillon, I.S., Heath Jr., R.W.: On the existence of equiangular tight frames. Linear Algebr. Appl. **426**(2–3), 619–635 (2007)

[Ste64] Steinberg, R.: Differential equations invariant under finite reflection groups. Trans. Am. Math. Soc. **112**, 392–400 (1964)

[Ste85] Steinberg, R.: Finite subgroups of SU_2, Dynkin diagrams and affine Coxeter elements. Pac. J. Math. **118**(2), 587–598 (1985)

[Str71] Stroud, A.H.: Approximate Calculation of Multiple Integrals. Prentice-Hall Series in Automatic Computation. Prentice-Hall, Englewood Cliffs (1971)

[Str08] Strohmer, T.: A note on equiangular tight frames. Linear Algebr. Appl. **429**(1), 326–330 (2008)

[Str11] Strawn, N.: Finite frame varieties: nonsingular points, tangent spaces, and explicit local parameterizations. J. Fourier Anal. Appl. **17**(5), 821–853 (2011)

[Str12] Strawn, N.: Optimization over finite frame varieties and structured dictionary design. Appl. Comput. Harmon. Anal. **32**(3), 413–434 (2012)

[SW70] Schweinler, H.C., Wigner, E.P.: Orthogonalization methods. J. Math. Phys. **11**, 1693–1694 (1970)

[SW71] Stein, E.M., Weiss, G.: Introduction to Fourier Analysis on Euclidean Spaces. Princeton Mathematical Series, vol. 32. Princeton University Press, Princeton (1971)

[SW09] Sloan, I.H., Womersley, R.S.: A variational characterisation of spherical designs. J. Approx. Theory **159**(2), 308–318 (2009)

[SZ84] Seymour, P.D., Zaslavsky, T.: Averaging sets: a generalization of mean values and spherical designs. Adv. Math. **52**(3), 213–240 (1984)

[Szö10] Szöllősi, F.: A two-parameter family of complex Hadamard matrices of order 6 induced by hypocycloids. Proc. Am. Math. Soc. **138**(3), 921–928 (2010)

[Szö12] Szöllősi, F.: Complex Hadamard matrices of order 6: a four-parameter family. J. Lond. Math. Soc. (2) **85**(3), 616–632 (2012)

[Szö13] Szöllősi, F.: Complex Hadamard matrices and equiangular tight frames. Linear Algebr. Appl. **438**(4), 1962–1967 (2013)

[Szö14] Szöllősi, F.: All complex equiangular tight frames in dimension 3 (2014)

[Szö17] Szöllősi, F.: A remark on a construction of D.S. Asche (2017)

[Tay72] Taylor, D.E.: Some topics in the theory of finite groups. Ph.D. thesis, University of Oxford, 1 (1972)

[Tre08] Tremain, J.C.: Concrete constructions of real equiangular line sets (2008)

[Tur65] Turyn, R.J.: Character sums and difference sets. Pac. J. Math. **15**, 319–346 (1965)

[TZ06] Tadej, W., Życzkowski, K.: A concise guide to complex Hadamard matrices. Open Syst. Inf. Dyn. **13**(2), 133–177 (2006)

[Veg00] Vegter, G.: The apolar bilinear form in geometric modeling. Math. Comp. **69**(230), 691–720 (2000)

[VW05] Vale, R., Waldron, S.: Tight frames and their symmetries. Constr. Approx. **21**(1), 83–112 (2005)

[VW08] Vale, R., Waldron, S.: Tight frames generated by finite nonabelian groups. Numer. Algorithms **48**(1–3), 11–27 (2008)

[VW10] Vale, R., Waldron, S.: The symmetry group of a finite frame. Linear Algebr. Appl. **433**(1), 248–262 (2010)

[VW16] Vale, R., Waldron, S.: The construction of G-invariant finite tight frames. J. Fourier Anal. Appl. **22**(5), 1097–1120 (2016)

[Wal03] Waldron, S.: Generalized Welch bound equality sequences are tight frames. IEEE Trans. Inf. Theory **49**(9), 2307–2309 (2003)

[Wal06] Waldron, S.: On the Bernstein-Bézier form of Jacobi polynomials on a simplex. J. Approx. Theory **140**(1), 86–99 (2006)

[Wal09] Waldron, S.: Continuous and discrete tight frames of orthogonal polynomials for a radially symmetric weight. Constr. Approx. **30**(1), 33–52 (2009)

[Wal11] Waldron, S.: Frames for vector spaces and affine spaces. Linear Algebr. Appl. **435**(1), 77–94 (2011)

[Wal13] Waldron, S.: Group frames. Finite Frames. Applied and Numerical Harmonic Analysis, pp. 171–191. Birkhäuser/Springer, New York (2013)

[Wal16] Waldron, S.: A sharpening of the Welch bounds and the existence of real and complex spherical t-designs. IEEE Trans. Info. Theory **63**(11), 6849–6857 (2017)

[WC14] Waldron, S., Chien, T.-Y.: The projective symmetry group of a finite frame. submitted, 12 (2014)

[WC15] Waldron, S., Chien, T.-Y.: Nice error frames, canonical abstract error groups and the construction of sics. preprint, 12 (2015)

[Wel74] Welch, L.R.: Lower bounds on the maximum cross correlation of signals. IEEE Trans. Inf. Theory **20**, 397–399 (1974)

[WF89] Wootters, W.K., Fields, B.D.: Optimal state-determination by mutually unbiased measurements. Ann. Phys. **191**(2), 363–381 (1989)

[WH06] Waldron, S., Hay, N.: On computing all harmonic frames of n vectors in $\mathbb{C}^d$. Appl. Comput. Harmon. Anal. **21**(2), 168–181 (2006)

[Whi57] Whitney, H.: Elementary structure of real algebraic varieties. Ann. Math. **2**(66), 545–556 (1957)

[XW01] Xu, Y., Waldron, S.: Tight frames of Jacobi polynomials on a simplex. preprint, 1 (2001)

[XZG05] Xia, P., Zhou, S., Giannakis, G.B.: Achieving the Welch bound with difference sets. IEEE Trans. Inf. Theory **51**(5), 1900–1907 (2005)

[YCD06] Yuan, J., Carlet, C., Ding, C.: The weight distribution of a class of linear codes from perfect nonlinear functions. IEEE Trans. Inf. Theory **52**(2), 712–717 (2006)

[You01] Young, R.M.: An Introduction to Nonharmonic Fourier Series, 1st edn. Academic Press Inc., San Diego (2001)

[Zau99] Zauner, G.: Quantendesigns – Grundzüge einer nichtkommutativen Designtheorie. Ph.D. thesis, University of Vienna, 1 (1999). http://www.gerhardzauner.at/qdmye.html

[Zau10] Zauner, G.: Quantum designs: foundations of a non-commutative design theory. Ph.D. thesis, University of Vienna, 1 (2010). English translation of 1999 Doctorial thesis including a new preface

[Zhu10] Zhu, H.: SIC POVMs and Clifford groups in prime dimensions. J. Phys. A **43**(30), 305305 (2010)

[Zhu12] Zhu, H.: Quantum state estimation and symmetric informationally complete POMs. Ph.D. thesis, National University of Singapore, 1 (2012)

[Zhu15] Zhu, H.: Super-symmetric informationally complete measurements. Ann. Phys. **362**, 311–326 (2015)

[Zim01] Zimmermann, G.: Normalized tight frames in finite dimensions. Recent Progress in Multivariate Approximation (Witten-Bommerholz, 2000). International Series of Numerical Mathematics, vol. 137, pp. 249–252. Birkhäuser, Basel (2001)

Index

Symbols

$(1,1)$-design, 126
$(\tilde{f}_j)$, 31, 32
(f_j^{can}), 35
(n,d,λ)-difference set, 269
1-uniform, 132
2-optimal frame, 268
24-cell, 22
$A = A_\Phi$, $B = B_\Phi$, 32
$A^\dagger$, 61
$B_\alpha^\Theta = B_\alpha$, 92
$B_{n,\Theta} = B_n$, 94
$C : \mathbb{C}^d \to \mathbb{C}^d; z \mapsto \overline{z}$, 382
$C : v \to \overline{v}$, 190
D square–free part of $(d-3)(d+1)$, 412
D_3, 213, 219, 331, 335
D_4, 231
$D_v f$, 93
F, 369
$F_A(a,b,c;x)$, 436
F_f, 413
$F_g = F_{g,j} \in ESL_2(\mathbb{Z}_{d'})$, 406
F_z, F_a, F_b, F_c, F_d, F_e, 389
G-equivariant map, 225
G-frame, 210
G-frame of multivariate orthogonal polynomials, 231
G-frame, characterisation, 214
G-frame, complement, 215
G-frames for infinite groups, 463
G-invariant frame, 209, 234
G-invariant polynomials, 239
G-invariant set, 210
G-invariant subspace, 210
G-map, 225
G-matrix, 332
G-module, 210
G-morphism, 225
G-orbit, 209
$G(m,p,d)$, 345
$G_g = G_{g,j} \in GL_2(\mathbb{Z}_{d'})$, 406
H, $\hat{H}$ Heisenberg group, 366
$H(p,q)$, 461
$H_V(W)$, 225
$H_g \in GL_2(\mathbb{Z}_{d'})$, 403
$L_\sigma f_j = f_{\sigma j}$, 191
M, 369
$M_1 = a_{F_a}$, 389
M_n^ν, 434
$N(\Pi) = N(S_0(\Pi)) = N(S(\Pi))$, 407
P_Φ, 72
P_σ, 369
R, 369
$S(\Pi) \subset ESL_2(\mathbb{Z}_{d'})$, 407
$S = S(v)$ symmetries of a fiducial v, 385
$S_0(\Pi) \subset ESL_2(\mathbb{Z}_{d'})$, 405
S_J, S_n, 190
S_Π symmetries of a SIC fiducial, 404
T_j Chebyshev polynomials of the first kind, 413
$V, V^*, S = VV^*$, 12
Z, W_1, W_2, 395
$Z_\xi^{(k)}$, 446
$\mathrm{Adj}(\Gamma)$, 291
$\mathrm{Ang}(\Phi)$, 253
$\mathrm{Aut}(\Gamma)$, 224
$\mathrm{C}(d)$ Clifford group, 368
$\mathbb{C}G$, 332
$\mathbb{C}G$-module, 227
$\mathbb{C}P^{d-1}$, 129
$\mathrm{EC}(d)$ extended Clifford group, 368

S.F.D. Waldron, *An Introduction to Finite Tight Frames*, Applied and Numerical Harmonic Analysis, https://doi.org/10.1007/978-0-8176-4815-2

EC(d) group of unitary and antiunitary maps, 368
$\mathbb{E}_0 = \mathbb{E}_0^{\Pi}$ fixed field of $\mathscr{G}_0$, 404
$\mathbb{E}_1$ fixed field of $\mathscr{G}_1$, 406
$\mathbb{E}_{\max}$ maximal SIC field, 410
$\mathbb{E}_{\min}$ minimal SIC field, 410
$EGL(\mathscr{H})$, 190
$EU(\mathscr{H})$, 190
$\mathbb{F}G$-homomorphism, 225
$\mathbb{F}G$-isomorphism, 225
$\mathbb{F}G$-module, 210
$\mathbb{F}G$-submodule, 210
$\mathbb{F} \subset \mathbb{C}$, 72
$\mathbb{F}_q$ the finite field of order q, 297
FF(a,b) frame force, 135, 138
FP(Φ) frame potential, 115
Gram(Φ), 13
$\mathrm{Harm}_k(\mathbb{R}^d) = \mathscr{H}_k(\mathbb{R}^d)$, 117
$\mathrm{Hom}_j(\mathbb{R}^d)$, 116
$\mathrm{Hom}_{\mathbb{F}G}(V,W)$, 228
Ind(d), 374
$\mathbb{K} := \mathbb{Q}(\sqrt{(d-3)(d+1)})$, 401
$O(d)$-invariant subspaces of $\mathscr{P}_n$, 451
PC(d) := C$(d)/[I]$ projective Clifford group, 368
PEC(d) := EC$(d)/[I]$ extended projective Clifford group, 368
$PU(\mathbb{C}^d)$, 365
$\Phi = (\phi_g)_{g\in G}$, 213
$\Phi \cup \Psi$, 99
$\Phi \hat{+} \Psi$, 106
$\Phi \oplus \Psi$, 100
$\Phi \otimes \Psi$, 108
Φ^{can}, 35
Π^G, 240
Π_+^G, 240
$\Pi_G = \Pi/\mathscr{I}_G$, 240
$\Pi_t^{\circ}(\mathbb{R}^d)$, 117
$\Pi_{t,r}^{\circ}(\mathbb{C}^d)$, 145
$\Pi_{t,t}^{\circ}(\mathbb{R}^d) = \Pi_{2t}^{\circ}(\mathbb{R}^d)$, 122
Ψ_a, 370
$\mathbb{Q}(\omega)$, 81
$SL_2(\mathbb{Z}_d)$, 370
$\mathbb{S} = \mathbb{S}^d$, 445
Seid(Γ), 291
$\mathrm{C}_{\mathrm{Sp}}(d) := \langle F, R, [I]\rangle$, 372
Sym(Φ), 191
Sym(μ), 231
$\mathrm{Sym}_{\mathrm{E}}(\Phi)$, 191
$\mathrm{Sym}_{\mathrm{EP}}(\Phi)$, 195
$\mathrm{Sym}_{\mathrm{P}}(\Phi)$, 195
$\mathrm{Sym}^t(\mathscr{H})$, 120
$\mathbb{T}$, 366
$\mathbb{Z}_4$-Kerdock codes, 325
area$(\mathbb{S})$, 446
$\mathscr{F}(f_1,\ldots,f_n)$, 113
$\mathscr{F}_{n,\mathbb{F}^d}$, 153
$\mathscr{G} = \mathrm{Gal}(\mathbb{E}/\mathbb{Q})$ Galois group of the SIC field, 402
$\mathscr{G}_0 = \mathscr{G}_0^{\Pi}$, 404
$\mathscr{G}_1 = \langle g_1\rangle$, 406
$\mathscr{H}$, 7
$\mathscr{H}^{\mathbb{C}} \cong \mathscr{H} \oplus i\mathscr{H}$, 227
$\mathscr{H}_k$, $\mathscr{H}_k(\mathbb{R}^d)$, $\mathscr{H}_k(\mathbb{S})$, 445
$\mathscr{I}_G$, 240
$\mathscr{L}_t(\mathscr{H},\mathbb{F})$, 120
$\mathscr{M}(d)$, 22
$\mathscr{N}_{n,\mathbb{F}^d}$, 152
$\mathscr{O}_1,\ldots,\mathscr{O}_\ell$, 406
$\mathscr{P}_k^{\mathrm{V}}$, 431
$\mathscr{P}_n$, 451
$\mathscr{P}_s^{\mathrm{V},n}$, 432
$\chi_p^{\Pi} := \mathrm{trace}(\Pi\hat{D}_p)$ overlap, 402
$\mathrm{tr}_{\mathrm{C}} : \mathrm{C}(d) \to \mathbb{Z}_d$, 392
dep(Φ), 72
$\Delta(v_{j_1}, v_{j_2}, \ldots, v_{j_m})$, 167
$\Delta_C(v_{j_1}, v_{j_2}, \ldots, v_{j_m})$, 176
$\Theta : GL_2(\mathbb{Z}_{d'}) \to GL_2(\mathbb{Z}_{d'})$, 407
$\alpha : SL_2(\mathbb{Z}_{d'}) \to \mathrm{C}_{\mathrm{Sp}}(d)/[I]$, 378
α-partition frame, 20
ω primitive d–th root of unity, 366
∇f, 136
$\sigma_1 = \sigma_x$, $\sigma_2 = \sigma_y$, $\sigma_3 = \sigma_z$, 353
σ_A, 370
$\theta : SL_2(\mathbb{Z}_{d'}) \ltimes \mathbb{Z}_d^2 \to \mathrm{Ind}(d)$, 376
$\hat{G}$, 133, 246
$\hat{D}_p$ displacement operator, 374
$\hat{\psi} : \mathrm{C}_{\mathrm{Sp}}(d)/[I] \to SL_2(\mathbb{Z}_d)$, 378
$\hat{\mathrm{FP}}(\Phi)$ normalised frame potential, 115
$\langle\cdot,\cdot\rangle_\Phi$, 80
$\langle\cdot,\cdot\rangle_\circ$, 120
$\int_{\mathbb{S}} f\,d\sigma$, 116
$\mathbb{E}$, 400
μ primitive $2d$–th root of unity, 366
$\mu(n)$ Möbius function, 81
mult$(W,\mathscr{H})$, 238
$\pi_\Phi : \mathrm{Sym}(\Phi) \to GL(\mathscr{H}) : \sigma \mapsto L_\sigma$, 191
$\psi : \mathrm{C}(d) \to SL_2(\mathbb{Z}_d)$, 372
$\sqrt{i} = e^{\frac{2\pi i}{8}}$, 85
srg(v,k,λ,μ), 291
θ-isogonal configuration, 212
$\varphi(n)$ Euler phi function, 81
$\mathrm{vol}_d(T)$, 430
$\xi^{\Theta}(x) = (\xi_j(x))$, 87
ξ^{α}, 92
$a * b$, 83

$a \circ b$, 183
$c^{\Phi}(f)$, 74
$c_t = c_t(d,\mathbb{F})$, 122
d-polytope-configuration, 344
d-polytope-configuration, flag of, 344
d-polytope-configuration, regular, 344
d-polytope-configuration, symmetry group of, 344
$f : SL_2(\mathbb{Z}_{d'}) \ltimes \mathbb{Z}_d^2 \to \mathrm{C}(d)/[I]$ Appleby indexing homomorphism, 375
$f_E : ESL_2(\mathbb{Z}_{d'}) \ltimes \mathbb{Z}_d^2 \to \mathrm{EC}(d)/[I]$, 382
$g_1 := gg_cg^{-1}$, 405
g_c complex conjugation, 400, 402
$h_{n,d}$, $h^{\mathrm{u}}_{p,d}$, $h^{\mathrm{l}}_{p,d}$, 258
j-faces, 344
k-flag, 201
$k_g \in \mathbb{Z}^*_{d'}$, 403
m-distance set, 309
m-distance tight frame, 310
m-products, 165, 167
m-th root signature matrix, 312
$m_d : SL_2(\mathbb{Z}_{d'}) \to SL_2(\mathbb{Z}_d)$, 378
n-th roots of unity, 263
p-groups, 256
$p(\partial)$, 145
q-Paley graph, 297
s-angular frame, 318
$z\psi_Z$, m_1, m_2, 394
z_a, 370
${}_2F_1$, 429

POVM, 364

A
abelian difference set, 269
abelian field extension, 412
absolute bound (on number of equiangular lines), 268
absolutely irreducible, $\mathbb{F}G$-module, representation, 228
ADE classification of the finite subgroups of $SL_2(\mathbb{C})$, 357
adjacency matrix of a graph, 224, 291
adjacent vertices of a graph, 291
affine generalised barycentric coordinates, 87
affine group, 261
affine independence, 98
affine map, 91
affinely equivalent subsets of a finite abelian group, 261
affinely independent points, 98
algebra of group matrices, 332
algebraic equations for tight complex equiangular lines, 316
algebraic variety, 151
algebraic variety of SIC fiducials, 414
alternate dual, 54
analysis operator, 12, 27
analysis operator of a generalised frame, 443
analytic SIC, 361
angle between subspaces/vectors, 266
angle multiset, 253
anti equivalent, 11
antilinear map, 11, 190
antisymmetry, 191
antiunitary map, 190
antiunitary symplectic, 382
apolar inner product, 120
Appell polynomials, 438
Appell's biorthogonal system, 429
Appleby index, 375
Appleby index, extended, 382
Appleby indexing of the Clifford operations, 374
approximate inverse, 53, 66
approximate left inverse, 66
approximately dual frames, 56
association schemes, 22
automorphism group of a graph, 224
automorphism of a graph, 224

B
balanced frame, 104
Bargmann invariants, 167
barycentric coordinates, 87, 430
basis, 73
Bernstein coefficients, 433
Bernstein frame, 93
Bernstein frame, properties, 93
Bernstein operator, 94
Bernstein polynomials, 431
Bernstein–Bézier coefficients, 431
Bernstein–Durrmeyer operator, 429, 434
Bessel identity, 8
Bessel identity, generalised, 121, 123
bilateral shift, 469
binary icosahedral group, 243
biorthogonal system, 33
Bloch sphere, 363
block designs, 275
block matrix calculation, 316
blocks, 275

Bombieri inner product, 120
Bombieri's inequality, 149
boring strongly regular graphs, 293
bra $\langle w|$, 364
bra–ket notation, 364
Bruck–Ryser–Chowla theorem, 269

C

canonical m-products, 176
canonical coordinates, 71, 74, 215
canonical coordinates and linear maps, 79
canonical coordinates for cyclotomic fields, 81
canonical coordinates, characterisation, 76
canonical coordinates, properties, 78
canonical copy of a tight frame, 15, 17
canonical dual frame, 32, 54, 56
canonical dual functionals, 74
canonical expansion, 74
canonical factorisation, 74
canonical Gramian, 54, 72, 74
canonical Gramian of a generalised frame, 443
canonical inner product, 80, 98
canonical inner product, properties, 83
canonical isomorphism between a vector space and its bidual, 76
canonical matrix, 75
canonical oblique dual, 59, 68
canonical order 3 Clifford operation, 394
canonical order 3 symplectic unitaries, conjugates of, 398
canonical tight frame, 31, 35, 62, 63, 186
canonical tight frame of a G-frame, 211
canonical tight frame of a generalised frame, 442
Cauchy–Schwarz inequality, 38, 114, 142, 149
Cayley graph, 309
Cayley transform, 155
Cayley transform, truncated, 155
CDMA systems, 113
central G-frame, 230, 336
central force, 135
central group frame, 230
central tight G-frames, classification, 337
centrally symmetric, 116
centred SIC fiducial, 405
character group, 133, 246
character of a representation, 334
character table of a finite abelian group, 246
characters of a finite abelian group, 246
Chebyshev method, 53
Chebyshev polynomials, 430
Chebyshev polynomials of the first kind, 413
Chinese remainder theorem, 398
chordal graph, 174
Chu–Vandermonde identity, 435
circulant graph, 303
circulant matrix, 213
class function, 336
Clifford action, 386
Clifford action of the extended Clifford group, 385
Clifford action on SICs, 386
Clifford group, 361, 368
Clifford group, generators for, 372
Clifford group, monomial representation of, 391
Clifford operation, 368
Clifford trace, 392
close frames, 51
closeness bound, 51, 66
code division multiple access systems, 113
coinvariant space, 240
coinvariants of a finite reflection group, 240
coisometry, 13, 25
column orthogonality (of irreducible characters), 246
column, of a linear map on $\ell_2(J)$, 12
commutativity relation for S and Ω, 366
commutativity relations for V,S,G, 62
complement of a G-frame, 215
complement of a difference set, 269
complement of a frames, 105
complement of a harmonic frame, 250
complement of a tight frame, 18
complementary tight frame, 18
complete frame graph, 168
complex (t,t)–design, 126
complex t–design, 126
complex conjugate equivalence, 11
complex conjugate of a Hilbert space, 26
complex conjugation map, 11, 190
complex equiangular tight frames, 310
complex frame, 41, 63
complex Hadamard matrix, 281
complex polytope, 343
complex projective sphere, 129
complex reflection, 240, 343
complex reflection group, 240
complex tight frame, 21
complex unit sphere, 122
complexification, 227

compressed sensing, 4
condition number of a frame, 46
condition number of the frame operator, 47
conductor, of a ray class field, 412
conference graph, 296
conference matrix, 296, 314, 327
conjugate gradient method, 53
conjugates of the canonical order 3 symplectic unitaries, 398
conjugation map, 26
connected components of a graph, 168
connection set of a graph, 309
constant diagonal Hadamard matrix, 282
continuous frame, 442
continuous generalised frame, 441
continuous tight frame, 2
continuous tight frame expansion, 441
continuous tight frame of zonal harmonics, 446
continuous tight frames for $\mathscr{P}_n$, 454
contraction map, 66
coordinate functionals, 33
coordinate star, 15
coordinates, 12
coordinates with respect to a tight frame, 29
Coulomb force, 135
cross in $\mathbb{R}^n$, 16, 346
cross–correlation, 3, 22, 113
cubature formula, 119, 122
cubature formula, existence of, 119
cubature rule, 119, 122, 123
cubature rule for the sphere, 459
cube, 222, 252, 346
cube roots, 283
cube, vertices of, 107, 248
cuboctahedron, 346
cycle space of a finite graph, 171
cyclic convolution, 83
cyclic difference set, 269
cyclic frame, 247
cyclic group, 210, 213
cyclic harmonic frame, 247
cyclic harmonic frames for $\mathbb{C}^1$, 255
cyclic harmonic frames for $\mathbb{C}^2$, 255
cyclic shift, 81
cyclic shift matrix, 366
cyclic shift operator S, 85
cyclic vector, 463
cyclotomic field, 80, 81, 374, 401
cyclotomic field $\mathbb{Q}(\omega)$, 81

D
de la Vallée–Poussin mean, 434
decomposition of frames, 103, 110
degree elevation of the Bernstein operator, 429
degree of a character, 334
degree raising operator, 431
degree reducing, 95, 96
degree reducing operator, 434
demicube, 347
density matrix, 364
density of rational tight frames, 156
dependencies, 72
dependencies dep(Φ), 72
determining set of m-products, 167
DFT matrix, 369
diagram vectors, 9, 25
dicyclic group, 357
difference set, 269
difference sets, infinite families of, 273
differentially 1–uniform, 132
dihedral group, 193, 213
dilation, 2
dimension towers, 413
Diophantine equation, 269
Dirac notation, 364
direct sum, 102
direct sum of frames, 100
direct sum of harmonic frames, 250
directional derivative, 93
discrete Fourier transform, 25
discrete Fourier transform matrix, 369
discrete frame, 442
discrete generalised frame, 441
discrete Heisenberg group, 366
discrete inner product, 251
disjoint union of frames, 99
disjointness, 101
displacement operation, 372
displacement operator, 374
distance between frames, 47
division ring, 228
dodecahedron, 222, 252
dual basis, 33, 75
dual frame, 31, 32
dual frame of a G-frame, 211
dual frame of a generalised frame, 442
dual frames, 54
dual Riesz basis, 45
dual sequences, 98
dual signed frame, 186

E
effective frame force, 138, 148

eigensteps, 154
elementary abelian p-group, 250
elliptic curve, 401
empty frame graph, 168
ENPTF, 23
equal-norm frame, 9
equal-norm tight frame, 27
equally spaced unit vectors in $\mathbb{R}^2$, 199
equally spaced vectors, 251
equations for SIC fiducials, 414
equiangular lines, 174, 265, 266
equiangular tight frame, 27, 73
equiangular tight frames from block designs, 275
equiangular vectors, 266
equiangularity, 3, 4
equispaced equal-norm frame, 263
equispacing, 4
equivalence of tight frames, 23
equivalence, unitary, 10
equivalent difference sets, 271
equivalent group representations, 210
equivalent Hadamard matrices, 321
erasures, 262
error operator basis, 353
Eulerian function, 254
Eulerian subgraph, 171
eutactic star, 16
existence of equalnorm tight frames, 157
expressible by radicals, 400, 401
extended Appleby index, 382
extended Clifford group, 368
extended projective Clifford group, 368
extended projective symmetry group of a frame, 195
extended projective symmetry group of a sequence of vectors, 195
extended symmetry group of a frame, 191
extended symmetry group of a sequence of vectors, 191

F

facts, about SICs, 401
faithful representation, 210
Fano plane, 273
FF–critical, 138
fiducial vector, 361, 365
fiducial vector, real, 417
field trace, 325
finite abelian extensions of $\mathbb{Q}$, 401
finite field of order q, 297
finite Fourier methods, 414
finite reflection group, 343
finite reflection groups, classification of, 345
finite tight frame, 7
finite tight frame, early examples, 2
finite tight frame, prototypical example, 1
finite tight frame, second prototypical example, 3
finite tight frames for $\mathscr{P}_n$, 459
Fisher inner product, 120
five vectors in $\mathbb{C}^3$, 199
flag of a d-polytope-configuration, 344
forward cyclic shift, 81
Fourier matrix, 9, 245, 246, 324, 369
Fourier transform, 443
Fourier transform matrix, 25
frame, 32
frame algorithm, 53
frame bound, 7
frame bounds, 32, 46, 64
frame force, 115, 135
frame force critical sequence, 138
frame graph, 100, 165
frame homotopy problem, 153
frame operator, 12, 27, 32, 62
frame potential, 115
frame potential, global minimisers, 136
frame potential, local minimisers, 136
Frame Research Centre, 186
frame theory, history, 2
frame transform operator, 12
frame with s angles, 318
frame with maximal projective symmetry, 198
frame, continuous, 442
frame, discrete, 442
frame, fusion, 9
frame, generalised, 442
frame, semicritical, 110
frames as orthogonal projections, 43
frames invariant under the unitary action of an abelian group, 236
full octahedral group, 346
functions on the complex sphere, 461
fundamental cycle basis, 171
fundamental cycle of an edge, 170
fusion frame, 9, 179
fusion frame operator, 180
fusion frame system, 180

G

Gabor system, 2, 5, 209, 366

Galois automorphism, 400
Galois correspondence of SIC subfields, 410
Galois field, 325
Galois group of the SIC field, 425
Galois multiplet, 403
Galois symmetries of a SIC, 361
Galois theory, 400
Gamma function, 430
Gauss sum, 381
Gauss sums, 330
Gegenbauer polynomials, 447, 453
generalised barycentric coordinates, 87
generalised barycentric coordinates, properties, 88
generalised Bernstein operator, 94
generalised cross, 346
generalised frame, 441, 442
generalised Gauss sums, 330
generalised Gegenbauer polynomials, 453
generalised Hermite polynomials, 453
generalised Pauli group, 366
generalised quaternian group, 357
generalised resolution of the identity, 28
generating set of m-products, 167
generic SIC, 401, 424
geometrically uniform frame, 247
Gerzon bound, 268
Givens rotation, 159
global field, 412
golden ratio, 206, 243, 327
Gram matrix, 13
Gram–Schmidt algorithm, 68, 72
Gramian, 13, 27, 62, 102
Gramian of a generalised frame, 443
Grassmann manifold of d-planes in $\mathbb{F}^n$, 153
Grassmannian, 42
Grassmannian frame, 268
Grassmannian packing problem, 268
Grassmannian space, 268
group algebra, 210, 215
group algebra over $\mathbb{C}$, 332
group covariance, 361, 365
group frame, 209, 210
group matrix, G-matrix, 213
group of rotations, 2
growing a flag set, 201

H

Haar measure, 463
Haar measure on $SO(d)$, 451
Hadamard conjecture, 25
Hadamard difference set, 273
Hadamard matrix, 9, 25, 278
Hadamard matrix of (v_j) and (w_j), 320
Hadamard product of matrices, 183
Hadamard type difference set, 273
Hahn polynomials, 429, 433
Hall's Eulerian function, 254
Hamiltonian circulant graph, 297
harmonic frame, 9, 229, 245, 247
harmonic frame with distinct vectors, 249
harmonic frame with real vectors, 249
harmonic function, 445
harmonic polynomials, 117
harmonic upper bound, 304
Heisenberg frame, 366
Heisenberg group, 366
Heisenberg group, normaliser of, 368
Heisenberg operation, 372
Hermite polynomials, 453
Hermitian complex Hadamard matrix, 281
Hessian matrix, 141, 149
hexadecachoron, 346
highly symmetric tight frame, 223, 349
highly symmetric tight frames, construction, 342
Hilbert class field, 412
Hilberts 12–th problem, 401
Hilbert–Schmidt inner product, 29
Hoggar lines, 365, 367, 399
homogeneous G-frame, 226
homogeneous components, 225
homogeneous polynomials, 117
homogeneous polynomials on a Hilbert space, 120
Householder transformation, 155, 159
hyperplane, 343

I

icosahedron, 116, 222, 252, 288
imprimitive finite reflection group, 345
index group of the Clifford operations, 374
index group, of a nice error frame, 353
index map $[a] \mapsto (\psi_a, z_a)$ of the Clifford operations, 374
inner direct sum of frames, 100
inner product preserving map, 27
inner product, apolar, 120
inner product, Bombieri, 120
inner product, Fisher, 120
intermediate SIC field, 410
invariant polynomials, 239
invariant subspace, 44

irreducibility, 85
irreducible G-frame, 218
irreducible character, 334
irreducible characters of a finite abelian group, 246
irreducible group representation/action, 218
irreducible tensors, 108
isogonal configuration, 41, 63, 212, 265, 309
isolated point (of a graph), 294
isometric embedding, 131
isometry, 8, 13, 25, 27

J
Jacobi inner product, 430
Jacobi measure, 431
Jacobi polynomials, 429, 433, 437
Jordan-Hölder theorem, 226

K
ket $|v\rangle$, 364
Kirkman frame, 312
Kronecker product, 252
Kronecker–Weber theorem, 401

L
Löwdin orthogonalisation, 35, 60, 68
La Jolla Difference Set Repository, 273
Laplaces equation, 445
Laplacian, 147
Lauricella function of type A, 429, 435
least squares solution, 60
Legendre polynomials, 429, 430, 437, 453
Legendre polynomials on a square, 231
Legendre symbol, 399
Lie group, 152, 163
lift, 110
lift of a frame, 104
lifted equally spaced vectors, 105
lifted roots of unity, 105
lifting a frame, 104
linear action, 210
linear dependencies, 72
linear map, 79
linear reproduction, 97
LLL (Lenstra–Lenstra–Lovász) algorithm, 404
local compact (topological) group, 463
local frame, 180
Lucas–Fibonnacci SIC, 413

M
Möbius function, 81
Magma, 342, 347
majorisation, 157
Maschkes theorem, 225, 227
matrices, 75
matrices with respect to a normalised tight frame, 29
matrix with respect to a tight frame, 29
maximal abelian $*$-algebra, 322
maximal number of MUBs, 174
maximal SIC field, 410
maximally symmetric frame, 198
maximally symmetric tight frame, 198
maximum cross-correlation, 266
maximum size $g(d)$ of a spherical two-distance set, 304
Mercedes–Benz frame, 3, 216, 288
minimal angle, 266
minimal SIC field, 410
minor of a graph, 316
modulation, 2
modulation matrix, 366
monomial representations of the Clifford group, 391
monomials, 121
mother wavelet, 366
MUB, 22, 127, 174, 209, 318
MUB problem, 22, 325
MUBs, 22
multi-indices, 92
multinomial identity, 96
multiple orthogonal polynomials, 438
multiplet, 403
multiplicative equivalence, 254
multiplicatively equivalent subsets of a group, 254
multiresolution analysis, 2
multivariate continuous Hahn polynomials, 439
multivariate Hahn polynomials, 439
multivariate Jacobi polynomials, 429
multivariate orthogonal polynomials, 231, 438
multivariate shifted factorial, 95
mutually unbiased bases, 22,174, 318
MWBE codebook, 271

N
Naĭmark's theorem, 15
near frames, 51
neighbourhood of a switching class of graphs, 294

nested roots, 400
nested SIC fields, 413
nice (unitary) error basis, 353
nice (unitary) error frame, 353
nice error bases, 221, 353
nice error basis, 221
nice error frames, 353
non-adjacent vertices of a graph, 291
noncanonical dual frame, 54
noncyclic harmonic frame, 248, 256
normal equations, 183
normal field extension, 401
normalised eutactic star, 16
normalised frame, 48
normalised frame potential, 115
normalised surface area measure, 122
normalised tight frame, 7, 8
normalised tight frames and linear mappings, 29
normalised tight frames, algebraic variety of, 151, 152
normalising a frame, 47
number of angles, 318
number of generators of a G-invariant frame, 234
number of harmonic frames, 258
numerical SIC, 361

O

oblique dual frames, 57
oblique projection, 58
octahedron, 222, 252, 346
octaplex, 346
optimal frame bounds, 32, 49, 50, 52, 61, 64
optimal Grassmannian frame, 268
order of a difference set, 269
orthogonal basis, 100
orthogonal compression, 101
orthogonal decomposition into homogeneous components, 226
orthogonal dilation, 101
orthogonal frames, 42
orthogonal polynomials, 69, 231
orthogonal polynomials of Appell, 437
orthogonal polynomials of Proriol, 437
orthogonal polynomials on the simplex, tight frames for, 429
orthogonal polynomials on the triangle, 429
orthogonal projection, 25
orthogonal projection formula, 24, 62
orthogonal projection, one-dimensional, 267
orthogonal resolution of the identity, 28
orthogonality (of irreducible characters), 246
orthogonality of frames, 101
orthonormal bases, 248, 267, 271
orthonormal basis, 9, 24, 189
outer direct sum of frames, 100
overlaps, 402
overlaps phases, 402

P

Paley graph, 297, 303
Paley-type difference set, 273
parabolic subgroup of a finite reflection group, 344
Parseval frame, 8, 23
Parseval identity, 8, 13, 24
partial difference set, 308
partial isometry, 8, 25, 36, 61
partition frame, 20, 105
partition of unity, 92
path graph, 290
pathconnected, 153
pathconnectivity, 153
Pauli matrices, 353
Pell's equation, 412
perfectly tight frame, 143
permutation matrices, 193
permutation matrices in the Clifford group, 380
permutations, 190
perturbation of a frame, 68
perturbation of a normalised tight frame, 61
phase-permutation representation, 391
planar difference set, 273
Plancherel identity, 8, 24
Plancherel identity, generalised, 121, 123
plane wave, 145, 146
Platonic solid, 209, 222, 252
Platonic solid, symmetry group, 222
Pochhammer symbol, 122, 431
Pohlke normal star, 15
point evaluation, 78, 187
pointwise product of matrices, 183
Poisson kernel $P(x, \xi)$, 450, 467
polar decomposition, 36
polarisation identity, 7, 24
pole, 446
pole of a zonal function, 446
polynomials on projective spaces, 148
Pontryagin duality, 246, 270
Pontryagin duality map, 133, 246

positive operator, 434
positive operator valued measure, 364
potential, 135
pre-frame operator, 12
precision bumping algorithm, 404, 410
Prime Power Conjecture, 273
primitive finite reflection group, 345
primitive roots of unity, 81
primitive strongly regular graphs, 293
principal submatrix of a matrix, 316
probability distribution, 364
projections of normalised tight frames, 25
projective, 9
projective (t,t)-designs, 130
projective Clifford group, 368
projective equivalence, 23
projective invariants, 165
projective linear group, 351
projective plane, 273
projective representation, 351
projective similarity, 176
projective similarity of vectors, 165
projective symmetry group of a frame, 189, 190, 195
projective symmetry group of a harmonic frame, 262
projective unitary equivalence, 11, 26, 165, 166
projective unitary equivalence of harmonic frames, 260
projective unitary group, 351, 365
projectively equivalent, 11
projectively repeated, 262
projectively similar, 176
projectively unitarily equivalent, 11
Proriol polynomials, 438
Proriols orthogonal basis, 429
prototypical example, first, 1
prototypical example, second, 3
pruning a set of flags, 201
pseudodual frames, 56
pseudodual, 57
pseudoinverse, 31, 60
pseudoreflection, 343
PSLQ algorithm, 404

Q
quadratic distance (between frames), 51
quadratic reciprocity law, 418
quadratic residue, 297
quadrature rule for the sphere, 459
quantum measurements, 364
quantum state determination, 22
quantum system, 364

R
radial function, 447
radially symmetric weights, 439
ramification, of a ray class field, 412
rank-one orthogonal projection, 31
rank-one projection, 60
rational tight frames, density of, 155
ray class conjecture, 412
ray class field, 401, 412
ray class group, 412
real algebraic variety, 152
real cyclic harmonic frame, 251
real fiducial vectors, 417
real frame, 41, 63
real harmonic frame, 251
real MUBs, 22
real orthogonal group, 189
real projective space, 129
real projective sphere, 129
real spherical t–design, 116
real spherical half–design, 116
real tight frame, 21
real unit sphere, 122
reconstruction operator, 12
rectified tesseract, 346
reduced Hadamard matrix, 322
reduced signature matrix of an equiangular frame, 284
redundancy, 7, 47
reflection group, 343
regular, 308
regular d-polytope-configuration, 344
regular character (of an abelian group), 270
regular polygon, 90
regular two-graph, 294
reindexing a frame, 254
relative bound (on number of equiangular lines), 287
reordering, 10
reordering a frame, 254
representation of a group, 210
representation theory, 209, 210
reproducing kernel, 118, 444
reproducing kernel for $\mathscr{P}_n$, 457
reproducing kernel Hilbert space, 444
reproducing kernel tight frame, 444
ridge function, 146, 447
ridge polynomial, 184, 442
Riesz basis, 45, 51

Riesz representer, 31, 120
ring of coinvariants, 240
robust signal transmission, 248
robustness to erasures, 1, 3, 4
roots of unity, 107
rotation, 2, 10
row construction of a tight frame, 17

S

sampling, 441
scalable frame, 182
scaling to a tight frame, 182
Schurs lemma, 226, 228
SchurHorn majorisation, 157
Schweinler–Wigner orthogonalisation, 60
Scott–Grassl numerical SIC, 385
seeding, 15
Seidel matrix, 289
Seidel spectrum, 290
semi-regular complex polytopes, 345
semicritical frame, 110
semidirect product, 375
sensor networks, 180
Shephard–Todd classification, 345, 347
shift invariant tight frame, 86
shift matrix, 366
shifted factorial, 95
short-time Fourier transform, 443
SIC, 3, 22, 127, 174, 209, 363, 364
SIC facts, 401
SIC fiducials, algebraic variety of, 414
SIC field $\mathbb{E} = \mathbb{Q}(\Pi, \mu)$, 400
SIC field of a multiplet, 410
SIC field, minimal, intermediate, maximal, 410
SIC problem, 22, 361
SIC problem, addictiveness, 361
SIC-POVM, 363, 364
signal transmission with erasures, 248
signal transmission with quantization, 248
signature of a signed frame, 181, 186
signed frame, 181, 186
signed frame operator, 186
similar frames, 41, 42
similarity, 42, 73, 101, 191
simple lift, 110
simple lift of a frame, 104
simplex, 73, 286, 290
simplex, vertices of, 271
Singer difference sets, 273
singular value decomposition, 62, 65
Sixteen equiangular lines in $\mathbb{R}^6$, 205
size of a Galois multiplet, 403
skew Hermitian matrix, 155
Sobolev dual, 67
solid spherical harmonics, 445
spanning set, 31, 32, 61, 73
spanning tree for a graph, 170
sparsness, 3
special unitary group, 152
sphere in $\mathbb{R}^d$, 445
spherical (t,t)-designs, numerical construction, 139
spherical 2-design, 117
spherical m-distance set, 309
spherical t-design, 116, 119
spherical t-designs, characterisations, 117
spherical design of harmonic index t, 118, 119
spherical half-design, 116
spherical half-design of order t, 119
spherical harmonics, 445
spherical two-distance set, 304
spherical two-distance tight frame, 304
sporadic SIC, 401, 416, 424
square free integer, 257
stability, 3
stable isogon, 359
stabliser, 342
standard m-distance tight frame, 310, 346
standard terminology for frames, 23
state, 364
state, of a quantum system, 364
Steinbergs fixed point theorem, 345
Steiner equiangular tight frame, 277
Steiner system, 275
STFT, 443
strongly centred SIC fiducial, 406
strongly disjoint frames, 42
strongly regular graph, 291
structured frames, 209
subsimplicial equiangular tight frame, 275
subsimplicial frame, 275
sum of frames, 106
sum of frames, direct, 100
sum of harmonic frames, 250
surface area measure, 122
surface area measure, invariance of, 123
surface spherical harmonics, 445
switching, 290
switching class (two-graph), 290
switching equivalent graphs, 290
symmetric t-linear maps, 120
symmetric block design, 269
symmetric Gram–Schmidt algorithm, 35

symmetric group on a set, 190
symmetric informationally complete positive operator valued measure, 363
symmetric orthogonalisation, 35, 60
symmetric tensors, 120
symmetries of a basis, 192
symmetries of a SIC fiducial, 404
symmetries of a SIC fiducial vector, 385, 390
symmetries of the vertices of a simplex, 192
symmetries of the Weyl–Heisenberg SICs, 389
symmetries, of a fiducial, 385
symmetry, 191
symmetry group of n equally spaced vectors, 193
symmetry group of a d-polytope-configuration, 344
symmetry group of a complementary frame, 196
symmetry group of a frame, 189–191
symmetry group of a harmonic frame, 250
symmetry group of a measure μ , 231
symmetry group of a Platonic solid, 222
symmetry group of the square, 233
symmetry group of the triangle, 233
symplectic index, 379
symplectic index, table of, 381
symplectic operation, 372
symplectic spreads, 325
symplectic unitaries, 372, 378
symplectic unitaries of order 3, 392
synthesis operator, 12, 27, 62
synthesis operator of a generalised frame, 443
Szegö kernel, 468

T
Tammes problem, 200, 269
Teichmüller set, 325
tensor product integration formula, 123
tensor product of frames, 108
tensor product of harmonic frames, 250
tensor product of Hilbert spaces, 108
tensor product of nice error frames, 355
tensor product of vector spaces, 108
tesseract, 346
tetrahedron, 222, 252
tight G-frames as idempotents of the group algebra, 333
tight equiangular lines, 265
tight frame, 7
tight frame, canonical, 35
tight frame, canonical copy, 15
tight frame, continuous, 2
tight frame, copy of, 15
tight frame, early examples, 2
tight frame, equal-norm, 9
tight frame, repeated vectors, 8
tight frame, unit-norm, 9
tight frames of orthogonal polynomials on the simplex, 429
tight frames, construction of, 158
tight frames, existence of, 157
tight fusion frame, 179
tight generalised frame, 441
tight signed frame, 181
tightness, 47
Tikhonov regularization (of the pseudoinverse), 484
time–frequency localisation, 1
topologically irreducible, 463
torsion point, 401
total potential, 135
trace formula, 13, 26
translate, 260
translation, 2, 260
Tremain equiangular tight frame, 278
Tremain frame, 278
triangular graph, 301, 310
triangulated graph, 174
trigonal bipyramid, 199, 343
trigonometric polynomials, 251, 463
triple products, 165, 167, 174, 284, 365
trivial difference sets, 271
two-distance set, 304
two-graph (switching class), 290

U
ultraspherical polynomials, 447
unfaithful action, 190
unimodular simplex, 278
union of frames, 99
unit sphere, 122
unit sphere in $\mathbb{R}^d$, 445
unit sphere, complex, 122
unit sphere, real, 122
unit-norm frame, 9
unital equiangular tight frame, 312
unitarily equivalent frames, 40
unitarily equivalent tight frames, 10
unitarily equivalent tight frames up to reordering, 10
unitarily equivalent via an automorphism, 254

unitarily inequivalent tight D_3-frames, 331
unitary action, 211
unitary continuous representation, 463
unitary equivalence, 23
unitary equivalence of harmonic frames, 253
unitary equivalence of tight frames, 10
unitary equivalence of tight frames up to reordering, 10
unitary equivalence up to reordering, 10
unitary group, 152
unitary group, parametrisations of, 155
unitary group, projective, 365
unitary images of tight frames, 24
unitary map, 211
unitary matrix, 152
univariate Jacobi polynomials, 430
univariate orthogonal polynomials, 68, 69

V
variational characterisation of tight frames, 114
variational formula, 13
vertex-transitive graphs, 224
vertices of the cube, 107, 248, 264
vertices of the simplex, 19, 20, 248
Von Neumann algebra, 465

W
Waring formula, 116
wavelet system, 2, 5, 209
wavelets, 1
WBE sequences, 113
Wedderburns Theorem, 238
weighted (t,t)-design, 128
Welch bound, 113
Welch bound equality sequences, 113
Weyl displacement operation, 372
Weyl–Heisenberg SIC, 366, 368
Weyl–Heisenberg SIC, symmetries of, 389
Wirtinger calculus, 508
Wirtinger complex differential operators, 146

Z
Zak transform, 465
Zauner matrix Z, 384
Zauners conjecture, 22, 361
Zauners conjecture, stronger form, 385
zonal function, 447
zonal harmonic, 446

ANHA Back Matter

Applied and Numerical Harmonic Analysis (82 volumes)

A. Saichev and W.A. Woyczyński: *Distributions in the Physical and Engineering Sciences* (ISBN 978-0-8176-3924-2)

C.E. D'Attellis and E.M. Fernandez-Berdaguer: *Wavelet Theory and Harmonic Analysis in Applied Sciences* (ISBN 978-0-8176-3953-2)

H.G. Feichtinger and T. Strohmer: *Gabor Analysis and Algorithms* (ISBN 978-0-8176-3959-4)

R. Tolimieri and M. An: *Time-Frequency Representations* (ISBN 978-0-8176-3918-1)

T.M. Peters and J.C. Williams: *The Fourier Transform in Biomedical Engineering* (ISBN 978-0-8176-3941-9)

G.T. Herman: *Geometry of Digital Spaces* (ISBN 978-0-8176-3897-9)

A. Teolis: *Computational Signal Processing with Wavelets* (ISBN 978-0-8176-3909-9)

J. Ramanathan: *Methods of Applied Fourier Analysis* (ISBN 978-0-8176-3963-1)

J.M. Cooper: *Introduction to Partial Differential Equations with MATLAB* (ISBN 978-0-8176-3967-9)

A. Procházka, N.G. Kingsbury, P.J. Payner, and J. Uhlir: *Signal Analysis and Prediction* (ISBN 978-0-8176-4042-2)

W. Bray and C. Stanojevic: *Analysis of Divergence* (ISBN 978-1-4612-7467-4)

G.T. Herman and A. Kuba: *Discrete Tomography* (ISBN 978-0-8176-4101-6)

K. Gröchenig: *Foundations of Time-Frequency Analysis* (ISBN 978-0-8176-4022-4)

L. Debnath: *Wavelet Transforms and Time-Frequency Signal Analysis* (ISBN 978-0-8176-4104-7)

S.F.D. Waldron, *An Introduction to Finite Tight Frames*, Applied and Numerical Harmonic Analysis, https://doi.org/10.1007/978-0-8176-4815-2

J.J. Benedetto and P.J.S.G. Ferreira: *Modern Sampling Theory* (ISBN 978-0-8176-4023-1)

D.F. Walnut: *An Introduction to Wavelet Analysis* (ISBN 978-0-8176-3962-4)

A. Abbate, C. DeCusatis, and P.K. Das: *Wavelets and Subbands* (ISBN 978-0-8176-4136-8)

O. Bratteli, P. Jorgensen, and B. Treadway: *Wavelets Through a Looking Glass* (ISBN 978-0-8176-4280-80)

H.G. Feichtinger and T. Strohmer: *Advances in Gabor Analysis* (ISBN 978-0-8176-4239-6)

O. Christensen: *An Introduction to Frames and Riesz Bases* (ISBN 978-0-8176-4295-2)

L. Debnath: *Wavelets and Signal Processing* (ISBN 978-0-8176-4235-8)

G. Bi and Y. Zeng: *Transforms and Fast Algorithms for Signal Analysis and Representations* (ISBN 978-0-8176-4279-2)

J.H. Davis: *Methods of Applied Mathematics with a MATLAB Overview* (ISBN 978-0-8176-4331-7)

J.J. Benedetto and A.I. Zayed: *Modern Sampling Theory* (ISBN 978-0-8176-4023-1)

E. Prestini: *The Evolution of Applied Harmonic Analysis* (ISBN 978-0-8176-4125-2)

L. Brandolini, L. Colzani, A. Iosevich, and G. Travaglini: *Fourier Analysis and Convexity* (ISBN 978-0-8176-3263-2)

W. Freeden and V. Michel: *Multiscale Potential Theory* (ISBN 978-0-8176-4105-4)

O. Christensen and K.L. Christensen: *Approximation Theory* (ISBN 978-0-8176-3600-5)

O. Calin and D.-C. Chang: *Geometric Mechanics on Riemannian Manifolds* (ISBN 978-0-8176-4354-6)

J.A. Hogan: *Time–Frequency and Time–Scale Methods* (ISBN 978-0-8176-4276-1)

C. Heil: *Harmonic Analysis and Applications* (ISBN 978-0-8176-3778-1)

K. Borre, D.M. Akos, N. Bertelsen, P. Rinder, and S.H. Jensen: *A Software-Defined GPS and Galileo Receiver* (ISBN 978-0-8176-4390-4)

T. Qian, M.I. Vai, and Y. Xu: *Wavelet Analysis and Applications* (ISBN 978-3-7643-7777-9)

G.T. Herman and A. Kuba: *Advances in Discrete Tomography and Its Applications* (ISBN 978-0-8176-3614-2)

M.C. Fu, R.A. Jarrow, J.-Y. Yen, and R.J. Elliott: *Advances in Mathematical Finance* (ISBN 978-0-8176-4544-1)

O. Christensen: *Frames and Bases* (ISBN 978-0-8176-4677-6)

P.E.T. Jorgensen, J.D. Merrill, and J.A. Packer: *Representations, Wavelets, and Frames* (ISBN 978-0-8176-4682-0)

M. An, A.K. Brodzik, and R. Tolimieri: *Ideal Sequence Design in Time-Frequency Space* (ISBN 978-0-8176-4737-7)

S.G. Krantz: *Explorations in Harmonic Analysis* (ISBN 978-0-8176-4668-4)

B. Luong: *Fourier Analysis on Finite Abelian Groups* (ISBN 978-0-8176-4915-9)

G.S. Chirikjian: *Stochastic Models, Information Theory, and Lie Groups, Volume 1* (ISBN 978-0-8176-4802-2)

C. Cabrelli and J.L. Torrea: *Recent Developments in Real and Harmonic Analysis* (ISBN 978-0-8176-4531-1)

M.V. Wickerhauser: *Mathematics for Multimedia* (ISBN 978-0-8176-4879-4)

B. Forster, P. Massopust, O. Christensen, K. Gröchenig, D. Labate, P. Vandergheynst, G. Weiss, and Y. Wiaux: *Four Short Courses on Harmonic Analysis* (ISBN 978-0-8176-4890-9)

O. Christensen: *Functions, Spaces, and Expansions* (ISBN 978-0-8176-4979-1)

J. Barral and S. Seuret: *Recent Developments in Fractals and Related Fields* (ISBN 978-0-8176-4887-9)

O. Calin, D.-C. Chang, and K. Furutani, and C. Iwasaki: *Heat Kernels for Elliptic and Sub-elliptic Operators* (ISBN 978-0-8176-4994-4)

C. Heil: *A Basis Theory Primer* (ISBN 978-0-8176-4686-8)

J.R. Klauder: *A Modern Approach to Functional Integration* (ISBN 978-0-8176-4790-2)

J. Cohen and A.I. Zayed: *Wavelets and Multiscale Analysis* (ISBN 978-0-8176-8094-7)

D. Joyner and J.-L. Kim: *Selected Unsolved Problems in Coding Theory* (ISBN 978-0-8176-8255-2)

G.S. Chirikjian: *Stochastic Models, Information Theory, and Lie Groups, Volume 2* (ISBN 978-0-8176-4943-2)

J.A. Hogan and J.D. Lakey: *Duration and Bandwidth Limiting* (ISBN 978-0-8176-8306-1)

G. Kutyniok and D. Labate: *Shearlets* (ISBN 978-0-8176-8315-3)

P.G. Casazza and P. Kutyniok: *Finite Frames* (ISBN 978-0-8176-8372-6)

V. Michel: *Lectures on Constructive Approximation* (ISBN 978-0-8176-8402-0)

D. Mitrea, I. Mitrea, M. Mitrea, and S. Monniaux: *Groupoid Metrization Theory* (ISBN 978-0-8176-8396-2)

T.D. Andrews, R. Balan, J.J. Benedetto, W. Czaja, and K.A. Okoudjou: *Excursions in Harmonic Analysis, Volume 1* (ISBN 978-0-8176-8375-7)

T.D. Andrews, R. Balan, J.J. Benedetto, W. Czaja, and K.A. Okoudjou: *Excursions in Harmonic Analysis, Volume 2* (ISBN 978-0-8176-8378-8)

D.V. Cruz-Uribe and A. Fiorenza: *Variable Lebesgue Spaces* (ISBN 978-3-0348-0547-6)

W. Freeden and M. Gutting: *Special Functions of Mathematical (Geo-)Physics* (ISBN 978-3-0348-0562-9)

A. Saichev and W.A. Woyczyński: *Distributions in the Physical and Engineering Sciences, Volume 2: Linear and Nonlinear Dynamics of Continuous Media* (ISBN 978-0-8176-3942-6)

S. Foucart and H. Rauhut: *A Mathematical Introduction to Compressive Sensing* (ISBN 978-0-8176-4947-0)

G. Herman and J. Frank: *Computational Methods for Three-Dimensional Microscopy Reconstruction* (ISBN 978-1-4614-9520-8)

A. Paprotny and M. Thess: *Realtime Data Mining: Self-Learning Techniques for Recommendation Engines* (ISBN 978-3-319-01320-6)

A. Zayed and G. Schmeisser: *New Perspectives on Approximation and Sampling Theory: Festschrift in Honor of Paul Butzer's 85th Birthday* (978-3-319-08800-6)

R. Balan, M. Begue, J. Benedetto, W. Czaja, and K.A Okoudjou: *Excursions in Harmonic Analysis, Volume 3* (ISBN 978-3-319-13229-7)

H. Boche, R. Calderbank, G. Kutyniok, J. Vybiral: *Compressed Sensing and its Applications* (ISBN 978-3-319-16041-2)

S. Dahlke, F. De Mari, P. Grohs, and D. Labate: *Harmonic and Applied Analysis: From Groups to Signals* (ISBN 978-3-319-18862-1)

G. Pfander: *Sampling Theory, a Renaissance* (ISBN 978-3-319-19748-7)

R. Balan, M. Begue, J. Benedetto, W. Czaja, and K.A Okoudjou: *Excursions in Harmonic Analysis, Volume 4* (ISBN 978-3-319-20187-0)

O. Christensen: *An Introduction to Frames and Riesz Bases, Second Edition* (ISBN 978-3-319-25611-5)

E. Prestini: *The Evolution of Applied Harmonic Analysis: Models of the Real World, Second Edition* (ISBN 978-1-4899-7987-2)

J.H. Davis: *Methods of Applied Mathematics with a Software Overview, Second Edition* (ISBN 978-3-319-43369-1)

M. Gilman, E. M. Smith, S. M. Tsynkov: *Transionospheric Synthetic Aperture Imaging* (ISBN 978-3-319-52125-1)

S. Chanillo, B. Franchi, G. Lu, C. Perez, E.T. Sawyer: *Harmonic Analysis, Partial Differential Equations and Applications* (ISBN 978-3-319-52741-3)

R. Balan, J. Benedetto, W. Czaja, M. Dellatorre, and K.A Okoudjou: *Excursions in Harmonic Analysis, Volume 5* (ISBN 978-3-319-54710-7)

I. Pesenson, Q.T. Le Gia, A. Mayeli, H. Mhaskar, D.X. Zhou: *Frames and Other Bases in Abstract and Function Spaces: Novel Methods in Harmonic Analysis, Volume 1* (ISBN 978-3-319-55549-2)

I. Pesenson, Q.T. Le Gia, A. Mayeli, H. Mhaskar, D.X. Zhou: *Recent Applications of Harmonic Analysis to Function Spaces, Differential Equations, and Data Science: Novel Methods in Harmonic Analysis, Volume 2* (ISBN 978-3-319-55555-3)

F. Weisz: *Convergence and Summability of Fourier Transforms and Hardy Spaces* (ISBN 978-3-319-56813-3)

C. Heil: *A Short Introduction to Metric, Banach, and Hilbert Spaces: With Operator Theory* (ISBN 978-3-319-65321-1)

S. Waldron: *An Introduction to Finite Tight Frames: Theory and Applications* (ISBN 978-0-8176-4814-5)

For an up-to-date list of ANHA titles, please visit http://www.springer.com/series/4968

Printed in the United States
By Bookmasters